6/03

130.00

BUILDING BLOCKS OF MATTER

EDITORIAL BOARD

BUILDING BLOCKS OF MATTER

A Supplement to the
MACMILLAN
ENCYCLOPEDIA OF PHYSICS

John S. Rigden

Editor in Chief

MACMILLAN
REFERENCE
USA™

THOMSON

GALE

New York • Detroit • San Diego • San Francisco • Cleveland • New Haven, Conn. • Waterville, Maine • London • Munich

Building Blocks of Matter: A Supplement to the Macmillan Encyclopedia of Physics
John S. Rigden, Editor in Chief

Library of Congress Cataloging-in-Publication Data

Building blocks of matter : a supplement to the Macmillan encyclopedia of physics / edited by John S. Rigden.
    p. cm.
Includes bibliographical references and index.
ISBN 0-02-865703-9 (hardcover : alk. paper)
1. Particles (Nuclear physics) I. Rigden, John S. II. Macmillan encyclopedia of physics.
QC793.2 .B85 2003

539.7'2—dc21
2002013396

Printed in the United States of America
10 9 8 7 6 5 4 3 2 1

CONTENTS

EDITORIAL AND PRODUCTION STAFF

Deirdre Graves, Brigham Narins *(Project Editors)*
Shawn Beall *(Editorial Support)*
Patti Brecht, Joseph Pomerance *(Copy Editors)*
Carol Roberts *(Indexer)*
Robyn Young *(Project Manager, Imaging and Multimedia Content)*
Pam Galbreath *(Art Director)*
GGS Information Services *(Typesetter)*
Mary Beth Trimper *(Composition Manager)*
Evi Seoud *(Assistant Production Manager)*
Rhonda Williams *(Buyer)*

Macmillan Reference USA
Frank Menchaca *(Vice President)*
Hélène G. Potter *(Director of New Product Development)*
Jill Lectka *(Director of Publishing)*

PREFACE

The concepts and ideas of elementary particle physics are abstract, and they are typically expressed in the language of mathematics. However, the goal of elementary particle physics is very simple, and all the efforts of elementary particle physicists are directed toward that simple goal: to identify the basic building blocks of matter and to understand how they interact to produce the material world we observe.

This encyclopedia contains articles intended for a broad audience of general readers and is designed to edify and give readers an appreciation for one of the most active and productive areas of physics throughout the twentieth century and to the present time. On the one hand, most of the articles have been written in ordinary language and provide a solid base in particle physics concepts and history for those who are new to the field. On the other hand, some topics in particle physics are difficult to express in everyday words, and in the articles on such topics, symbols appear and even an occasional equation. Even these articles, however, are written so that the reader with little physics background can capture a general sense of the topic covered.

Several features of the encyclopedia are designed to help the general reader navigate the language of physics and mathematics included in the articles on the more complex topics. A glossary in the back of the book provides definitions for terms that may be unknown to the reader, both in the field of physics and in related sciences. A list of common abbreviations and acronyms at the beginning of the book is included to aid readers unfamiliar with those used in the book. Numerous tables, figures, illustrations, and photographs supplement the information contained within the articles and provide visual tools to better understand the material presented.

Entries are arranged alphabetically and include extensive cross-references to refer the reader to additional discussions of related topics. In each article, a bibliography directs the reader to books, articles, and Web sites that provide additional sources of information. The articles themselves focus on particular topics that, taken together, make up the intellectual framework called elementary particle physics. Articles such as those on accelerators, quarks, leptons, antimatter, and particle identification provide a working base for the study of particle physics. Articles such as those on quantum chromodynamics, neutrino oscillations, electroweak symmetry breaking, and string theory bring readers to subjects that fill the conversations of contemporary particle physicists. Finally, articles such as those on the cosmological constant and dark energy, supersymmetry, and unified theories discuss the key topics replete with many exciting questions left to be answered.

Articles also detail the history of particle physics, including the discovery of specific particles, such as the antiproton and the electron. In addition to the historical articles, a time line is included to provide an overview of the development of the field of particle physics. This time line of research and development in what is now called particle physics extends back almost three millennia. The time line demonstrates the commanding grip that the desire to identify the basic building blocks of matter has had on the minds of past and present scientists. Biographical articles of physicists who have made seminal contributions to our understanding of the material world complete the encyclopedia's coverage of the history of particle physics. The selection of physicists for the biographies was based on the desire to provide a historical background for the topics presented in this encyclopedia, and so no living physicist was included.

Since experimentation is a vital part of particle physics, detailed articles discuss the technologies used to discover particles, including current accelerator types and subsystems. Articles also profile the international laboratories that house these accelerators, describing experiments, both historic and current, conducted at these labs. Articles on case studies are included to provide the reader with more in-depth information as to how these technologies contribute to the past and continuing search for particles.

Particle physics both affects and is affected by other sciences as well as by the political and philosophical environment. Articles discuss the interaction of particle physics and cosmology, astrophysics, philosophy, culture, and metaphysics. Also included are articles describing the spin-off technologies created in the search for particles as well as the funding of this research.

A reader's guide in the beginning of the encyclopedia arranges the topics into broad categories and thereby helps organize the array of individual entries into a comprehensive field of study. Additionally, the article on elementary particle physics provides an overview of the field and its current questions.

The authors of the articles contained in this encyclopedia work in the top particle physics laboratories and are professors at renowned colleges and universities. Not only does this encyclopedia provide a comprehensive coverage of the field of particle physics, but it also brings together articles from the top members of the physics and scientific community.

This collection of articles would not have been possible without the effort of those who contributed, and I thank each of the authors. Jonathan Rosner, University of Chicago, has responded to personal requests I made of him, and I thank him. Also, I am grateful to both editors, Jonathan Bagger, Johns Hopkins University, and Roger H. Stuewer, University of Minnesota, for their work and advice. Lastly, the Macmillan editor, Deirdre Graves, has been devoted in her assistance throughout the project. We, the editors, thank her.

John S. Rigden

INTRODUCTION

Physicists distinguish between classical and modern physics. The classical era began in the Scientific Revolution of the seventeenth century and extended throughout the eighteenth and most of the nineteenth centuries. By then there were rumblings among some prominent physicists that their subject was complete, that no more basic physics remained to be discovered. Then, in 1895, Wilhelm Conrad Röntgen discovered X rays, and abruptly, although perhaps unknowingly, the modern era of physics began. During the following year Henri Becquerel discovered radioactivity, and in 1897 the work of several physicists culminated in the discovery of the electron, which is generally credited to J. J. Thomson. With the first subatomic particle, the electron, to account for, physicists knew that a new era was under way.

The idea of basic building blocks of matter is at least 2,600 years old. In the sixth century B.C.E. Thales proposed that all things reduced to water, and, coming out of the Greek-Roman eras and for centuries to come, the four basic elements were thought to be earth, water, fire, and air. The atomic hypothesis, originating in the fifth century B.C.E., lingered in the background for centuries until experimental support, through the work of eighteenth- and nineteenth-century chemists, brought atoms to the fore as the basic building blocks of matter. By the early years of the nineteenth century, quantitative measurements had established that hydrogen was the least massive of the chemical elements, and in 1815 William Prout proposed that hydrogen was the building block of all the chemical elements. Prout's idea had supporters through the nineteenth century, but it was finally discredited with the discovery of isotopes early in the twentieth century.

One of the major themes of twentieth-century physics, a spectacular period in the history of physics, has been the continuation, although greatly intensified, of the ancient quest to identify and understand the fundamental constituents of matter. The electron, discovered in 1897, was the first elementary particle, and, after a century that saw "elementary" particles come and go with great profusion, the electron was and remains truly elementary.

What makes a particle elementary? Simply put, it contains no parts. The electron has no hidden constituents. The electron is elementary. The proton, long considered to be an elementary particle, does have parts—three quarks. The proton is not elementary. There are currently twelve elementary particles that physicists believe make up the observable matter throughout the universe: six quarks—up, down, charm, strange, top, and bottom—and six leptons—electron, electron neutrino, muon, muon neutrino, tau, and tau neutrino—all of which fit nicely into three groups, called generations, each consisting of two quarks and two leptons. The first generation consists of the four lightest particles—the up and down quarks and the electron and the electron neutrino—which are the particles responsible for ordinary matter as we currently know it. The composition of dark matter remains a mystery. The particles of the second and third generations are successively more massive, and these heavier particles are believed to have played roles during the moments following the Big Bang. The twelve elementary particles make up the Standard Model.

The electron and proton were discovered by experimental set-ups built on a small table. By contrast, quarks were discovered by means of vast accelerators with dimensions measured in miles and with subsystems that dwarfed the physicists walking among them. The century's trend toward larger and larger accelerators was necessitated by the need for higher and higher energies. In turn, higher energies were required to probe the innards of particles such as the proton as well as to create new particles with substantial masses such as the W and Z as well as the top quark.

The objective of elementary particle physics is twofold: to establish the identity of all the elementary particles of nature and to determine the means by which the elementary particles interact so as to give rise to our material world. Four basic interactions, or forces, have been identified: gravitational, electromagnetic, weak, and strong. Each of these four forces is transmitted between particles by the exchange of a force-carrying particle; the photon transmits the electromagnetic force, W and Z particles the weak force, and gluons the strong force. The graviton, which has not been established experimentally, is assumed to transmit the gravitational force. With the twelve "matter" particles and the four "interaction" particles, the behavior of all the observed matter in the universe can be described.

The ability to describe ordinary matter in terms of a few basic entities is a triumph of contemporary physics. In this remarkable process, however, physicists have moved toward a new threshold that portends stunning insights into the physical world—insights whose outlines can be observed, but only dimly. As is always true, good science raises profound questions. Is space three-dimensional or are there hidden di-

mensions hovering within our intellectual and experimental reach? Dark matter is a reality, but what is it? Dark matter pulls our universe together, but dark energy pushes it apart. What is dark energy? Will the expansive effect of dark energy override the contractive effect of dark matter? Why do the elementary particles have their particular masses? Will the Higgs boson bring understanding to this question? Gravitation remains to be unified with the other basic interactions. What will be required to accomplish this unification? The answers to such questions may transform the conceptual landscape of physics and, in the process, fundamentally alter the way humans view their world.

During the past two decades, nature's extremes have been linked. At one extreme are the elementary particles with their infinitesimal sizes and masses; at the other extreme is the universe with its incomprehensibly immense size and mass. The detailed knowledge of elementary particles accumulated over the past century has illuminated events immediately following the Big Bang and has provided a reasonable explanation of how the universe evolved from the zero-of-time to its current state fifteen billion years later. The physics of elementary particles has joined hands with cosmology, and together they have brought knowledge and understanding to a level that could not have been imagined when the electron was first observed in 1897. Of course, many questions, major questions, await answers; and many details, significant details, await elaboration. Good science begets good questions.

At a practical level, particle physics has dramatically changed contemporary culture. Many of the electronic methods that drive modern societies and many of the computer powers that are now omnipresent were developed to meet the stringent demands of detecting and following events in the unseen domain where the elementary particles blink in and out of existence. The international character of elementary physics, with team members located in laboratories around the globe, required new and efficient ways of communication. The World Wide Web was invented by elementary particle physicists at CERN, the accelerator laboratory in Switzerland, to exchange information quickly and accurately. Many other contributions to society have their origins in accelerator laboratories.

Particle physics has had a profound influence on scientific explanation. For much of the twentieth century, explanations have been sought by reducing complex systems to their simplest parts. Although no one can deny the fruitfulness of this approach and the great appeal of its explanations, it remains an open question whether the simple parts can meet the challenges ahead. Do new phenomena emerge with complexity that cannot be understood in terms of the basic interactions between nature's simplest particles? Indeed, all material systems consist of elementary particles, but as systems move up the ladder of complexity, are there threshold rungs that break the explanatory line of logic back down to the particles? Only further scientific experimentation will provide the answer.

John S. Rigden

LIST OF ARTICLES

LIST OF CONTRIBUTORS

Kazuo Abe
Japanese High-Energy Accelerator Research Organization
Japanese High-Energy Accelerator Research Organization, KEK

Peter Arnold
University of Virginia, Charlottesville
Electroweak Phase Transition

Robert G. Arns
University of Vermont, Burlington
Reines, Frederick

Neil Ashby
University of Colorado, Boulder
Case Study: Gravitational Wave Detection, LIGO

Gordon J. Aubrecht II
Ohio State University
Lawrence, Ernest Orlando

Lawrence Badash
University of California, Santa Barbara
Radioactivity, Discovery of
Rutherford, Ernest

Jonathan Bagger
Johns Hopkins University
Planck Scale

Vernon Barger
University of Wisconsin, Madison
Gauge Theory
Grand Unification

William A. Barletta
Lawrence Berkeley National Laboratory
Devices, Accelerating

Katharina Baur
Stanford Synchrotron Radiation Laboratory
Radiation, Synchrotron

Benjamin Bayman
University of Minnesota, Minneapolis
Cyclotron
Radioactivity

Karl Berkelman
Cornell University
Cornell Laboratory for Elementary Particle Physics

William K. Brooks Jr.
Thomas Jefferson National Accelerator Facility
Accelerators, Fixed-Target: Electron

Laurie M. Brown
Northwestern University
Neutrino, Discovery of
Pauli, Wolfgang
Tomonaga, Sin-itiro
Yukawa, Hideki

Nina Byers
University of California, Los Angeles
Noether, Emmy

Lawrence S. Cardman
Thomas Jefferson National Accelerator Facility and University of Virginia
Thomas Jefferson National Accelerator Facility

R. Sekhar Chivukula
Boston University
Electroweak Symmetry Breaking

Lawrence A. Coleman
University of Arkansas at Little Rock
Momentum
Quantum Tunneling

Janet Conrad
Columbia University
Lepton

Robert P. Crease
State University of New York, Stony Brook
Brookhaven National Laboratory

Sally Dawson
Brookhaven National Laboratory
Boson, Gauge
Standard Model

Michael Dine
University of California, Santa Cruz
Particle
Symmetry Principles

Gabor Domokos
Johns Hopkins University
Resonances

John F. Donoghue
University of Massachusetts, Amherst
Broken Symmetry
Renormalization

Gerald F. Dugan
Cornell University
Accelerator

Guy T. Emery
Bowdoin College
Atom

William E. Evenson
Brigham Young University
Energy
Energy, Center-of-Mass
Energy, Rest

Isobel Falconer
Open University, UK
Electron, Discovery of
Thomson, Joseph John

Adam F. Falk
Johns Hopkins University
Hadron, Heavy

Jonathan L. Feng
University of California, Irvine
Supersymmetry

Kenneth W. Ford
American Institute of Physics (retired)
Conservation Laws

Gordon Fraser
Accelerators, Colliding Beams: Hadron

Wendy L. Freedman
Carnegie Observatories, Pasadena, CA
Hubble Constant

Robert Garisto
Physical Review Letters
Virtual Processes

Marcelo Gleiser
Dartmouth College
Phase Transitions

Charles Goebel
University of Wisconsin, Madison
Gauge Theory

M.C. Gonzalez-Garcia
European Laboratory for Particle Physics (CERN)
Neutrino Oscillations

Howard A. Gordon
Brookhaven National Laboratory
Case Study: LHC Collider Detectors, ATLAS and CMS

Paul Grannis
State University of New York, Stony Brook
Detectors and Subsystems

Benjamin Grinstein
University of California, San Diego
Flavor Symmetry

Lee Grodzins
Massachusetts Institute of Technology
Kendall, Henry

David Gross
University of California, Santa Barbara
Unified Theories

Howard E. Haber
University of California, Santa Cruz
Boson, Higgs

Francis Halzen
University of Wisconsin, Madison
Neutrino Oscillations

Frederick A. Harris
University of Hawaii, Honolulu
Beijing Accelerator Laboratory

Donald Hartill
Cornell University
Injector System

Wick C. Haxton
University of Washington, Seattle
Neutrino, Solar

Kenneth J. Heller
University of Minnesota, Minneapolis
Particle Physics, Elementary

JoAnne Hewett
Stanford Linear Accelerator Center
CKM Matrix
Scattering

Christopher T. Hill
Fermi National Accelerator Laboratory
Higgs Phenomenon

David Hitlin
California Institute of Technology
Detectors, Collider

L. Donald Isenhower
Abilene Christian University
Detectors, Particle

Maurice Jacob
European Laboratory for Particle Physics (CERN)
CERN (European Laboratory for Particle Physics)
International Nature of Particle Physics

Michel Janssen
University of Minnesota, Minneapolis
Einstein, Albert

Elizabeth Jenkins
University of California, San Diego
SU(3)

T. W. B. Kibble
Imperial College, London
Salam, Abdus

Chung W. Kim
Johns Hopkins University and *Korea Institute for Advanced Study, Seoul, Korea*
Neutrino

Robert P. Kirshner
Harvard-Smithsonian Center for Astrophysics, Cambridge, MA
Supernovae

Adrienne W. Kolb
Fermi National Accelerator Laboratory
Fermilab

Noemie Benczer Koller
Rutgers University
Wu, Chien-Shiung

Helge Kragh
University of Aarhus, Denmark
Cosmology
Dirac, Paul

Lawrence M. Krauss
Case Western Reserve University
Cosmological Constant and Dark Energy

Graham Kribs
University of Wisconsin, Madison
Grand Unification

G. Peter Lepage
Cornell University
Lattice Gauge Theory

Harry J. Lipkin
Weismann Institute of Science, Rehovot, Israel
Quarks, Discovery of

Raphael Littauer
Cornell University
Accelerators, Colliding Beams: Electron-Positron

Byron G. Lundberg
Fermi National Accelerator Laboratory
Experiment: Discovery of the Tau Neutrino

Robert H. March
University of Wisconsin, Madison
Muon, Discovery of

William J. Marciano
Brookhaven National Laboratory
Quantum Electrodynamics

John Marriner
Fermi National Accelerator Laboratory
Accelerators, Fixed-Target: Proton
Cooling, Particle
Extraction Systems

Boyce D. McDaniel
Cornell University
Wilson, Robert R.

Kevin McFarland
University of Rochester
Detectors, Fixed-Target

Stephen G. Naculich
Bowdoin College
Cosmic Strings, Domain Walls

Meenakshi Narain
Boston University
Experiment: Discovery of the Top Quark

Dwight E. Neuenschwander
Southern Nazarene University
Antimatter

Keith Olive
University of Minnesota, Minneapolis
Dark Matter

Mark J. Oreglia
University of Chicago
Charmonium

Wolfgang K. H. Panofsky
Stanford University
Funding of Particle Physics

Elizabeth Paris
Massachusetts Institute of Technology
Antiproton, Discovery of

Roberto Peccei
University of California, Los Angeles
Basic Interactions and Fundamental Forces

Nan Phinney
Stanford Linear Accelerator Center
Z Factory

William H. Pickering
California Institute of Technology (emeritus)
Anderson, Carl D.

Joseph Polchinski
University of California, Santa Barbara
String Theory

John Polkinghorne
Queens College, Cambridge, UK
Culture and Particle Physics
Metaphysics

Stephen Pordes
Fermi National Accelerator Laboratory
Detectors

Richard H. Price
University of Utah, Salt Lake City
Relativity

Helen Quinn
Stanford Linear Accelerator Center
J/ψ
SLAC (Stanford Linear Accelerator Center)

David Rainwater
Fermi National Accelerator Laboratory
Experiment: Search for the Higgs Boson

Krishna Rajagopal
Massachusetts Institute of Technology
Quark-Gluon Plasma

Regina Rameika
Fermi National Accelerator Laboratory
Experiment: Discovery of the Tau Neutrino

Pierre Ramond
University of Florida, Gainesville
Family

Blair N. Ratcliff
Stanford Linear Accelerator Center
Radiation, Cherenkov

Rashmi Ray
Physical Review Letters
Virtual Processes

Michael L. G. Redhead
University of London, UK
Philosophy and Particle Physics

David Rice
Cornell University
Accelerators, Colliding Beams: Electron-Positron

Steven Ritz
NASA Goddard Space Flight Center
Detectors, Astrophysical

B. Lee Roberts
Boston University
Experiment: g−2 Measurement of the Muon

Natalie Roe
Lawrence Berkeley National Laboratory
B Factory

Xavier Roqué
Universitat Autònoma Barcelona, Bellaterra, Spain
Positron, Discovery of

Jonathan L. Rosner
University of Chicago
CP Symmetry Violation
Eightfold Way

Lewis Ryder
University of Kent, Canterbury, UK
Annihilation and Creation
Feynman Diagrams
Parity, Nonconservation of

David H. Saxon
University of Glasgow, UK
Accelerators, Colliding Beams: Electron-Proton
Influence on Science
Particle Identification

Silvan S. Schweber
Brandeis University
Feynman, Richard
Schwinger, Julian

Robert W. Seidel
University of Minnesota, Twin Cities
Accelerators, Early

Ramamurti Shankar
Yale University
Quantum Field Theory

Pierre Sikivie
University of Florida, Gainesville
Axion

Joseph I. Silk
University of Oxford, UK
Big Bang

Albert Silverman
Cornell University
Wilson, Robert R.

Elizabeth H. Simmons
Boston University
Electroweak Symmetry Breaking

Alexander N. Skrinsky
Budker Institute of Nuclear Physics
Budker Institute of Nuclear Physics

Henry W. Sobel
University of California, Irvine
Case Study: Super-Kamiokande and the Discovery of Neutrino Oscillations

Paul Söding
Deutsches Elektronen-Synchrotron Laboratory
DESY (Deutsches Elektronen-Synchrotron Laboratory)

Suzanne T. Staggs
Princeton University
Cosmic Microwave Background Radiation

George Sterman
State University of New York, Stony Brook
Asymptotic Freedom
Jets and Fragmentation
Quantum Chromodynamics

Roger H. Stuewer
University of Minnesota, Minneapolis
Chadwick, James
Neutron, Discovery of

Daniel F. Styer
Oberlin College
Quantum Mechanics

Michael J. Syphers
Fermi National Accelerator Laboratory
Beam Transport

John Terning
Los Alamos National Laboratory
Technicolor

Alvin V. Tollestrup
Fermi National Accelerator Laboratory
Quarks

Virginia Trimble
University of California, Irvine
Astrophysics

Neil G. Turok
University of Cambridge, UK
Inflation

Roger K. Ulrich
University of California, Los Angeles
Big Bang Nucleosynthesis

Erich Vogt
University of British Columbia, Vancouver, Canada
Wigner, Eugene

C. Jake Waddington
University of Minnesota, Minneapolis
Cosmic Rays

Terry P. Walker
Ohio State University
Universe

Albert Wattenberg
University of Illinois, Urbana-Champaign
Fermi, Enrico

Bebo White
Stanford Linear Accelerator Center
Computing

Frank Wilczek
Massachusetts Institute of Technology
Benefits of Particle Physics to Society
Quantum Statistics

Edmund J. N. Wilson
European Laboratory for Particle Physics (CERN)
SSC

Bruce Winstein
University of Chicago
Outlook

Stanley G. Wojcicki
Stanford University
Case Study: Long Baseline Neutrino
Detectors, K2K, MINOS, and OPERA

Zhipeng Zheng
Institute of High Energy Physics, Beijing, China
Beijing Accelerator Laboratory

COMMON ABBREVIATIONS AND ACRONYMS

AGS . . . . Alternating Gradient Synchrotron (BNL)

ALEPH . . . . . Apparatus for LEP Physics (CERN)

ALICE . . . . . . A Large Ion Collider Experiment (CERN)

AMANDA . . . Antarctic Muon and Neutrino Detector Array (South Pole)

ATLAS . . . . . A Toroidal LHC Apparatus (CERN)

BEPC . . . . . . Beijing Electron-Positron Collider (IHEP)

BES . . . . Beijing Spectrometer (IHEP)

BINP . . . . . . Budker Institute of Nuclear Physics

BNL . . . . Brookhaven National Laboratory

c . . . . speed of light

CDF . . . . Collider Detector at Fermilab (FNAL)

CDM . . . . cold dark matter

CEBAF . . . . . Continuous Electron Beam Accelerator Facility (JLAB)

CERN . . . . . . European Laboratory for Particle Physics (Conseil Européen de la Recherche Nucléaire)

CESR . . . . . . Cornell Electron Storage Ring (LEPP)

CHAOS . . . . Canadian High Acceptance Orbit Spectrometer (University of Regina, Canada)

cm . . . . centimeter

cm^3 . . . . cubic centimeter

CMS . . . . Compact Muon Spectrometer (CERN)

COMPASS . . Common Muon and Proton Apparatus for Structure and Spectroscopy (CERN)

cos . . . . cosine

CPS . . . . CERN Proton Synchrotron (CERN)

CREST . . . . . Cryogenic Rare Events Search with Superconducting Thermometers (Gran Sasso Laboratory, Italy)

DAFNE, ... Double Annular Factory for Nice
   DAPHNE   Experiments (LEPP)

dc . . . . direct current

DELPHI . . . . Detector with Lepton Photon and Hadron Identification (CERN)

DESY . . . . . . Deutsches Elektronen-Synchrotron Laboratory

DOE . . . . Department of Energy (United States)

DONUT . . . . Direct Observation of the Nu Tau (FNAL)

DORIS . . . . . Double Ring Store (DESY)

e . . . . electronic charge

ELSA . . . . Electron Stretcher Accelerator (Bonn University)

esu	electrostatic unit
eV	electron volt
FNAL, ... Fermilab	Fermi National Accelerator Laboratory
ft	foot
g	acceleration of free fall
g	gram
G	gravitational constant
GeV	giga electron volt
GHz	gigahertz
GLAST	Gamma Ray Large Area Space Telescope
GUT	grand unified theory
h	Planck's constant
HERA	Hadron Electron Ring Accelerator (DESY)
Hz	hertz
ICARUS	Imaging Cosmic and Rare Underground Signal (Gran Sasso Laboratory, Italy)
IHEP	Institute of High Energy Physics
in	inch
ISR	Intersecting Storage Rings (CERN)
J	joule
JLAB	Thomas Jefferson National Accelerator Facility
J/s	joules per second
K	degrees Kelvin
KEK	Japanese High-Energy Accelerator Research Organization
kg	kilogram
km	kilometer
LANL	Los Alamos National Laboratory
LBNL	Lawrence Berkeley National Laboratory
LEAR	Low Energy Antiproton Ring (CERN)
LEP	Large Electron Positron Collider (CERN)
LEPP	Laboratory for Elementary Particle Physics (Cornell)
LHC	Large Hadron Collider (CERN)
LIGO	Laser Interferometer Gravitational-Wave Observatory (California Institute of Technology)
linac	linear accelerator
m	meter
MeV	mega electron volt

MHz	megahertz
MINOS	Main Injector Neutrino Oscillation Search (FNAL)
ml	milliliter
mm	millimeter
Mpc	megaparsec
mph	miles per hour
ms	millisecond
MV	megavolt
nb	nanobarn
NIST	National Institute of Standards and Technology
NOMAD	Neutrino Oscillation Magnetic Detector (CERN)
ns, nsec	nanosecond
n/s	neutrons per second
n/s cm^2	neutrons per second per a square centimeter
NSF	National Science Foundation
ns/m	nanoseconds per meter
NuMI	Neutrinos at the Main Injector (FNAL)
OPAL	Omni Purpose Apparatus for LEP (CERN)
PDG	Particle Data Group
PEP	Positron Electron Project (SLAC)
PETRA	Positron Electron Tandem Ring Accelerator (DESY)
PMT	photomultiplier tube
PS	Proton Synchrotron (CERN)
QCD	quantum chromodynamics
QED	quantum electrodynamics
R&D	research and development
rf	radio frequency
RHIC	Relativistic Heavy-Ion Collider (BNL)
s, sec	second
SLAC	Stanford Linear Accelerator Center
SLC	Stanford Linear Collider (SLAC)
SNO	Sudbury Neutrino Observatory (Queen's University, Canada)
SPEAR	Stanford Positron Electron Accelerating Ring (SLAC)
SPS	Super Proton Synchrotron (CERN)
SSC	Superconducting Super Collider Project

SSRL. . . . . . . Stanford Synchrotron Radiation Laboratory

TASSO . . . . . Two Arm Spectrometer Solenoid (DESY)

TESLA . . . . . TeV Energy Superconducting Linear Accelerator (DESY)

TeV. . . . . tera electron volt

TRT . . . . Transition Radiation Tracker

VLHC. . . . . . Very Large Hadron Collider (FNAL)

VLPC . . . . . . Visible Light Photon Counter

W. . . . . watt

A

ACCELERATOR

Particle accelerators are scientific instruments used to accelerate elementary particles to very high energies. They are of paramount importance for the study of elementary particle physics because the fundamental structure of matter is most clearly revealed by observing reactions of elementary particles at the highest possible energies. Historically, the development of elementary particle physics has been strongly coupled to advances in the physics and technology of particle accelerators. The first modern particle accelerators were developed in the 1930s and led to fundamental discoveries in nuclear physics. From 1930 to 1990, the energies attainable in particle accelerators have increased at an exponential rate, with an average doubling time of about two years. This progress has been due to a remarkable synergy between accelerator physics concepts (such as resonant acceleration, alternating gradient focusing, and colliding beams) and accelerator technology developments (such as microwave cavities, superconducting magnets, and broadband feedback systems). The consequence has been enormous progress in our understanding of the fundamental forces and constituents of matter.

Types of Accelerators

The large varieties of high-energy accelerators all share two basic common features. The first feature is the way that they accelerate the collection of moving charged particles within the accelerator (which is called the beam). In all accelerators, the energy of the beam is increased by passing it through electric fields, which exert a force on the beam parallel to its direction of motion. This force causes the beam's energy to increase. The second common feature is

the method of controlling the direction of motion of the beam. All accelerators do this by the use of magnetic fields, which exert a force perpendicular to the direction of motion of the beam.

Accelerators can be usefully classified according to their geometry. A linear accelerator is a straight-line arrangement of many electric fields, with a few magnetic fields interspersed between the electric fields to focus the beam. A circular accelerator typically has only a few electric fields. Many magnetic fields bend the orbit of the beam into a closed, roughly circular path, as the beam particles pass through the electric fields once each revolution. Over many revolutions, the energy of the beam increases. As explained below, the magnetic field strength required to deflect a particle beam through a given angle is proportional to the momentum of the beam. In a synchrotron (the most common form of circular accelerator), the strength of the magnetic field is increased with the beam energy to maintain a constant radius orbit.

Accelerators may also be distinguished according to the species of particle that they accelerate: electrons or heavier particles such as protons (also called hadrons). One of the features of circular electron accelerators is the production of large amounts of electromagnetic radiation due to the centripetal acceleration of the electrons. This radiation, called synchrotron radiation, complicates the design of circular electron accelerators, since the radiated energy must be restored to the beam particles, increasing the requirements on the accelerator's electric fields. However, the radiation (a highly directional source of X rays) has been found very useful for applications in condensed matter physics, chemistry, and biology. Many accelerators (called synchrotron radiation sources) have been built whose sole purpose is the production of such radiation. For a fixed-radius accelerator, the power dissipated in synchrotron radiation increases as the fourth power of the beam energy, placing a very severe limit on the ultimate energy of circular electron accelerators. To achieve very high-energy electron beams, linear electron accelerators are required.

In hadron accelerators, protons or heavier ions are accelerated. Because of their larger mass, the synchrotron radiation of protons in circular accelerators is much weaker than that of electrons. Consequently, much higher energies are possible in circular proton accelerators than in circular electron accelerators. However, unlike the electron, the proton is not a true elementary particle: it is a composite system of three quarks and multiple gluons. The energy carried by a proton is shared among the quarks and gluons, so the energy of a single quark is much lower than the proton beam energy.

Accelerators may also be classified in terms of the final use of the accelerated beam. In accelerators prior to the 1960s, the high-energy beam struck a stationary target, in which the reactions to be observed took place. This was done either by placing the target within the accelerator or by manipulating the orbit of the accelerated beam so that it emerged from the accelerator (a process called extraction) and struck the target. In either case, an accelerator that is used in this way is called a fixed-target accelerator. The energy E_R available for a reaction in a target is given by

$$E_R \cong \sqrt{2E_b mc^2}$$

in which E_b is the total beam energy, m is the rest mass of the target atom, and c is the speed of light.

Starting in the 1970s, circular accelerators were developed in which two counter-rotating beams were made to collide with reactions occurring at the collision point. Such an accelerator is called a collider. If both beams share orbits controlled by a single set of magnetic fields, one of the beams must be composed of the antimatter partner of the other (e.g., protons and antiprotons, or positrons and electrons). The advantage of a collider lies in the fact that the energy available for a reaction is given in this case by

$$E_R = 2E_b.$$

Since typically $E_b \gg mc^2$, the energy available for a reaction is much larger than in a fixed-target accelerator. Colliders may also be built using two separate accelerators, which share a small overlap region where collisions take place; in this case, antimatter is not required. All current and planned accelerators operating at the energy frontier are colliders.

Circular colliders often utilize a special type of accelerator called a storage ring. This is a circular accelerator in which the beam simply circulates at a fixed energy. Collisions take place during the storage time of the beam, which is usually in the range of several hours. During this time, the beams may undergo billions of collisions. Nevertheless, the number of particles in the beam is diminished very slowly, since the probability of a high-energy reaction occurring in a single collision is very low.

Very high-energy electron circular colliders are not feasible due to excessive synchrotron radiation. To obtain very high energies in the collisions of electron beams, it is necessary to collide the beams from two opposing electron linear accelerators. Such a machine is called a linear collider.

Although the beam energy of a collider is a key measure of its usefulness for the study of elementary particle reactions, it is not the only figure of merit. Equally important is a measure of the rate at which reactions will occur: this measure is called the luminosity. For a collider, the luminosity is proportional to the density of the beams at the collision point and to the rate at which collisions take place. The design of a high-energy collider is often dominated by the need to attain sufficient luminosity to permit the observation of an adequate number of high-energy reactions.

Injector

The injector is the source of the particles for an accelerator. The injector is required to deliver to the accelerator a beam of a specified quality and energy. The quality of a beam is a measure of the beam's intensity and size: a high-quality beam will typically have a large number of particles (perhaps 10^{10}) and a relatively small transverse size (ranging from millimeters to nanometers, depending on its energy and its location within the accelerator). For low-energy accelerators, the injector may be a small device, such as a hot-filament electron source or a discharge ion source. For high-energy accelerators, the injector is itself a complex arrangement of lower-energy accelerators. For hadron colliders, the luminosity is influenced heavily by the beam quality delivered by the injector.

Colliders that utilize beams of antimatter require very specialized injectors that can efficiently collect antimatter. The antimatter is typically produced in a fairly diffuse, low quality state from a target illuminated by the beam of an auxiliary fixed-target accelerator. The quality of the antimatter beam must be increased by orders of magnitude, in a process called beam cooling. For electrons and positrons, specially designed storage rings, called damping rings, are used, in which the process of synchrotron radiation reduces the size and increases the density of the beam. For antiprotons, an artificial process (called stochastic cooling) involving sophisticated microwave signal processing is often employed. After sufficient cooling has occurred, the injectors can deliver high-quality antimatter beams to a collider.

Acceleration System

For accelerators used in elementary particle physics, the acceleration system is a set of resonant cavities or waveguides carrying time-varying electromagnetic fields. The beam passes through the cavities, and the electric fields increase the energy of the beam. The frequency of the electromagnetic cavity fields can range from below 50 MHz to above 30 GHz. The electric field strengths can range from below 5 MV/m to above 100 MV/m. The beam is accelerated in "bunches" whose length is related to the wavelength of the cavity fields, ranging from meters (for accelerators using 50 MHz fields) to fractions of a millimeter (for high-frequency accelerators).

A key concept in an acceleration system is that of resonant acceleration. This requires that each bunch arrive at each cavity at about the same phase of the electromagnetic field, so that each bunch always receives roughly the same energy gain. The cavity spacing and the field's frequency must be appropriately matched to the beam velocity to achieve resonant acceleration. An important feature of the beam dynamics is called phase stability. This guarantees that, under the appropriate circumstances, the beam is stable under small deviations from the resonant acceleration condition (that is, if displaced from the resonant condition, the beam will oscillate stably about it, rather than continue to deviate further from it).

Orbit Control System

The orbit control system in an accelerator is a set of magnets placed along the beam's trajectory. The magnets do not change the energy of the beam but exert forces on the beam that define its orbit. The magnets are most often electromagnets, with fields that are either constant in time (in storage rings) or which increase in strength as the beam's energy is increased (in synchrotrons). Permanent magnets, with fixed magnetic fields, may also be used in storage rings. The most common types of magnets used in an accelerator are dipole magnets and quadrupole magnets.

Dipole magnets produce a spatially uniform magnetic field and are used to deflect the orbits of all particles in the beam by the same amount. The fundamental orbit control system in a circular accelerator is a series of dipole magnets that bend the orbit of the beam into a roughly circular path. The Lorentz force exerted by the dipole's field provides the centripetal force required for circular motion. This leads to the following relation between the momentum of the beam particle p, the magnetic field B, the beam particle's charge q, and the beam's orbit radius R:

$$p = qBR.$$

This equation shows that for a high-energy beam, with a large value of p, either a large magnetic field or a large orbit radius is required. The need to limit the accelerator's size, for economic reasons, puts a great premium on the use of high magnetic fields for high-energy circular accelerators. Very high magnetic fields can be generated without excessive power dissipation through the use of magnets whose conductors are made from superconducting materials. This is why today's largest high-energy circular accelerators rely on superconducting magnet technology for their orbit control system.

Quadrupole magnets are used to focus the beam. A useful analogy may be made between the orbits of charged particles in an accelerator and the paths of light rays in an optical system. Prisms deflect all the rays in a monochromatic light beam by the same amount in the same way that dipole magnets deflect all the orbits in a monoenergetic charged particle beam by the same amount. Optical lenses focus light beams by providing a deflection of a light ray that is proportional to the distance of the ray from the lens' axis. Similarly, charged particle beams are focused using quadrupole magnets, which have a magnetic field strength that is proportional to the distance from the magnet's axis. The use of quadrupole magnets is essential to the operation of all types of accelerators. Their focusing properties ensure that the beam will oscillate stably about the ideal orbit if displaced from it.

Optical lenses are cylindrically symmetric and can focus simultaneously in both transverse planes. Unfortunately, the equations of electrodynamics do not allow this for quadrupole magnets: if they focus in one transverse plane, they must defocus in the other. Nevertheless, it is possible to construct a system of alternating focusing and defocusing magnets whose net effect is focusing. This is called the principle of alternating gradient focusing. Accelerators with a focusing system based on this principle were first developed in the 1950s, and since then all accelerators make use of this feature in their orbit control system.

Final Use Systems

In a fixed-target accelerator, the high-energy beam is usually extracted from the accelerator prior to its use in the creation of high-energy reactions. Extraction is very simple from a linear accelerator. Extraction from a circular accelerator can be more challenging. It is usually not desirable to extract the entire beam from the accelerator in one revolution, as the resulting instantaneous rate of reactions in the target may be too high to be useful. Generally, the beam must be extracted "slowly," over many thousands of revolutions. Such a procedure often relies on the generation of small nonlinear disturbances in the accelerator's magnetic fields, which slowly divert the beam from its stable orbits. The location of these disturbances must be carefully controlled to ensure that the entire beam emerges from the accelerator at a single location from which it can be transported by a linear array of quadrupole and dipole magnets (called a beam line) to the target.

In a collider, the beams do not need to be extracted but must be tailored to have very specific features at the collision point. Since the luminosity is proportional to the density of the beams at the collision point, the beams must be focused very tightly to as small an area as possible. A system of very strong quadrupole magnets, placed within the accelerator very close to the collision point, provide this focusing. When the high-density beams collide, the electromagnetic fields of one beam can strongly perturb the motion of the other beam. This beam-beam interaction is one of the fundamental limitations on the achievable beam density, and hence luminosity, in a circular collider. In a linear collider, the beams interact only once, and so the density can be made much higher. Nevertheless, the luminosity is comparable to that in a circular collider because the rate at which collisions occur is much lower in a linear collider.

To record the results of the colliding beam reactions, a system of high-energy particle detectors is installed surrounding the collision point. These particle detectors often have their own magnetic fields, which can influence the orbits of the colliding beams, and must be considered in the design of the accelerator. Conversely, background reactions from stray particles in the beam can severely comprise the performance of the particle detector. The need for careful and close integration of the particle detector and the accelerator is an important feature of a collider.

See also: ACCELERATORS, COLLIDING BEAMS: ELECTRON-POSITRON; ACCELERATORS, COLLIDING BEAMS: ELECTRON-PROTON; ACCELERATORS, COLLIDING BEAMS: HADRON; ACCELERATORS, EARLY; ACCELERATORS, FIXED-TARGET: ELECTRON; ACCELERATORS, FIXED-TARGET: PROTON; BEAM TRANSPORT; DETECTORS; EXTRACTION SYSTEMS; INJECTOR SYSTEM

Bibliography

Bryant, P. J., and Johnsen, K. *The Principles of Circular Accelerators and Storage Rings* (Cambridge University Press, New York, 1993).

Conte, M., and MacKay, W. W. *An Introduction to the Physics of Particle Accelerators* (World Scientific, Singapore, 1991).

Edwards, D. A., and Syphers, M. J. *An Introduction to the Physics of High Energy Particle Accelerators* (Wiley, New York, 1993).

Lawson, J. D. *The Physics of Charged-Particle Beams* (Oxford University Press, New York, 1988).

Livingood, J. J. *Principles of Cyclic Particle Accelerators* (Van Nostrand, Princeton, NJ, 1964).

Livingston, M. S., and Blewett, J. P. *Particle Accelerators* (McGraw-Hill, New York, 1962).

Reiser, M. *Theory and Design of Charged Particle Beams* (Wiley, New York, 1994).

Wangler, T. P. *Principles of RF Linear Accelerators* (Wiley, New York, 1998).

Wiedemann, H. *Particle Accelerator Physics I: Basic Principles and Linear Beam Dynamics,* 2nd ed. (Springer-Verlag, New York, 1998).

Gerald F. Dugan

ACCELERATORS, COLLIDING BEAMS: ELECTRON-POSITRON

Astronomy, cosmology, and space travel have expanded our frontiers outward to the limits of the universe and to its earliest moments. In the opposite direction, atomic, nuclear, and particle physics have pushed inward toward the ultimate constituents of matter. Both frontiers offer thrilling adventure and great triumphs. Both call for impressively large and expensive machines. And, surprisingly, the discoveries on the innermost scale shed light also on the grandest event of cosmology: the Big Bang—a veritable cauldron of elementary particles.

To explore them deep inside, atoms are bombarded with beams of particles brought to high energy in an accelerator: the higher the projectile's energy, the deeper it can probe into an atom and its nucleus. More spectacularly, as a consequence of relativity, a collision with enough energy can also create new particles. (Conservation laws may call for pair creation, particle plus antiparticle, to balance the books.) The required energy is the equivalent of the total mass created.

The rest energies, $E_0 = m_0c^2$, of some interesting particles are given in Table 1. The energy stakes in this game can be high, well above the rest energy of the projectiles themselves—especially if the projectiles are electrons, whose rest energy is only 0.00051 GeV. For example, a 5.1-GeV electron has 10,000 times the energy it had at rest; equivalently, its mass is 10, 000 times its original rest mass. Such an electron is ultrarelativistic.

It is not efficient to shoot a massive particle at a stationary target (as does a fixed-target accelerator).

TABLE 1

Rest Energies of Selected Particles

Particle		Rest Energy
Electron (e)	Stable	0.00051 GeV
Proton (p)		0.94 GeV
Muon (μ)		0.11 GeV
pi-zero meson (π^0)		0.14 GeV
Omega meson (ω)		0.78 GeV
Tau lepton (τ)		1.8 GeV
J/psi meson (J/ψ)	Unstable	3.1 GeV
Upsilon meson (Υ)		9.5 GeV
W boson		79 GeV
Z^0 boson		91 GeV
Top quark (t)		170 GeV

CREDIT: Courtesy of Raphael Littauer.

When a massive projectile strikes a light target, it flies on almost undisturbed, retaining most of its energy—like a truck that has hit a mosquito. A projectile gives up energy only if it is slowed down. That can happen if two beams of particles are aimed at each other: in a head-on collision, both particles may slam to a stop and release all their combined energies. Unfortunately, compared to a slab of stationary matter, an oncoming particle beam makes a frustratingly elusive target. It took single-minded optimism and dedication to overcome this problem. Nevertheless, since the 1960s, colliding beam accelerators have become the dominant tool for particle research.

Event Rate: Cross Section and Luminosity

On the subatomic scale, hitting a target is a matter of chance, rather like shooting into a swarm of mosquitoes. However, large mosquitoes are hit more often than small ones: they present more frontal area. By analogy, the probability of hitting a particle, producing a specified type of outcome, can be represented as an effective frontal area. This is called the production cross section σ (sigma). That is, if a target particle is somewhere within an area A, and one projectile is shot into this area, the chance of obtaining an event of the type specified is σ/A. Shooting N_1 projectiles f times per second at N_2 targets in an area A will produce, on average, $fN_1N_2(\sigma/A)$ events per second. (For particles, a convenient unit for σ is the nanobarn (nb); 1 nb $= 10^{-9}$ barn $= 10^{-33}$ cm^2. Barn is the name jokingly given to a cross section of 10^{-24} cm^2, as easy to hit as the side of a barn!)

The luminosity $£$ of the collider is defined as the factor that multiplies σ; that is, event rate = $£\sigma$. In the situation just described, $£ = fN_1N_2/A$. For example, if a collider produces events of cross section one nb at an average rate of 1 per second, its luminosity is $£ = 1/$nb/s (or 10^{33} cm^{-2} s^{-1}).

Most colliders use storage rings to keep bunches of particles circulating in opposite directions, passing through each other on every turn at one or more interaction points (IP). In principle, the particles continue to circulate until they finally collide; in practice, there are other losses. To increase the luminosity, the bunches are focused into a very small spot at the IP by a low-beta insertion, a set of strong lenses that act like back-to-back burning glasses. (Beta is an optical parameter related to the size of the bunch; its value at the IP also indicates the maximum bunch length that can be accommodated given the diverging bunch profile on either side of the focus.)

A storage ring fulfills two other functions: particles can be accumulated in each bunch from many injection cycles to increase N_1 and N_2 above what is directly available from an injector (particle source plus preaccelerator). Also, when accumulation is complete, the particles can (if necessary) be accelerated to the desired collision energy while they circulate in the ring.

Choice of Particle

Colliders using electrons (e^-) and their antiparticles, positrons (e^+) represent one of several types of colliding beam accelerators. Electrons—used generically, the term includes both e^- and e^+—are distinguished by the type of physics information they reveal and also by the technical aspects of their storage:

- Electrons are truly elementary: they have no internal components. Their collisions produce pristine, precisely controllable conditions. (By contrast, the quarks and gluons that make up a proton can lead to very complicated scenarios.) Moreover, an electron and a positron can annihilate each other when they collide, surrendering all their energy to the collision products. The energy of the collider can then be set with almost surgical precision to match a desired final situation. The advantages of this annihilation mode are so compelling that they far outweigh the difficulty of first having to create the positrons. Because of their opposite charges, e^- and e^+ can circulate in opposite directions in a single ring.

- Electrons at collider energies are ultrarelativistic; when forced to circulate in a ring, they emit strong synchrotron radiation. This energy loss is a major burden for electron storage; however, because it damps particle oscillations, it also has beneficial effects. (The "waste" radiation was soon exploited for an impressive range of research and industrial applications. Specialized synchrotron light sources have since proliferated.)

Figure 1 schematically illustrates the main components of a storage-ring collider. Figure 2 is a view inside the tunnel for the Cornell Electron Storage Ring (CESR).

Source of Particles

Electrons are readily emitted from a heated metal, as in a TV picture tube. By contrast, positrons must first be created. This is done by bombarding a converter—a slab of heavy metal—with high-energy electrons from a linear accelerator (linac). Near the heavy nuclei of the converter a cascade of processes develops: electrons radiate some of their energy as photons, and photons, in turn, produce electron-positron pairs. Emerging from the converter is a cloud of electrons, positrons, and photons, from which positrons are directed by magnetic lenses into another linac. For injection into the storage ring, the particles (e^+ or e^-) are brought to high energy in one or more boosters (linac or synchrotron).

Injection and Storage

A storage ring is a special-purpose synchrotron. The particles circulate in a vacuum chamber placed in a magnetic guide field, with quadrupoles (magnetic lenses) keeping them close to the desired trajectory. Energy lost by radiation is replaced as the particles traverse one or more radio frequency (rf)

FIGURE 1

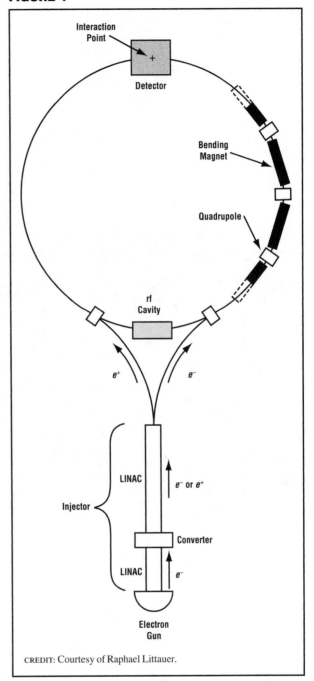

CREDIT: Courtesy of Raphael Littauer.

Schematic layout of a storage-ring collider. The chain of bending magnets and quadrupoles continues all the way round the ring; only a few are shown.

cavities—hollow metal structures in which a strong oscillating electric field is maintained. Conveniently, this time-dependent field gathers the particles into

FIGURE 2

View inside the tunnel for CESR. The booster synchrotron is on the left, the storage ring on the right. CREDIT: COURTESY OF CORNELL UNIVERSITY.

short, synchronized bunches by the mechanism of phase stability: particles arriving early or late receive different energy increments that return them toward the bunch center.

Because of the high intensity of the stored bunches and the long storage times, very stringent stability conditions must be met by the components of a synchrotron used in storage mode. Also, to avoid derailing the particles already stored, new ones must be injected on a displaced path that weaves about the central orbit. Fortunately, synchrotron radiation damps these injection oscillations, so the new particles soon coalesce with the older bunch.

When accumulation (and final acceleration, if any) is complete, the circulating bunches are steered to meet head-on at an interaction point (IP), around which the detector is placed. This consists of sophisticated equipment to track and analyze the fragments emerging from a collision, often identifying special patterns in as few as one in a million cases. (In terms of complexity and expense, detectors may rival the collider itself.) After an experimental run is initiated, the bunches may circulate for an hour or more, passing through each other many millions of times. Ultra-high vacuum is maintained in the beam chamber to reduce collisions with residual gas, which would shorten their lifetime and also cause background in the detector.

Synchrotron Radiation (SR)

As they circulate in a ring, continually steered inward, electrons emit synchrotron radiation (SR), a broad spectrum of electromagnetic waves reaching typically into the ultraviolet and X-ray region. The most dramatic feature of SR is its steep rise with beam energy E: the energy radiated per turn is proportional to E^4. ($E \times 10 \rightarrow$ SR $\times 10,000!$) To maintain the beams, the radiated energy must be resupplied continuously by the rf cavities. The required power, sometimes tens of megawatts, can become prohibitive; to lower it, the ring radius is made large (SR power is inversely proportional to the radius squared). At the highest energies, SR ultimately becomes prohibitive for electron storage rings, forcing a retreat to linear colliders (discussed below).

SR is emitted in a narrow forward cone, like light from a car's headlights. A particle traveling at an angle to the ideal trajectory emits SR at that angle; this carries off some of the transverse momentum. Since the rf cavities resupply purely forward momentum, transverse oscillations are gradually damped. (The effect—analogous to friction steadying a pendulum—is used specifically in damping rings to form compact particle bunches.)

SR is not emitted continuously but instead in individual quanta (photons), each of which jolts the electron with a step in energy. This gives the bunch an energy spread; also, because off-energy particles want to travel at different radii, it excites transverse oscillations in the plane of orbit. The ultimate bunch dimensions are governed by equilibrium between quantum excitation and radiation damping; typically, a bunch comprising upward of 10^{11} particles may be several millimeters wide, a fraction of a millimeter high, and some tens of millimeters long.

Intensity Limitations

Short bunches of many particles represent very large instantaneous beam currents—often several hundreds of amperes—accompanied by strong electromagnetic pulses. These wake fields echo around the vacuum chamber and can react back on the bunch (or subsequent bunches) causing instability. The beam's environment (chamber, rf cavities, and auxiliary apparatus) must be carefully controlled to raise the usable intensity. In addition, feedback devices can detect incipient oscillations and, within limits, act to suppress their growth.

Unfortunately, as the particles pass through the electromagnetic field of the opposing bunch, they are deflected by an amount that varies strongly across the bunch. This unavoidable beam-beam interaction (BBI) dilutes bunch density and limits the maximum usable intensity per bunch. The ensuing ceiling on luminosity is raised by tighter focusing at the IP (lower beta), but here the limit is set by the bunch length. Further increases ultimately result only from raising the number of bunches in each beam.

When B bunches circulate in each of the two beams, they make $2B$ encounters around the ring, at each of which the BBI must be controlled. With only a few bunches, each crossing point can be configured with a low-beta insertion as a usable IP. Many colliders have done this, but only at the cost of exacerbating the BBI. For more bunches, multiple meeting points must be avoided by separating the bunches with electric fields. Even so, residual BBI makes it progressively harder to raise the number of bunches. CESR represents an extreme case: with up to forty-five bunches each of e^+ and e^- it produces a luminosity ten times that of other single-ring colliders (Table 2).

The highest luminosities, achieved in colliders ambitiously known as particle factories, are obtained with two separate rings, where the trajectories are separated except near an IP.

Asymmetrical Colliders

Use of equal-energy colliding beams is motivated by the energy yield achieved in head-on collisions. However, the available energy is not much reduced if the two beams have somewhat unequal energy. The collision products are then carried forward in the direction of the higher-energy beam, which makes the decay points of short-lived collision products visible by moving them away from the IP. Knowing how long an unstable particle survived is important in some experiments, such as those looking for particle-antiparticle asymmetry in the decay of B and $\bar{B}$ mesons. (This information could shed light on how the universe evolved from the Big Bang to a state where matter dominates over antimatter.)

Physics Results from Electron Colliders

Some major electron-positron colliders are listed in Table 2. There has been dramatic progress on both frontiers: energy and luminosity. Because a collider yields peak performance over only a relatively narrow energy span, many different colliders are in service. The largest ring, LEP, about 27 kilometers in circumference, reached an energy (100 + 100 GeV) still far short of the energies possible with proton rings. (Protons, 2,000 times more massive than electrons, are less relativistic for a given energy and emit only an insignificant amount of synchrotron radiation. On the other hand, in comparing effective collision energies, one must consider that the real projectiles and targets—the quarks within the protons—each carry only a fraction of the proton's energy as a whole.)

Energy alone is not enough for a collider; there must also be sufficient luminosity to yield an acceptable event rate. To place this in perspective, Figure 3 shows how the cross section σ varies with total energy

E. (Note that, to do justice to the very wide range of values, the scales on this graph are logarithmic.) The dominant feature is the steep decrease of σ—in proportion to $1/E^2$. (Every time E increases tenfold, σ is divided by 100.) This trend underlies all collision processes that start with e^+e^- annihilation, which is the dominant mode in the region up to approximately 100 GeV. The cross section for the production of a lepton pair—e, μ, or τ, involving no strong forces, only quantum electrodynamics (QED)—is shown as a broken line. Measurements at successive colliders have checked the theoretical prediction to great accuracy, verifying that leptons are indeed pointlike particles, down to a scale of 10^{-16} cm.

In the late 1960s, when ADONE came into operation, it was a pleasant surprise to many how readily e^+e^- collisions yielded hadrons (strongly interacting particles). This was interpreted as being initiated by production of a quark pair and helped quarks gain acceptance as likely constituents of matter. The solid curve in Figure 3, with its dotted extension, shows that the cross section for quark-pair processes is a constant multiple of the lepton-pair cross section; the numerical ratio confirms a fundamental tenet of the Standard Model, namely, that quarks come in three "colors."

TABLE 2

Selected Electron-Positron Colliders

Name	Location	Maximum Energy (GeV)	Circumference (m)	Dates	Luminosity (events/nb/s)*
ACO	Orsay, France	0.6 + 0.6	22	1967–1974	0.0001
ADONE	Frascati, Italy	1.5 + 1.5	105	1969–1995	0.0006
SPEAR	Stanford, California, USA	4.1 + 4.1	234	1972–1990	0.02
VEPP-2M	Novosibirsk, Russia	0.9 + 0.9	18	1975–2001	0.005
PETRA	Hamburg, Germany	22 + 22	2,304	1978–1987	0.02
CESR	Ithaca, New York, USA	8 + 8	768	1979–	1.3
DORIS–II	Hamburg, Germany	5.5 + 5.5	288	1979–	0.03
VEPP-4M	Novosibirsk, Russia	5.5 + 5.5	365	1979–	0.006
PEP	Stanford, California, USA	15 + 15	2,200	1980–1995	0.03
TRISTAN	Tsukuba, Japan	15 + 15	3,016	1987–1995	0.014
BEPC	Beijing, China	2.5 + 2.5	240	1988–	0.01
LEP	Geneva, Switzerland	100 + 100	26,659	1989–2001	0.1
SLC	Stanford, California, USA	50 + 50	n/a (linear)	1989–1998	0.003
PEP-II	Stanford, California, USA	9 × 3 (10.4)	2,200	1998–	4
DAFNE	Frascati, Italy	0.51 + 0.51	98	1999–	0.05
KEK-B	Tsukuba, Japan	8 × 3.5 (10.6)	3,016	1999–	7

*See text for this unit; 1/nb/s = 1×10^{33} cm^{-2}s^{-1}. Quoted luminosities are values achieved by time of writing (May 2002).

CREDIT: Courtesy of Raphael Littauer.

FIGURE 3

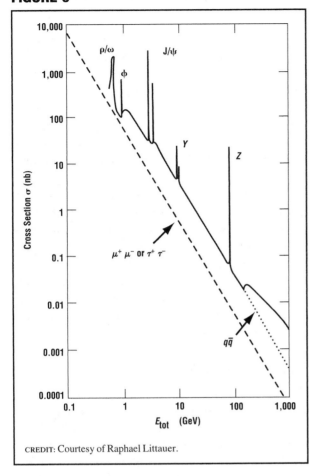

CREDIT: Courtesy of Raphael Littauer.

Electron-positron annihilation cross section as a function of total collision energy.

The tall, narrow peaks of Figure 3 superimposed on the sloping curve are resonances that occur when the collision energy matches the rest energy of a specific final-state particle. Finding such a resonance can be dramatic. For example, in 1974 the event rate at SPEAR went up a hundredfold when the energy coincided with the rest energy of the J/ψ meson (consisting of a pair of charmed quarks), but the collider's energy had to be correct to within 0.00005 GeV! Once found, such resonances are a cornucopia of information about the particle, its lifetime and decay patterns, and the properties of the secondary particles in their turn. Studying rare decay modes requires a large number of "raw" events; even with a relatively large resonance production cross section, there is always a call for higher luminosity.

At the energies first reached by PETRA, particles often emerge from the collision clustered in two or three "jets." These trace back to individual collision products, confirming the presence of quarks and providing the first evidence of gluons, the essential carriers of the strong force.

The last peak on the graph, for the Z^0 intermediate boson, comes from LEP. Measuring the lifetime of the Z^0 (via the width in energy of its resonance) indicated that there are three, and only three, generations of light neutrinos, an important piece of information for the Standard Model.

The τ lepton—a member of the third generation in the Standard Model—was discovered at SPEAR in 1976.

Toward Higher Energy

To reduce the burden of synchrotron radiation, it is tempting to think about colliding leptons more massive than electrons (and thus not so extremely relativistic). Unfortunately, the next best candidate, the muon, is unstable, with a lifetime of only 2.2 microseconds. Relativistic time dilation helps: for example, a 100-GeV muon lives for 2 milliseconds, but this is still very marginal. Creating muons and antimuons at a sufficient rate to make up for their decay is a major challenge.

Less speculative is the use of a linear collider, avoiding rings entirely—but giving up their advantages, too. In a linear colliders there is no accumulation, ramping up to final energy, or recycling of particles: it is like single-shot injection at full energy. To recoup luminosity, the bunches must be made to cross in an extremely small spot, perhaps a few nanometers in diameter. At such densities, the beam-beam interaction becomes so strong that it is better called disruption. The bunches also radiate energy as they cross, which degrades the energy definition (and will ultimately limit the maximum energy of e^+e^- colliders).

In a dramatic drive for quick progress, the Stanford Linear Collider (SLC) was built at the Stanford Linear Accelerator Center (SLAC), using just their single linac. Bunches of electrons and positrons were accelerated to 50 GeV in close succession and were then steered into head-on collisions by two guide-field arcs, through which they needed to pass only once. Although the complexities of bending and focusing the very small bunches limited luminosity, both high-energy and accelerator physics results were noteworthy.

Several linear collider projects are currently under development. NLC in the United States and JLC in Japan both use room-temperature copper accelerating rf cavities; TESLA in Germany uses superconducting niobium. These machines would stretch over 30 kilometers in total length and likely be deep underground. A facility of this size will, of necessity, be an international project. Several laboratories are conducting research on more potent (higher field) accelerating structures for a second generation of linear colliders.

See also: ACCELERATOR; BEAM TRANSPORT; EXTRACTION SYSTEMS; INJECTOR SYSTEM

Bibliography

CERN. "Hands-on CERN." <http://hands-on-cern.physto.se>.

Edwards, D. A., and Syphers, M. J. *Introduction to the Physics of High-Energy Accelerators* (Wiley, New York, 1993).

Horgan, J. "Particle Metaphysics." *Scientific American* **270** (2), 96–106 (1994).

Kane, G. *The Particle Garden* (Addison-Wesley, Reading, MA, 1995).

Myers, S., and Picasso, E. "The LEP Collider." *Scientific American* **263** (1), 54–61 (1990).

The Particle Data Group. "The Particle Adventure." <http://particleadventure.org>.

Quinn, H. R., and Witherell, M. S. "The Asymmetry Between Matter and Antimatter." *Scientific American* **279** (4), 76–81 (1998).

Riordan, M. *The Hunting of the Quark* (Simon & Schuster, New York, 1987).

Wilson, R. R. "The Next Generation of Particle Accelerators." *Scientific American* **242** (1), 42–57 (1980).

Raphael Littauer
David Rice

ACCELERATORS, COLLIDING BEAMS: ELECTRON-PROTON

The electron and the proton, along with the neutron (a close relative of the proton), are the particles that make up all of the observed universe. The study of their structure and properties is therefore fundamental.

One of the consequences of relativity is that a truly elementary particle, that is, one with no internal structure, must be of zero size. The electron satisfies this criterion as far as we can tell. Measurements at the highest energies probe the smallest distances and show that the electron is certainly smaller than 10^{-18} m across. The proton, by contrast, is 10^{-15} m across and has internal structure. It consists of three quarks, which give it its properties, plus a fluctuating sea of quark-antiquark pairs, strongly bound together by gluons, which are the carriers of the strong force, which binds protons and neutrons together into the nuclei of atoms. The very strength of the strong force makes it hard to study: it is hard to isolate one feature of it at a time. Any disturbance one makes in a strongly interacting system has massive side effects from the collision that can confuse the features under investigation.

The Need for High Energies

One way to open up the strong force to study is to use the highest possible energies. This allows us to single out one interaction at a time while the effect of the underlying, less energetic interactions can be ignored. The key to high energies is to use colliding beams. Early experiments to create high-energy collisions in the laboratory used a single beam of high-energy particles hitting a stationary target. This has the disadvantage that a large fraction of the energy of the beam is wasted in uselessly imparting momentum to the target particle that is struck. The law of conservation of momentum makes this loss unavoidable and severe. The need to add momentum to the target soaks up precious beam energy.

When the energies concerned become large compared to the masses of the particles involved (using $E = mc^2$ as a comparator), the useful collision energy grows only very slowly as the beam energy is increased. Thus, if a proton beam is used on a fixed proton target, a beam kinetic energy six times the proton mass produces a useful collision energy twice the proton mass. Worse, an energy 800 times the proton mass produces a useful energy less than 40 times the proton mass. The process of putting energy into reactions therefore becomes very inefficient.

Benefits of Colliding Beams

Colliding beams are much more efficient. Here two beams of particles collide head on. In the case where the two beams consist of identical particles and have equal energies, they will have equal and opposite momenta. The two momenta then cancel out, no energy is needed to set anything in motion, and 100 percent of the energy is available for reaction. One must be careful when asserting this when one of the beams is made of composite particles, such as protons. The constituent quarks, antiquarks, and gluons each take only a modest fraction of the momentum of the moving proton. If one thinks of an electron-proton collider as an electron-quark collider (with the remainder of the proton constituents acting as spectators), then it is the balance between the electron and quark momenta that matters.

Colliding Beam Options

Practical colliding beams in use in 2002 use recirculating beams of electrically charged particles. One can compare electron-positron colliders (the positron is the antiparticle of the electron) with proton-antiproton and electron-proton (or positron-proton) colliders. Properties of selected colliders are shown in Table 1. The collision energy listed in the table is the total energy available for interactions after allowing for the effects of momentum conservation. The typical energy is the energy available in a quark collision. (Recall the typical quark carries only a small fraction of the proton energy). For some purposes, such as searching for new particles, the collision energy may be more important than the typical energy.

Electron-positron colliders create new matter out of the energy of the collisions they create and address many questions, but since there is no proton present, they do not directly touch on the question of proton structure. Proton-antiproton collisions (or proton-proton collisions planned for the European Laboratory for Particle Physics [CERN] Large Hadron Collider) reach the highest attainable energies but involve colliding two internally complex objects, and so as far as the structure of the proton is concerned, are harder to interpret. By contrast, electron-proton colliders use the pointlike electron as a scalpel to dissect the proton. Nature holds the quarks strongly inside the proton. One can never see a free quark but only a "jet" of pions and other particles produced from electron-proton collisions.

Figure 1 shows an example of an electron-proton interaction. The single electron track is turned almost around by the vigor of the collision, and a jet of particles is produced by the quark that is struck.

Features of a Colliding Beam Facility

A colliding beam accelerator operates in the following way. One starts with a source of electrons (a hot wire in a vacuum under high voltage, for example) and protons (simply the nuclei of the atoms in a cylinder of hydrogen gas). (If positrons are required, the procedure is more complex: a beam of electrons with an energy of several million volts hits a target, producing many matter-antimatter electron-positron pairs. Positrons are then accumulated in a small storage ring

TABLE 1

Selected Colliders and Their Properties

Facility	Location	Beam 1	Beam 2	Energy 1 (GeV)	Energy 2 (GeV)	Collision Energy (GeV)	Typical Energy (GeV)	Year
LEP	Geneva	Electron	Positron	104	104	208	208	2000
Tevatron	Chicago	Proton	Antiproton	980	980	1,960	60	2001
HERA	Hamburg	Electron/ positron	Proton	27.5	920	320	55	2000
LHC	Geneva	Proton	Proton	7,000	7,000	14,000	420	2006

CREDIT: Courtesy of David H. Saxon.

FIGURE 1

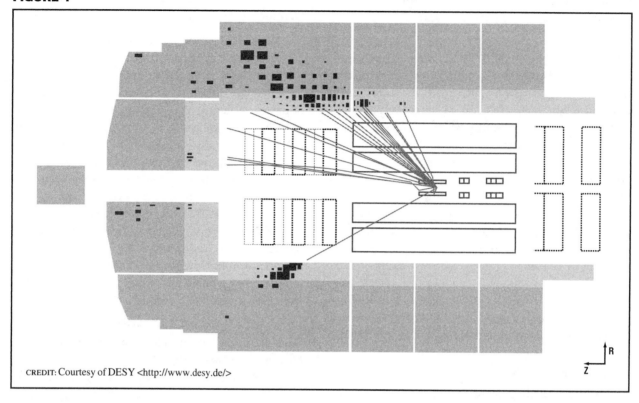

Reconstruction of a high-energy electron-proton interaction seen in the H1 detector at HERA. The electron beam is incident along the detector axis from the left and the proton beam from the right. The single outgoing track in the lower portion is the outgoing electron. The proton is broken up, giving a jet of particles traveling into the top portion of the detector. Lines show particle tracks, and squares indicate energy deposited in the detector.

until enough have been made.) Then the particles are accelerated to high energy using a synchrotron.

The basis of acceleration is the addition of energy to an electrically charged particle by passing it through a voltage difference. The trick is that this voltage difference is oscillating inside a metal cavity at very high frequency, switching sign a billion times per second. If a localized bunch of particles arrives at just the right moment, their energies are increased. The particles are then recycled in a circular path within a narrow vacuum pipe (using a ring of magnets to steer and focus them) and passed repeatedly through the same cavities, gaining energy at each revolution. The magnetic field must be increased in step with the energy in order to keep the particles on the same circular path. (Synchronization of the magnetic field to the beam energy at each step in the acceleration process gives the synchrotron its name.) In practice, a set of accelerator rings is used, each ring stepping up the energy a certain amount until the large storage ring is filled with a train of bunches of electrons and protons (some 200 bunches in the case of HERA, the storage ring at the Deutsches Elektronen-Synchotron Laboratory [DESY] in Hamburg, Germany). The process of filling and accelerating the two separate electron and proton beams to the required energies normally takes up to an hour. The magnet rings are then kept filled at the maximum energy while the beams make repeated orbits, with the possibility of reactions occurring at each collision of an electron bunch with a proton bunch.

The need to recycle the particles during acceleration and storage governs the size of the machine. Using superconducting magnet technology, the HERA magnets (built in 1992) achieve a maximum magnetic field of over 5 Tesla around the circular arcs of the accelerator (more than twice that attainable using copper conductors). For a given proton beam energy, this dictates the radius of the arcs and hence the circumference of the machine. To achieve a smaller arc radius, a higher magnetic field would be required. The Large Hadron Collider magnets will run at 8.3 Tesla in 2007.

The HERA Collider

Figure 2 gives a schematic of the HERA facility showing the succession of booster rings. The electron (or positron) and proton beams circulate in opposite directions in separate magnet rings of 6.4 km

FIGURE 2

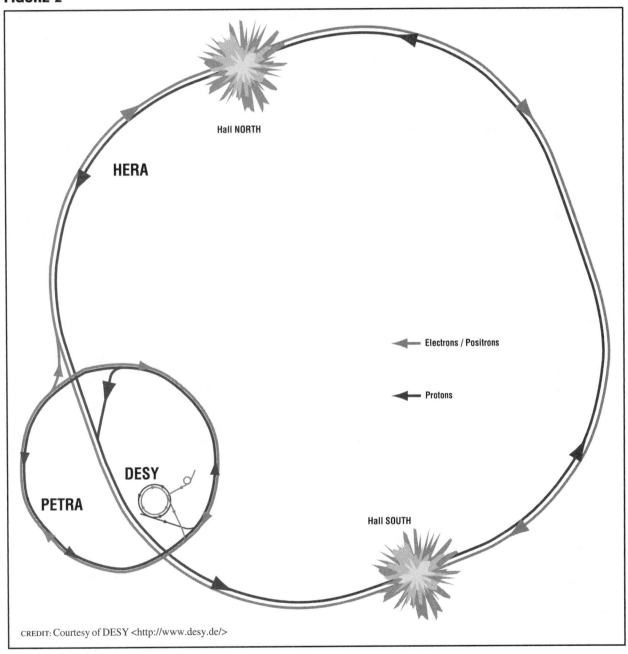

CREDIT: Courtesy of DESY <http://www.desy.de/>

Schematic of the HERA collider at DESY. All particles start in the small linear collider and are accelerated to progressively greater energies in the DESY, PETRA, and HERA rings. Electrons or positrons travel clockwise in the HERA ring, and protons travel counterclockwise. The two beams are brought into collision in the North and South experimental halls.

circumference. Each ring has circular arcs of magnets (for the acceleration and recycling of the beams) and straight sections containing the electrical cavities and focusing elements. The electron and proton beams are brought into collision at two points north and south of the ring where the collisions can be studied.

Colliding beams offer a big advantage over fixed targets in terms of energy but are much more problematic in terms of intensity. Protons and electrons are accumulated individually: the total mass of protons circulating at any time is only 10^{-11} g, so precious few are available for the electrons to hit. The key lies in focusing, to increase the local density at the collision point, and recycling. At the interaction points the electron and proton beams are focused down to a spot only 0.05 mm by 0.2 mm across. Each individual particle, traveling close to the speed of light, makes some 47,000 circuits per second. In a typical 8 hour running period, each proton has 2,700,000,000 opportunities to interact collide and interact. Scientists select about five interactions per second where interesting processes have occurred for detailed study.

See also: ACCELERATOR; BEAM TRANSPORT; COOLING, PARTICLE; EXTRACTION SYSTEMS; INJECTOR SYSTEM

Bibliography

Fraser, G. *The Particle Century* (IoP Publishing, Bristol, UK, 1998).

Wille, K. *The Physics of Particle Detectors, an Introduction* (Oxford University Press, Oxford, UK, 2000).

Wolf, G. *Proceedings International School of Subnuclear Physics, Erice,* Vol. 37, edited by A. Zichichi (World Scientific, Singapore, 1999).

David H. Saxon

ACCELERATORS, COLLIDING BEAMS: HADRON

Accelerators providing colliding beams of hadrons have become one of the major tools of high-energy physics research. Hadrons, from the Greek word for "thick," or "heavy," are those elementary particles which feel the strong nuclear interaction. The proton and its antimatter counterpart, the antiproton, are the only stable hadrons. All other hadrons ultimately decay, yielding protons and other, nonhadronic, particles. Because protons are electrically charged and stable, they can be made into beams (beam optics) and controlled by electric and magnetic fields, for example, in particle accelerators such as the cyclotron, synchro-cyclotron, and synchrotron. Although they are held together by the strong nuclear force, atomic nuclei (positively charged ions) strictly speaking are not hadrons but are stable and can be controlled and accelerated in the same way as protons.

Fixed Targets versus Colliding-Beam Machines

In particle accelerators, beams of particles such as protons are taken to high energy by suitable arrangements of electric and magnetic fields. When the resultant beams hit a fixed target (a block of material), energy is released in the collisions between the incident particles and the constituent protons and neutrons of the target nuclei. According to Einstein's equation $E = mc^2$, this released energy can produce additional particles that were not present initially. From the late 1940s, these fixed-target accelerators enabled physicists to discover many new kinds of elementary particles. The greater the energy of the incoming particle, the greater the energy that can be liberated for making new particles. To calculate how much collision energy is thus made available means viewing the collision in a reference frame which moves with the center of mass of the colliding particles. When a high-energy proton collides with a stationary proton, the resultant collision energy E_{cm} in the center-of-mass frame is

$$E_{cm} = mc^2 \sqrt{2 + 2E/mc^2}$$

where m is the proton mass, E the energy of the moving particle, and c the velocity of light. If E is very large, this can be approximated to

$$E_{cm} = mc^2 \sqrt{2E/mc^2}.$$

This shows that the collision energy which can be exploited for making new particles increases only as the square root of the energy of the incident proton. The majority of the incident energy is "lost" simply in making the target particle recoil.

This energy loss can be avoided by colliding two particle beams together, when a greater share of the energy of the two colliding particles becomes available for the production of new particles. In 1943, Rolf Wideröe, who built the first linear accelerator in 1927 and subsequently developed the betatron, a machine for accelerating electrons, proposed such a colliding beam idea, which he patented in 1953. In 1956 Gerald O'Neill of Princeton took the idea further by proposing the use of a standard synchrotron to accelerate the particles and then hold them in two rings which met at a common tangent, where the stored beams would collide. Soon such electron-electron colliders were built at Stanford, California, by a Stanford-Princeton collaboration, and in Novosibirsk, Russia. These developments soon led to the first electron-positron colliding beam accelerators, in which a beam of electrons and a beam of positrons (the antimatter counterpart of electrons) can be held in adjacent orbits in a single ring before being made to collide inside the ring.

As charged particles are accelerated in a ring, any particles with velocities directed out of the beam oscillate around the stable beam orbit, and these oscillations have to be controlled if the circulating beam is to remain stable. Being very light particles, electrons and positrons when accelerated radiate a considerable amount of electromagnetic energy in the form of "synchrotron radiation." This radiation emission naturally damps the oscillations of electrons and positrons, ensuring that the circulating beams are narrow. It also means that a storage ring for electrons and positrons has to replenish continuously the energy of the circulating beams to compensate for these energy losses.

For protons, several major obstacles had to be overcome before the colliding beam idea could be put to work. Beams of protons are nowhere near as dense as even the lightest of natural materials, so most of the time the colliding beams would simply slide past each other, with no collisions taking place. To overcome this, the particle beams would have to be held in storage rings, where they could cross and hopefully collide over and over again. In these rings, the particle beams would have to circulate stably, so that they could be held over much longer time periods than a normal fixed-target machine. But protons, being heavy, do not lose much energy through

FIGURE 1

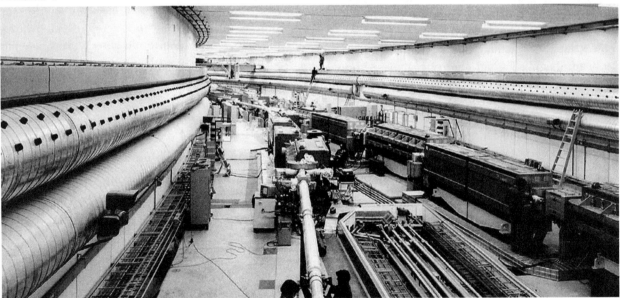

The CERN Intersecting Storage Rings (ISR) commissioned in 1971. The ISR was the first Hadron collider. One of the machine's eight proton beam intersections can be seen in the background. CREDIT: COURTESY OF CERN (EUROPEAN ORGANIZATION FOR NUCLEAR RESEARCH). REPRODUCED BY PERMISSSION.

synchrotron radiation, and other techniques are required to ensure that the beams become sufficiently concentrated.

This can be achieved using the idea of beam stacking, invented in 1956 by the Mid-Western Universities Research Association (MURA), based in Madison, Wisconsin. For beam stacking, the injected particles are stored over as wide a kinematic range in the machine as possible. At the time, CERN, the European Laboratory for Particle Physics in Geneva, Switzerland, was well on the way to building its first high-energy synchrotron, the 25-GeV (giga electron volt) Proton Synchrotron (PS), which supplied its first beams in 1959. But CERN was already looking further ahead and wanted to use the PS to feed new physics facilities and reach higher collision energies. Protons, unrestricted by synchrotron radiation losses, attain much higher collision energies than an electron-positron collider, and in much smaller rings.

At CERN, one possibility was to build two rings to hold contra-rotating 25-GeV proton beams from the PS and bring them into collision, thereby exploiting the full 50 GeV of two colliding particles. This scheme, the Intersecting Storage Rings (ISR), with two interlaced rings 300 meters in diameter crossing at eight different points, was approved for construction in 1965. Mindful that its performance would depend on tight control of the proton beams, tight tolerances were imposed everywhere—magnets, power supplies, vacuum. Special attention was paid to beam diagnostics and control systems. Although it was designed primarily as storage rings, the ISR had its own radio frequency system, to stack the proton pulses arriving from the PS, and this radio frequency power later allowed the ISR protons to be accelerated to 31 GeV.

Many pessimists predicted that the ISR would not work. Unlike the spectacularly successful electron-positron colliders, proton machines at these energies are not subject to synchrotron radiation. This threatened that every ripple and field error would impress itself on the circulating particle beams, and these effects would accumulate, destroying the beam. But no such thing happened. Beams in the ISR did grow slowly, mainly due to scattering on residual gas, and to minimize these effects the ISR pioneered new ultra-

high-vacuum techniques, reaching 10^{-12} torr. The ISR produced its first collisions on January 24, 1971, and the machine subsequently ran for thirteen years. Beams were routinely stored for physics runs of fifty hours or more without replenishment.

In 1968, even before the ISR was running, Simon van der Meer at CERN proposed an idea for beam cooling, controlling the spread of the transverse beam momentum in a beam's particles. Van der Meer's stochastic beam cooling scheme would monitor the fluctuations of a circulating beam and transmit a suitable correction signal across the diameter of the ring to meet the same particles as they came round. Over a period of time, this would beat down the statistical fluctuations and increase the beam density. Stochastic cooling was first demonstrated at the ISR in 1974.

Meanwhile another technique for beam cooling had appeared. At Novosibirsk, where one of the first electron colliders had been built, Gersh Budker had the vision of building a proton-antiproton collider. Instead of having two rings, like CERN's ISR, such a machine could hold protons and antiprotons in a single ring before colliding them together, analogous to an electron-positron collider. But antiprotons are difficult to produce and then even more difficult to control. To handle such unruly beams, Budker proposed the idea of electron cooling, surrounding the particles with a sleeve of well-behaved (cold) electrons which would absorb transverse motion from the enclosed beam. In this way, the core beam would become better behaved. Electron cooling was first demonstrated at Novosibirsk by Alexander Skrinsky, a colleague of Budker, in 1974.

Proton-Antiproton Colliders

In particle-antiparticle colliders, the colliding particles, with their mutually opposite quantum numbers, can annihilate. In this way, all the shared collision energy becomes available for producing completely new kinds of particle. This was soon exploited in electron-positron colliding beam machines. One of the earliest of such machines, SPEAR, at the Stanford Linear Accelerator Center (SLAC), transformed electrons and positrons into new varieties of quark-antiquark bound states in 1974.

As well as quark-antiquark bound states, particle-antiparticle collisions could in principle also furnish other particles, such as the tau lepton discovered at SLAC in 1975. The hope had been that electron-positron colliders could also find the long-awaited W and Z particles, respectively the electrically charged and neutral carriers of the weak interaction. But in the mid-1970s, experimental results from a new generation of synchrotrons at Fermilab, near Chicago, and at CERN began to suggest values for the hitherto unknown masses of the W and Z particles. These masses, about 100 GeV, were out of reach of any contemporary electron-positron collider.

In 1976, David Cline, Peter McIntyre, and Carlo Rubbia proposed building a proton-antiproton collider, using the new beam cooling techniques to control the antiprotons. The higher energies available from proton-antiproton colliders with multihundred-GeV beams could be used to look for the W and Z particles. At this time, Fermilab's new 500-GeV synchrotron was furnishing its first beams, and CERN's new 400-GeV Super Proton Synchrotron (SPS) was nearing completion. Fermilab was not immediately interested in the proton-antiproton collider proposal, being already committed to building a more powerful synchrotron, the Tevatron, using superconducting magnets cooled to 4.2 K.

After constructing a special ring to test the new electron cooling and stochastic beam cooling techniques, CERN saw that the latter route was best suited for high-energy antiprotons, and in 1979 CERN decided to convert its SPS synchrotron into a proton-antiproton collider, building special experiments around the beam collision points. This first meant constructing an ambitious antiproton supply system using stochastic beam cooling, and with this in place the new SPS proton-antiproton collider duly delivered its first collisions of 270-GeV protons and 270-GeV antiprotons in the summer of 1981. In 1983, the experiments at the collider discovered the W and Z particles, with masses near 80 and 90 GeV, respectively. The following year, Rubbia, who had pushed the proton-antiproton collider scheme from the beginning and had led the major experiment, and van der Meer, the architect of stochastic beam cooling, were awarded the Nobel Prize in Physics.

Bigger Colliders

Even while these developments were taking place, physicists were looking further ahead at the subsequent generation of machines to explore higher collision energies. In 1962, the United States decided to begin work on a "Super-ISR" at Brookhaven, near New York City, where the Alternating Gradient Synchrotron (AGS) would supply 30-GeV proton beams to a new two-ring arrangement, called ISABELLE, to collide together 200-GeV proton beams in a 3.8-kilometer circumference tunnel. Meanwhile Fermilab was pushing ahead with its superconducting Tevatron project, to supply 1-TeV (1,000 GeV) proton beams. Once the Tevatron was available, then this could be adapted, CERN SPS style, into a proton-antiproton collider, but at higher energies than had been available at CERN. The Tevatron went on to provide its first proton-antiproton collisions in 1985, and in 1995 experiments there discovered the sixth (top) quark. The Tevatron collider initially supplied two 800-GeV beams, and this beam energy was subsequently increased to 980 GeV. The Tevatron pioneered the use of superconducting magnets in a hadron collider, making it possible to achieve collision energies in the TeV range.

CERN's discovery of the W and Z particles in its proton-antiproton collider wrested the crown of particle physics from the United States, which had monopolized particle physics discoveries since the post–World War II introduction of high-energy accelerators. To re-establish its position, the United States proposed an ambitious new machine, the Superconducting Supercollider (SSC). A plan emerged for an 84-kilometer oval racetrack at a totally new laboratory at Waxahachie, Texas, to collide 20-TeV proton beams. With such a large machine, the advantages of using protons and antiprotons in the same ring were outweighed by the difficulties of pumping as many particles as possible into the ring. Even with beam cooling, the antiproton supply was necessarily limited. The SSC was therefore designed as a two-ring proton-proton collider. In the initial SSC approval package, one condition had been that work would stop on Brookhaven's ISABELLE, which meanwhile had been renamed the Colliding Beam Accelerator (CBA).

SSC construction began in Texas in the early 1990s, but in the decade since the project's initial approval, the financial climate had changed. In 1992, SSC funding was drastically reduced, and in 1993 the decision came from Washington, D.C., to scrap the project completely. Of what would have been the world's largest hadron collider, there remained a few huge superconducting magnets and 23 kilometers of empty tunnel in Texas. In the wake of the SSC cancellation, the superconducting CBA scheme at Brookhaven was resurrected and transformed into the Relativistic Heavy Ion Collider (RHIC). Instead of protons, RHIC's initial aim was to collide beams of heavy ions, such as gold nuclei, at energies of up to 200 GeV per nucleon. RHIC went on to collide its first ion beams in 2000. RHIC also collides beams of polarized (spin-oriented) protons.

Several laboratories had occasionally proposed colliding protons with electrons. At the Deutsches Elekronen-Synchrotron Laboratory (DESY) in Hamburg, Germany, a large electron-collider, HERA, was built. Using 30-GeV electrons and 850-GeV protons in a 6.3-kilometer tunnel, using superconducting magnets for the proton ring, HERA provided the world's first high-energy electron-proton colliding beams in 1991.

The LHC

Even while CERN was still building its proton-antiproton collider, a longer-term plan was emerging for a 27-kilometer-circumference tunnel housing the LEP electron-positron collider, to mass-produce Z particles. LEP construction began in 1983, and the machine began operating in 1989. However, with a 27-kilometer-circumference ring available, a hadron collider—either for protons on protons or for protons on antiprotons—eventually could be built in the LEP tunnel.

Plans for this machine, the Large Hadron Collider (LHC), initially emerged in the mid-1980s in parallel with the United States's SSC scheme. Like the SSC, the LHC would be a proton-proton collider, but with its smaller circumference it needed powerful magnets to hold its 7-TeV proton beams. The LHC design thus uses superfluid helium at 1.9 K as the superconducting cooling medium. To optimize the difficult cryogenics, the LHC design evolved into a scheme with the two separate proton rings held inside a single, common cryostat.

The LHC beams will emit small, but measurable, quantities of synchrotron radiation. This is useful for beam diagnostics but insufficient to damp beam oscillations, so traditional beam stacking procedures still have to be relied on. But for a highly cryogenic machine, even a small amount of heating produced by synchrotron radiation has to be kept under tight control. LEP was decommissioned in 2000 so that LHC construction work could begin. The LHC is scheduled to produce its first proton-proton collisions in 2007. It will also sometimes be used to collide beams of heavy nuclei. The large-scale experiments at the LHC attract research physicists from all over the world.

While LEP used 50- to 100-GeV electron and positron beams, the LHC will operate with 7-TeV protons in the same 27-kilometer circumference tunnel, illustrating well the different constraints on accelerating and storing protons and electrons in circular machines. With heavy particles such as protons, the challenge is to provide a strong magnetic field to enable the particles to reach as high an energy as possible in the ring, while with electrons, which are very light particles, the beams have to be gently constrained to minimize synchrotron radiation losses in the ring.

In broad terms, hadron colliders provide high collision energies and are best suited for an initial exploration of a new physics regime. Protons contain three valence quarks together with an attendant cloud of quarks, antiquarks, and gluons, and the collision energy is smeared out among these constituent particles. Electrons and positrons, on the other hand, contain no constituents, and in electron-positron colliders the resultant collision energy can be more sharply focused.

The experiments built for the initial generation of hadron colliders (the ISR and the SPS collider at CERN) revolutionized the design of large-scale particle detectors, with a powerful magnet and with tracking and energy measurement (calorimetry) surrounding the point where the beams collide so as to intercept as many as possible of the emerging particles. This approach is used in all major experiments at colliders. Hadron colliders and planetary systems share the distinction of being the only large-scale systems that have stable orbits for 10^{12} and more revolutions.

TABLE 1

World Hadron Colliders

Machine	Circumference	Beam 1 (max)	Beam 2 (max)	1st operation	Closed
ISR, CERN, Geneva	1 km	31 GeV protons	31 GeV protons	1971	1984
SPS, CERN	7 km	315 GeV protons	315 GeV antiprotons	1981	1991
Tevatron, Fermilab	6.4 km	980 GeV protons	980 GeV antiprotons	1985	
RHIC, Brookhaven	3.8 km	200 GeV/nucleon ions	200 GeV/nucleon ions	2000	
SSC, Texas	84 km	20 TeV protons	20 TeV protons	cancelled	
LHC, CERN	27 km	7 TeV protons	7 TeV protons	2007 (scheduled)	

CREDIT: Courtesy of Gordon Fraser.

See also: ACCELERATOR; BEAM TRANSPORT; COOLING, PARTICLE; EXTRACTION SYSTEMS; INJECTOR SYSTEM

Bibliography

Bryant, P. J., and Johnsen, K. *The Principles of Circular Accelerators and Storage Rings* (Cambridge University Press, Cambridge, UK, 1993).

Fraser, G. *The Quark Machines* (Institute of Physics Publishing, Bristol and Philadelphia, 1997).

Johnsen, K. "The CERN Intersecting Storage Rings: The Leap into the Hadron Collider Era" in *The Rise of the Standard Model, Particle Physics in the 1960s and 1970s*, edited by L. Hoddeson, L. Brown, M. Riordan, and M. Dresden (Cambridge University Press, Cambridge, UK, 1997).

Richter, B. "The Rise of Colliding Beams" in *The Rise of the Standard Model, Particle Physics in the 1960s and 1970s*, edited L. Hoddeson, L. Brown, M. Riordan, and M. Dresden (Cambridge University Press, Cambridge, UK, 1997).

Gordon Fraser

ACCELERATORS, EARLY

The study of particle physics has involved artificial means of accelerating particles since the discovery of the electron, which required the manipulation of cathode rays by an electromagnetic field in an evacuated tube. The need for more powerful sources of accelerated particles was articulated by Ernest Rutherford, the discoverer of the atomic nucleus, in 1927. The development of electron acceleration for X-ray tubes in order to provide high-voltage X rays provided a more practical rationale for the development of early accelerators. Discovered in 1895, X rays were in common medical use within a very short time, although it was not until the First World War that tubes with reliable output were manufactured by W. D. Coolidge at General Electric.

At the California Institute of Technology (Caltech), Charles Lauritsen and his colleagues capitalized on a million-volt testing laboratory built by Southern California Edison Company to devise a million-volt accelerating tube for a high-voltage X-ray machine that was designed to provide deep therapeutic X rays to treat cancer.

It is not surprising that physicists turned to electrical engineers for high voltages in order to conduct nuclear investigations in the 1920s. In addition to the Caltech high-voltage tube, they turned to sources such as Tesla coils, electrostatic generators, and even lightning as power sources for their accelerating tubes. Like Benjamin Franklin's early experiments, these had fatal consequences. More generally, the problems attending the insulation and regulation of high voltage made their straightforward application problematic.

Robert J. Van de Graaff developed the electrostatic accelerator. While working on his Ph.D. at Oxford University in 1926, he conceived of a vacuum-insulated high-voltage generator composed of concentric Faraday cages. He subsequently became a National Research Fellow at Princeton University working with Karl T. Compton, who encouraged him to pursue the idea. When Compton became President of the Massachusetts Institute of Technology (MIT) in 1929, he invited Van de Graaff to be a Research Associate, and it was agreed that MIT would have a half-interest in any patents acquired for a source of extremely penetrating X rays as well as for an electrostatic motor and the artificial transmutation of elements. Compton acquired a dirigible hanger at the Round Hill Estate of railroad magnate Edward Howland Robinson Green in South Dartmouth, Massachusetts, to perfect the tube.

At Round Hill, Van de Graaff built a pair of accelerators that could be used to double the potential along an accelerating tube. Although this made more than a million volts available, it did not enable him to achieve the "transmutation" of the atom that he sought. This was done by Rutherford's students, John Douglas Cockcroft and Thomas Sinton Walton, in the Cavendish Laboratory in Cambridge, UK. Cockcroft and Walton shared a background in electrical engineering and an interest in nuclear physics and used a tube developed by T. E. Allibone to achieve the transmutation of the atom. They benefited from the theoretical calculations of George Gamow, then at the Niels Bohr Institute in Copenhagen, that showed that the nucleus could be penetrated by particles with energies below the Coulomb potential through a quantum mechanical tunneling effect. Gamow came to the Cavendish in early 1929, where he discussed his theory with Cockcroft and Walton. Their accelerator was at first a transformer coupled to a vacuum-tube rectifier built by Allibone that produced 300 kilovolts (kV). When the transformer failed, Cockcroft conceived a voltage-multiplying circuit to produce a high-voltage direct current from the transformer's alternating current. This voltage was applied along an accelerating tube made of two glass cylinders and evacuated by oil pumps producing over 700,000-volt protons. Using one of Rutherford's scintillation detectors, they observed the disintegration of the lithium nucleus into two alpha particles on April 14, 1932. The reaction of lithium and hydrogen nuclei produced two alpha particles and energy corresponding to the difference in masses between the reactants and products according to Einstein's famous equation $E = mc^2$. The accelerator served to disintegrate many other elements and has become a staple in more complex accelerators where it serves as the first stage. Cockcroft and Walton received the Nobel Prize in Physics in 1951 for transmutation of atomic nuclei by artificially accelerated atomic particles.

Cockcroft and Walton also invented this accelerator before competing American physicist Ernest Orlando Lawrence had developed a means to avoid the use of high voltages in accelerating particles to high energies by reusing the same potential in a series of accelerating gaps through which the particles passed in circular orbits in resonance with the radio frequency of the voltage. Resonance acceleration had been proposed by Swedish physicist Gustaf Ising and experimentally demonstrated by Rolf Wideröe in a thesis written at the Aachen Technische Hochscule in 1927. Although linear accelerators for mercury and other heavy ions could be developed using these frequencies, the acceleration of light particles required acceleration in a radial direction by a magnetic field so that the particles traveled in a spiral as they were accelerated. Lawrence learned of the latter's work in 1929 and pursued both techniques in 1930 and 1931 with the assistance of N. E. Edlefsen, who built a 10-centimeter prototype, and M. Stanley Livingston, who demonstrated resonance acceleration in his Ph.D. thesis at the University of California, Berkeley in 1931 with a 4-inch-diameter chamber. During the years of 1931 and 1932, Livingston and Lawrence built a magnet with 10-inch-diameter pole faces and a brass chamber that achieved 1.2-million-electron-volt (MeV) protons in early 1932.

By increasing the diameter of the magnetic field, one could accelerate particles to higher energies, and Lawrence was already working on a machine with magnetic poles of 27 inches in diameter when he learned of Cockcroft and Walton's success. Ironically, he might have anticipated them with a machine 10 inches in diameter that had been built by his graduate student, M. Stanley Livingston, had he recognized the quantum mechanical implications of George Gamow's work, which had been done independently by Ronald W. Gurney and Edward U. Condon in the United States. Moreover, he had not constructed the detectors required; they were not installed until the summer of 1932 when Yale physicists Donald Cooksey and Franz Kurie introduced them to the Radiation Laboratory that Lawrence had built to house his larger cyclotron.

Although the Van de Graaff, Cockcroft-Walton, and cyclotron accelerators dominated the field of particle acceleration in the 1930s, their contributions to the development of nuclear and particle physics were eclipsed by more conventional techniques. The positron, the first antiparticle predicted by Paul Dirac, was detected in cosmic rays by Carl Anderson at Caltech, and the neutron was established as a nuclear constituent by Rutherford's associate, James Chadwick, both of whom used natural sources of particles. The meson, predicted by the Japanese physicist Hidekei Yukawa, was also discovered in cosmic rays by Anderson.

The new field of artificial radioactivity was opened by others using traditional techniques, especially Frédéric and Irène Joliot-Curie in 1934. Like nuclear disintegration, this discovery rested upon the availability of suitable detectors, and the recognition that radioactivity persisted after the original source was removed. In 1934 and 1935 Enrico Fermi and his associates in Rome used neutrons slowed by light elements to induce radioactivity in a variety of elements. The nuclear reactions that produced these unstable isotopes had been unsuspected by the accelerator builders, who were able to duplicate them almost immediately.

Ernest Lawrence pursued the art with his cyclotrons that were capable of much higher energies and produced isotopes that he and his sponsors hoped to market for medical purposes, as they had marketed high-voltage X-ray tubes developed in his laboratory. Because of the higher energies available to them, Lawrence and his Radiation Laboratory at the University of California were responsible for the discovery of the majority of the reactions producing artificially radioactive substances (radioisotopes) in the 1930s. In 1936, he won funding for a Medical Cyclotron as well as for the Crocker Radiation Laboratory, which would house a 60-inch cyclotron to be used for medical research and for experimental therapy with neutrons produced by the cyclotron. Samuel Ruben and Martin Kamen, in a search for biologically useful radioisotopes, found carbon-14 in 1940.

Emilio Segrè used parts of the 27-inch cyclotron supplied to him in Sicily to discover a new element, technetium, in 1937. After he left Italy, Segrè joined the Radiation Laboratory in Berkeley. In 1940, Edwin M. McMillan, investigating the newly discovered fission of uranium, found that it could be transmuted into another new element, heavier than uranium, which he called neptunium. Emilio Segrè and Glenn Seaborg picked up this work and discovered the next of the "transuranic" elements, plutonium, in 1940. Seaborg and McMillan received the Nobel Prize in Chemistry in 1951 for their discoveries of the transuranium elements. Seaborg and Albert Ghiorso continued this program at Berkeley after World War II, discovering the elements 95 (Americium) and 96 (Curium) in 1946, elements 97 (Berkelium) and 98 (Californium) in 1950, element 102 (Nobelium) in 1958, and element 103 (Lawrencium) in 1961.

The development of early particle accelerators depended upon a variety of historical factors, some of which were remote from the interest in subatomic particles, although it was regarded as "pure science" by physicists at the time. While participants have focused upon the experimental utility of particle accelerators in describing the atomic nucleus, it is clear that entrepreneurs like Lawrence were successful in presenting these machines as a variety of X-ray equipment as well as production machines for radioisotopes at a time when nuclear reactors were not yet available. The perceived increase in the incidence of cancer in the interwar period enhanced the market for technological cures promised by radiologists and physicists who built their giant machines. The public demonstration of the use of radioisotopes using Geiger counters to detect the circulation of radioactive sodium in the blood was a standard marketing tool used by Lawrence, while universities saw medical applications as an easily understood rationale for the support of physics research with accelerators. The National Cancer Institute encouraged experiments with accelerators as did a number of medical philanthropies. Lawrence even went so far as to postulate the production of nuclear energy using the cyclotron, although Rutherford felt compelled to brand this as "moonshine." Nevertheless, the Cavendish Laboratory acquired a cyclotron.

In addition to Great Britain and the United States, cyclotrons found homes in Denmark, France, Sweden, the Soviet Union, and Japan in the 1930s. The construction of these cyclotrons was based upon Lawrence's designs, which he shared freely with physicists elsewhere. In most cases, physicists who had worked on the original machines at the Radiation

Laboratory in Berkeley were among those who built the accelerators, since the problems related to creating and maintaining a suitable vacuum in the larger cyclotrons as well as the focusing of magnetic fields in all such machines had not been reduced to engineering practice but was, at the University of California and elsewhere, still tacit knowledge.

Engineering and theoretical understanding of cyclotron behavior was developed only in the late 1930s by William Brobeck and Robert R. Wilson of the Radiation Laboratory. Before World War II, however, empirical techniques of shimming magnets and leak prevention remained an important part of cyclotron construction and operation.

Van de Graaff electrostatic generators were also widespread in Europe and the United States. Here the interest extended to the generation of electrical power as well as cancer therapy and nuclear physics, as might be expected given Van de Graaff's early experience in electrical power generation. The steady currents available from the accelerators made them more reliable for nuclear experiments in the 1 to 10 MeV range, and Merle Tuve at the Carnegie Institution of Washington, William Fowler's group at Caltech, Ray Herb's group at the University of Wisconsin, and John Williams's group at the University of Minnesota made significant contributions to nuclear physics and Van de Graaff machine design in the 1930s.

At the University of Illinois, Donald M. Kerst, aided by Radiation Laboratory veteran Robert Serber, built a magnetic induction electron accelerator, the betatron, just before World War II. The scheme, which had been proposed by Wideröe and Walton, was made to work by shimming of the magnetic field in much the same way that the cyclotron had been.

World War II brought an end to particle accelerator development as physicists turned to work on radar and the atomic bomb. Lawrence's cyclotrons were rebuilt as calutrons to separate the isotopes of uranium for the bomb, and a large Oak Ridge facility was created to house these machines, which processed the uranium-235 that was used in the Hiroshima bomb. The war did not prevent physicists from thinking about particle accelerators, however, and the thoughts of Luis Alvarez and Edwin M.

McMillan bore postwar fruit in a new generation of accelerators. McMillan, who had been assigned to the new Los Alamos Laboratory of the University of California where the first nuclear weapons were constructed, conceived of a means of escaping the energy limits on conventional cyclotrons, which was caused by the increase in mass of accelerated particles as they approached the speed of light with their increasing mass. By changing the frequency as the particles were accelerated, it was possible to keep them in synchrony. This principle, called phase stability, is the basis of modern proton synchrotrons. It was first demonstrated in the 184-inch cyclotron in 1946 and was applied to an electron synchrotron at Berkeley subsequently. William Brobeck designed a very large proton synchrotron to provide protons with energies of 10 billion electron volts, which became the basis of the design of the first two American machines, the Cosmotron at the new Brookhaven National Laboratory and the Bevatron (named for its billion electron volt energies) at Berkeley. These machines, completed in 1951 and 1954, respectively, were the first to produce particles with energies approaching those in cosmic rays. The 184-inch cyclotron produced the first human-made mesons in 1948. The Cosmotron produced a series of strange particles, such as the K-meson, so named because of their unexpectedly long lifetimes, and demonstrated the principle of associated production. The Bevatron produced an entirely new particle, the antiproton, in 1955, an accomplishment for which Emilio Segrè and Owen Chamberlain won the Nobel Prize in Physics. Marcus Oliphant built a smaller proton synchrotron.

Luis Alvarez pursued the development of linear proton accelerators using microwave-frequency generators similar to those used in wartime radar and built a 40-foot accelerator at Berkeley after World War II. The linear accelerator (LINAC), as it was called, produced 40-MeV protons, and a larger accelerator, the materials testing accelerator, was built at Livermore, California, to produce fissile and other neutron-enriched elements to supply America's need for nuclear explosives. It was abandoned after two years of development when the discovery of natural sources of uranium made it economically unfeasible. The Alvarez linear accelerator, like the Cockcroft-Walton machine, is often used to preaccelerate particles fed into high-energy synchrotrons. One of Alvarez's associates, Wolfgang Panofsky, applied the principle to the acceleration of electrons in the Stanford Linear Accelerator.

With the liquid hydrogen bubble chamber, a 6-foot-long detector built in the 1950s, Alvarez discovered many new subatomic particles, providing the empirical basis for the Standard Model. He was awarded the Nobel Prize in Physics in 1967 for this work.

The Bevatron and subsequent proton synchrotrons were the principal instruments of particle physics, replacing cosmic rays as sources of particles and, after the discovery of strong focusing at Brookhaven in 1953, produced increasingly higher energies to probe the nucleus.

See also: CYCLOTRON; LAWRENCE, ERNEST ORLANDO

Bibliography

Crease, R. P. *Making Physics : A Biography of Brookhaven National Laboratory, 1946–1972* (University of Chicago Press, Chicago, 1999).

Hartcup, G., and Allibone, T. E. *Cockcroft and the Atom* (A. Hilger, Bristol, UK, 1984).

Heilbron, J. L., and Seidel, R. W. *Lawrence and his Laboratory* (University of California, Berkeley, 1989).

Livingston, M. S. *Particle Accelerators; A Brief History* (Harvard University Press, Cambridge, MA, 1969).

Perkowitz, S. "Brother, Can You Spare a Cyclotron? Physics Research During the Great Depression." *MIT's Technology Review* **100**, 45–50 (1997).

Wilson, E. *An Introduction to Particle Accelerators* (Clarendon Press, Oxford, 2001).

Wilson, R. R., and Littauer, R. *Accelerators: Machines of Nuclear Physics* (Anchor Books, Garden City, NY, 1960).

Robert W. Seidel

ACCELERATORS, FIXED-TARGET: ELECTRON

A tremendous amount of scientific insight has been garnered over the past half-century by using particle accelerators to study physical systems of subatomic dimensions. These giant instruments begin with particles at rest, then greatly increase their energy of motion, forming a narrow trajectory or beam of particles. In fixed-target accelerators, the particle beam impacts upon a stationary sample or target that contains or produces the subatomic system being studied. This is in distinction to colliders, where two beams are produced and are steered into each other so that their constituent particles can collide.

The acceleration process always relies on the particle being accelerated having an electric charge; however, both the details of producing the beam and the classes of scientific investigations possible vary widely with the specific type of particle being accelerated. Fixed-target accelerators produce beams of electrons, the lightest charged particle.

As detailed below, the beam energy has a close connection with the size of the physical system studied. Here a useful unit of energy is the giga electron volt (GeV). (One GeV, the energy an electron would have if accelerated through a billion volts, is equal to 1.6×10^{-10} joules.) To study systems on a distance scale much smaller than an atomic nucleus requires beam energies ranging from a few GeV up to hundreds of GeV and more.

A correct description of the accelerated electrons' motion requires Einstein's theory of special relativity, because their speed is close to the speed of light. For example, an electron with only 0.01 GeV of energy is already traveling at 99.9 percent the speed of light. This simplifies the acceleration scheme because no matter what energy the beam has at a given stage of acceleration, the speed of its particles is nearly constant.

All high-energy accelerators use a rapidly alternating high voltage to accelerate charged particles. (Constant high voltages, which can also accelerate particles, are not practical for particle acceleration above 0.1 GeV because of material property limitations.) The geometry of the accelerating structure is periodic, and the arrival times of the particles are synchronized so that as they are transported along they feel a "push" where the accelerating electric field is large and pointing in the direction of motion but no "pull" where the electric field switches back and becomes small. One consequence is that the particles must be accelerated in separate groups or bunches. If the beam bunches arrive at the target separated by a short time interval (such as a few nanoseconds), then the beam is considered to be "continuous" compared to the overall detector response, which is usually much slower.

The structures most commonly used to provide the accelerating voltage are electromagnetically resonating cavities that allow very high voltages to be developed at a particular frequency. The alternating high voltage is conventionally referred to as rf from the historical use of "radio frequencies" for this purpose. Linear accelerators, or linacs, consist of an arrangement of numerous resonating cavities in a line, through which the beam passes. (An alternative type of circular accelerator, the synchrotron, is not as well suited for high-energy electrons unless a very-large-circumference path is used. The beam particles emit a light called synchrotron radiation when they are forced into a curving path. This removes energy from the beam, decreasing its overall quality.) In rf cavities, fabricated from ordinary electrical conductors such as copper, the large surface currents generated by the alternating high voltage create a significant amount of heat in the cavity material. An operational impact of this is that such high-field cavities have to be operated at a reduced duty factor, that is, they are repeatedly switched off to cool after operating a short time; duty factors of 1 percent or less are common, that is, the cavities are cooling off 99 percent of the time, during which no beam is accelerated. A major recent development in accelerator science has been the successful implementation of rf cavities fabricated from superconducting materials, in which the electrical resistance is greatly reduced. Although incurring the need for a helium cryogenic facility for cooling the cavities, this development has provided not only a means of achieving significant electrical power savings, it has also resulted in greatly improved beam quality. Small beam diameters and divergences (together referred to as the beam emittance), an extremely well-defined beam momentum, and a duty factor of 100 percent are noteworthy benefits of superconducting rf cavities. (An accelerator with a duty factor of 100 percent is conventionally referred to as a cw machine, an abbreviation for "continuous wave.") These new capabilities permit whole new classes of important measurements to be performed.

Fixed Target vs. Colliders

There are a number of differences between fixed-target accelerators and colliders. One advan-

FIGURE 1

A photograph of a seven cell acceleration cavity fabricated from niobium, which is a superconductor at liquid helium temperatures. The cavity is approximately 1 meter in length. CREDIT: COURTESY OF UNITED STATES DEPARTMENT OF ENERGY'S JEFFERSON LAB. REPRODUCED BY PERMISSION.

tage of fixed-target mode is that, unlike colliders, it provides great flexibility in the selection of targets. For example, sophisticated cryogenic liquid targets that can absorb nearly a kilowatt of power without boiling have been developed at fixed-target facilities. Increasing the thickness of a given target (and consequently the reaction rate) is usually a simple matter, whereas for colliders the equivalent action of increasing the density of collisions is a complex undertaking. This is because highly focused colliding beams tend to become less dense when they overlap, due to very strong electromagnetic forces. The density of collisions is quantified by the luminosity, which for fixed-target accelerators is the product of the target thickness and the beam intensity. For a particular process, the reaction rate is simply proportional to the luminosity. The potential limitations on the luminosity achievable include both the beam current (current density for a collider) that the accelerator can produce, and the rate of outgoing particles which the particle detectors can accept without malfunctioning. Collider luminosities are typically several orders of magnitude smaller than those achieved in comparable fixed-target accelerators.

The great advantage of colliders is that the amount of energy available for the reactions of interest is maximal. For colliding beams of equal momentum, all of the kinetic energy of the beams is available for the interaction. By contrast, for fixed-target accelerators only a fraction of the kinetic energy of the beam is available, the rest constituting the energy of the center of mass. For example, for electron-proton scattering with a 4-GeV electron beam, approximately half of the kinetic energy is available for the reaction; at 50 GeV, only 17 percent is available, and the fraction continues to decrease as $E^{-1/2}$, where E is the electron beam energy. Since the costs of building an accelerator generally increase as the beam energy increases, a collider is clearly to be preferred at the highest beam energies for experiments that can tolerate their limitations. An additional feature of colliders is that the particles emerging from the collision tend to be spread out over a larger range of angles than in fixed target accelerators. This means that good detector acceptance (the range of particle angles and momenta to which the detector is sensitive) for particles scattered

at small angles (relative to the initial beam direction) is more challenging to achieve in fixed-target accelerators than in colliders.

Storage rings can be operated as a special category of fixed-target accelerators. In these devices the beam circulates in a ring-shaped path for an extended time, often minutes or hours. The beam is usually added, or injected, into the ring, from another accelerator. The targets can be very thin layers of solid material or gases that are located at a point on the ring; the same beam particles pass through these internal targets many times. In storage rings it is possible to make a high-purity internal gas target with high polarization, an important property related to the magnetic character of the target particle. External targets can also be used by extracting the beam from the ring and transporting it to the target.

Scientific Topics

A wide variety of scientific topics can be addressed at fixed-target electron accelerators. Of the three fundamental forces in the universe (the gravitational force, electroweak force, and strong force), the electrons which scatter from the targets are only capable of interacting via the electroweak force. This force is much weaker than the strong force at small distance scales, and therefore using the electron as a probe does not disturb the system under study as much as strongly interacting probes. A further advantage is that the electroweak interaction is well understood. Studies with electron beams are complementary to those performed with other particle beam types (such as proton beams); some experimental quantities can only be efficiently accessed by one particular type of beam. Historically, the weakness of the electroweak force was a disadvantage because it meant that the rate at which interactions take place is much smaller than for strongly interacting beams, and therefore the experiments were of long duration and collected relatively few events. This disadvantage has been overcome by technological advances such as developing cw accelerators with high luminosities.

At subatomic distance scales, forces between particles are transmitted through the exchange of other particles. When the exchanged particle is transmitting the force, it is referred to as a virtual particle; it exists for a very brief time and does not emerge from the interaction region. In the case of electron scattering, the dominant process is to exchange a single virtual photon. (Photons are the particles that make up visible light, as well as all other electromagnetic fields such as microwaves or X rays.)

The characteristics of the virtual photon exchanged in an interaction are determined by measuring the characteristics of the electron scattered from the target. The energy of the scattered electron, and the angle at which it emerges (compared to the initial beam direction), yield the properties of the virtual photon. The energy transferred to the target can be determined, as can the distance scale at which the interaction takes place, which is determined by the virtual photon's momentum. (Accessing smaller distance scales requires larger momenta, which in turn require larger beam energies.) Using this information, many types of subatomic systems can be studied. For instance, basic properties of the proton and neutron can be determined, such as their distributions of charge and magnetization, which yield information on the nature and distribution of their constituent particles. Systems consisting of unstable particles that decay rapidly, such as highly excited protons or heavier mesons, can be produced and characterized as well.

An example of a line of scientific inquiry is the study of the structure of protons. Protons are thought to be made up of three particles called quarks, bound together by forces transmitted via the exchange of particles called gluons. These forces become extremely strong when the quarks are further apart and are very weak when they are close together. A "cloud" of short-lived virtual particles surrounds these stable quarks. The virtual particles include other quarks (called sea quarks) and gluons. Most of this complex picture of proton structure has been derived from fixed-target electron accelerators.

Other particles besides the electron can also be detected, and, as might be expected, detecting two or more particles in coincidence yields significantly more information on the structure of subatomic systems. High-quality coincidence experiments place higher demands on the performance of the accelerator and detector, however. Important ingredients in coincidence experiments include high duty factor, high luminosity, and large acceptance detection. The high duty factor is essential for most coincidence experiments; all other things being equal, a given experiment would take one hundred times longer at an accelerator with a 1 percent duty factor than it would with a cw accelerator. The other two ingredi-

FIGURE 2

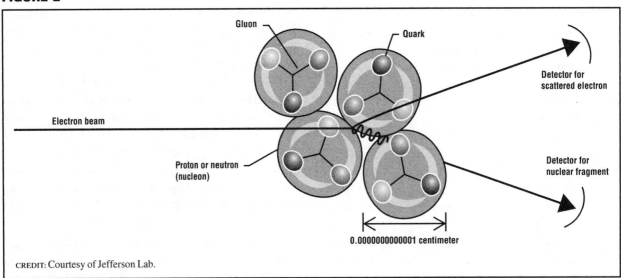

CREDIT: Courtesy of Jefferson Lab.

A conceptual illustration showing an electron from the beam exchanging a virtual photon, which interacts with a quark contained in a helium nucleus.

ents, luminosity and acceptance, can compensate for each other. For instance, designing a magnet-detector combination, or spectrometer, to measure a particle's momentum with very high precision will often require sacrificing detector acceptance. The lower acceptance can be compensated for by increasing the luminosity, if permitted by the accelerator-detector combination.

Existing Facilities

Historically, there have been many fixed-target electron accelerators devoted to studies in nuclear and particle physics. Over time, there have been changes in both the scientific focus of the fields served by these accelerators and the state-of-the-art in accelerator technology. This has lead to the adaptation of many of the older machines to other purposes. For instance, some accelerators formerly used in particle physics as a primary accelerator now serve as injectors to larger accelerators, often for colliders with much higher energies.

Of the fixed-target electron accelerators devoted to nuclear or particle physics, the most recently constructed is the Continuous Electron Beam Accelerator Facility (CEBAF) at the Thomas Jefferson National Accelerator Facility in Virginia, a U.S. Department of Energy laboratory for nuclear physics. CEBAF consists of two superconducting linacs connected together by two magnetic arcs. The beam accelerated through the first linac is directed via an arc into the second linac and further accelerated. The beam can be recirculated up to a total of five passes through each linac to achieve the maximum energy. Each time the beam emerges from the second linac, a fraction of it can be extracted and sent into an experimental area. The accelerator produces electron beams ranging from 0.4 to 6 GeV in energy, with a beam current of up to 200 microamperes. Very high beam polarization is available at high current, and the beam can be delivered to multiple experimental areas so that up to three experiments can be performed simultaneously with three different energies, each with a cw beam. An excellent emittance produces tiny beam spot sizes of only a few hundred microns in diameter.

Accelerators with lower maximum beam energies include the MIT-Bates Linear Accelerator in Massachusetts and the Mainz Microtron (MAMI) facility in Mainz, Germany, both of which are capable of producing close to 1-GeV electron beams. The MAMI facility employs a three-stage specialized accelerator (called a microtron) to produce a cw beam of up to 100 microamperes, with high polarization and a small beam spot size. The Bates facility consists of a pulsed, normal-conducting linac injecting into a storage ring with an internal target, along with external target capabilities. Another facility utilizing a storage ring at higher energies is the Electron Stretcher Accelerator (ELSA) in Bonn, Germany. By injecting high-intensity current into a stretcher ring, a high duty factor is achieved at low beam currents for up to 3.5-GeV electrons.

High-energy electron and positron beams of up to nearly 30 GeV are used in combination with gas targets in the HERMES experiment, which uses the HERA storage ring at the Deutsches Elektronen-Synchrotron Laboratory (DESY) in Hamburg, Germany. These ultrapure gas targets include polarized hydrogen, deuterium, and helium-3, as well as other unpolarized gases.

The highest electron energies available for fixed-target experiments are found at the Stanford Linear Accelerator Center (SLAC) in California. This facility consists of a 2-mile-long, normal-conducting linac.

FIGURE 3

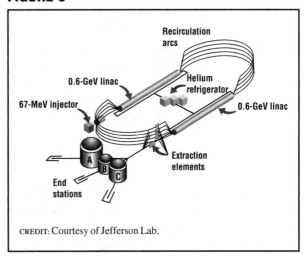

CREDIT: Courtesy of Jefferson Lab.

The layout of the Continuous Electron Beam Accelerator Facility at Jefferson Lab, showing the linacs, recirculating arcs, and experimental areas (labeled A, B, and C).

Although the duty factor is typically quite low (for example, 0.03 percent), electrons of up to 50 GeV can be produced without beam quality degradation due to synchrotron radiation.

See also: ACCELERATOR; BEAM TRANSPORT; EXTRACTION SYSTEMS; INJECTOR SYSTEM

Bibliography

Deutsches Elektronen-Synchrotron Laboratory. <http://www.desy.de>.

Fernow, R. *Introduction to Experimental Particle Physics* (Cambridge University Press, Cambridge, UK, 1989).

Leemann, C.; Douglas, D. R.; and Krafft, G. A. "The Continuous Electron Beam Accelerator Facility: CEBAF at the Jefferson Laboratory" in *Annual Review of Nuclear and Particle Science,* Vol. 51 (Annual Reviews, Palo Alto, CA, 2001).

Perkins, D. H. *Introduction to High Energy Physics* (Addison-Wesley, Reading, MA, 1987).

Stanford Linear Accelerator Center. <http://www.slac.stanford.edu>.

Thomas Jefferson National Accelerator Facility. <http://www.jlab.org>.

William K. Brooks Jr.

ACCELERATORS, FIXED-TARGET: PROTON

Particle accelerators invariably accelerate either protons or electrons because they are the only stable, electrically charged particles. Copious synchrotron radiation of electron beams at high energies tends to make the design of proton and electron accelerators very different, and a given accelerator is normally only used for one type of particle. Many proton accelerators are, however, capable of accelerating heavy nuclei.

Proton Accelerators

Proton accelerators begin with an ion source. The most commonly used sources produce either protons or H^- ions. An H^- ion consists of a proton with two bound electrons. Proton sources can result in higher currents, but the advantage of H^- sources is that they enable the use of a special technique known as "multi-turn injection." While the pulse of protons that can be injected into a circular accelerator is limited to one turn (i.e., the time it takes for a particle to traverse the circumference), an H^- pulse can be injected for many turns. Multi-turn injection is achieved by passing the H^- beam through a thin foil inside the circular accelerator to strip off the two electrons and thereby create a proton beam. A variety of techniques are available to produce heavy atoms in various states of ionization. The ions emitted by a source may be accelerated initially by a static electric field. However, it is impractical to obtain static fields with the billions of volts that are required to produce high-energy beams.

In modern, high-energy accelerators, the first step toward high energy takes place in a linear accelerator (linac). A linac stores the energy required for acceleration in radio frequency (rf) cavities. In these cavities the electrical field changes sign at a rate determined by the cavity frequency. Particles arriving at the correct time are accelerated while particles arriving half an rf cycle later are decelerated. The beam is bunched so that all particles are accelerated; no beam passes through the cavities during the decelerating cycle. If no countermeasures were taken, the beam would expand its transverse dimensions indefinitely, like the beam from an ordinary flashlight. In practice, the beam is continually refocused by a series of quadrupole magnets that are used to bend the diverging beam inward, resulting in a beam whose size oscillates between somewhat larger and smaller sizes but does not grow over distance. The linac technique can be continued indefinitely, but the amount of acceleration is proportional to the number of rf cavities, which are a dominant component of the cost.

Circular accelerators are used to overcome the limitations of proton linacs. The principle of operation is similar to the linac except that dipole magnets are used to bend the beam in a circle. The beam passes through the rf cavities thousands or even millions of times, dramatically reducing the number of rf cavities required. Circular accelerators tend to be dominated by the need for a large number of high-field magnets to bend the beam; the rf cavities, because they are used more efficiently, typically occupy only a small fraction of the circumference.

A large number of proton accelerators have been built. Table 1 lists some of the higher-energy proton accelerators that are used for fixed-target experiments. The beam energy is given in terms of the total voltage of the static electrical field that would be required to achieve the beam energy measured in billions of volts (giga electron volts or GeV).

Comparison of Fixed Targets and Colliding Beams

High-energy physics creates new forms of matter by colliding particles at extremely high energy using two basic techniques. The first technique, the colliding beam technique, involves colliding one beam of particles with another. The second involves the interaction of a particle beam with a fixed target made of gaseous, liquid, or solid material. A major difference between these techniques is that the conservation of momentum significantly affects the amount of collision energy that can be converted into mass. For example, a 1,000-GeV proton beam colliding head-on with another 1,000-GeV proton (or antiproton) beam can produce a state with masses up to $2,000 \text{ GeV}/c^2$ because the momenta of the two beams are equal and opposite. The same proton beam colliding with a stationary target nucleon can create at most $43 \text{ GeV}/c^2$ of mass. For this reason, experiments intending to produce high-mass particles (such as the recently discovered top meson) are performed with colliding beam accelerators.

Fixed-target experiments have the advantage of producing higher interaction rates. In a proton-proton collider, a proton beam might circulate for hours and still not be exhausted. In a fixed-target experiment, the same beam can be produced repetitively every few seconds, each pulse being completely exhausted in a dense target. Even more important is the possibility of creating a wide variety of secondary beams from interactions of protons in the target. The higher interaction rates with fixed targets is crucial in making secondary beams, but, ironically, the need to conserve the momentum of the primary proton beam is an advantage in this case: the secondary particles are produced mainly along the initial direction of the proton beam.

Types of Targets

A variety of targets are used. Some experiments (where the protons are used to produce a secondary beam, for example) do not require a particular nuclear

TABLE 1

Some of the High-Energy Accelerators Used for Fixed-Target Experiments				
Accelerator	**Location**	**Beam Energy (GeV=10⁹ eV)**	**Beam Intensity (protons/pulse)**	**Cycle Time (sec)**
Alternating Gradient Synchrotron (AGS)	Brookhaven National Laboratory (BNL) Upton, New York, USA	28	7×10^{13}	2.5
CERN Proton Synchrotron (CPS)	European Laboratory for Particle Physics (CERN) Geneva, Switzerland	26	2×10^{13}	2
Super Proton Synchrotron (SPS)	European Laboratory for Particle Physics (CERN) Geneva, Switzerland	400	4×10^{13}	10
Proton Synchrotron (PS)	National Laboratory for High Energy Physics (KEK) Tsukuba, Japan	26	5×10^{12}	2.5
Proton Synchrotron (U70)	Institute of High Energy Physics (IHEP) Protvino, Russia	70	2×10^{13}	10
Main Injector	Fermi National Accelerator Laboratory (FNAL) Batavia, Illinois, USA	120	3×10^{13}	2

CREDIT: Courtesy of John Marriner.

composition. In these cases, a copper target might be used because of its attractive mechanical and thermal properties. Occasionally, targets are chosen because of the number of protons in the target: beryllium is often used when a nucleus with a small number of protons ($Z = 4$) is desired. Lead ($Z = 82$) or tungsten ($Z = 74$) may be used when a large number of protons per nucleus or high material density is desired. Many experiments want the simplest nuclear target possible, namely, hydrogen atoms, whose nucleus consists of a single proton. In order to make a high-density target, the hydrogen is usually cooled to cryogenic temperatures ($-255°C$). When a neutron target is desired, deuterium is used. The deuterium nucleus, known as "heavy hydrogen," consists of a proton and a neutron. Comparison of hydrogen and deuterium targets allows experimenters to deduce the characteristics of beam interactions with neutrons.

Nuclei possess an intrinsic angular momentum known as "spin." When the nuclear spins point in a common direction, a material is said to be "polarized." Since interactions generally depend on spin, it is an advantage to be able to control the target polarization. Special targets, using special materials, have been built that allow creation of significant polarization. The techniques are not generally applicable to all nuclei, and practical considerations restrict polarized target usage to those experiments that require them.

Types of Beams

Protons interact with nuclei via the "strong" force and are therefore very effective in producing a wide variety of particles. Particles that are stable except for decays mediated by the weak force can generally form beams suitable for experiments. The very massive bottom (and top) mesons, however, are rarely produced and have lifetimes that are too short (at currently available energies) to make beams. Particle beams that decay via weak interactions can also be used to produce tertiary products (like neutrinos) that are not produced in strong interactions.

Secondary beams produced by protons generally consist of more than one particle type. A bending magnet is used almost universally with charged particle beams to select a specific momentum range. Often this simple technique produces a beam of the correct momentum and adequate purity, but it may be possible to restrict beam particles to a particular velocity. When combined with momentum selection, the velocity selection can be restrictive enough to specify the particle mass, uniquely identifying the particle type. More commonly the velocity (or some other property) of the particle is measured, but undesired particles are not removed from the beam; the information is used later in the analysis of the interactions.

Primary Proton Beams

The simplest fixed-target experiment at a proton accelerator involves steering the primary beam onto a fixed target. Proton beams can be used to study elastic and total cross-sections of proton-nuclei interactions. The rates of production of various particles can be studied, and particles produced by proton interactions can be used to form secondary beams. In many cases, the nature of neither the target nor the projectile is important: only the amount of energy that can be converted into mass is important. For these experiments, the high-intensity primary beam can be an overwhelming advantage in studying rare processes. The challenge is to design an experiment that is sensitive to the rarely produced state but able to ignore more common states.

Charged Meson Beams

When a high-energy proton beam hits a target, the most prolifically produced charged particles are pi-mesons (π^+ and π^-), and K mesons (K^+ and K^-). These mesons have long lifetimes ($\tau_\pi = 2.6 \times 10^{-8}$ second and $\tau_K = 1.2 \times 10^{-8}$ second), so they can be focused into beams and transported a hundred meters at high energy, where the lifetime is stretched by the relativistic effect known as time dilatation.

Meson beams have been used extensively to analyze meson production at a wide variety of momenta and from a wide variety of nuclei. Meson beams have also been used to study interactions with nuclei, including total and elastic cross-sections. Interactions with electrons in the target have been used to measure the distribution of quarks in mesons. The decays of mesons have been studied extensively, including extremely rare processes, as infrequent as one in many billion decays.

Antiprotons

Antiprotons are produced in the same way as mesons, but at much lower rates. Since antiprotons are stable, they can be collected in specially designed storage rings. These storage rings have been used mainly to collect antiprotons for proton-antiproton colliding beam accelerators but have also been used with internal, gas jet targets to produce charmonium states (bound states of a charm quark and an anticharm quark) and to form antihydrogen atoms (the bound state of an antiproton and a positron). Antiproton beams have been extracted from storage rings at controlled rates, and their interactions have been studied. Antiproton beams have also been produced at high energy for scattering experiments similar to those performed with π and K mesons.

Hyperon Beams

Proton beams can be used to produce hyperon beams. Hyperons are baryons (protonlike particles) with the strangeness quantum number equal to -1 (Σ^+, Σ^-, Λ^0), -2 (Ξ^- and Ξ^0), or -3 (Ω^-). The Σ^0 is unstable against electromagnetic interactions (decaying $\Sigma^0 \to \Lambda^0 \gamma$) and has a lifetime that is too short (7.4×10^{-20} second) to form a beam. Hyperon lifetimes (typically 10^{-10} second) are much shorter than those of π and K mesons (about 10^{-8} second), so hyperon beam lines tend to be much shorter.

Hyperon beams have been used to study decays and hyperon properties including precise measurements of magnetic moments. Hyperon beams can also be used to study hyperon-nuclei interactions.

Neutral Beams

A proton beam impinging on a target will produce neutral particles, such as π^0's, K^0's, and neutrons. The charged particles produced by proton interactions can be swept out of the beam by a magnetic field, leaving only the neutrals. The π^0 is unstable against electromagnetic decays ($\pi^0 \to \gamma\gamma$), and its lifetime is too short (8×10^{-17} sec) to make a beam. The π^0 decays, however, can be used to produce a photon (γ) beam. The γ's can also be used to produce a tertiary electron beam via the reaction $\gamma N \to e^+ e^- N$. While it might seem much eas-

ier to produce electrons with an electron accelerator, proton accelerators tend to have higher energies, and it has been attractive to produce high-energy (albeit low-intensity) electron beams at proton accelerators.

Photon beams interact electromagnetically with nuclei; at high energies they probe the nuclear quark distribution. Since the electromagnetic interaction is probably the most studied, and best understood, of all interactions, it is a great advantage to study nuclei with this well-understood probe.

Neutral K mesons have been a subject of intense study, because the neutral K meson system is one of the few ways that a property known as CP violation can be studied. The K-meson system consists of two particles: K_s^0 (K short) and K_L^0 (K long). The K_s^0 has a short lifetime (0.9×10^{-10} second) and decays rapidly, leaving a beam of primarily (K_L^0 lifetime 5.2×10^{-8} second). However, the K_s^0 can be "regenerated" by the interaction of a K_L^0 beam in a nuclear target.

Muon and Neutrino Beams

Neutrinos are particles that interact only weakly and are not produced directly in proton nuclei interactions. A muon neutrino beam can, however, be produced by a decay of a meson beam via $\pi^+ \to \mu^+ \nu_\mu$ and $K^+ \to \mu^+ \nu_\mu$. Some electron neutrinos are also produced via decays such as $\mu^+ \to \bar{\nu}_\mu e^+ \nu_e$ and $K^+ \to \pi^0 e^+ \nu_e$. Tau neutrinos can be produced at low rates from short-lived, heavy meson decays. Antineutrino beams can be produced from mesons of the opposite sign.

Neutrinos are an especially good probe of nuclei since they mainly interact with individual quarks via the weak interaction. A key disadvantage of neutrino beams is that massive targets are required (perhaps 1,000 tons) to produce reasonable interaction rates. Neutrino beams have been used to establish the nature of the weak interaction and to measure the distribution of quarks in the nucleus.

The decays of mesons can also be used to produce beams of muons. Muon beams have been used to study the distribution of quarks in nuclei, and static properties of the muon, such as its magnetic moment, have also been measured. Low-energy muon beams can be brought to rest in an absorber. Stopped muons have proven to be a sensitive way to search for rare decays.

Polarized Beams and Ion Beams

Polarized beams must be produced by a special ion source. There is no practical method for developing polarization of nuclei after they leave the ion source, so the polarization must be meticulously preserved as the beam is accelerated. Proton beams are polarized most often, but sources have been developed to polarize other nuclei. Polarized beams have been used in conjunction with polarized targets to determine the spin-dependent properties of proton-proton interactions.

Ion beams are also produced by special sources, but in general, the ions are not polarized. Before acceleration to high energy, ions invariably have all their electrons stripped off the nucleus although some electrons are typically present at intermediate stages between the source and the first circular accelerator. Ion beams are mainly used to study nucleus-nucleus interactions.

See also: ACCELERATOR; BEAM TRANSPORT; EXTRACTION SYSTEMS; INJECTOR SYSTEM

Bibliography

Lee, S. Y. *Accelerator Physics* (World Scientific, Singapore, 1999).

Wilson, R.R. "The Batavia Accelerator." *Scientific American* **230** (2), 72–83 (1974).

John Marriner

ALPHA DECAY

See RADIOACTIVITY

ALTERNATING GRADIENT SYNCHROTRON, AGS

See ACCELERATORS, FIXED-TARGET: PROTON

ANDERSON, CARL D.

Carl David Anderson, winner of the Nobel Prize in Physics in 1936 at age 31, was born in New York City on September 3, 1905. He was the only son of Swedish immigrant parents.

The family moved to Los Angeles in 1912 where Anderson attended local public schools. In 1923 he entered the newly established California Institute of Technology (Caltech), intending to study electrical engineering. He was an outstanding student, and in 1926 he was awarded the Junior Travel Prize, which was a grant sufficient for him to spend six months traveling in Europe. During his travels, he met the eminent physicists Hendrik A. Lorentz and Heike Kamerlingh-Onnes.

In 1927 Anderson graduated with a B.S. degree in physics engineering. He continued with graduate studies in physics and received his Ph.D. magna cum laude in 1930. His thesis was on the spatial distribution of electrons ejected from gases by X rays.

Robert A. Millikan, Nobel Prize in Physics winner in 1923 and Chief Executive of Caltech, was Anderson's graduate advisor. He recommended that Anderson broaden his experience by applying for a National Research Council fellowship. Anderson approached Arthur H. Compton at the University of Chicago and was offered a position. However, Millikan had changed his mind and persuaded Anderson to return to Pasadena to work on the cosmic ray research team he was setting up.

Millikan had become very interested in cosmic radiation when he realized that this was a very-high-energy radiation striking the Earth from outer space. To support his interest he established three research groups using different techniques to observe the radiation, namely, electroscopes to study ionizing effects, Geiger counters to count cosmic ray particles directly, and cloud chambers to photograph incoming particle tracks. Anderson was responsible for the cloud chamber program.

The cloud chamber, invented by Charles T. R. Wilson, is a device consisting of a chamber, usually with a glass front and back, which contains moist air. The chamber is designed so that the pressure inside

American physicist Carl D. Anderson (1905–1991) won the 1936 Nobel Prize in Physics for his discovery of the positron. CREDIT: CORBIS. REPRODUCED BY PERMISSION

can be suddenly dropped. This cools the air, and cloud droplets will form on any suitable nucleus. If a charged particle has passed through the chamber shortly before the pressure drop, it will leave a trail of ions that will appear as a trail of cloud droplets. By photographing the chamber at this time, the path of the particle is made visible. This device proved to be one of the most useful tools for the study of radiation phenomena.

Anderson set up his chamber in the Guggenheim Aeronautics building at Caltech, where an ample supply of electricity was available. He needed the power to operate a large electromagnet to develop a magnetic field in the chamber. Charged particles travers-

ing the chamber would have their tracks curved by this field. By measuring the curvature, the energy of the particles could be calculated. At the time he was doing the experiments, only two elementary particles were known, namely, the electron and the proton. These differed in charge and in mass so that their tracks in the chamber could be easily distinguished.

By adding Geiger counters above and below the chamber, he was able to trigger the chamber by a pulse from the counter and thus ensure that a particle had indeed passed through the chamber. Thus almost every picture contained one or more tracks.

To his surprise Anderson found tracks that had the ion density he expected from electrons but were curved by the magnetic field as though they had a positive charge. Could these particles be previously undiscovered positively charged electrons? The alternative explanation that they were traveling in the opposite direction was of course a possibility.

To settle the matter Anderson placed a lead plate in his chamber so that the particles would have to pass through the plate. In so doing they would lose energy, and therefore the radius of curvature of the track after passing through the plate would be less than it was on entering the plate. This observation would give a definite answer to the question of the direction of travel of the particle. In 1932 he recorded the historic photograph of a track that had to be made by a positively charged electron. Surprisingly, the particle was traveling upward through the chamber.

Within a few years physicists were overwhelmed with additional new particles, constituting a veritable "zoo" of particles. Anderson contributed to this zoo by taking his cloud chamber to the top of Pikes Peak in Colorado. The trip to the top of Pike's Peak was a challenge to the experimenters and their old truck. However, they made it and obtained many photographs with their cloud chamber. These photos contained evidence for a short-lived particle with a mass that had a value that was in between the mass of electron and the mass of the proton. They suggested the name mesotron, but the particle became known as the muon.

During World War II, Anderson was associated with the Caltech project to develop barrage rockets for the Navy. He was primarily concerned with the problem of launching these rockets from aircraft. His work took him to the Normandy beachhead shortly after the invasion. He assisted in installing the rockets on various aircraft.

Data collected by Anderson and others in the 1930s showed that cosmic rays had enormous energy. For many years these rays were the only source of such radiation available for the study of high-energy particle physics. In recent years, however, the experimenters have been building ever higher energy machines, and these have now become the tools for particle research.

Carl Anderson married Lorraine Bergman in 1946. It was her second marriage, and her son Marshall was adopted by Anderson. A second son David was born in 1949.

All of his professional life Anderson was at Caltech. He became professor of Physics in 1939 and retired in 1970. He received many honors and awards in addition to the Nobel Prize. These include member of the National Academy of Sciences 1938, Chair of its physics section 1963–1966, the Elliott Cresson medal of the Franklin Institute 1937, and the John Ericsson medal of the American Society of Swedish Engineers 1960. He received honorary degrees from three colleges: Colgate University in 1937, Temple University in 1949, and Gustavus Adolphus College of St. Peter in 1963.

Carl Anderson was a first class experimental physicist, and his discovery of the positron was one of the major discoveries in particle physics of the twentieth century. He died on January 11, 1991.

See also: ANTIMATTER; DIRAC, PAUL; MUON, DISCOVERY OF; POSITRON, DISCOVERY OF

Bibliography

Anderson, C. D. *The Discovery of Anti-matter,* edited by R. J. Weiss (World Scientific, Singapore, 1999).

Anderson, C. D., and Neddermeyer, S. H. "Cloud Chamber Observations of Cosmic Rays at 4300 Meter Elevation and Near Sea Level." *Physical Review* **50**, 263 (1936).

Anderson, C. D., and Neddermeyer, S. H. "Nature of Cosmic Ray Particles." *Review of Modern Physics* **11**, 191 (1939).

Nobel Foundation. "C. D. Anderson—Nobel Lecture." <http://www.nobel.se/physics/laureates/1936/anderson-lecture.html>.

William H. Pickering

ANNIHILATION AND CREATION

In Newtonian mechanics mass is conserved—it can neither be created nor destroyed. Energy is also conserved. In Einstein's relativistic mechanics, however, these two conservation laws are replaced by one law only: mass-energy is conserved. It is possible, in Einstein's theory, for mass to be changed into energy and vice versa; the formula giving the equivalence between them is of course $E = mc^2$ (c is the speed of light). Thus relativity sets the scene for creation of particles of matter from energy alone, and their annihilation into energy alone.

Modern particle physics relies heavily on this phenomenon; in fact, virtually *every* reaction in particle physics involves the conversion of mass into energy or energy into mass. For example, an electron e^- and its antiparticle the positron e^+ can annihilate each other into photons—quanta of electromagnetic radiation

$$e^- + e^+ \rightarrow \gamma + \gamma. \qquad (1)$$

The restmass m_e of an electron has an energy equivalent $m_e c^2 = 0.5$ MeV. So, if in the above reaction e^- and e^+ have very little kinetic energy (that is, if they are moving very slowly), then the total energy available is about 1 MeV, which the photons divide between themselves, moving off in opposite directions in the center-of-mass frame.

An important series of experiments involving e^+ e^- annihilation was undertaken at the Stanford Linear Accelerator Center (SLAC) from the late 1960s, in which electrons and positrons were accelerated to very high energies—5 GeV (= 5,000 MeV) and beyond. At such energies new, heavy particles may be produced, a typical example being the J/ψ particle:

$$e^- + e^+ \rightarrow J/\psi + X. \qquad (2)$$

J/ψ has a mass of 3.1 GeV/c², and its significance lies in the fact that it is a bound state of the charm quark c and its antiparticle $\bar{c}$. (In the above reaction, X simply stands for any other particle or particles which may be produced; for the purposes of the experiment they are not of interest.) The annihilation process (1) is an example of mass being converted into energy, whereas (2) exemplifies energy being converted into mass—the creation of new particles from energy alone (in this case, the kinetic energy of e^- and e^+). Relativity allows reactions like these to happen, whereas Newtonian mechanics does not.

A final example of historical importance is the discovery of the W boson, the field quantum of the weak field. It was first found in the reaction

$$p + \bar{p} \rightarrow W^+ + X^-$$

in which proton and antiproton, at very high energy, annihilate and produce a W (and other stuff X). The proton restmass is $m_p\, c^2 = 0.98$ GeV, and the W mass $m_W c^2 = 80.6$ GeV—almost all the mass of the W comes from the p and $\bar{p}$ kinetic energy. The W particle, predicted by the electroweak theory of Sheldon Lee Glashow, Steven Weinberg, and Abdus Salam, was discovered in the above reaction at the European Laboratory for Particle Physics (CERN) in 1983.

See also: FEYNMAN DIAGRAMS; QUANTUM FIELD THEORY; QUANTUM MECHANICS; QUANTUM STATISTICS; RELATIVITY; RESONANCES; SCATTERING; VIRTUAL PROCESSES

Bibliography

Alldar, J. *Quarks, Leptons, and the Big Bang* (IoP Publishing, Philadelphia, 1998).

Brehm, J. J., and Mullin, W. J. *Introduction to the Structure of Matter* (Wiley, New York, 1989).

Okun, L. B. α, β, γ . . . Z; *A Primer in Particle Physics* (Harwood Academic Publishers, Chur, Switzerland, 1987).

Taylor, J. C. *Hidden Unity in Nature's Laws* (Cambridge University Press, Cambridge, UK, 2001).

Tipler, P. A. *Physics for Scientists and Engineers,* 4th ed. (W. H. Freeman, New York, 1999).

Lewis Ryder

ANTIMATTER

The proton in a hydrogen atom carries one unit of positive charge, the electron one unit of negative charge. Imagine reversing the signs of the charges, producing a negative proton and a positive electron. One would have a sample of antimatter, in this case an atom of antihydrogen.

Concept and Confirmation

The existence of antimatter was predicted by Paul Dirac in 1928. While attempting to make quantum mechanics consistent with special relativity, Dirac encountered negative-energy solutions. In special relativity, when a free particle of mass m moves with momentum $\mathbf{p}$, its energy E is given by

$$E^2 = (mc^2)^2 + (pc)^2$$

where c denotes the speed of light in a vacuum. The square root yields a spectrum of positive *and* negative energies (Figure 1):

$$E = \pm[(mc^2)^2 + (pc)^2]^{1/2}$$

FIGURE 1

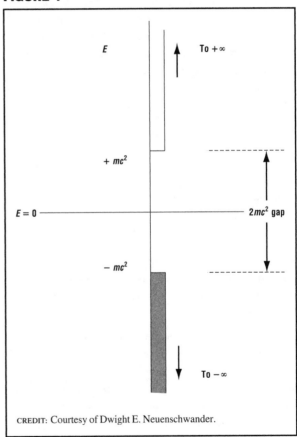

CREDIT: Courtesy of Dwight E. Neuenschwander.

Energy spectrum for a fermion of mass m: $E = \pm[(mc^2)^2 + (pc)^2]^{1/2}$, where $p \geq 0$. The vertical bands denote the energies allowed to the particle. Note the presence of the $2mc^2$ gap between the positive- and negative-energy states and also that the allowed energies go to positive and negative infinity as $p \rightarrow \infty$. In the vacuum state, the negative-energy states are filled (filled band), and the positive-energy states are unoccupied (unfilled band).

With no ground state, the negative energies offered no immediate physical interpretation. But for spin-1/2 particles (fermions) such as the electron, which are constrained by the Pauli exclusion principle, Dirac conceptualized a way to make sense of these states, with observable consequences. Dirac supposed that the vacuum consists of no positive-energy particles and the negative-energy states filled with electrons. The exclusion principle prohibits the negative-energy electrons from dropping further into the bottomless energy well. This vacuum thus carries infinite negative charge and energy, but Dirac postulated that only *departures* from this situation are observable.

Suppose a negative-energy electron receives energy from an external source, such as an electromagnetic field, sufficient to jump the $2mc^2$ energy gap. This boosts the electron into a positive-energy state, leaving a hole in a negative-energy state (Figure 2). The *absence* of a negative-energy electron (the hole) would be observed in the laboratory as the *presence* of a particle of positive energy and positive charge. In the laboratory, this process is observed as pair production:

$$Electromagnetic\ Energy \rightarrow e^- + e^+$$

where e^- denotes the electron and e^+ the antielectron (the positron). The electron will eventually drop back into the hole in pair annihilation:

$$e^- + e^+ \rightarrow Electromagnetic\ Energy$$

The electromagnetic energy is carried by gamma-ray photons γ. The positron was first observed in cloud chamber tracks of cosmic rays, photographed by Carl Anderson in 1932.

Pair production and annihilation respect the conservation of energy, linear and angular momentum, and electric charge. Charge conservation requires particle-antiparticle pairs in production and annihilation, and momentum conservation requires more than one gamma ray.

Because the proton is a fermion, Dirac's argument requires the antiproton to exist. In 1955 Emilio Segrè and others demonstrated its existence with an accelerator at Berkeley. Since the proton is a bound state of constituent fermions called quarks, antiquarks evidently exist as well. Today antiprotons are

FIGURE 2

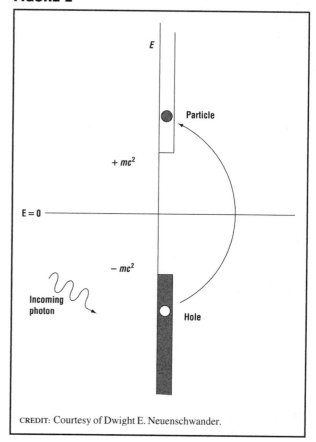

CREDIT: Courtesy of Dwight E. Neuenschwander.

Particle-hole production. An incoming photon's energy is absorbed by a negative-energy fermion, elevating it into a positive-energy state. This process leaves a positive-energy particle plus a hole in the negative-energy state. The hole is interpreted as a positive-energy antiparticle because positive energy from an outside source (the photon) was deposited there to cancel out the original negative energy.

produced in the accelerators of Fermilab in Batavia, Illinois, and the European Laboratory for Particle Physics (CERN) in Geneva, Switzerland, in collisions that convert kinetic energy into mass, such as

$$p + p \rightarrow p + p + p + \bar{p}$$

where $\bar{p}$ denotes the antiproton. The antiprotons are stored and then collided head-on with a proton beam. The pair annihilations result in gamma rays and other final states because the proton and antiproton participate in the weak and strong interactions, in addition to the electromagnetic interction.

The exclusion principle does not apply to integer spin particles (bosons) such as photons or pions.

In 1948 Richard Feynman created another way to think of the negative energy states that applies to fermions *and* bosons. In Feynman's version of relativistic quantum electrodynamics, the negative-energy particles move backward in time. This concept is interpreted *physically* as a positive-energy antiparticle moving forward in time. This interpretation is analogous to the description of electric current. In an electric circuit, positive charges moving from *A* to *B* (the so-called conventional current) and the same amount of negative charge moving from *B* to *A* produce identical changes in the charge distribution. Likewise, when the relativistic quantum mechanics equations have a negative-energy particle moving backward in time, the phenomena show a positive-energy particle *of the opposite charge* moving ahead in time—the antiparticle (Figures 3a and 3b).

For every particle species *P*, there is a corresponding antiparticle $\bar{P}$. The particle and antiparticle

FIGURE 3(a)

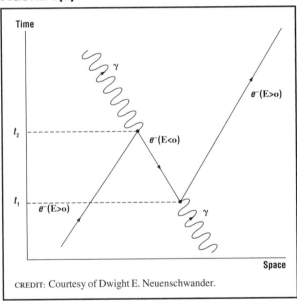

CREDIT: Courtesy of Dwight E. Neuenschwander.

Particles with negative energy propagate backward in time. Before time t_1, a positive-energy particle (e.g., an electron, represented by the solid line) and a photon (wavy line) move through space and advance through time. At time t_2, in emitting another photon, the electron acquires negative energy and recoils *backward* in time. At time t_1, the original photon collides with the negative energy, backward-propagating electron, and the negative-energy electron absorbs the photon and recoils forward in time. The overall reaction describes one mechanism for photon–electron collisions, because one photon and one electron exist in the initial and final states.

FIGURE 3(b)

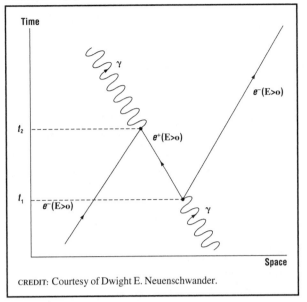

CREDIT: Courtesy of Dwight E. Neuenschwander.

The physical situation corresponding to the events of Figure 3(a). Before time t_1, an electron and photon move through space and advance through time. At time t_1, the photon decays into an electron-positron pair (pair production). At time t_2, the original electron collides with the positron, and the pair annihilation produces a new photon. Back in Figure 3(a), note the continuous electron-positron line throughout the process; thus, in Feynman's approach, electrons and positrons are treated as the same mathematical entity.

have identical mass but opposite charge, opposite magnetic dipole moment, and (if applicable) opposite signs of all other conserved quantum numbers (such as strangeness or baryon number that extend the charge concept to nonelectromagnetic interactions). What about uncharged bosons such as the photon or the Z^0? For these species the particle and antiparticle are indistinguishable—they are their own antiparticles.

What Happened to All the Antimatter?

In high-energy collisions, where kinetic energies exceed particle masses, photons and particle-antiparticle pairs readily change into one another. The ultimate high-energy "laboratory" was the early universe immediately following the Big Bang. That ultra-relativistic environment witnessed dynamic equilibrium between the photons and particle-antiparticle pairs (here denoted generically as $P\bar{P}$) through the reaction

$$P + \bar{P} \leftrightarrow \gamma + \gamma$$

As the universe expanded and cooled, the production of particles and antiparticles of mass m was quenched when the photon energy dropped below mc^2. This occurred for the protons and neutrons (collectively called baryons) about 10^{-6} s after the Big Bang. However, pair annihilations continued (until the onset of nucleosynthesis, about 3 minutes later) without pair production to replenish the supply of baryon-antibaryon pairs. Since particles and antiparticles appear and annihilate as pairs, why did not all baryons annihilate with antibaryons, leaving the universe a gas of photons? That the universe today consists of matter with little antimatter suggests a lack of symmetry in their reactions. The decay channels available to antiparticles differ slightly from those for the corresponding particles. For example, suppose that, left to themselves, particle P is stable, but its antiparticle $\bar{P}$ can decay, with a small probability δ, into something else: $\bar{P} \to a + b$. The probability of $\bar{P}$ *not* decaying is $1 - \delta$. Every $\bar{P}$ that decays is unavailable to annihilate with a P particle, so when all the $\bar{P}$ antiparticles have disappeared through decay or annihilation, some P particles are left over. Such reactions are indeed observed in the laboratory, as in the decay of the kaons. They demonstrate matter-antimatter asymmetry, a mechanism for producing a universe dominated by matter.

See also: ANNIHILATION AND CREATION; ANTIPROTON, DISCOVERY OF; BIG BANG NUCLEOSYNTHESIS; POSITRON, DISCOVERY OF

Bibliography

Berstein, J.; Fishbane, P. M.; and Gasiorowicz, S. *Modern Physics* (Prentice Hall, Upper Saddle River, NJ, 2000).

Feynman, R. P. *Quantum Electrodynamics* (W. A. Benjamin, Reading, MA, 1962).

Feynman, R. P.; Leighton, R. B.; and Sands, M. *The Feynman Lectures on Physics* (Addison-Wesley, Reading, MA, 1965).

Hawking, S. *A Brief History of Time: From the Big Bang to Black Holes* (Bantam, New York, 1988).

Segrè, E. *From X-Rays to Quarks: Modern Physicists and Their Discoveries* (W.H. Freeman, San Francisco, CA, 1980).

Weinberg, S. *The First Three Minutes: A Modern View of the Origin of the Universe* (Basic Books, New York, 1988).

Wilczek, F. "The Cosmic Asymmetry between Matter and Antimatter." *Scientific American* **243**, 82–90 (1980).

Dwight E. Neuenschwander

ANTIPARTICLE

See ANTIMATTER

ANTIPROTON, DISCOVERY OF

The notion of the existence of antimatter in general, and of antiprotons in particular, can be traced at least as far back as the 1930s. The first unambiguous identification of the antiproton, however, did not occur until September 1955 at the University of California, Berkeley's Radiation Laboratory (later renamed the Lawrence Berkeley National Laboratory). In an article published in the journal *Physical Review*, Owen Chamberlain, Emilio Segrè, Clyde Wiegand, and Thomas Ypsilantis described how they employed Berkeley's new proton accelerator, the Bevatron, to record the presence of these negatively charged pieces of antimatter, whose mass is identical to that of the positive proton. In 1959 the Royal Swedish Academy of Sciences awarded Chamberlain and Segrè the Nobel Prize in Physics for their efforts.

Theoretical Context

As early as 1928, British theoretical physicist Paul Dirac (1902–1984) realized that solutions to his equations—which described behaviors of negatively charged electrons quite successfully—contained a puzzling feature. The equations allowed particles of negative energy to exist in addition to their positive energy counterparts. According to the equations, such particles would have a positive charge. Few knew what to make of this strange property. During the next four years, many, including Dirac, speculated that the well-known and positively charged proton might somehow account for these odd solutions. However, the significantly greater mass of the proton (two thousand times that of the electron), among other things, cast doubt on such assumptions. In 1931, Dirac proposed "a new kind of particle, unknown to experimental physics, having the same mass and opposite charge to an electron," the "anti-electron" (Dirac, p. 61). Such a particle would be rare in nature since it would tend to recombine into a state of pure energy with any of the many electrons present, but it should otherwise be stable. Observed experimentally in 1932, the antielectron soon became known as the positron; its presence suggested that other fundamental particles, like the proton, might also have antimatter counterparts. Some theoreticians proceeded to find a place for further antimatter in their equations, and some experimentalists proceeded to seek further antimatter in their observations.

Early Observations

The tendency of antiparticles to recombine with their particle counterparts presented a challenge to any experimental physicist wishing to record the presence of these elusive tidbits. The relatively high energy necessary to produce a proton-antiproton pair meant that physicists in the 1930s and 1940s could only expect to find an antiproton during cosmic ray observations. These measurements utilized the high-energy particles bombarding the Earth daily from extraterrestrial sources. In performing the experiments, scientists used cloud chambers and nuclear emulsions to visually record the paths of the incoming particles and their collisions. By analyzing pictures of particle movements in a magnetic field, observers made claims about the masses, energies, and charges of particles that had passed through these devices. From 1946 through 1955, cosmic ray experimenters suggested that one or another event might be traced to an antiproton, but they never completely convinced themselves or their colleagues. Copious production of antiprotons would have to wait for an artificial source, namely, a particle accelerator. Before 1954, however, no accelerator in the world was able to achieve more than half of the minimum energy predicted to be necessary.

The Accelerator and the Experiment

In 1946 and 1947 physicists at both the Radiation Laboratory at the University of California, Berkeley and the Brookhaven National Laboratory on Long Island, New York, lobbied for larger and more expensive accelerators to probe the finer structure of matter and energy. Given political and funding constraints, the laboratories had to moderate their requests. Brookhaven agreed to the quicker construction of a machine optimized for lower-energy particles, the Cosmotron. Berkeley embraced the longer-term task of building a machine whose particle energies could be expected to produce enough antiprotons for reliable measurements.

Berkeley's Bevatron beam, however, would also produce much larger amounts of another negatively charged particle, the negative pion. Therefore, the challenge of detecting antiprotons from an accelerator focused on convincingly identifying the presence of antiprotons from among the formidable noise of the pion background. The physicists needed to invent a scheme whereby the lighter pions would be excluded from their detectors and only negatively charged particles of a mass equal to the proton would be selected. To accomplish this, they set up a series of magnetic fields and detectors which indicated only particles whose charge, momentum, and velocity all matched the readings expected for antiprotons.

The selection was performed as follows (see Figure 1): the Bevatron's proton beam was steered into a copper target. The negatively charged particles from the ensuing collision passed through a magnet, M1, which was designed to bend the path of the particles with the appropriate momentum through focusing magnets, Q1, and through shielding into the first of three scintillation counters, S1. The particles then continued along through a second set of focusing magnets, Q2, and a second bending magnet, M2, until they reached the second scintillation counter, S2. If the particles made it through both M1 and M2, then they possessed the desired momentum. However, the heavier antiprotons would travel more slowly between S1 and S2 (51 billionths of a second compared to the pions' 40 billionths), so only particles with this longer time-of-flight would be recorded. The Cerenkov counter, C2, double-checked the velocity by registering only when triggered by the slower particles. (Slower, in this case, meant about 75 percent rather than 99 percent of the speed of light.) The remaining counters ensured that particles made it all the way through the apparatus and that two pions traveling some distance apart were not mistaken for a single, slower antiproton. The apparatus could also be adjusted to select for other particles as a test of its effectiveness. When tuned for antiprotons, it registered negatively charged particles with a mass identical to that of the proton. Thus, the first convincing

FIGURE 1

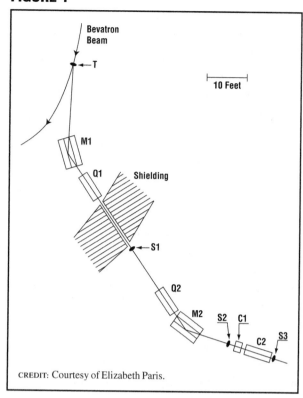

Bevatron
Beam

T

10 Feet

M1

Q1

Shielding

S1

Q2

M2   S2   C1

C2   S3

CREDIT: Courtesy of Elizabeth Paris.

Diagram of the experimental arrangement used in 1955 to detect antiprotons.

Chamberlain, O.; Segrè, E.; Wiegand, C.; and Ypsilantis, T. "Observation of Antiprotons." *Physical Review* **100**, 947–950 (1955).

Dirac, P. A. M. "The Quantum Theory of the Electron." *Proceedings of the Royal Society of London, Series A* **117**, 610–624 (1928).

Dirac, P. A. M. "Quantised Singularities in the Electromagnetic Field." *Proceedings of the Royal Society of London, Series A* **133**, 60–72 (1931).

Galison, P. *Image and Logic* (University of Chicago Press, Chicago, 1997).

Heilbron, J. L. "The Detection of the Antiproton." in *Proceedings of the International Conference on the Restructuring of Physical Sciences in Europe and the United States, 1945–1960, Universitá "La Sapienza," Rome, Italy, 19–23 September 1988*, edited by M. De Maria, M. Grilli, and F. Sebastiani (World Scientific, Singapore, 1989).

Pais, A. *Inward Bound* (Oxford University Press, New York, 1986).

Seidel, R. W. "Accelerating Science: The Postwar Transformation of the Lawrence Radiation Laboratory." *Historical Studies in the Physical Sciences* **13** (2), 375–400 (1983).

Elizabeth Paris

evidence for the antiproton was recorded electronically rather than visually and originated from an artificial source. Later, some of the earlier cosmic ray and nuclear emulsion observations were confirmed, and nuclear emulsion images were created in conjunction with the electronic apparatus.

Disputed Credit

In 1972 Italian émigré Oreste Piccioni brought a well-publicized lawsuit against Chamberlain and Segrè, claiming that he had originated the idea for the apparatus and had communicated it to them at a meeting in Berkeley in December 1954. The suit was dismissed on technicalities in 1973.

See also: ANTIMATTER; CONSERVATION LAWS; DIRAC, PAUL

Bibliography

Brown, L. M.; Dresden, M.; and Hoddeson, L., eds. *Pions to Quarks: Particle Physics in the 1950s* (Cambridge University Press, Cambridge, UK, 1989).

ASTROPHYSICS

Astrophysics is the branch of physics that attempts to understand the structure and evolution, appearance and behavior, of astronomical objects, especially those outside the solar system, including stars and galaxies, assemblages of these, and the material between them. Study of the universe as a whole is sometimes included but is more often given the separate name cosmology. Subdisciplines include (1) nuclear astrophysics, focused on nuclear reactions as energy sources in stars and as synthesizers of the chemical elements, nearly all of which are made in stars and stellar explosions; (2) high-energy astrophysics, which studies objects such as supernovae, pulsars, quasars, X-ray sources, radio galaxies, and astrophysical black holes (where the energy per particle or photon or the total energy is much larger than in typical stars and galaxies, and where strong magnetic fields, motions at close to the speed of light, and strong gravitational fields often occur); (3) particle astrophysics, whose main topic is the evidence for the existence and behavior of particles other than protons, neutrons, electrons, and photons (light) in stars, galaxies, and the universe; and (4) plasma astrophysics, which is concerned with low-density, very hot gases in the coronae of our sun and other stars, in the ejecta from supernovae and other cosmic explosions, and in interplanetary, interstellar, and intergalactic medium.

Stars and Nucleosynthesis

Stars, of which the Sun is quite typical, derive most of their energy from nuclear reactions, predominately the fusion of hydrogen to helium. They have lifetimes ranging from a few million years, for the biggest and brightest, to about 10 billion years for solar type stars, and up to 100 billion years or more for the tiniest and faintest. Most stars, indeed most of the matter in the universe, is, by weight, about three-quarters hydrogen and one-quarter helium, with only 1 or 2 percent of all the other elements. Oxygen, carbon, neon, magnesium, silicon, iron, sulfur, and nitrogen are the most abundant (notice they are also the biologically important ones), and uranium, thorium, and tantalum are among the rarest. The vast majority of stars are smaller and fainter than the Sun. The ones seen on a dark night are not a fair sample and include many of the rare, very bright stars, which are far away from the Earth.

Stars form, usually in groups or clusters, from clouds of gas and dust in the interstellar medium of galaxies. The gas is mostly molecular hydrogen, and the dust is mostly carbon compounds, silicates, and ices. The clouds are turbulent and pervaded by magnetic fields. The details of star formation involve complex interactions among rotation, gas flow, turbulence, and magnetic fields and are not completely understood.

In contrast, theoretical understanding of the structure and evolution of stars, once they have formed, is on quite solid ground, based in laboratory measurements of nuclear reaction rates and of how atoms and molecules behave when light shines on them. The structure is expressed in a set of differential equations, which can be solved numerically and integrated forward in time to describe changes in the brightness and colors of stars as hydrogen is gradually transformed to helium. The Sun is currently about 30 percent brighter than when it formed and began hydrogen fusion or burning, and about

five billion years in the future, it will become still brighter but redder (a red giant). It will end by fusing helium to carbon (C) and oxygen (O) and puffing off its outer layer to leave a dense core of C and O, called a white dwarf.

More massive stars continue on to the fusion of carbon, neon, oxygen, and silicon, producing cores of nickel, iron, cobalt, and other intermediate mass elements, which then collapse, leaving a neutron star (so-called because it is made mostly of neutrons, with only about 1 percent protons and electrons) or a black hole (so-called because its escape velocity exceeds the speed of light c, and light cannot get out).

The most abundant elements, hydrogen and helium, are (mostly) left from the early universe. Those from carbon to germanium in the periodic table are made by fusion reactions from which evolved, massive stars derive some of their energy. Lithium, beryllium, and boron are produced when cosmic rays strike atoms of carbon, nitrogen, and oxygen in the interstellar medium. And all of the elements heavier than iron and its neighbors in the periodic table come from the capture of neutrons by iron "seeds," followed by beta decays, late in the lives of stars and in their explosions. Energy is absorbed in these reactions, and the elements so made are all rather rare (hence the price of gold).

Stellar astrophysics is, for the most part, a consumer of information from atomic, nuclear, and particle physics, though it has been a donor in the past. For instance, it showed that fusion reactions must be possible, and it also showed that atoms like carbon have particular properties, before these properties had been measured in the laboratory. During the first half of the twentieth century, several elementary particles were recognized among cosmic ray secondaries before they had been produced on Earth. Close to the end of the century, it became clear that the deficit of neutrinos coming to the Earth from the Sun, relative to the numbers calculated, had been trying for more than 30 years say something about neutrinos, rather than about the Sun.

White Dwarfs, Neutron Stars, and Stellar Black Holes

White dwarfs, neutron stars, and black holes are the three expected end points of stellar evolution.

Table 1 compares some of their properties. There is observational evidence for all of them, including single white dwarfs and neutron stars, and ones with normal stars as binary companions, bound to them by gravitation, and for black holes in binary systems. Single black holes must surely exist, but unless one came very close to the solar system or passed across our line of sight to a normal star and bent the light rays coming to us, we would not be aware of them. There have been tentative detections of a few such gravitational lensing events, implying black holes with masses about six times that of the Sun, close to what is found in the black hole binaries.

Single neutron stars are most conspicuous when they are young and so have strong magnetic fields and short rotation periods. These cause beams of radio waves (and sometimes visible light and X rays) to sweep around like searchlight beams, which sometimes intersect the Earth. They are called pulsars. The best known is part of the remnant of a supernova explosion, seen in 1054 C.E., and called the Crab Nebula.

When white dwarfs, neutron stars, or black holes have binary companions, their masses can be mea-

sured, using Newton's laws of gravity, the same way as ordinary stars. Indeed having a mass too large for a neutron star (more than about three times that of the Sun) is part of the "signature" of a black hole. In addition, because the three are all very compact, material falling down onto or into them gets very hot and can radiate brightly, often at X-ray wavelengths.

A binary companion can supply gas to be accreted, some of which may also undergo nuclear reactions. The combination of accretion energy and nuclear energy is responsible for the many kinds of astronomical events and sources that occur in these binaries. Examples are nova explosions, X-ray binaries, some kinds of supernovae, and (probably) gamma-ray bursters. All of these are bright enough to be studied throughout our own galaxy and in those nearby; the supernovae and gamma-ray bursters can be seen even when they are in very distant galaxies (but they are rare).

The measured masses, sizes, rotation periods, and surface temperatures of some neutron stars are very close to the maximum or minimum possible according to calculations. It has been suggested that

TABLE 1

Properties of White Dwarfs, Neutron Stars, and Black Holes

Property	White Dwarfs	Neutron Stars	Black Holes (Stellar)
Masses (solar units)	0.4–1.4	1.2–2.2	6–10
Interior composition	He, C + O, O + Ne + Mg	99 percent neutrons	Cannot tell
Progenitor star masses	$\leq$8–10	$\geq$8–10	Very large or otherwise unusual
Example or evidence	Sirius B	Pulsars, e.g., CP 0529 in Crab	Some MACHO events (gravitational lensing)
Binary phenomena	Novae and other cataclysmic variables	X-ray binaries, e.g., Cen X-1	X-ray binaries, e.g., Cyg X-1
Luminosity solar units	10^{-4}–10	Up to 10^4	Up to 10^5
Rotation periods	Minutes to centuries	1.55 msec to hours	None (but disks msec to minutes)
Surface magnetic fields	$\leq 10^{-4}$–$10^{8.5}$ G	10^8–10^{14} G	None (but fields attached to disks)
Gravitational redshifts from surface	1–3×10^{-4}	0.25	Infinite
Escape velocity	~4,000 km/sec	200,000 km/sec	$\geq c$

CREDIT: Courtesy of Virginia Trimble.

the more extreme cases may be made not of pure neutrons but, at least at the centers, primarily of pions or even strange quarks.

Galaxies and Dark Matter

Galaxies as now seen are assemblages of stars—anything from a million to 10^{12} or more—often with 5 to 25 percent residual gas that is still forming stars. Galaxies come in a couple of characteristic shapes, called elliptical and spiral, and many others are irregular in appearance, either because there has not been time for gravitational processes to smooth them out or because they have been involved in a collision or near miss with another galaxy in the past few hundred million years. Both sorts of irregular galaxies are seen to be more common when looking back into the past by looking far away. This and other evidence indicate that galaxy formation was largely complete billions of years ago, though interactions (and star formation) are still going on. Small elliptical galaxies (called dwarf spheroidals) are the commonest sort of galaxy. Large spirals are the prettiest and so the kind most often shown in pictures.

Our galaxy is the Milky Way, a large spiral galaxy. It is part of a group including another large spiral (the Andromeda Nebula) and about three dozen smaller galaxies, called the Local Group (LG). The LG, in turn, is on the outskirts of a supercluster, whose center is a cluster of more than 1,000 large galaxies. It can be seen by looking through the star pattern of Virgo and so is called the Virgo Cluster. Hierarchical structure of galaxies, groups, clusters, and superclusters is typical. There seem to be no isolated galaxies (nor, for that matter, no more than a very few isolated stars, probably kicked out of their galaxies).

The largest structures have sizes about ten times our distance from the Virgo cluster (100 to 200 megaparsecs in the units actually used in astrophysics or 300 to 600 light-years in the units of science fiction and some well-meaning introductory books). The large-scale distribution of galaxies and clusters is more like sheets and filaments in a honeycomb, sponge, or foam than like matzoh balls in chicken soup. The space between the galaxies is very empty, with a density less than that of the space between the stars by a factor of 100,000 or more. On the other hand, while stars in the Milky Way are separated by distances that are millions of times their own sizes, galaxies in clusters are separated by only about ten times their diameters.

The process of galaxy formation is much less well understood than is star formation because there is a critical part of the physics about which very little is known. From 1922 onward, astronomers have gradually become aware, and forced themselves to accept, that the total masses of galaxies, clusters, and superclusters are very much larger than would have been supposed by adding up the masses of the stars and gas that contribute to their emission of light, radio waves, and X rays. This so-called dark matter reveals itself by exerting gravitational forces on the stars and galaxies we see, on the gas between them, and even on the very light rays that pass through clusters of galaxies (a process called gravitational lensing). The evidence therefore comes from measuring the motions of stars and galaxies, from the temperatures of X-ray emitting gas in elliptical galaxies and clusters, and from images of lensed galaxies and quasars (Figure 1).

The evidence is both internally consistent and quite strong for quasars. Alternative interpretations of the data have been attempted, but involve significant changes in the behavior of gravitation, which would have to differ from the familiar Newtonian and Einsteinian (general relativistic) versions very profoundly and in ways that probably disagree with other kinds of astrophysical data. The gradual increase in ratio of dark to luminous material as one looks on larger and larger scales, from cores of galaxies to their outskirts and to pairs, groups, clusters, and superclusters, means that the dark stuff is spread more uniformly through space than is the luminous stuff. Indeed, it would have been reasonable to guess that there would be still more dark matter entirely outside the largest structures seen. This does not seem to be the case.

Strangely, although the evidence for dark matter is very compelling, very little is known of its nature or behavior. This is a major stumbling block in trying to model the formation of galaxies and larger-scale structures, since these must arise by gravitational amplification of subtle variations in the density of matter when the universe was young. And most of the gravitational

FIGURE 1

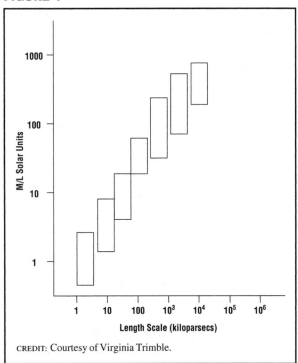

CREDIT: Courtesy of Virginia Trimble.

Plot of mass-to-luminosity ratio, which is roughly the ratio of dark matter to luminous matter, in units of solar mass per solar luminosity, as a function of the size of the system for which the measurement is made (in kiloparsecs). The data for small scales come from motions of stars and gas in our own and other individual galaxies. The intermediate scales represent orbits of star clusters and small satellite galaxies around large ones and of pairs of galaxies orbiting each other. The largest scales are whole clusters and superclusters of galaxies.

attraction will come from the dark matter, whatever it is, because there is more of it than there is of ordinary matter. Thus it is possible to provide a table of dark matter candidates (Table 2), but no actual laboratory examples of dark matter, though searches for various sorts have been under way since about 1980.

Of course old white dwarfs and brown dwarfs (like stars but too small in mass for nuclear reactions to light them up) also exist, but there are too few of them to constitute most of the dark matter. Similarly, the evidence for small, but nonzero, rest mass for neutrinos is quite strong, but they are also not most of the dark matter. This leaves the field for the "cold" dark matter candidates (axions or WIMPs, also called neutralinos, lowest-mass-supersymmetric-particles, etc.), topological defects like monopoles, and the cosmological constant (also called quintessence).

Quasars, Other Active Galaxies, and Galactic Black Holes

The light of normal galaxies comes largely from stars and so is emitted by the whole body of the galaxy. About 1 percent of galaxies (at present, but much larger billions of years ago) display much larger luminosities in light, X rays, radio waves, or some combination, that either comes directly from a tiny core or nucleus of the galaxy or is clearly powered by streams of fast-moving particles from the nucleus. These are the active galaxies.

The main types of active galaxies are radio galaxies (ellipticals with blobs of radio-emitting plasma far outside their visible limits, connected to the nuclei by jets), quasars (also strong radio sources but with cores emitting more visible light than all the rest of the galaxy by a factor of 100 or more), quasi-stellar objects (QSO) (like quasars but without radio emission), blazars (QSOs with very rapid changes in luminosity and structure), and Seyfert galaxies (spirals with cores about as bright as the rest of the galaxy).

When the radio galaxies and quasars were first recognized between 1954 and 1963, it was suggested that they might be collisions between normal galaxies and galaxies made of antimatter. This would have been very important as evidence about the symmetry between matter and antimatter, but it was the wrong answer because it predicted strong emission in gamma rays, which is not seen. The rapid firing of many supernova explosions in galactic centers also did not fit all the observations. What has turned out to work is the presence of a massive black hole at the center of an active galaxy, accreting stars and gas from its surroundings and using some of the energy that is released by the accretion to accelerate relativistic particles and amplify magnetic fields. These, in turn, radiate the photons we see.

Evidence for the black holes and their masses (from about a million to perhaps as much as 10 billion solar masses) comes from (1) the large luminosities, (2) the rapid variability, (3) the strong concentration of light in a cusp at the very center, and (4) the large velocities of stars and gas in that central cusp. Simple calculations show that a quasar can last for at most 1 percent of the age of the universe, and observations show that they were much

TABLE 2

Dark Matter Candidates

Type	Properties	Comments
Old white dwarfs brown dwarfs	Baryonic	Known to exist, ≈ 5 percent of total
Neutrinos with $m \neq 0$	Hot dark matter (large velocities when galaxies form)	Almost certainly exist, = 5 percent of total
Axions	Particle m ≪ electron mass. Cold (i.e., small velocities when galaxies form)	Predicted by theories beyond the Standard Model; could be 20–35 percent of total
WIMPs, neutralinos lowest mass supersymmetric partner, inos	Particle m ≈ proton mass Cold.	Predicted by theories beyond the Standard Model; could be 20–35 percent of total
Monopoles, strings domain walls textures	Seeds for galaxy formation	Predicted by symmetry breaking at phase changes; small fraction of total
Cosmological constant, quintessence, or dark energy	Pressure negative	Could be 60–79 percent of total; some observational evidence in favor

CREDIT: Courtesy of Virginia Trimble.

commoner in the past. Thus "dead quasars" (that is, black holes that are not currently accreting very much and so are not very bright) are expected inside many normal galaxies. The strongest evidence comes from the Milky Way, where the motions of the stars and gas near the center indicate that there is a central core of about three million solar masses, too compact to be anything except a single, massive black hole.

See also: BIG BANG; COSMIC MICROWAVE BACKGROUND RADIATION; COSMIC RAYS; COSMOLOGICAL CONSTANT; COSMOLOGY; HUBBLE CONSTANT; INFLATION; NEUTRINO, SOLAR; SUPERNOVAE

Bibliography

Cox, A. N. *Allen's Astrophysical Quantities,* 4th ed. (Springer-Verlag, New York, 2000).

Lang, K. R. *Astrophysical Data,* 2nd ed. (Springer-Verlag, New York, 1992).

Lang, K. R. *Astrophysical Formulae,* 3rd ed. (Springer-Verlag, New York, 1999).

Trimble, V. "Existence and Nature of Dark Matter in the Universe." *Annual Review of Astronomy and Astrophysics* 25, 425–472 (1987).

Virginia Trimble

ASYMPTOTIC FREEDOM

Asymptotic freedom is a characteristic property of quantum chromodynamics (QCD), the component of the Standard Model that describes the strong interactions. Asymptotic freedom ensures that when QCD is probed over short enough distances and times, it is well described by weakly interacting quarks and gluons. The idea that the strong interactions should somehow involve weakly interacting quarks may seem a bit paradoxical, but it is one of the triumphs of the Standard Model.

The strength of the color force is quantified by the QCD coupling α_s, the analog of the fine-structure constant of quantum electrodynamics (QED). There are a number of ways to conceptualize α_s, but perhaps the most intuitive is to define it in terms of the potential energy U_{QCD} between a quark and an antiquark held apart by a distance r through the relation:

$$U_{QCD}(r) = \frac{\alpha_s \hbar c}{r} \rightarrow \alpha_s = \left(\frac{r}{\hbar c} \right) U_{QCD}(r). \qquad (1)$$

Here c is the speed of light, and $\hbar = h/2\pi$, with h being Planck's constant. In Equation (1), α_s is itself a function of the distance. This dependence is a quantum mechanical effect called the running of the coupling. Asymptotic freedom is the property that α_s decreases as r decreases. It states that the color charge on a quark grows weaker when measured at shorter distances and stronger at longer distances.

In quantum field theory, the origin of Equation (1) is gluon exchange, shown in Figure 1, for a quark and antiquark separated by distance r. The quark spontaneously emits a gluon, which travels to the antiquark (or vice versa). This violates energy conservation by the energy of the gluon E_{gluon} for the period of time that it takes the gluon, which travels at the same speed as light, to arrive at the antiquark:

$$\Delta t = \frac{r}{c}. \qquad (2)$$

In quantum mechanics, such a process is allowed, so long as it satisfies the time-energy uncertainty relation

$$E_{gluon} \leq \frac{\hbar}{\Delta t} = \frac{\hbar c}{r} \qquad (3)$$

where Equation (2) has been used in the second step. Given this inequality, the potential energy $U_{QCD}(r)$ in Equation (1) can be thought of as the maximum energy of a gluon that the quark and antiquark can exchange at distance r times the probability that, if one were to look, such a gluon would be present. This probability is identified with $\alpha_s(r)$. Such an interpretation is only approximate, however, because the direct exchange of a single gluon is only the first of an endless set of possibilities that generalize Figure 1.

The next most important scenarios are shown in Figure 2. In one, the gluon is emitted as above, but before it arrives at the antiquark, it splits for a while into a quark-antiquark pair, then reforms, and next is absorbed. Similarly, the gluon sometimes splits into two gluons for part of its journey.

The running of the coupling is due to the mixture of the processes in Figures 1 and 2. Suppose that the potential energy is measured at some fixed distance r_0, and $\alpha_s(r_0)$ is defined by Equation (1), $\alpha_s(r_0) = r_0 U_{QCD}(r_0)/(\hbar c)$. Then one changes the distance between the quark and antiquark to $r_0 - \Delta r$, with Δr being much less than r, and measures again. The change in α_s is entirely due to Figure 2 and is pro-

FIGURE 1

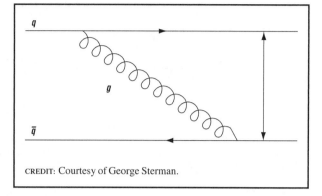

CREDIT: Courtesy of George Sterman.

Representation of gluon g exchange between a quark q and an antiquark $\bar{q}$ held a distance r apart.

portional to the fractional change in distance, $\Delta r / r_0$. The extra splitting in Figure 2 can be thought of as an extra emission and absorption so that the change in probability is also proportional to $\alpha_s^2(r_0)$. In summary, the change in the coupling is of the form

$$\alpha_s(r_0 - \Delta r) = \alpha_s(r_0) - b_0 \alpha_s^2(r_0) \frac{\Delta r}{r_0}. \qquad (4)$$

Determining the constant b_0 requires careful calculation, but the answer is quite simple: $b = (1/2\pi)(11 - 2n_f/3)$, where n_f is the number of quarks whose rest energies mc^2 are less than the energy of the gluon. The number "11" is the contribution of the graph with only gluons in Figure 2, and $2n_f/3$ is the contribution of the graph with a quark pair. In the Standard Model, n_f is always less than 6, so that b_0 is always positive. This means that the coupling decreases as r decreases. This is asymptotic freedom.

How did this come about? When the gluon in Figure 2 spends time as a pair of quarks, the charges of the quark pair tend to screen the charge of the original quark. The larger r is, the longer the screening goes on, which tends to decrease the charge as r increases, just the opposite of asymptotic freedom. On the other hand, when the gluon spends part of its time as two gluons, the charges of the produced gluons tend to enhance, or antiscreen, the original quark charge. There is a close analogy between the screening of color and diamagnetism, in which a material establishes an in-

FIGURE 2

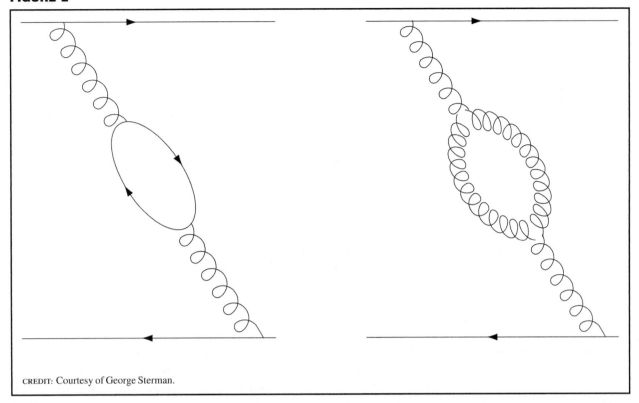

CREDIT: Courtesy of George Sterman.

Additional processes that contribute to the quark-antiquark potential.

ternal magnetic field that opposes an applied field. Correspondingly, the antiscreening of color may be compared to paramagnetism, in which a medium tends to enhance the applied field. In the case of QCD, the external field is the color field of a quark or gluon, and the medium is the QCD vacuum itself, capable of excitations to virtual states. The running of the coupling in QCD results from the competition of these two effects, and antiscreening wins out.

As long as $\alpha_s(r)$ is much less than 1, the following explicit formula is equivalent to Equation (4):

$$\alpha_s(r) = \frac{1}{b_0 \ln\left(\dfrac{hc}{r\Lambda}\right)} , \qquad (5)$$

where b_0 is the same constant as above, and Λ is an energy that must be determined from experiment. In practice, Λ turns out to be of order 2×10^8 electron volts (200 MeV). One can easily verify that Equation

(5) is an approximate solution to Equation (4). To do so, substitute $r_0 - \Delta r$ for r in (5). Then denote $x = r_0 \Lambda / hc$ and $\delta x = \delta r_0 \Lambda / hc$, and use the relation

$$\frac{1}{\ln\left(\dfrac{1}{x - \delta x}\right)} \approx \frac{1}{\ln\left(\dfrac{1}{x}\right)} - \frac{\delta x}{x \ln^2\left(\dfrac{1}{x}\right)} , \qquad (6)$$

which is accurate when δx is much smaller than x.

α_s is measured experimentally by accelerating quarks. This can be done by colliding them with leptons, or with other quarks. For example, in high-energy jet production, a quark whose energy is in the range of 10^{11} electron volts radiates a gluon of comparable energy (E_{gluon}) about one-tenth of the time. This can be thought of as the likelihood that the quark was struck just at the time it had emitted such a virtual gluon, as in Figure 1. However, this probability is $\alpha_s(r)$, $r = \hbar c / E_{gluon}$, which implies that $\alpha_s \approx 0.1$ for $r = \hbar c / (10^{11} \text{ eV}) \approx 2 \times 10^{-16}$cm. Given one such result, Equation (5) enables one to determine

FIGURE 3

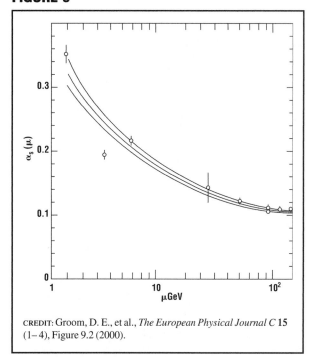

CREDIT: Groom, D. E., et al., *The European Physical Journal C* **15** (1–4), Figure 9.2 (2000).

Measurements of the QCD running coupling as a function of energy. The points at higher energy μ are from jet production, those at lower energy from lepton-proton scattering and particle decays. The curves are fits to $\alpha_s(\mu)$ similar to Equation (5).

Λ and then to extrapolate to any value of r, all the way up to $r = (hc)/\Lambda \approx 1$ fm $= 10^{-13}$cm. At this distance scale, α_s in Equation (5) becomes undefined because the logarithm of 1 is zero. This means that the probability of finding a gluon at such an r is too large for this method of calculating $\alpha_s(r)$ to be valid. Such distances are in the realm where the strong interactions become truly strong. It leaves, however, a wide range in which asymptotic freedom applies.

Experimental results on α_s are shown in Figure 3, compiled from a variety of sources, plotted against a scale of energy μ, and related to distances by $\mu = hc/r$. The running of the coupling and its asymptotic freedom are more than evident; the change in α_s is greater than a factor of 2. Only because of asymptotic freedom is it possible to understand jet cross sections and the scaling phenomenon in electron-proton scattering through which quarks were first observed directly. The discovery of asymptotic freedom lifted the veil from the strong interactions making possible their consistent description in the Standard Model.

See also: QUANTUM CHROMODYNAMICS; QUARKS; STANDARD MODEL

Bibliography

Johnson, G. *Strange Beauty* (Vintage, New York, 1999).

Kane, G. *Modern Elementary Particle Physics* (Perseus, Cambridge, MA, 1993).

Zee, A. *Fearful Symmetry* (Princeton University Press, Princeton, NJ, 1999).

George Sterman

ATOM

Ordinary physical and chemical matter is made of atoms. An atom is defined by its atomic number Z. A neutral atom has Z electrons, whose negative electric charge is matched by an equal positive charge on a small but massive nucleus. The structure of the atom is determined by the electrostatic attraction of each electron to the nucleus, by the electrostatic repulsion between the electrons, by the laws of quantum dynamics that apply to matter on small scales, and, finally, by the feature of quantum physics called the Pauli principle, which limits the degree to which a group of electrons may be compressed. An atom can exist in many states, each with a definite energy E and angular momentum J. Free atoms in their ground state are approximately spherical, and, although they do not have a sharp boundary, when in interaction with other atoms, they have an effective diameter between approximately 0.04 and 0.28 nanometers. Atomic ions are structurally like neutral atoms, except the number of electrons is different from that required to balance the nuclear electric charge.

Atomic structure is a rich subject, studied experimentally by a wide variety of methods, among which optical, X-ray, and radiofrequency (rf) spectroscopy are especially important. Atomic spectra are affected by external magnetic fields (the Zeeman effect) and electric fields (the Stark effect). The coupling of nuclear and atomic degrees of freedom leads to small effects, such as electric and magnetic hyperfine structure and isotope shift, but only in such fine details does atomic structure depend on the atomic mass A (or equivalently, the neutron number N). The electron capture mechanism of nuclear beta decay, and the internal conversion that parallels nuclear gamma decay, are other kinds of atom-nuclear interaction.

In the early 1980s Hans Dehmelt showed that single atoms can be isolated and studied in an electromagnetic trap. In a similar way, with the help of laser beams, small-scale gases of identical atoms (e.g., a few to millions of atoms) can also be produced, cooled to low temperature, and studied. At sufficiently low temperature the effects of the Pauli principle can be observed, leading to a filled Fermi sea or to Bose-Einstein condensation, depending on whether the total number of fermions (electrons and nucleons) is odd or even, respectively. For the even case, coherent (laserlike) beams of atoms can be produced through these same techniques.

Charged particles moving through matter lose energy by ionizing atoms, and the rate of energy loss (e.g., MeV/cm) depends on the velocity and charge of the particle, as well as on the atomic number of the atom. Many important detectors of high-energy particles use these properties of the energy loss mechanism to help identify the velocity and mass of the particles. In a similar way the absorption and scattering of photons depend on both the photon energy and the atomic number Z of the material. Experimental particle physics relies strongly on these ways in which energetic particles interact with atoms. Carefully chosen atoms also serve as minilaboratories for the study of the nonconservation of parity and of time reversal.

Atomlike systems can be formed by a variety of combinations of oppositely charged particles. Such systems obey the same laws of quantum mechanics as ordinary atoms and thus have very similar atomic properties, appropriately scaled by charge and mass. Positronium is an atom composed of an electron and its antiparticle, a positron. Muonic atoms (a negative muon and a nucleus) have been useful in measuring the charge distribution in the nucleus. Pionic, kaonic, antiprotonic, etc., atoms allow study of the additional complications of the strong interaction; the combination $\mu^+ e^-$ has been studied. Exotic atoms is a general name used to describe all these systems.

The idea that matter consists of small, indivisible objects, moving in otherwise empty space, is thought to have arisen in Greece during the period from 480 to 400 B.C.E. The objects were called atoms (in Greek, $\alpha\tau o\mu o\sigma$, uncuttable), and the earliest writings on them come from Democritus. According to his version, atoms were small but solid, and varied in shape and size. He thought everything was composed of atoms and void, and the properties of materials resulted from the arrangements of the atoms as well as their nature.

In the years around 1800 the concept of chemical elements became firmer. In the early 1800s John Dalton took the approximately integer ratios of combining volumes (of gases) and of combining weights (of condensed matter) to indicate that each chemical element was composed of atoms specific to that element and that all the atoms of an element were essentially the same. Also during the 1800s advances in the molecular picture of gases, especially the kinetic theory of gases, gave additional evidence that matter was composed of small units and even gave rough estimates of the sizes of these units (atoms, molecules).

H. A. Lorentz's interpretation of the Zeeman effect in 1896 strongly indicated that what were later called electrons were constituents of atoms, an idea that several physicists, especially J. J. Thomson, placed on a secure experimental foundation in 1897. A more definite picture of the positively charged part of the atom came with the discovery of the nucleus by Ernest Rutherford and his students in the years 1911 through 1913. Niels Bohr then pictured the atom as having a nucleus surrounded by electrons moving around it with motions like those of planets around a sun. His explanation had puzzling aspects (and the idea of definite orbits was in most respects misleading), but it accounted with amazing precision for many features of atomic hydrogen and qualitatively for observations of other elements.

The structure of the atom became much better understood with the discovery of quantum mechanics by Werner Heisenberg, Erwin Schrödinger, and others during 1925 and 1926. According to Schrö-dinger, electrons are described by wave functions (complex-number functions of position) from which the measurable properties of the atom can be calculated. For example, the absolute value squared of an electron wave function gives the distribution in space of that electron. Electrons are pointlike, as far as is known, but they have an intrinsic angular momentum, or spin, whose magnitude is $\hbar/2$, and whose projection on any spatial direction must equal either $+\hbar/2$ or $-\hbar/2$. All electrons are alike, indistinguishable from each other, except as they differ in their projection of angular momentum. Electron spin was discovered by Samuel Goudsmit and George Uhlenbeck in 1925.

An essential feature of the structure of atoms is that electrons obey the symmetry and statistical relations described by Enrico Fermi and Paul Dirac. As Wolfgang Pauli showed in 1926, the wave function of a two-electron combination must change sign if the coordinates of the electrons are interchanged. An important result is that two electrons of the same spin direction can never occupy the same position in space or have the same wave function. It is this fact, that electrons are Fermi-Dirac particles, that keeps atoms from falling into a state where all the electrons are in the same innermost orbital; the result is that atoms of all elements have radii falling within a limited range, rather than increasing with atomic number as Z^2.

The original quantum mechanics was not consistent with the requirements of relativity, but an appropriate relativistic wave equation was found by Dirac in 1927. There were still difficulties with a relativistic description of the coupling of electrons to radiation fields, but these were overcome during the period of 1946 to 1948, leading to a fully relativistic quantum electrodynamics (QED). QED allows many properties of simple atoms to be correctly calculated to a very high degree of precision.

The mid-1800s through the 1930s was a golden age for the exploration of atomic structure. Optical and X-ray spectroscopic work was intimately linked to the understanding of chemical properties, the periodic table, and the assignment of Z values. Series of spectral lines, fine structure, multiplicity, and intensity patterns all played important roles in the de-velopment of a theoretical picture of the microscopic world. Atomic spectroscopy became a widely used tool in practical applications and in other developing areas of physics. Especially important was the strong relationship developed by Henry Rowland, Henry Norris Russell, and others between laboratory spectroscopy and the spectra of the stars, which was responsible for great advances in astrophysics as well as in atomic physics and particle physics. The application of the laser to atomic spectroscopy, starting in the 1970s, has perhaps led to a new golden age of the atom.

Matter is made of atoms. Atoms are made of nuclei and electrons. Nuclei are made of nucleons (protons and neutrons). Nucleons, and other strongly interacting particles, are made of quarks. Searching for the structure of a physical system—looking for the units it is made of and how they are combined—has been one of the major themes of modern physics.

See also: ELECTRON, DISCOVERY OF; NEUTRINO, DISCOVERY OF; NEUTRON, DISCOVERY OF; POSITRON, DISCOVERY OF; QUARKS, DISCOVERY OF; RADIOACTIVITY, DISCOVERY OF

Bibliography

Boorse, H. A., and Motz, L., eds. *The World of the Atom,* 2 vols. (Basic Books, New York, 1966).

Drake, G. W. F., ed. *Atomic, Molecular, & Optical Physics Handbook* (AIP Press, Woodbury, NY, 1996).

Ketterle, W. "Experimental Studies of Bose-Einstein Condensation." *Physics Today* 30–35 (December 1999).

Wieman, C. E.; Pritchard, D. E.; and Wineland, D. J. "Atom Cooling, Trapping, and Quantum Manipulation." *Reviews of Modern Physics* **71**, S253–S262 (1999).

Guy T. Emery

AXION

According to the Standard Model of elementary particles, the strong interactions are described by quantum chromodynamics (QCD). QCD is similar to quantum electrodynamics (QED), except that electrons are replaced by quarks, photons by gluons, and the U(1) gauge group by SU(3). The physics of QCD and QED differ significantly, however. One important difference is that the coupling constant of QCD becomes stronger at long distances. Another is that QCD depends on an extra parameter θ that arises because of nonperturbative effects associated with QCD instantons. The dependence of QCD on θ causes a difficulty that the existence of an axion solves.

When θ differs from zero, QCD violates the discrete symmetries P and CP. P stands for parity and CP for the product of charge conjugation invariance and parity. Because the strong interactions obey P and CP in the laboratory, QCD can only describe them well if θ is very small, of order 10^{-9} or less. However, in the Standard Model, there is no reason why θ should be small. The theory must violate P and CP because these symmetries are broken by the weak interactions. The P and CP violation in the weak interactions feeds into the strong interactions in such a manner that θ is expected to be of order unity. The inability of the Standard Model to account for P and CP conservation by the strong interactions is called the strong CP problem.

A solution to the strong CP problem can be achieved by modifying the theory in such a way that θ becomes a dynamical field. In these models, the previously mentioned nonperturbative QCD instanton effects produce an effective potential for the θ field. Because the minimum energy state occurs at $\theta = 0$, the θ field relaxes to zero, and the strong CP problem is solved. This solution predicts the existence of a new particle, called the axion. The axion is the quantum of oscillation of the θ field. It has zero spin, zero electric charge, and negative intrinsic parity. It is coupled to all other particles with a strength proportional to its mass.

The mass m_a of the axion is not known *a priori*. Indeed, the axion solves the strong CP problem regardless of the value of its mass. However, masses larger than 50 keV are ruled out by searches for the axion in high-energy and nuclear physics experiments. Also, the masses between 300 keV and 3 milli-eV are ruled out by stellar evolution considerations, specifically the ages of red giants and the duration of the observed neutrino pulse from Supernova 1987a. Finally, masses less than approximately 1 micro-eV are ruled out because axions that light are so abundantly produced in the early universe, they would exceed the closure density. The only remaining window of allowed axion masses is $10^{-6} < m_a < 3 \times 10^{-3}$ eV.

In that window, axions make an important contribution to the present cosmological energy density. In fact, axions are one of the leading candidates to constitute the dark matter that appears clustered in halos around galaxies and also appears to be present on a larger scale within galactic clusters and the universe as a whole. Dark matter axions can be searched for on Earth by stimulating their conversion to microwave photons in an electromagnetic cavity permeated by a strong magnetic field. Searches of this type are presently underway in the United States and in Japan. Existing detectors are able to detect dark matter axions if axions constitute all of the halo matter and if their coupling is favorable. Future detectors will be sensitive to even a fraction of the halo density in a model-independent way.

See also: CP SYMMETRY VIOLATION; DARK MATTER; QUANTUM CHROMODYNAMICS; QUARKS

Bibliography

Raffelt, G. G. "Astrophysical Methods to Constrain Axions and Other Novel Particle Phenomena." *Physics Reports* **198**, 1–113 (1990).

Sikivie, P. "Axion Searches." *Nuclear Physics B* (Proceedings Supplements) **87**, 41 (2000).

Turner, M. S. "Windows on the Axion." *Physics Reports* **197**, 67 (1990).

Pierre Sikivie

B

B FACTORY

A B factory is a particle collider dedicated to producing B mesons. A meson is a bound state of a quark and an antiquark. A B meson contains a heavy bottom antiquark together with a light up or down quark while a $\overline{B}$ meson contains a bottom quark and light antiquark. Whereas up and down quarks are stable, bottom quarks decay with a lifetime of just 1.5 picoseconds (1.5 trillionths of a second). B mesons can decay in many different ways. Measurements of the pattern of decay rates have led physicists to a much deeper understanding of the fundamental properties of quarks and of the electroweak force.

B mesons also possess the remarkable ability to change, or oscillate, from matter to antimatter. This happens when a B^0 meson, the bound state of a b antiquark and a d quark, turns into a $\overline{B^0}$ meson, consisting of a b quark and a d antiquark. B^0-$\overline{B^0}$ oscillations were first observed in 1984, and this property together with the relatively long lifetime of the B meson led to the realization that it might be possible to observe an asymmetry between matter and antimatter in B meson decays. This asymmetry, known as CP violation, is required to explain how the universe evolved from a matter-antimatter symmetric configuration just after the Big Bang into the present matter-dominated state. CP violation was first observed in 1974 by James Cronin and Val Fitch in decays of K^0 mesons. However, further experiments with K^0 mesons did not reveal the source of CP violation, and thus the possibility of studying it in another, heavier, quark system was very attractive.

The first B factory was the Cornell Electron-Positron Storage Ring (CESR). CESR began operation in 1979, following the discovery of the b quark by Leon Lederman and colleagues at Fermilab in Batavia, Illinois, in 1977. A storage ring in Germany called DORIS was upgraded in the early 1980s so it could also produce B mesons. At CESR and DORIS, equal-energy positron and electron beams were collided at a center-of-mass energy of 10.58 GeV/c^2. At this energy, B meson production is resonantly enhanced, occurring in a quarter of all interactions. Another advantage is that B mesons are pair-produced with no additional particles to complicate the event. A disadvantage of the symmetric B Factory is that the B mesons are almost at rest and travel only 0.03 mm on average before decaying. This distance is too short to be measured with present detector technology, making it impossible to study time-dependent effects, such as lifetimes, at a symmetric B factory.

CP violation appears as a time-dependent difference in the B^0 and $\overline{B^0}$ decay rates to special final states known as CP eigenstates. This time dependence occurs because of a special quantum coherence between a pair of *B* mesons that are resonantly produced. The CP asymmetry may only be observed if one can measure the time between the two *B* meson decays and determine whether the CP eigenstate decay occurred first. The other *B*, called the tagging *B*, is identified as a B^0 or $\overline{B^0}$. The time difference between the two *B* decays is plotted separately for B^0 and $\overline{B^0}$ tagged events, and a shift between the two distributions is evidence of CP violation (Figure 1). Because the overall CP asymmetry will average to zero if the time between the decays of the two *B* mesons is not measured, this interesting phenomenon could not be studied at symmetric *B* factories.

In 1987 Piermaria Oddone of the Lawrence Berkeley National Laboratory proposed the asymmetric *B* factory (ABF) as a means to combine the copious and clean *B* meson production of an *e+ e−* *B* factory with the lifetime information required for CP violation studies. In an ABF, the electron and positron beams have unequal energy, but the total energy available in the center of mass is tuned to

10.58 GeV for resonant *B* meson production. The resulting *B* mesons move in the laboratory frame along the direction of the more energetic beam and thereby travel a measurable distance before they decay. A typical decay distance is on the order of one-fourth of a millimeter and can be measured using highly precise silicon-strip vertex detectors.

The construction of two ABFs started in 1994 and both were completed in 1999. The PEP-II ABF was built at the Stanford Linear Accelerator Center (SLAC) in California. In Japan the KEK-B ABF was built at the Japanese High-Energy Research Organization (KEK) in Tskuba, Japan. PEP-II and KEK-B are very similar in design. To achieve high interaction rates, both require very-high-intensity electron and positron beams, typically 1 ampere or more of stored current for each beam. Both reused existing electron-positron storage rings that had formerly operated as symmetric rings with much higher energies and lower currents, and both required the construction of an additional ring in order to store the beam of lower energy. The designs differed somewhat in how the beams are brought into collision. KEK-B employs a small crossing angle between the two beams to separate them after colliding, while in PEP-II the beams are collided head-on and separated by means of strong permanent dipole magnets located close to the interaction point.

PEP-II and KEK-B began operations in 1999 and soon achieved very high interaction rates. In the first year of operation, PEP-II produced approximately 20 million pairs of *B* mesons, whereas KEK-B produced about 10 million. In the future both machines expect to reach a rate of 100 million *B* meson pairs per year. The *B* meson decay products are recorded with large detectors that surround the electron-positron collision point. The PEP-II detector is called BaBar and the KEK-B detector is called Belle; both detectors were built by large international teams of physicists and are designed to record the charged tracks as well as the photons produced in *B* meson decays. The data are analyzed to reconstruct the *B* mesons from their decay products. The decay rates of B^0 mesons are compared to the corresponding $\overline{B^0}$ decay rates as a function of the time between decays to look for evidence of a time-dependent CP-violating asymmetry. In the summer of 2001 the BaBar and

FIGURE 1

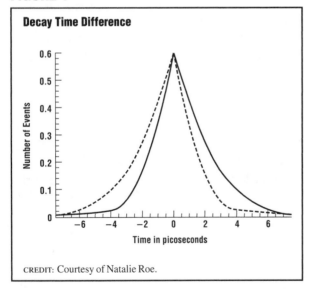

Decay Time Difference

CREDIT: Courtesy of Natalie Roe.

The time between the decays of the two *B* mesons shows a time-dependent shift depending on whether the event was tagged with a *B* meson (solid line) or a $\overline{B}$ meson (dashed line). The difference between the two distributions provides experimental evidence for a difference between matter and antimatter, and establishes CP violation in *B* meson decays.

Belle collaborations both reported significant evidence of CP violation in *B* mesons.

The 2002 results from BaBar and Belle are in agreement with the predictions of the Standard Model. However, current theories of baryogenesis cannot explain the amount of matter in the universe within the context of Standard Model CP violation. The future goal of the BaBar and Belle experiments is to measure CP violation in *B* decays more accurately and in a number of different decays in an effort to find an inconsistency that will point to new physics beyond the Standard Model.

See also: ACCELERATORS, COLLIDING BEAMS: ELECTRON-POSITRON

Bibliography

Quinn, H. R., and Witherell, M. S. "The Asymmetry between Matter and Antimatter." *Scientific American* **279** (4), 76–81.

Natalie Roe

BARYON NUMBER

See CONSERVATION LAWS

BASIC INTERACTIONS AND FUNDAMENTAL FORCES

One of the remarkable features of the fundamental forces governing the interactions among the constituents of matter is that they appear to act at a distance. For example, the force of gravity and the electrostatic force between two charged objects are both inversely proportional to the square of the distance between the bodies in question. Even in the case of the short-range forces between nucleons, the nuclear force appears to act in a nonlocal fashion. That is, a nucleon located in a given position exerts a force on another nucleon a distance away. In the case of the nuclear force, the strength of this interaction decreases exponentially with distance, whereas for gravity and electromagnetism, as well as for the effective van der Waal forces in molecules, the strength of the force only decreases as a power of the distance.

It is possible to understand the origin of these apparently nonlocal forces on a deeper level and view them as arising from local interactions between the constituents of matter- and force-carrying fields. Thus, for example, the electrostatic force between two charges arises as a result of the local interaction of each charge and the electrostatic potential established by the other charge at the location of the charge in question. Similarly, the gravitational forces of attraction between two massive bodies is the result of the local interaction between the energy density of each body and the gravitational potential set up by the other body.

These concepts are embedded in James Maxwell's theory for electromagnetic interactions and in Albert Einstein's theory of gravitation. According to these theories, electric charges and masses of particles are the sources of electromagnetic and gravitational fields, respectively. These fields permeate all space, and it is their local interactions with other charges or masses that result in the apparent action at-a-distance behavior of the electromagnetic and gravitational forces observed in nature. Similar but slightly more complex arguments can be made for other, less fundamental action at-a-distance forces. For instance, the van der Waal forces among the atoms in a molecule are the result of partially screening some of the electrostatic forces between the positively charged nuclei and the negatively charged electrons in the molecule.

These classical concepts may be naturally extended to the quantum domain. Indeed, in this context, they lead to an appealing physical interpretation, first advanced by Richard Feynman within the specific context of quantum electrodynamics (QED). All known matter appears to be made up of a few fundamental constituents, or elementary particles that are characterized by a set of intrinsic properties, such as their mass, electric charge, and possibly other (quantized) charges. These fundamental constituents have local interactions with force-carrying fields, each of which couples to one of these intrinsic attributes. It is also possible to associate particlelike excitations with each of these force-carrying fields, so that the local interactions between the fundamental constituents and the force-carrying fields can be viewed simply as an interaction between the fundamental constituents and the particles associated with the force-carrying fields. In this picture then, the forces between the fundamental constituents of matter result from the interchange of the force-carrying particles.

This concept is best illustrated by several examples. The electromagnetic force between two electrons can be viewed as resulting from the exchange of photons (the force-carrying particles associated with the electromagnetic field). The source of the electromagnetic force is the local interaction of photons with the electric charge of the electrons. One can show that to reproduce the inverse square dependence of the electrostatic force with the distance between the two charges, it is necessary that the mass of the exchanged photon be precisely zero. Any finite mass would lead to a force with an exponential dependence on the distance, with the range being inversely proportional to the force-carrying particle's mass.

Hideki Yukawa's inference of the existence of π mesons as mediators of the nuclear force, with masses of the order of one-tenth the proton mass, originated precisely with this kind of argument, which was based on the known range of the nuclear force. Although this particle-exchange example for the nuclear force is essentially correct, it is nonetheless incomplete. The exchange of π mesons explains only a part of the nuclear force, since this force, like the van der Waal forces, results from a more complex residual interaction—in this case, between the quarks in each nucleon. Quarks have local interactions with gluons, the carriers of the strong force. These interactions are responsible for producing neutrons and protons, as well as π mesons, as quark bound states. They are also directly responsible for the nuclear force, with π meson exchange representing the simplest form of quark-antiquark exchange, leading to the formation of nuclei as nucleon bound states.

At the fundamental level, there appear to be only four forces: gravity, electromagnetism, the strong force, and the weak force. In fact, strong evidence exists that electromagnetism and the weak force responsible for weak decays—such as that of the neutron when it decays into a proton, an electron, and an electron-antineutrino—are part of a unified electroweak force. Furthermore, hints exist that at very high energies the strong and electroweak forces themselves may unify into a single grand unified force. Finally, through string theories, where point particles are replaced by vibrational modes of strings, it may be possible to put gravity on the same footing as the other forces and thereby achieve a unification of all forces.

Each of these four forces couples to certain attributes of matter, and each force has an associated force-carrying particle. For example, the carrier of the electromagnetic force, the photon, couples to electromagnetic current. Remarkably, the specific form of this coupling is dictated by a symmetry principle: the freedom of being able to make local phase transformations of the fields associated with charged particles, in conjunction with an associated gauge transformation of the electromagnetic field. Electromagnetic interactions arise precisely by requiring that nature be invariant under these local $U(1)_{em}$ transformations, which requires only one parameter.

The carriers of the strong force, the gluons, and their interactions can be understood in the same manner as the electromagnetic interactions. The details are a bit more complicated for the carriers of the strong force, but the principles involved are quite similar. There are, in fact, eight gluons because the strong interactions, instead of being invariant under $U(1)_{em}$ transformations, are invariant under an $SU(3)$ symmetry group of transformations, which requires more than one parameter. To be invariant under $SU(3)$ symmetry transformations, eight separate force carriers, one for each independent parameter of the $SU(3)$ group of transformations, are needed. Similar considerations are involved for the weak force that has three carriers: the W^+, W^-, and Z. These three particles, along with the photon, are needed to be able to also associate the electroweak force with an underlying symmetry group of transformations—in this case, the four-parameter group $SU(2) \times U(1)$. Finally, the gravitational force is more akin to electromagnetism with only one carrier of the gravitational force, called the graviton.

At its most basic level, matter appears to be composed of two kinds of excitations: quarks and leptons. Quarks feel all four forces, whereas leptons (of which the electron and the electron neutrino are examples) are not subject to the strong force. The specific interactions of quarks and leptons with the strong and electroweak force carriers are dictated by the way these states transform under the full symmetry group connected with these interactions: SU(3) × SU(2) × U(1). For example, quarks are triplets [3] under SU(3), whereas antiquarks are antitriplets [$\bar{3}$]. The interactions giving rise to the strong force must respect the SU(3) group of transformations, since this is the symmetry of the theory. This is achieved by coupling the octet current of quarks and antiquarks [3 × 3 = 8 + 1] in an invariant manner to the eight gluons. The resulting theory, known as quantum chromodynamics (QCD), is a generalization of quantum electrodynamics in which the electron-positron electromagnetic current couples to the photon.

The weak force is more complex. First, the weak bosons couple asymmetrically to matter states, depending on their polarization. Thus, for example, only the left-handed component of electrons and electron neutrinos couples to the SU(2) force-carrying particles. Second, only the electromagnetic piece of the electroweak force is associated with a massless excitation (the photon) and gives rise to long-range forces. The remaining excitations of the electroweak interactions (the W^+, W^-, and Z weak bosons) actually acquire heavy masses—of order 100 times the proton mass. This arises as the result of the spontaneous breakdown of the SU(2) × U(1) symmetry to the phase symmetry connected to electromagnetism, which is associated with another Abelian symmetry group U(1)$_{em}$. Hence, weak interactions are very short-range. This is the reason why, at first, there did not seem to be any connection between the weak forces and electromagnetism. However, at energies much above the masses of the W and Z bosons, the distinction between electromagnetism and the weak interactions ceases to be important and the underlying SU(2) × U(1) symmetry is clear. If in nature, indeed, a grand unified theory does exist, it may well be that the distinction between the strong and electroweak forces, so apparent now, may also, in fact, disappear at superhigh energies.

The particles associated with the strong and electroweak forces all carry spin 1, whereas the graviton has spin 2. In the static limit, it is easy to show that the exchange of a spin-2 particle gives rise to an attractive force, while spin-1 exchange leads to a repulsive force. Since gravitons couple to mass, and this is always positive, one understands why gravity is attractive. In contrast, photon exchange is only attractive among charges of opposite sign but is repulsive for charges of the same sign. Because the gluons and the weak bosons carry a non-Abelian charge, the channels that are attractive or repulsive will depend in detail on the group structure of the matter states. Furthermore, because these force-carrying particles have nontrivial quantum numbers under the SU(3) × SU(2) × U(1) group, they themselves will feel these forces. So, in effect, for gluons and weak bosons the distinction between matter states and force-carrying particles blurs, since these states both carry and feel the forces.

Perhaps the most far-reaching and important feature of the basic interactions is that their form is fixed purely by symmetry considerations. As discussed, the theories that describe these interactions are invariant under a group of symmetry transformations. In fact, the invariance exists under local transformations of the symmetry group, in which each point in space-time can be subject to an independent transformation of the symmetry group. This is a much stronger requirement, and to achieve this invariance, which is known as gauge invariance, it is necessary to introduce compensating fields at each space-time point. These compensating fields are precisely those associated with the fundamental forces, and their interaction with the matter constituents is entirely fixed by the requirement of gauge invariance. Thus, remarkably, symmetry principles fully determine the forces of nature.

See also: BOSON, GAUGE; BROKEN SYMMETRY; ELECTROWEAK SYMMETRY BREAKING; FLAVOR SYMMETRY; GAUGE THEORY; GRAND UNIFICATION; SU(3); SYMMETRY PRINCIPLES

Bibliography

Fermi National Accelerator Laboratory. "The Science of Matter, Space and Time." <http://www.fnal.gov/pub/inquiring/matter>.

Quigg, C. "Elementary Particles and Forces." *Scientific American* **252** (4), 84 (1985).

t' Hooft, G. "Gauge Theories of the Forces between Elementary Particles." *Scientific American* **242** (6), 104 (1980).

Roberto Peccei

BEAM TRANSPORT

The trajectories of moving charged particles can be altered through the use of electromagnetic fields. In this manner, particle beams are guided around a circular accelerator, such as a cyclotron or synchrotron, for repetitive encounters with an accelerating cavity. Likewise, beams of particles can be transported from one accelerator to another or from an accelerator toward an experimental target, or even to a patient in medical applications. In each case, this is accomplished by an arrangement of appropriate electromagnetic fields.

Armed with an understanding of the motion of charged particles in magnetic fields, one can imagine a system of electromagnets used to guide a beam of charged particles. Steering a particle from one point to another is only one issue; keeping a stream of charged particles focused along the central trajectory is a vital concern for any beam transport system. Dipole and quadrupole magnets are used for these two purposes. The ability to perform fine adjustments to the particle beam's trajectory and focusing characteristics must also be included in any beam transport system design. For example, smaller, adjustable electromagnets can be used to adjust the position of a particle beam to a small fraction of a millimeter in a particle accelerator that may be many kilometers in circumference. Transport systems can be built to guide and focus particle beams of very high energies to great precision.

Charged Particle Motion in Electromagnetic Fields

The Lorentz force governs the motion of a charged particle,

$$\vec{F} = q(\vec{E} + \vec{v} \times \vec{B}),$$

where E and B are the electric and magnetic field strengths, and q and v are the particle's charge and speed, respectively. The arrows indicate that the direction of the fields and of the particle's motion dictates the direction of the resulting force.

Imagine a positively charged particle entering a localized region of electric field, where the field lines are perpendicular to the initial direction of motion. The particle's trajectory will be deflected in the direction of the electric field lines. The trajectory through this region is a parabola.

Now, suppose a particle enters a localized region of magnetic field where, again, the field lines are perpendicular to the initial direction of motion. The particle's trajectory will be deflected in a direction perpendicular to the particle's velocity and perpendicular to the direction of the magnetic field. Through this region the trajectory will be an arc of a circle.

These two cases are illustrated in Figure 1. By using localized regions of electric and magnetic fields, charged particles can be steered in any general direction, transporting them toward an experimental apparatus, or from one particle accelerator to another, or toward the phosphorescent screen of a television set!

From the Lorentz equation above, it can be seen that particles with very low speeds are more easily governed by electrical forces, whereas particles at very high speeds—say, approaching the speed of light—are more easily affected by magnetic forces.

FIGURE 1

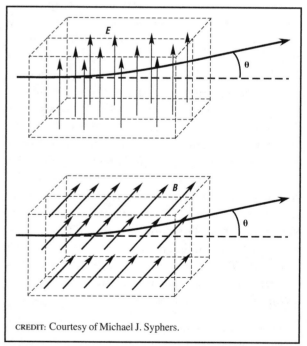

CREDIT: Courtesy of Michael J. Syphers.

Charged particle trajectories in electrical and magnetic fields.

For comparison, consider a charged particle moving near the speed of light (3×10^8 m/s). If it encounters a magnetic field of strength 1 Tesla (T), which is typical of an iron electromagnet, then the product of these two quantities is 3×10^8 T-m/s, or equivalently, 3×10^8 volt/m. Thus, to generate the same force with an electric field, the electric field strength would have to generate 3 million volts over a distance of 1 cm—an extremely large voltage over a relatively short distance! For particles with high momentum, therefore, magnetic fields are used in beam transport systems.

The ratio of a particle's charge to its momentum is called the magnetic rigidity of the particle and has units of Tesla-meters (T-m). Since the common unit of energy of an elementary particle is the electron volt (eV), and the unit of momentum is written in terms of electron volts divided by the speed of light (eV/c), then a particle's magnetic rigidity can be conveniently approximated as

$$\frac{10}{3} \, p_{GeV/c} \text{ Tesla-meters}$$

where $p_{GeV/c}$ is the particle's momentum in units of GeV/c. (*Note:* 1 GeV = 10^9 eV.) The approximation lies in the 3 of the denominator, which comes from the approximation that the speed of light is 3×10^8 m/s. As an example, a particle with a momentum of 3 GeV/c will have a magnetic rigidity of 10 T-m. This says that if the particle is in a magnetic field of 2 T, then it will move in a circular path of radius 5 m. Going one step further, if the particle is in this 2-T field over a distance of only 10 cm, then its trajectory will be deflected through an angle of 10 cm/ 5 m = 0.02 radians (1.15 degrees). The above discussion illustrates how one can build a system of magnetic elements of given lengths and field strengths to steer the paths of particles of a given energy.

The Need for Transverse Focusing

In the design of a beam transport system, an ideal trajectory of an ideal particle is laid out. The ideal particle is one with the design energy or momentum and with a required initial trajectory (i.e., initially headed in the right direction). Magnetic elements are then arranged to guide this ideal particle to its final destination. However, particle accelerators and beam transport systems usually handle streams of many particles, typically billions at a time. Such beams will have a distribution of particles with an average energy that might be ideal but that has a spread in energy about this average. Likewise, not all particles will be headed along exactly the same trajectory but will have trajectories that start nearby. Therefore, beam transport systems need to control more than the "ideal" particle trajectory: they must appropriately control surrounding trajectories as well.

Static electric and magnetic fields, as shown above, can be used to control the trajectory of an ideal particle. However, a nearby particle with a slightly different trajectory needs to be guided back toward the ideal path. As a particle begins to deviate from the ideal course, one would like for it to be forced back toward its nominal position, much like what happens when a spring pushes and pulls a mass back toward its equilibrium location. Following this analogy, imagine a mass hanging motionless on a spring. When the mass is pulled and released, the spring exerts a force on the mass that is proportional to the displacement of the mass from its equilibrium point. The mass oscillates about the equilibrium point, undergoing simple harmonic motion.

A magnetic field whose strength is zero at one point and becomes stronger proportionally to the distance from the center is a quadrupole field. Such a field, derived from a quadrupole magnet, is depicted in Figure 2. In this figure, imagine a positively charged particle coming toward the reader. A particle moving down the center of the magnet will experience no force. However, particles displaced farther horizontally from the center will experience stronger forces directed back toward the center. One problem, however, is that particles that are displaced vertically from the center will experience forces that deflect them away from the center. This is a consequence of the fact that magnetic fields in free space are irrotational. Since one quadrupole field will focus the particle beam in one degree of freedom (horizontally, say) and defocus in the other degree of freedom (vertically, say), then one needs to examine arrangements of magnetic elements carefully to provide proper guiding (focusing) in both degrees of

FIGURE 2

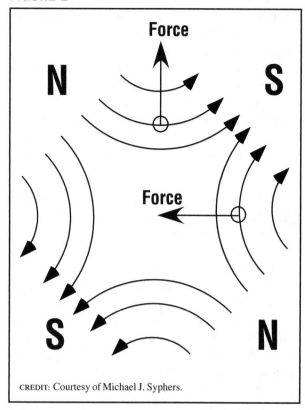

CREDIT: Courtesy of Michael J. Syphers.

A quadrupole magnetic field.

FIGURE 3

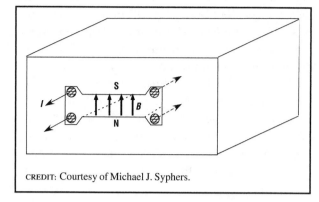

CREDIT: Courtesy of Michael J. Syphers.

A dipole magnet.

height of the pole gap, and μ_0 is the permeability of free space. The length of the magnet times the field generated inside the gap will determine the deflection of a particle's trajectory.

For focusing particle beams, a quadrupole magnet is typically used. A sketch of a quadrupole magnet design is provided in Figure 4. The vertical magnetic field in the gap is proportional to the horizontal displacement from the center, and the horizontal field is proportional to the vertical displacement from the cen-

freedom simultaneously. This topic is discussed further below.

Magnetic Elements

Uniform magnetic fields used for guiding particle beams are typically generated by electromagnets with two poles (north and south) as depicted in Figure 3. In this traditional iron magnet, electric current is carried through the body of the magnet in copper conductors, and the lines of magnetic flux circulate around the copper, through the iron yoke, and into the magnet gap. The field in the gap, typically expressed in units of Tesla, is given by the equation

$$B = \frac{2\mu_0 NI}{d}$$

where I is the current in the conductor, N is the number of conductor windings around each pole, d is the

FIGURE 4

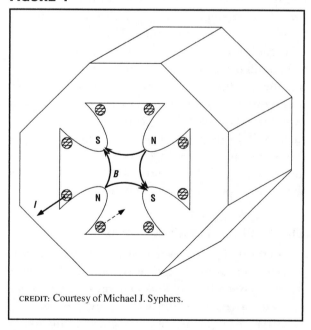

CREDIT: Courtesy of Michael J. Syphers.

A quadrupole magnet.

ter. In each case, the constant of proportionality, called the quadrupole gradient, is given by

$$G = \frac{2\mu_0 NI}{a^2}$$

where $2a$ is the distance between opposite pole tips. The gradient is typically expressed in units of Tesla/meter. To generate the desired quadrupole field, the iron poles are machined to an appropriate hyperbolic shape.

Likewise, the actual deflection of a particle trajectory due to a quadrupole field will depend on the length of the magnet. Since the deflection also depends on the displacement of the particle from the center of the quadrupole field, the magnet acts basically as a lens. If one considers the trajectory of a light ray passing through a simple lens, as depicted in Figure 5, one can see that the corresponding focal length of a quadrupole magnetic lens is given by

$$\frac{1}{f} = \frac{qGL}{p}$$

where p is the particle's momentum, q is its charge, L is the length of the magnet, and G is the quadrupole gradient defined above. As long as the length of the magnet is short compared to the focal length, the magnet can be treated as a "thin lens," and the focusing characteristics of the beam transport system can be understood by the standard rules of thin lens optics. One must be careful to note, however, that a lens which focuses in one degree of freedom will defocus in the other degree of freedom. Thus, vertical

FIGURE 5

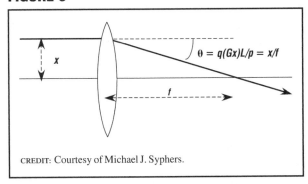

θ = q(Gx)L/p = x/f

CREDIT: Courtesy of Michael J. Syphers.

Definition of focal length.

and horizontal particle motion must be examined simultaneously in the magnetic optical system.

Also, it should be pointed out that other types of electromagnetic designs—for example, the coil configurations found in superconducting magnets—exist in addition to the simple examples cited above.

Beam Lines and Circular Accelerators

A beam transport system that delivers particles from one point to another is often referred to as a beam line. Such a system must transport the particles along the ideal trajectory, and the maximum displacement from the ideal trajectory of any given particle needs to lie within the physical aperture of the system. Any one focusing element along the way affects each degree of freedom differently; thus, the particle motion in each degree of freedom must be examined simultaneously. To analyze such a system's design, extreme initial conditions of possible particle trajectories can be traced through the system to be sure that they meet the final conditions required at the end of the beam line.

Tracking a particle's trajectory once around a circular accelerator, however, is not necessarily sufficient to determine the functionality of an accelerator's design. Circular accelerators take particles from one point back to the same point again, and again, and again. Thus, such a system must be "stable" under repetitive traversals. Further analysis of the basic magnetic system is therefore required.

Weak and Strong Focusing

Imagine a uniform magnetic field that is used to guide a particle in an ideal circular trajectory. If the field lines are pointed vertically, for instance, then horizontal motion is stable in this system. That is, if a particle begins its trajectory near but displaced slightly horizontally from the ideal circle, it will simply be guided around in a circular trajectory of the same radius as the ideal circle, but slightly offset. It will oscillate around the ideal trajectory with one oscillation per revolution. However, if the particle is given any vertical momentum, it will spiral around the vertical magnetic field lines and gain vertical displacement until it reaches the vacuum chamber walls; as shown in Figure 6.

FIGURE 6

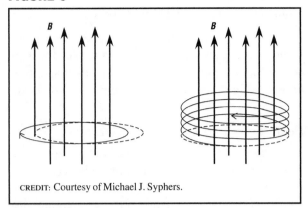

Particle motion in uniform magnetic field with and without vertical particle momentum.

Focusing in both the horizontal and vertical directions can be restored by widening the gap of the guiding magnet's pole face at its outside edge, as shown in Figure 7. The magnetic field lines are made to curve, and the magnetic field will become weaker near the radial outside and stronger near the radial center of the guiding magnets. Although this gives slightly less horizontal focusing, it provides vertical focusing. Since the force is always perpendicular to the field direction, a particle whose trajectory deviates in the vertical direction will now experience a force guiding it back toward the center of the magnet gap. So long as the wedge angle of the gap is not too steep, both horizontal and vertical motions will be stable and a particle will oscillate about the ideal circular trajectory. This form of focusing is called weak focusing and was commonly used in the design of particle accelerators until the mid-1950s.

FIGURE 7

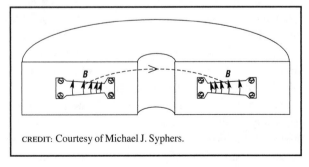

A weak focusing accelerator magnet.

Strong focusing can be understood by noting that a combination of two thin lenses, one focusing and one defocusing, separated by some appropriate distance will form a system that is itself focusing. In this way, a series of quadrupole magnets with their gradients alternating in sign (alternately focusing and defocusing in the horizontal direction, for instance) can create a system that focuses in both degrees of freedom simultaneously. As an example, motion through a repetitive system of equally spaced thin lens quadrupoles is stable as long as the lenses alternate between positive and negative focal lengths, and the absolute value of the focal length must be larger than half the distance between the lenses. This focusing structure, sometimes referred to as a FODO cell, is commonly used in long beam lines and large-circumference particle accelerators. The beam size can be kept arbitrarily small by focusing frequently enough and is not dependent on the overall length of the beam line or accelerator.

As the bending radii of weak focusing accelerators became larger and larger to accommodate higher and higher particle energies, their magnet apertures became larger accordingly and the amount of steel and copper required to generate the necessary magnetic fields made these devices very expensive to build. The invention of strong focusing in 1952 (Courant and Snyder) decoupled the focusing characteristics of the circular accelerator from its bend field requirements and thus allowed for accelerators of very large circumferences to be designed and built.

Beam Control

In an accelerator or beam line, dipole magnets, typically much smaller in strength than the main bending magnets, are used for the fine adjustment of the particle beam trajectory. By placing these steering magnets advantageously around the accelerator, the position and angle of the beam at important locations can be readily adjusted. Such uses might center the beam within a particle detector, as in a colliding beams experiment, or adjust the beam trajectory coming into the accelerator and onto a desired orbit. Independent control is often required over the particle beam position and its slope at a certain location. To perform a transverse position adjustment that is localized to the point in question, three steering magnets are necessary. With this so-called three bump system, the first magnet defines the trajectory leading to the desired position adjustment, and the second and third magnets bring the trajectory (position and slope) back to its original value.

While a three bump system can control a specific position in the beam line, independent control of both position and slope at a location requires two steering magnets upstream of the location in question and two downstream. Together, the two upstream magnets can be adjusted simultaneously to produce any desired position and slope, and then the two downstream magnets are adjusted to bring the trajectory back to its original state downstream of the system. Most large accelerators and beam lines are constructed with many such correctors to allow for adjustments to beam trajectories at arbitrary locations.

In addition to control of the beam trajectory, focusing adjustments are commonly required in beam transport systems as well. One example of their use would be to make fine adjustments to the number of oscillations a particle makes about the ideal trajectory in a circular accelerator. A small quadrupole magnet can be used to adjust the focusing characteristics of the horizontal oscillations, for example. However, the same magnet will also alter the vertical oscillations as well. Thus, two such quadrupole magnets are required for independent control of the oscillations in both degrees of freedom. For fine adjustments of the oscillation frequency, "families" of many small correction quadrupoles are typically located around the accelerator at favorable positions and connected in series electrically. Two independent families will control the horizontal and vertical motion independently. Having many such correctors reduces the necessary strength of the correction magnets and also serves to reduce the perturbations that the magnets make to the primary focusing structure of the system.

See also: ACCELERATOR; ACCELERATORS, COLLIDING BEAM: ELECTRON-POSITRON; ACCELERATORS, COLLIDING BEAM: ELECTRON-PROTON; ACCELERATORS, COLLIDING BEAM: HADRON; ACCELERATOR, FIXED-TARGET: ELECTRON; ACCELERATOR: FIXED TARGET: PROTON; EXTRACTION SYSTEM; INJECTOR SYSTEM

Bibliography

Chao, A. W., and Tigner, M., eds. *Handbook of Accelerator Physics and Engineering* (World Scientific, Singapore, 1999).

Conte, M., and MacKay, W. W. *An Introduction to the Physics of Particle Accelerators* (World Scientific, Singapore, 1991).

Courant, E. D., and Snyder, H. S. "Theory of the Alternating Gradient Synchrotron." *Annals of Physics* **3** (1), 1–48 (1958).

Edwards, D. A., and Syphers, M. J. *An Introduction to the Physics of High Energy Accelerators* (Wiley, New York, 1993).

Wilson, E. J. N. *An Introduction to Particle Accelerators* (Oxford University Press, Oxford, 2001).

Michael J. Syphers

BEIJING ACCELERATOR LABORATORY

Beijing Electron-Positron Collider

The Beijing Electron-Positron Collider (BEPC) is located 12 kilometers west of the center of Beijing, China, at the Institute for High Energy Physics. The purpose of the laboratory is to conduct research in elementary particle physics and to serve as a source of synchrotron radiation, which is a special form of light created when electrons' paths bend in a strong magnetic field. The laboratory is composed of three main components: a collider (BEPC), a magnetic spectrometer (Beijing Spectrometer [BES]), and synchrotron radiation facilities.

In a collider, bunches of electrons and positrons circle in opposite directions in an evacuated tube, shaped like a doughnut, and are held in their orbits by strong magnetic fields. They can collide head-on, annihilate, and form new particles. Since 1990, BEPC has been the only collider in the world operating in the 2 to 5 billion electron volt energy range. This is a very interesting energy region because this is the threshold region for the production of tau leptons (heavy versions of the electron) and many particles made from charmed quarks.

Origin

The 1970s were an exciting period in high-energy physics. The J/ψ particle had been discovered simultaneously in 1974 in different experiments by groups led by Burton Richter at the Stanford Linear

Accelerator Center (SLAC) in Palo Alto, California, and by Chinese-American Samuel C. C. Ting at the Brookhaven National Laboratory in Long Island, New York. The discovery at SLAC was made in a new electron-positron collider, called SPEAR. The J/ψ particle, more than three times heavier than a proton, is composed of a charmed quark and an anti-charmed quark held together by the strong nuclear force. Richter and Ting shared the Nobel Prize in Physics in 1976 for their discoveries.

In 1981, the director of SLAC, Wolfgang K. H. Panofsky, and Chinese-American Nobel Prize winner Tsung Dao Lee suggested that China build a 5 billion electron volt collider. This suggestion received strong support from Chinese physicists, and a proposal was made to build a machine with a similar energy range as SPEAR but with higher interaction rates, allowing more detailed studies in this region. Premier Deng Xiaoping officially approved the BEPC project in 1982 and attended the groundbreaking on October 7, 1984. After four years, on October 22, 1988, he attended the celebration of the first electron-positron collisions.

Beijing Electron-Positron Collider

Figure 1 shows the BEPC accelerator complex. The straight section on the right is the injector, which is a 202-meter-long linear accelerator. An electron gun, which is the source of electrons, is at the beginning of the injector. Electrons are accelerated to an energy of 1.1 to 1.4 billion electron volts and are transferred into the storage ring on the left. The storage ring has a circumference of 240 meters and is shaped like a racetrack. To make positrons, electrons strike a moveable tungsten target located partway down the linear accelerator and create both positrons and electrons. Some of the positrons formed are accelerated through the remaining portion of the linear accelerator and injected in the opposite direction into the collider. The beams are separated vertically except where they collide at the interaction point.

The Beijing Spectrometer Detector

Many different kinds of particles can be formed in the collisions. In most cases, the particles exist for a very short time before they decay into other longer-lived particles. To record what happens, a large magnetic detector named the Beijing Spectrometer

(BES) is used. Its main function is to measure the parameters of the particles and to identify them. BES is cylindrical in shape, approximately 4.7 meters long and 4.9 meters in diameter, and is composed of many subsystems. Close to the beam pipe is a vertex detector to measure the positions of the charged particles passing through it. Next is the main drift chamber, which measures the trajectories of charged particles. Because the chamber is located in a large magnetic field, the trajectories curve, and the amount of curvature determines the momentum of each particle.

Particles next pass through the time of flight system, which determines the time of travel from the interaction point for each particle and thereby the particle's velocity. By measuring the velocity and momentum, the mass and type of particle can be determined.

Further out is the shower counter, which measures the energies of photons (particles of light) coming from the interaction. Photons are electrically neutral and are not detected in the main drift chamber. Outside the shower counter is the coil of the magnet that supplies the magnetic field in BES. All the signals from the various subsystems of BES are read into a computer using special electronics and are written to tape for later analysis.

Physics Results

BES began taking data in 1990. In 1991, physicists from the United States joined the effort, making the BES experiment an international collaboration. In 2001, the collaboration included about 150 physicists from China, Japan, Korea, and the United States.

BES has published papers in English and Chinese journals on many physics topics. One of the most important measurements was the measurement of the mass of the tau lepton. In the early 1990s, the values of the tau mass and tau lifetime, measured by previous experiments, implied disagreement with the theory of lepton universality, part of the Standard Model of high-energy physics. According to this theory, all leptons (electrons, muons, and taus) should have the same behavior.

In 1992, BES measured the tau mass with a precision of 0.02 percent by carefully measuring the tau production rate near threshold. The measured value was

FIGURE 1

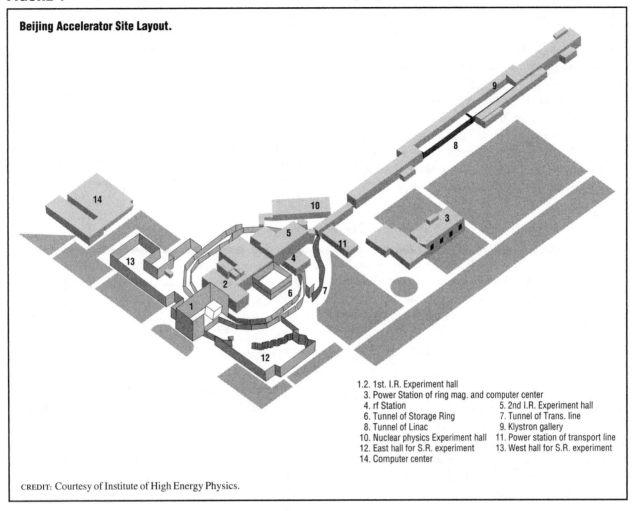

Beijing Accelerator Site Layout.

1.2. 1st. I.R. Experiment hall
3. Power Station of ring mag. and computer center
4. rf Station   5. 2nd I.R. Experiment hall
6. Tunnel of Storage Ring   7. Tunnel of Trans. line
8. Tunnel of Linac   9. Klystron gallery
10. Nuclear physics Experiment hall   11. Power station of transport line
12. East hall for S.R. experiment   13. West hall for S.R. experiment
14. Computer center

CREDIT: Courtesy of Institute of High Energy Physics.

lower than the previous value, bringing it into very good agreement with the theory of lepton universality.

In 1998 and 1999, BES measured the electron-positron interaction rate at ninety-one different energies between 2 and 5 billion electron volts. The uncertainties on these values were less than one-half the previous ones. These improved measurements are important for precision tests of the Standard Model and have been used to improve the predicted mass of the Higgs meson.

See also: INTERNATIONAL NATURE OF PARTICLE PHYSICS

Bibliography

Drell, S. "Electron-Positron Annihilation and the New Particles." *Scientific American* **232**, 50–62 (1975).

Riordin, M. *The Hunting of the Quark* (Simon & Schuster, New York, 1987).

Zheng. Z. P. "The Present and Future of China's Particle Physics Research." *Science* **262**, 368 (1993).

Frederick A. Harris
Zhipeng Zheng

BENEFITS OF PARTICLE PHYSICS TO SOCIETY

Compared to most scientific endeavors, though not to space exploration or to some defense-related technology research, high-energy physics is an expensive enterprise. Modern accelerator facilities capable of expanding the high-energy frontier, such as Fermilab or the European Laboratory for Particle Physics (CERN) Large Hadron Collider (LHC) project, are big science, involving the concerted efforts of thousands of people and costing several billions of dollars. High-energy physics has been supported almost entirely by government agencies and thus ultimately by taxpayers. It is entirely appropriate that scientists who promote these expenditures should be expected to justify this investment by society as a whole, by explaining its benefits to society as a whole.

The primary aim of research in high-energy physics is easily stated. It is, simply, to produce a better understanding of fundamental physical law by following a reductionist strategy. That is, scientists attempt to understand the behavior of matter in general by working up from profound understanding of the properties and interactions of its elementary constituents.

This strategy has proven remarkably fruitful and successful, especially over the course of the twentieth century. We have discovered that strange but precise and elegant mathematical laws, summarized in the so-called Standard Model, govern the laws of physics on subatomic scales. There is every reason to think that these laws, as presently formulated, are adequate to serve as the foundation for materials science, chemistry (including biochemistry), and most of astrophysics.

One must be careful in interpreting this sort of statement, which superficially might appear quite arrogant. Chemists in pursuit of their profession are rarely, if ever, concerned with the equations of quantum chromodynamics (QCD). They take the existence and basic properties of atomic nuclei as given. For most chemical purposes it is adequate to approximate nuclei as pointlike concentrations of charge and mass. In a few applications nuclear spin also plays a role, but rarely any other aspect of nuclear structure. So in saying that QCD provides part of the "foundation" for chemistry, one means no more (and no less) than that it provides equations which in principle should allow one to derive the existence of nuclei, and to calculate a few of their properties, from a few proven properties of their constituent quarks and gluons. It does not thereby

directly solve, or even address, any properly chemical problems. In the same spirit, it might be said that acoustics provides the foundation for music or lexicography the foundation for literature.

As the inner frontier of the reductionist program has moved from explaining matter in terms of atoms to explaining atoms in terms of electrons and nuclei and then from explaining nuclei in terms of protons and neutrons to explaining protons and neutrons in terms of quarks and gluons, the models it creates have become ever more accurate and more broadly applicable. But with this progression, the domain of phenomena in which the new models provide qualitatively new insights, as opposed to better foundations, has grown increasingly remote from everyday life. Subatomic physics allowed us to understand and refine the basic principles of chemistry and to design materials with desired electric and magnetic properties; nuclear physics allowed us to understand the energy source of stars and the relative abundance of the elements; quark-gluon physics allowed us to understand the behavior of matter in the very early universe. Future developments may help us to penetrate more deeply into the early moments of the Big Bang or to recognize and understand yet undiscovered extreme astronomical environments, but apart from this, it is hard to anticipate their direct application to the natural world. It would be quite disingenuous to hold out the promise of economically significant new technologies based on future discoveries in high-energy physics.

From a broader perspective, however, the picture looks quite different. Over recent history, again and again fundamental, curiosity-driven research has led to unexpected developments and spin-offs whose economic value far exceeds the cost of the investments that spawned them. Sometimes the payoff was delayed by many decades and came from directions that no one remotely anticipated. The whole world of radio and wireless communication grew from Michael Faraday's vision of empty space as a dynamical medium and the experiments it inspired. Lasers and digital cameras grew from struggles of Max Planck and Albert Einstein to understand the strange wave-particle dualism of light-photons. Modern microelectronics, with all its ramifications, grew out of J. J. Thompson's discovery of electrons and the revolutionary insights of Niels Bohr, Werner Heisenberg, and Erwin Schrödinger in quantum theory.

Nor do we lack examples closer to the present, recognizably belonging to the modern era of high-energy physics. The central tool of the field, the accelerator, has become a ubiquitous medical device. The simplest and most familiar incarnation, perhaps, is the X-ray machine, but other particle beams are used in cancer therapy and for diagnosis. Who would have thought that reconciling quantum theory with special relativity would lead to important clinical technologies? Yet Paul Dirac's theory predicated antimatter, and positron emission tomography (PET scans) has become a powerful tool for looking inside the brain. Another fascinating application of accelerators is mass spectroscopy. The ability this technique supplies, to analyze accurately the chemical and isotopic composition of very small samples and thereby to characterize and date them, has supported significant contributions to geology, archaeology, and art history.

At this moment, synchotron light sources are providing new, cutting-edge tools for investigations in structural biology and chemical dynamics. For high-energy physics the production of synchotron radiation as an inevitable accompaniment of charged particle acceleration was regarded as a nuisance, draining energy from the particles of interest. But it turns out that this "waste product" allows scientists to look at molecules with unprecedented resolution in space and time. So now special accelerators are designed specifically to be sources of synchotron radiation. The new windows they are opening will undoubtedly reveal extraordinary new vistas.

Besides its direct impact, the development of high-energy accelerators has also spurred progress in a number of supporting technologies. Notably, these accelerators require large powerful magnets to guide the particle orbits. Such magnets have become the workhorse of magnetic resonance imaging (MRI), another major medical technology.

A completely unanticipated, quite recent spin-off may become the most important of all. Modern high-energy physics experiments typically involve many tens or even hundreds of collaborators, who must share their data and their analyses. It was to facilitate this process that Tim Berners-Lee, a software engineer working at CERN, developed the concept of the World Wide Web and the first browser-editor, thus initiating the Internet revolution. Many other innovations in high-speed electronics, less well known but central to commercial computing and communication technology, were developed in response to the challenges of guiding vast numbers of particles moving at velocities very close to the speed of light and interpreting the complicated results their collisions produce.

More difficult to identify specifically, but also important, are spin-offs from conceptual developments in high-energy physics. Quantum field theory was developed as the rigorous language of elementary processes but also turns out to be the appropriate tool to understand superconductivity. The renormalization group, first developed as a technical tool within quantum field theory, turns out to be the key to understanding phase transitions and is playing a dominant role in the emerging theories of pattern formation, chaos, and turbulence.

Why do such valuable surprises occur so regularly? Can more be anticipated in the future? There is a simple, yet profound explanation. In essence, it was put forward by William James, who spoke of "the moral equivalent of war." It is the fact that human beings can be inspired by difficult problems and challenges to work very hard and selflessly and to find more in themselves than they knew existed. Especially in youth, they even seem driven to seek—or manufacture!—such problems. Perhaps evolution selected this ability to rise to the occasion partly in response to the pressures of human conflict. In any case, we should cherish the opportunity to direct it into constructive channels.

Certainly, high-energy physics offers an abundance of tough challenges. Ultimate questions about the closure of fundamental dynamical laws and the origin of the observed universe begin to seem accessible. Tantalizing hints point toward new worlds of phenomena involving supersymmetry and unified field theory, but present ideas contain many loose ends and unsatisfactory details. The great challenge of reconciling general relativity with quantum mechanics might be met with superstring theory, but as yet this is far from reaching fruition in specific world-models. And the great embarrassment of the cosmological term, whose measured value is many orders of magnitude smaller than current theories suggest, threatens to upset the whole applecart. On the experimental side the challenges are more tangible and no less awesome. The next generation of accelerators will be engineering projects of grandeur, both in their size and in their precision. They will be modern civilization's answer to the pyramids of Egypt, but nobler, built to improve our understanding rather than to appease superstition and tyrannical theocracy. We must learn how to handle the tremendous flow of data these accelerators will generate. The ATLAS experiment already planned for CERN's Large Hadron Collider is expected to collect 10^{15} bytes/year—equivalent to a million human genomes. Amidst this torrent we must identify the fraction, probably a mere trickle, which does not fit the Standard Model. New ultrafast methods of communication and computation will need to be developed. It would be surprising if the effort of rising to these challenges did not produce some spectacular by-products.

In short, the economic fruits of fundamental investigation, though unpredictable in detail, have arrived with wonderful reliability and have been reliably wonderful. Investment in this area is ultimately an investment in people, specifically in the power of great problems to inspire great efforts.

In this connection, it is appropriate to emphasize that the human effects of big scientific projects ramify far beyond their immediate research community. Construction of a modern high-energy accelerator, its detectors, and its information infrastructure brings engineers into intimate contact with exotic frontiers of technology and with problems of a quite different nature from those they would ordinarily encounter. Also, most of the young people going into these projects will not find permanent academic employment. They enter this life with open eyes, foregoing security for the opportunity to participate in something great. When these engineers and researchers return to the outside world, they bring with them unique skills and experience.

Finally, the visible commitment of society to high-profile scientific endeavors sends an important message to young people considering what career to enter, encouraging them in scientific and technological directions. This is important, since our soci-

ety needs capable scientists and engineers, and they are always in great demand.

In addition to spin-offs and indirect benefits, there is also the intrinsic worth of the prospective knowledge. There are several identifiable questions that seem ripe for progress.

Universal Condensate and the Origin of Mass

The theory of the weak and electromagnetic interactions postulates that what is ordinarily regarded as empty space is in fact filled with a pervasive condensate. It is only by interacting with this condensate that many particles, notably including the W and Z bosons, which mediate the weak interaction, acquire their mass. Although the theory is extremely successful, this central aspect has not been tested directly. Physicists hope to excite the condensate, either producing so-called Higgs particles, or revealing some more complex structure.

Unification of the Theory of Matter

The Standard Model, containing both the theory of weak and electromagnetic interactions and QCD (the theory of the strong interaction) provides a remarkably complete theory of the behavior of matter. The different pieces of the Standard Model have related mathematical structures, embodying various symmetries, and it is tempting to speculate that there is a master symmetry encompassing them all. There appears to be a compelling candidate for such a "grand unified" symmetry. Will it hold up to further scrutiny?

Supersymmetry

The unification mentioned above, when pursued quantitatively, requires another important addition: supersymmetry. This idea postulates the addition of extra quantum-mechanical dimensions. Motion of particles in these dimensions will make them appear to be particles with quite different, but broadly predictable, properties. So far none has been found, but according to theory they must begin to show up in higher-energy collisions.

The Arrow of Time

A few exceptional microscopic processes that exhibit a preferred direction in time (that look different when run backward) have been observed. This phenomenon is vital to understanding how the cosmic asymmetry between matter and antimatter arose. To understand it properly, we need to see more examples of how it works, especially at high energy.

Unification with Gravity

Gravity is not deeply integrated into the Standard Model or even its unified extensions. But there are bold ideas for how a completely unified theory, including both the Theory of Matter and gravity, might be constructed. Some of these ideas lead to predictions of new particles, and patterns among their masses, that could be observed. In this way, we might for the first time acquire empirical information on the nature of quantum gravity, or indications of the existence of extra curled-up spatial dimensions.

Transcending whatever specific answers it supplies, continued pursuit of the reductionist program expresses society's commitment to some of the deepest ideals of our scientific culture: to pursue the truth wherever it leads; to ground our working picture of nature in empirical realities and challenge that picture; and to see whether the marvelous simplicity and mathematical beauty of the description that has emerged from previous investigations can be refined further, or whether it reaches some limit.

See also: CULTURE AND PARTICLE PHYSICS; INFLUENCE ON SCIENCE; INTERNATIONAL NATURE OF PARTICLE PHYSICS; PHILOSOPHY AND PARTICLE PHYSICS

Bibliography

Berners-Lee, T. *Weaving the Web* (Harper, San Francisco, 2000).

DESY. "Conceptual Design Report for the TESLA Test Facility Free-Electron Laser." <http://tesla.desy.de/TTF_Report>.

Winstein, B., et al. *Elementary-Particle Physics: Revealing the Secrets of Matter* (National Academy Press, Washington, D.C., 1998).

Frank Wilczek

BETA DECAY

See RADIOACITIVITY

BIG BANG

The starting point for the Big Bang theory is Einstein's theory of general relativity in combination with the cosmological principle. General relativity is a metric theory of gravity, now verified to high precision via observations of the binary pulsar. The cosmological principle has also been verified to exquisite precision, as far as this can be achieved.

The cosmological principle asserts that the universe is statistically isotropic and homogeneous for any observer. The cosmic microwave background has demonstrated isotropy to a level of order 1 part in 100,000. There is no preferred direction such as might be associated with the geometrical center of the universe. Homogeneity has been demonstrated by galaxy redshift surveys that provide three-dimensional maps of the universe, given the strong empirical correlation discovered by Edwin Hubble in 1929 between redshift and distance. As one probes deeper and deeper into the universe, to distances as great as several gigaparsecs (1 parsec = 3.2 light-years), the density of galaxies is found to be uniform. Humankind definitely does not inhabit a fractal universe of vanishingly low density in the mean, as some have argued. One can set a limit on any large-scale nonuniformities of approximately 10 percent; otherwise, excessive perturbations would be induced in the Hubble diagram of galaxy redshift versus distance.

Friedmann-Lemaître Cosmology

The cosmological principle applied to the Einstein gravitational field equations led to a remarkable simplification. In 1917 Einstein found a static cosmological model that could only be prevented from collapsing under the relentless tug of gravity by invoking a repulsion force. This force was enshrined as the cosmological constant, a term that has no counterpart and no effect in Newtonian gravity, but is important only on cosmological scales. In modern parlance, one identifies the cosmological constant with the energy of the vacuum, and its introduction leads to the Einstein static universe.

In fact, Einstein had overlooked the only true cosmological solution to the field equations that satisfied the cosmological principle and that did not require the introduction of a cosmological constant. His mistake was soon rectified by Alexander Friedmann in 1924 and independently by Georges Lemaître in 1927, who discovered the expanding universe cosmological models.

The expanding universe was much later dubbed the Big Bang for the simple reason that it expanded from a pointlike singularity of infinite density. It was realized at the outset that this singularity was a mathematical artifact indicative of missing physics that was only supplied half a century later. Space itself was uniform, unbounded, and expanding. There was no center and no edge to space.

The Big Bang theory indeed predicted the expansion of the universe, a result that many scientists in the early decades of the twentieth century, including Hubble himself, found too radical to accept. The equations that describe the evolution of the universe come from Einstein's theory of general relativity. To describe expansion, one introduces the scale factor $a(t)$. Physical distance is $d = ra(t)$, where r is coordinate or comoving distance, just a fixed number conventionally evaluated with respect to a mass-scale at present. The Einstein equations are

$$G_{\mu\nu} = 8\pi G T_{\mu\nu} + \Lambda g_{\mu\nu}.$$

Here, $g_{\mu\nu}$ is the metric of the universe, which is incorporated into the Einstein tensor $G_{\mu\nu}$ to describe gravity, and $T_{\mu\nu}$ is the energy-momentum tensor that describes the matter and radiation content of the universe that acts as the source of gravity. Another important source of gravity that corresponds to the density of the vacuum is the cosmological constant Λ.

Under the cosmological principle, equivalent to spherical symmetry about every point, the Friedmann-Lemaître equation of cosmology is obtained. This can be cast in the form of a cosmic energy equation:

$$\dot{a}^2 - \frac{1}{3}(8\pi G\rho + \Lambda)a^2 = -k.$$

The first term on the left describes the kinetic energy of a shell of matter, and the second term is its gravitational potential energy. There are three distinct solutions in the absence of the cosmological constant term. These are conveniently described by the curvature of space that in the Newtonian limit corresponds to the total energy of an arbitrarily placed expanding shell of matter. The shell may have either zero total energy, in which case space is flat; negative energy, in which case space is positively curved like a spheroidal surface; or positive energy, which results in space being negatively curved like the surface of a hyperboloid. The constant $-k$ represents the total energy of the shell per unit mass and may be -1, 0, or -1, corresponding to a universe of negative, zero, or positive energy. The flat and negatively curved spaces are infinite, and only the positively curved space is finite.

One can discriminate between the three solutions of the Friedmann-Lemaître equation by introducing the critical density. This is the density of the flat or Einstein–de Sitter universe and is equal to $\frac{3H_0^2}{8\pi G}$. Here, H_0 is Hubble's constant. Because of the matter content, a universe without a cosmological constant is decelerating. At early times, the three spaces are indistinguishable. Only at late times do they deviate from one another, the negative-energy spatially closed model decelerating more strongly than the other models before reaching its maximum extent and then recollapsing to a future singularity.

To further examine the deceleration, one may apply conservation of mass-energy that leads to

$$\dot{\rho} = -(\rho + p)\,\frac{\dot{a}}{a},$$

and an equation can now be derived for the deceleration of the universe:

$$\frac{\ddot{a}}{a} = \frac{-4\pi G(\rho + 3p)}{3 + \dfrac{\Lambda}{3}}.$$

From the deceleration equation, one learns that the universe actually accelerates if a negative energy condition is satisfied, $\rho + 3p \leq 0$. Indeed, the cosmological constant satisfies $\rho + 3p = 0$, and the solution is $a(t) \propto \exp\dfrac{\sqrt{\Lambda t}}{3}$. In this solution, the density is constant, equal to $\Omega_\Lambda \equiv \Lambda/H^2$, where Ω_Λ is the vacuum density corresponding to the cosmological constant relative to that of the $k = 0$ universe with density $\dfrac{3H_0^2}{8\pi G}$, and H_0 is Hubble's constant, $\dfrac{\dot{a}}{a}$ at present.

The Friedmann-Lemaître equation today reduces to

$$\Omega_{tot} \equiv \Omega_{mat} + \Omega_{rad} + \Omega_\Lambda - 1 = k/a^2 H^2$$

where the mass density of matter $\rho_{mat} \propto a^{-3}$ and radiation $\rho_{rad} \propto a^{-4}$. Three possibilities for the pressure content of the universe are $p = p_m \ll \rho c^2$, applicable since the epoch of matter-radiation decoupling; $p = \rho_{rad} c^2 / 3$, applicable prior to matter-radiation decoupling; and $p = w\rho$, a generalization of the cosmological constant to an arbitrary equation of state. In this latter case, $\rho \propto a^{-3(1+w)}$, with $w = -1$ corresponding to the case of the cosmological constant. The universe is radiation-dominated prior to $a/a_0 = \Omega_{rad}/\Omega_{mat}$ or at redshift larger than

$$1 + z_{eq} = a_0/a(t_{eq}) = 3{,}000(\Omega_{mat}/0.3).$$

The Distance Scale

Lemaître formulated one of the greatest predictions of modern physics, that the universe should be expanding, into a relation that expressed the proportionality between the recession velocity of a distant galaxy and its distance. In 1929 Hubble verified the redshift-distance relation, which became known as Hubble's law, $v = H_0 d$, where v is recession velocity and d is luminosity distance. The latter is measured by identifying a class of luminous variable stars, Cepheids, that were used to establish the size of the Milky Way galaxy and, more recently, the distances to its nearest neighbor galaxies. Hubble used the brightest stars in more distant galaxies as his basic distance indicators. He explored a region that extends to the Virgo cluster of galaxies. With hindsight, one knows that Hubble's distance indicators were erroneous, since he could not distinguish HII regions from stars. It is also known that the region between the Earth and the Virgo cluster, where Hubble's galaxies were located, is dominated by random motions. The uniformity of the universe only becomes manifest beyond Virgo. Nevertheless, Hubble in 1929 announced his discovery of the redshift-distance law.

The redshift was produced by the Doppler effect and resulted in a systematic displacement toward longer wavelengths for a receding galaxy. Blueshifts would be indicative of approach; only a few of the nearest galaxies have blue-shifted spectra.

The prevalence of galaxy redshifts had been discovered in the first decades of the twentieth century by Vesto Slipher. The fainter the galaxy, on the average, the larger its redshift. However, the observers who tried to understand the relation between distance and redshifts paid too much attention to the theoretical cosmologists, who only knew about the possibility of redshifts in the de Sitter universe. The de Sitter cosmological model was a strange beast. It was an empty universe in which the distance depended exponentially on redshift. The nearby galaxies in this model displayed a quadratic dependence of distance on redshift. To his credit, Hubble did not care a great deal about theory. He reevaluated distances more precisely than his predecessors had done and inferred the linear relationship that is known as Hubble's law. To his dying day, Hubble could not accept that the universe was expanding, despite the prediction of Lemaître and Friedmann before him. Within a year of Hubble's announcement, most of the cosmological community seized on Hubble's law to infer that space was expanding.

It is difficult in retrospect to understand how Hubble inferred a linear law, given the enormous uncertainties in galaxy distances and the fact that Hubble only initially sampled such a small volume of the universe. Hubble's constant is measured in units of velocity per unit distance, in effect an inverse time. Hubble inferred a value of 600 km/s/Mpc. The modern value of H_0 is smaller by an order of magnitude, amounting to 70 km/s/Mpc with an uncertainty of about 15 percent. The inferred timescale $1/H_0$ is a measure of the age of the universe if no deceleration (or acceleration) has occurred. The age inferred from Hubble's measurement was approximately 1.5 billion years and far less than the known age of the Earth. Hence, many astronomers were at first reluctant to accept the expanding universe interpretation.

What changed? First, the cosmologists were very ingenious. Under the influence of Lemaître, the cosmological constant, first introduced by Einstein to make the universe static, was reintroduced. Eddington and Lemaître advocated a universe that began from a static phase that would last as long as necessary before beginning to expand. Lemaître showed that galaxies could form in such a universe. A variant was an expanding universe that underwent an extended coasting phase as a consequence of the effect of the cosmological constant, with expansion eventually taking over. Such approaches greatly extended the age of the universe.

Most significantly, however, the observers revised the distance scale. This came about in part by recognition of Hubble's significant error in confusing the brightest stars with giant HII regions. Alan Sandage from 1960 onward was primarily responsible for developing a new distance calibrator that made use of the brightest galaxies in clusters as standard candles. This enabled him to probe the universe to great distances and to reduce Hubble's constant to 200 km/s/Mpc. A major breakthrough occurred when Walter Baade recognized that there are two types of Cepheid variable stars. The confusion between the two types only dissipated when Baade succeeded in identifying populations I and II in the Andromeda galaxy and realized that there were two types of Cepheids which differed appreciably in luminosity. He was able to double the distance scale. The remaining improvements happened more slowly. For nearly 40 years, cosmologists debated Hubble's constant within the range of 50 to 100 km/s/Mpc. Resolution came when the Hubble Space Telescope was able to resolve Cepheids in several galaxies outside our Local Group, in which supernovae were also found. The supernovae are of a type associated with the merger of a pair of white dwarfs that explode catastrophically once the Chandrasekhar mass limit on a white dwarf is exceeded. These SNeIa are found in old stellar populations in both spiral and elliptical galaxies and are luminous enough to be detectable at the edge of the observable universe and also to be reliable distance indicators. Type Ia supernovae seem to have identical luminosities, amounting to the light from a billion suns at maximum light and fading away after a year. The light curve is interpreted as resulting from the radioactive decay of 0.6 solar mass of Ni^{56} produced in core collapse and provides the energy source for an ideal standard candle.

The Age of the Universe

Type Ia supernovae have been detected out to a look-back time of half the present age of the universe, from when light was redshifted by a factor of 2 in wavelength. The distance measurements are precise enough (to 15%) that acceleration of the universe has now been confirmed. Deviations from Hubble's linear law are found for the most distant supernovae. The measured age of the universe, inferred from Hubble's constant and the measured acceleration, is 15 billion years.

There are two completely independent measures of the age of the universe. Radioactive dating via thorium and uranium isotope measurements is applied to the abundances in old halo stars. Both $thorium^{232}$ with a half-life of 14 Gyr and $uranium^{238}$ with a half-life of 4.5 Gyr have been detected in two halo stars, measured with the world's largest telescopes. Nuclear astrophysics theory provides an estimate of the initial abundances relative to iron. The observed ratio provides an estimate of the age of the universe since the supernova synthesized these elements and ejected them in the debris that eventually was incorporated into the molecular clouds from which stars such as the Sun formed.

Another age determination comes from application of the theory of stellar evolution to globular clusters. Globular star clusters are systems of millions of stars that predate our galaxy. One knows they are old because the abundances of metals as measured in stellar spectra are low compared to those in the Sun. Hence, the globular clusters must have formed long before the Sun. As stars radiate energy by thermonuclear burning of hydrogen into helium, they evolve in luminosity, becoming brighter as the fossil fuel is gradually exhausted and the central temperature rises. Heavier elements are burnt, first helium, then carbon, to provide the central temperature and pressure. Once the nuclear fuel supply of hydrogen, helium, and carbon is exhausted, the star soon runs out of fuel.

If the star initially weighed less than 8 solar masses, its final fate as the core heats up is that its envelope swells. The star becomes a luminous supergiant. The outer shell is expelled to become visible as a planetary nebula. The ejecta slowly, after some 10^4 years, fade away, and a white dwarf is all that remains. If the star initially weighs more than 8 solar masses, its central pressure builds up to a level that the star core implodes via neutron capture and neutrino emission. A neutron star forms in the core, and the release of binding energy drives a supernova explosion of Type II.

In a globular cluster, the stars formed coevally, and so one has a snapshot of stars of different masses that have reached differing evolutionary points. One can thereby infer the age of the globular cluster from comparison with models of stellar evolution, the best estimate being 13 billion years. To this must be added about 1 billion years for the delay between the Big Bang and the formation of the globular cluster, to give an age for the universe of some 14 billion years. Remarkably, this agrees well with the independently determined ages from cosmology and uranium or thorium decays.

Cosmic Acceleration and Dark Energy

What causes the acceleration? Cosmologists have gone full circle, ending up with a value of the cosmological constant about 30 percent smaller than Einstein originally introduced for the static universe. One can interpret Λ, the cosmological constant, as a constant energy density of the vacuum that has only recently begun to dominate the mass density of the universe. One does not observe any such energy directly. Hence, it is often referred to as dark energy. The matter density decreases as the universe expands. When the universe was about one-quarter of its present size, at redshift 4, the dark energy first became comparable to the matter density. One consequence is that the universe switched from deceleration under the influence of the gravitational attraction of matter to acceleration under the influence of the gravitational repulsion of the dark energy. The universe began to accelerate.

Dark energy produces acceleration because it has a large negative pressure, indeed $p = -\rho c^2$, where p is the dark energy pressure, and ρ is the dark energy density. In a normal gas, pressure is positive, and Einstein's theory of relativity predicts that its contribution to gravity is attractive. Ordinary gas pressure acts as a source of gravity.

The ultimate fate of black hole formation by a collapsing massive star cannot be avoided by the ac-

tion of gas pressure; in fact, it is enhanced. In the expanding universe, positive pressure produces deceleration, as does matter. As the universe expands, ordinary pressure does less and less work and produces less and less heat energy. However, negative pressure has the opposite effect. An elastic string when expanded gains energy. More energy means that the pressure of an elastic string is negative. In the expanding universe, negative pressure accordingly acts in the opposite way to positive pressure: more and more work is done as the universe expands. This is what drives acceleration. Negative pressure acts like antigravity: it is repulsive.

Dark energy accounts for two-thirds of the mass-energy density of the universe. There is no explanation for dark energy; it can simply be regarded as a contribution to the energy of the vacuum. Dark energy is completely uniform and does not cluster under the effect of gravity as does ordinary matter. It is only detectable via its effect on the acceleration of the expansion of the universe. In terms of fundamental units, the energy density associated with the cosmological constant is remarkably small, amounting to $10^{-121}\, m_{pl}^4$, where the Planck mass m_{pl} is 1.2×10^{19} GeV. In conventional units, where the cosmological constant is an inverse square length, its magnitude is naturally the inverse Hubble length squared, or $10^{-56}\mathrm{cm}^{-2}$ or $10^{-121}\, \lambda_{pl}^{-2}$.

Dark Matter

Dark matter, in contrast, is detectable. And it amounts to about a third of the total mass-energy density of the universe. The cosmic mass budget is best expressed with respect to the critical density for a universe that is spatially flat, the Einstein–de Sitter model, namely $\dfrac{3H_0^2}{8\pi G}$. This can be expressed as $3 \times 10^{11} h \mathrm{M}_\odot\, \mathrm{Mpc}^{-3}$. The luminosity density of the universe is measured to be $2 \times 10^8 h\, \mathrm{L}_\odot\, \mathrm{Mpc}^{-3}$. The mass-to-light ratio for closure is therefore $1{,}500 h \mathrm{M}_\odot/\mathrm{L}_\odot$. This is a clear prediction for closure of the universe.

What is actually measured is far less. Galaxy clusters gave the first indication of the prevalence of dark matter on large scales as early as 1933. The first reliable values, however, came from galaxy rotation curves, which provided proof of dark matter dominance in ordinary galaxies and, in particular, in the

Milky Way galaxy. The rotation curves for large spiral galaxies are generally flat at large distances, indicating that far from the Keplerian expectation, if mass traces light, the mass in fact increases with increasing galactocentric radius, $M(<r) \propto r$. Typical values of the mass-to-light ratio are $100h\text{M}_\odot/\text{L}_\odot$, whereas within the half-light radius, one finds a value of approximately $10h\text{M}_\odot/\text{L}_\odot$, the actual value depending slightly on the type of galaxy. Galaxy rotation curves are measured at low resolution via radio techniques using the 21 cm of atomic H and at high resolution in the optical band by $H\alpha$ emission lines. Consistent results are obtained, and dark matter is found to be ubiquitous on scales of up to 100 kpc.

In galaxy clusters, great progress has been made since the early determinations that used the virial theorem applied to the optically measured dispersion of radial velocities of cluster galaxies. Two independent techniques confirm the dynamical measurements. One utilizes X-ray measurements of the hot intracluster gas that is assumed to be in hydrostatic equilibrium, and another makes use of the gravitational lensing by the cluster of remote background galaxies and the consequent image distortions. All three methods consistently yield a value of $300h\text{M}_\odot/\text{L}_\odot$. The scale probed is 1 Mpc.

On larger scales, there are no equilibrium gravitationally bound structures that can be reliably probed. One method utilizes the infall motion of galaxies into the Virgo Supercluster. This probes the dark matter density on scales of up to 20 Mpc. Another probe of the dark matter density on even larger scales, up to 100 Mpc, makes use of the variance in the counts of galaxies obtained in large-scale galaxy redshift surveys. The clustering of the galaxy distribution on large scales is measured by fluctuations in the galaxy counts averaged over randomly placed spheres. The matter on large enough scales must be correlated with the light. The fluctuations inferred in the matter density provide a gravitational source that induces perturbations in the Hubble flow, observable as random motions of galaxies and of galaxy clusters. The observed Hubble flow dispersion requires a value of the mass-to-light ratio that is equivalent to $\Omega_m \approx 0.3$, in agreement with the mass-to-light ratio inferred for galaxy clusters. Were the universe at critical density, much larger

Hubble flow distortions and galaxy peculiar velocities and cluster streaming motions would be observed amounting to 1,000 or more km/s. The observed random motions of galaxies amount to approximately 300 km/s. This method probes the dark matter out to 100 Mpc.

Similar conclusions are reached from studies of the peculiar velocity pattern of galaxies in the Virgo Supercluster, from large-area weak lensing of high redshift galaxies, and from the redshift evolution of the number density of clusters. The rich cluster abundance above a given mass is observed to only increase slowly as the universe expands. The theory of cluster formation predicts a rapid increase of the massive cluster abundance in a critical density universe, due to the growth of density fluctuations driven by gravitational instability, and this effect is systematically suppressed if the density of the universe is below the critical value.

Cosmic Blackbody Radiation

Only about 10 percent of the dark matter in the universe is baryonic. Nucleosynthesis of the light elements was predicted by George Gamow and his collaborators in the 1940s. This necessitated a hot origin to the universe and led, in turn, to the prediction of the cosmic radiation background by Ralph Alpher, George Gamow, and Robert Herman in 1950. The blackbody, and hence microwave nature, of the cosmological background radiation was first appreciated by Andrei Doroshkevich and Igor Novikov in 1964, and by Robert Dicke and his collaborators in 1965. The search by the latter group was overtaken by the contemporaneous discovery by Arno Penzias and Robert Wilson of the Cosmic Microwave Background (CMB) in 1965.

In 1990 the COBE satellite confirmed the blackbody nature of the CMB to remarkable precision. No deviations to the blackbody spectrum are found to within a fraction of a percent. This provides eloquent testimony to a hot origin for the universe when matter and radiation were in thermal equilibrium. The blackbody temperature is measured to be 2.728 K, with an uncertainty of only 0.004 K. Gamow had already laid down the key ingredient of a hot universe. The present-day universe is cold and dominated by matter. But, the observed radiation density, while today only amounting to $\Omega_{rad} = 10^{-5}$ in mass density, would have dominated in the past as a consequence of the redshifting of the photon energy during expansion.

Baryon Density

Modern determinations of the abundances of He^4, He^3, D, and Li^7 are found to be consistent with a Big Bang origin, and, now that the CMB blackbody temperature is measured, they provide an accurate accounting of the primordial baryon abundance. One finds that $\Omega_{baryon}h^2 = 0.02$, with an uncertainty of only 10 percent. Independent confirmation of the baryon fraction in the universe comes from studies of the intergalactic medium at two distinct epochs. At high redshift one sees intergalactic neutral hydrogen in absorption in the spectra of quasars. The gas exists in vast numbers of clouds and filaments, and one needs to apply the ionizing photon flux, measured directly via the quasar emission spectra, to infer the total amount of intergalactic gas. At low redshift, the hot intracluster gas in galaxy clusters is measured via its X-ray emission flux to be approximately 10 percent of the total cluster mass. Since clusters are considered to be sufficiently massive to have preserved their original baryon content, one can also deduce the baryon content of the nearby universe. Both methods agree. There is a problem, however. One can only account for about half of the predicted baryon fraction today in known sources such as stars and diffuse intergalactic gas. There is also a dark baryonic matter problem.

Thermal History

The discovery of the CMB led to some remarkable insights into the beginning of the universe. The Big Bang was once a fireball. Only after redshift Ω_m/Ω_{rad}, about 3×10^4, did the universe become matter-dominated. Only in a universe dominated by matter could the density fluctuations be gravitationally unstable and grow in strength. Moreover, Stephen Hawking and Roger Penrose derived a theorem which proved that as a consequence of the dense past of the universe (inferred in order for the CMB to have thermalized and been isotropized by the photons scattering off free electrons, then under classi-

cal general relativity) the universe must inevitably have undergone a past singularity.

One could now begin to reconstruct the thermal history of the universe. Quantum gravity supplants general relativity on the Planck scale at an epoch of 10^{-43} s or at a temperature of 10^{19} GeV. This is where unification of the four fundamental forces—electromagnetic, weak nuclear, strong nuclear, and gravity—occurred. There is as yet no preferred theory for this regime, although higher-dimensional theories of quantum gravity include models in which Planck scale physics is manifest at TeV energy scales. As the universe expanded and cooled below the Planck scale, the ensuing evolution can be sketched as follows.

Above 10^{16} GeV, the electromagnetic, weak and strong nuclear forces were indistinguishable and of equal strength. This was the grand unification (GUT) era. As the temperature dropped below 10^{16} GeV, the symmetry of grand unification was spontaneously broken. The resulting change in phase of the matter content of the universe involved the transient appearance of a scalar energy field that was responsible for the inflation of the universe. The universe expanded exponentially as long as this so-called inflaton field was the dominant source of energy density. The universe is then 10^{-36} s old. The inflaton is similar to the cosmological constant, except that its energy density was larger by about 120 factors of 10. The potential energy of the inflaton field dominates the kinetic energy, and this provides the constant energy density that drives the universe to inflate. The potential energy drops (by design), and inflation ends by about 10^{-35} s. The enormous kinetic energy thermalizes, or turns into heat, and one is now again in the conventional hot Big Bang phase, initially dominated by radiation and relativistic particles.

At 100 GeV, the electroweak forces decouple, and the fundamental force strengths subsequently resemble those observed today, with the nuclear forces being strong and short-range compared to the feebler electromagnetic force and the vastly weaker gravitational interaction. At this epoch, the change in phase of the universe helps generate a small asymmetry in the baryon number, the number of particles minus antiparticles. The baryon number is expressed in dimensionless form as $(N - \bar{N})/(N +$

$\bar{N}$) and is only 10^{-9}. However, as the temperature drops further, all the strongly interacting particles annihilate into radiation. The radiation redshifts to become the CMB. There are 10^9 CMB photons per proton in the universe. The relic particles freeze out of thermal equilibrium once the temperature drops below a fraction of the particle mass. Very few $p\bar{p}$ pairs survive, because of the strong interactions that annihilate almost all the pairs. The baryon excess, however, means that the baryons, which have no anti-baryon counterparts, do survive to become the present matter content of the universe. The observed universe consists almost exclusively of matter: the antimatter content is less than a hundredth of a percent, otherwise, one would see gamma rays from matter-antimatter annihilations.

If stable weakly interacting particles were present, these would freeze out in substantial numbers, regardless of whether there was any primordial asymmetry. The lightest supersymmetric particle or neutralino is such a possible stable relic. Its abundance is determined by its annihilation cross section, so that $\Omega_\chi \approx 10^{-38} \text{cm}^2/<\sigma v>$. For typical values of the weak cross section, the neutralino is a viable candidate for the nonbaryonic dark matter in the universe. Typical predicted mass scales are of order 0.1 to 1 TeV, the preferred supersymmetry energy scale.

The universe is now a soup of quarks, gluons, electron-positron pairs, neutrinos, and photons. At about 200 MeV, another phase transition occurs when the quarks and gluons form hadrons. The universe now contains protons and neutrons in thermal equilibrium with $n_n/n_p \approx e^{-\Delta m/kT}$, or about 0.1, where δm is the mass difference between proton and neutron. Once the temperature drops below 1 MeV, the neutron-producing reactions stop, and neutrons freeze out. At 0.5 MeV, e^+e^- pairs annihilate and neutrinos freeze out. The stage is now set for nucleosynthesis of the light elements that commences at 0.1 Mev or 10^9 K, when deuterium nuclei can first form. Subsequent reactions produce He^3, He^4, D, Li^7, all of which are generated in abundances that are measurable today in primordial environments. Lack of stable nuclei at masses 5 and 8 means that nucleosynthesis peters out after He^4 is synthesized. The predicted primordial He^4 simply incorporates all the neutrinos: $Y = 2n_n(n_p - n_n)^{-1} \approx 0.25$ by mass.

One expects to find primordial helium in such unprocessed environments as the intergalactic medium, the outermost parts of galaxies, metal-poor galaxies, and even with suitable extrapolation, meteorites, and the atmosphere of Jupiter. All abundances are consistent with a universal baryon fraction $\Omega_b h^2 = 0.02$ to within 10 percent. The universe remained dense and hot enough for the thermal equilibrium of matter and radiation to be maintained until an epoch of about 1 month. This was when the cosmic blackbody radiation was effectively generated. Any spectral distortions would probe the physics of the universe at this epoch.

The temperature continued to drop. The hydrogen is ionized and the radiation scatters frequently. There are 10^9 photons for every baryon, and these suffice to keep the hydrogen fully ionized until the temperature drops below 0.2 eV. At this point, there are too few photons with energy above the hydrogen ionization threshold of 13.6 eV to keep the hydrogen fully ionized. The protons and electrons combine to form hydrogen atoms. Unlike free electrons, these are very poor scatterers of electromagnetic radiation. Scattering of the photons abruptly stops. The universe is now transparent to the CMB, from a redshift of 1,000 or 300,000 years after the Big Bang, to the present.

CMB Fluctuations

Measurement of the 2.728 Kelvin blackbody radiation spectrum confirms the thermal history of the universe back to an epoch of a few days after the Big Bang. Detection of temperature fluctuations at a level of $\delta T/T \sim 10^{-5}$ has revealed the irregularities at the epoch of last scattering that trace the density fluctuations from which large-scale structure evolved. The primary temperature anisotropies that emerge from last scattering are measured at angular scales that range from the dipole (180 degrees), associated with motion relative to the CMB frame, to a few minutes of arc, which are induced at the moment of last scattering.

The density fluctuations prior to last scattering are like sound waves in a medium with a sound velocity approaching that appropriate to that of a relativistic plasma, $c/\sqrt{3}$. After last scattering, the radiation thermally decouples, and the sound speed drops to that of a gas at a few thousand degrees Kelvin. This means that the density fluctuations, which previously were pressure-driven sound waves, now respond only to gravity, the pressure being completely unimportant at least for fluctuations that contain the masses of even the smallest galaxies. Indeed, the minimum size for gravity to dominate, and thus for the first self-gravitating gas clouds to form, is about a million solar masses. As time proceeds, the clouds build up in mass by clustering together under the action of gravity to form a galaxy and eventually cluster mass clouds. The galaxy mass clouds are able to cool and fragment into stars. One ends up with galaxies and clusters of galaxies, the latter containing large amounts of gas that is too hot to have cooled.

The sound waves leave a remarkable imprint on the CMB. Inflation, or some equivalent theory, generates these waves that just begin to undergo their first compression peak when they enter the horizon. The wavelength simply spans the distance traveled by light since the Big Bang. Such waves that are cresting at last scattering for the first time have the largest amplitude. They produce a peak in the CMB fluctuations at an angular scale corresponding to the horizon scale at last scattering, about 1 degree. Shorter waves that are cresting for the second time at last scattering are amplified less and leave a smaller angular scale peak. Waves undergoing their first rarefaction also leave a peak on an intermediate scale, since rarefactions are measured in quadrature as fluctuations that are either negative or positive, the density field being random. There are a series of peaks predicted to be of decreasing strength until one reaches wavelengths that are so inefficient at scattering the radiation that there are no further fluctuations. It is then said the fluctuations are damped out, and this occurs at a physical scale corresponding to the thickness of the last scattering epoch, the distance a primordial sound wave could travel over the time the universe undergoes the transition from ionized to neutral. This amounts to approximately 30,000 years, so the smallest surviving primary fluctuations are on a scale of about one-tenth of a degree.

A series of peaks have been measured in the CMB temperature fluctuations. The first, second, and third peaks have been detected. The angular position of the peaks is sensitive to the curvature of the universe. If one lived, for example, in an open universe with hyperbolic geometry, the peaks are shifted to smaller angular scales, the universe acting like a giant concave lens. This effect is not observed: the universe is found to be flat to within an accuracy of 10 percent, in terms of the critical energy density, $\Omega_m + \Omega_\Lambda \approx 1$.

The detection of the acoustic peaks is another independent confirmation of the dominance of nonbaryonic dark matter in the universe; the peaks are produced by baryons, scattering by electrons. From their strength, a value $\Omega_b \approx 0.04$ is independently inferred. $\Omega_m \approx 0.3$ is required in order to have enough fluctuation growth in the early universe to make the fluctuations as small as they are observed. From the locations of the peaks, the equation of state is also measured, and one infers from both large-scale structure and cosmic microwave background observations that w is less than approximately -0.5, not far from the value corresponding to the cosmological constant. Hence, an independent confirmation of Λ holds: for the universe to be flat, $\Omega_\Lambda \approx 0.7$. This constitutes the concordance model of the Big Bang.

See also: ASTROPHYSICS; BIG BANG NUCLEOSYNTHESIS; COSMOLOGY; HUBBLE CONSTANT; INFLATION

Bibliography

Hu, Wayne. "The Physics of Microwave Background Anisotropies." <http://background.uchicago.edu/>.

NASA. "Cosmology: The Study of the Universe." <http://map.gsfc.nasa.gov/m_uni.html>.

"The N-Body Site." <http://star-www.dur.ac.uk/~moore/>.

"Ned Wright's Cosmology Tutorial." <http://www.astro.ucla.edu/~wright/cosmo_01.htm>.

Raine, D., and Thomas, E. *An Intro to the Science of Cosmology* (IoP, Philadelphia, 2001).

Scott, Douglas. "The Cosmic Microwave Background." <http://www.astro.ubc.ca/people/scott/cmb.html>.

Silk, J. *The Big Bang* (W. H. Freeman, New York, 2001).

"What Is Theoretical Cosmology?" <http://astron.berkeley.edu/~jcohn/tcosmo.html>.

Joseph I. Silk

BIG BANG NUCLEOSYNTHESIS

The search for the origin of our universe and its contents, including the Earth and its living organisms, is a fundamental object of human curiosity. Following the discovery by Hubble that the galaxies of the universe all recede from each other, a simple projection back in time allowed Gamow to estimate that the universe must have originated from a very dense and hot condition that allowed the formation of the chemical elements out of more elementary constituents. By turning the problem around to deduce conditions in the early universe, the investigation of Big Bang nucleosynthesis provides one of the most powerful probes of the origin of the universe.

In Big Bang models there is a time of precisely zero when the scale of the universe becomes zero. Although this zero point itself is outside the domain of the physical model, arbitrarily small times near but not equal to zero are within the scope of the model. For these earliest times the density of matter and energy as well as the temperature become arbitrarily high. The words "arbitrarily near zero" mean that there can be as many zeros between the decimal point and the first nonzero number as one may sensibly describe. With time measured in seconds, the limit of physical theories is now at what is called Planck time, with forty-two zeros preceding the first digit. Quantifying the evolution of the universe from this early time until the present is a central goal of modern cosmology.

The fundamental forces of nature in the present universe include gravity, which binds matter into planets; the weak interaction, which allows the creation of electrons and neutrinos when a neutron decays into a proton; the electromagnetic force, which binds electrons and atomic nuclei into atoms and molecules as well as the creation of photons from moving electrons; and the strong force, which holds together nuclei. Associated with these forces are different classes of particles. The strongly interacting particles are composed of two or three quarks and are known as hadrons. The weakly interacting particles are known as leptons and include the electrons, muons, tau particles, and their associated neutrinos. The hadrons include the baryon subgroup that includes protons and neutrons.

At high density and temperature, the universe is filled with particles of many unfamiliar types. With the extreme temperature and energy, the forces of nature lose some of their distinct properties and combine into a more unified form called a Grand Unified Theory (GUT) or still further a theory of everything

(TOE) if gravity is included. So far combinations beyond weak forces and electromagnetic fields are only a goal of particle physics and cosmology. As the expansion continues, more familiar particles like protons, neutrons, electrons, and photons begin to appear. Initially, there are both matter and antimatter particles. The matter particles are found today, whereas the antiparticles were annihilated by matter prior to the time of nucleosynthesis. At present just the excess of matter particles over antimatter particles remains in existence. One of the first questions that must be confronted in understanding the formation of the elements is as follows: "Why is there an excess of matter over antimatter?" There is no evidence that any regions of the universe contain pockets of antimatter, so some asymmetry in the conservation laws governing the early universe must favor that form of matter that prevails today. Either form of matter could have been favored by this asymmetry, and naturally one refers to the form one is not made of "antimatter." The study of this question is called barygenesis.

A successful barygenesis model must include forces that favor particles instead of antiparticles. This, however, is not enough and the forces must include other nonsymmetric aspects. All particles are described by sets of numbers that specify their properties. An example of such a property is the charge. Another property is called the parity, and it depends on the handedness of a particle. A normal corkscrew goes down into the cork when it is turned clockwise as viewed from above. An anticorkscrew would have to be turned counterclockwise to penetrate the cork. The forces between particles depend on their parities and charges. If the force depends on these properties in a nonsymmetric manner, the force is said to violate CP symmetry. Andrei Sakharov pointed out in 1967 that in order for a process favoring matter over antimatter to succeed in leaving our present universe with the observed matter excess, there must be an asymmetry between the forces on the particles and antiparticles and there must be a force that violates CP symmetry. Models that have both these features are not developed to the level where they can reproduce the observed baryon density in the universe, and Big Bang nucleosynthesis beyond barygenesis treats the baryon density as a free parameter from which the relative abundances of the light elements are deduced.

FIGURE 1

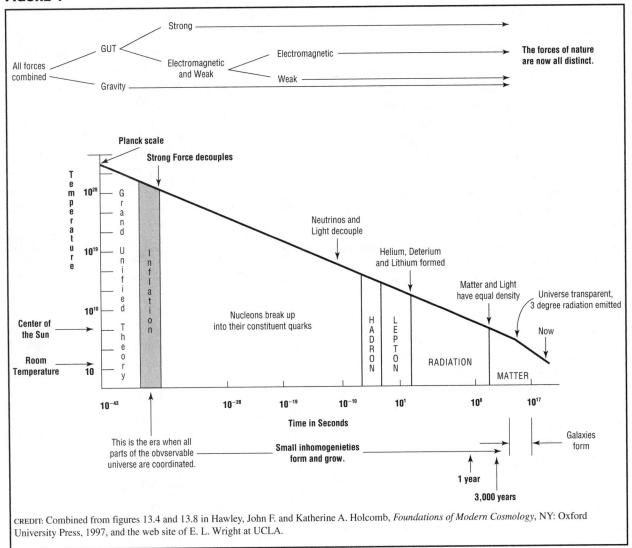

The sequence of key events in the expansion of the universe.

As one follows the expansion of the universe past the time when the baryon density is established, there are two general principles that govern the unfolding of the universe's content: (1) The more matter and energy the universe contains per volume at a given time, the more rapidly it expands, and (2) many constituents of the universe are not in balance with other constituents. The first point is true because the universe is in the reverse of a free-fall collapse—a free expansion. Neither free fall nor free expansion involve frictional processes, and so these two processes are time-reversed versions of each other. It is easy to see that the more mass attracting an object in free

fall, the more rapidly the object will accelerate. Because of this effect, if the energy content in one model of the universe is larger than that of a second model of the universe at a moment of time, its rate of expansion will also be larger. The second point says that the relative abundances of the elements need not be in thermodynamic equilibrium with each other. During the key period of the universe expansion, various isotopic species have abundances that differ from those characteristic of a steady thermodynamic equilibrium. When the temperature changes more rapidly than the forward and backward reaction rates can follow, the abundances become fixed near values appropriate to this last equilibrium temperature. Different isotopes are characterized by different last equilibrium temperatures. This process is described as the freeze-out of particle species.

In the time prior to element building, neutrinos are formed in equilibrium with their associated electrons, muons, and tau particles. The number of distinct leptons governs the number of neutrino families that can be created during the lepton era. This, in turn, governs the energy density of the universe since a larger number of distinct neutrino types increases the energy in the form of neutrinos. Because of the free expansion character of the early evolution of the universe, a larger number of neutrino types increases the rate at which the universe expands and reduces the time available for element building. Consequently, the abundances predicted by the Big Bang nucleosynthesis models constrain the number of independent neutrino families.

The temperature drops during the universe expansion until the neutrons and protons have frozen out. The era of element building is somewhat later than the neutron/proton freeze-out, but the time interval is short enough that the decay of the free neutrons has no major impact on nucleosynthesis. Although reactions involving leptons and the neutron-proton conversions are generally slow, their reaction rates are very temperature-dependent and initially the neutron/proton ratio is the thermodynamic equilibrium value. This ratio depends on temperature since the mass energy of the neutron at rest is larger that of the proton—a higher temperature gives a higher neutron/proton ratio. For a more rapid expansion the freeze-out temperature is higher so that a more rapidly expanding universe has a greater neutron abundance and ultimately a greater ^{4}He abundance.

After the neutron/proton ratio has become frozen, the building of the light isotopic species ^{2}D, ^{3}He, ^{4}He, and ^{7}Li can take place. If one is able to determine the present abundances of these species and be sure that no other process has contributed to their production, these abundances may be used to learn about conditions during the early universe. As a complication, nucleosynthesis also occurs in stellar cores, and the production or destruction of species can alter their present abundance.

To distinguish between stellar and Big Bang nucleosynthesis one can use the fact that the early universe differs from stellar cores in two important respects: (1) The product of matter density and the time available for nucleosynthesis is very small compared to conditions in stellar cores, and (2) with few exceptions there are no free neutrons in stellar cores. These two differences have consequences that permit the identification of nuclei that have been produced exclusively by Big Bang nucleosynthesis and those that have been altered by star-based nucleosynthesis. In particular, ^{12}C, ^{13}C, ^{14}N, and ^{16}O as well as most other heavy isotopes require longer times and higher densities than are available during the Big Bang; they are considered to be the products of nucleosynthesis in star cores or supernovae. In contrast, the light isotopes except for ^{4}He are destroyed in stellar interiors, and ^{4}He is generally not ejected back into space from stars with a higher abundance than is found in the interstellar gas. Thus, the light isotope abundances are the best evidence about conditions during Big Bang nucleosynthesis.

Big Bang nucleosynthesis begins with the individual baryons—the protons and neutrons. The neutrons are unstable as free particles, but due to the shortness of time during the nucleosynthesis era of the Big Bang, their abundance is only slightly reduced by this decay. In order to build up multiple baryon isotopes, pairs of the individual baryons must combine. In the absence of the neutrons, the first step would have to combine two protons to form a product described as ^{2}He. This combination cannot exist even briefly unless one of the protons is converted into a neutron to produce deuterium. However, this conversion involves a weak interaction and is too slow to occur during the Big Bang. Consequently, the presence of the neutrons at the beginning of Big Bang nucleosynthesis is a critical requirement for the formation of the light isotopes. With neutrons starting the sequence, the nuclei with more than two nucleons are easily created by adding either protons or neutrons until ^{4}He is reached as the dominant product. Heavier nuclei are more difficult to build due to the absence of nuclei having five or eight nucleons. The combination of ^{4}He with one of the lighter intermediate species permits the mass 5 gap to be bridged but at the expense of a reduced abundance of the product. Big Bang nucleosynthesis is effective in the production of isotopes up to ^{11}B. Of the light isotopes useful observational constraints are available for ^{2}D, ^{4}He, and ^{7}Li. Observed abundances are also available for ^{3}He, but these are not as useful because of possible alterations by stellar processing.

The comparison between models of Big Bang nucleosynthesis and observations requires both good observations and good model calculations. The input parameters to the model calculation include the density of baryons relative to the cosmic background radiation, the number of neutrino families, and a set of nuclear reaction rates derived from laboratory observations. Figure 2 shows the output abundances provided by one of these model calculations as re-

FIGURE 2

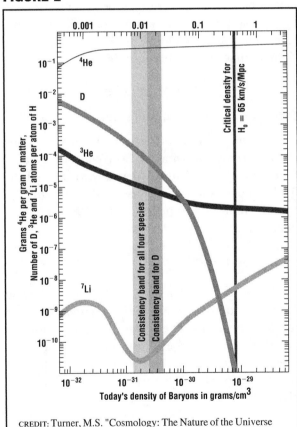

CREDIT: Turner, M.S. "Cosmology: The Nature of the Universe Debate, Cosmology Solved? Quite Possibly!" *Publications of the Astronomical Society of the Pacific* **111,** 264 (1999).

Ratio of the baryon density to the critical density for a Hubble expansion rate of 100 km/s/Mpc.

ported by M. S. Turner in 1999. Compared to the models and shown as the cross-hatched bands are recent estimates of the abundances and their uncertainties reported by K. A. Olive, G. Steigman, and T. P. Walker in 2000 (OSW[2000] in Figure 2) and S. Burles, K. M. Nollett, and M. S. Turner in 2001 (BNT[2001] in Figure 2). The vertical grey band indicates the density range where there is good agreement between the model and the observations of ^{2}D. The observations of ^{2}D in the interstellar gas along lines of sight to distant quasars provide the most precise constraint on the baryon density. The abundance of ^{4}He restricts the number of independent lepton families to three or possibly four. The final species shown in Figure 2 is ^{7}Li. Although there are stellar processes that form and destroy ^{7}Li, there is a large set of measurements for stars believed to be members of an older population, and the distribution of the abundances shows a plateau that is interpreted as the primordial abundance. This interpretation is subject to systematic uncertainty, for which an estimate is included in the plotted abundance range.

The application of Big Bang nucleosynthesis provides three important results:

(1) The widely distributed isotopic species of ^{2}D, ^{4}He, and ^{7}Li can be produced from a fully self-consistent model.

(2) There are no more than three or possibly four independent families of leptons and their associated neutrinos.

(3) The baryons can only account for about 8 percent of the mass needed to achieve a closed universe. Other methods of determining the amount of matter in the universe show that the baryons represent only a small fraction of the mass density.

See also: BIG BANG; CP SYMMETRY VIOLATION; SYMMETRY PRINCIPLES

Bibliography

Alpher, R. A.; Bethe, H., and Gamow, G. "The Origin of the Chemical Elements." *Physical Review* **73**, 803–804 (1948).

Burles, S.; Nollett, K. M.; and Turner, M. S. "What Is the Big-Bang Nucleosynthesis Prediction for ht Baryon Density and How Reliable Is It?" *Physical Review D* **63**, 063512-1–063512-5 (2001).

Collins, P. D. B.; Martin, A. D.; and Squires, E. J. *Particle Physics and Cosmology* (Wiley, New York, 1989).

Copi, C.; Schramm, D. N.; and Turner, M. S. "Big-Bang Nucleosynthesis and the Baryon Density of the Universe." *Science* **267**, 192–199.

Gamow, G. "Expanding Universe and the Origin of Elements." *Physical Review* **70**, 572–573 (1945).

Olive, K. A.; Steigman, G.; and Walker, T. P. "Primordial Nucleosynthesis: Theory and Observations." *Physics Reports* **333–334**, 389–407 (2000).

Sakharov, A. D. "CP Violation and Baryonic Asymmetry of the Universe." *Journal of Experimental and Theoretical Physics Letters* **5**, 24-27 (1967).

Schramm, D. N., and Turner, M. S. "Big-Bang Nucleosynthesis Enters the Precision Era." *Reviews of Modern Physics* **70**, 303–318 (1998).

Steigman, G.; Schramm, D. N.; and Gunn, J. E. "Cosmological Limits to the Number of Massive Leptons." *Physics Letters* **66B**, 202–204 (1977).

Trodden, M. "Electroweak Baryogenesis." *Reviews of Modern Physics* **71**, 1463–1499 (1999).

Turner, M. S. "Cosmology: The Nature of the Universe Debate, Cosmology Solved? Quite Possibly!" *Publications of the Astronomical Society of the Pacific* **111**, 264–273 (1999).

Roger K. Ulrich

BOSON, GAUGE

The gauge principle is used to understand the interactions between fundamental particles. According to this principle, the weak, electromagnetic, and strong forces are all described by the interactions of spin-1 gauge bosons with the quarks and leptons. Each of the gauge bosons is associated with an underlying symmetry. The electromagnetic force is mediated by the photon, the strong force by the gluons, and the weak forces by the charged W^+ and W^- and the neutral Z bosons.

Basics

A quantum mechanical state is described by a wave function $\psi(x)$ where x is the space and time coordinate. Then all physical observables are described by the interactions of operators O with the wave function of the system:

$$\langle O \rangle = \int \psi^*(x)\, O \psi(x)\, dx.$$

The only physical observable is the expectation value $\langle O \rangle$ which is unchanged by changes in the phase of $\psi(x)$:

$$\psi(x) \rightarrow e^{i\vartheta}\psi(x)$$
$$\psi^*(x) \rightarrow e^{-i\vartheta}\psi(x)$$

where ϑ is a constant at every space and time point x. The wave function itself cannot be measured; the only measurable quantity is the expectation value. The invariance of the expectation value under phase changes implies that the phase of the wave function has no physical significance and so also can never be measured in an experiment.

The set of all such global phase transformations (change of the wave function by a constant phase) forms a U(1) (Abelian) symmetry group.

Since ϑ has no physical importance, one would like to be able to choose ϑ to be different for different space and time locations x. If this were the case, the system would be invariant under phase changes that were different in different places:

$$\psi(x) \rightarrow e^{i\vartheta(x)}\psi(x).$$

This is known as a local gauge transformation.

The interactions of particles in quantum mechanics (using the Dirac or Schrödinger equation, for example) always involve derivatives acting on the fields. Under a local phase change, the derivative operating on the wave function changes the wave function by a factor ($\partial_\mu = \partial/\partial x^\mu$):

$$\partial_\mu \psi(x) \rightarrow e^{i\theta(x)}\left(\partial_\mu \psi(x) + i\partial_\mu \theta(x)\psi(x)\right)$$

In this equation, $\mu = 0, 1, 2, 3$, with x^0 being the time coordinate and x^1, x^2, x^3 representing the spatial dimensions. The second term, proportional to $\partial_\mu \theta(x)$, destroys the invariance under the local gauge transformation. The local gauge invariance can be restored, however, if the derivative is replaced by

$$\partial_\mu \rightarrow D_\mu = \partial_\mu + igA_\mu(x).$$

D_μ is called the gauge covariant derivative, whereas the field $A_\mu(x)$ is called a gauge field and must change under local phase transformations as

$$A_\mu(x) \to A_\mu(x) - \frac{1}{g} \partial_\mu \theta(x)$$

The parameter g describes the strength of the coupling of the gauge field to other particles, such as the electron.

Invariance of the laws of physics under local gauge transformations therefore requires the introduction of a massless gauge field $A_\mu(x)$ and the replacement of all derivatives by gauge covariant derivatives. The simplest example of a gauge theory constructed according to this principle is quantum electrodynamics, describing the interactions of the photon with the electron.

Abelian Gauge Bosons

The electromagnetic field $A_\mu(x)$ describing the interactions of the photon is an example of an Abelian gauge field. The interactions of the photon are described by a U(1) gauge symmetry. This symmetry requires that the interactions be invariant under local phase transformations that depend on the space-time point as explained in the previous section. The self-interactions of the photon are contained in the Lagrangian:

$$L = -\frac{1}{4} F^{\mu\nu} F_{\mu\nu}$$

where $F_{\mu\nu} = \partial_\mu A_\nu(x) - \partial_\nu A_\mu(x)$. This interaction is clearly unchanged by the shift

$$A_\mu(x) \to A_\mu(x) - \frac{1}{e} \partial_\mu \vartheta(x)$$

where e is the charge of the electron. A mass term for the photon would have the form

$$L = \frac{1}{2} m^2 A^\mu A_\mu.$$

It is easy to see that this interaction violates the local gauge invariance, and so local gauge invariance requires that the photon be massless. Massless gauge bosons such as the photon have spin 1 and two transverse degrees of freedom, with the spin of a transverse photon being perpendicular to the photon's direction of motion.

The interactions of the photon with fermion fields Ψ such as the electron are restricted by the requirements of local gauge invariance and described by the Dirac equation

$$L = \overline{\Psi}(i\gamma^\mu D_\mu - m_e)\Psi$$

where m_e is the mass of the electron and γ^μ are 4×4 Dirac matrices. Since $D_\mu = \partial_\mu + ieA_\mu(x)$, the Dirac equation represents a coupling between the photon and the fermion field with strength e. There are no free parameters in the Dirac theory since it depends only on the mass and charge of the electron, both of which are measured experimentally.

Non-Abelian Gauge Boson

A gauge theory described by a special unitary group SU(N) is termed a non-Abelian gauge theory. An SU(N) gauge theory has $N^2 - 1$ gauge bosons that interact in a manner exactly specified by the gauge theory. The simplest example of a non-Abelian gauge theory is the SU(2) gauge theory describing the electroweak interactions. This theory was first written down by Chen Ning Yang and Robert Mills. In SU(2) gauge theory, the interactions are invariant under the local gauge transformations:

$$\Psi(x) \to e^{-i\Sigma_i \sigma_i \theta_i(x)}\Psi(x)$$

where σ_i, $i = 1, 2, 3$ are the 2×2 Pauli matrices, and θ_i, $i = 1, 2, 3$ are three real parameters that can depend on the space-time point x. The Pauli matrices can be written as

$$\sigma_1 = \begin{pmatrix} 0 & 1 \\ 1 & 0 \end{pmatrix}, \sigma_2 = \begin{pmatrix} 0 & -i \\ i & 0 \end{pmatrix}, \sigma_3 = \begin{pmatrix} 1 & 0 \\ 0 & -1 \end{pmatrix}$$

An SU(2) gauge group has three massless gauge bosons, $W_{i\mu}$, i=1, 2, 3. (Each gauge boson has four components, corresponding to the energy of the boson and the three spatial directions). In order to maintain the local gauge invariance, derivatives acting on Ψ must be replaced by covariant derivatives:

$$\partial_\mu \to \partial_\mu + ig \sum_i \frac{\sigma_i}{2} W_{i\mu}.$$

The strength of the gauge coupling is represented by the parameter g, and the self-interactions of the gauge bosons are given by the square of the field strength tensor:

$$L = -\frac{1}{4} \sum_{i,\mu,\nu} F_{i\mu\nu} F^{i\mu\nu}$$

$$F_{i\mu\nu} = \partial_\nu W_{i\mu} - \partial_\mu W_{i\nu} + g \sum_{j,k} \varepsilon_{ijk} W_{j\mu} W_{k\nu}$$

where ε_{ijk} changes sign under the exchange of any two of its indices. The non-Abelian gauge bosons have self-interactions between two and three gauge bosons, unlike the Abelian gauge bosons of quantum electrodynamics. Because of the self-interactions of the gauge bosons, the strength of the non-Abelian gauge boson self-interactions decreases at high energy (corresponding to short distances) and increases at low energy (large distances). This property is known as asymptotic freedom.

The strong interactions are described by an $SU(3)_c$ gauge theory called quantum chromodynamics. This theory contains eight massless gauge bosons termed gluons that provide the interactions between quarks. Since the theory is non-Abelian, the strength of the coupling between the quarks and gluons increases with large distances and so provides the force that confines quarks into hadrons such as the proton.

Spontaneously Broken Gauge Theories

An unbroken non-Abelian gauge theory contains only massless gauge bosons. The Standard Model of electroweak interactions consists of a product group, SU(2) × U(1), which contains a spontaneously broken gauge symmetry. A spontaneously broken gauge symmetry has at least one scalar field, termed a Higgs field. This scalar field is used to break the gauge symmetry, while maintaining the gauge invariance of the interactions. When the gauge symmetry of the SU(2) × U(1) electroweak theory is broken, three of the gauge bosons receive masses, while one remains as the massless photon of quantum electrodynamics. The massive bosons are linear combinations of the SU(2) gauge bosons $W_{i\mu}$ ($i = 1, 2, 3$) and the U(1) gauge boson B_μ:

$$W^\pm = \frac{W_{1\mu} \pm i W_{2\mu}}{\sqrt{2}}$$

$$Z_\mu = W_{3\mu} \cos \theta_W - B_\mu \sin \theta_W.$$

The angle θ_W is called the weak mixing angle and is experimentally measured to be $\sin^2 \theta_W = .23$. The weak mixing angle is a measure of the mixing between the SU(2) gauge bosons and the U(1) gauge boson. The remaining combination of neutral gauge bosons remains massless after the spontaneous symmetry breaking and is identified with the photon of quantum electrodynamics.

The massive gauge bosons contain three degrees of freedom: two are the transverse polarizations described in the previous section for the photon, and the third is the longitudinal polarization in which the spin of the gauge boson is parallel to the direction of motion of the gauge boson.

Experimental Successes of Gauge Theories

The predictions of quantum electrodynamics have been spectacularly confirmed by atomic physics measurements, such as the Lamb shift, and by high-energy measurements, such as the anomalous magnetic moments of the electron and the muon. These measurements leave no doubt that quantum electrodynamics describes the interactions of the photon with fermions.

The SU(2) × U(1) gauge theory of electroweak interactions has also received substantial experimental confirmation. The masses of the electroweak gauge bosons, $W^\pm$ and Z, are predicted in terms of the weak mixing angle and the Fermi coupling of beta decay as $M_W = 81$ GeV and $M_Z = 91$ GeV. These masses were predicted before the experimental discoveries of the gauge bosons and have been verified by measurements at the Fermilab Tevatron and the European Laboratory for Particle Physics (CERN) LEP collider. The interactions of the quarks and leptons with the gauge bosons are completely specified in terms of the gauge coupling constants. Many of these interactions, particularly those of the quarks and leptons with the Z boson, have been precisely measured, with most measurements agreeing with the predictions to within a percent.

See also: BASIC INTERACTIONS AND FUNDAMENTAL FORCES; GAUGE THEORY; QUANTUM FIELD THEORY; RENORMALIZATION; STANDARD MODEL

Bibliography

Abers, E., and Lee, B. "Gauge Theories." *Physics Reports* **9**, 1–143 (1973).

Quigg, C. *Gauge Theories of the Strong, Weak, and Electromagnetic Interactions* (Benjamin-Cummings, Menlo Park, CA, 1983).

Yang, C., and Mills, R. "Conservation of Isotopic Spin and Isotopic Gauge Invariance." *Physical Review* **96**, 191–195 (1954).

Sally Dawson

BOSON, HIGGS

Much of contemporary research in elementary particle physics is focused on the search for a particle called the Higgs boson. This particle is a critical missing piece of the present theoretical understanding of the fundamental forces of nature, which describe the interactions of the elementary constituents of matter. The forces include gravity and electromagnetism as well as two additional forces—the strong and weak nuclear forces. The nuclear forces are short-ranged and can be felt only over extremely small subatomic distance scales.

The discovery of nuclear forces and the development of a comprehensive theory to explain them have been one of the profound achievements of twentieth-century physics. The strong nuclear force is responsible for the binding of protons and neutrons inside the nucleus. The weak nuclear force is a little more mysterious. Two consequences of the weak force are beta decay (a form of natural radioactivity) and hydrogen fusion (which ultimately is responsible for the energy received from the Sun). However, initial attempts to arrive at a sound theoretical description of this force ran into trouble. Eventually, it became clear that the existence of the weak force demanded the existence of new elementary particles, which had not yet been observed in atomic experiments.

Several times in the development of the theory of the weak force and associated phenomena, theoretical physicists "invented" new elementary particles that were later discovered in the laboratory. Wolfgang Pauli invented the neutrino in 1930 in order to explain certain anomalies in beta-decay radioactivity. Twenty-six years after his bold prediction, the neutrino was discovered in the laboratory by Frederick Reines and Clyde Cowen. By 1961 a theory of the weak force had been formulated by Sheldon Glashow (and others), which invoked the existence of a new set of fundamental particles, called W and Z bosons. Indeed, the W and Z were detected for the first time in high-energy particle collisions in 1982, and their theoretically predicted properties were verified. The term boson describes a class of particles whose interactions with ordinary matter transmit a force of attraction or repulsion. For example, the electromagnetic force between charged particles is transmitted by the photon (the quantum of light). Likewise, the W and Z transmit the weak force, which is responsible for beta decay. However, unlike the photon (which has no mass), the W and Z must be very massive in order to explain the short-ranged nature of the weak force. This means that producing the W and Z in the laboratory requires very high-energy colliding particle beams. In the collision process, energy is converted to mass (as predicted by Einstein's relativity theory that asserted the equivalence of mass and energy) with sufficient collision energy to create the heavy W and Z bosons.

One aspect of Glashow's theory was troubling. The photon is massless because a deep theoretical principle, called gauge invariance, that underlies the theory of electromagnetism. Glashow's theory of the weak interactions was constructed from the same set of principles, and so it seemed to require that the W and Z should also be massless. This was inconsistent with the short-ranged nature of the weak force, which requires the W and Z to be very massive (as experimentally observed). Thus, Glashow's theory of the weak force was incomplete, since it did not provide an explanation for the masses of the W and Z particles. It was not possible to modify the theory by simply adding masses "by hand" for the W and Z. Such modifications either violate Einstein's principle of relativity or lead to nonsense predictions such as negative probabilities for scattering processes.

The key to overcoming this dilemma was found independently in 1964 by Peter Higgs; by Tom Kibble, Gerald Guralnik, and C. Richard Hagen; and by Robert Brout and Francois Englert. These physicists showed that the physics of electromagnetic fields in superconductors, as clarified by Yochiro Nambu, could be generalized to address the problem of mass generation for the carriers of forces. In a superconductor, pairs of electrons condense and organize themselves macroscopically. The superconducting metal then repels the magnetic field. The mechanism can be described mathematically as resulting from the generation of a mass for the photons that propagate within the superconducting material. Higgs and others showed that this mechanism can lead to a sensible relativistic theory with massive W and Z particles. To construct a realistic model, it was necessary to postulate further new particles to play the role of the electron condensate of the superconductor. At a minimum, one new particle is required. It is the Higgs boson.

In 1967 Steven Weinberg and Abdus Salam constructed a theory of weak interactions based on the Higgs mechanism. The model incorporated Glashow's theory and added a Higgs boson. In doing so, they combined the theory of the electromagnetic and weak forces into a unified description, called the electroweak theory. They showed that in this theory, masses are generated for the W and Z, but the photon remains massless, exactly as required. That is, the symmetry of the gauge boson masses (which are all zero prior to invoking the Higgs mechanism) has been broken. In this case, it is said that the Higgs boson is responsible for electroweak symmetry breaking.

Further examination of the electroweak theory showed that the Higgs boson has just the right properties such that its condensate can also give mass to the electron (and other charged leptons). A subsequent generalization of the model (to incorporate the weak interactions of quarks) showed that quark masses could also be generated in a similar fashion.

It is tempting to suppose that all mass is ultimately due to the interactions of the Higgs bosons. However, that is incorrect. For example, most of the mass of the proton results from the interaction energy of the strong force among its constituent quarks. Perhaps the Higgs boson may be dispensed with entirely by generating mass for the leptons, quarks, and the W and Z bosons in a similar manner, say, by inventing a new strong subnuclear force (not yet discovered). Many theorists have tried to do this, but the results so far have been unsatisfactory.

In particular, any theory of electroweak symmetry breaking is significantly constrained by experimental data that provide precision tests of electroweak phenomena. These data are in very good agreement with the simplest theory of electroweak symmetry breaking in which a single Higgs boson is added to the known fundamental particles—if the mass of the Higgs boson is less than approximately twice the mass of the Z boson. To confirm or refute this theory, one must determine whether or not the Higgs boson exists.

The most comprehensive search for the Higgs boson has been undertaken at the large electron-positron (LEP) collider at the European Laboratory for Particle Physics (CERN) in Geneva, Switzerland, with collisions above 200 billion volts of energy. If the mass of the Higgs boson were less than 1.25 times the mass of the Z, then it would have been possible to create Higgs bosons at LEP. This would have been achieved by colliding electrons and positrons, which annihilate into pure energy and then materialize as a Z boson and Higgs boson. Both the Z and the Higgs bosons are unstable, and both decay almost instantaneously into lighter elementary particles with probabilities that can be predicted from the electroweak theory. The theory of Z decay has been tested and verified to high precision at the LEP collider. After a dedicated search for the Higgs boson, the experimental collaborations at LEP announced that there was no definitive evidence of Higgs boson production in their data.

Two colliders now take aim at the potential discovery of the Higgs boson. The Fermilab Tevatron is a proton-antiproton collider, with collisions of 2 trillion volts of energy. If the Tevatron can achieve a sufficient number of collisions between 2002 and 2007, then calculations show that it may be possible to discover the Higgs boson at the Tevatron if the Higgs boson mass lies in its expected mass range. Otherwise, for a definitive discovery, physicists must wait for the Large Hadron Collider (LHC) now under construction at CERN, which is expected to begin operations in 2007. The LHC is a proton-proton collider, which will operate with collisions of 14 trillion volts of energy. When two protons collide, probability exists that some of the constituents of the two protons will annihilate into a Higgs boson. In this way, the Higgs boson will be prolifically produced—perhaps one million Higgs bosons per year! However, its discovery will not be easy.

Although produced in great numbers, each Higgs boson decays immediately into lighter elementary particles. To prove that the Higgs boson has been produced, one must reconstruct its presence from the debris it has left behind. This is not an impossible task; nevertheless, it requires particle detectors of a very specialized nature as well as extremely sophisticated data analyses. Much work has already been devoted to developing the tools and techniques necessary for this task. For example, if the Higgs boson mass is up to 1.5 times the mass of the Z (but beyond the reach of the LEP collider), then the following technique will be employed. The electroweak theory predicts that roughly one time out of a thousand, the Higgs boson (in this mass range) will decay into two photons. This would be a very distinctive event, in which the two photons have a definite and reproducible total mass equal to that of the Higgs boson. However, one must statistically differentiate such events from other more mundane (so-called "background") events in which photons are produced from the interactions of ordinary matter. Simulations have been performed suggesting that with one year of data collection at the LHC, it should be possible to discover the Higgs boson in this way. Other techniques have also been developed if the Higgs boson turns out to be heavier. Ultimately, it will be possible to discover the Higgs boson at the LHC if its mass is less than approximately ten times the mass of the Z.

Does the Higgs boson exist? Or, does the existence of mass require fundamentally new phenomena that await discovery in future experiments? The answers, although not yet known, will be discovered during the first decade of the twenty-first century.

See also: ELECTROWEAK SYMMETRY BREAKING; EXPERIMENT: SEARCH FOR THE HIGGS BOSON; HIGGS PHENOMENON; PARTICLE; STANDARD MODEL; TECHNICOLOR

Bibliography

Lederman, L., and Teresi, D. *The God Particle* (Houghton Mifflin, Boston, 1993).

Gunion, J.; Dawson, S.; Kane, G; and Haber, H. *The Higgs Hunter's Guide* (Addison-Wesley, Redwood City, CA, 1990).

't Hooft, G. "Gauge Theories of Forces between Elementary Particles." *Scientific American* **242**, 104–138 (1980).

Veltman, M. "The Higgs Boson." *Scientific American* **255**, 76–84 (1986).

Howard E. Haber

BOTTOM

See QUARKS

BRANES

See STRING THEORY

BROKEN SYMMETRY

Symmetry principles have turned out to be very important in the theory of fundamental interactions. In some cases the symmetries are exact within the limits of present knowledge. However, equally important and interesting are the situations involving symmetries that are broken or hidden in some manner. In fact, symmetry techniques are often more useful in these cases. There are many possible fates for a symmetry. It may be exact, explicitly broken, or dynamically or spontaneously broken.

Symmetries arise when a theory has an invariance under some transformation of the basic fields of the theory. This invariance may be either discrete or continuous. Continuous symmetries involve transformations where the magnitude of the transformation can take on a continuous range of values, in particular it can be close to zero (i.e., no change). An example of this is a transformation that shifts the position of an object, that is, translation invariance. In contrast, discrete symmetries involve finite noncontinuous transformations. An example of a discrete symmetry is parity invariance, which involves changing all the spatial coordinates into the negative of their value, $x \rightarrow -x$.

It is possible that a symmetry is clearly realized in all states seen in nature. This is referred to as an exact symmetry. An example is the invariance under electromagnetic gauge transformations. This symmetry allows a redefinition of the phase of the electron's wavefunction, different at each point in space, as long as one simultaneously makes a change in the scalar and vector potentials of electromagnetism, leaving all physics invariant. The existence of this symmetry predicts that electric charge is conserved. As far as is known, all particles and interactions respect this symmetry, and electric charge is conserved.

Another possibility is that a symmetry may be explicitly broken. This occurs if the theory contains an interaction that does not obey the proposed symmetry. The symmetry would be valid if this interaction vanished. Even though the symmetry is not fully present, the use of symmetry techniques could still be useful if the interaction that breaks the symmetry is in some sense small. In that case, in a first approximation one can analyze the theory in the limit where the symmetry is valid and then treat the breaking interaction as a perturbation. An example of this is isospin symmetry, which in the Standard Model would reflect the invariance of transformations among linear combinations of up and down quarks—a continuous symmetry. A consequence of this symmetry would be the equality of the masses of the neutron and the proton, which are made of the up and down quarks. However, the electromagnetic interaction spoils this symmetry, as it is different for the up and down quarks because of their different charge. The different masses of the up and down quarks also explicitly break the isospin symmetry. However, both electromagnetism and the quark mass differences have only small effects on the masses of the nucleons. This can be seen from the mass difference of the neutron and proton, which is only 1 percent of their average mass. Isospin symmetry also predicts other regularities in the interactions of particles, and these are generally valid predictions at the level of a few percent.

The most subtle case concerns dynamically or spontaneously broken symmetry. This situation occurs when the symmetry reflects a continuous invariance of the underlying theory, yet the observed spectrum of particles does not display such a symmetry. The most important state to consider is the ground state. In these theories, the ground state is not unique, and there is a continuous family of possible ground states. A common visual analogy is the lowest energy state of a classical particle in a vertical wine bottle. The particle could be at rest anywhere on the circle that defines the bottom of the bottle. The different ground states would be the different positions around the bottom of the bottle, and the symmetry would reflect invariance of the physics under rotations around the circle. A consequence of the symmetry is that each of the possible ground states possesses the same energy. Nevertheless, despite this symmetry, only one ground state can exist at any time, and any one ground state breaks the symmetry by choosing a preferred direction. A similar situation occurs in quantum field theories. In this case, the ground state is defined by some configurations of the fields, and there could be a continuous family of configurations related to each other by the symmetry—all with the same energy. However, any one of these ground states would break the symmetry by itself.

Whenever this phenomenon occurs, a massless particle generally exists as a consequence. This can be seen from the initial premise that there are many different states with the same energy. Once one of these states is selected as the ground state, the other configurations would be other states with the same energy. In field theory all excitations are described as particles, and only massless particles can be excited with no cost in energy (assuming they also carry zero momentum). This requirement of massless particles is called Goldstone's theorem, and the particles themselves are often referred to as Goldstone particles or Goldstone bosons.

The archetypal case involves a spinless (scalar) field that is allowed to take on both real and imaginary values, that is, it is a complex field. The symmetry involves the transformation of the field by any complex phase. This is an invariance of the theory if only the absolute value of the field enters the theory. Such a theory could result in a symmetry that is either exact or broken. If the ground state of the theory was a state where the value of the field was zero, such a state would be invariant under a change of phase, and the symmetry would be preserved. However, if the energetics of the theory favored a nonzero value of the field in the state of lowest energy, then the symmetry would be spontaneously broken. The initial invariance tells us that the ground state could occur for any value of the phase (much like the particle at the bottom of the wine bottle), but once a specific value for the phase occurs the symmetry is broken. A complex field has two components, that is, real and imaginary parts. Only one combination is fixed by the ground state condition, and it is the orthogonal combination that becomes the Goldstone boson. Which of these two outcomes occurs depends on the nature of the potential energy for the theory in question, but commonly either scenario could result for different values of the parameters of such a theory.

The only exception to the Goldstone's theorem occurs through what is called the Higgs mechanism. When a gauge symmetry is broken in this fashion, instead of obtaining a massless Goldstone particle, the gauge bosons of the theory acquire a mass. Instead of the two spin states of a massless gauge boson (like the photon), the massive one has three spin states. The degree of freedom that would have been the Goldstone boson has transformed into this extra component of the gauge boson.

The terms spontaneous symmetry breaking and dynamical symmetry breaking both refer to the phenomenon previously discussed, in which the theory has a continuous invariance but the ground state does not. The phrases are not quite identical, with spontaneous symmetry breaking being used most often to refer to the situation where a scalar field is responsible for the symmetry breaking and dynamical most often used when there are no fundamental scalar fields involved.

The Standard Model reveals all forms of broken symmetries. The theory involves an $SU(2) \times U(1)$ gauge symmetry that describes the interactions of quarks and leptons with the gauge bosons (the W and Z bosons, and the photon). The $SU(2)$ portion gauge symmetry is spontaneously broken by the vacuum state of the Higgs scalar field, leading to massive W and Z fields through the Higgs mechanism. There is a residual exact symmetry, that of the electromagnetic gauge symmetry. If the up and down quarks had the same mass and charge, the isospin

symmetry mentioned above would exist: it is an example of a useful explicitly broken symmetry. There is also an example of a dynamically broken symmetry: chiral symmetry. This symmetry is an extension of isospin symmetry—if the up and down quark masses were both equal to zero, an independent isospin invariance of each of the two spin states of these quarks would exist. The Goldstone bosons would be the pions. Because the quark masses are not exactly zero, this symmetry is also explicitly broken. The pions are then not strictly massless, but they are, in fact, the lightest of the observed strongly interacting particles. The rich and varied symmetry of the Standard Model is one of the reasons that symmetry techniques have been so fruitful in exploring fundamental interactions in the physical universe.

See also: SYMMETRY PRINCIPLES

Bibliography

Donoghue, J. F.; Golowich, E.; and Holstein, B. R. *Dynamics of the Standard Model* (Cambridge University Press, Cambridge, UK, 1992).

Kane, G. L. *Modern Elementary Particle Physics* (Addison Wesley, Redwood City, CA, 1987).

Zee, A. *Fearful Symmetry: The Search for Beauty in Modern Physics* (Macmillan, New York, 1986).

John F. Donoghue

BROOKHAVEN NATIONAL LABORATORY

Funded by the U.S. Department of Energy (DOE) and one of ten national laboratories, Brookhaven National Laboratory (BNL) carries out research and development in four main areas: basic science and applied technology, environmental quality, national security, and energy resources. The lab accomplishes this research and development (R&D) mission by designing, constructing, and operating some of the world's largest and most sophisticated research facilities for scientists across the country and around the world; by carrying out long-term programs of basic and applied research; by advancing technology and transferring it to industry; and by educating future scientists.

Brookhaven National Laboratory. The circle at the top is the Relativistic Heavy Ion Collider, which is 2.4 miles in circumference. CREDIT: COURTESY OF BROOKHAVEN NATIONAL LABORATORY. REPRODUCED BY PERMISSION.

In July 1946, a consortium of nine universities in the northeastern United States banded together to form Associated Universities, Inc. (AUI), which contracted with the Atomic Energy Commission (AEC) to build and operate BNL for the scientific community as a national educational and scientific resource. This unique contractual arrangement was later copied by other national and international laboratories in the United States and abroad. The lab opened in 1947. Its first major generation of large instruments included the Cosmotron (1952-1966), which for a while was the world's most powerful particle accelerator; the Brookhaven Graphite Research Reactor (BGRR, 1950-1969), the first reactor built specifically for peacetime research; and the Brookhaven Medical Research Reactor (BMRR, 1959-2000), the first reactor built specifically for medical research. The lab's second generation of large instruments included the Alternating Gradient Synchrotron (AGS, completed 1960) and the High Flux Beam Reactor (HFBR, 1965-2000). Starting in 1972, BNL attempted to build a new large particle accelerator, ISABELLE, but problems with the superconducting magnets delayed the project, and in 1983 the DOE (the AEC's successor agency) canceled the project—the laboratory's single biggest failure. In 1997, following discovery of radioactive contamination from the HFBR, the DOE fired AUI as the lab's contractor and replaced it with Brookhaven Science Associates, a new company established by the Research Foundation of the State University of New York (on behalf of SUNY Stony Brook) and Battelle Memorial Institute.

A world-renowned scientific leader and national resource since its inception, Brookhaven is home to four Nobel Prize–winning discoveries in physics: 1957, for the 1956 theory of parity nonconservation, which explains the difference between the real world and its mirror opposite; 1976, for the co-discovery, in 1974, of the J/ψ particle, the first known particle to contain a charmed quark; 1980, for the 1964 discovery of CP violation, which accounts for the predominance of matter over antimatter in the universe; and 1988, for the 1962 discovery of the muon neutrino, as distinguished from the already known electron neutrino.

In the life sciences, Brookhaven has pioneered research in positron emission tomography (PET), exploring the link between dopamine and addiction thanks to a Brookhaven-developed medical tracer used worldwide to diagnose cancer, brain disease, psychiatric illnesses, and heart disease. Other tracers developed at BNL include technitium-99m, the medical radiotracer employed in 85 percent of the world's nuclear medicine procedures; and thallium-201, used worldwide in stress tests of the human heart. Brookhaven's medical breakthroughs also include establishment of the quantitative connection between salt and hypertension, which resulted in the elimination of salt in baby foods and recommendations restricting salt intake; use of L-dopa for relief of Parkinson's disease symptoms; and synthesis of the first human insulin for use by diabetics, replacing the use of animal insulin.

Still other important Brookhaven discoveries involved machine design: the alternating gradient principle, used in all modern high-energy accelerators; the undermoderated core used in all high-flux reactors; and the magnet arrangement or "lattice" used in all modern synchrotron sources. Brookhaven scientists developed the first video game as a toy for visitors (1958) and were awarded a patent for magnetically levitated, or "maglev," trains (1968).

Brookhaven draws about $450 million in federal dollars to New York State and attracts over 4,000 scientists per year to its facilities. These include the Relativistic Heavy Ion Collider (RHIC), commissioned in

2000 as the world's newest and largest accelerator for nuclear physics research into the structure of matter that existed moments after the Big Bang; the National Synchrotron Light Source, commissioned in 1982, which has become one of the world's most widely used facilities due to its ability to provide very bright beams of X rays and other light to look at molecular structure and function in physical and biological materials; the AGS, the accelerator that produced three of the lab's four Nobel Prizes and the only U.S. heavy-ion accelerator used in experiments to determine the biological effects of space travel; and the Scanning Transmission Electron Microscope, one of three in the world to image individual heavy atoms. RHIC has recently been upgraded to conduct polarized proton research. Future programs at Brookhaven include a nanoscience center.

See also: BENEFITS OF PARTICLE PHYSICS TO SOCIETY; FUNDING OF PARTICLE PHYSICS

Bibliography

Brookhaven National Laboratory. <http://www.bnl.gov>.

Crease, R. P. *Making Physics: A Biography of Brookhaven National Laboratory 1946–1972* (University of Chicago Press, Chicago, 1999).

Robert P. Crease

BUDKER INSTITUTE OF NUCLEAR PHYSICS

The Budker Institute of Nuclear Physics (BINP), located in Novosibirsk, Russia, was founded in 1958. It originated from Gersh (Andrey) Budker's Laboratory of New Acceleration Methods at the Institute of Atomic Energy, headed by Igor Kurcharov. Until his death in 1977, academician Gersh Budker was director of the institute. Since then, academician Alexander Skrinsky has served as its director. Scientific and economic policy is controlled by "The Round Table"—the Scientific Council of BINP.

One of the main scientific objectives of BINP is to study elementary particles. The existing scheme to describe elementary particles, the Standard Model, considers as elementary six quarks, six leptons, and four carriers of the fundamental interactions—the

photon, the $W^{\pm}$ and the Z bosons, and the gluon. All other particles are composite: for example, a proton consists of three quarks (uud) whereas all mesons are quark-antiquark particles. BINP contributed substantially to the development of this picture.

Since the mid-1960s, BINP has studied elementary particles using electron and electron-positron colliders—the most important method in modern elementary particle physics. The institute made many pioneering contributions to the development of this method and to the research in this field, including the work of Gersh Budker, Alexey Naumov, Veniamin Sidorov, and Alexander Skrinsky. From 1965 to 1967 the electron-electron collider VEP-1, simultaneously with the Princeton-Stanford rings, was used to test the Coulomb law at small distances, and it was shown that the electron size does not exceed 10^{-14} cm. In the world's first annihilation experiments in electron-positron collisions, carried out at the VEPP-2 collider in 1967, rho-meson parameters were measured. In the experiments performed at VEPP-2 until 1970, the main parameters of vector mesons were studied, and two-photon processes and multiple production of hadrons were discovered. The latter process is an important confirmation of the existence of quarks.

Between 1974 and 2000 the electron-positron collider VEPP-2M, with a productivity that was a hundred times larger, provided much physical information on rare decays of vector mesons. For example, the first evidence for the existence of exotic mesons (possibly four-quark states) was obtained. In these experiments the total cross section of e^+e^- annihilation was measured with record accuracy in the energy range of 0.36 to 1.4 GeV. At VEPP-4, another e^+e^- collider operating at BINP, the hadronic cross section was precisely measured between 7.2 and 10.3 GeV. Precise knowledge of the total cross section is important for accurately determining fundamental physical parameters such as the anomalous magnetic moment of the muon and the electromagnetic fine structure constant at high energies. Such experiments will be continued at the new e^+e^- colliders—the VEPP-4M, which is currently operating, and the VEPP-2000, which is under construction.

The method of resonant depolarization developed at BINP with contributions by Lev Barkov, Lery Kurdadze, Alexey Onuchin, Igor Protopopov, Veniamin Sidorov, Vladimir Smakhtin, Yuri Shatunov, Alexandr Skrinsky, Yuri Tikhonov, and German Tumaykin was successfully applied to establish with very high accuracy (about 10^{-5}) the absolute mass scale of elementary particles from 1 to 100 GeV. Most of the experiments were performed in Novosibirsk from 1975 through 1984; in 1994 the Z boson mass was measured at the European Laboratory for Particle Physics (CERN).

Parity nonconservation (PNC) in atoms was discovered at BINP. In 1974 Iosif Khriplovich (BINP) proposed an experiment to look for the rotation of the plane of polarization of light passing through atomic bismuth vapor. Simultaneously, this proposal was made at Oxford, UK, and Seattle, Washington. The effect was discovered experimentally by Lev Barkov and Mark Zolotorev in February 1978. It was a vivid demonstration of parity nonconservation, that is, the absence of symmetry between right and left (the plane of polarization of light prefers, say, left rotation to right). In this experiment PNC was first observed as a macroscopic coherent effect. A new kind of weak interaction between electrons and nucleons, resulting from the so-called neutral currents, was first discovered at BINP. It was one of the first decisive confirmations of the unified theory of electroweak interactions.

In 1980 Victor Flambaum and Iosif Khriplovich of BINP predicted that the PNC effects in atoms, which depend on nuclear spin, were due mainly to the so-called nuclear anapole moment (AM). AM corresponds to a special configuration of the electromagnetic field, of the type produced by the current in a toroidal winding. The AM of the cesium nucleus was discovered in an optical experiment in Boulder, Colorado, in 1997.

One of the most frequently cited works in the world is the article on sum rules in quantum chromodynamics by Arkady Vainshtein (BINP) and his Moscow coauthors (1979). Additionally, the most popular model to describe hadron scattering at high energies is the so-called BFKL equation, proposed by Victor Fadin and Eduard Kuraev of BINP

in 1975 with his coauthor from Leningrad (now St. Petersburg).

The theoretical discovery of asymptotic freedom was anticipated at BINP. In 1968 Iosif Khriplovich was the first to correctly calculate charge renormalization in the Yang-Mills theory. He pointed out the unusual sign of the effect and gave a simple, intuitive explanation for it.

The electron cooling of the beams of heavy particles was suggested by Gersh Budker in 1966 and realized and developed between 1966 and 1985 at the Budker Institute of Nuclear Physics largely through the efforts of Nikolay Dikansky, Igor Meshkov, Vassli Parkhomchuk, Dmitri Pestrikov, Rustam Salimov, Alexander Skrinsky, and Boris Sukhina. The cooling of beams of charged particles creates a decrease of the phase space occupied by the particles in the storage ring. Cooling substantially increases the particle density in the phase space, compresses the beam, and decreases the spread of particle velocities. This allows one to apply multiple injection to store more and more particles in the phase space sites that become free after cooling.

The electron cooling of the beams of heavy particles is based on the interaction of the beam to be cooled with the cold electron beam. To this end, in one of the straight sections of the storage ring, an electron beam with a small spread of velocities is passed through a circulating beam of heavy particles with the same average velocity. Because of the Coulomb interaction between "cold" electrons and "hot" heavy particles, an intensive heat exchange takes place resulting in the cooling of the heavy particles. The cooling decrements grow proportionally to electron density and decrease rapidly when the angular spread in the ion beam and its energy increase.

The equilibrium of this spread is determined by the equality of the temperatures of electrons and heavy particles:

$$\theta_i = \theta_e \sqrt{\frac{m_e}{M_i}}$$

Because of the large mass difference (m_e and M_i are the electron and ion mass, respectively), the angular spread in the beam of the heavy particles θ_i is much smaller than in the cooling electron beam θ_e.

The longitudinal magnetic field applied for the beam transport further strengthens the cooling action of the electron beam: the transverse thermal motion of electrons is frozen (heavy particles flying far enough away from the electron do not feel their fast rotation in the magnetic field along the Larmor orbits), and the temperature of the longitudinal motion of electrons is often much smaller than the transverse one.

Experiments with electron cooling, even at BINP's first installation, NAP-M, allowed the cooling of the proton beam with an energy of 65 MeV to a temperature of $T \sim 1°K$ in the time $\tau \sim 50$ ms. Electron cooling is one of the most important techniques in the experimental physics of nuclei and elementary particles, and it is used in laboratories all over the world.

There have been many other important achievements at BINP, for example, the pioneering of a physically self-consistent project of linear colliders able to reach interaction energies ten times higher for electrons, positrons, and photons; and the proposal to reach muon-muon collisions of even higher energy and high luminosity using ionization cooling followed by muon acceleration and storing.

See also: BENEFITS OF PARTICLE PHYSICS TO SOCIETY; FUNDING OF PARTICLE PHYSICS; INTERNATIONAL NATURE OF PARTICLE PHYSICS

Bibliography

Auslender, V., et al. "The Study of Ro-meson Resonance in Electron-Positron Annihilation." *Physics Letters* **25B**, 433 (1967).

Barkov, L. M., and Zolotarev, M. "Observation of Parity Nonconservation in Atomic Transitions." *Soviet Physics: JETP Letters* **27**, 357 (1978).

Budker, G. I., et al. "Check on Quantum Electrodynamics by Electron-Electron Scattering." *Soviet Journal of Nuclear Physics* **6** (6), 889–892 (June 1968).

Budker Institute of Nuclear Physics. <http://www.inp.nsk.su/>.

Fadin, V. S.; Kuraev, E. A.; and Lipatov, L. N. "On the Pomeranchuk Singularity in Asymptotically Free Theories." *Physics Letters* **B60**, 50–52 (1975).

Khriplovich, I. B. "Feasibility of Observing Parity Nonconservation in Atomic Transitions." *Soviet Physics: JETP Letters* **20**, 315 (1974).

Shifman, M. A.; Vainstein, A. I.; and Zakharov, V. I. "QCD and Resonance Physics. Sum Rules." *Nuclear Physics* **B147**, 385–447 (1979).

Skrinsky, A. N., and Shatunov, Y. M. "High Precision Measurements of Elementary Particles Masses using Colliders with Polarized Beams." *Soviet Physics USPEKHI* **32** (6), 548–554 (1989).

Alexander N. Skrinsky

CASE STUDY: GRAVITATIONAL WAVE DETECTION, LIGO

LIGO is an acronym for Laser Interferometer Gravitational-Wave Observatory. The LIGO observatory is funded by the National Science Foundation and is managed jointly by the California Institute of Technology and the Massachusetts Institute of Technology. Several hundred scientists collaborate on observatory design/construction, on predictions of gravitational waveforms, and on methods of analysis of the gravitational wave signals whose detection is expected in the near future. Two LIGO detectors, consisting of Michelson-type interferometers with arms 4 km long, are nearing completion at Hanford, Washington, and Livingston, Louisiana. A third detector with 2-km-long arms is also under construction at the Livingston site.

LIGO is one of several gravitational wave detectors being constructed around the world. The objective of these efforts is not only to detect gravitational waves directly for the first time but also to use the signals as a tool for observational astronomy. Processes that give rise to gravitational radiation typically involve large masses undergoing rapid motions, as in supernova collapse or in the coalescence of massive black holes. Observations of gravi-

tational signals from such events provide details about the dynamics of motion of huge masses deep inside an evolving system; the gravitational radiation produced is usually low in frequency and easily escapes from the system because of the weakness of gravitational interactions. By contrast, electromagnetic radiation comes from individual atomic-size particles, is of high frequency, and is easily obscured by dust or other material at the surface of stars.

Gravitational waves from distant sources have flat (plane) wave fronts by the time they reach the Earth. These waves are actually ripples in space-time that exert forces on matter in their path. Understanding these forces suggests various ways of detecting such waves. Figure 1 illustrates the force fields that are established in the path of a plane gravitational wave. In Figure 1 the propagation direction occurs at right angles to the page. The waves can be fully described in terms of two so-called polarizations. Consider first the × polarization in the left-most part of Figure 1. Relative to a reference mass at the origin, a free particle placed anywhere in the same plane will experience a time-varying force—hence an acceleration—in the direction of the arrows. The force reverses every half-period. The strength of the force is greater the greater the distance of the free particle from the origin; this is indicated in the figure by the increasing density of the lines of force away from the center.

FIGURE 1

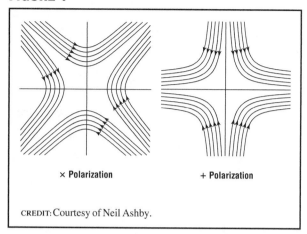

× Polarization   + Polarization

CREDIT: Courtesy of Neil Ashby.

Two independent polarizations of plane gravitational waves. The waves are propagating perpendicular to the page. A free particle placed on one of the lines of force will experience acceleration in the direction of the arrow, relative to a reference mass of origin. The density of the field lines, hence the force, increases as the distance from the reference mass increases. For a wave of frequency f or period $T = 1/f$, the fields vary harmonically and reverse direction each half-cycle. The + and × polarization fields occur everywhere at right angles to each other. One polarization diagram can be obtained from the other by a 45° rotation.

The + wave polarization is illustrated in the rightmost part of Figure 1; it is basically the × figure turned 45°. The force fields of the two polarizations are at right angles to each other. An elastic bar lying along the vertical axis of the + wave would be squeezed, while an elastic bar placed along the horizontal axis would be stretched at the same time.

This alternate squeezing and stretching effect is the fundamental basis for interferometric detectors such as LIGO. An interferometer detector is diagrammed in Figure 2. Consider the + polarization. Imagine a reference mass at the origin upon which a beam splitter is placed. Mirrors are placed at the top and at the right of the figure so that light sent through the beam splitter and out along the axes is reflected by the mirrors back to a photodetector near the origin. As the gravitational wave passes through the interferometer, the distance of one of the mirrors from the beam splitter increases (the stretched arm) while the distance of the other mirror decreases (the squeezed arm). This will change the interference conditions on the two beams when they recombine, and the change in the interference pattern

can be measured. The structure of such an interferometer, which was invented by A. A. Michelson in an attempt to observe the Earth's absolute motion, is thus very well suited to detect mirror motions in the presence of gravitational waves.

Gravitational wave effects are proportional to a unitless amplitude h called the wave strain. (A unitless strain in a compressed rod, for example, is the fractional change in length, or the change in length divided by starting length.) The force fields of Figure 1 correspond to a plane wave strain that is the same everywhere in the plane; this means that the movement of a reflecting mirror is proportional to the distance of the mirror from the beam splitter, whereas the wavelength of the light propagating in the interferometer arm does not change. The longer the interferometer arms, the greater the change in the interference pattern and the more sensitive the detector will be. Arm lengths of a few kilometers are a practical limit for interferometers placed on the Earth's surface, in part because the entire apparatus must be placed in high vacuum. Plans are under way for an interferometer in orbit, in the vacuum of space, with arms in a triangular configuration about 5 million km on one side (LISA).

Theoretical calculations of the wave strain amplitude are usually based on the theory of general relativity. Such calculations are mathematically very challenging, but for stellar sources of known configuration—such as stellar binaries—they lead to predictions of the wave strain amplitude in which one can place a great deal of confidence. The strains are predicted to be in the neighborhood of 10^{-20} or smaller. This means that for an interferometer arm with a length of 4 km, the motion of one of the mirrors is only approximately $.4 \times 10^{-15}$ m, which is smaller than the size of an atomic nucleus. That measurements of such small motions can be seriously contemplated is incredible. Only by venturing beyond the frontiers of technology, inspired by a series of brilliant ideas, have these measurements become possible. The LIGO detector has been designed in such a way that as new ideas are proposed and technology advances, improvements can be incorporated into the instrument so that its sensitivity will improve.

One method of increasing the sensitivity of the interferometer detector is to reflect the light beam

FIGURE 2

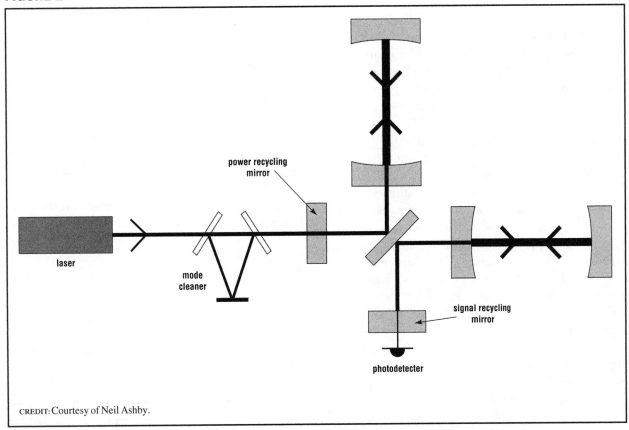

CREDIT: Courtesy of Neil Ashby.

Simplified diagram of initial LIGO detector, a Michelson-type interferometer with very long arms, with additional optical components designed to overcome noise sources that limit detector sensitivity. The view is from above; optical components are suspended with wires attached to supports (not shown). A + polarized gravitational wave passing through the interferometer in a direction perpendicular to the plane of the paper will compress one arm while it stretches the other, causing a shift in the interference pattern produced at the detector after the beams recombine. Power recycling mirrors, mode cleaners, and signal recycling mirrors are indicated.

in each arm back upon itself with additional mirrors, so that the light bounces back and forth in the arms many times before escaping. Such mirrors are shown between the beam splitter and the end mirrors in Figure 2. This effectively increases the arm length and transforms each arm into a resonant Fabry-Perot cavity. If the arm length becomes too large, however, this advantage is lost as the forces resulting from the wave would reverse and average out while the light circulates. There is thus a storage time limit for photons in the interferometer arms beyond which the sensitivity does not improve. The storage time limit is approximately half the period of the gravitational wave; during this time, the forces act generally to push the mirrors in a consistent direction. For a wave of frequency 500 Hz the storage time limit is

0.001 s, and in an arm with a length of 4 km there should not be more than about 75 bounces. The sensitivity of a 4-km interferometer can be pushed to a lower frequency (corresponding to a larger number of bounces) if the mirrors are extremely smooth and made of ultra-low-loss material.

Another improvement in sensitivity can be achieved within a limited frequency range by placing a partially transmitting signal recycling mirror between the beam splitter and the photodetector. This transforms the interferometer into a cavity that resonates and thus gives improved sensitivity within a frequency range controlled by mirror reflectivity and position. This technique is expected to be useful when searching for continuous signals of definite frequencies, such as those from rotating neutron stars.

One of the most serious difficulties to overcome is that of isolating the beam splitter and mirrors from the Earth's seismic vibrations. (Interferometers in space are not subject to seismic noise.) In order to allow the mirrors to respond to a gravitational wave, freedom of motion parallel to the interferometer arm is necessary. This is made possible, while also providing some isolation from seismic vibrations, by suspending the beam splitter and mirrors like pendulums, from supports which are further isolated from vibrations with carefully designed stacks of absorbing material or heavy masses. A pendulum that is free to swing parallel to the arm can be set into motion by a seismic vibration of the support in this direction. However, if the seismic vibration frequency f is large compared to the pendulum frequency f_0, the mirror motion will be smaller than the support motion by a factor of $(f_0/f)^2$. Several stages of such isolation can be considered, but the pendulum support limits the low-frequency sensitivity of the detector. In future detectors, external equipment may sense seismic motions and actively push the detectors to compensate for seismic disturbance.

Other kinds of vibrations must also be reduced. Vertical motions of the support can couple to horizontal motions, for example, because the mirrors at the ends of very long interferometer arms are actually not parallel—the suspension wires tend to point toward the Earth's center. Also, there may be cross-couplings of vibrations resulting from mechanical misalignments. The suspending wires can vibrate like strings on a violin; all such mechanical oscillations must be severely suppressed. Even if all such vibrations could be completely eliminated, there would still be gravity gradient noise—time-varying gravitational forces on the mirrors arising from people and vehicles moving in the vicinity—that cannot be shielded from the apparatus and that will limit the sensitivity at low frequencies.

There are many other sources of random noise that can cause spurious mirror motions. One of the most important of these is due to random fluctuations in the number of photons that are circulating in the interferometer arms, called shot noise. If the number of circulating photons can be significantly

increased, however, the sensitivity of the detector can be improved. The apparatus is most sensitive to mirror motion when there is destructive interference of the recombined light beams at the photodetector; in other words, the photodetector is at the dark port. Then most of the light energy goes back through the beam splitter toward the light source. Another mirror, called a power-recycling mirror, can be placed in the path of this light to reflect it back into the interferometer. This can significantly increase the power circulating in the arms; fluctuations resulting from shot noise then play less of a role. There is a tradeoff, however, because if the power in the arms is too great, there can be undesirable thermal heating and distortion of the mirrors, causing losses of light, as well as fluctuations in radiation pressure on the mirrors. The placement of the power-recycling mirror is shown in front of the laser in Figure 2.

A disadvantage of having high power circulating in the arms is that radiation pressure on the mirrors fluctuates, causing unwanted mirror motions. The effort to reduce shot noise by increasing power in the arms is thus thwarted by radiation pressure. The two effects depend on frequency differently with shot noise tending to increase at frequencies of 200 Hz and higher, and radiation pressure increasing at lower frequencies. Interferometer designers must find a compromise between these two effects—see the discussion of Figure 3 below.

Heat can excite vibration modes of the mirror masses or of the suspension systems, and these can couple into resonances of the system such as pendulum modes and cause unwanted mirror motions. Reducing energy losses in the suspension materials can reduce such effects. Thermal motions have amplitudes that depend on the temperature, and making the interferometer arms long can reduce the harmful effects of thermal excitations.

Other significant sources of noise are instabilities in frequency and power coming from the laser. Usually, the interferometer arms cannot be made exactly equal in length. Then if the frequency of the laser light fluctuates, the interference conditions at the dark port will change, and a spurious signal will be detected. To overcome this, a mode cleaner is inserted between the laser output and the power-recycling mirror. (See Figure 2.) A mode cleaner is an optical assembly that acts like a narrow band filter so that light of unwanted frequency cannot pass through. Power fluctuations can also cause signals at the dark port; partly for this reason, lasers such as Nd:YAG that are easier to control will be used in LIGO.

There are numerous other sources of noise that must be dealt with in order to reduce unwanted signals at the dark port. These include light scattered out of the main light beams and then scattered back in, for example, from the walls of the vacuum chamber. The phase of this scattered light will not agree with the phase of the main beam and so will contaminate the signals. A system of baffles to absorb scattered light within the vacuum system is therefore required. Residual gas in the vacuum chamber can cause fluctuations in the index of refraction that disturb the beams, and gas particles can also bounce off the mirrors, causing slight displacements. Motions from any source (such as seismic disturbances) in the optical systems (as in the mode cleaner) cause the beam position and direction to fluctuate and hence cause some noise at the dark port. Stray electric and magnetic fields can affect the mirrors. For example, if there is stray parasitic charge on any of the optical surfaces, electric fields can exert forces on these elements. Magnetic fields can interact with actuator magnets bonded to the mirrors (such magnets are used to control mirror orientation) and exert forces on them.

The construction of the first-generation LIGO I detectors has been completed (April 2002) and rigorous testing of the many systems—optical, vacuum, laser, control, data analysis—is under way. Figure 3 shows the design sensitivity of LIGO I (curve marked I) and of the advanced LIGO (curve marked II) that features enhanced sensitivity due to the incorporation of many features made possible by developing technology. (Gravitational wave sensitivities h are usually quoted with units $1/\sqrt{\text{Hz}}$. The squared strain noise in a narrow range of frequencies Δf is then $h^2 \Delta f$ and is dimensionless.) LIGO I's sensitivity lies in the frequency range of 30 to 1,200 Hz. LIGO II should have much improved sensitivity over a wider frequency range.

Interferometer detectors are most sensitive to waves propagating perpendicular to the interferometer plane. However, as the Earth rotates, this direction sweeps across the sky, sampling waves from different directions. The Hanford and Livingston facilities are sufficiently far apart on the Earth's surface that seismic disturbances are likely to be uncorrelated; also, the detectors are oriented differently. A gravitational signal burst would generally arrive at different times at the two sites. Looking for delayed coincidences between signals arriving at the two sites can determine the direction of the source as well as yield better detection sensitivity and eliminate false alarms. The growing worldwide network of gravitational wave observatories (LIGO in the United States, VIRGO and GEO600 in Europe, and TAMA300 in Japan) is working out agreements whereby different projects may save and exchange data. It will thus be possible to operate all of them together in order to observe common coincidences.

There are a few astrophysical sources whose gravitational wave strain amplitudes are large enough to be detected by LIGO I. Estimates of the radiation emitted by the pulsar remnant from the Crab supernova, which occurred in 1054, show that the signal strength may be large enough to be observed. Astronomers have been surprised many times by observations of previously unknown objects when new instruments came online, so it is possible that sources of as yet unknown types exist, which could be observed. LIGO I serves as a proving ground from which much is being learned about noise sources and the methods of detecting gravitational signals. Because signals are weak and many noise sources are present, detection involves continuous observation over many months, so that the noise can be averaged down. Data are analyzed using the technique of matched filtering—correlation of the data with calculated "templates" of waveforms expected from the source.

There are many additional astrophysical sources, however, that should be detectable with LIGO II. These usually involve violent motions of large masses. A few such sources are indicated with heavy lines in Figure 3. For example, the upper heavy line, sloping downward as the frequency increases, corresponds to theoretical estimates of observable signals from the interaction of two ten-solar-mass black holes, at a distance less than 100 Megaparsecs. (1 Megaparsec [Mpc] = 3.3 million light-years. The Milky Way galaxy is about 0.03 Mpc in diameter; there are about

FIGURE 3

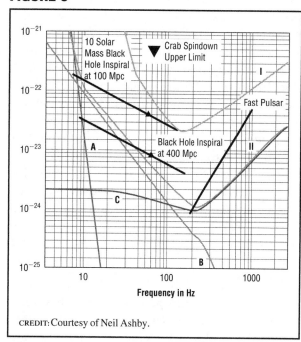

CREDIT: Courtesy of Neil Ashby.

Sensitivity of the first LIGO detector I and the advanced LIGO detector II. The most important noise sources that limit LIGO II are indicated. These are seismic noise A, radiation pressure B, and shot noise C. The amplitudes of a few of the radiation sources are also indicated with heavy lines on the graph. The inspiral of two black holes, each of ten solar masses, should be observable by LIGO II if the distance from Earth is less than about 100 Mpc.

twenty other galaxies containing many potential sources within approximately 1 Mpc of the Milky Way.) As the two black holes orbit around each other, they lose energy because of the emission of gravitational radiation, and spiral inward toward each other. The radiation frequency increases as they come closer, until finally the black holes collide, merge, and probably vibrate violently until the system settles down to a rotating black hole. Each stage of such a process involves different signal frequencies and amplitudes. If the binary system is closer than 100 Mpc, the line will rise on the figure; LIGO I might then be able to detect such signals.

The lower black line at the left of Figure 3 corresponds to black hole–black hole inspiral at distances less than 400 Mpc. The heavy black line toward the right side of Figure 3 shows the signal amplitudes from fast-spinning pulsars of unknown frequencies but at distances less than 0.01 Mpc with an averaging time of three months, if one assumes the pulsars are slightly nonspherical.

Signal recycling can improve LIGO II's sensitivity significantly beyond what is plotted in Figure 3, within a narrow frequency band near approximately 700 Hz (not shown in Figure 3). In this frequency range there are many low-mass X-ray–emitting binary systems that should be observable by LIGO II in a narrow-frequency-band operation mode.

Other binary systems such as neutron star binaries, or neutron star–black hole binaries, are possible sources. In many such sources, the underlying physics is poorly understood; theoretical estimates of the signal amplitudes of waveforms are, at best, approximate. The actual observation of signals would have a very significant impact on understanding the dynamics of such processes.

Heavy stars that exhaust their stores of nuclear energy after many millions of years may undergo catastrophic collapse and subsequently produce a supernova explosion, emitting strong bursts of gravitational radiation. Chunks of matter ejected from such systems could remain in close orbit and radiate at frequencies equal to twice their orbital periods.

Another potential source, not shown in Figure 3, is random gravitational waves arising perhaps from the Big Bang or from events during early stages of expansion, such as transitions from domination of the universe by one type of matter to another.

Clearly, there are many interesting astrophysical processes that serve as potential candidates for gravitational wave detection. Studies of such sources are actively underway by a large number of researchers. At the same time, improvements in detector sensitivity are being actively pursued on many fronts, such as in development of ultrasmooth mirror surfaces and coatings, better suspensions and attachments, actuation of optical components, data analysis, and many other areas.

The LIGO collaboration consists of hundreds of researchers, from all over the world, organized into working groups that concentrate actively on specialized aspects of the overall effort. These efforts are expected to result in the successful observation of gravitational waves, opening up a new and exciting field of astronomy using a new tool that can observe the inner details of large-scale, fast astrophysical processes that cannot be observed by any other technique.

See also: ASTROPHYSICS; BASIC INTERACTIONS AND FUNDAMENTAL FORCES; BOSON, GAUGE; COSMOLOGY; RELATIVITY

Bibliography

Robertson, N. "Laser Interferometric Gravitational Wave Detectors." *Classical and Quantum Gravity* 17 (15), R19–R40 (2000).

Neil Ashby

CASE STUDY: LHC COLLIDER DETECTORS, ATLAS AND CMS

Thousands of scientists from many countries all over the world are collaborating to find the answer to the fundamental question in elementary particle physics: What gives particles their mass? Physicists have noticed that some particles are light, whereas others are heavy. The effort to understand why this is so will move significantly ahead at two enormous detectors called the A Toroidal LHC Apparatus (AT-LAS) and the Compact Muon Spectrometer (CMS) located at the Large Hadron Collider (LHC). These detectors are being built at the European Laboratory for Particle Physics (CERN) in Geneva, Switzerland. The detectors operate in large underground caverns hundreds of feet below the surface, since the LHC is located underground. These detectors have evolved from a series of earlier collider detectors, which made many earlier discoveries at lower energies. However, each of these detectors is so large in scale that they may be thought of more appropriately as "laboratories" since there are thousands of physicists there conducting a large number of diverse precision measurements and searches for new phenomena.

What Questions Do the LHC Detectors Address?

The LHC detectors are intended to explore the source of electroweak symmetry breaking (EWSB): why the massless photon is in the same family as the massive *W* and *Z* bosons. Scientists know that some new physics relating to EWSB must appear at the TeV

mass scale (10^{12} electron volts [eV]). If there were no such new physics, including no Higgs boson, then the Standard Model would become inconsistent. In particular, the cross section for longitudinally polarized *W*'s would exceed the so-called unitarity bound. The simplest possibility is the single Standard Model Higgs boson. However, it is known that the Standard Model is incomplete because it does not include gravity. In this simplest possibility, one would not know why the Higgs mass is ~100 GeV (10^{11} eV vs. 10^{28} eV associated with gravity). There are three classes of models that have been widely discussed to resolve this question: (1) Extend the Standard Model to a larger symmetry called supersymmetry; (2) replace the Higgs boson with a dynamical condensate (technicolor is the prototype); or (3) extend the four-dimensional space-time at short distances to include extra dimensions.

In order to determine whether one or another of these possibilities exists, experiments will be performed to search for new particles produced in proton-proton collisions. The two large detectors must be sufficiently versatile to detect and identify what is produced when the protons collide and interact.

Although it is not known exactly what the new physics will be, there is confidence that measuring the products of the collisions will lead to the discovery and measurement of the properties of the new physics. In particular, each detector must be capable of measuring the momenta and directions of constituents of the proton (quarks and gluons), electrons, muons, taus, and photons and be sensitive to energy carried off by weakly interacting particles such as neutrinos that cannot be directly detected.

The new particles being sought are likely to be heavy themselves—in the 1 tera electron volt (TeV), or 10^{12} eV, range—so the LHC must have sufficient energy to produce these particles. The LHC is a proton-proton colliding beam accelerator. Two counterrotating beams of discrete bunches of protons of 7 TeV traveling in almost circular orbits collide at four locations around the 16-mile ring. The energy of the proton-proton collisions is 14 TeV. When the protons collide, the hard scattering of the constituents of the protons, the quarks and gluons, becomes of interest. Those constituents have only a small fraction of the momentum of each 7-TeV proton; however, masses of several TeV can be easily produced. Collisions of the proton bunches will occur every 25 ns (nanosecond = 10^{-9} s) with an average of twenty-five collisions in every bunch crossing when the accelerator reaches its design goals. This means that in a year of data accumulation, each of these detectors would witness 10^{16} collisions, while only a tiny fraction of those collisions would provide evidence of the search for the Higgs, for example. In order to reach the physics goals in this challenging environment of unprecedented energy and collision rates, these detectors are larger and more complex than previous detectors. Both experiments are under construction and expect the first collisions to occur in 2007.

How Does an LHC Detector Work?

Figure 1 shows schematically how different aspects of a collider detector are used to measure the

FIGURE 1

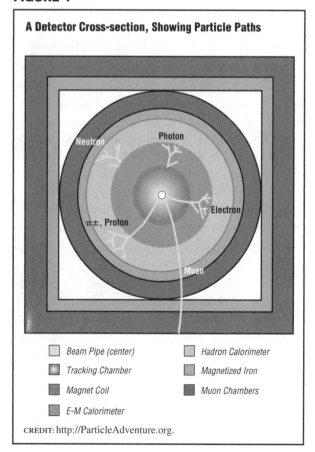

How different products of a proton-proton collision appear in a detector.

products of the collision. In a collision of the two protons of interest, hundreds of electrically charged and neutral particles are produced at the collision point and travel outward through the detector. The detectors is seen schematically as having many separate layers, one outside the next, each with different capabilities. All charged tracks are measured in the tracking chamber consisting of several layers. A critical part of the tracking detectors is the innermost pixel layers, which can identify b quarks (the next to heaviest of the six flavors of quarks). The presence of b quarks is crucial in certain scenarios. The pixels are made from tiny rectangles of silicon, which identify the position of a charged track very close to the collision point. The entire tracking volume is immersed in an axial magnetic field so that the charged particles bend as they emerge from the collision point and traverse the tracking detectors. The tracking detectors accurately measure the position of the charged particles, which pass through them so that the curvature and direction of the particles can be measured. The curvature is proportional to the momentum of the particle. The electromagnetic (EM) calorimeter identifies and measures the energy and direction of the electrons and photons. Both electrons and photons produce localized energy deposits in the EM calorimeter. However, photons have no charged track aiming at the energy deposit, while electrons have a charged track, with the momentum matching the energy seen in the EM calorimeter. The hadron calorimeter measures the direction and energy of the hadrons. The muons penetrate these inner layers and are identified as a charged track outside of the calorimeters. The momentum of the muon is measured by determining the curvature of the muons in the outer muon detectors, which are located in a magnetic field. The collisions also produce neutral particles such as neutrinos, which are only inferred by taking measurements of all the other particles in the collision and calculating what is missing.

ATLAS (see Figure 2) and CMS (see Figure 3) use some similar and some different detector technologies to accomplish these functions. ATLAS will be as tall as a seven-story building and weigh about 5,500 tons (note the relative size of the people in the figure!). Both have silicon pixel and strip detectors as the first element of the tracking region. The pix-

FIGURE 2

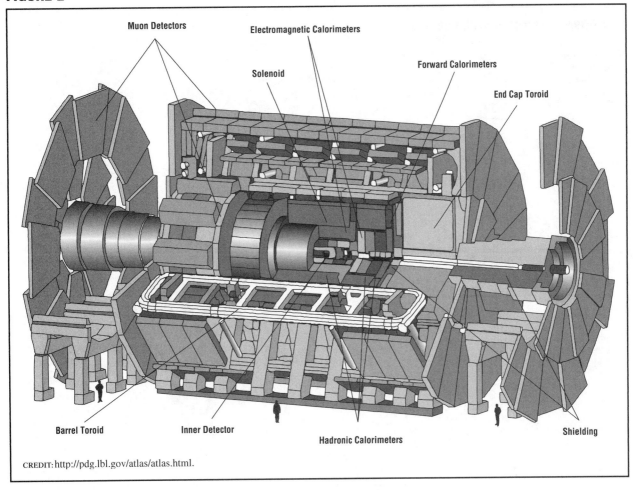

CREDIT: http://pdg.lbl.gov/atlas/atlas.html.

The ATLAS Detector

els in the case of ATLAS are about 10^8, $50 \times 400 \ \mu m^2$, arranged in cylinders and disks all within a radius of about 8 inches from the beam axis. These small dimensions are used to search for charged tracks from heavy quarks (*b* or *c* quarks) that result from secondary vertices very close to the primary vertex or collision point. In CMS the entire tracking volume outside the pixel layers is composed of silicon-strip detectors, whereas ATLAS uses silicon-strip detectors to a radius of 22 inches. ATLAS then employs a Transition Radiation Tracker (TRT) beginning at the outer radius of the silicon-strip cylinders and disks to a radius of about 42 inches. The TRT not only measures the position and curvature of charged tracks (adding to the information obtained from the silicon layers) but also can identify electrons by the transi-

tion radiation (X-ray photons) they produce when traversing layers of differing indices of refraction.

A primary difference between the two experiments is the EM calorimetry. CMS uses a large array of lead tungstate crystals that produce light proportional to the EM energy. ATLAS uses a liquid argon calorimeter consisting of radial layers of accordion-shaped stainless-steel-coated lead plates separated by thin layers of liquid argon and electrodes. The movement of the free electrons produced by the electromagnetic shower in the liquid argon is measured as a current on the electrodes. Each technique has advantages and disadvantages. The energy resolution is superior for the crystals, whereas the liquid argon has a more stable response. Also, the liquid argon

FIGURE 3

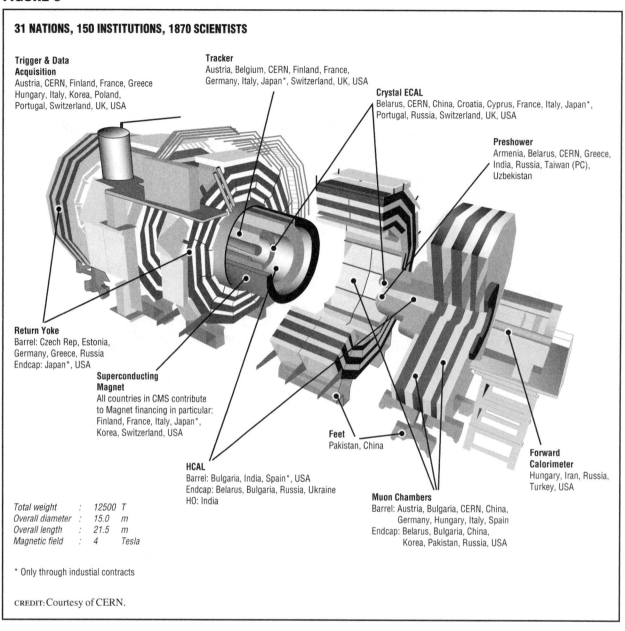

31 NATIONS, 150 INSTITUTIONS, 1870 SCIENTISTS

Trigger & Data Acquisition
Austria, CERN, Finland, France, Greece
Hungary, Italy, Korea, Poland,
Portugal, Switzerland, UK, USA

Tracker
Austria, Belgium, CERN, Finland, France,
Germany, Italy, Japan*, Switzerland, UK, USA

Crystal ECAL
Belarus, CERN, China, Croatia, Cyprus, France, Italy, Japan*,
Portugal, Russia, Switzerland, UK, USA

Preshower
Armenia, Belarus, CERN, Greece,
India, Russia, Taiwan (PC),
Uzbekistan

Return Yoke
Barrel: Czech Rep, Estonia,
Germany, Greece, Russia
Endcap: Japan*, USA

Superconducting Magnet
All countries in CMS contribute
to Magnet financing in particular:
Finland, France, Italy, Japan*,
Korea, Switzerland, USA

Feet
Pakistan, China

Forward Calorimeter
Hungary, Iran, Russia,
Turkey, USA

HCAL
Barrel: Bulgaria, India, Spain*, USA
Endcap: Belarus, Bulgaria, Russia, Ukraine
HO: India

Muon Chambers
Barrel: Austria, Bulgaria, CERN, China,
Germany, Hungary, Italy, Spain
Endcap: Belarus, Bulgaria, China,
Korea, Pakistan, Russia, USA

Total weight	:	12500	T
Overall diameter	:	15.0	m
Overall length	:	21.5	m
Magnetic field	:	4	Tesla

* Only through industial contracts

CREDIT: Courtesy of CERN.

The CMS Detector.

calorimeter is subdivided into more transverse and longitudinal layers to allow better photon and electron identification. Both detectors use scintillator tile hadronic calorimeters. These tiles produce light when particles pass through them. When the hadrons interact in the absorber material of the calorimeter, the light produced is proportional to the energy of the hadron. This light is channeled to a photomultiplier tube, which transforms the light into an electrical signal.

The names of the two detectors refer to the different methods used to identify and measure muons. CMS has a larger central field and measures the momentum of the muon in the inner tracking volume while using the outer muon layers for identification

FIGURE 4

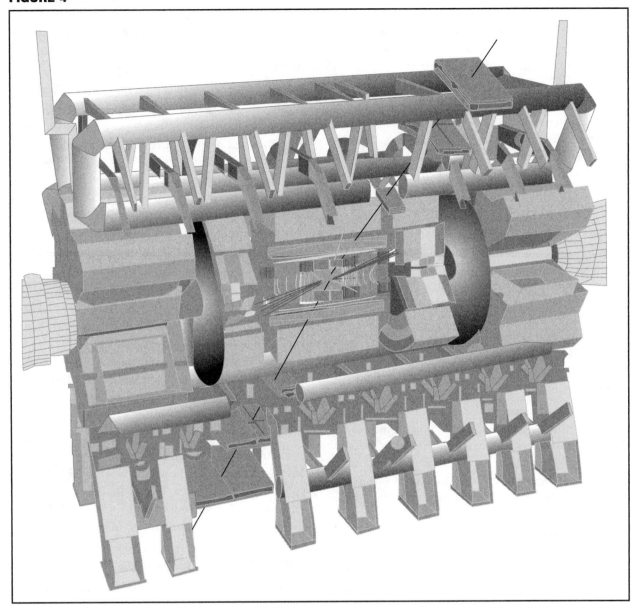

A Higgs event decaying into two electrons (grey showers in the EM calorimeter) and two muons (tracks going through to the outer muon chambers) in the ATLAS Detector. CREDIT: COURTESY OF LAWRENCE BERKLEY NATIONAL LABORATORY.

and triggering. ATLAS employs a large air-core toroidal magnet system to measure muons independently outside of the inner tracking volume. Both detectors have sophisticated trigger systems. The purpose of the trigger is to make a fast selection (in a few microseconds) as to whether the particular collision is likely to contain an event that may be a signature of one of the physics quests. For example, events with two EM energy depositions above 30 GeV with no charged track pointing to them would be a candidate for a Higgs particle. Another event with three or four muons with momentum greater than 10 GeV would also be a Higgs candidate decaying into two photons. An example of how one particular event appears in the ATLAS detector can be seen in Figure 4.

Much of the capability of these detectors follows from advances in technology. For example, both detectors use superconducting magnets, which allow a higher magnetic field than conventional magnets. The miniaturization of electronics allows the subdivision of the detector into very small parts to allow measurement of the complex collisions.

The main reason to have two detectors with different technical approaches is to ensure that when a potential discovery is observed, it is not the result of an instrumental effect. An independent experimental method and team can verify the discovery or reject the data.

How These Collaborations Work

The ATLAS collaboration includes about 1,850 physicists and engineers from 175 institutes in 34 countries. CMS has a similar list of participants often from the same countries, but it does not completely overlap. Most institutes participating in the LHC have joined only one of these two collaborations. Each institute has specific responsibilities as formalized in a "Memorandum of Understanding." Financial support comes from the respective governmental funding agencies.

A heavily documented process exists for first establishing the objectives of each detector and then developing detailed technical specifications. Design is next completed with a full prototyping of each component. Testing, installation, commissioning, and operations follow fabrication. Many committees are asked to review this project at each step of the way. Approvals are required before progressing to the next step.

Groups that have specific expertise in some aspect of detector technology form each of these collaborations. For example, one group may have expertise in growing crystals, another in testing the uniformity of light output of the crystals, another in the readout of the light from the crystals, etc. Each of the collaborations has an overall management and is then organized into subsystems. A subsystem may include the overall tracking or only the pixel detectors. Each subsystem has a series of subgroups. An example is the Liquid Argon group for ATLAS, which has subgroups for barrel mechanics, endcap mechanics, cryogenics and feedthroughs, front-end electronics, control of voltages and temperatures, etc. The central leadership for each collaboration includes a technical coordination group that addresses the overall configuration and integration (Do all the detectors fit together?), the routing of services (electric power, cooling, and signals), and installation and access of the detectors for repair.

These collaborations have organized meetings to resolve specific design issues and to divide the work. These meetings may occur all over the world, often over the telephone or through video conferencing, but they are mostly held at CERN. Decisions and technical specifications are documented in technical design reports, drawings, and other documents available on the World Wide Web (which was, in fact, invented at CERN by particle physicists).

These state-of-the-art detectors rely on advanced technology that can withstand a harsh radiation environment, which is highest at a small radius and in the forward and backward directions. Every component of the detector must be able to survive in the radiation levels predicted for its location. The scientists designing these experiments have worked closely with industry to use existing electronic technologies as well as develop new ones that will operate in the inner regions of the detector.

Leading industrial companies from all over the world fabricate the components of the detector. Many of these components are assembled at various collaborating institutes. Final installation and commissioning of each component occur at CERN with the participation of the collaborating teams.

Computing and Data Analysis

Once a collision occurs, the data from each detector must be held in a buffer until the trigger logic can decide whether to keep the event for future study. A complete event in ATLAS may contain 1 megabyte of information. If there were no trigger, one would need to analyze 10^{15} collisions times 10^6 bytes = 10^{21} bytes per year! This amount of data would be impossible to deal with and so a selection—the trigger—must be made. The purpose of the trigger is to reject the "normal" events while retaining with good efficiency those events that may contain rare new physics signatures such as the Higgs. The multistep trigger reduces the data sample to about 100 events per second, which must be recorded for

FIGURE 5

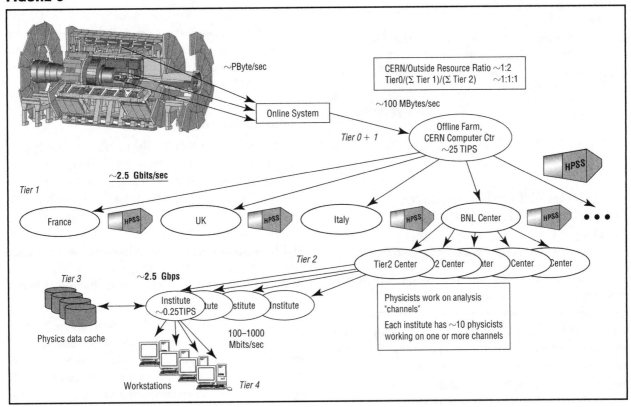

Hierarchical computing plan for ATLAS.

further analysis. One hundred events recorded per second times 1 megabyte per event equals 100 megabytes per second. This means that 1 petabyte (10^{15} bytes) will be produced by each experiment for each year of 10^7 seconds of data taking.

The computing power needed to analyze this huge amount of data is larger than what is available. These experiments are actively participating in the development of a data grid to facilitate such analysis. The word "grid" as used here is analogous to the power grid: users who need computing resources, data, or computational power will be able to request this from their home institutions, which will in turn be connected to all the collaborating institutions within a tiered structure (see Figure 5).

International groups will analyze the data. Members of these will be in constant electronic contact. When an analysis reveals something new and exciting, presentations will be made to the group and eventually to the entire collaboration. When the col-

laboration is sure that the new piece of physics is convincingly demonstrated to exist, a scientific paper will be written and published. No one knows for sure where this new understanding of the universe will lead, but one has only to look back 100 years to the discovery of the electron to realize how much of everyday life changed with the electron's discovery.

See also: CERN (EUROPEAN LABORATORY FOR PARTICLE PHYSICS); DETECTORS AND SUBSYSTEMS

Bibliography

Adams, S. *Particle Physics* (Heinemann Educational, Oxford, UK, 1998).

ATLAS. <http://pdg.lbl.gov/atlas.html>.

Barnett, R. M.; Muehry, H.; and Quinn, H. *The Charm of Strange Quarks: Mysteries and Revolutions of Particle Physics* (Springer-Verlag, New York, 2000).

Close, F.; Marten, M.; and Sutton, C. *The Particle Explosion* (Oxford University Press, New York, 1987).

CMS. <http://cmsdoc.cern.ch/cms/outreach.html>.

Fraser, G. *The Quark Machines: How Europe Fought the Particle Physics War* (Institute of Physics Publishing, Bristol, UK, 1997).

Kane, G. *The Particle Garden: Our Universe As Understood by Particle Physicists* (Addison-Wesley, New York, 1996).

Weinberg, S. "A Unified Physics by 2050?" *Scientific American* **281**, 68–75 (1999).

Howard A. Gordon

CASE STUDY: LONG BASELINE NEUTRINO DETECTORS, K2K, MINOS, AND OPERA

Neutrinos are elementary particles, which are still very poorly understood, mainly because their interactions with matter are very weak and thus difficult to study. Neutrinos come in at least three different species, electron, muon, and tau. If neutrinos have mass, they can oscillate between different species as they propagate through vacuum or matter, that is, a beam of initially pure muon neutrinos can develop an electron or tau neutrino component as it travels through space. The distance over which this happens, that is, characteristic wavelength of oscillations, is inversely proportional to difference in mass squared between different kinds of neutrinos and directly proportional to the neutrino energy.

There is evidence for the possible existence of oscillations from studies of neutrinos produced in the Sun and of neutrinos produced in the Earth's atmosphere. Efforts are under way to study both of these phenomena in more detail and under more controlled conditions. The phenomena observed in solar neutrinos appear to have mass squared difference such that they are best studied with reactors, which produce low-energy neutrinos. The atmospheric neutrino phenomena appear to have a mass squared difference such that they can be best studied with accelerator neutrinos and source to detector distance of the order of several hundred kilometers. They are generally referred to as long baseline neutrino oscillation experiments.

Currently there are three accelerator experiments, running or in preparation, designed to study atmospheric neutrino phenomena. They are K2K in Japan, OPERA in Europe, and MINOS in United States. Even though their general goal—study of oscillations in the atmospheric neutrino region—is the same, their specific goals are different, and thus they require different beams and detectors.

Neutrino Production

Neutrino production in an accelerator begins with acceleration of protons. All three experiments under discussion use a circular accelerator: KEK Proton Synchrotron (PS) in Tsukuba, Japan, with a peak energy of 12 GeV; Super Proton Synchrotron (SPS) in the European Laboratory for Particle Physics (CERN), spanning the French-Swiss border region near Geneva, with a peak energy of 400 GeV; and the Main Injector (MI) at Fermilab in Batavia, Illinois, with a peak energy of 120 GeV.

Once the peak energy is reached, the proton beam is extracted from the accelerator and allowed to impinge on a target. The resulting proton interactions with the nuclei in the target produce secondary particles, the most numerous of which are pi mesons, and they are the parents of most of the neutrinos used in the experiments. Most of the secondaries produced travel in a rather wide cone in a forward direction, and thus if nothing were done to contain them, they would disperse over a large area. To avoid this, a focusing system is constructed for the pions with energy of interest, which compresses those with energy of interest into a much narrower cone, much like the parabolic mirror in a flashlight focuses the beam of light from its lightbulb. The focusing system consists of magnetic horns, which are specially shaped conductors through which a current is pulsed when the beam arrives. The resulting magnetic field bends most of the desired pions into a relatively parallel beam.

The pion beam is then allowed to enter a decay volume, an evacuated pipe with length of the order of 100 to 1,000 meters. As the pions travel through this pipe, some of them decay into a muon and a muon neutrino. The neutrino is emitted into a very narrow cone along the pion direction. The length of the decay pipe is generally proportional to the chosen pion energy, which in turn is related to the wanted neutrino energy: the very forward decay neutrinos take 42 percent of the pion energy. The decay volume is generally followed by earth shielding to stop the undecayed pions and the decay muons. Neutrinos, being very weakly interacting, pass easily through the earth shield and continue on to the detector. Besides the far detector, located several hundred kilometers away, long baseline experiments also frequently have a near detector, located just downstream of the absorber shield. Its purpose is to measure the properties of the neutrino beam before the neutrinos had a chance to oscillate.

Detectors

Neutrino detectors tend to be very massive so as to obtain a sufficiently large sample of neutrino interactions in spite of their very weak interactions. Large size is even more important in long baseline experiments since their detectors are far from the source, and thus the neutrino beam has diverged significantly before it arrives there. The detailed design of the detector depends on the specific goal of the experiment and sometimes on special circumstances.

All three experiments under discussion here locate the detectors deep underground, that is, on the order of a kilometer or more below the surface. The reason for this is to suppress the cosmic ray background, which on the surface is about 100 particles/s m^2. Because there will be relatively few neutrino interactions, one must suppress all possible sources of background as much as possible. Additional background suppression is obtained from the directionality of the events observed and the time at which they occur, since the accelerator beam is pulsed and on only a fraction of the time.

Detectors are built by the collaborations doing the experiment. Once its general structure is agreed on and the design complete, different institutions assume responsibilities for different subsystems of the detector. Frequently those subsystems are built and tested at home institutions far away from the detector site. The lifetime of a typical neutrino experiment is long, several years or more, and this required longevity is one of the factors influencing the design.

Collaborations

The formation and operation of collaborations performing long baseline neutrino oscillation experiments is qualitatively very similar to what happens in other large-scale particle physics experiments. These experiments are complex and of long duration; it takes a large group of scientists and engineers to perform them. A collaboration typically involves 100 to 200 people coming from fifteen to forty different institutions, primarily national laboratories and universities. The collaborations tend to be international in scope with the majority of institutions from the general region where the experiment is being performed.

The experiments tend to be initiated by a smaller number of scientists brought together by a common scientific interest. Frequently these people have worked together before or are working on a common experiment at the time of the new proposal. This group will generally be too small to construct and carry out the experiment but will be large enough to do the initial preliminary design to establish viability of the experiment. It may then take several years to obtain scientific approval and the required financial resources from the host laboratory and/or the appropriate funding agencies. These steps are usually somewhat different in different regions (United States, Western Europe, or Japan) and are frequently influenced by the potential existence of some relevant infrastructure, for example, beam line or detector.

Once this initial phase is completed, the collaboration will grow to the required size. Other groups may be invited to join since they may possess required expertise for some subsystem of the detector; alternatively, groups can express interest in participation on their own. In parallel, a detailed design of the required apparatus and software is made and responsibilities for different subsystems are assigned to specific individuals and/or groups. During the next phase (research and development [R&D], prototyping, testing, and construction), the work is done at home institutions. The different groups then deliver their hardware to the beam or detector site where the whole system is put together with the participation of members of various institutions. The checkout and subsequent data taking follows afterward and is a collaborative effort with subsystem experts playing a prominent role.

Early in its life the collaboration adopts a constitution defining governance and decision making.

This will vary in different collaborations, but generally there will be one or more spokespersons who act as representatives of the collaboration vis-à-vis the outside world and as CEOs of the collaboration, a policy-making board composed of senior members of the collaboration, and an institutional board where all institutions have a voice. The role of these groups, the method of their selection, and their term of office are spelled out in the constitution. In addition, different individuals are given formal responsibilities for different technical components, and the overall technical direction may reside in a group of individuals forming a technical board.

Frequently there is a parallel structure with fiduciary and management responsibilities, which is appointed by and reports to the host laboratory that provides the funding. This organization is headed by a project manager, frequently reporting directly to the laboratory director. The project manager is responsible for the appointment of individuals to head the work on different subsystems. The project thus relieves the collaboration as an organization from most of the management responsibilities in the technical area even though the individuals assigned these responsibilities will be members of the collaboration and come both from the host laboratory and other participating institutions. The project organization is generally formed during the latter part of the design phase and exists for the duration of the construction of the experiment.

The communication between members of the collaboration occurs in a variety of ways: collaboration meetings that generally occur several times per year as well as meetings, conferences, and workshops organized by smaller subgroups of the Collaboration. They can be face-to-face meetings or via telephone or videoconferences. The subgroups tend to be organized around specific technical subsystems during the construction phase and around a physics or software topic during the data taking and analysis phase. The Internet plays an important role in collaboration communication.

K2K Experiment

The K2K experiment uses the neutrino beam created by the KEK (National Laboratory for Particle Physics in Japan) proton synchrotron and the Su-

per-Kamiokande detector about 250 kilometers away. The latter is located in a working zinc mine with about 1,000 meters overburden of rock and earth above it. Super-Kamiokande consists of a tank filled with 50 kilotons of purified water, covered on its inside surface with about 11,146 twenty-inch photomultipliers (PMTs). Neutrino interactions produce charged particles, most of which emit Cerenkov light in a cone around the trajectory of the particle. This light is detected by the photomultipliers, and the nature of the event is subsequently deduced from the pattern of the PMT hits.

The origin of this experiment is somewhat different from those of MINOS and OPERA insofar that the detector already existed, having been built previously for other reasons: the study of solar and atmospheric neutrinos, the detection of neutrinos from future supernovas, and the search for nucleon decay. The additional elements required were a neutrino beam line and a near detector, requiring significantly less resources than the Super-Kamiokande demanded originally. Thus the experiment is quite cost effective. The K2K Collaboration, composed mainly of Japanese and U.S. groups, has a very strong overlap with the Super-Kamiokande collaboration. Thus, many of the steps generally required for initiating an experiment were avoided in this case since they had been taken earlier for the Super-Kamiokande experiment.

The main goal of the experiment is verification of the existence of oscillations in the atmospheric neutrinos with a well-controlled accelerator experiment. The atmospheric neutrino studies indicate that muon neutrinos, produced at the top of the atmosphere, are depleted as they travel through the Earth and/or atmosphere. The dependence of the effect on the zenith angle of the neutrinos, equivalent to the total pathlength traveled, strongly favors the oscillation interpretation. The K2K beam is essentially a pure muon neutrino beam, whose characteristics, such as intensity and energy, can be calculated and also verified in a near detector. The researchers try to see whether the observed neutrino interaction rate is different from the no-oscillation prediction and if it is, to determine the energy dependence of the effect.

The projected event rate is modest, about two hundred observed events in a four-year run without oscillations, less if oscillations exist. The modest rate is due to the relatively low intensity of the KEK accelerator and low energy of the neutrino beam, peaking around 1 GeV. The experiment commenced taking data in 1999 and was scheduled to run through 2003. Toward the end of 2001, an implosion occurred in the Super-Kamiokande tank that destroyed over 50 percent of the photomultiplier tubes used in the detector. Restoration of the experiment, with a less dense photomultiplier coverage, is underway. Data taking is scheduled to resume at the beginning of 2003.

OPERA Experiment

The OPERA experiment involves a neutrino beam produced by the CERN SPS and a detector to be located in the Gran Sasso Laboratory in Italy, 732 kilometers away. Both the beam and detector are being constructed specially for this experiment.

The goal of the OPERA experiment is the observation of interactions of tau neutrinos, which are expected to be the main end product of muon neutrino oscillations. In other words, a relatively pure muon neutrino beam produced at CERN slowly develops a tau neutrino component as it travels through the Earth. The main challenge for the experiment lies in the fact that tau neutrino interactions are difficult to identify. The unambiguous signature of a tau neutrino interaction is the production and decay of a tau lepton. Because the tau lepton is very short lived—in the OPERA experiment its typical length is of the order of 1 millimeter—the detector has to have very good spatial resolution. At the same time it has to be massive to observe a significant number of events. Accomplishment of these two goals simultaneously is difficult.

The basic elements of the OPERA detector are modules composed of sandwiches of sheets of iron and photographic emulsion coated plastic. The iron provides target material; the emulsion provides a detecting medium with one micron resolution and hence the capability of observing tau events. The experiment presents a number of challenges of which the most formidable are identification of the interaction within a small volume (of the order of a few cm^3) and the need to process—scan and measure—very large volumes of emulsion. The former is han-

dled by interleaving electronic detectors in between the layers of iron/emulsion modules, information which allows one to locate the vertex. The ability to handle the second challenge is the result of many years of development of automated emulsion scanning and measuring techniques, principally by a group in Nagoya, Japan.

A neutrino interaction is identified by the electronic detectors and in most cases can be localized to a given iron/emulsion module. Periodically, roughly once a day, the modules with neutrino interactions are pulled out of the detector and developed underground so as not to contaminate the emulsion with cosmic rays. The emulsion in the vicinity of the identified vertex is then scanned quickly to search for possible evidence of tau production and decay. The potential tau candidates are subsequently subjected to additional and more sophisticated analysis.

Simulations show that tau events can be identified with negligible background from other, non-tau, neutrino interactions. This is essential to the success of the experiment since the tau neutrino production rate and detection efficiency is such that for the oscillation parameters suggested by the Super-Kamiokande results, one can expect only about twenty observed and identified events in five years of running. It is hoped that the experiment will start data taking in 2005, but the financial situation at CERN may necessitate delay. The collaboration consists of thirty-three groups as of 2002, mainly from Europe and Japan.

MINOS Experiment

The MINOS experiment uses the neutrino beam from the Main Injector accelerator at Fermilab and a detector in the former iron mine in Soudan, Minnesota, 700 meters underground and 735 kilometers away. The mine is currently run as a state park, and the experiment relies on the infrastructure provided by the park; an additional cavern was excavated to house the MINOS detector.

The main goal of the experiment is to measure the oscillation parameters by studying the disappearance of muon neutrinos. A nearby detector on the Fermilab site is used to measure the properties of the neutrino beam. Those measurements allow prediction of what the beam will be at the Soudan location. Deviations from the predicted intensity and energy spectrum would be evidence for oscillations, and their quantitative study would determine oscillation parameters (for example, difference in mass squared of different neutrino states) with a precision of about 10 percent.

The neutrino beam is designed so that it can provide beams with different energies by changing the location of the target and the focusing elements. The energy range that can be covered extends from 1 to 20 GeV. The initial running will be with a low-energy beam configuration that has a spectrum centered at about three GeV, since that configuration appears to be best for the measurement of oscillation parameters suggested by the Super-Kamiokande experiment. One expects to see about 1,000 neutrino interactions per a year.

The 5.4-kiloton far detector will be composed of 486 alternating layers of 1-inch steel plates and scintillator planes in the shape of 8-meter wide octagons. The scintillator planes consist of 4.1-centimeter wide strips and enable both positional information and energy information to be obtained. The iron will be magnetized to enable the muons produced in the neutrino interactions to have their energies measured by curvature.

The detector assembly commenced in July 2001 and should be completed in 2003. The conventional construction at Fermilab should be finished sometime in 2003 at which point the installation of the beam technical components and the near detector will start. The first beam is expected toward the beginning of 2005.

The MINOS Collaboration is composed of about 175 scientists and engineers from thirty institutions in five countries. The majority of the institutions are in the United States and in the United Kingdom.

Data Analysis

In most particle physics experiments the data arrive as electronic signals. After some processing and filtering online, the data are stored on some kind of mass storage device for offline analysis. Generally the host laboratory acts as a repository for the data, but the data are readily available to the collaborating institutions. This pattern will apply to both K2K and MINOS experiments. Because the number of events involved will be much less than for a typical particle physics experiment, the data handling issues here are relatively simple.

In the OPERA experiment the situation is more complex because the essential information consists not only of the digital data from electronics but also of the pattern of developed grains on the photographic emulsion. Thus the data analysis will involve a considerable amount of scanning and measuring of emulsions before all the data can be reduced into digital format. The scanning and measuring phase will take place at the home location of several of the collaborating institutions.

The collaborations as a whole organize the data analysis with specific responsibilities assigned to different institutions and/or individuals. This division is generally organized around physics topics, but in the initial stages different groups focus on work needed to understand different subsystems and features of the detector, such as efficiency, resolution, systematic uncertainties, etc. Graduate students generally assume responsibility for some physics topic, which then becomes the subject of their Ph.D. thesis.

See also: DETECTORS AND SUBSYSTEMS; NEUTRINO

Bibliography

Ahmad, Q. R., et al. "Measurement of the Rate of ν_e + d → p + p + e$^-$ Interactions Produced by ^{8}B Solar Neutrinos at the Sudbury Neutrino Observatory." *Physical Review Letters* **87**, 071301-1–071301-6 (2001).

Apollonio, M. et al. "Limits on Neutrino Oscillations from the CHOOZ Experiment." *Physics Letters*, **B466**, 415–430 (1999).

Bahcall, J. N.; Calaprice, F.; McDonald, A. B.; and Totsuka, Y. "Solar Neutrino Experiments; The Next Generation." *Physics Today* **49**(4), 30–36 (1996).

Boehm, F., and Vogel, P. *Physics of Massive Neutrinos* (Cambridge University Press, Cambridge, UK, 1992).

Fisher, P.; Kayser, B.; and McFarland, K. S. "Neutrino Mass and Oscillation." *Annual Review of Nuclear and Particle Science* **49**, 481–527 (1999).

Fukuda, S., et al. "Solar ^{8}B and hep Neutrino Measurements from 1258 Days of Super-Kamiokande Data." *Physical Review Letters* **86**, 5651–5655 (2001).

Fukuda, Y., et al. "Evidence for Oscillation of Atmospheric Neutrinos." *Physical Review Letters* **81**, 1562–1567 (1998).

Kayser, B.; Gibrat-Debu, F.; and Perrier, F. *The Physics of Massive Neutrinos* (World Scientific, Singapore, 1989).

Kim, C. W., and Pevsner, A. *Neutrinos in Physics and Astrophysics* (Harwood Academic Publishers, Langhorne, PA, 1993).

Klapdor-Kleingrothaus, H. V., and Staudt, A. "Neutrino Os-cillations" in *Non-accelerator Particle Physics* (Institute of Physics Publishing, Philadelphia, 1995).

Nakamura, K. "Status of K2K." *Nuclear Physics B (Proc. Suppl.)* **91**, 203–204 (2001).

Rubbia, A. "ICANOE and OPERA Experiments at the LNGS/ CNGS." *Nuclear Physics B (Proc. Suppl.)* **91**, 223–229 (2001).

Wojcicki, S. "Status of the MINOS Experiment." *Nuclear Physics B (Proc. Suppl.)* **91**, 216–222 (2001).

Stanley G. Wojcicki

CASE STUDY: SUPER-KAMIOKANDE AND THE DISCOVERY OF NEUTRINO OSCILLATIONS

In the 1970s theoretical physicists were hot on the trail of what they believed would be a Grand Unified Theory. They hoped that this theory would join three of the four forces of nature together into one theoretical construct. It would provide a unified explanation for the electromagnetic, strong, and weak interactions. This theory, named SU(5) after the mathematical group that it contained, predicted that the proton, until then thought to be an absolutely stable particle, would decay into other particles. The observation of proton decay would therefore be evidence for grand unification, and several experimental groups were formed to attempt this search. Unfortunately, this search would not be an easy one, as the predicted lifetime of the proton (or bound neutron) was 10^{28-30} years!

Luckily, proton decay is a statistical process. If one were to watch a single proton and wait for it to decay, one would expect to wait a very long time, but if a large number of protons were watched simultaneously, say, 10^{30} protons, one of these protons would be expected to decay in a year. Thus, the search for nucleon decay requires massive detectors. A search with a sensitivity of 10^{33} years requires a detector with approximately 10^{33} nucleons. Since there are 6×10^{29} nucleons per ton of material, this implies detectors of kiloton scale. Over the past two decades, there have been two types of nucleon decay detectors: water Cherenkov detectors such as IMB, Kamiokande, and Super-Kamiokande; and fine grain sampling detec-

tors such as NUSEX, Frejus, and Soudan. The largest and most sensitive of these detectors is the Super-Kamiokande water Cherenkov detector, which was completed and began taking data in 1996.

Water Cherenkov Detectors

A water Cherenkov detector consists of a large volume of purified water viewed by photomultiplier tubes (very sensitive photocells that convert light into an electrical signal). These massive detectors employ water both as an inexpensive target material and the detection medium. When a charged particle passes through a transparent material with a speed faster than the speed of light in that material, it emits light (Cherenkov light). The light is given off at a constant angle to the particle track. The angle is determined by the particle speed and the index of refraction of the medium. For a charged particle traveling in water at a speed close to that of light in vacuum, this angle is about 42°. In the case of a track that begins and ends in the detector, a ring of Cherenkov light is formed on the detector wall. By measuring the time of arrival of each photon at a photomultiplier tube, and by knowing the angle that the photons made with the particle track, it is possible to fully reconstruct the particle track, including its direction and its beginning (vertex) and ending points.

The Super-Kamiokande detector (see Figure 1) is a 50,000 metric ton (55,000 ton), water Cherenkov detector located at a depth of 1,000 m (3,300 ft) of rock in the Mozumi zinc mine in central Japan. It consists of a cylindrical stainless steel tank, 39 m (128 ft) in diameter and 41 m (135 ft) high, filled with purified water. The detector is optically segmented into an inner volume, 34 m (112 ft) in diameter and 36 m (118 ft) in height, and an outer (anticoincidence) region, 2.5 m (8 ft) thick, on the top, bottom, and sides of the inner volume. The inner detector is viewed by 11,146 photomultiplier tubes, 50 cm (20 in.) in diameter, uniformly distributed on the inner boundary. The total mass of water inside the surface of the inner detector defined by the photomultiplier tubes is 32,000 metric tons (35,000 tons). The fiducial mass, defined to be 2 m (6.5 ft) inside the photomultiplier tube plane, is 22,500 metric tons (25,000 tons).

The outer annulus of the detector is called an anticoincidence region as it is used to tag entering charged particles as well as to attenuate low-energy

FIGURE 1

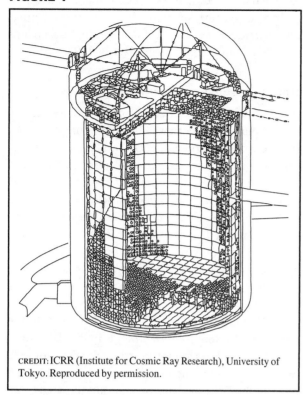

CREDIT: ICRR (Institute for Cosmic Ray Research), University of Tokyo. Reproduced by permission.

A sketch of the Super-Kamiokande detector showing the inner and outer regions.

gamma rays and neutrons, which cause undesirable background events in the sensitive volume. The outer annulus also complements the measurement of particle energy in the inner detector by measuring the energy loss due to exiting particles. This outer detector region is viewed by 1,885 photomultiplier tubes, 20 cm (8 in.) in diameter, with additional light collection afforded by attached plastic plates. The walls of the anticoincidence region are reflective to enhance light collection. The photomultiplier tubes are mounted facing outward on the same superstructure as the 50 cm (20 in.) photomultiplier tubes of the inner volume. Also, an optical barrier is mounted on the same structure to separate the inner and outer regions.

Results on Proton Decay

Background for nucleon decay arises from interactions between the target material and muons and neutrinos produced by cosmic-ray interactions in the upper atmosphere. By locating the detectors underground, experimenters can reduce cosmic-ray muons to a manageable level, but since neutrinos penetrate large thicknesses of material easily, some neutrino background is unavoidable. The vast majority of atmospheric neutrino interactions bear little resemblance to nucleon decay, but a small fraction are indistinguishable (based on topology and kinematic parameters) from the signal. At present, Super-Kamiokande has found no candidates that can be unambiguously ascribed to baryon-number-violating nucleon decay, but as described below, the study of the neutrino backgrounds led to a very important discovery.

Grand unified theories continue to predict a broad range of possible proton lifetimes. There is evidence that the fundamental approach to unification

FIGURE 2

The Super-Kamiokande neutrino detector contains 50,000 tons of ultra-pure water surrounded by thousands of phototubes that catch flashes of light from neutrino interactions in the water. CREDIT: ICRR (INSTITUTE FOR COSMIC RAY RESEARCH), THE UNIVERSITY OF TOKYO. REPRODUCED BY PERMISSION.

is sound, and nucleon decay is one of the few accessible regimes where grand unified theories can be directly confronted with experimental data. Further progress in the detection of this unique process may be crucial to the future development of particle physics; this strong possibility indicates that the search for evidence of nucleon decay should be pursued with renewed vigor.

Atmospheric Neutrinos: From Background to Signal

The search for nucleon decay naturally yields a sample of atmospheric neutrino interactions fully contained in Super-Kamiokande's inner detector volume. Although these atmospheric neutrinos present a background to some physics searches (e.g., proton decay and extraterrestrial neutrino astronomy), they are a rich source of physics in their own right. They provide, for example, a unique opportunity for the study of the neutrinos themselves. In fact, the story of Super-Kamiokande is a classic example of that old bit of wisdom in experimental science which states: One person's background is another person's signal.

Atmospheric neutrinos are generated in cosmic-ray interactions in the atmosphere: a cosmic ray (either proton or nucleus) interacts, producing a cascade of secondaries; the secondaries (typically pions and kaons) may either interact again or decay. This cascade continues until the primary particle's energy is dissipated in the atmosphere, primarily by ionization. In such a cascade, the number of positively charged secondaries will be nearly equal to the number of negatively charged secondaries. Thus, if one assumes that all low-energy pions decay and that their muon secondaries also decay, simple counting of the daughters allows the following conclusion: there will be twice as many muon neutrinos as electron neutrinos. These estimates are borne out by sophisticated Monte Carlo calculations of the atmospheric flux. The atmospheric flux covers a broad range of neutrino energies from 10s of MeVs to 10s of TeVs and beyond.

Super-Kamiokande is the largest existing underground detector of atmospheric neutrinos. During its first 1,489 days of operation, it collected about 12,000 events in which a neutrino interacted in the water and the resulting charged particle was fully captured and identified in the detector. This is by far the world's largest sample of atmospheric neutrino interactions and constitutes a valuable resource—the interpretation of these events has given new insight into the question of neutrino oscillations.

In addition, the detector records a vast number of entering (meaning some light is seen in the anti-coincidence volume of the detector) particles, about three each second. Of these, a small fraction, about one every other day, enters from below the horizon. These are of special interest since they are mainly muons produced in muon neutrino interactions in the rock surrounding the detector. Such neutrinos travel large distances, up to 12,000 km between production and interaction. As shall be seen in the next section, the distance neutrinos have traveled since production is of central importance.

Neutrino Oscillations

There are three known types of neutrinos: the electron neutrino, the muon neutrino, and the tau neutrino. Although neutrinos are one of the fundamental constituents of matter, until recently, whether or not they had any mass was unknown. If neutrinos do have mass, then it is possible for one type to spontaneously change into another and then back again as it travels through space. This process is called neutrino oscillations. The parameters describing this process involve the difference in the square of the individual neutrino masses (δm^2) and the probability of the oscillation occurring (known as the mixing probability). The mixing probability is written as the square of the sine of twice the mixing angle, $\sin^2 2\,\theta$. Full mixing would be $\sin^2 2\,\theta = 1.0$. The mathematics of this process predicts that the oscillation length (or wavelength of the oscillation) should be proportional to the energy of the neutrino divided by δm^2. This means that a very small δm^2 produces a very long oscillation length.

As previously mentioned, atmospheric neutrinos are produced in the interactions of primary cosmic rays in the atmosphere. They penetrate the Earth from all directions, arriving almost isotropically at the detector. On their way to the detector, they travel distances from about 10 km (when coming directly from above) to about 12,000 km (when they come from below the detector). Such large oscillation distances are not available with other, humanmade neutrino beams in this energy range and for these flavors of neutrinos. The large distances together with the low mean energy of the neutrinos provides an opportunity to study oscillations with δm^2 as low as a few times 10^{-5} eV^2.

The flux is composed of roughly equal proportions of electron and muon neutrinos and antineutrinos. Since it is believed that lepton flavor is conserved in near-instantaneous particle interactions, the flavor of an incoming neutrino can be determined by identifying the lepton type produced in the detector. In Super-Kamiokande, this is accomplished by examining in detail the nature of the particle tracks.

Electron tracks are separated from muon tracks by taking advantage of a difference in patterns of illuminated phototubes caused by the difference in the mechanisms of energy deposition by these two classes of particles. An electron track is a showering track that result from the cascade shower of electrons and positrons formed as the electron passes through the water. This showering track produces a fuzzy ring in the detector. A muon, on the other hand, does not shower and produces a sharply defined Cherenkov ring pattern.

The energy spectrum of the incoming atmospheric neutrinos is mostly below the threshold needed for the production of a tau, so any oscillation to tau neutrinos manifests itself predominantly as a loss of muon neutrinos.

The most accurately predicted feature of the atmospheric neutrinos ($\pm 5\%$) and historically the first to be studied is the ratio of the muon neutrino to the electron neutrino flux. If oscillations of any kind take place, this ratio could change. The IMB and Kamiokande detectors did find a significant deficit of muon neutrinos while recording approximately the predicted number of electron neutrinos. Taken at face value, this suggested neutrino oscillations, but, of course, such an interpretation relied on the accuracy of the Monte Carlo simulations of neutrino production in the atmosphere. Super-Kamiokande was in an ideal position to decide this issue. The enormous size of Super-Kamiokande yielded a very large number of atmospheric neutrinos to study. If neutrino os-

cillations were the reason for the muon neutrino deficit, it would be possible to see a difference in the number of muon neutrinos coming from above the detector, where the path length of the neutrino since production was small, from those coming from beneath the detector, where the path was through the entire Earth. This comparison essentially eliminates reliance on atmospheric simulations since it compares the downward-traveling events with the upward-traveling ones. Conclusive evidence of the oscillation was observed and announced in 1998. A significant, angular dependent deficit of muon-type neutrinos was observed. As the number of electron-type neutrinos showed no such angular dependence, it was clear at this point that the muon-type neutrinos were turning into either tau neutrinos or some previously undiscovered (sterile) form of neutrino. Subsequent studies of Super-Kamiokande's data confirmed that the muon neutrinos were predominantly converting into tau neutrinos, whereas the existence of a sterile neutrino has become increasingly unlikely.

As mentioned earlier, another source of information for oscillation studies is provided by the neutrinos that interact in the rock surrounding the detector. These interactions mainly produce muons that enter the detector from below. The observed rate of these upward-traveling muons is compared with computer simulations based on a variety of models for atmospheric neutrino fluxes, neutrino interactions, and the transport of interaction products through the rock. Once again, the observations are significantly below the predictions and are consistent with the contained neutrino interaction sample interpretation, namely, muon neutrinos turning into tau neutrinos. Since this analysis is dependent on atmospheric simulations, the sensitivity to mixing angle is limited mainly by the 20 percent systematic uncertainty in the flux. In order to overcome this limitation, the ratio of the rate of up-going tracks that stop in the detector and the rate of those that traverse the detector and exit were studied. This ratio is flux-independent. Since the median energy of stopping muons is much smaller than that of the exiting ones, the ratio is sensitive to oscillations, and the same oscillation parameters as with the other two analyses are obtained, thus confirming the oscillation interpretation and hence the existence of massive neutrinos with an entirely different event sample.

Continuing Studies of the Neutrino

Super-Kamiokande's atmospheric and solar neutrino results, along with those of other atmospheric and solar neutrino experiments, have conclusively established the reality of neutrino mass and mixing. This discovery is among the most fundamental in particle physics within the past several decades, requiring the first significant modification in the long history of the Standard Model. The further study of neutrino mass splittings and mixing angles, an entirely new and unexplored sector of physics, has just begun but has already revealed surprises. The mixing angle between the second and third generations (inferred from atmospheric data) is apparently near maximal, whereas solar neutrino data also favor large mixing between the first and second generations. Clearly, these leptonic mixings bear little resemblance to those of the quarks—a new and fundamental revelation. New experiments have begun and others are in the construction and planning stages to explore this new and exciting avenue of study.

See also: JAPANESE HIGH-ENERGY ACCELERATOR RESEARCH ORGANIZATION, KEK; LEPTON; NEUTRINO; NEUTRINO, DISCOVERY OF

Bibliography

Aglietta, M., et al. "Experimental Study of Atmospheric Neutrino Flux in the NUSEX Experiment." *Europhysics Letters* **8,** 611–614 (1989).

Agrawal, V., et al. "Atmospheric Neutrino Flux above 1 GeV." *Physical Review D* **53,** 1314–1323 (1996).

Allison, W. W. M., et al. "Measurement of the Atmospheric Neutrino Flavour Composition in Soudan 2." *Physics Letters B* **391,** 491–500 (1997).

Casper, D., et al. "Measurement of Atmospheric Neutrino Composition with the IMB-3 Detector." *Physical Review Letters* **66,** 2561–2564 (1991).

Daum, K., et al. "Determination of the Atmospheric Neutrino Spectra with the Frejus Detector." *Zeitschrift für Physik C* **66,** 417–428 (1995).

Fukuda, Y., et al. "Evidence for Oscillation of Atmospheric Neutrinos." *Physical Review Letters* **81,** 1562–1567 (1998).

Fukuda, S., et al. "Tau Neutrinos Favored over Sterile Neutrinos in Atmospheric Muon Neutrino Oscillations." *Physical Review Letters* **85,** 3999–4003 (2000).

Hirata, K. S., et al. "Experimental Study of the Atmospheric Neutrino Flux." *Physics Letters B* **205,** 416–420 (1988).

Henry W. Sobel

CERN (EUROPEAN LABORATORY FOR PARTICLE PHYSICS)

CERN is the everyday name of the European Laboratory for Particle Physics located near Geneva in Switzerland, but the acronym actually stands for Conseil Européen de la Recherche Nucléaire (European Council for Nuclear Research). This Council was set up in 1952 to establish the European Organization for Nuclear Research that came into being in 1954 when the agreement establishing CERN was signed by the participating member states. The Council stressed the "necessity to build a laboratory for the high energy study of elementary particles in western Europe," and its advice was followed.

Whereas the first model for such a laboratory was the Brookhaven National Laboratory in the United States, only its academic (high-energy physics) part was deemed appropriate for an international venture. CERN provides scientists from its member states with experimental facilities for the study of high-energy physics. At the end of the twentieth century, CERN included twenty member states: Austria, Belgium, Bulgaria, the Czech Republic, Denmark, Finland, France, Germany, Greece, Hungary, Italy, the Netherlands, Norway, Poland, Portugal, the Slovak Republic, Spain, Sweden, Switzerland, and the United Kingdom. All countries agree on a yearly budget, which was approximately $600 million in 2001 and which is shared by the member states in proportion to their respective net national income.

Belgium, Denmark, France, Germany, Greece, Italy, Norway, the Netherlands, Sweden, Switzerland, the United Kingdom, and Yugoslavia created CERN in 1954. Austria quickly joined, but Yugoslavia withdrew for financial reasons. The core group of twelve member states throughout the sixties and seventies was eventually joined by Spain and Portugal in the eighties and by the Central European countries in the early nineties, soon after the fall of the Berlin Wall. Finland joined at the same time and was followed by Bulgaria, the latest member. CERN was also the first international organization in which postwar Germany participated. Besides providing access to unique facilities and making it possible to participate in great scientific achievements, member states participate in a highly successful European organization conducive to creating better links between European nations.

As it grew in size, CERN became international in another aspect. Although the official location of the organization is in Switzerland, the Laboratory now extends well into neighboring France, and two of its machines straddle the Franco-Swiss border, with accelerated particles crossing it several thousand times per second.

The Goal of CERN, Its Machines, and Its Users

World War II had left continental Europe in a state of scientific inferiority. The creation of CERN responded to the desire of European physicists to establish in Europe, as had been established in the United States, a frontier domain of research. However, there were technical, financial, and political roadblocks to overcome in creating such a large international scientific laboratory. The latter were removed after the encouraging address of Isidor I. Rabi at the UNESCO conference of 1950. Pierre Auger and Eduardo Amaldi took care of the former. These three scientists are considered the founding fathers of the organization.

CERN developed rather quickly into a large laboratory with first-class resources, some of them unique in the world, but it took many years to reach research parity with the United States. This was achieved in the early eighties, and since then there have been more American physicists using CERN than European particle physicists using American facilities (in particular, Brookhaven, the Stanford linear accelerator, and Fermilab).

The goal of CERN was to study the deep structure of matter, an academic domain requiring large facilities in which cooperation and collaboration could easily prevail over competition. The term "nuclear," which appears in the name of the organization, comes from the fact that nucleons were still considered the basic building blocks of matter in the 1950s. However, research was quickly oriented primarily toward high-energy particle physics, and the first big CERN accelerator to be built was the proton synchrotron (PS), which followed a smaller machine, the

synchro-cyclotron (SC). The PS was commissioned in 1960 shortly before the Brookhaven Alternating Gradient Synchrotron (AGS). Both accelerators used a new strong focusing technique and thus reached energies over 25 giga electron volts (GeV).

CERN's Intersecting Storage Rings (ISR) were completed in the early seventies. This machine, unique in the world, provided head-on collisions between protons accelerated in the two rings to energies up to 31 GeV. The resulting collision energy was equivalent to that of a 2,000 GeV machine using a fixed target but with a very low intensity. The CERN Super Proton Synchrotron (SPS) came into operation in 1977, five years after its American counterpart at Fermilab. Its transformation into a proton-antiproton collider in the early 1980s resulted in the attainment of the highest energies accessible at that time, with colliding beams of 300 GeV. The next big CERN machine was the Large Electron Positron Collider (LEP), an electron-positron collider with 100 GeV, and later 200 GeV, of collision energy. It operated from 1989 to 2000. The Large Hadron Collider (LHC) is under construction and will begin operation in 2007. It will collide protons at 14 tera electron volts (TeV) and will be the highest-energy machine in the world. The SPS and LEP are, respectively, 7 and 27 kilometers in circumference and are both located in underground tunnels.

CERN does not carry out research on behalf of its member states. It builds and operates facilities that are open to external users from universities and research laboratories. Physicists represent a minority (about 10%) of CERN's 2,500 employees, and most of the CERN physicists hold nonpermanent positions. The number of scientific users has steadily climbed from 1,500 at the commission of the SPS in the mid-seventies to over 6,500 in 2002. There were very few scientific users from nonmember states in the sixties, but their number quickly increased to about a third of the present total. In particular, the uniqueness of certain facilities (the ISR, the proton-antiproton version of the SPS, LEP, and now the preparation of the LHC experiments) has attracted many American groups. From an initially primarily European center, CERN has now grown into a laboratory serving a worldwide community. In 2002 half of the high-energy physicists in the world were CERN users.

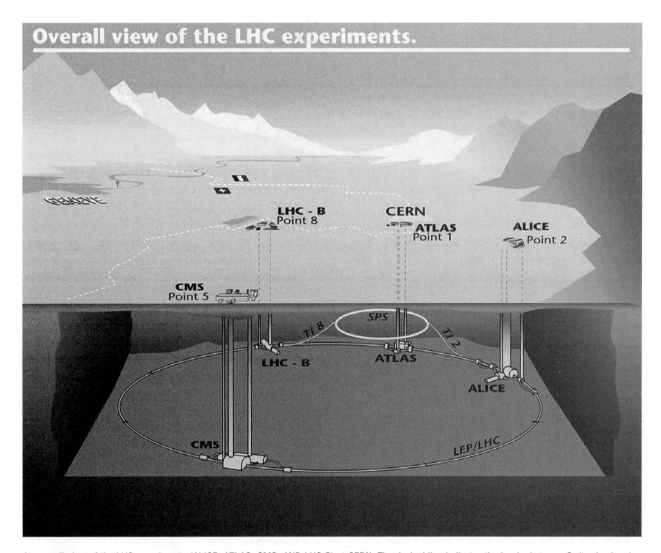

An overall view of the LHC experiments (ALICE, ATLAS, CMS, AND LHC-B) at CERN. The dashed line indicates the border between Switzerland and France. CERN is officially located in Switzerland but extends into neighboring France. The LHC is being constructed using the existing LEP cryogenics infrastructure and 27-km underground tunnel to help reduce costs. CREDIT: COURTESY OF CERN (EUROPEAN ORGANIZATION FOR NUCLEAR RESEARCH). REPRODUCED BY PERMISSION.

Nuclear physicists have been attracted by the heavy ion and muon beams as well as by the Low Energy Antiproton Ring (LEAR), a facility providing beams of low-energy antiprotons, and by ISOLDE with its on-line study of radioactive isotopes. The advent of colliders (with beams accelerated in opposite directions and clashing in special zones), as opposed to accelerators using fixed targets, has imposed the use of very big detectors dwarfing even the giant bubble chambers of the sixties and seventies. Detectors are built by large collaborations (hundreds of physicists for LEP and over a thousand for the LHC), and the dominant fraction of their cost (around 80 per-

cent) is borne by different funding agencies. The detectors are assembled at CERN from components built all over the world.

Research at CERN

Research at CERN has long paralleled research in the United States. Thanks to CERN, European scientists did well in the 1960s and 1970s, hunting new particles, studying hadron and neutrino interactions, elucidating the nature of weak and strong interactions, and collecting evidence for the quark structure of matter; however, CERN was following rather than leading. European physicists contributed much to these experiments, but the Nobel Prize discoveries (the two neutrinos, the track chamber hunt of hadron resonances, the discovery of CP violation and that of the surprising new J/ψ particle, the quark structure of the proton as first seen in electron scattering, and the discovery of the tau lepton) came from America.

The first significant CERN discovery in this competitive race was the electron decay mode of the pion, which had resisted discovery efforts in the United States. Results at the ISR did not reach the Nobel Prize level, but the 1973 discovery of the weak interaction with neutral currents could have. The research group working on the big heavy liquid bubble chamber, Gargamelle, was led by André Lagarrigue, who sadly passed away soon afterward. However, the major discovery of W and Z, the vector bosons of the weak interaction, was made possible with the use of the CERN proton-antiproton collider. This research brought the 1984 Nobel Prize in Physics to Carlo Rubbia, the project leader, and to Simon van der Meer, who had designed a clever way to build an intense antiproton beam, which was later used at the Fermilab Tevatron. These two discoveries, made a decade apart, vindicated the theory of the electroweak interaction and provided clear evidence for hadronic jets in hadronic collisions, which had been predicted by quantum chromodynamics (QCD). With these results, CERN could claim some world supremacy in the field of high-energy physics.

LEP was built to test with precision the Standard Model of fundamental particles and fundamental interactions, which combines the electroweak interaction and chromodynamics, thus bringing unity and simplicity at the quark-lepton level to the structure of matter. LEP succeeded very well but without bringing the hoped-for surprises that could have heralded a departure from the predictions of the Standard Model. Research at LEP stopped with only a hint that the Higgs boson (the missing element in the Standard Model) could be there at the very limit of the energy available on the machine. The top quark was discovered at Fermilab, but its mass could be predicted from the very precise LEP data. Even though the LEP energy was not sufficient to produce a top-antitop pair (which was possible using the Fermilab

Tevatron), the presence and mass of the top quark could be inferred from the precision measurement of other processes.

LEP was shut down for the construction of the LHC, which will explore fertile new ground. The heavy-ion program, pursued since the mid-1980s, could provide evidence for the formation of a new state of matter at very high temperatures where quarks are no longer confined within hadrons but can briefly roam freely over the volume of colliding ions. This research, now being pursued at the Relativistic Heavy-ion Collider (RHIC) at Brookhaven, will continue at the LHC, at increased energies, when the LHC is completed. This transition between free to bound quarks played a key role in the evolution of the universe 10 microseconds after the Big Bang.

Opening Up to the World

CERN was built to bring European scientists together, working on a common endeavor, soon after they had been on opposing sides during World War II. This it did remarkably well, and several other European scientific organizations, modeled after CERN, were created dealing with astronomy, space research, and molecular biology. There is no experiment at CERN that does not involve scientists from several nations.

CERN opened, maintained, and encouraged cooperation with the East, in particular with the international laboratory at Dubna, near Moscow, and with the Soviet laboratory at Serpukhov. Exchanges of scientists quickly developed, and this had a noticeable and highly positive impact as mutual understanding and appreciation prevailed over lack of knowledge and mistrust. It was very important to maintain this scientific collaboration during the Cold War, and this contributed to the eventual thaw in the political climate. It was quite natural for the Central European countries to join CERN in the early 1990s, basing their new membership on already existing collaborations. CERN has many users from the new independent states of the former Soviet Union.

The laboratory opened beyond Europe with many scientists coming from the United States, Japan, China, and India. They all work together, learn physics from each other, and also learn each other's cultural ways. Financial contributions for the construction of the machines, besides those naturally associated with the detectors, come from the United States, Japan, and other countries. In the case of the United States, this tallies surprisingly well with the European contribution to the Hubble telescope.

The success of CERN taking a leading role in particle physics worldwide is largely due to the fact that the laboratory grew up in one place. Although this resulted from the impossibility of reaching an agreement on a new site for the SPS in the 1970s and from the advent of affordable tunnel boring techniques that allowed big machines to be built in suburban areas, this fixed location paid strong dividends as the construction of new machines exploited previous ones, and an imposing, stable infrastructure was built up, along with a highly competent staff. A unique network of interconnected machines, each one feeding the next, made it possible to build LEP and then the LHC without an increase of the annual budget. The PS and the SPS will serve as the injector for the LHC, which will use the LEP tunnel and the LEP cryogenics infrastructure. The cost of the LHC is therefore half of what it would have been in a new location.

A New Style of Work

Particle physics at CERN has gradually shifted to the use of very large detectors built and operated by imposing collaborations. This is a new style in physics research. It has its drawbacks when compared to the smaller congenial and flexible teams of the past. Nevertheless, there are three positive elements. First, despite the size of the collaboration, it is always possible to see who has done what so that individual recognition is possible within the large collective effort. Research could not work otherwise. Second, working on such large and highly sophisticated facilities, with tight schedules and in an international atmosphere, has a very high training value much appreciated in industry. Indeed, half of the new Ph.D.s working on LEP for their degree have turned to industry where their acquired skills are appreciated. This is a very important spin-off of this academic research. The age distribution of CERN users, which has been stable over the LEP decade, shows a peak at 28, a 10-year expanse in width, followed by a

plateau at about half the peak height and extending until retirement age. This shows explicitly that training through research has become as important as training for research even at this advanced level. This is welcome since this research with large facilities requires far more young people than academia can eventually absorb. Finally, this concentration of skills in large collaborations, with a critical assessment of new ideas and an urge to achieve what is needed for the research, is highly conducive to new technical developments. It is not a surprise that the World Wide Web was born at CERN.

See also: BENEFITS OF PARTICLE PHYSICS TO SOCIETY; FUNDING OF PARTICLE PHYSICS; INFLUENCE ON SCIENCE; INTERNATIONAL NATURE OF PARTICLE PHYSICS; UNIVERSE

Bibliography

CERN. <http://www.cern.ch>.

Hermann, A.; Krige, J.; Mersits, U.; Pestre, D.; and Belloni, L. *History of CERN,* 3 vols. (North-Holland, Amsterdam, 1987–1990).

Maurice Jacob

CHADWICK, JAMES

James Chadwick, who won the Nobel Prize in Physics in 1935 for his discovery of the neutron, was born in Bollington, south of Manchester, England, on October 20, 1891. He was the first of four children whose parents were so poor that they had to send him to live in nearby Macclesfield with his grandmother, who saw to his primary education. Thereafter he attended the Manchester Municipal Secondary School, where he excelled in his studies and gained a scholarship to enter the University of Manchester in 1908, all the while living at home and skipping lunch because he could not afford it. He intended to study mathematics, but prior to his matriculation he was interviewed by mistake as a potential physics student and was too immature and shy to correct the error. That turned out to be fortunate, however, because during his first year his physics lecturer had to go to London for a few weeks and his place was taken by the new professor

English physicist James Chadwick (1891–1974) received the Nobel Prize in Physics in 1935 for his discovery of the neutron. CREDIT: COURTESY OF BETTMANN/CORBIS. REPRODUCED BY CORBIS CORPORATION.

at Manchester, Ernest Rutherford, whose infectious enthusiasm for physics captivated the young student. Later, in his third and final year as an undergraduate, Chadwick began to carry out some research under Rutherford's direction. He received his B.A. degree with First Class Honors in Physics in 1911, embarked on further research, received his M.Sc. degree in 1912, and gained fellowship support for the academic year 1912–1913 to continue his research under Rutherford's direction.

In the fall of 1913, on Rutherford's recommendation, Chadwick was awarded an 1851 Exhibition Senior Research Studentship to pursue research under Hans Geiger, who in 1908 had invented the particle counter named after him while working as Rutherford's research assistant in Manchester, and who in 1912 had returned to Germany to become the director of his own laboratory in the Physikalisch-Technische Reichsanstalt (Imperial Physical-Technical Institute) in Charlottenburg, a suburb of Berlin. There Chadwick mastered the German language and

came into contact with Albert Einstein, Otto Hahn, Lise Meitner, Walther Bothe, and other prominent physicists and chemists. In the spring of 1914, he made an important and perplexing discovery, that of the continuous spectrum of beta rays emitted by certain radioactive elements, whose correct explanation would have to wait until the end of 1930 when Wolfgang Pauli postulated the existence of the neutrino to preserve the laws of conservation of energy and momentum in the beta-decay process.

Soon after Chadwick made that discovery, on August 4, 1914, war was declared. As a British subject, he was interned in Ruhleben, near Spandau on the western outskirts of Berlin, where he and around four thousand other foreigners were billeted in unused racehorse stables for the duration of the Great War. Conditions were harsh, especially during the winter famine of 1916–1917, but even then he managed to carry out some research with other interned scientists using primitive equipment. After the Armistice on November 11, 1918, he returned to Manchester, his digestion permanently impaired by his wartime ordeal. Around six months later, when Rutherford moved from Manchester to Cambridge to succeed J. J. Thomson as Cavendish Professor of Experimental Physics, he was able to take Chadwick along with him because Gonville and Caius College awarded him a Wollaston Studentship and a small supplement for teaching, just enough to sustain his meager needs. In 1921 he was elected as a Fellow of Gonville and Caius College, which gave him an assured higher income plus college rooms and meals. In 1924 he was formally appointed as Assistant Director of Research of the Cavendish Laboratory. In 1925 he married Aileen Stewart-Brown, the daughter of a prominent Liverpool stockbroker; two years later they became the parents of twin daughters, Joanna and Judy. In 1927 Chadwick was elected as a Fellow of the Royal Society of London, the highest distinction apart from the Nobel Prize that a British scientist can receive.

At the Cavendish Laboratory, Chadwick introduced and taught each year his "Attic Course," an introductory six-week course on radioactivity, glass blowing, the production of high vacua, and the like, from which generation after generation of entering research students learned basic experimental skills.

His own research throughout the 1920s, most of which he carried out with Rutherford, focused on following up Rutherford's 1918 discovery of artificial disintegration, that alpha particles from radioactive elements can disintegrate light nuclei with the emission of protons. Chadwick, like Rutherford, was a brilliant experimentalist, and the two worked together closely, elucidating new phenomena and defending their work successfully when it was challenged, especially by researchers working in Vienna. Rutherford and Chadwick gave a full account of their work in their book, *Radiations from Radioactive Substances,* which they published jointly with Charles D. Ellis in 1930, and which became the standard work from which numerous others learned about the emerging field of experimental nuclear physics. Chadwick's experiments and discussions with Rutherford during their long and close association in the Cavendish Laboratory laid the groundwork for his discovery of the neutron in February 1932.

Chadwick left Cambridge in 1935 to accept the Lyon Jones Chair of Physics at the University of Liverpool, knowing that he would never be able to overcome Rutherford's opposition to building a cyclotron in the Cavendish Laboratory, and being firmly convinced that future advances in nuclear physics would require the use of one. He realized that dream in Liverpool, where he and his coworkers constructed and brought a 36-inch cyclotron into operation in July 1939, just before the outbreak of war in Europe, which would end his career as a research scientist and open up a new one as a scientific administrator and statesman.

Thus, following the death of Rutherford in October 1937, Chadwick was the most prominent nuclear physicist in Britain, and as such he became deeply involved with the development of nuclear energy for military purposes during the second world war. He became a member of the so-called MAUD committee, which was established to look into the possible military use of uranium, and he coordinated the work on nuclear fission at British universities until the British decided to abandon that work and join the American atomic bomb project. In 1943 he moved to the United States to take charge of the British effort on the bomb project, becoming the only British scientist who was granted complete access to all aspects of the Manhattan Project. His scientific acumen and diplomatic skills earned him the full respect and confidence of its military head, General Leslie R. Groves, as well as the head of the Los Alamos laboratory, J. Robert Oppenheimer, and other leading scientists. In 1945 he was knighted, becoming Sir James Chadwick. Many other honors also were bestowed upon him.

After the war Chadwick returned first to Liverpool and then, in 1948, he left to become Master of Gonville and Caius College in Cambridge. A shy and reserved man, nervous and anxious in front of any audience, he had little taste for controversy, and when one developed that promised to sap his limited energy, he suddenly resigned the Mastership in 1958 and retired to a cottage in North Wales. In 1969 he and his wife, Lady Chadwick, returned to Cambridge to be near their two daughters. He died there on July 24, 1974, at the age of 82.

See also: NEUTRON, DISCOVERY OF

Bibliography

Brown, A. *The Neutron and the Bomb: A Biography of Sir James Chadwick* (Oxford University Press, Oxford, UK, 1997).

Massey, Sir H., and Feather, N. "James Chadwick 20 October 1891–24 July 1974." *Biographical Memoirs of Fellows of the Royal Society* **22**, 11–70 (1976).

Stuewer, R. H. "Artificial Disintegration and the Cambridge-Vienna Controversy" in *Observation, Experiment, and Hypothesis in Modern Physical Science,* edited by P. Achinstein and O. Hannaway (MIT Press, Cambridge, MA, 1985).

Roger H. Stuewer

CHARGE CONJUGATION, C

See SYMMETRY PRINCIPLES

CHARM

See QUARKS

CHARMONIUM

Charmonium refers to a class of composite particles formed when a charm quark binds to a charm antiquark. The study of this system offers a unique probe into the force between quarks, known as the strong force and also into quantum chromodynamics.

Any type of quark can bind together with any type of antiquark to make the composite particles known as mesons. However, only the charm (c) and bottom (b) quarks form bound states having spectra and energy levels reminiscent of atoms. The charm quark, discovered in 1974, has a mass of approximately 1,500 MeV/c^2, or 2.7 10^{-27} kg.

The hydrogen atom is formed from a proton and an electron bound together by the electric field; charmonium's quarks are bound together by the strong interquark force. The relative motion of the particles gives rise to orbital angular momentum, measured in whole-number multiples of Planck's constant h, which contributes to the total energy of the system. The quarks, like the proton and electron, also possess some intrinsic angular momentum called spin.

Quantization of these angular momenta results in a unique set of allowed energy levels for the system. The spacings between these energy levels are determined by the particle masses and angular momenta and depend on how the force between the bound particles varies with their separation. The energy levels can be inferred by measuring the energies of light emitted when the atomic system relaxes from a higher-energy state E_n to a lower state E_m; the emitted photon has energy $E_n - E_m$.

Figure 1 shows the spectrum of photon energies observed when one of the excited energy states of charmonium (the so-called ψ' particle) relaxes to lower-energy levels. The ψ' charmonium state has a mass of 3,684 MeV/c^2. There are higher-mass charmonium states, such as ψ'' (3770), which prefer to decay into a pair of D (1865) particles (the meson formed from a c quark and a u antiquark) rather than de-excite by photon emission. This type of decay is not possible for ψ' because its mass is less than $2m_D$; it must decay into lower-energy states by radiating one or more photons.

FIGURE 1

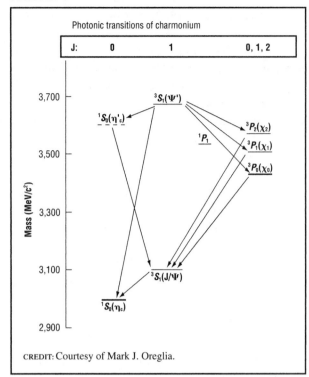

CREDIT: Courtesy of Mark J. Oreglia.

Masses of the charmonium states and the radiative transitions. States are denoted by spectroscopic notation $^{2s+1}L_J$ and also by the historical name (e.g., J/ψ). A state predicted but not yet observed is indicated by the dashed line; the $^1P_1$ state cannot be reached from $^3S_1$ by emission of a photon.

The energy levels of charmonium are governed by three quantum numbers. The radial quantum number (integer n) describes the successive quantum energy steps of a given configuration (e.g., the difference between the $n = 2$ ψ' and the $n = 1$ ψ). Next, there is the total spin of the two c quarks. Each quark has an intrinsic angular momentum of $\pm\frac{1}{2}h$, resulting in a total spin angular momentum s of 0 or 1h. Finally, there is the orbital angular momentum L, in whole-number multiples of h. Spectroscopic notation (from atomic physics) is used to label the charmonium states shown in Figure 1; the nomenclature is $^{2s+1}L_J$, where J is the total angular momentum and L is denoted by the historical notation S for $L = 0$, P for $L = 1h$, and so on. In quantum mechanics the various angular momenta add vectorially. For instance, with $s = 1$ and $L = 1$, J can take on the values 0, 1, or 2; in the figure this is evident for the $s = 1$ P states ($L = 1h$) that split into three energy levels.

The observation of the charmonium energy levels has verified the spin of the c quark and, more importantly, provided a better idea of the behavior of the strong force. Although the strong force increases linearly with quark separation for distances greater than 10^{-15} meters (and this is why one never sees single quarks!), charmonium tells us that the behavior is like that of the electric force for distances between 10^{-14} and 10^{-15} meters.

See also: J/Ψ; QUARKS

Bibliography

Bloom, E. D., and Feldman, G. J. "Quarkonium." *Scientific American* **246** (5), 66–77 (1982).

Mark J. Oreglia

CKM MATRIX

The CKM matrix contains information on all quark flavor transitions and is the source of the violation of CP symmetry in the Standard Model.

Before electroweak symmetry breaking occurs, the Standard Model fermions are massless, and they obey the symmetries of the Standard Model. The quarks and leptons appear in pairs known as doublets under the left-handed weak interactions. A charge $+\frac{2}{3}$ and a charge $-\frac{1}{3}$ quark compose a left-handed weak doublet, and three such doublets, known as families or generations, are known to exist. When the electroweak symmetry is spontaneously broken by the Higgs mechanism, masses are generated for the fermions via their interactions with the Higgs field. These massive states are the physical particles that are observed in the laboratory.

The quark-Higgs interactions are not diagonal under the weak interactions. This means that quark states that carry the weak charge are mixtures of the physical quark states. If the quarks were massless (or had equal masses), they would not mix. By convention, this mixing is mathematically ascribed to the charged $-\frac{1}{3}$ quark states: down, strange, and bottom. The wave-

functions of the weak flavor quark states, denoted here by primes, are then expressed in terms of the physical quark wavefunctions (unprimed) by

$$
\begin{aligned}
d' &= V_{ud}d + V_{us}s + V_{ub}b, \\
s' &= V_{cd}d + V_{cs}s + V_{cb}b, \\
b' &= V_{td}d + V_{ts}s + V_{tb}b.
\end{aligned}
\tag{1}
$$

The nine quantities V_{ij} are numbers, which are in principle complex. They describe the weighting of each physical quark wavefunction in the mixture and are labeled by the quark states that they link. These V_{ij} are the elements of the 3×3 Cabibbo-Kobayashi-Maskawa (CKM) matrix.

This matrix is named after Nicola Cabibbo, who first developed these ideas in 1963 when only three quark flavors (up, down, and strange) were known, and Makoto Kobayashi and Toshihide Maskawa, who extended it to the six quark flavors that are now known to exist. Although only three quark flavors were known in 1973 when Kobayashi and Maskawa hypothesized this extension, they realized that CP violation could be naturally accommodated in the Standard Model if one incorporated six quarks within this framework.

The CKM matrix is unitary, meaning that the product of the matrix with its complex conjugate must be the unit matrix. This implies that the values of the elements are interconnected. For example, the sum of the squares of the elements in a row or column equals unity

$$
|V_{ud}|^2 + |V_{us}|^2 + |V_{ub}|^2 = 1,
\tag{2}
$$

for three generations. In addition, the multiplication of two different rows or columns must equal zero:

$$
V_{ub}^*V_{ud} + V_{cb}^*V_{cd} + V_{tb}^*V_{td} = 0,
\tag{3}
$$

where the asterisk denotes the complex conjugate of the element. This last relationship can be represented as a triangle in the complex plane (see Figure 1), where the length of the sides of the unitary triangle is given by the magnitudes of the elements, and the angles of the triangle are related to the phases in the matrix. A 2×2 mixing matrix consists of only a single parameter: one mixing angle that is known as the Cabibbo angle and that mixes the down and strange quarks. The 3×3 CKM matrix can be

FIGURE 1

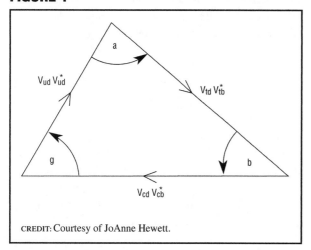

CREDIT: Courtesy of JoAnne Hewett.

described in terms of four independent parameters, three mixing angles and one phase that gives rise to complex entries. It is the existence of this phase which is responsible for CP violation.

This mixing of quark flavors, as well as the generation of quark masses through the Higgs field, has no fundamental explanation in the Standard Model. The values of the quark masses and mixing angles are arbitrary and not predicted. Numerous theoretical attempts have been made to derive these quantities from a basic theory of flavor, but as of 2002 no compelling model exists. The numerical values of the quark masses and mixings are determined from laboratory measurements.

The values of the individual CKM matrix elements determine the rates of transitions, or weak decays, of the quarks. For example, a heavy quark of charge $-\frac{1}{3}$, such as the strange quark, can decay to a lighter quark of charge $+\frac{2}{3}$, such as an up quark, by emitting a W^- particle. The strength of this sample transition is specified by the value of the quantity V_{us}. There are nine such transitions, corresponding to the nine elements of the CKM matrix. These transitions are known as charged current decays of the quarks, as the charged weak gauge bosons $W^\pm$ are emitted. Since the $W^\pm$ particles are heavier than all the quarks except for the top quark, they subsequently decay to a pair of lighter quarks or leptons after being emitted. Neutral current couplings, that is, quark couplings to the neutral weak gauge boson Z^0, are flavor diagonal in the Standard Model. Flavor changing neutral current decays, such

as $b \rightarrow s\gamma$, only occur when quantum corrections to the Standard Model are included.

The most accurately measured CKM matrix element is V_{ud}, which is determined from comparing superallowed nuclear β decay to muon decay. Its value is close to unity. The elements $V_{us,ub,cd,cs,cb}$ are all measured in three-body semileptonic decays (where the emitted $W^\pm$ boson decays into a lepton pair) of the heavier of the linked quarks. For example, the transition $b \rightarrow cl^- \bar{\nu}_1$ determines V_{cb}. $V_{cd,cs}$ can also be determined in neutrino nucleon collisions, $\nu_\mu N \rightarrow \nu + c + X$. All these measurements are subject to theoretical uncertainties (in addition to experimental errors) introduced by the fact that the quarks are not free particles but are bound inside of hadrons. The quantity V_{tb} is measured directly in the two-body decay $t \rightarrow bW^+$ since the top quark is massive enough to decay into a W^+ boson. Its value is also close to unity. The elements $V_{td,ts}$ have yet to be measured directly, but a range for their values can be inferred from the unitarity properties of the CKM matrix.

Present knowledge of the CKM matrix elements is given at the web site of the Particle Data Group, which summarizes the measurements of particle properties. In 2002 the absolute values of the matrix elements lie in the following ranges:

$$
\begin{pmatrix} V_{ud} & V_{us} & V_{ub} \\ V_{cd} & V_{cs} & V_{cb} \\ V_{td} & V_{ts} & V_{tb} \end{pmatrix} =
$$

$$
\begin{pmatrix} 0.9741\text{--}0.9756 & 0.219\text{--}0.226 & 0.0025\text{--}0.0048 \\ 0.219\text{--}0.225 & 0.9734\text{--}0.9748 & 0.038\text{--}0.044 \\ 0.004\text{--}0.014 & 0.037\text{--}0.044 & 0.9990\text{--}0.9993 \end{pmatrix} \quad (4)
$$

This includes the constraints imposed from unitarity. Note that the matrix is nearly diagonal and that first-second generation mixing is strong, whereas the first-third generation mixing is very weak. The explanation of CP violation in the Standard Model requires a nonvanishing value of V_{ub}, which is satisfied by the data.

An active research topic within high-energy physics is improving the determination of the CKM matrix elements. This is being accomplished via better experimental precision with larger data samples and refined experimental resolution with modern detectors, and with more precise theoretical calculations. An accurate determination of the fundamental parameters of the Standard Model will hopefully provide insight to the underlying theory of flavor.

See also: CP SYMMETRY VIOLATION; FAMILY; QUARKS; STANDARD MODEL

Bibliography

Cabibbo, N. "Unitary Symmetry and Leptonic Decays." *Physical Review Letters* **10**, 531 (1963).

Hagiwara, K., et al. "CP Violation in the Renormalizable Theory of Weak Interaction." *Physical Review* **D66**, 010001 (2002).

Kobayashi, N., and Maskawa, T. "The Review of Particle Physics." *Progress of Theoretical Physics* **49**, 652 (1973).

JoAnne Hewett

COLLIDER

See ACCELERATORS, COLLIDING BEAMS: ELECTRON-POSITRON; ACCELERATORS, COLLIDING BEAMS: ELECTRON-PROTRON; ACCELERATORS, COLLIDING BEAMS: HADRON

COMPACTIFICATION

See STRING THEORY

COMPUTING

Applications relevant to elementary particle and high-energy physics (HEP) computing can be categorized as follows:

1. Triggering and data acquisition

2. Simulation

3. Data handling and storage

4. Commodity hardware and software

5. Data analysis and visualization

6. Control and monitoring systems

7. Information systems and multimedia

Triggering and Data Acquisition

In addition to their specialized detection and measurement systems (for example, calorimeters, drift chambers, etc.), the detectors in high-energy physics experiments are, in fact, sophisticated computing systems. Without reliable triggering and data acquisition (DAQ) computing systems present in these detectors, all other experimental computing is of little consequence. Triggering and DAQ systems ensure that the physics events occurring in the detector are observed, measured, and accurately transformed into analyzable data. In a typical experiment, the first level trigger, implemented in hardware, initiates the data collection. Data from the front-end electronics are digitized and collected with electronic data modules. A readout computer reads experiment conditions from the control system, reads event fragments from the data modules over the local network, and builds events from fragments. These event data are then written to buffer storage and/or transmitted via a local network to archival storage for additional processing and eventual analysis.

The scale of triggering and DAQ systems can be seen in the design of the ALICE experiment at the Large Hadron Collider (LHC) at the European Laboratory for Particle Physics (CERN) in Geneva, Switzerland. The ALICE detector will measure up to 20,000 particles in a single interaction event resulting in a data collection rate of approximately seventy five million bytes per event. The event rate is limited by the bandwidth of the data storage system. Higher rates are possible by selecting interesting events and subevents or efficient data compression.

Simulation

A computer simulation (sometimes referred to as a Monte Carlo simulation) of particle interactions in the experimental configuration is essential to most HEP experiments. Computer software providing these simulations plays a fundamental role in the design of detectors and shielding components, in the investigations of the physics capabilities of the experiment, and in the evaluation of background (non-experimental, for example, cosmic and/or terrestrial radiation) data. Simulation software must be complete and capable of generating simulated experimental data comparable in scope to genuine experimental data. The simulation software must support a description of the experimental detector from the point of view of the materials used and the geometry adopted, both for the structural and the active event-detecting components. The configurations adopted for the data output and the logic of DAQ on the quality of the physics results are also modeled in order to evaluate their impact on the overall performance of the detector. The simulation must be able to describe the properties and the physics processes of the particles involved both in the expected signal/output and in the background. Especially important is the capability to handle physics processes across a wide energy range, which in such experiments simulation may span several orders of magnitude. An ideal simulation system is also flexible and open to evolution and to the integration of external tools. This is particularly important since a number of software tools are already commonly used within the scientific community where a particular experimental environment may require the ability to extend the simulation functionalities, for instance, to include the ability to deal with peculiar physical processes. One of the most powerful and widely used simulation toolkits is GEANT4 developed at CERN.

Data Handling and Storage

Particle physics experiments generate enormous amounts of data. For example, the BaBar experiment at the Stanford Linear Accelerator Center (SLAC) is designed to accommodate 200 terabytes (200 million million bytes) of data per year. As of April 2002, the BaBar database contained over 500 terabytes of data in approximately 290,000 files. This database is the largest known in the world (with the possible exception of some with military/government content). Such data rates and database sizes push the limits of state-of-the-art data handling and database technologies.

In order to handle such large volumes of data, experiment data handling and storage systems/database must be designed to

- provide reliable and robust storage of the raw detector data, simulation data, and other derived data;

- keep up with production processing; be able to process raw data files within minutes of writing them to tape;

- provide easy, rapid, and intuitive access to data on a variety of systems at a wide variety of locations where processing and data storage resources are available to physicists;

- provide accurate detailed information on the processing steps that transformed event data—from the trigger through reconstruction and all the way to the creation of individual or group datasets;

- provide mechanisms for policy-based allocation and use of disk, central processing unit (CPU), network, and tape drive resources.

Commodity Hardware and Software

Commodity hardware and software refers to the hardware and software architectures and configurations used to accomplish off-line batch and interactive data processing. In the past such processing was often accomplished by large mainframe computers. In recent years, large (200 or more) compute farms of inexpensive computers have become a common replacement for these mainframe systems. These compute farms are fundamentally groups of networked desktop systems (without monitors and keyboards) that are housed in a single location and which function as a single entity. A computer farm streamlines internal processes by distributing the workload between the individual components of the farm and expedites computing processes by harnessing the power of multiple CPUs. The farms rely on load-balancing software that accomplishes such tasks as tracking demand for processing power from different machines, prioritizing the tasks, and scheduling and rescheduling them depending on priority and demand that users put on the network. When one computer in the farm fails, another can step in as a backup. Combining servers and processing power into a single entity has been relatively common for many years in research and academic institutions. Compute farms provide an effective mechanism for handling the enormous amount of computerization of tasks and services that HEP experiments require. Farms of Intel-based computers running the Linux operating system (OS) have become common at many HEP institutions.

The computing grid is the next generation of compute farms. A grid is a distributed system of computing resources (a cyberinfrastructure) in which computers, processor farms, disks, major databases, software, information, collaborative tools, and people are linked by a high-speed network. The term grid was coined as a result of the analogy with an electrical power distribution system. Grid resources are made available transparently to a distributed community of users through a set of new middleware that facilitates distributed collaborative working in new ways. The nine-institution Particle Physics Data Grid collaboration—consisting of Fermi National Laboratory, SLAC, Lawrence Berkeley Laboratory, Argonne National Laboratory, Brookhaven National Laboratory, Jefferson National Laboratory, CalTech, the University of Wisconsin, and the University of California at San Diego—will develop the distributed computing concept for particle physics experiments at the major U.S. high-energy physics research facilities.

Data Analysis and Visualization

Analysis systems are often at the core of an experiment's physics efforts, and the constraints imposed by those systems can heavily influence the physics event reconstruction and analysis framework. Conversely, an analysis system which lacks key features (or worse, implements them incorrectly) can be a serious handicap. Physicists are constantly searching for new and interesting ways to extract physical information through two-dimensional and three-dimensional computer visualization/modeling, animation, histogram plotting, etc. Key also is the development of techniques for data interactivity—methods for interacting with a program or data. These techniques often include graphical user interfaces (GUIs) but also scripting, browsing and other technologies. There have even been some attempts to utilize virtual reality techniques wherein a physicist becomes "immersed" in experimental data. Development of data analysis and visualization tools has been the subject of numerous international collaborations. The result has been the creation of specialized software libraries used, supported, and maintained by these collaborations but generally available to all physicists.

Control and Monitoring Systems

The infrastructure surrounding experiment detectors is highly complex. The hardware devices used in detectors and the systems of experiments consist of commercial devices used in industry, specific devices used in physics experiments, and custom devices designed for unique application. The control and monitoring system must insure that these devices interface correctly with one another by providing testing and error diagnostic functionality. The administrative component of a control and monitoring system provides access to the control of an experiment often distributed between supervision and process control functions.

Information Systems and Multimedia

The World Wide Web (WWW) is the best example of how the requirements of physics research and the need for experimental collaboration have led to developments in information systems and multimedia. The Web was developed to allow physicists in international collaborations to access data and information easily, quickly, and in a device-independent (i.e., computer and operating system) manner. There has been increasing use of collaborative environments supporting point-to-point and multipoint videoconferencing, document, and application sharing across both local and wide area networks; video on demand (broadcast and playback); and interactive text facilities. Resources such as the HEP preprints database at SLAC and the Los Alamos National Laboratory electronic preprint server, officially known as the e-Print Archive, support physicist research and authoring. The first U.S. web server, at SLAC, was installed to provide access to the pre-prints database. Other information systems and multimedia applications include electronic logbooks used to improve and replace paper logbooks, and streaming media servers to provide widespread access to seminars and lectures.

See also: DETECTORS, COLLIDER

Bibliography

Foster, I., and Kesselman, C., eds. *The GRID: Blueprint for a New Computing Infrastructure* (Morgan Kaufmann, San Francisco, 1999).

GEANT4. <http://wwwinfo.cern.ch/asd/geant4/geant4.html>.

Particle Physics Data Grid. <http://www.ppdg.net>.

White, B. "The World Wide Web and High-Energy Physics." *Physics Today* **51** (11), 30–36 (1998).

Bebo White

CONSERVATION LAWS

A conservation law is a statement of constancy in nature. One quantity (the conserved quantity) remains constant while other quantities may change. For example, in a collision of elementary particles, the total momentum is conserved (it is the same after the collision as before) while other quantities, such as the speeds and directions of the particles, and even the number of particles and their masses, may change.

The world of particles, like the larger-scale world around us, is a world of incessant change. Probing the small-scale world, one might not be surprised to find that *everything* changes, that nothing is constant. Yet scientists have, in fact, identified a limited number of conserved quantities. These quantities have special significance because constancy is an idea of such power and simplicity and because conserved quantities are related to symmetry principles in nature.

Conservation laws are not just keystones of theory, they are practical tools of analysis. They can be applied to processes whose complex details are beyond any capability of measurement or calculation. In a particle collision, conserved quantities such as energy, momentum, angular momentum, and electric charge are the same after the collision as before, even though incredible complexity, with countless interactions, may attend the process.

Conservation laws are tested not so much by measuring some quantity before and after an interaction to see if it is the same as by looking for, and *not* finding, evidence of a process, which, if it occurred, would violate the conservation law. For example, the decay of an electron into lighter neutral particles has never been observed. A process consistent with all known conservation laws except charge conservation is the decay of an electron into a neutrino of the electron type (called an electron neutrino) and a photon (or gamma ray), indicated by

$$e^- \not\rightarrow \nu_e + \gamma.$$

Energy, momentum, and angular momentum could all be conserved in this process, but not charge. The slash through the arrow means that the process does *not* occur—or, more accurately stated, has never been seen. We can say that the electron is stabilized by charge conservation. Since the electron is the lightest charged particle, its decay would necessarily violate the law of charge conservation. Experiment puts the lifetime of the electron at more than 4×10^{24} years. This means, roughly, that in the lifetime of the universe, no more than one out of a million billion electrons could have decayed. It is on this basis (together with a theoretical underpinning) that we call charge conservation an absolute conservation law.

Conservation laws apply to isolated systems, those for which external influences are absent or too small to be significant. Particle processes are almost always isolated in this sense. Gravity, electric fields, magnetic fields, and neighboring atoms have no appreciable effect during the brief moment of a particle collision. (There are certain examples of nuclear gamma decay within an atom in a crystal where neighboring atoms do have an effect.)

Absolute and Partial Conservation Laws

Conservation laws may be absolute or partial. An absolute conservation law is one for which no confirmed violation has ever been seen and which is believed to be valid under all circumstances. By this definition, momentum, energy, angular momentum, and charge are absolutely conserved, as is the color charge of the strong interaction. (Although color charge is an attribute of strongly interacting particles only, its conservation can be considered universal because leptons and their associated bosons have zero color—they are "colorless.") Another absolutely conserved quantity is the combined symmetry called TCP, standing for time reversal, charge conjugation (or particle-antiparticle inversion), and parity (or left-right inversion). (The individual symmetries T, C, and P are only partially conserved.)

Baryon conservation occupies a special and ambiguous place. Baryons are a class of heavy particles (those made up of three quarks) that include the proton and the neutron. The law of baryon conservation, which states that the number of baryons minus the number of antibaryons never changes, is valid experimentally. Its most stringent test is the absence of proton decay. Since the proton is the lightest baryon, its decay would imply a change of baryon number, a violation of the law of baryon conservation. Experiment puts the lifetime of the proton at greater than 10^{33} years. This leaves room for only an incredibly tiny probability of decay (much less even than the limit on electron decay probability), yet theorists are reluctant to call baryon conservation an absolute law. The law has no known theoretical basis, and indeed there is theoretical reason to expect a tiny but nonzero probability of proton decay. Searches for it continue.

Like baryon conservation, lepton conservation is absolute so far as experiment is concerned. The law of lepton conservation states that the total number of leptons of all types (electron type, muon type, and tau type) minus the number of antileptons never changes. No violation of this law has ever been reported. Yet there is no known theoretical reason for it, and physicists expect that in the end it will prove to be partial, not absolute. For many years, the numbers of leptons of the three individual types appeared to be separately conserved (from which it followed, of course, that the total of all lepton types is conserved). Because of recent evidence for the "oscillation" of one type of neutrino into another type, the conservation laws of individual lepton types are now recognized to be partial, not absolute (although the transformation of one type of charged lepton into another has yet to be seen).

What is a partial conservation law? At first, it sounds like a contradiction in terms, like partial pregnancy. A conservation law is called partial if the quantity it governs is conserved when certain interactions are at work but not conserved for all interactions. Stated differently, a partial conservation law is one obeyed by one or more kinds of interaction and violated by at least one kind of interaction. For example, conservation of the number of quarks of a particular "flavor" (up, down, strange, charm, top, or bottom—in each case counting particles minus antiparticles) is obeyed by the strong and electromagnetic interactions but not by the weak interaction. The weak interaction can cause one type of quark to turn into another type.

The weak interaction is recognized as a "symmetry-breaker" that prevents several conservation laws from being absolute. Another symmetry-breaker is the "Higgs interaction," an interaction between every particle and the Higgs boson. The Higgs boson (named after one of its inventors,

Scotland's Peter Higgs) is the quantum of a still-hypothetical field believed to permeate all space and to account, through its interactions, for particle masses. The Higgs interaction, along with the strong and electroweak interactions, is incorporated into the so-called Standard Model of particles. Whether gravity, which lies outside the Standard Model, is an additional symmetry-breaker is unknown. If it is, its effects in the particle world will surely be tiny and hard to detect. The present status of conservation laws is shown in Table 1.

TABLE 1

Status of Conservation Laws and Invariance Principles	
Conserved or Invariant Quantity	Comment
Energy	Believed to be absolutely conserved
Momentum	Believed to be absolutely conserved
Angular momentum (orbital + spin)	Believed to be absolutely conserved
Electric charge	Believed to be absolutely conserved
Time inversion, or time reversal (T)	T and the combination CP are violated, presumably by the Higgs interaction
Particle-antiparticle inversion, or charge conjugation (C)	Violated by the weak interaction
Space inversion, or mirror inversion, called parity (P)	Violated by the weak interaction
Combined inversions, TCP	Believed to be absolutely conserved
Isospin (charge independence of interactions)	Violated by the electromagnetic and Higgs interactions (masses depend on charge)
"Color"	Believed to be absolutely conserved. (Quarks and gluons have color. Leptons, photons, and W and Z bosons are "colorless.")
Individual quark "flavors," i.e., "upness," "downness," "charm," "strangeness," "topness," "bottomness"	Violated by the electroweak interaction and ultimately the Higgs interaction
Baryon conservation (combined number of all quark types)	Experiment consistent with absolute conservation, but theorists predict a very weak violation that would be evidenced by proton decay
Electron-family number, Muon-family number, Tau-family number	Observed violation for neutrinos and predicted minute violation for charged leptons via the Higgs mechanism
Combined lepton number (e-family + mu-family + tau-family)	Experiment consistent with absolute conservation, but violation by the Higgs interaction is predicted

Note: Except for the lepton-number rules, which are not relevant to this generalization, all of the conservation laws and invariance principles enumerated above are believed to be valid principles for the strong interactions. Violations occur through the electroweak interaction and/or the Higgs interaction. Gravity's role, if any, is unknown.

CREDIT: Courtesy of Kenneth W. Ford.

Conservation Laws and Feynman Diagrams

Quantum theory replaces smooth change with explosive change. Every interaction is believed to be driven ultimately by the creation and annihilation of particles. Simple space-time diagrams, or Feynman diagrams (Figure 1), illustrate this idea and show how conservation laws work at the most fundamental level.

For all interaction types, Feynman diagrams have the same "three-prong" structure. At a space-time "vertex," one fermion world line ends, another fermion world line begins, and a boson world line begins or ends. No particle survives an interaction. (If an electron enters an interaction event and an electron leaves it, theory treats them as different particles, one being annihilated and one created.) What do survive an interaction are conserved quantities—whichever ones are preserved by the interaction in question (with a subtlety involving energy and momentum, to be discussed below). So conservation laws apply not just to the before and after of a process in the laboratory. They are believed to apply to the before and after of every separate interaction. From that base, they reach out to all that happens, at whatever scale of size.

It turns out that each of the simple Feynman diagrams shown in Figure 1 represents only a "virtual," not a real, process because energy and momentum cannot be simultaneously conserved across the vertex. So real processes that conserve all relevant quantities involve a "daisy chain" of at least two such interaction vertices (and perhaps millions of them). Illustrating this point is a diagram for a real process, the decay of a negative muon, shown in Figure 2. The time sequence is from the bottom to the top of the diagram. In this process, muon family number, electron family number, and electric charge are conserved.

Examples of Absolute Conservation Laws

Momentum is a vector (directed) quantity. Following a particle collision, the vector sum of the momenta has the same direction and the same magnitude as before the collision. Nonrelativistically, the momentum of an object is its mass times its velocity ($\mathbf{p} = m\mathbf{v}$). Relativistically, the definition is different but the conservation law remains valid. Momentum conservation is related to the homogeneity of space (that the laws of physics are the same at every point).

FIGURE 1

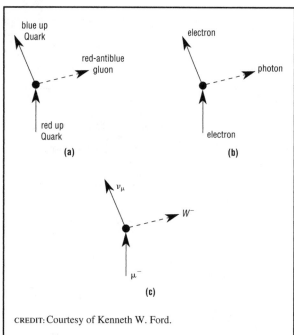

CREDIT: Courtesy of Kenneth W. Ford.

Space-time (or Feynman) diagrams representative of three fundamental interactions. (a) Strong quark-gluon interaction. Color charge, electric charge, and quark type are conserved. (b) Electromagnetic interaction linking charged particles and a photon. Electric charge and particle type are conserved. (c) Weak interaction linking leptons and a W boson. Electric charge and particle family number are conserved.

FIGURE 2

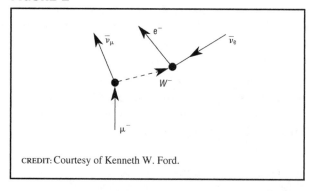

CREDIT: Courtesy of Kenneth W. Ford.

Feynman diagram for negative muon decay involving two fundamental interaction events. Muon family number, electron family number, and charge are conserved, as well as energy, momentum, and angular momentum. (The backward pointing arrow designates an antiparticle. A forward-in-time antiparticle is equivalent to a backward-in-time particle.)

Energy is a scalar (numerical) quantity that takes many forms. Its relevant forms in the particle world are kinetic energy (energy of motion) and mass. The energy locked up in mass is $E = mc^2$, where c is the speed of light. Initial kinetic energy is required if new mass is to be created. If mass decreases, as in a decay process, kinetic energy increases. A simple implication of energy conservation is that all spontaneous decay processes must be "downhill" in mass—that is, the total mass of the products must be less than the mass of the initial particle. Energy and momentum conservation taken together prevent one particle from decaying into another single particle. If the initial particle is at rest, the final particle would have to be at rest, too, to preserve zero momentum, but then energy would not be conserved. And if the final particle moves in such a way as to conserve energy, momentum is not conserved. So every decay process results in at least two particles. Energy conservation is related to the homogeneity of time (nature's laws are the same at one time as another).

Angular momentum, like momentum, is a vector quantity. It measures the strength of rotational motion and is directed along the axis of rotation. Particles may have *spin* angular momentum and *orbital* angular momentum, roughly analogous to the spin of the Earth about its axis and the orbiting of the Earth around the Sun. Remarkably, a particle may possess spin even if it has no spatial extent. In units of Planck's constant h divided by 2π, written $\hbar$, fermions have half-odd-integer spin ($\frac{1}{2}$, $\frac{3}{2}$, etc.) and bosons have integer spin (0, 1, 2, etc.). The spins of the fundamental particles are as follows: quarks and leptons, $\frac{1}{2}$; gluons, photon, and the W and Z bosons, 1; the Higgs boson, 0. Orbital angular momentum is always integral (0, 1, 2, etc.). One consequence of angular momentum conservation is that if the number of fermions before a reaction is even, the number afterward must also be even, and if the number before is odd, the number after must be odd (Figure 2, for example, shows a process where one fermion decays into three fermions.) Angular momentum is related to the isotropy of space (nature's laws don't depend on direction).

The law of charge conservation goes back to Benjamin Franklin in the eighteenth century. It has withstood the test of time. It is related now to the

masslessness of the photon through a principle called gauge invariance. In units of the proton charge e, leptons have charges 0 and -1 (antileptons 0 and $+1$); quarks have charges $+\frac{2}{3}$ and $-\frac{1}{3}$ (antiquarks $-\frac{2}{3}$ and $+\frac{1}{3}$); gluons, the photon, the Z boson, and the Higgs boson have charge 0; and the W boson has charge -1 (its antiparticle $+1$). As noted above, the most salutary effect of charge conservation is to prevent the decay of the electron.

The conservation of "color" (or "color charge") is much like the conservation of electric charge. Both color charge and electric charge are quantized properties of particles that are preserved in every interaction, even when no particle survives the interaction. (Think of runners in a relay race as particles and the baton they carry as color charge or electric charge. When one runner stops and another starts, the baton continues on.) Quarks may have any one of three "colors," arbitrarily called red, green, and blue, so there are really eighteen different quarks, not six. There are no "colorless" quarks. Antiquarks are said to have anticolor. Gluons carry a color-anticolor mixture such as red-antiblue, of which there are eight independent combinations. Figure 1a shows an example of a color-conserving quark-gluon interaction.

Lepton conservation means that the number of leptons minus the number of antileptons is preserved in every interaction. Figure 2 illustrates the principle with a decay process in which one lepton turns into two leptons and an antilepton. Another example is the decay of the neutron into a proton, an electron, and an antineutrino,

$$n \rightarrow p^+ + e^- + \bar{\nu}_e.$$

The lepton number is zero before and after. This process illustrates several other conservation laws as well: charge conservation (0 to 0), baryon conservation (1 to 1), energy conservation ("downhill" in mass), and angular momentum conservation (odd number of fermions to an odd number of fermions). This process and the one in Figure 2 also show conservation of the individual lepton families (electron family and muon family), for which there are no known exceptions involving charged leptons.

Examples of Partial Conservation Laws

The concept of isospin was introduced in the 1930s to describe the similarities of the proton and neutron—they have nearly the same mass and appeared to have the same strong interaction. The proton and neutron came to be regarded as two states of a single underlying particle, the nucleon, to which the mathematics of a spin-$\frac{1}{2}$ particle with its two orientations of spin could be applied (thus the name isospin, which otherwise has nothing to do with spin). Later other particle "multiplets" were found, such as the pion triplet and the xi-particle doublet (and some singlets such as the lambda particle). The law of isospin conservation states that the strong interaction is identically the same for the members of each multiplet. Isospin conservation is clearly violated by the electromagnetic interaction because particles within a given multiplet don't have the same charge (and also differ slightly in mass).

Isospin conservation is now recognized to be a consequence of "flavor invariance." The six quark flavors are called up, down, charm, strange, top, and bottom; flavor invariance means that the strong interactions do not distinguish among flavors. Thus replacing an up quark by a down quark (which could mean changing a proton to a neutron), or even replacing a strange quark by a top quark, does not change any property of the strong interaction. The electroweak interaction does, however, depend on flavor, or quark type—that is, it violates the law of flavor invariance. This is attributed, ultimately, to effects of the Higgs boson.

The strong interaction goes further and conserves each quark flavor separately. If the conservation of quark flavor were an absolute law, there would be no transformations among different quark types. An observed decay process that violates the law of flavor conservation is the decay of a lambda particle into a proton and a negative pion,

$$\Lambda^0 \rightarrow p^+ + \pi^-.$$

Using u for up, d for down, and s for strange, this process is represented in terms of its quark constituents by

$$uds \rightarrow uud + \bar{u}d.$$

Balancing the books here requires that a strange quark be transformed into a down quark. This is a process forbidden by the strong interaction but allowed by the weak interaction. The mean life of the lambda is many orders of magnitude greater than if the process occurred through the strong interaction, confirming that the flavor violation is provided by the weak interaction.

According to the reductionist view that dominates modern science, conservation laws in the large-scale world result from the action of such laws in the particle world. There is extensive evidence that this is the case. But how fascinating it will be if future discovery reveals large-scale regularity not attributable to small-scale laws.

See also: ENERGY; MOMENTUM; NOETHER, EMMY; SYMMETRY PRINCIPLES

Bibliography

Bernstein, J. *The Tenth Dimension* (McGraw-Hill, New York, 1989).

Bernstein, J.; Fishbane, P.; and Gasiorowicz, S. *Modern Physics* (Prentice Hall, Upper Saddle River, NJ, 2000).

Feynman, R. *The Character of Physical Law* (MIT Press, Cambridge, MA, 1967).

Pagels, H. *The Cosmic Code: Quantum Physics as the Language of Nature* (Simon & Schuster, New York, 1982).

Kenneth W. Ford

COOLING, PARTICLE

Beam Cooling

In a particle accelerator or storage ring, particle motion consists of oscillations around the nominal orbit. Normally, the amplitude of the oscillations is constant in time. However, special techniques, known collectively as "beam cooling," can be used to reduce the average oscillation amplitudes of beam particles. A cooled beam will have a reduced momentum spread, a smaller physical size, a reduced angular divergence, or some combination of these characteristics. Beam cooling is conceptually similar to cooling of ordinary matter, which involves a reduction in the amplitude of the random motion of the constituent atoms.

Beam cooling techniques have proven essential for achieving high interaction rates at electron-positron and proton-antiproton colliding beam facilities, largely because of the need to create a dense beam of antiparticles. Beam cooling is not required for the Large Hadron Collider (LHC) project, a proton-proton colliding beam facility currently under construction at the European Laboratory for Particle Physics (CERN) in Geneva, Switzerland.

Electron Cooling

Electron cooling occurs when a beam particle, say a proton, moves slowly through a beam of electrons of uniform velocity. As it bumps into the electrons, the proton loses energy until it is *at rest* relative to the electrons. Thus, protons slower than the electron beam speed up and protons faster than the electron beam slow down. The motion of particles transverse to the electron velocity is decreased for the same reason. The process results in the transfer of energy from the hot proton beam to the cold electron beam. The electron beam is normally generated continuously and discarded after a single interaction with the hot beam.

Electron cooling has been implemented at several accelerator facilities. A typical application involves the creation of a dense, low-energy, low-momentum spread beam that can be used for high-precision spectroscopy experiments. The applicability of the method has been restricted to low energies because the cooling rates tend to be slower at high energy and because it is difficult to produce a high-energy electron beam that is sufficiently uniform in electron velocity.

Synchrotron Radiation Damping

Synchrotron radiation is the spontaneous radiation of charged particles subjected to a strong magnetic field. The radiation is proportional to the fourth power of the energy to mass ratio, $(E/m)^4$, and, as a consequence, is important mainly for electrons at high energy. The radiation makes it difficult to accelerate or store electrons at high energy in circular accelerators and is a dominant consideration in the design of these accelerators. Special rings, known as synchrotron light sources, utilize synchrotron radiation to produce intense light beams that are scattered from a variety of experimental samples. Synchrotron radiation also produces beam cooling effects that are utilized in electron-positron colliding beam accelerators and special storage rings known as damping rings.

Synchrotron radiation cooling works primarily by providing a braking force antiparallel to the particle velocity. The reduction of the transverse component of the beam momentum results in smaller transverse oscillation amplitudes, but the loss of longitudinal momentum does not significantly change the momentum spread, as is required for cooling. Momentum cooling is achieved primarily by exploiting the coupling between the longitudinal and horizontal particle oscillations. Proper design of the focusing properties of the storage ring can result in a beam that is cooled in all three dimensions. The rapid loss of longitudinal beam momentum invariably requires that it be restored by electromagnetic fields, which are created in high-power, radio frequency cavities.

A number of circular electron-positron colliding beam machines have been designed and built to take advantage of the cooling properties of synchrotron radiation. Damping rings have been used to cool electrons and positrons at the Stanford Linear Collider (SLC), the only linear colliding beam facility that has been built. A linear accelerator avoids the copious synchrotron radiation associated with the magnetic fields required to bend the particles in a circle but requires damping rings utilizing synchrotron radiation to cool the beam to a small size to produce a high particle collision rate.

Ionization or Muon Cooling

Ionization cooling is similar to electron cooling except that the electrons are part of a medium (usually liquid hydrogen). The energy loss creates a braking force (similar to synchrotron radiation) resulting in transverse cooling and the need to replace the average longitudinal momentum that is lost in the medium. The strong momentum cooling effect present in electron cooling is not present in ionization cooling because the beam velocity is necessarily higher than that of the electrons in the medium, which are at rest.

This technique is practical only for muons: proton beams are too severely disrupted by nuclear interactions in the medium, and electron beams are disrupted by electromagnetic interactions. Muon beams are subject to electromagnetic interactions, but the magnitude is greatly reduced because the muon mass is larger than the electron mass. Ionization cooling has not been utilized nor even demonstrated in a practical system but has been extensively studied for possible future use in neutrino sources based on muon beams and for muon colliding beam facilities.

Stochastic Cooling

Stochastic cooling is a technique that requires the ability to measure fluctuations in the average beam motion about the nominal orbit. If it were possible to measure each particle individually, its motion could be corrected to coincide with the nominal orbit. In practice, only large groups of particles can be measured and corrected as a group. In any sample of particles, some have a position that is, say, lower than the nominal orbit and some have a higher position. The average position is close to the nominal orbit, but sometimes there are more high particles and sometimes there are more low particles. If the motion of the beam sample is corrected according to its average, the individual particle oscillation amplitudes are reduced after many sample positions are measured and corrected. The effectiveness of the technique requires the samples to consist of a continuously changing population of particles. Stochastic cooling systems are normally designed to cool one beam coordinate at a time, but multiple systems can be used to achieve cooling in three dimensions.

Stochastic cooling has been used most notably to cool antiproton beams. Antiprotons are produced in high-energy collisions but with a very low density compared to particle accelerator beams. Antiprotons are collected and transported into a storage ring where they are cooled in a succession of steps. The transverse oscillations are reduced about five-fold, but the main goal is to increase the beam intensity without increasing the momentum spread. In this process, known as "stochastic stacking," a pulse of antiprotons is added to an existing stack by placing it at a slightly higher momentum than the previously stacked beam. The cooling system then reduces the momentum spread to the size of the previously stacked beam and a new pulse is added. The cycle can be repeated thousands of times to produce a high intensity antiproton beam.

Other Cooling Techniques

Other cooling techniques are available to cool ion beams, very low energy particles (captured in particle traps), and atoms stored in traps. These techniques are not used for high-energy particle physics.

See also: ACCELERATOR; ACCELERATORS, COLLIDING BEAM: ELECTRON-POSITRON; ACCELERATORS, COLLIDING BEAM: ELECTRON-PROTON; ACCELERATORS, COLLIDING BEAM: HADRON; ACCELERATORS, EARLY; ACCELERATORS, FIXED-TARGET: ELECTRON; ACCELERATORS: FIXED-TARGET: PROTON; BEAM TRANSPORT; EXTRACTION SYSTEMS; INJECTOR SYSTEM

Bibliography

Poth, H. "Electron Cooling: Theory, Experiment, Application." *Physics Reports* **196**, 135–297 (1990).

van der Meer, S. "Stochastic Cooling and the Accumulation of Antiprotons." *Reviews of Modern Physics* **57** (3), 689 (1985).

John Marriner

CORNELL LABORATORY FOR ELEMENTARY PARTICLE PHYSICS

The Cornell Laboratory for Elementary Particle Physics (formerly called the Cornell Laboratory of Nuclear Studies) was established in 1946 by the Trustees of the University as a research unit within the Physics Department. Its mission then was "to investigate the particles of which atomic nuclei are composed and to discover more about the nature of the forces which hold these particles together." Over the next two decades under the leadership of Robert R. Wilson the Laboratory designed, built, and operated a series of electron synchrotrons of successively higher energies.

In a synchrotron a beam of electrons runs around in a ring-shaped vacuum chamber, guided by electromagnets and accelerated on each turn by microwave cavities. The 10-GeV synchrotron was completed in 1968, when Wilson left Cornell to become director of Fermilab. It was housed in a ring tunnel one-half mile in circumference and 12 meters under the Cornell campus. Experimenters from Cornell and collaborating universities used the electron and photon beams extracted tangentially from the synchrotrons to map the internal structure of the proton and neutron, study the production of pi and K mesons, and test the theory of quantum electrodynamics.

The Cornell Electron Storage Ring

By the late 1970s it became clear that it was more economical to reach higher energies by colliding oppositely circulating electron and positron beams head on than by bombarding targets with electrons. Positrons are the positively charged antiparticles of the negatively charged electrons. Electron-positron pairs can be created by running a high-energy electron beam into a target. When an electron and positron annihilate in a colliding beam ring, all of their energy can go into the production of new states of matter, and none of it has to be wasted in conserving the net forward momentum of a beam-target collision.

So, under the leadership of Boyce McDaniel, the next laboratory director after Wilson, and Maury Tigner, the chief designer and project manager, and with the support of the National Science Foundation, the Laboratory constructed the Cornell Electron Storage Ring (CESR) in the same tunnel alongside the existing synchrotron. After electrons or positrons reach their maximum energy in a fraction of a second in the synchrotron, a pulsed electromagnet kicks them into a transfer channel that takes them into the storage ring where they can coast around and around for an hour or more. A storage ring is similar to a synchrotron except that the beam energy is held fixed; that is, the microwave cavities provide just enough push to make up for the energy that is lost by radiation. In CESR the two beams orbit at energies up to six billion electron volts per particle, that is, at velocities that are within 1.1 m/s of the 299,792,458 m/s speed of light.

The CLEO Experiment

The beams in CESR are configured so that the electrons and positrons collide at one point in the ring. Surrounding that point is a large, sophisticated, multipurpose apparatus designed to detect and identify all the particles that are produced in the electron-positron annihilation. It is built and operated by a collaboration of about 200 faculty, staff, and graduate students from about twenty universities. This collaboration, called CLEO, has been studying the products of the high-energy electron-positron collisions since CESR began operating in 1979. The CLEO detection apparatus records the dozen or so particles that are produced in each of the millions of electron-positron collisions.

Present understanding of the basic constituents of matter and the laws that govern their strong, electromagnetic, and weak interactions is embodied in the Standard Model. It explains how quarks make up the protons and neutrons in atomic nuclei and how the nuclei and the electrons (the lightest leptons) combine to make atoms, molecules, and bulk matter. It explains radioactivity, radio waves, chemistry—all sorts of phenomena familiar and unfamiliar. Although the model has had many successes in correlating the properties and behavior of the fundamental quarks, leptons, and bosons that make up the universe, it is incomplete. Basic questions are unanswered. Why are there six quarks and six leptons? Why do they have the masses and coupling strengths that they do? How did the symmetry between particles and antiparticles get broken to give the very asymmetric abundances in the present universe? The mission of the Cornell Laboratory for Elementary Particle Physics has over the years evolved from an original concern with nuclei to a quest for understanding the basis of the Standard Model of quarks and leptons.

An electron-positron collision in CESR is like a miniature version of the Big Bang and is an ideal way to create the more exotic heavier quarks and leptons that have decayed away since the Big Bang. CESR is particularly well adapted to the production of the charm and bottom quarks and the tau lepton. Since the discovery of the B meson (a bottom quark and a light antiquark) and the first measurement of its mass at CESR in 1980, CLEO has measured rates for over a hundred different decay modes of the B

The superimposed oval shows the location of the Cornell Electron-Positron Storage Ring in a tunnel about 40 feet underground. The brick building at the near side of the oval is the Wilson Laboratory building, housing most of the laboratory facilities. CREDIT: COURTESY OF KARL BERKELMAN, CORNELL UNIVERSITY. REPRODUCED BY PERMISSION.

meson, including many that are sensitive to violation of particle-antiparticle symmetry. The radiative decay of the bottom quark to the strange quark is particularly sensitive to the top quark and to the possible existence of hypothetical particles too massive to be directly produced in any present-day accelerator, and the CLEO measurement of the decay rate has been used to constrain numerous theoretical speculations on possible extensions of the Standard Model. CLEO has also been the leader in mapping out the spectroscopy of the particle states formed by the charm quark. Most of the known charmed baryon states (a charm quark bound to two other quarks) were discovered by CLEO.

Research in Accelerator Physics and Technology at Cornell

In spite of the advantages of electron-positron collisions for the study of heavy quarks, there is an important drawback; the annihilation probability is very small; the electrons and positrons usually pass by each other without interacting. Progress in particle research is therefore limited by the achievable beam-beam luminosity, a measure of the rate of collisions. This motivated efforts to raise the beam currents and focus them more tightly where they intersect. Instead of circulating a single bunch of electrons and a single bunch of positrons as in the original design, CESR now has forty-five bunches in each beam, with electric fields separating the orbits so that the bunches can collide at only one point. Thanks to these and other innovations, CESR held through the 1990s the world's luminosity record for colliding beam machines. These tricks have been copied in the design of later storage rings, surpassing CESR's record. CESR physicists have pioneered in the application of superconductivity to microwave particle acceleration cavities.

Other Research

The circulating electron and positron beams in CESR emit X rays tangentially to the beam orbit. This by-product radiation is thousands of times brighter than normal laboratory X-ray sources. Nine experimental stations are administered by a separate National Science Foundation (NSF) supported organization, called CHESS (for Cornell High Energy

Synchrotron Source). Hundreds of X-ray experiments have been carried out in material science, molecular biology, medicine, and other fields by scientists from Cornell, other universities, government laboratories, and industry.

The theory group at the Laboratory of Nuclear Studies has a distinguished history, marked by Nobel Prizes for Hans Bethe in 1967 and for Kenneth Wilson in 1982. Current work covers a wide range, from the physics of supernovae, through quantum electrodynamics and lattice quantum chromodynamics, to superstring theory and relativistic astrophysics.

See also: BENEFITS OF PARTICLE PHYSICS TO SOCIETY; FUNDING OF PARTICLE PHYSICS

Bibliography

Berkelman, K. "Upgrading CESR." *Beam Line* **27** (2), 18 (1997).

Besson, D., and Skwarnicki, T. "Upsilon Spectroscopy." *Annual Review of Nuclear and Particle Science* **43**, 333 (1993).

"CESR Set to Bow out of B-particle Business." *CERN Courier* **41** (2), 10 (2001).

Cornell Laboratory for Elementary Particle Physics. <http://www.lns.cornell.edu>.

McDaniel, B. D., and Silverman, A. "The 10-GeV Synchrotron at Cornell." *Physics Today* **21**, 29 (October 1968).

McDaniel, B. D., et al., "CESR Design Report." *IEEE Transactions on Nuclear Science* **NS-28**, 1984 (1981).

"New Meson Physics Pushes on in 2000." *CERN Courier* **39** (10), 11 (1999).

Richman, J. D., and Burchat, P. R. "Leptonic and Semileptonic Decays of Charm and Bottom Hadrons." *Reviews of Modern Physics* **67** (4), 893 (1995).

Stone, S. *B Decays,* Revised 2nd ed. (World Scientific, Singapore, 1993).

Karl Berkelman

COSMIC MICROWAVE BACKGROUND RADIATION

The cosmic microwave background radiation (CMBR) comprises the remnant photons from an early period after the Big Bang in which the electrons, protons, and photons constituted a hot plasma filling the universe. The CMBR has the spectral form of blackbody radiation. The expansion of the universe stretches the wavelengths of the CMBR photons, reducing their energy and thus cooling off the radiation. The present temperature of the CMBR is 2.728 K, and so it is sometimes called the 3K background. The intensity of this blackbody radiation peaks at a wavelength of about 3 mm; in the microwave range, the CMBR is the dominant source of signal observed by telescopes looking up through the disk of our galaxy and away from known point sources (like supernova remnants). Other notable features of the CMBR are its isotropy (meaning that in all directions on the sky, its temperature is measured to be the same to within a fraction of a percent) and its lack of any sizable polarization. The existence of such relic radiation is one of the foundations of modern cosmology. Any theory attempting to address the large-scale disposition and history of the universe must satisfactorily explain such radiation, including its relative isotropy and resemblance to a perfect blackbody.

Discovery

The CMBR was discovered serendipitously in 1965 by Robert W. Wilson and Arno A. Penzias of Bell Laboratories in Holmdel, New Jersey. In 1978 Penzias and Wilson were awarded the Nobel Prize in Physics for their discovery. The work that lead to the discovery stands as an excellent example of careful scientific method. Penzias and Wilson made their discovery in the course of characterizing an antenna-receiver system designed to calibrate the absolute intensity of several astronomical sources known to emit radio waves. Absolute calibration of an antenna-receiver system requires understanding every source of thermal noise to which it responds. Noise sources include the room temperature (300 K) radiation from the Earth and from the metal comprising the antenna itself, thermal emission from the atmosphere, and noise from the amplifier. Through a laborious series of tests, Penzias and Wilson characterized their system to better than 1 K, so that the "excess antenna temperature" of 3 K was undeniably external to their apparatus. As they mulled over the import of their result, Penzias and Wilson became aware of ongoing theoretical work at nearby Princeton University. There, P. James E. Peebles and Robert H. Dicke had just deduced that in a universe evolving from a big bang, relic thermal radiation should be detectable. In fact, David Wilkin-

son and Peter G. Roll had already begun building an experiment to try to measure it. The two groups collaborated and published back-to-back papers announcing the discovery and offering a cosmological interpretation for it.

Origin

The universe is presently observed to be expanding; therefore, at earlier times it was smaller than it is today and thus much denser. At some point, its contents included electrons, protons, and photons, among other things. (Other constituents included neutral particles like neutrons and neutrinos, as well as dark matter, which does not interact with the plasma.) Initially, the universe was so hot that the electrons and protons were completely ionized. This plasma was opaque; photons did not travel in straight lines for long, as they were continuously interacting with (or scattering off) the charged particles. However, as the universe expanded and cooled, it was eventually no longer hot enough to sustain the plasma. Photons no longer had enough energy to ionize hydrogen when it chanced to form, and eventually all the electrons and protons were bound up in hydrogen. Suddenly, the universe became transparent to the photons, since they quit scattering off particles. This epoch in the history of the universe is sometimes referred to as the time of last scattering for that reason. Another name is decoupling, since the neutralization of the plasma decoupled the photons from the matter. This epoch occurred a few hundred thousand years after the Big Bang. The universe is 10 to 20 billion years old, so the CMBR photons reaching the Earth today have been traveling straight toward the planet for the last 10 to 20 billion years.

Nature

The most noticeable aspect of the CMBR is its lack of features. The absolute temperature of the CMBR has been measured from wavelengths of 50 cm to 0.5 mm; across this broad range, no significant deviations from a blackbody shape have been observed. The Far Infrared Absolute Spectrometer (FIRAS) on board the COBE satellite limited any distortions at wavelengths between 0.5 and 5 mm to be smaller than 0.1 percent. The CMBR is isotropic to one part in a thousand. At that level, it exhibits a dipole anisotropy. (Anisotropy means lack of isotropy; a related term is inhomogeneity.) The dipole arises because the solar system is moving with respect to the CMBR. In the direction the solar system is moving, the CMBR appears blue-shifted (to hotter temperatures) by the Doppler effect. In the opposite direction, the CMBR is red-shifted. Primordial anisotropy in the CMBR (meaning anisotropies observed today that are interpreted as reflecting conditions at the time of last scattering) was first measured by the Differential Microwave Receiver (DMR), also on board COBE. Detection of the anisotropy was announced in 1992. These measurements were very hard to make because (aside from the dipole) the anisotropy is tiny: ten parts in a million.

Revelations

The extreme isotropy of the CMBR (prior to the DMR result) puzzled scientists for two reasons. The first is that the rest of the universe today is not at all isotropic. How could the present lumpy matter distribution (planets, stars, galaxies, and so on) have evolved from a completely isotropic beginning? The answer, of course, is that the beginning was not completely isotropic. Study of the small anisotropies of the CMBR, and how the amplitudes of such anisotropies vary with spatial scale, reveals features of the early universe. In the decade after the COBE result, numerous ground-based and balloon-borne experiments made measurements of CMBR anisotropy at many spatial scales. In the summer of 2002, the MAP satellite was launched with the goal of measuring the CMBR anisotropy on many scales with unprecedented accuracy.

The second problem involves causality. At the time of last scattering, the universe was only a few hundred thousand years old, so light could only have traveled a distance R of a few hundred thousand light-years at most. Since information cannot be transmitted faster than light can travel, causality arguments dictate that regions in the universe farther apart than R cannot be in thermal equilibrium. Today, R subtends an angle a bit smaller than a degree, and yet the CMBR is isotropic over much larger regions than a degree. The solution to this conundrum was inflation, an epoch in which regions that were originally in causal contact moved apart much more

rapidly than the speed of light because of an extremely rapid period of expansion of the universe.

See also: ASTROPHYSICS; BIG BANG; COSMOLOGICAL CONSTANT AND DARK ENERGY; HUBBLE CONSTANT; INFLATION

Bibliography

Dicke, R. H.; Peebles, P. J. E.; Roll, P. G.; Wilkinson, D. T. "Cosmic Black-body Radiation." *Astrophysical Journal* **142**, 414–419 (1965).

Fixsen, D. J.; Cheng, E. S.; Gales, J. M.; Mather, J. C.; Shafer, R. A.; and Wright, E. L. "The Cosmic Microwave Background Spectrum from the Full COBE FIRAS Data Set." *Astrophysical Journal* **473**, 576–587 (1996).

Peebles, P. J. E. *Principles of Physical Cosmology* (Princeton University Press, Princeton, NJ, 1993).

Penzias, A. A., and Wilson, R. W. "A Measurement of Excess Antenna Temperature at 4080 Mc/s." *Astrophysical Journal* **142**, 419–421 (1965).

Silk, J. *The Big Bang* (W. H. Freeman, New York, 1996).

Smoot, G. F., et al. "Structure in the COBE Differential Microwave Radiometer First-year Maps." *Astrophysical Journal* **396**, L1–L5 (1992).

Wilson, R. W. "The Cosmic Microwave Background Radiation." *Science* **205**, 866–874 (1979).

Suzanne T. Staggs

COSMIC RAYS

The idea that there was some type of unknown energetic radiation falling on the Earth from space arose from studies of radioactivity that began in the 1890s. As instruments and understanding of radioactivity improved, the presence of a residual background of radiation that could not be accounted for became more troubling and significant. As early as 1900, C. T. R. Wilson suggested that there could be some form of "cosmic" radiation. However, it was not until 1912 that Victor Hess, in a classic and daring series of manned balloon flights, proved that indeed some form of energetic radiation was penetrating the atmosphere from space.

Initially this cosmic radiation was assumed, by analogy, to be similar to the most penetrating form of radiation produced by radioactivity, gamma radiation. In the 1930s, it was found that the cosmic radiation was influenced by the Earth's magnetic field and hence had to be made up of charged particles, presumed to be electrons. When further studies showed that these particles were positively charged, the assumption was that these particles were positively charged electrons, positrons, examples of which had recently been discovered in the cosmic rays. By 1940 it was becoming apparent that the majority of the incident particles were not electronic but nuclear, presumably protons. It is now known that although most of the particles are indeed protons, there are also energetic nuclei of all the elements in the periodic table present with roughly the same abundance as those found in the solar system. These particles originate from outside the solar system, although, in addition, the sun sometimes produces bursts of copious but relatively low energy particles, solar energetic particles (SEP). The energies of the cosmic ray particles cover an enormous range, from values typically found in radioactive decays, 10^6 electron volts (eV), to greater than 10^{21} eV, energies that are greatly above any that can be produced in the largest accelerators made by humans.

Few, if any, of the primary particles that enter the top of the earth's atmosphere reach the ground unaffected. Instead they typically undergo nuclear interactions with the nuclei in the atmosphere and produce showers of secondary particles. If the initial energy is high enough, this process may continue for many generations of interactions, resulting in a large burst of secondary particles reaching sea level. While the majority of these secondary particles are electrons, there are also present many of the unstable elementary particles that are produced in high-energy nuclear interactions. The lower-energy particles will lose their energy by ionization in the atmosphere even if they do not interact. As a result, the number of particles in the atmosphere reaches a maximum at an altitude of about 20,000 meters and declines nearer the surface.

Until about 1954, primary and secondary cosmic ray particles were the only source of available high-energy particles that could be used to study the physics of nuclear interactions at energies greater than those typical of radioactive decays. Many of the fundamental processes of the production and existence of elementary particles were first studied by looking at the secondary cosmic ray particles.

Pair Production and the Positron

The first new particle discovered in the secondary cosmic rays was the positively charged electron, the positron, the first example of antimatter proved to exist. In 1932 Carl Anderson used a cloud chamber to look at the tracks produced by cosmic ray particles passing through the chamber. A magnetic field was applied to the chamber so that the sign of the charge of each particle could be determined from the curvature in the field. A lead plate was placed in the chamber so that the direction of motion could be determined from the loss of energy in the lead. Anderson found an example of a particle with the mass of an electron but with a positive charge. This discovery was rapidly confirmed by pictures in cloud chambers of the tracks of pairs of electrons with opposite curvature, examples of the pair production of electrons and positrons by gamma rays. The discovery of these positrons confirmed the theoretical predictions of Paul Dirac (1928) of negative energy states of the electron.

Light Mesons

The presence of unstable particles in the secondary cosmic rays with masses intermediate between that of the proton mass of 1836 electron masses, m_e, and that of the electron, was established as the result of a wide range of experiments. In 1937 particles were observed which had both positive and negative charges, did not lose energy as fast as electrons, but were not as massive as a proton. These particles, originally called "mesotrons," then "mesons," appeared at first to resemble those predicted in 1935 by Hideki Yukawa to explain the forces within the nucleus. The observed mesons were found to have a mass some 200 times that of the electron and to be unstable. However, they had a half-life of 2.1×10^{-6} second, which was some twenty times longer than predicted, and did not interact strongly with matter, as the Yukawa particle should.

After World War II rapid advances in experimental techniques and in particular the development of sensitive nuclear photographic emulsions by Cecil Powell and his group solved the problem of the discrepancies between the predicted and observed properties of these mesons. Emulsions exposed on mountaintops and on high-altitude balloons showed that there were two types of light meson. Pi-mesons, with a mass of about 273 m_e, were observed coming to rest in the emulsion and decaying into a lighter mu-meson with a mass of about 207 m_e. This mu-meson then also came to rest and decayed into an electron. Now we know that charged pi-mesons decay to mu-mesons and a neutrino. The mu meson then decays into an electron and two more neutrinos. The lifetime of the pi meson is 2.5×10^{-8} second, much less than that of the mu meson. In addition, negatively charged pi mesons, which are not repelled by the positive nuclei in atoms, tend to be captured and interact with nuclei before they can decay, proving that they are strongly interacting particles of the sort predicted by Yukawa. Mu mesons, on the other hand, do not interact but decay even though they have a much longer lifetime. These pi mesons are now known to exist with positive, negative, and neutral charges, with the neutral pi mesons having very short lifetimes (8.4×10^{-17} second) and generally decaying into two energetic gamma rays.

Heavier Mesons and Hyperons

Although a light meson was predicted theoretically, the existence of additional mesons heavier than the pi meson was not anticipated. However, as early as 1944 there was a cosmic ray report of the observation of a particle with a mass about 1,000 m_e. Further isolated examples were found over the next few years. Examples were seen of particles that came to rest in emulsion and decayed into three pi-mesons. Other events showed decays, both in flight and at rest, of heavy mesons into various lighter particles. It became clear from observations in cloud chambers and emulsions that there were a number of different modes of decay of so called "K mesons" with masses around 1,000 m_e being produced in high-energy nuclear interactions. By 1954 there had been observations of at least six different modes of decay of charged K mesons and several modes of decay of neutral K mesons.

At the same time there were also observations of unstable particles with masses between that of the proton and deuteron. These "hyperons" were initially regarded as being excited states of the nucleon. They could be charged or neutral, decaying into a proton and a charged or neutral pi-meson.

In every case only a few examples of each type of particle were observed. In many cases there was limited information on the properties of the incident particles and the decay products. Only when artificial accelerators were built with sufficient energy to create these particles and study them under controlled conditions was it possible to begin to understand the underlying physics of nuclei and their forces.

Current Status

The studies using cosmic ray particles only just touched on the problem of these unstable particles. The full complexity of the possible unstable forms of matter that could be created in high-energy nuclear interactions was not unraveled until extensive experiments using high-energy particles from artificial accelerators led to the development of the quark theory of matter. In this theory strongly interacting particles known as hadrons consisted either of baryons (nucleons and hyperons) composed of three quarks or mesons composed of two quarks. Hence, unlike the earlier assumption in the cosmic ray studies, mesons could include particles with masses greater than that of a nucleon. Hyperons also were found to be more than just excited nucleons.

See also: ASTROPHYSICS

Bibliography

Dirac, P. A. M. "The Quantum Theory of the Electron." *Proceedings of the Royal Society of London* **A117**, 610–624 (1928).

Friedlander, M. W. *Cosmic Rays* (Harvard University Press, Cambridge, MA, 1989).

Millikan, R. A. *Cosmic Rays* (Cambridge University Press, Cambridge, UK, 1939).

Montgomery, D. J. X. *Cosmic Ray Physics* (Princeton University Press, Princeton, NJ 1949).

Sekido, Y., and Elliot, H., eds. *Early History of Cosmic Ray Studies* (D. Reidel Publishing Company, Dordrecht, Netherlands, 1983).

Wilson, J. G. *Cosmic Rays* (Springer-Verlag, New York, 1976).

C. Jake Waddington

COSMIC STRINGS, DOMAIN WALLS

Certain models of elementary particle physics predict the existence of extended objects, such as cosmic strings or domain walls, in addition to the usual pointlike particles, such as quarks, leptons, and gauge bosons. Cosmic strings and domain walls are filamentlike and sheetlike structures, respectively, of microscopic thickness, typically less than 10^{-18} meters, but of arbitrary, and possibly astronomical, length (and width). No definitive experimental evidence for either of these objects currently exists, but, if detected, their properties could help determine the correct theory of elementary particles. They might also play an important role in structure formation in the early universe. Cosmic strings and domain walls are examples of topological defects, so called because their existence is determined by the topology of the set of ground states of the theory.

Spontaneous Symmetry Breaking

Topological defects can only arise in theories with a feature called spontaneously broken symmetry. The ground state (state of lowest energy) in such theories is different at high and low temperatures, with the high-temperature ground state having a greater degree of symmetry than the low-temperature ground state. The early universe was much hotter than the current universe, and so was in a stable, symmetric phase. As the universe expanded and cooled, this symmetric phase became unstable, and the universe made a phase transition to a state of reduced symmetry. This process of spontaneous symmetry breaking can lead to the formation of topological defects, as explained below.

The Standard Model of elementary particle physics, which describes the strong, weak, and electromagnetic forces, has a symmetry that is spontaneously broken at a critical temperature of $T = 10^{15}$ Kelvin; nevertheless, this model predicts neither cosmic strings nor domain walls. The Standard Model, however, is not believed to be a complete description of reality but only part of a larger theory. Some candidates for this enveloping theory, called grand unified theories, have additional symmetries broken at much higher temperatures, typically around $T = 10^{29}$ Kelvin. Depending on their symmetries and how these are broken, grand unified models may predict cosmic strings and/or domain walls, whose thickness would typically be about 10^{-32} meters.

The Higgs Field

In particle physics models, spontaneous symmetry breaking is usually caused by a Higgs field, described here through an analogy. Consider a pencil suspended from its tip. Its gravitational potential energy is minimized when it hangs straight down. The vertical pencil is rotationally symmetric, since rotating the configuration does not change it. This maximum-symmetry, minimum-energy configuration is analogous to the ground state of the Higgs field above the critical temperature.

Next consider a pencil balancing vertically on its tip on a table. This configuration also has rotational symmetry but is unstable because the potential energy is a maximum. A configuration of minimum energy, with the pencil lying on its side, breaks the rotational symmetry. Moreover, because the pencil could fall in any direction, a continuum of minimum energy configurations exists. These broken-symmetry, minimum-energy states are analogous to the ground states of the Higgs field below the critical temperature.

Domain Walls

The existence of more than one ground state with broken symmetry is the key feature of spontaneous symmetry breaking that allows the possibility of topological defects. To see why, consider an infinite two-dimensional array of pencils balancing vertically on a table, with adjacent pencils connected by springs. Assume that the tips of the pencils are hinged so that each can only fall in either of two directions: to the left (L) or to the right (R). Let all the pencils be released simultaneously. If one of the pencils begins to fall to the right, the springs will cause nearby pencils to fall in the same direction, creating a region of the plane (or domain) in which all the pencils have fallen to the right. In another region of the plane, all the pencils might fall to the left. The plane is thus divided up into L and R domains. At the boundary between an L domain and an R domain will be a swath of standing or leaning pencils, supported by the springs connecting them and interpolating between the left-pointing and right-pointing fallen pencils. This swath of pencils separating two domains is termed a domain wall; it characterizes a region in space where the potential energy is not minimized.

If one tries to reduce the energy further by pushing the standing pencils down in one direction or the other, the springs will force nearby pencils to pop up; the domain wall will move. Thus, a domain wall represents trapped energy density, which can move but cannot spread out or dissipate.

The above analogy describes a theory with discrete symmetry breaking, one in which the Higgs field has a finite number of ground states (two, in this case: L and R) below the critical temperature. In such a theory, when the universe cools below the critical temperature (corresponding to the release of the pencils), the Higgs field at each point evolves from the symmetric state to one of the ground states; this is called a cosmological phase transition. The gradient energy of the Higgs field causes the transitions at nearby points to be correlated (just as the springs cause nearby pencils to fall in the same direction). Because the symmetry breaking occurs simultaneously and randomly, and because information can travel no faster than the speed of light, distantly separated regions will not necessarily be in the same ground state. Since the Higgs field fills space, the L and R domains are three-dimensional, so the boundaries separating them are two-dimensional domain walls. The mismatch of the ground states in different domains forces the Higgs field at the boundaries to remain in the higher-energy symmetric state. This trapped energy represents the mass of the domain wall.

Cosmological phase transitions are studied using numerical simulations, in which each spatial region is randomly assigned one of the broken-symmetry states. Typically, a wall that spans the universe will form. Such a wall would disrupt the observed homogeneity of the cosmic microwave background radiation unless the symmetry-breaking scale is much less than $T = 10^{15}$ Kelvin. This constraint rules out grand unified theories that predict domain walls (unless an era of inflation occurs during or after domain wall formation).

Cosmic strings

Domain walls can only form in theories with discrete symmetry breaking. Theories with a continuously broken symmetry, those in which the set of ground states below the critical temperature forms a connected continuum, do not give rise to domain walls, but can host other types of topological defects, for example, cosmic strings.

A cosmic string can be described by another analogy. Again imagine a two-dimensional array of pencils standing on a table and connected by springs, but now suppose that the pencils may fall in any direction. Each pencil now has a continuum of ground states, characterized by the angle θ of the fallen pencil (measured counterclockwise from the east). Because of this multiplicity of ground states, pencils in different regions of the plane will randomly fall in different directions when released. The springs cause nearby pencils to fall in nearly the same direction, and because the ground states are continuously connected, some realignment of the pencils can occur after they have fallen to minimize the stretching of the springs. Suppose, however, that the pencils all fall outward from some arbitrary origin: The pencils to the east fall in the direction $\theta = 0$, those to the north in the direction $\theta = 90$, those to the west in the direction $\theta = 180$, and those to the south in the direction $\theta = 270$, with the pencils at intermediate points of the compass interpolating smoothly between these angles. Since the springs do not allow the directions of nearby pencils to differ greatly, the pencils near the origin of the configuration must remain standing or partially standing. This core of standing pencils constitutes a region of stable, trapped potential energy density; although it can be moved around, no amount of realignment can eliminate it.

This analogy describes a theory with a continuously broken symmetry. Since the Higgs field fills three dimensions, the core of trapped energy density extends along one dimension and is called a cosmic string. The two-dimensional configuration of pencils above represents a cross section of the string (see Figure 1). A cosmic string can curve or form loops, and it can move in space.

As with domain walls, the formation and evolution of cosmic strings are studied using numerical simulations. These simulations show that when the universe cools below the critical temperature, the majority of cosmic strings that form stretch across the universe, with the remainder being closed loops. When two strings intersect, the ends can break and reconnect differently. If a single string intersects itself, it can reconnect so that a loop breaks off from

FIGURE 1

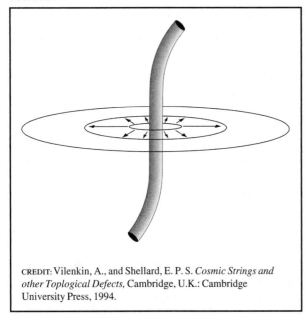

CREDIT: Vilenkin, A., and Shellard, E. P. S. *Cosmic Strings and other Toplogical Defects,* Cambridge, U.K.: Cambridge University Press, 1994.

Cosmic string showing a cross section of the Higgs field.

the rest of the string. Closed loops can lose energy through gravitational or other radiation, shrinking away to nothing. Simulations show, however, that in theories that predict cosmic strings, some of the strings would persist to the present day.

Because of their mass (typically 10^{21} kg/m), cosmic strings would curve space-time and act as gravitational lenses. They would also cause anisotropy in the cosmic microwave background but typically not so much as to conflict with observations. The distribution of matter was extremely homogeneous in the early universe. Through gravitational attraction, cosmic strings may have acted as seeds for matter to clump onto, leading to the formation of galaxies or quasars.

String Theory

Cosmic strings should not be confused with the fundamental strings of superstring theory, an ambitious framework for describing all the fundamental forces of nature, including gravity. If superstring theory is true, fundamental strings are the building blocks of all particles and fields; they have no thickness and typically microscopic length. Cosmic strings, on the other hand, are made out of the Higgs field, with their thickness depending on the details

of the theory and with typically astronomical lengths. Superstring theories, however, can also predict the existence of cosmic strings and domain walls, as well as other types of extended objects called D-branes.

See also: COSMOLOGY; ELECTROWEAK SYMMETRY BREAKING; INFLATION

Bibliography

Vilenkin, A. "Cosmic Strings." *Scientific American* **257** (6), 94–100 (1987).

Vilenkin, A., and Shellard, E. P. S. *Cosmic Strings and other Topological Defects* (Cambridge University Press, Cambridge, U.K., 1994).

Stephen G. Naculich

COSMOLOGICAL CONSTANT AND DARK ENERGY

Shortly after the development of his theory of general relativity, Albert Einstein recognized a potential problem with his new theory. General relativity was the first theory to describe not only the dynamics of objects within space-time but also the dynamics of space-time itself. As such, the theory offered the possibility of providing a first-principles understanding of the evolution of the universe itself. However, the fact that it reproduced Newtonian gravity in weak gravitational fields meant that no stable static cosmological solution of Einstein's equations existed involving merely matter and radiation. Since the gravitational attraction of all such sources of energy is universally attractive, any initially static system of mass points, such as galaxies, will inevitably collapse inward. In 1917, however, it appeared that the universe on large scales was indeed static.

In an effort to resolve this problem, Einstein recognized that he could preserve the symmetries that led him to develop the theory of general relativity by modifying his equation with the addition of an extra term, which he dubbed the "cosmological term." Such a term could produce, on large scales, a small repulsive force throughout the universe that might serve to counterbalance the standard gravitational attraction of distant masses, while leading to no observable effects on terrestrial scales that might disagree with existing observations.

Unfortunately, however, almost immediately problems arose with Einstein's idea. The physicist Willem de Sitter demonstrated that a consistent cosmological solution existed in the presence of a cosmological constant in which an otherwise empty universe could continue to expand forever. Einstein found such a possibility distasteful. But, a far more serious concern arose as it was recognized over the next decade, largely due to the work of the astronomer Edwin Hubble, that the universe is not static but is in fact expanding. In such a case, no additional repulsive force would be needed to counterbalance conventional gravity. In an expanding universe gravity could merely work to slow the expansion. Whether the observed expansion would stop was thus a simple question of initial conditions. The attempt to determine the expansion rate of the universe and its mass density—the two factors needed to ascertain the ultimate fate of the Universe—then became the central focus of cosmology for much of the rest of the twentieth century, while the cosmological constant faded from interest.

The question of the possible existence of a cosmological constant emerged again, however, following World War II as theoretical physicists began to grapple with the quantum mechanical properties of elementary particles. It was soon recognized that empty space need not be precisely empty. Virtual particle-antiparticle pairs could spontaneously appear and disappear again, as long as they did so in a time interval short enough so that no direct observations of the violation of the conservation of energy and momentum could be observed. With this recognition came the recognition that quantum mechanically, at least, one would in general expect the vacuum state of nature to possess energy. An examination of the form that this energy would take, dictated by the fundamental symmetry called Lorentz invariance, demonstrated that such vacuum energy produced an additional contribution to Einstein's equations identical in form to Einstein's cosmological term.

A new challenge then arose, which has since been termed the cosmological constant problem. The lack of any observed repulsive force in nature governing the expansion of the universe placed very strong constraints on the size of any possible vacuum energy density today. When compared with naïve theoretical estimates based on extrapolating our current knowledge of elementary particle physics to scales as small as the Planck length, the observational upper limit on the cosmological constant is about 125 orders of magnitude smaller. This is perhaps the worst prediction in all of physics!

It was generally assumed by particle physicists that an ultimate resolution of the cosmological constant problem would involve some mechanism, perhaps based on an unknown symmetry argument that required the ultimate vacuum energy density of the universe to be precisely zero. Only in this case could one hope to gain accord with observations while not requiring some unprecedented fine-tuning mechanism that might reduce the vacuum energy density by 125 orders of magnitude.

Here the situation remained until the last decade of the twentieth century. A host of new observational data began to expose some inconsistencies in the standard model of cosmology, which involved a flat universe dominated by some new form of nonbaryonic matter, conventionally called cold-dark matter. The problems arose from three separate fronts: (1) The age of a flat-matter-dominated universe was predicted to be less than about 10 billion years old, given the measured expansion rate of the universe. However, a determination of the age of the oldest stars in our own galaxy suggested that these stars were at least 12 to 20 billion years old. (2) Observations of the clustering properties of matter on the largest observable scales gave some estimate of the overall density of matter, and these estimates repeatedly began to suggest that there was not sufficient mass density in the universe to result in its being spatially flat. (3) Direct probes of the matter density in clusters of galaxies, based on X-ray measurements of the hot gas contained there, and on their evolution as a function of time, both put upper limits on the total matter density that fell far short of the amount needed to result in a flat universe today.

In order to resolve the latter two problems, some additional source of unclustered, spatially uniform energy would have to exist in the universe in order to provide the extra energy necessary for a flat universe. If such energy was unclustered, then both

X-ray and gravitational probes on the scale of galaxy clusters would not be sensitive to it. One possible source might be vacuum energy density. Such energy would have the additional impact of causing a net acceleration to the expansion of the universe that could resolve the apparent discrepancy with the expansion age of the universe compared to stellar ages. If the universe had expanded at a rate slower than that by which it is now measured to be expanding, then it would have taken longer for galaxies observed a certain distance away from Earth to have achieved that separation. Thus, the expansion age could be longer than it would otherwise be if vacuum energy plays a dynamical role today.

Between 1998 and 2000 these indirect arguments were bolstered by two significant developments from cosmology. First, observations of distant supernovae allowed a comparison of the relationship between physical distance and redshift that could probe the temporal evolution of the expansion. Much to the surprise of a large segment of the scientific community, evidence was obtained that the expansion of the universe is indeed accelerating, precisely at the amount required if vacuum energy is the dominant energy density of the universe today and leads to a flat universe. Concurrently, in 1999, observations of small fluctuations in the temperature of the cosmic microwave background (CMB) radiation that permeates space as a remnant of the Big Bang allowed the direct measurement of the large-scale geometry of space-time. As the geometry of the universe changes from open to closed, the angular scale corresponding to a fixed physical location observed from a distance changes. By comparing the angular scale of anisotropies in the CMB radiation, which has traveled unimpeded since matter became neutral about 300,000 years into the Big Bang, with model predictions, three different experiments confirmed with good accuracy that the universe is flat.

The combination of these two observations definitely suggests that some energy appears to be associated with empty space, indeed, this is the dominant energy density in the universe today! All cosmological observations are now consistent with the notion of a flat universe, in which about 30 percent of the total energy density results from matter and about 70 percent results from some form of dark energy.

The term dark energy is used to describe the energy associated with empty space because existing observations cannot distinguish between a true cosmological constant and some other unclustered form of energy permeating all of space. Much of the inquiry in present-day cosmology focuses on trying to find techniques that might distinguish a cosmological constant from something else. This observational activity will be very challenging, however.

The existence of dark energy presents one of the greatest puzzles in all of physics. As of yet scientists have no good theoretical grasp of what causes this energy, nor do they understand why it has its apparent value. Moreover, the presence of such dark energy has completely altered perceptions of the possible future evolution of the universe. Such energy violates a condition in general relativity called the strong energy condition that allows a one-to-one connection between geometry and the ultimate fate of an expanding universe. As a result, although one of the central goals of cosmology in the twentieth century, to determine the geometry of the universe, has now been achieved, the future is unfortunately far less certain than expected. Allowing for the presence of dark energy, a closed universe can expand forever, and an open or flat universe can recollapse. If one is to ever unambiguously be able to predict the future evolution of the universe, an understanding of the nature of dark energy arising from the fundamental theories of particle physics will be required.

See also: BIG BANG; COSMOLOGY; INFLATION

Bibliography

Krauss, L. M. "Cosmological Antigravity." *Scientific American* **280** (1), 34–41 (1999).

Krauss, L. M. *Quintessence* (Basic Books, New York, 2000).

Weinberg, S. "The Cosmological Constant Problem." *Reviews of Modern Physics* **61**, 1–24 (1989).

Lawrence M. Krauss

COSMOLOGY

Although astronomers and natural philosophers have always been interested in cosmology, it is only in the twentieth century that a science of the universe has become a reality. In particular, from about 1950 cosmological theory began to interact ever more strongly with nuclear and particle physics. This development has accelerated in the later part of the 20th century with the result that parts of cosmology (especially the early big bang theory) have become thoroughly integrated with new theories of elementary particles and fundamental interactions. Not only have theories of microphysics greatly advanced cosmological knowledge, but the early universe has also proved a valuable testing ground for fundamental particle physics.

Early Cosmology

Modern cosmology dates essentially from 1917, when Albert Einstein suggested a cosmological model based on his new theory of gravitation, the general theory of relativity. Einstein's universe was static, spatially finite, and filled with dilute matter. The same year, 1917, Willem de Sitter showed that the theory of relativity allowed another cosmological solution, corresponding to an empty universe. Both models made use of the cosmological constant that Einstein had introduced in his field equations as a parameter of a hypothetical antigravity force.

Nonstatic or evolutionary models were first investigated in 1922, by the Russian physicist Alexander Friedmann, but his work was ignored. Five years later, in 1927, the Belgian Georges Lemaître rediscovered the expanding universe and suggested a model in which the universe expands steadily from an Einstein world. The theories of Friedmann and Lemaître only became known in 1930, after Edwin Hubble had found observational evidence for the expanding universe. From that time onward, most physicists and astronomers accepted that the universe is in a state of expansion. At the same time, the cosmological constant fell in discredit and was often seen as an unnecessary complication.

A further step was taken by Lemaître in 1931, when he suggested for the first time what would later be called a Big Bang model, that is, a universe of finite age that has originated from an ultradense state. Lemaître called this state the primeval atom and suggested that its explosion was a quantum process, like in radioactive decay. The Einstein–de Sitter model of 1932, an early Big Bang model, assumed that there is just enough matter to keep the universe spatially flat. This matter density, which depends on the value of the Hubble constant, is called the critical density.

When Lemaître suggested his primeval atom hypothesis, the proton and the electron were the only known (massive) elementary particles. With the emergence of nuclear physics in 1932, new possibilities were offered for astrophysics and a more physical approach to cosmology. The entry of nuclear and particle physics into cosmology was largely a result of astrophysicists' attempts to understand the energy production of stars and the formation of heavier elements out of protons and neutrons. In a 1938 paper, Carl Friedrich von Weizsäcker sought to infer the state of the early universe by assuming that the chemical elements had been formed under conditions that would reproduce their present individual abundance. Von Weizsäcker was led to a kind of Big Bang universe, starting in an extremely hot and dense state, but he did not develop his ideas further. Such development was left to George Gamow, the Russian-American nuclear physicist and pioneer of cosmology.

Gamow's Physical Cosmology

Von Weizsäcker's methodology—to use the cosmic distribution of chemical elements as evidence for the physical state of the early universe—is sometimes known as nuclear archaeology. It was this method that guided Gamow's development of physical Big Bang cosmology in the years between 1946 and 1956. Together with his collaborators Ralph Alpher and Robert Herman, he was led to consider the primordial universe as a hot, dense gas of neutronic matter. Right after the Big Bang, neutrons would decay into protons and electrons, and some of the protons would combine with neutrons to form deuterons. The essential process in the building-up of higher nuclei was believed to be the capture of neutrons. Between 1948 and 1950 it was realized that this picture was too primitive and that the assumption of a matter-dominated early universe was untenable. Gamow and Alpher argued that during the first phase of the expansion, the energy content of the universe would be governed almost entirely by electromagnetic radiation (that is, photons). Only at a later stage, when the universe had become colder, would matter begin to dominate over radiation.

Gamow's program of Big Bang cosmology culminated in a 1953 paper, which can be considered the first example of particle cosmology. Written by Ralph Alpher, Robert Herman and James Follin, the paper provided a detailed analysis of the early universe, now thought to consist mainly of photons, neutrinos, and electrons (including positrons), but also with small amounts of nucleons and muons. The Alpher-Herman-Follin theory made full use of the most recent progress in particle physics, including the theory of weak interactions as applied to beta decay and processes involving electrons, neutrinos, and muons. The three physicists found the present abundance of helium to be about 30 percent and also calculated the neutron-to-proton ratio at the time nucleosynthesis started. However, the wider aim of the Gamow program—to account for the formation of all chemical elements—failed. Gamow and his collaborators were unable to build up elements heavier than helium, and this was widely considered a failure of the Big Bang theory itself.

Only in 1965 did Big Bang cosmology experience a renaissance, this time irreversibly. The discovery of the cosmic microwave background radiation of temperature about 3K provided an important parameter for new and improved calculations of the synthesis of the lightest elements. In 1966, James Peebles calculated from the Big Bang theory a helium abundance between 26 and 28 percent, in excellent agreement with observations. Subsequent calculations of the light elements (including deuterium and lithium) only improved the fit between theory and experimental data. The standard cosmological model accounts satisfactorily for the primordial abundance not only of helium-4, but also of deuterium, helium-3, and lithium-7. (The heavier elements are produced in stars and novae, not cosmologically.) The success of the predictions strongly suggests that the hot Big Bang theory is accurate all the way back to 1 second after the universe came into origin. Moreover, theories of nucleosynthesis give a good estimate of one of cosmology's most important parameters, the density of baryons (essentially protons and neutrons). Calculations imply that the density of baryons is somewhere

between 1 and 10 percent of the critical density, a result that has important implications for cosmology.

Antimatter and Baryogenesis

Antiparticles, predicted by Paul Dirac in 1931, have played an important role in the development of cosmology. As early as 1933, Dirac suggested the existence of entire antistars (made up of positrons and antiprotons), and in 1956 Maurice Goldhaber even speculated about an "anticosmos" symmetric to the cosmos in which we live. Neither these nor other speculations about abundant masses of antimatter have received observational support.

On the contrary, observations strongly indicate that there is only very little antimatter in the universe. If the universe were initially symmetric, annihilation between matter and antimatter would have resulted in a present state almost completely dominated by photons and with only trace amounts of baryons and antibaryons. This annihilation catastrophe obviously has not occurred, which means that the early universe must have possessed a slight excess of matter over antimatter. Until about 1970 the only explanation was to postulate a charge asymmetry in the very early universe, a slight predominance of quarks over antiquarks. Why would there be, for every 300 million quarks, just 299 million antiquarks? This question relates critically to the possibility of creation of baryons—baryogenesis—out of processes in which the number of baryons is not conserved.

The earliest attempt to explain how a baryon excess could be generated was published by Andrei Sakharov in a 1967 paper, but his suggestion did not attract much interest at the time. To explain the baryon asymmetry, Sakharov assumed violation of charge conjugation and parity (CP) conservation, a process which in 1964 had been detected in the decay of neutral kaons. More speculatively, he postulated an interaction that violated baryon conservation. With the first versions of Grand Unified Theories (GUTs), the idea of baryogenesis received theoretical support, and Sakharov's speculations were reconsidered. According to the early GUTs, as developed by Howard Georgi, Sheldon Glashow, Steven Weinberg, and others in 1974–1975, transitions between quarks and leptons are possible, that is, the baryon number is not precisely conserved. In 1978, Motohiko Yoshimura used the new GUTs to predict a baryon-antibaryon asymmetry caused by primordial vector bosons, and since then baryogenesis has attracted great interest. Within the GUT framework there is no need to postulate a baryon excess, for the observed baryon number could have been created by baryon number nonconserving processes.

During the years 1978–1980, Yoshimura, Weinberg, Frank Wilczek, and others used GUT to solve another cosmological problem. The number of photons in the observable part of the universe is around 10^{88} and that of baryons is around 10^{79}. The photon-to-baryon ratio is considered a fundamental quantity, because theory prescribes it to be constant in time. Why are there one billion times as many photons than baryons? The answer, according to the GUT theorists, is that this has not always been the case but was the result of the slight asymmetry in quark-antiquark annihilation processes in the early universe. Although the GUT-based theory of baryogenesis has still no direct experimental support, it is considered compelling by many particle physicists who see it as support for an intimate relationship between cosmology and particle physics.

Nucleosynthesis and baryogenesis are examples of particle physics applied to cosmology. Conversely, cosmology has also been used as a probe of fundamental physics, to gain knowledge of physics at very high energies. For example, in 1977 Gary Steigman, David Schramm, and James Gunn showed that the number of neutrino types could not be larger than four if the hot Big Bang theory were correct. Further refinements led to a lower limit of three types or families of neutrinos. At that time, only two neutrino types were known (the electron and muon neutrinos), but in 1993 evidence for a third (tau) neutrino was produced in accelerator experiments, which was seen as a brilliant confirmation of the cosmological prediction. Moreover, calculations showed that the primordial nucleosynthesis required the mass of the tau neutrino to be less than 0.5 MeV. This constraint agreed with, but was finer than, the one obtained experimentally and thus afforded another test of the big bang scenario. No wonder that Schramm concluded that "the marriage between particle physics and cosmology had indeed been consummated" (Schramm 1996, p. xvii).

Inflation Models

Grand unified theories also led to the first inflation models, introduced in the early 1980s by Alan Guth, Andrei Linde, and others. According to the inflation model, the very early universe underwent an extreme phase transition and approached a state known as a false vacuum. In Guth's original model, the energy density of the false vacuum was attributed to a mechanism derived from GUT, spontaneous symmetry breaking. The mechanism of inflation can be imagined as the vacuum energy effectively acting as a cosmological constant that boosts the expansion of the universe exponentially until the vacuum energy is converted into heat and the universe enters its epoch of slow expansion. However, most later versions of the inflation model do not depend critically on GUT or relate to the spontaneous symmetry breaking of particle physics. The essential feature is the false vacuum. Guth's paper of 1981 inspired a massive influx of particle physicists into cosmology and a burst of theoretical activity. By 1997, more than 3,000 papers had been published on inflation theory, most by particle physicists.

Among the many attractive features of the inflation model is that it offers an explanation of how the energy of the universe came into existence, namely, when the huge energy stored in the inflated false vacuum was released at the end of the brief inflation era. Moreover, it avoids the problem of the magnetic monopoles, particles that have never been detected. Most GUTs predict an abundance of primordial monopoles, many of which should still exist, but inflation takes care of the problem.

The confidence that many particle physicists and cosmologists have in the inflation model (in one of its several versions) is related to one of cosmology's most exciting problems, the problem of dark matter. According to the inflation model, the mass (or energy) density of the universe should be critical. However, observations show that there is far from enough ordinary (baryonic) matter to produce a critical density. It is known that most of the matter in the universe must be in an exotic, dark form. Particle physics suggests a number of dark matter candidates, from massive neutrinos to hypothetical particles predicted by certain fundamental theories of the GUT type. However, none of these particles has been detected, and the problem of dark matter is thus still unsolved. Nonetheless, there is a growing consensus that dark matter particles are "cold," that is, slowly moving. Particles called axions and neutralinos are examples of such cold dark matter, but, even if they exist, they may not be the most abundant form of energy in the universe.

An Accelerating Universe

Observations of supernovae in 1998 indicated that the expansion rate of the universe is much greater than hitherto assumed. Subsequent observations have substantiated the result, and it is now generally accepted that the universe is in a state of acceleration rather than deceleration. According to the standard big bang theory, an accelerated universe contains less matter than given by the critical density, but things look different if a form of energy with negative pressure is admitted. Such strange forms of energy were studied prior to the inflation model, and in 1974 Linde argued that a vacuum energy with negative pressure acts as an effective cosmological constant. (The same insight can be found as early as 1934, in a paper by Lemaître.) However, Linde did not realize that this effective cosmological constant may greatly influence the initial stage of the evolution of the universe. This was an insight of the inflation theory in which the cosmological constant has a natural interpretation.

Today, it is widely assumed that the vacuum energy of the accelerated universe must be attributed to a positive cosmological constant which is responsible for as much as two-thirds of the total energy content of the universe. The symbol for the cosmological constant is Λ, and cold dark matter is abbreviated CDM. For this reason, physicists sometimes speak of the ΛCDM scenario or, because of the connection to inflation theory, the inflation $+$CDM scenario. Now, in the beginning of the twenty-first century, many cosmologists explore both this and alternative theories of the early universe.

See also: ASTROPHYSICS; BIG BANG; BIG BANG NUCLEOSYNTHESIS; COSMIC STRINGS, DOMAIN WALLS; COSMOLOGICAL CON-STANT AND DARK MATTER; EINSTEIN, ALBERT; HUBBLE CONSTANT; INFLATION; UNIVERSE

Bibliography

Gribben, J. *Companion to the Cosmos* (Phoenix Giant, London, 1997).

Guth, A. *The Inflationary Universe* (Addison-Wesley, Reading, MA ,1997).

Kolb, E. W., and Turner, M. S. *The Early Universe* (Addison-Wesley, Reading, MA, 1993).

Kragh, H. *Cosmology and Controversy: The Historical Development of Two Theories of the Universe* (Princeton University Press, Princeton, NJ, 1996).

Lightman, A., and Brawer, R. *Origins: The Lives and Worlds of Modern Cosmology* (Harvard University Press, Cambridge, MA, 1990).

Schramm, D. N. *The Big Bang and Other Explosions in Nuclear and Particle Astrophysics* (World Scientific, Singapore, 1996).

Turok, N., ed. *Critical Dialogues in Cosmology* (World Scientific, Singapore, 1997).

Helge Kragh

CP SYMMETRY VIOLATION

The symmetry known as CP is a fundamental relation between matter and antimatter. The discovery of its violation by James Christenson, James Cronin, Val Fitch, and René Turlay (1964) has given us important insights into the structure of particle interactions and into why the universe appears to contain more matter than antimatter.

In 1928, Paul Dirac predicted that every particle has a corresponding antiparticle. If the particle has quantum numbers (intrinsic properties), such as electric charge, the antiparticle will have opposite quantum numbers. Thus, an electron, with charge $-|e|$, has as its antiparticle a positron with charge $+|e|$ and the same mass and spin. Some neutral particles, such as the photon, the quantum of radiation, are their own antiparticles. Others, like the neutron, have distinct antiparticles; the neutron carries a quantum number known as baryon number B $= 1$, and the antineutron has B $= -1$. (The prefix bary- is Greek for heavy.) The operation of charge reversal, or C, carries a particle into its antiparticle.

Many laws of physics are invariant under the C operation; that is, they do not change their form, and, consequently, one cannot tell whether one lives in a world made of matter or one made of antimatter. Many equations are also invariant under two other important symmetries: space reflection, or parity, denoted by P, which reverses the direction of all spatial coordinates, and time reversal, denoted by T, which reverses the arrow of time. By observing systems governed by these equations, one cannot tell whether the world is reflected in a mirror or in which direction its clock is running. Maxwell's equations of electromagnetism and the equations of classical mechanics, for example, are invariant separately under P and T.

Originally it was thought that *all* elementary particle interactions were unchanged by C, P, and T individually. In 1957, however, it was discovered that a certain class known as the weak interactions (for example, those governing the decay of the neutron) were not invariant under P or C. However, they appeared to be invariant under the product CP and also under T. (Invariance under the product CPT is a very general feature of elementary particle theories.) Thus, it was thought that one could not distinguish between our world and a mirror reflected world made of antimatter, or our world and one in which clocks ran backward.

Murray Gell-Mann and Abraham Pais (1955) used an argument based on C invariance (recast in 1957 in terms of CP invariance) to discuss the production and decay of a particle known as the neutral K meson, or K^0. This particle, according to a theory by Gell-Mann and Kazuo Nishijima, carried a quantum number called strangeness, with $S(K^0) = +1$, and so there should exist a neutral anti-K meson, called $\bar{K}^0$, with $S(\bar{K}^0) = -1$. The theory demanded that strangeness by conserved in K meson production but violated in its decay. Both the K^0 and the $\bar{K}^0$ should be able to decay to a pair of π mesons (e.g., $\pi^+\pi^-$). How, then, would one tell them apart?

Gell-Mann and Pais solved this problem by applying a basic idea of quantum mechanics: The particle decaying $\pi^+\pi^-$ would have to have the same behavior under C (in 1957, under CP) as the final $\pi^+\pi^-$ combination, which has CP = +1. (That is, its quantum-mechanical state is taken into itself under the CP operation.) A quantum-mechanical combi-

nation of K^0 and $\bar{K}^0$ with this property was called K_1^0. There should then exist another combination of K^0 and $\bar{K}^0$ with CP = -1 (i.e., its quantum-mechanical state is changed in sign under the CP operation). This particle was called K_2^0. (The subscripts 1 and 2 were used simply to distinguish the two particles from one another.) The K_2^0 would be forbidden by CP invariance from decaying to $\pi^+\pi^-$ and thus, being required to decay to three-body final states, would be much longer-lived. This predicted particle was discovered in 1956.

Christenson, Cronin, Fitch, and Turlay performed their historic experiment in the early 1960s at Brookhaven National Laboratory to see if the long-lived neutral K meson could occasionally decay to $\pi^+\pi^-$. They found that indeed it did but only once every 500 decays. For this discovery Cronin and Fitch were awarded the 1980 Nobel Prize in Physics.

The short-lived neutral K meson was renamed K_S and the long-lived one K_L. The K_L lives nearly 600 times as long as the K_S. The discovery of its decay to $\pi^+\pi^-$ was the first evidence for violation of CP symmetry. The K_S is mainly CP-even, while the K_L is mainly CP-odd. Within any of the current interaction theories, which conserve the product CPT, the violation of CP invariance then also implies T-invariance violation.

Shortly after CP violation was detected, Andrei Sakharov (1967) proposed that it was a key ingredient in understanding why the universe is composed of more matter than antimatter. Another ingredient in his theory was the need for baryon number (the quantum number possessed by neutrons and protons) to be violated, implying that the proton should not live forever. The search for proton decay is an ongoing topic of current experiments.

CP violation can also occur in quantum chromodynamics (QCD), the theory of the strong interactions, through solutions that violate both P and T. However, this form of CP violation appears to be extremely feeble, less than a part in ten billion; otherwise it would have contributed to detectable effects such as electric dipole moments of neutrons. It is not known why this form of CP violation is so weak; proposed solutions to the puzzle include the existence of a light neutral particle known as the axion.

The leading theory of CP violation was posed by Makoto Kobayashi and Toshihide Maskawa in 1973. Weak coupling constants of quarks (the subunits of matter first postulated in 1964 by Gell-Mann and George Zweig) can have both real and imaginary parts. These complex phases lead not only to the observed magnitude of CP violation discovered by Christenson et al., but also to small differences in the ratios of K_S and K_L decays to pairs of charged and neutral π mesons (confirmed by experiments at the European Laboratory for Particle Physics [CERN] and Fermilab), and to differences in decays of neutral B mesons and their antiparticles. Experiments at the Stanford Linear Accelerator Center (SLAC) using the BaBar detector (named after the character in the children's book) and at the National Laboratory for High Energy Physics in Japan (KEK) using the Belle detector have recently reported convincing evidence for this last effect (Aubert et al., 2001; Abe et al., 2001). At a deeper level, however, both the origin of the matter-antimatter asymmetry of the universe discussed by Sakharov and the source of the complex phases of Kobayashi and Maskawa remain a mystery, perhaps stemming from some common source.

See also: BIG BANG NUCLEOSYNTHESIS; CKM MATRIX; STANDARD MODEL; SYMMETRY PRINCIPLES

Bibliography

Abe, K., et al. "Observation of Large CP Violation in the Neutral B Meson System." *Physical Review Letters* **87**, 091802-1-7 (2001).

Aubert, B., et al. "Observation of CP Violation in the B^0 Meson System." *Physical Review Letters* **87**, 091801-1-8 (2001).

Bigi, I. I., and Sanda, A. I. *CP Violation* (Cambridge University Press, Cambridge, 2000).

Branco, G. C.; Lavoura, L; and Silva, J. P. *CP Violation* (Clarendon Press, Oxford, 1999).

Christenson, J. H.; Cronin, J. W., Vitch, V. L., and Turlay, R. E. "Evidence for the 2π Decay of the K_2^0 Meson." *Physical Review Letters* **13**, 138–140 (1964).

Cronin, J. W. "CP Symmetry Violation: The Search for its Origin." *Reviews of Modern Physics* **53**, 373–383 (1981).

Fitch, V. L. "The Discovery of Charge Conjugation Parity Asymmetry." *Reviews of Modern Physics* **53**, 367–371 (1981).

Gell-Mann, M., and Pais, A. "Behavior of Neutral Particles Under Charge Conjugation." *Physical Review* **97**, 1387–1389 (1955).

Kobayashi, M., and Maskawa, T. "CP Violation in the Renormalizable Theory of Weak Interaction." *Progress of Theoretical Physics* **49**, 652–657 (1973).

Sakharov, A. D. "Violation of CP Invariance, C Asymmetry, and Baryon Asymmetry of the Universe." *Soviet Physics—JETP Letters* **5**, 24–27 (1967).

Jonathan L. Rosner

CREATION

See ANNIHILATION AND CREATION

CULTURE AND PARTICLE PHYSICS

To seek to understand the basic constituents out of which the great variety of the physical world is made is an intellectual quest of deep significance. The enterprise is as old as the speculations of the presocratic philosophers, such as Thales and Anaximenes, who suggested that there might be a limited number of kinds of fundamental stuff out of which the world around us is made. Two and a half millennia after this kind of thinking began, we have found that the answer is very much more interesting than the simple possibilities, such as fire, earth, air, or water, that occurred to those early thinkers. Any educated person imbued with intellectual curiosity should want to learn about the discoveries that have been made concerning the constitution of matter. Yet some people seem content simply to suppose that "atoms" is an adequate answer, despite those atoms being, from the point of view of people who know about quarks and gluons, systems that are large and composite. Such an attitude of indifference amounts to self-imposed cultural deprivation. It also poses problems for particle physicists, who have to depend upon the good will of the taxpayer for the large sums of money necessary for the pursuit of their research.

History

In 1945, at the end of World War II, particle physics began to move into its Big Science mode, as accelerators and detectors of increasingly larger size and greater cost began to become indispensable means for further advance. The leaders of the experimental community had learnt the value of teamwork from such wartime cooperative experiences as the Manhattan Project, and they had also earned the gratitude of government for their substantial contributions to the war effort. Initially, therefore, funding was comparatively easily obtained but, as the sums requested began to rise steeply, this period of financial honeymoon came to an end. Eventually it became necessary to work collaboratively on an international scale, if sufficient resources were to be available. The consortium of European nations that has made the European Laboratory for Particle Physics (CERN) possible is an outstanding example of the fruitfulness of enterprises of this kind, demonstrating to the world that collaborations of such wide scope can be successful and need not succumb to the disrupting effects of national rivalries.

In their appeal to the taxpayer for finance, particle physicists can point to the pragmatic usefulness of some of the spin-offs from their work. The construction of precision engineering devices operating on kilometer scales has required the development of techniques that have proved to have applications beyond the control system of a synchrotron or storage rings. At the same time, many talented young people have served scientific apprenticeships within the high-energy physics community and gone on to make use of the talents, skills, and experience they acquired in other fields of application. Yet these are merely collateral arguments for the support of particle physics.

The fundamental justification for the activity of the subject lies in the intrinsic intellectual value of uncovering and understanding the substructure of the physical world. As the American accelerator builder, Robert Wilson, said, when questioned by a senator about what the latest project for which he was seeking funding would do for the defense of the United States, "Nothing—but it will help to make the United States worth defending!" The heart of the cultural case for particle physics is that it affords us knowledge of a kind that, in its depth and fundamentality, is the source of its own intrinsic value. Yet particle physicists face a number of problems in conveying this truth to the general educated public.

Cultural Problems

One problem is the lack of pictorial appeal in the visual material of particle physics. The contrast with the cosmologists—those other fundamental scientists who are operating at the other end of the length scale—is striking. Astronomical pictures, such as the deep space photographs taken with the Hubble space telescope, are often breathtakingly beautiful and immediately seize the public imagination. On the other hand, the average bubble chamber photograph, or a diagram of spark discharges in a detector, is frequently messy and unappealing. Only when these weird patterns are interpreted do they reveal their fascination and stimulate the viewer's interest. Few outsiders, however, will persevere in penetrating far enough into the subject to be able to make this discovery.

Another problem arises simply from the minute scale on which the phenomena of particle physics take place. Everyone has looked up at the starry heavens and wondered what is going on there. Few people are disposed to look inward at the structure of matter in a similar way. Genes in the case of living beings, and atoms in the case of inanimate entities, is about as small as many are prepared to go.

A third difficulty, endemic in physics as a whole but particularly acute in the case of particle physics, is the essential role of mathematical thinking. Paul Dirac expressed his conviction that the fundamental laws of physics are expressed in beautiful equations, illustrating this fact in striking fashion by his discovery of the relativistic equation of the electron. Yet mathematical beauty is a rarified aesthetic experience which comparatively few are privileged to share. The intellectual attractions of gauge field theories, let alone the rarified delights of string theory, are inaccessible to those who have to rely on words alone to receive the message.

Perhaps a fourth difficulty has arisen from the inevitable development of huge experimental groups, involving literally hundreds of Ph.D. physicists in their activities. The work is necessarily fragmented and distributed to many subgroups, inhibiting the possibility of telling a story of bold simplicity centering on the work of a single vivid personality. This prevalence of group activity has also encouraged the suspicions of contemporary sociologists of knowledge. The so-called science wars have been about whether science is discovery at all, but rather it is asserted that theories are simply the tacit agreement of a like-minded in-group (the invisible college of particle physicists) to see things this way. On the contrary, scientists, not the least of which particle physicists, are very conscious of how surprising nature frequently turns out to be, frustrating initial expectations and forcing conclusions that are more intellectually satisfying than the scientists could have anticipated beforehand. This experience seems to physicists only to be credibly explained in terms of discovery and not mere construction. If postmodern culture were to succeed in casting doubt on the attainability of a verisimilitudinous mapping of the physical world by science, this would have serious consequences for the health of particle physics and for an honorable interest in its activities. After all, what would be the point of so much expenditure of effort, talent, and money if it were not telling what matter is really like? The physical regimes created in high-energy experiments are very extreme, far from situations likely to be of direct relevance to the processes of everyday life. Their significance is fundamental and not simply pragmatic.

Cosmic Questions

The regimes investigated in modern accelerator experiments are relevant, however, to the state of matter at the beginning of the observable universe. A fusion of cosmology and particle physics is necessary for the discussion of the very early cosmos and this has enabled particle physics to acquire some of the glamour associated with cosmology. One of the most astonishing impacts of science upon general culture observed in recent years has been the multi-million-copy sales of Stephen Hawking's *A Brief History of Time* (1988). Partly, no doubt, this has been due to respect for the remarkable character of its author, but this success was also clearly fueled by a feeling that to know something about the fundamental history and nature of the physical universe is to know something of real value and significance. Yet one must also acknowledge that all theories of quantum cosmology are not only doubly difficult to expound, combining the need to explain both quantum theory and general relativity, but also, pending the full reconcilia-

tion of these two great discoveries, dependent on a precarious degree of intellectual conjecture.

A profound cultural consequence of particle physics is the recognition that the universe is both rationally transparent to human inquiry (in that people are able to penetrate the secrets of the subatomic world, despite its nature being so counterintuitive when compared with the world of everyday experience) and also rationally beautiful to an astonishing degree (those beautiful equations). The fact that particle physics is possible at all is surely a highly significant fact about reality. Human thought has proved able to access and understand processes taking place on extremely short length scales and, when that understanding has been gained, it has proved to be of a kind that excites wonder in those privileged to participate in it. A particle physicist will instinctively doubt a theory that does not display the recognizable, if abstract, character of mathematical beauty. Part of the current dissatisfaction with the Standard Model is that its possession of so many adjustable parameters denies it that elegance and economy that scientists have come to expect as the hallmark of a truly fundamental theory. Physicists believe that this stance is no mere attitude of mathematical aestheticism, for the use of these nonempirical criteria as techniques of discovery and as indicators of validity has proved itself time and again in the history of physics by the long term fruitfulness displayed by resulting theories of this kind. Dirac's discovery of the relativistic equation of the electron, and Albert Einstein's discovery of the equations of general relativity, are cases in point. These considerations, together with the intrinsic interest of the questions it addresses, give particle physics a secure and significant place in the cultural attainments of humankind.

See also: BENEFITS OF PARTICLE PHYSICS TO SOCIETY; INFLUENCE ON SCIENCE; METAPHYSICS; PHILOSOPHY AND PARTICLE PHYSICS; UNIVERSE

Bibliography

Barrow, J. D. *The World within the World* (Oxford University Press, Oxford 1988).

Brown, L. M., and Hoddeson, L., eds. *The Birth of Particle Physics* (Cambridge University Press, Cambridge, England, 1983).

Gallison, P. *How Experiments End* (University of Chicago Press, Chicago, 1987).

Hawking, S. W. *A Brief History of Time* (Bantam Press, London, 1988).

Pais, A. *Inward Bound: Of Matter and Forces in the Physical World* (Clarendon Press, Oxford, 1986).

Polkinghorne, J. C. *The Particle Play* (W. H. Freeman, Oxford, 1979).

Weinberg, S. *Dreams of a Final Theory: The Search for the Fundamental Laws of Nature* (Hutchinson Radius, London, 1993).

John Polkinghorne

CYCLOTRON

High-energy charged particles have many applications in fundamental and applied research in the physical and biological sciences. They are produced by starting with a source of low-energy charged particles, such as a plasma in a gaseous discharge, and then accelerating the particles to high energy. A cyclotron is one of the devices that can be used as a charged-particle accelerator.

To accelerate a particle, it is necessary to exert a force on it. Forces on charged particles can be exerted by electric fields and by magnetic fields. The force exerted by a magnetic field is always exactly perpendicular to the velocity of the particle. Such a force can change the direction in which a particle is moving, but it cannot change the energy of the particle. However, the force exerted by an electric field acts in the direction of the electric field. If this direction has a non zero component in the direction of the velocity of the particle, then the effect of the electric field will be to increase the speed, and hence the energy, of the particle. Thus any device that increases the energy of a charged particle must use electric fields (not magnetic fields) to produce the increase. In a cyclotron, the sideways forces produced by magnetic fields are used to keep the particles moving in approximately circular orbits, while electric fields at certain places around these orbits provide the increases in particle energy.

Figure 1 shows a simplified drawing of a cyclotron. The two shaded regions in Figure 1 are semicircular metal chambers, called "dees," which are shown in

FIGURE 1

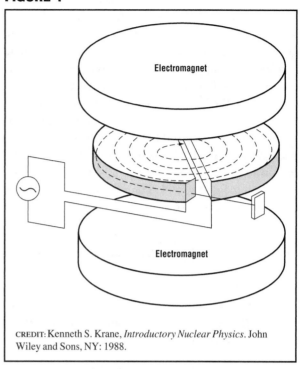

CREDIT: Kenneth S. Krane, *Introductory Nuclear Physics*. John Wiley and Sons, NY: 1988.

A simplified drawing of a cyclotron. The two shaded regions are semicircular metal chambers called dees. The beam spirals outward from the center and is accelerated each time it crosses the gap between the dees. Eventually, it is extracted and directed against a target.

more detail in the cross-section drawing in Figure 2. An alternating potential difference between the dees is produced by an oscillator. Because of this potential difference, there is an electric field between the dees which accelerates a particle as it moves from one dee to the other. Within a dee, there is only a uniform magnetic field perpendicular to the plane of the dee. The associated magnetic force provides the centripetal acceleration needed to keep the particle in a circular orbit as it moves around the dee. The dee is enclosed in an evacuated chamber, and the particles can pass unimpeded from one dee to another.

The source of low-energy particles is near the center of the cyclotron, in the space between the dees. The particles are accelerated across this space, into a dee, then move halfway around the cyclotron and reenter the electric field between the dees traveling in the opposite direction. In the time it takes the particle to move halfway around the dee, the sign of the potential difference between the dees must change so that the electric field is still in the direction needed to accelerate the particles. For example, if the particle is a proton, it must always move from the dee of high potential to the dee of low potential, and so the oscillator must ensure that this is always the situation when a proton is in the space between the dees. If the time required for a half-orbit were the same at all radii, acceleration would occur every time the particle crossed from one dee to another as it spiraled outward. This is true as long as the particle moves slowly compared to the speed of light, in a uniform magnetic field. Then the frequency of the oscillator must be given by

$$f \text{ (hertz)} = \frac{qB}{2\pi m},$$

where q is the charge of the particle in coulombs, m is its mass in kilograms, and B is the magnetic field in tesla. If the particle spirals out to a radius R, in meters, it will reach a kinetic energy of

$$\text{K. E. (Joules)} = \frac{q^2 B^2 R^2}{2m}.$$

For example, a proton cyclotron using a magnetic field of 0.1 tesla (1 kilogauss) requires an oscillator of frequency of 1.5 MHz (1.5×10^6 Hz). If the maximum orbit radius is 1 meter, then the proton will reach a kinetic energy of 7.7×10^{-14} joules = 0.48 MeV (0.48×10^6 eV).

The description above referred to a "standard" cyclotron, with fixed magnetic field and fixed fre-

quency. This device was developed by E. O. Lawrence and M. S. Livingston in the early 1930s. Its great merit was that particles could be accelerated to high speeds in repeated small increments, so that high voltages were not required. It produces a continuous beam of particles, whose energy can be varied by adjustment of the magnetic field and oscillator frequency. Much of the research in nuclear reactions done before about 1960 depended upon beams produced by standard, fixed-frequency cyclotrons.

If the speed of the particle being accelerated is not small compared to the speed of light, then the time taken for a half-orbit around the cyclotron ceases to be independent of the radius of the orbit. This prevents the use of a fixed-frequency cyclotron to produce very high-energy beams. One solution to this difficulty is provided by the synchrocyclotron. In this device, low-energy particles are injected into the space between the dees in short bursts (bunches). The magnetic field is still uniform, but the frequency with which the potential on the dees is alternated varies as the bunch spirals outward, in such a way that the electric field in the space between the dees always has the direction needed to accelerate the pulse passing through it. In a typical synchrocyclotron, each bunch lasts about 10^{-4} seconds, with about a 10^{-2}-second interval between bunches. Thus a target put in a synchrocyclotron beam would be bombarded for only about 1 percent of the total time it was exposed to the beam, whereas a standard cyclotron would produce continuous bombardment. For some purposes, such as the measurement of the lifetime of a state produced by bombardment, the bunching of the beam is an advantage.

In both the standard cyclotron and a synchrocyclotron, the radius of the orbit of the particle starts out small, and increases with each half-cycle, all the while immersed in a uniform magnetic field. To reach very high energy it would be necessary to reach very large radii, which would then require very large (and very expensive) magnets. A more economical solution would be to have each bunch move in an orbit of constant radius, while the strength of the magnetic field and the oscillator frequency are increased as the energy of the particles in the bunch increases. The beam is confined to an evacuated "beam pipe," containing one or more accelerating sections, with the

FIGURE 2

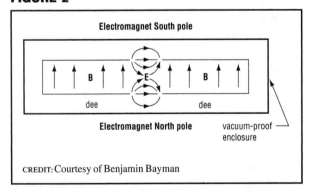

CREDIT: Courtesy of Benjamin Bayman

Within each dee there is a uniform magnetic field, perpendicular to the plane of the dee. The electric field is in the space between the dees.

magnets arranged around the pipe. This device, in which both the frequency and magnetic field strength are varied in order to produce acceleration at constant radius, is called a synchrotron. The Tevatron is a proton synchrotron at Fermilab, with a diameter of 2 kilometers. Superconducting magnets produce a maximum field strength of 4.2 tesla, about 15,000 times stronger than the Earth's magnetic field. It takes about 20 seconds for the magnetic field to rise from 0.66 to 3.54 tesla, while the proton energy increases from 150 to 850 GeV ($1 \text{ GeV} = 10^9 \text{ eV}$). The final beam energy is approximately 980 GeV. The world's most powerful accelerator, the Large Hadron Collider (LHC), is under construction at the European Laboratory for Particle Physics (CERN) in Geneva. It is designed to accelerate protons to an energy of 7 TeV ($7 \times 10^{12} \text{ eV}$). It is planned that it will begin operation in 2007.

There are many technical problems that must be overcome in the design and operation of a working particle accelerator. An especially important issue is the stability of the beam with respect to spatial and temporal fluctuations. This requires careful shaping of the magnetic field and electric fields, and very high vacuum in the beam pipe. Accelerator technology is a very important component of modern physical science.

See also: ACCELERATORS, EARLY; LAWRENCE, ERNEST ORLANDO

Bibliography

European Laboratory for Particle Physics (CERN). <http://www.cern.ch/>.

Fermi National Accelerator Laboratory (FNAL). <http://www.fnal.gov/>.

Fishbane, M. F.; Gasiorowicz, S.; and Thornton, S. T. *Physics for Scientists and Engineers* (Prentice-Hall, Upper Saddle River, NJ, 1993).

Krane, K. S. *Introductory Nuclear Physics* (Wiley, New York, 1988).

Lawrence, E. O., and Livingston, M. S. "The Production of High Speed Light Ions Without the Use of High Voltages." *Physical Review* **40**, 19 (1932).

Benjamin Bayman

D

DARK ENERGY

See Cosmological Constant and Dark Energy

DARK MATTER

Despite the enormous amount of progress made during the last century in physics and astronomy, scientists still can not identify over 90 percent of the overall composition of the universe. Indeed, what has been learned is that the visible matter forming planets, stars, and galaxies only makes up a relatively small fraction of the total mass-energy of the universe. In the past, this unseen component was called the missing matter or missing mass of the universe. However, many different astronomical observations have confirmed its presence. Most of the mass of the universe is not missing, it is just dark, and what it is remains unknown.

Observational Evidence

The presence of dark matter was suggested by Fritz Zwicky in 1933. Typical velocities v of galaxies in a cluster of galaxies can be related to the total mass M and size R of the system through the simple Newtonian relation

$$v^2 \sim G_N M/R$$

where G_N is Newton's constant. Measurements of velocities of galaxies in the Coma and Virgo clusters of galaxies indicated that there was significantly more matter than could be accounted for by the individual light-producing galaxies. This discrepancy is often accounted for by computing a mass-to-light ratio. A mass-to-light ratio is the mass of the system divided by the luminosity of the system and is usually expressed in solar units (i.e., it is compared to the mass and luminosity of the Sun). In the solar neighborhood, the mass-to-light ratio is about 2 $M_\odot/L_\odot$ The solar neighborhood, however, is not typical of either the galaxy or the universe as a whole. In the bright central parts of galaxies, the ratio is (10–20) h_0. For the Coma cluster, this ratio is about 400 h_0, indicating that there is significantly more matter on the large scales associated with clusters of galaxies than in galaxies themselves. Both of the mass-to-light ratios above carry an uncertainty as a result of the very large distance to other galaxies or clusters. The uncertainty is qualified by the Hubble parameter H (see discussion below). The quantity h_0 is a scaled value of the present Hubble parameter and is given by $h_0 = H_0/100$ km Mpc^{-1} s^{-1}.

The present value of the scaled parameter is $h_0 = 0.71 \pm 0.07$.

There are several modern astronomical techniques for establishing the existence of dark matter. The same Newtonian relation expressed above can be applied to measurements of the rotation velocities of the disks of spiral galaxies. Rotation velocities can be determined from the Doppler shifts of 21-cm hydrogen lines in the far outer reaches of spiral galaxies. At radii far beyond the point where most of the light is concentrated, one would expect that the velocities diminish as $v \propto 1/\sqrt{R}$. Instead, one finds the following surprising result: the velocities remain constant at large radii, producing what are called flat rotation curves and indicating that the mass of galaxies continues to increase as $M \propto R$. This provides direct evidence for the presence of dark matter.

Additional evidence is available from X-ray observations of elliptical galaxies and clusters of galaxies. The X rays are emitted by the hot hydrogen gas surrounding these systems. The atoms in a gas, with a temperature of order 10^6 K, have velocities that would enable them to escape the system unless there was a sufficient amount of dark matter to gravitationally bind the gas to the system.

It is also possible to infer the presence of dark matter in clusters of galaxies through gravitational lensing. The large mass associated with clusters causes the trajectories of light from background galaxies to be bent. The degree of bending is related to the amount of dark matter in the cluster.

There is also a strong theoretical argument for dark matter arising from the theory of structure formation. Galaxies and clusters of galaxies are thought to have grown from primordial density fluctuations produced in the very early universe. These fluctuations begin to grow as a result of gravitational collapse that occurs when the universe becomes dominated by nonrelativistic matter. Without dark matter and, more importantly, without nonbaryonic dark matter (see below), there is not enough time for these perturbations to grow into the structures observed today.

The Density of Matter in the Universe

It is often convenient to relate the overall density of matter in the universe to a critical density ρ_c

that is given in terms of Newton's constant and the Hubble parameter H:

$$\rho_c = \frac{3H^2}{8\pi G_n} = 1.88 \times 10^{-29} \, h_0^2 \text{ g cm}^{-3}.$$

The ratio $\Omega = \rho/\rho_c$ is called the cosmological density parameter and is related to the overall spatial geometry of the universe. If $\Omega > 1$, the universe is closed and finite in spatial extent; if $\Omega < 1$, the universe is open and infinite; and if $\Omega = 1$, the universe is spatially flat and infinite. The universe is observed to be expanding so that the distances to far-away objects are increasing. Furthermore, the expansion is uniform so that the increase in distance can be related to a common scale factor for the universe $a(t)$. The Hubble parameter expresses the rate of change of the scale factor and determines the velocities of distant objects:

$$H = \frac{\dot{a}}{a} = \frac{v}{d}$$

where $\dot{a}$ is the time rate of change of a, v is the velocity of a distant object, and d is the distance to that object. This is known as Hubble's law.

The density parameter may have several different types of contributions. Ordinary matter made up of neutrons and protons is referred to as baryonic matter. (Baryons are a class of particles composed of three quarks or three antiquarks. Neutrons and protons are the lightest- and longest-lived baryons.) Other forms of matter that may contribute to Ω can be collectively called nonbaryonic dark matter. The energy densities of both baryonic and nonbaryonic matter scale with the expansion of the universe in such a way that the density decreases inversely with the volume expansion, $\rho_m \sim a^{-3}$.

Baryonic Dark Matter

The total amount of baryonic matter can be determined from the observations of the light element abundances D, ^{3}He, ^{4}He, and ^{7}Li. These elements were produced within the first 3 minutes after the Big Bang. In particular, the abundance of deuterium is particularly sensitive to the baryon density. The value of the baryon density ρ_b is commonly expressed relative to the critical density. Therefore, one can define $\Omega_b = \rho_b/\rho_c$. A firm upper limit to the quantity $\Omega_b h_0^2$ is 0.03. By the same token, Big Bang nucleosynthesis also requires that $\Omega_b h_0^2 > 0.006$. The detailed spectrum of microwave background fluctuations can also determine the baryon density. Current estimates yield $\Omega_b h_0^2 = 0.021 \pm 0.004$. However, the *observed* baryon density is much smaller. By adding up the densities of baryons in stars (both living and dead) as well as the observed gas, it has only been possible to find a fraction $\Omega_b h_0^2 \sim 0.01$.

There is still the possibility that baryons contribute to some of the dark matter, particularly on the scale of galaxies. In general, baryons make poor dark matter candidates since they are typically associated with luminous objects such as stars or X-ray-producing gas. However, it is possible that dark baryons reside in the dead remnants of stars, such as black holes, neutron stars, or white dwarfs. Gravitational microlensing has been useful in limiting the amount of dark matter in these forms.

Particle Dark Matter

The data so far indicate that most of the dark matter in the universe is nonbaryonic. Various techniques, which include the study of the dynamics of galaxies and clusters, show that the total density of matter is roughly $\Omega_{matter} \sim 0.3 - 0.5$. The difference between this density and the baryon density yields the nonbaryonic component to the density of the universe. Fortunately, there are many potential candidates for dark matter in the well-studied models of particle physics.

From the standpoint of the Standard Model of particle physics, the simplest candidate for dark matter would be a neutrino. Originally, the theory of electroweak interactions was constructed so that the neutrino was massless. Recent data on solar neutrinos, and neutrinos produced by cosmic-ray collisions in the atmosphere, strongly indicate that neutrinos of different types oscillate and thereby have mass. As such, they could, in principle, contribute to the dark matter.

Even a small neutrino mass can make a large contribution to the overall density of matter. If m_v is the mass of neutrino, then so long as $m_v < 1$ MeV, its contribution to the density parameter is $\Omega_v h_0^2 =$ $m_v/(94$ eV$)$. However, a neutrino-dominated universe is strongly disfavored by the observed large-scale structure of the universe. Light neutrinos remain relativistic until relatively late times. By the time they come to dominate the mass density of the universe, they have traveled across immense distances, erasing the possibility of the growth of structures on smaller scales. Thus, a neutrino-dominated universe inevitably produces too much large-scale structure. Furthermore, it appears that the oscillation data require neutrino masses that are too small to make a dominating contribution to the overall mass density.

In contrast to light neutrinos, which are typically labeled hot dark matter because of their relativistic velocities, an ideal dark matter candidate should be almost at rest with respect to the cosmic expansion. These cold dark matter candidates lead to the formation of smaller structures (galaxy size and smaller) first and fit the observations reasonably well.

Beyond the Standard Model of particle physics there are many particle candidates for cold dark matter. One of the best-studied candidates is found in an extension of the Standard Model based on supersymmetry. Supersymmetry is a symmetry that relates particles of different spin (an internal quantum number assigned to all particles). The supersymmetric Standard Model contains many new particles and predicts that one of these will be stable. This new stable particle is normally a neutralino (it is the supersymmetric partner of the photon and Z gauge boson) and may have a mass of order 100 times the mass of the proton. Another potential candidate is called the axion, and it arises in a theoretical solution to what is known in particle physics as the strong CP problem. Fortunately (or not), many other particle candidates are found in theories extending the Standard Model.

Dark Energy

It is also possible that there is a component of Ω for which the energy density is constant (with respect to the expansion of the universe). This is called the cosmological constant Λ. It was originally introduced by Albert Einstein in order to cancel the expansion he found in his cosmological models. When the expansion of the universe was observed, he described the introduction of the cosmological constant as his biggest blunder. Nevertheless, its potential contribution must be considered.

The total density parameter has been established to be very close to unity by microwave background experiments, which indicate that $\Omega_{total} = 1.03 \pm 0.06$. As mentioned above, it appears that the contribution of Ω due to matter (both baryonic and nonbaryonic) is approximately 0.3–0.5. The difference between Ω_{total} and Ω_{matter} has been called dark energy. The existence of dark energy has also been suggested by recent supernovae observations, which detect that the universe may be accelerating (i.e., its expansion rate is increasing). The cosmological constant could play the role of dark energy as could any smoothly distributed energy associated with the vacuum.

See also: ASTROPHYSICS; COSMOLOGICAL CONSTANT AND DARK ENERGY; COSMOLOGY; OUTLOOK

Bibliography

Borner, G. *The Early Universe: Facts and Fiction* (Springer-Verlag, Heidelberg, Germany, 1993).

Krauss, L. M. *The Fifth Essence: The Search for Dark Matter in the Universe* (Basic Books, New York, 1989).

Riordan, M., and Schramm, D. N. *The Shadows of Creation: Dark Matter and the Structure of the Universe* (W. H. Freeman, New York, 1992).

Srednicki, M. *Particle Physics and Cosmology: Dark Matter* (North-Holland, Amsterdam, 1990).

Keith Olive

DESY (DEUTSCHES ELEKTRONEN-SYNCHROTRON LABORATORY)

DESY is a national research laboratory in Germany and, besides the European Laboratory for Particle Physics (CERN), the other major particle accelerator center in Europe. The name (German electron synchrotron) derives from the first accelerator constructed there. DESY is located in a suburban area of Hamburg. It has a staff of about 1,200 and a budget of about 150 M Euro.

The laboratory was established in 1959. Its principal funding agency is the federal ministry for education and research. The original mission was to design, construct, and run a high-energy particle accelerator, so that researchers from German universities and other research institutes, wishing to participate in the emerging field of particle physics, could conduct experiments there. Gradually, DESY attracted users from all over Europe and from the United States, Canada, Japan, and China. The laboratory, though formally remaining a national institution, factually became international.

From the beginning, DESY fostered an active program of synchrotron radiation applications. This work, including condensed matter physics and material science, geology, chemistry, life sciences, and medical applications has steadily grown in scope and importance and has gradually transformed the laboratory into an interdisciplinary research establishment.

The Past

The first accelerator constructed at DESY was an electron synchrotron of 7.5 GeV beam energy. Research interests included the structure of the nucleon investigated by elastic and inelastic electron-nucleon scattering, the production of hadrons and hadronic resonances by high-energy photons, and tests of quantum electrodynamics. The synchrotron was commissioned in 1964; experiments continued for a decade. Meanwhile DESY investigated the possibility of constructing electron-positron storage rings of high energy, building on the pioneering work at Stanford University and the Italian national laboratory in Frascati. The machine was completed in 1974 and was called DORIS (from the German Doppel-Ring-Speicher); it had a maximum energy of the colliding beams of initially 4.3 GeV, later upgraded to 5.6 GeV. It came into operation just after the discovery of the J/ψ resonance, the first charm-anticharm bound state, and was to become a significant player in unraveling the physics of the new quark states and of the τ lepton, discovered shortly afterward.

Notable observations at DORIS include the P-wave charm-anticharm states and the S-wave states of the bottom-antibottom quark states, the so-called upsilon resonances. In 1982 a comprehensive investigation of *B* mesons, the hadrons containing one heavy *b* quark or antiquark, was launched. In its course, the ARGUS collaboration discovered the quantum mechanical mixing of neutral *B* mesons with their antiparticles, the first such mixing case found since an analogous phenomenon had been discovered with K mesons more than two decades earlier. This observation led the way to exciting prospects in b quark research, in particular to the possibility of charge parity–parity (CP) violation by B mesons, a development that culminated in the construction of specialized *B* factories at the Stanford Linear Accelerator Center (SLAC) and the Japanese High-Energy Accelerator Research Organization (KEK). DORIS has been, since 1993, used as a dedicated synchrotron radiation source.

The next accelerator at DESY was an electron-positron collider as large as would fit on the site. The PETRA (Positron Electron Tandem Ring Accelerator) ring had a circumference of 2.3 kilometers and a collision energy in the center-of-mass of initially about 30 GeV. It was commissioned in 1978, and a few months after its start, the international Two-Arm Solenoid Spectrometer (TASSO) collaboration began to observe so-called three jet events. These are events in which out of the electron-positron collision, instead of just a lepton-antilepton or quark-antiquark pair, a quark-antiquark pair accompanied by an additional energetic "hard" gluon appeared, resulting in a third jet of hadrons. The interpretation followed readily, and the results were soon corroborated by the other three collaborations working at PETRA, each using their own detector. This was a crucial step toward establishing the theory of quantum chromodynamics.

PETRA was then gradually upgraded to 46 GeV energy in an attempt to find the top quark and/or supersymmetric particles. At the time, it was not clear how large the masses of these particles would be. On the way, many important observations were made at PETRA (and at the similar PEP machine at the Stanford Linear Accelerator Center). They concerned the radiation of hard gluons, the formation of jets of hadrons from quarks and gluons, and the properties of the gluons like spin and coupling strength to quarks; furthermore, they also concerned properties of *b* quarks such as the lifetime, the electroweak couplings of the heavy quarks and of the μ and τ lep-

tons, tests of quantum electrodynamics at very small distances, and more. These results went a long way toward establishing the Standard Model as a viable description of the particles and their interactions down to distances of the order of 10^{-15} cm. Experiments at PETRA were carried on until 1986. PETRA was then converted into an injector for HERA, the next big accelerator at DESY, and today (2002) it also serves as a synchrotron radiation source for hard X rays.

HERA

Meanwhile plans were made at DESY for a new type of accelerator—an electron-proton collider. This project, called HERA (Hadron Electron Ring Accelerator), apart from being a novelty in machine design, faced two further challenges. Since it would not fit on the DESY site, it was constructed below ground in an adjacent area, which was partly industrial and partly residential. And since both the human resources needed and the price tag were substantially higher than what the German funding agency was prepared to grant, it was funded and built in international collaboration. Remarkably, authorization to operate the accelerator 15 meters underground, directly under private homes, was obtained. The project was co-funded by the funding agencies of a number of foreign (that is, non-German) countries, and several foreign institutes sent substantial human resources to DESY to help construct HERA. It was completed in 1991.

HERA is housed in a 6.3-km-long tunnel, in which protons of up to 920 GeV energy and electrons of 28 GeV circulate. The electron energy is lower because of synchrotron radiation while the proton energy is limited by the magnetic guiding fields generated by superconducting magnets. Owing to the colliding beams, the lepton-nucleon interactions in HERA have more than ten times higher energy in the center of mass than in lepton-nucleon scattering experiments that employ fixed nuclear targets. The interactions are observed at two beam intersections where detectors called H1 and ZEUS have been set up by large international collaborations. As the scattering takes place in the unified electroweak region where the weak force is no longer small compared to the electromagnetic force, the finite range

An aerial view shows the layout of PETRA (small ring) and HERA (large ring) at DESY in Germany. On Hera's North side (N) is the H1 detector, while ZEUS is located on the South side (S). HERMES is located on the East side (E). HERA-B is located on the West (W), closest to PETRA. CREDIT: COURTESY OF DEUTSCHES ELEKTRONEN-SYNCHROTRON (DESY). REPRODUCED BY PERMISSION.

effects in the electron-quark interactions were seen for the first time. The analysis of these interactions is yielding precise information on the structure of the nucleon in terms of the quarks and gluons down to scales of 10^{-16} cm, about one thousandth of the nucleon size. A surprise was the very large density of "soft" gluons and quark-antiquark pairs found in the proton. Quantum chromodynamics interprets this as a consequence of a continuing emission and reabsorption of gluons by the quarks, with the gluons in turn radiating further gluons and generating quark-antiquark pairs which can again annihilate into gluons—a multistep process by which the interior of the nucleon becomes something like a permanently fluctuating, dense liquid of quarks, antiquarks, and gluons. This state of the interior of the nucleon has so far defied quantitative description in terms of quantum chromodynamics. It presumably is closely related to the phenomenon of quark confinement.

Measurements of similar nature, at lower energy but with longitudinally polarized electrons or positrons scattered on polarized nucleons, are made in the experiment HERMES (HERA Measurement of Spin). It uses the electron beam of HERA, observing scatterings on a target of polarized gas molecules in a spectrometer. The aim is to unravel the spin structure of the nucleon in terms of the angular momenta of the constituent quarks, antiquarks, and gluons. In agreement with other experiments it is found that the three valence quarks are carrying only about one third of the nucleon's spin. Present effort is directed toward finding the contribution of the gluons to the angular momentum. There is a fourth large detector at HERA, called HERA-b, in which the final states from scattering of the 920-GeV proton beam on a fixed target are being measured; b quarks are produced whose rare decay modes will be analyzed in the HERA-b spectrometer. About 1,200 physicists are participating in the experiments at HERA. It is planned that this experimental program will continue until at least 2006.

Other Activities

In addition to accelerator and detector construction and particle physics synchrotron radiation experiments, theoretical studies in particle physics and cosmology take place at DESY. Furthermore, DESY has a branch institute in Zeuthen near Berlin that has evolved from the former East German Institute of High Energy Physics. Besides collaborating in the general research and development (R&D) program at DESY, physicists in Zeuthen have a major interest in neutrino astrophysics, aimed at the detection of cosmic neutrinos in underwater and under-ice Cherenkov telescopes—in particular, participation in the Antarctic Muon and Neutrino Detector Array (AMANDA) and IceCube projects at the South Pole. Another activity pursued at DESY-Zeuthen is lattice gauge theory including the development, together with the Istituto Nazionale di Fisica Nucleare (INFN) in Italy, of parallel computers specially designed for the necessary calculations.

TESLA Project

Since 1992, DESY has been pursuing, on the initiative of its late director Bjorn Wiik and in the framework of an international collaboration, the design of a superconducting linear collider for electrons and positrons with center-of-mass energies of 500 to 800 GeV, named the Tera Electronvolt Superconducting Linear Accelerator (TESLA). Such a machine would be an ideal instrument for definitely establishing the Higgs mechanism and testing its various aspects. One needs to know whether the Higgs particle is standing alone, whether it is a supersymmetric Higgs, or whether it is the first sign of something completely new. To this end one must measure all of its interactions and decays precisely. If there is a light Higgs particle and if supersymmetric particles are found, the experiments at TESLA, combined with results from the LHC, may become for supersymmetry what optical spectroscopy was for quantum mechanics: the establishment of precision data from which the underlying theory can be developed. If no evidence for the Higgs mechanism is found at TESLA, it will be even more interesting to find out what takes its place.

TESLA would be ideally suited also as driver of a self-amplifying spontaneously emitting free electron laser which could serve as a source of hard, coherent X rays. An X-ray laser of this sort would be far more powerful than any currently available X-ray source, opening unprecedented opportunities for X-ray-based research and for X-ray applications in a wide variety of fields, from materials research to life sciences.

In 2001, DESY and the TESLA collaboration presented a detailed proposal to construct TESLA and an associated X-ray laser laboratory. The machine would be housed in a 33-km-long underground tunnel. A possible site near DESY has been identified. The project would have to be an international collaboration and would involve founding a new administrative structure in which DESY would act as host and as one of the partners.

See also: BENEFITS OF PARTICAL PHYSICS TO SOCIETY; FUNDING OF PARTICLE PHYSICS; INTERNATIONAL NATURE OF PARTICLE PHYSICS; UNIVERSE

Bibliography

DESY. <http://www.desy.de>.

Paul H. Söding

DETECTORS

The apparatus of particle physics has evolved from table-top experiments performed by a small group of people into detectors weighing thousands of tons, equipped with millions of channels of electronics and powerful computing systems, and staffed by collaborations of hundreds of scientists and engineers. As each generation of experiments brings new insights into the fundamental particles of matter, physicists strive to design more capable detectors to investigate the questions raised by new knowledge. This article is designed to give an introduction to some of the techniques involved.

The Detector as Camera

The apparatus for a modern experiment can be thought of as a digital camera taking pictures of individual interactions of elementary particles. The important questions about this camera are as follows: how accurate is the image, how well does it resolve objects that are close to each other, how long does it take to form the picture, and how quickly can it take consecutive pictures? An "intelligent" camera, moreover, would only take a picture when there was something interesting to be seen. Depending on the purpose of the experiment, the apparatus design concentrates on some or all of these aspects.

This camera, however, is not taking a picture of an object by reflected light; it is imaging the interactions of the tiniest elements of the subatomic world. The interaction could be the decay of a kaon into two pions, the annihilation of a high-energy electron with a positron, the collision of a 980-GeV proton with a 980-GeV antiproton, or the interaction of a neutrino with a proton. The experimenter typically wants to know the trajectories and energies of the particles produced in the interaction and to identify the type of particles (electron, muon, pion, etc.) produced in the interaction.

Experiments can be classified as colliding beam experiments where particles in counterrotating beams collide with each other or as fixed-target experiments where particles—from an accelerator, from the Sun, or from the depths of the universe—strike some material target. The detector challenges can be quite similar, although the geometry of the solutions may be rather different.

The Physical Basis of Detection Techniques

Ionization

The basis of most detection techniques is the fact that an energetic charged particle ionizes the material through which it passes, leaving a trail of positive ions and free electrons. Ionization chambers and silicon detectors measure the liberated charge directly. In proportional chambers, the liberated electrons seed a multiplication process in a strong electric field to produce a signal. In a scintillation counter, the molecules of the scintillator emit light as the electrons return to their ground state, and the light signal is then converted into an electrical signal by a photomultiplier. In bubble chambers and cloud chambers, the positive ions serve as nucleation centers for the formation of bubbles in the super-

heated liquid of the bubble chamber and for the formation of droplets in the supersaturated vapor of the cloud chamber. Charged particles also leave a trail in photographic emulsion.

Cerenkov and Transition Radiation

Cerenkov and transition radiation are two other phenomena exploited for particle detection. These processes produce free photons rather than electrons. Cerenkov radiation is emitted by a charged particle passing through a medium if the particle speed exceeds the speed of light in the medium; transition radiation is emitted by fast-moving particles as they cross the boundary between materials of different refractive index. In both cases, the radiation can be used as a sensitive indicator of a particle's speed and is often used in distinguishing different types of particles. How these basic phenomena are used in systems of detectors will be described in the following sections.

Tracking Detectors

Tracking detectors, as the name implies, are used to determine the paths of particles produced in the event. A particle's trajectory may be used to derive its momentum and to determine whether it emerged directly from the interaction or was born in the decay of some other particle produced in the event. The momentum is determined by placing the tracking device in a known magnetic field and measuring the track's curvature. Cloud chambers and photographic emulsion were the earliest tracking devices, but bubble chambers were the detectors that presented some of the wonders of the elementary particle world most directly.

Bubble chambers, however, have the disadvantage that they can only take one picture every few seconds. Most current experiments rely on electronic detectors that may have an exposure time of a microsecond or less and need no recovery time.

Wire Chambers

A standard device for tracking particles over a large area is the proportional wire chamber; this invention (for which Charpak received the Nobel Prize in Physics) and its derivatives have allowed a thousand-fold increase in the sensitivity of experiments over previous detectors. A proportional chamber contains a number (from tens to tens of thousands) of thin wires, typically 0.02 millimeters in diameter and spaced by a few millimeters, enclosed in a volume of an appropriate gas. The wires are maintained at a high positive voltage with respect to some cathode. A charged particle traversing the gas liberates electrons along its path, which then drift in the electric field to the closest wire. Near the wire, the electric field is so strong that the drift electrons gain enough energy to ionize the gas, and a multiplication occurs inducing a signal on the wire. Particles from the interactions pass by several sets of wires, and by recording which wires produce signals, one can infer the trajectories of the particles. In a drift chamber, the wires are spaced further apart, and the time of the signal on the wire is recorded. This time depends on how far the particle passed from the wire and allows its position to be determined to 200 microns (0.2 mm) or better. Wire chambers range in size from a few cubic centimeters to many cubic meters.

In a Time Projection Chamber, the wires, rather than being distributed throughout the volume as above, are placed on the end walls, and a large electric field is applied across the gas volume. The tracks then appear as "projections" on these walls (see Figure 1).

Silicon Strip and Pixel Detectors

While wire chambers track particles as they fly away from the interaction, physicists want to study what happens right at the interaction point, and silicon detectors have been developed to meet this need. These detectors are capable of making measurements with a precision of 10 microns (0.01 mm) and distinguishing tracks separated by a hundred microns. They are made from pieces of silicon, several centimeters on a side and typically 300 microns (0.3 mm) thick. Readout strips that run the length of the silicon are spaced by 50 microns (0.05 mm). When a particle passes through the silicon, it generates a signal on the closest strip. Arrays of these detectors are placed close to the interaction point to reconstruct the event vertex in great detail.

With such tiny spacing, it takes a large number of elements to cover a sensible area, and such detectors are presently only feasible where the location

FIGURE 1

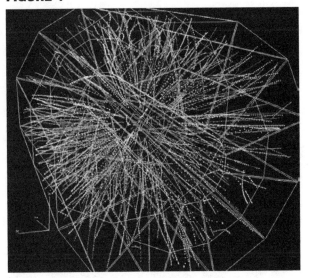

A perspective view of an event in the STAR TPC. CREDIT: COURTESY OF BROOKHAVEN NATIONAL LABORATORY. REPRODUCED BY PERMISSION.

of the interactions is known in advance, as in colliding beam experiments. Even then, the Babar detector at the Stanford Linear Accelerator Center (SLAC) has 150,000 channels, and the collider experiments at the Fermilab Tevatron in Batavia, Illinois, each have silicon systems with approximately 750,000 channels. Not to be deterred, physicists are exploiting new technologies in microelectronics to develop pixel detectors in which the individual element is not a strip 50 microns by several centimeters but instead is a rectangle typically 100 microns by 150 microns. Detectors with tens of millions of pixel elements are currently in the prototyping stage.

Scintillation Counters

The term "scintillation" was used at the turn of the twentieth century to describe the faint flashes of light visible under a microscope produced by alpha particles hitting a zinc-sulphide screen, and a scintillation "counter" was a scientist such as Ernest Rutherford looking through a microscope. A modern scintillation counter is made of a transparent material that emits light when a charged particles passes through it; the light is viewed by a photomultiplier that produces an electrical signal. Scintillation counters range in size from a few square centimeters to a square meter or more. Figure 2 shows a scintillator

FIGURE 2

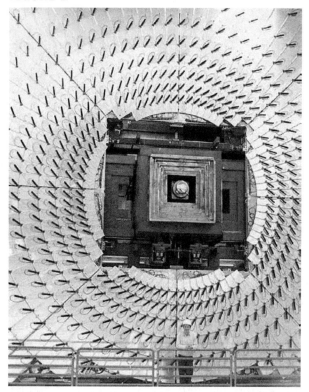

The Scintillator wall used for triggering on muons at the DZero Detector. There are approximately 700 counters in the wall. CREDIT: COURTESY OF FERMILAB PHOTO. REPRODUCED BY PERMISSION.

wall containing approximately 700 scintillation counters. Scintillation counters do not usually provide precise spatial information, but they can yield accurate information on the time of an event and are often used to indicate the occurrence of an interesting event. The development of plastic scintillating fibers 1 millimeter in diameter and smaller and of solid-state photomultipliers with 80 percent quantum efficiency has led to the use of scintillating material for tracking detectors.

Particle Identification

Particle identification is intended to determine the types of particles produced in the event. Nondestructive identification techniques depend on measuring both the speed of the particle and its energy or momentum, thus deriving its mass. The direct determination of speed by measuring the flight time over a known path is only useful at low ener-

gies. Another technique is to measure the ionization produced by the particle per unit distance—this is also only useful at low energies. For medium and high energies, the more powerful current technique measures the angle of the Cerenkov light produced as the particle passes through some suitable radiator and combines this with a momentum measurement to determine the particle type.

Calorimeters

A calorimeter is a device used to measure the energy of a particle or a set of particles. For neutral (uncharged) particles like photons and neutrons, calorimetry is the only direct way of measuring their energies and trajectories. Despite its name, the technique does not usually involve measuring a temperature rise, although it does mean absorbing the energy of the incident particle(s). The quantity measured is typically the total amount of ionization or light produced, with the assumption that this is proportional to the incident energy.

Calorimeters can be divided into electromagnetic calorimeters, used to measure the energies of electrons, positrons, and photons, and hadron calorimeters, used to measure the energies of hadrons (pions, kaons, protons, neutrons, etc.). The distinction arises because electrons and photons are fully absorbed in 30 centimeters of material of high atomic number, whereas it takes 2 meters or more of steel to fully absorb the energy of a hadronic particle.

The sequence of interactions produced when a particle strikes a calorimeter is called a cascade or shower. Sampling calorimeters use plates of a dense passive material that initiate and absorb most of the cascade energy, alternating with detectors between the plates that sample the cascade. The sampling detectors may be scintillation counters, ionization chambers filled with liquid-argon, or wire-chambers. Fully active calorimeters are sensitive to all the charged particles in the cascade.

Electromagnetic calorimeters use materials containing elements with high atomic number. Any photon, electron, or positron striking such a material produces a cascade of photons, electrons, and positrons. In sampling electromagnetic calorimeters, the passive material is a metal such as lead, tungsten, or uranium. Fully active calorimeters are made from scintillators such as cesium iodide, lead tungstate, or sodium iodide; from lead glass in which the electrons produce Cerenkov light; or from liquid krypton used as an ionization medium. Water and oil are also used as a Cerenkov radiator at lower energies.

Unlike electromagnetic calorimeters, hadron calorimeters extend a few meters in length to ensure that the incident hadron interacts and that the ensuing cascade of hadronic particles is fully absorbed. Because of the amount of material needed to absorb this cascade, hadron calorimeters are all of the sampling style. Uranium, brass, iron, and lead have all been used as the passive material.

Muon Identification and Measurement

The system for identification and measurement of muons is the last detector a particle may encounter. Muons play an important role in the search for new phenomena, and the muon system is an important aspect of many experiments. Muons are distinguished by their ability to penetrate meters of matter, and the standard technique for muon identification is to intersperse tracking detectors between plates of magnetized iron. Only muons will penetrate the iron, and their deflection (due to the magnetic field in the iron) can be used to estimate their momentum. In high-energy collider experiments, the muon identifier is the outermost and often the most massive system in the experiment. In neutrino experiments exposed to a muon-neutrino beam, the functions of target, hadron calorimeter, and muon identification are often combined (see Figure 3). In both types of experiment, the muon system may weigh thousands of tons.

When the discovery of the top quark, the most massive elementary particle ever observed, was announced, the experiments reporting findings on it were described as "5,000-ton three-story marvels of

FIGURE 3

A side view of the MINOS neutrino detector in the Soudan Mine (Minnesota) showing the iron plates and the readout of the scintillator sampling planes. CREDIT: COURTESY OF JERRY MEIER, UNIVERSITY OF MINNESOTA. REPRODUCED BY PERMISSION.

FIGURE 4

The tracking chamber being installed in the CDF detector for Run II.
CREDIT: COURTESY OF FERMILAB PHOTO. REPRODUCED BY PERMISSION.

sophisticated circuitry and engineering with tens of thousands of electronic channels to scan and record the products of tens of thousands of collisions every second." Figure 4 shows the CDF apparatus being assembled for the second run of the Tevatron. The structure of the experiment collaborations, however, was equally remarkable. Participating scientists came from fifteen different countries, detector components were developed and fabricated on four continents, and the experimental data were analyzed collectively, with no regard for national origin. Just like the detectors, this is a characteristic feature of the large experiments in particle physics. They are true international collaborations, united in the quest to learn more about the fundamental nature of matter.

See also: DETECTORS AND SUBSYSTEMS; DETECTORS, ASTRO-PHYSICAL; DETECTORS, COLLIDER; DETECTORS, FIXED-TARGET; DETECTORS, PARTICLE

Bibliography

Brookhaven National Laboratory. <http://www.bnl.gov>.

European Laboratory for Particle Physics. <http://www.cern.ch>.

Ferbel, T., ed. *Experimental Techniques in High Energy Nuclear and Particle Physics* (World Scientific, Singapore, 1991).

Fermi National Accelerator Laboratory. <http://www.fnal.gov>.

Leo, W. R. *Techniques for Nuclear and Particle Physics Experiments* (Springer-Verlag, New York, 1994).

Stanford Linear Accelerator Center. <http://www.slac.stanford.edu>.

Stephen Pordes

DETECTORS AND SUBSYSTEMS

The study of elementary particles and the forces between them is made possible by modern high-energy accelerators that bring beams of electrons or protons—or their antiparticles, positrons, and antiprotons—to nearly the speed of light and energies of hundreds or thousands of (GeV). (A GeV is the energy that an electron or proton would reach if accelerated across the terminals of a billion-volt battery).

Colliding beam accelerators focus two beams traveling in opposite directions onto each other; compared with the older technique of directing a single beam on a target of stationary nuclei, the energy available for creating new particles is much enhanced. The energy of modern colliders is larger than the energy associated with a particle's significant mass, $E = mc^2$, where c is the speed of light. The proton and antiproton have a mass energy of about 1 GeV; the electron and positron have approximately $\frac{1}{2}$ MeV, where 1 GeV = 1,000 MeV. Thus, when a collision takes place between two particles of the opposite beams, sufficient energy is brought to the system that hundreds of particles may be created in the event. The particles created fly outward from the collision point, with the imprint of what occurred in the original collision encoded in their energies and momenta. The momentum of a particle is its mass times its velocity; it carries directional information. The magnitude of the momentum for the highly relativistic particles typical of accelerator collisions is related to the energy by $E = cp$.

In contrast to colliding beam accelerators, collisions of particles in a single beam with nuclei in a fixed target occur at lower effective energy. However, added flexibility exists in that a wider range of beam particles can be used—not only the stable electron or proton but also unstable particles such as pions, kaons, or hyperons and even the very weakly interacting neutrinos. Thus for some specific studies, fixed-target experiments are preferred.

It is the job of the particle detector to record the momenta and energies of particles produced in a collision and, to the extent possible, determine the particle's identity. For example, when high-energy electrons or muons emerge from collisions of protons and antiprotons, it typically signals that something rare and interesting has occurred, and it is important to flag their presence to help distinguish the interesting events from the large background of less interesting collisions. A collider detector surrounds the accelerator pipes, centered on the collision point. A fixed-target detector is arranged downstream of the interaction target, as the produced particles are almost all thrown forward. Either type of detector consists of a set of subdetectors, each providing specific information; these subdetector types are the same for both collider and fixed-target use.

Tracking Detectors

The tracking subdetector system located closest to the collision point determines the momentum and direction of all the particles carrying electric charge. A strong magnetic field is imposed on the tracking region, usually by a solenoid magnet surrounding this detector in a collider experiment or by a dipole magnet in a fixed-target experiment. The charged particles bend in this field as a result of the Lorentz magnetic force. The bending radius is inversely proportional to the momentum. With care in design, a resolution of better than 0.1 percent can be achieved at low momentum, but for high-momentum particles the bending decreases and ultimately the resolution becomes very poor.

The tracking detectors sense the ionization trails left by the particle's disruption of atoms in their material. Several types of tracking detector may be used, often in concert. In the widely used drift chamber, wires with positive high voltage are arranged at regular intervals within a low-density gas environment. Electrons drift with constant velocity to these wire electrodes where they create detectable signals through an avalanche in the large electric field close to the wires. The time taken for the electrons to

reach the wire measures the coordinate along the drift direction with precision, typically a few hundred microns. Alternate wires may be stretched at small angles to each other, allowing the measurement of both coordinates perpendicular to the particle's direction. An alternate detector type employs thin crystalline silicon layers with conducting narrow (typically 50 μm wide) strips or pixels prepared on the surfaces. A voltage between the two surfaces causes ionization electrons or holes to drift to the strip electrodes, allowing spatial measurements with precision better than strip widths. Still other choices for track-sensitive detectors exist, and for special purposes these may be considered. Organic plastics called scintillators are available that deliver visible light from the deexcitation of the molecules of the plastic after disruption by ionizing particles. This light is delivered to photo-optic devices such as photomultipliers or avalanche photodiodes through optical waveguides. The fast time response of scintillators makes them attractive possibilities. Scintillating fibers with diameters less than a millimeter have recently been employed, giving good spatial resolution for tracking detectors.

Typically, the silicon strip and outer tracking detectors are used together, with the higher-resolution but more expensive silicon detectors arrayed nearest to the collision point, and the drift chamber or scintillating fibers starting approximately 20 cm from the beams. In this way, the presence of particles that live for a short time and travel only several hundred microns before their decay can be sensed in the high-resolution silicon detector, whereas the larger depth of the drift chamber can be used for a more accurate measurement of the bending radius and hence, particle momentum. The knowledge that there are short-lived particles present—such as those containing bottom or charm quarks, or the tau lepton—is often crucial for an experiment, as these particles signal the presence of interesting events that have produced new heavy states of matter. A general requirement for tracking detectors is that they do not significantly degrade the energy of the particles traversing them, and do not create new particles through interactions with the material. It is important that the directions of the charged particles not be seriously affected by the material, so the measurement of momentum from the bending in the magnetic field is not compromised.

Particle Identification

In some cases, the ability to identify specific particle types is of enough importance to the experiment that some of the tracking detector volume may be devoted to special particle identification detectors. These employ methods to differentiate particles of different mass but the same momentum. Cerenkov detectors record the light emitted when a particle exceeds the speed of light within a material medium and have a response that depends on particle velocity. Together with the momentum measurement in the standard tracking detectors, they allow determination of the particle mass. A transition radiation detector, containing many closely spaced layers of a dielectric material, detects the X rays emitted when a highly relativistic particle passes through it. Since the lightest of all particles, the electron, is the most relativistic for a given energy, the transition radiation detector can make a useful electron identifier. A final particle identification choice is a time of flight detector made from thick scintillation counters at the outside of the tracking region; with their exceptional timing response, such counters are able to distinguish the velocity of particles of differing mass at a given momentum. It is especially important to distinguish electrons and muons from other charged particles; this is primarily achieved using the calorimeter and muon detectors discussed below. The special particle identifying detectors typically require the use of valuable space, and unless there are strong reasons to include them, cost considerations tend to disfavor their inclusion in collider detectors. The special-purpose fixed-target experiments often place a higher premium on particle identification and more often incorporate such subdetectors.

In tracking detectors, charged particle energies are poorly measured and neutral particles (photons, neutrons, long-lived neutral K mesons) are not seen at all. Thus, it is necessary to have detectors outside the tracking region to measure particle energies. Such detectors are called calorimeters, based on the analogy with the energy-measuring calorimeters of chemistry experiments at a much-lower-energy scale. A calorimeter is based on the showering of particles

in a dense medium. A high-energy particle incident upon a material travels, on average, some characteristic interaction distance λ before suffering a collision where several new particles are produced. These daughter particles jointly carry the energy of the incident particle, so they are at lower energies. After a further distance of about λ, these too suffer collisions, resulting in the further multiplication of particles in the shower. This process continues until the particles in the shower are so low in energy that further multiplication is not possible, and the particles stop. The shower process is statistical; the locations of the collisions and the number of particles produced vary from shower to shower, but for high-energy incident particles the total number of particles (or their total travel distance in the medium) is, on average, proportional to the incident energy. Although particles that interact primarily through the electromagnetic interaction (electrons and photons) have a rather different scale λ from the hadrons that interact primarily by the strong nuclear interaction (protons, pions, K mesons, etc.), the principles are the same. Typically, calorimeters are subdivided into an initial first thinner section for electromagnetic particles and a thicker backing section for the hadrons.

In a calorimeter, the showering process is generated through collisions in a medium with large atomic number; lead, iron, or uranium are often used. Generally, the absorber material is made into sheets with interleaved gaps in which particle detectors are placed to record the signals left by the particles traversing it. The total signal from these 50 to 100 interleaved detectors samples the total number of particles in the shower; hence, it is proportional to the energy of the incident particle. Owing to the statistical nature of the showering process itself and the statistical sampling of its content, the relative energy resolution improves with energy like the square root of the energy. The active detectors are segmented in the directions perpendicular to the shower direction to differentiate the energy deposits from different particles in the event. Several choices for the active detector are possible; they differ in their stability, ease of calibration, ability to withstand radiation, ease of segmentation, and time response. A typical choice is scintillation counters, arranged in pads of a few centimeters across, with the light piped out to external photon detectors. Alternate choices are liquid argon or silicon wafers, in which the signal deposits are collected directly as electronic charges on transversely segmented electrodes at the surface of the detector. The energy resolutions achievable are in the range $\sigma/E = 10$ to 20 percent/$\sqrt{E}$ for electromagnetic particles and $\sigma/E = 40$ to 70 percent/$\sqrt{E}$ for hadrons, where σ is the standard deviation energy error, and E is measured in GeV. If better electromagnetic energy resolution is required, it is possible to make calorimeters with transparent heavy crystals such as cesium iodide or bismuth germanate. For these calorimeters, the absorber and active elements are combined, and the sampling fluctuations can be avoided. These more expensive options may be chosen for fixed-target experiments where the specific goals might require exceptional energy resolution. If improved spatial resolution is required, a separate, more finely segmented, section of the calorimeter called a preshower detector may be added before the main calorimeter.

Muon Detector

The final detector system of a modern collider or fixed-target detector measures the muon, which is capable of penetrating the calorimeter without causing showers and thus does not have its energy well measured. The muon is the only observable particle that can penetrate the material of the calorimeter without substantial degradation of its energy, so seeing a track after the calorimeter gives a clear signal for a muon. Its ionization trail is typically measured in detectors such as drift chambers or scintillation counters interleaved with magnetized iron plates. The magnetic field in the muon detector iron permits a second measurement of the muon's momentum, independent of that performed in the inner tracking. In a collider detector, the outermost muon system is very large, so there is a premium on cheap detectors.

Neutrino Detection

Neutrinos can be produced in the decays of several particles of particular interest, so it is important to know if they are present in a collider experiment event. Neutrinos have such small interaction probabilities that they pass through the detector without showing a detectable trace. However, their presence can be inferred from the balance of momentum in the event. Before a collision, the beam particles have no momentum components in the two directions perpendicular to the beams; momentum conservation ensures that after the collision, the sum of all momenta in these directions should also be zero. If a neutrino is present, it will result in an apparent imbalance in this transverse momentum and thus be revealed in the analyses of the event.

Event Selection

The rate at which collision events occur is very high—in excess of 1 million collisions per second at colliding beam accelerators using protons or antiprotons, and even higher in some fixed-target experiments. This rate is much too large to allow the recording of every event, so selection of only the most interesting must be made. This is the job of the trigger system. Information from each of the detectors is made available within a few microseconds after the collision to special electronics processors. These enable a quick, but somewhat crude, look at the pattern of activity in the detector. By finding evidence for particles such as electrons, muons, or short-lived particles, or by sensing the particularly interesting topologies of many particles, the first-level trigger can flag an event for more detailed scrutiny. On receipt of such a flag, more complete digitized information from subdetectors is collected, and a refined examination is made in dedicated microprocessors, leading to a decision on whether to transfer the full set of subdetector information to the on-line computer for logging the data to permanent storage. Typically, the rate of events saved for permanent archiving is about 100 per second in a collider experiment, where the total amount of information to be stored for each event is hundreds of thousands of bytes. In a fixed-target experiment where smaller event data sizes are typical, the rate of logged events may be even higher. The recorded events can then be analyzed off-line in great detail and studied in a multitude of specific physics analyses.

A full collider detector is extremely large and complex. Existing detectors at the 2,000 GeV Fermilab antiproton-proton collider in Batavia, Illinois, are

FIGURE 1

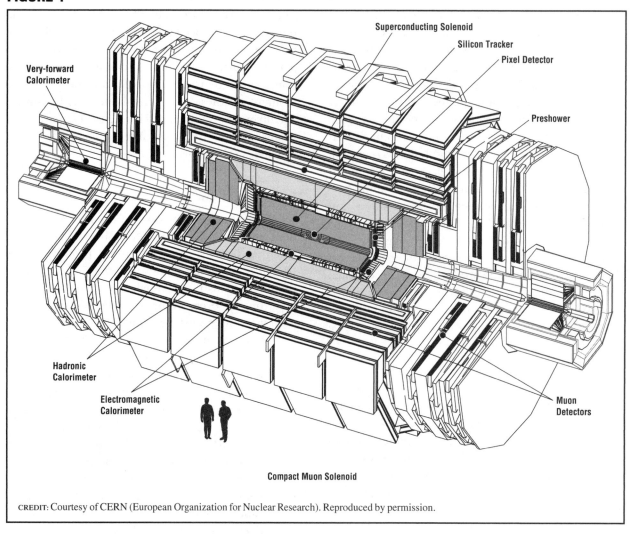

Superconducting Solenoid

Silicon Tracker

Pixel Detector

Very-forward Calorimeter

Preshower

Hadronic Calorimeter

Electromagnetic Calorimeter

Muon Detectors

Compact Muon Solenoid

CREDIT: Courtesy of CERN (European Organization for Nuclear Research). Reproduced by permission.

Cutaway view of the CMS detector being prepared for the CERN LHC.

about 15m high and wide, and 20m long. Future detectors for the 14,000-GeV large hadron collider (LHC) (see Figure 1) at the European Laboratory for Particle Physics (CERN) in Geneva, Switzerland, will be twice this size. There are well over a million channels of electronic readout for the various subdetectors. The collaborations of physicists and engineers who design, build, and operate these detectors number in the hundreds; these individuals come together from universities and laboratories across the world.

Further information on particle detectors can be found for experiments currently under way at major research laboratories around the world (D0 and CDF at Fermilab, BaBar at Stanford in Palo Alto, California, CLEO at Cornell in Ithaca, New York, and H1 and ZEUS at the Deutsches Elcktroncn-Synchrotron Laboratory [DESY] in Hamburg, Germany). A description of the detectors being planned for the LHC (ATLAS and CMS) include very accessible explanations of the language of particle physics experiments and Web-based tours of the operational principles of subdetectors.

See also: CASE STUDY: LHC COLLIDER DETECTORS, ATLAS AND CMS; CASE STUDY: LONG BASELINE NEUTRINO DETECTORS, K2K, MINOS, AND OPERA; DETECTORS, ASTROPHYSICAL; DETECTORS, COLLIDER; DETECTORS, FIXED-TARGET; DETECTORS, PARTICLE; PARTICLE IDENTIFICATION

Bibliography

Abachi, S., et.al. "The D0 detectors." *Nuclear Instruments and Methods* **A324**, 53 (1993).

Abe, F., et.al. "The CDF Detector: An Overview." *Nuclear Instruments and Methods* **A271**, 387 (1988).

Bell Collaboration. <http://belle. kek.jp>.

Center for European Laboratory for Particle Physics. "The CMS Detector."<http://cmsinfo.cern.ch/Welcome.html>.

Cornell Electron Storage Ring Laboratory. "The CLEO Detector." <http://w4.lns.cornell.edu/public/CLEO/>.

Deutsches Elektronen-Synchrotron Laboratory. "The H1 Detector." <http://www-h1.desy.de/>.

Deutsches Elektronen-Synchrotron Laboratory. "The ZEUS Detector." <http://www-zeus.desy.de/>.

European Laboratory for Particle Physics. "The ATLAS Detector." <http://pdf.lbl.gov/atlas/atlas.html>.

Fermilab. "The CDF Detector." <http://www-cdf.fnal.gov/>.

Fermilab. "The D0 Detector." <http://www-d0.fnal.gov/>.

Stanford Linear Accelerator Center. "The BaBar Detector." <http://www.slac.stanford.edu/BFROOT>.

Paul Grannis

DETECTORS, ASTROPHYSICAL

Particle physics is defined more by the questions addressed than by the techniques used. Although accelerator-based experiments have been, and will be, the primary experimental tool for particle physics research, nonaccelerator measurements are playing an increasingly important role. The intellectual connection between particle physics and other fields, particularly astrophysics and cosmology (the study of the history and evolution of the universe), has grown significantly over the past decade, as great progress has been made on many questions of common interest. In most cases, these detectors use the same measurement principles as in accelerator-based detectors, but they are adapted for a much wider variety of environments. The detector choices are motivated by the physics being investigated.

Connections and Physics Goals

Laboratory techniques for fundamental physics investigations have long been successfully applied to astrophysical measurements. In turn, making increasingly precise measurements of astrophysical phenomena, over immense distance scales and energy scales, allows one to address fundamental questions, test limits of physical law in the most extreme environments and over immense distances, and study the relationships between physical law and the evolution of the universe. Some of the goals of this research include

- Exploring black holes: The goal is to understand the acceleration mechanisms producing ultra-high-energy jets from supermassive (10^6 to 10^9 solar mass) black hole systems, which are nature's highest-energy accelerators, and study the characteristics of these remarkable objects. Black holes provide an important laboratory for testing theories of gravity, which is the least well-understood of the fundamental forces. The fact that many black hole systems shine brightly in high-energy particles makes them especially interesting to particle physicists and high-energy astrophysicists;

- Finding the origin(s) of the highest-energy cosmic rays (energetic particles propagating in outer space). Historically, particle physics grew out of cosmic ray physics (and nuclear physics), yet the origin of the cosmic rays remains an unsolved problem;

- Understanding gamma-ray bursts, which are brief and intense flashes of gamma-rays from far-away explosions that appear to be the most powerful since the Big Bang;

- Uncovering galactic dark matter, a hypothetical new form of matter that is required to explain a variety of observations of galaxies and clusters of galaxies;

- Studying the cosmic microwave background (CMB), an important and remarkable fossil from the early universe;

- Testing inflation, currently the most widely accepted paradigm for models of the earliest stages in the evolution of the universe;

- Searching for other Big Bang relics, which are clues to the puzzle of the early universe and which may provide information about physics at energy scales far higher than those achievable with artificial accelerators;

- Detecting gravity waves, a clear prediction of general relativity (the theory of gravity) and an entirely new window for astronomical observations;

- Confirming and studying the dark energy, the generic explanation for the recently discovered apparent *increase* in the expansion rate of the universe;

- Most importantly, discovering the unanticipated.

Many of the detectors that address these physics topics are surveyed below. For reasons of scope and space, a number of important categories are omitted, most notably optical, microwave, and X-ray detectors. These instruments have provided spectacular results in astrophysics and cosmology, and they are playing increasingly important roles in particle physics. In coming decades, particularly as dark energy and dark matter are better understood, these detectors may become part of major particle physics experiments. In what follows, energy is expressed in units of electron volts (eV), with $1\ keV = 10^3\ eV$, $1\ MeV = 10^6\ eV$, $1\ GeV = 10^9\ eV$, and $1\ TeV = 10^{12}\ eV$.

Gamma Rays

Gamma rays are the highest-energy photons, which are the quanta of electromagnetism. Gamma-ray photons are very similar to the photons of visible light humans see, except they are at least a million times more energetic. Astrophysical sources of gamma rays identify sites of extreme particle acceleration, such as neutron star and black hole systems, or signal decays of very massive particles that might have been produced in the early universe.

Gamma rays are so energetic that, when interacting with matter, pairs of new particles with opposite electric charge can be created when a portion of the photon energy is converted into the new particles' mass. The pair is usually an electron and its antimatter partner, the positron. If there is sufficient energy, this process of converting energy into matter continues and a shower of particles is produced. The number of particles in the shower is proportional to the kinetic energy of the initial particle. When a celestial gamma ray encounters the upper atmosphere, it interacts and is stopped by this process. If the gamma ray has very high energy (more than approximately 50 to 100 GeV), enough information about the shower reaches the ground, and the direction and energy of the initiating gamma ray can be measured by ground-based detectors. For lower-energy gamma rays, these measurements must be made above the atmosphere, in space. For both ground-based and space-based detectors, one of the most significant challenges is background identification and rejection. For every gamma-ray photon from the weakest astrophysical sources, there can be upward of 10,000 background particles (mostly charged cosmic rays). This fact has a strong influence on the instrument designs and sometimes sets the fundamental limitations on the scientific capabilities of a particular detector.

There are two basic types of ground-based gamma-ray detectors: airshower Cerenkov telescopes (ACTs) and extended airshower (EAS) array detectors. Cerenkov light is emitted when a charged particle travels through a medium at speeds faster than light propagates in that same medium. The emission pattern is somewhat like the wake made by a speedboat in water. ACTs detect the Cerenkov light emitted by the electrons and positrons in the gamma-ray-induced shower. Large reflectors collect the light and focus it onto an array of sensors (usually phototubes) that record an image of the shower. The

TABLE 1

Selected Astrophysical Detectors

Detector	Type/Purpose
Whipple, Cangaroo, HEGRA, CAT,STACEE, CELESTE, HESS, MAGIC, VERITAS, Milgro, ARGO	Present-day and future ground-based gamma-ray detectors
EGRET, GLAST, AGILE	Recent and future space-based gamma-ray detectors
AGASA, HiRes, Auger, OWL	Present and future ultra-high-energy cosmic ray detectors
AMS, PAMELA, BESS, CAPRICE	Cosmic ray antimatter detectors
Super-Kamiokande, SNO, Soudan II, Borexino, KamLAND, ICARUS, Homestake,Gallex, SAGE, MACRO, AMANDA, ICECUBE, Baikal, NESTOR, ANTARES	Present and future neutrino detectors
LIGO, VIRGO, GEO, TAMA,AIGO, LISA	Gravity wave interferometers
CRESST, DAMA, CDMS, UKDMC, AXION	Dark matter detectors

CREDIT: Courtesy of Steven Ritz.

size, shape, and orientation of the image identify the shower and provide a measure of the gamma-ray energy and direction. The Whipple observatory, located on Mount Hopkins in Arizona, is one of the pioneering ACTs. The detector consists of a 10-meter reflector dish, comprising 248 mirror segments, and a camera array of approximately 500 phototubes. Other ACT observatories include CANGAROO (Australia), HEGRA (Canary Islands), and CAT (France). Two other experiments, STACEE in New Mexico and CELESTE in France, use large arrays of solar energy collectors for extended sensitivity to lower gamma-ray energy: during the daytime, these facilities are solar energy power plants, and during the nighttime, they detect gamma rays from space. Future detectors in planning or under construction use arrays of ACTs that can work together or independently. These include HESS and VERITAS. Note that gas Cerenkov detectors are also used in accelerator-based detectors for particle identification but in very limited volumes; in contrast, ground-based gamma-ray ACTs use the atmosphere as the detection medium, and they therefore have enormous collecting areas capable of detecting faint gamma-ray sources.

ACTs can only operate on cloudless, moonless nights, so EAS detectors are also being used for ground-based gamma-ray astrophysics. When the gamma-ray energy is high enough (typically greater than 500 GeV), the resulting air shower particles can reach the ground, and EAS detectors directly sense the passage of these particles. The number and geographic distribution of the particles gives a rough measure of the gamma-ray energy, and the relative arrival times of the particles across the array give a measure of the gamma-ray direction. Relative to ACTs, EAS detectors have much higher operating efficiency because they can make measurements day and night, and they also have a much larger field of view; however, they also typically have much worse measurement precision. The two techniques are therefore complementary. An example of a gamma-ray EAS is the Milagro detector, located in the Jemez mountains of New Mexico. The detector is an enclosed pool of water, about the size of a football field, with an array of over 700 phototubes. When airshower particles pass through the water, they emit Cerenkov light that is detected by the phototubes.

Space-based high-energy gamma-ray detectors use the pair conversion process even more directly. The gamma ray converts inside the active volume of the instrument, and the electron and positron trajectories are detected using the same kind of charged particle tracking detectors found in accelerator-based experiments. The combined tracks give the gamma ray direction. The energy is measured in a calorimeter placed behind the tracking detectors. Space-based calorimeters are quite similar to accelerator detector calorimeters, except they are far smaller and less massive due to launch volume and weight limitations. Other significant differences are that space-based detectors must withstand the accelerations and acoustic shocks from the launch vehicle; they must operate on far less power; they must work reliably without any possibility of repair (though access inside immense accelerator-based experiments is becoming equally difficult); and, of course, they must operate in the environment of space. The most successful space-based high-energy gamma-ray instrument was EGRET, aboard the Compton Gamma Ray Observatory (GRO) that was launched in 1991. EGRET had a gas-based spark chamber tracking detector and a single calorimeter. GLAST, which is planned for launch in 2006, uses modern particle physics techniques (precision silicon strip tracking detectors and a segmented calorimeter) and will provide much greater sensitivity. The Italian mission, AGILE, which is a miniature version of GLAST using similar technology and with sensitivity comparable to EGRET, is scheduled to launch in 2003. Since GRO was deorbited in 2001, AGILE will fill an important gap in time until GLAST launches.

Ground-based and space-based gamma-ray detectors have complementary capabilities, and results from each are even more scientifically significant when combined together. The next generation of instruments will, for the first time, have significant overlap in energy coverage, providing new opportunities for important crosschecks and comparisons of results.

Cosmic Rays

Cosmic rays are high-energy particles that propagate through space. The vast majority of cosmic rays are protons, and their observed energies range over more than twelve orders of magnitude to more than

10^{20} eV. There are many important mysteries to solve in this area, but the question of most direct interest to particle physicists is the origin of the highest-energy cosmic rays. At present, the acceleration mechanism is not understood, and it is possible that some new physics at very high energy scales is required. Compounding this mystery is the fact that, while the universe is mostly transparent to most protons traveling through interstellar space, protons with energy greater than a particular cutoff energy, called the Greisen-Zatsepin-Kuzmin (GZK) energy, collide and interact with the pervasive CMB photons, causing them to lose energy. (Below the GZK cutoff energy, there is not enough energy to cause the reaction with the CMB photons.) One would therefore expect a reduction in the number of detected particles with energy greater than the GZK energy. Surprisingly, there is evidence that the number actually increases. However, the statistical evidence is not conclusive, and this must be confirmed by more than one experiment with greater statistical significance. Since the flux of these highest-energy particles is very low (approximately one particle per square kilometer per century!), gathering useful numbers of events within a reasonable amount of time requires instruments with enormous collecting areas.

As with ground-based gamma-ray detectors, the enormous collecting area is achieved by using the atmosphere as the principal medium. Showers of particles in the atmosphere, initiated by the cosmic ray particle, are detected either directly with EAS arrays (see the Gamma-ray Section above) or indirectly via collection of the light emitted by the shower particles. For these measurements, the light comes from atmospheric nitrogen that is excited by the developing airshower. Unlike Cerenkov light, this nitrogen fluorescence light is emitted isotropically, which greatly increases the effective collecting area of the sensors.

The world's largest cosmic ray EAS array is the AGASA array in Akeno, Japan. It has 111 particle detectors spaced approximately 1 kilometer apart, covering a total area of about 100 square kilometers. The AGASA array has detected the largest number of cosmic ray events with estimated energy greater than 10^{20} eV. The HiRes experiment, in Utah, is a nitrogen fluorescence detector consisting of a set of collecting mirrors and phototube cameras. By imaging the light from two different perspectives, the properties of the airshower (and therefore the primary particle) can be better determined. HiRes has an effective collecting area that is larger than that of AGASA, but it has a lower observing efficiency. The Pierre Auger Observatory, under construction in Argentina, will combine the two techniques by using both types of detectors. Some of the airshowers will be observed by both types of instruments, allowing important cross-checks and a combined analysis. Auger will have a much larger collecting area, consisting of 1,600 EAS particle detectors covering 3,000 square kilometers, along with 24 fluorescence detector telescopes. A second Auger installation in the Northern Hemisphere is also being planned. Pushing the nitrogen fluorescence technique to its ultimate implementation, OWL, a space-based light collector that will view a large fraction of the entire atmosphere, is being considered. OWL would have an effective aperture approximately 1,000 times larger than that of AGASA.

Another important puzzle is the asymmetry between matter and antimatter in the universe. If models of the early universe are correct, matter and antimatter were produced in equal amounts, so why does matter apparently dominate today? The AMS experiment, which will fly on the International Space Station, is being built to search for an anomalous flux of antimatter particles in space. AMS has silicon strip charged particle tracking detectors and a strong magnet to measure the momentum and electric charge of particles passing through it. Combined with other sensors in the experiment, the particle type can be identified. AMS, and related experiments such as PAMELA, will carry into space investigations that have been done on high-altitude balloons by experiments such as BESS and CAPRICE.

Neutrinos

There are several astrophysical sources of neutrinos, including solar neutrinos, which are produced in the core of the sun as a fusion reaction byproduct; atmospheric neutrinos, which are produced in airshowers initiated by cosmic rays; neutrinos from supernovae within our galaxy; and extragalactic neutrinos, which are expected from supermassive black hole accelerator systems and gamma-ray bursts. Neutrinos would also be produced by decays of most hypothetical massive relic states from the Big Bang and early universe.

Neutrinos interact only via the weak nuclear force (and, of course, gravity), so detecting them usually requires very massive instruments with which the neutrinos have a practical probability of interacting. There are two categories of weak nuclear interactions, called neutral current (NC) and charged current (CC), and the neutrino can participate in both of them. In a NC interaction, the neutrino scatters off a particle in the detector (nucleus or electron), imparting some of the neutrino's energy and momentum to that particle. After the collision, the neutrino escapes the detector. In a CC interaction, the neutrino is converted into a charged lepton (electron, muon, or tau, depending on the type of neutrino), which can be detected directly, and the nucleus is also converted to that of a neighboring element in the periodic table or is broken up if enough energy is transferred to it. By measuring the trajectory of the emerging charged lepton, the initial direction of the neutrino can be inferred. There are many different kinds of neutrino detectors, but they all rely on NC and/or CC interactions in some manner. The main distinction is the method of detecting the deposited energy and emerging lepton. In addition, the detector setup is optimized to study a particular source of neutrinos. All of these detectors operate deep underground to minimize backgrounds.

Proton decay detectors, which are very large-volume instruments that can detect small energy releases within those volumes, are often also excellent neutrino observatories. These detectors have made important observations of neutrinos from a supernova, atmospheric neutrinos, and solar neutrinos. An example is the Japanese Super-Kamiokande detector, a huge, 50-ton, underground imaging water Cerenkov detector. The inner portion of the detector is viewed by 11,146 phototubes. The cone of Cerenkov light projects a ring onto the array of phototubes, and the character, location, size, and shape of the ring tell the particle type, energy, and direction. In addition to neutrino interactions within the volume of the detector, the instrument can also detect CC interactions that occur nearby in the surrounding rock, effectively instrumenting the Earth.

Airshower neutrinos from the other side of the Earth travel through the Earth and can produce upward-going muons that are seen in the detector. Analysis of these events has provided important evidence for neutrino oscillations. The Sudbury Neutrino Observatory (SNO) in Canada is a similar type of detector, but the active medium is heavy water (D_2O) instead of ordinary water. This allows distinct measurements of the CC and NC interactions and provides other important cross-checks for neutrino oscillation measurements. The Soudan II detector, located in an underground laboratory in Minnesota, has very similar physics goals but uses a different detection principle. The heart of the detector is a 960-ton iron calorimeter with gaseous charged particle sensors. Borexino, at the Gran Sasso laboratory in central Italy, has as its main emphasis the study of solar neutrinos. The detector consists of 300 tons of liquid scintillator, viewed by 2,200 phototubes. ICARUS is a large-volume charged particle tracking detector, called a time projection chamber (TPC), whose target and detection medium is not gas but rather liquid argon.

Isotope experiments, rather than viewing the interactions immediately when they happen, detect the presence of small numbers of converted element nuclei indicating neutrino CC interactions after the fact. The Homestake detector, located in a gold mine in South Dakota, was constructed to measure the flux of solar neutrinos. Six hundred tons of tetrachloroethylene served as a target for the neutrinos. A CC interaction converted the ^{37}Cl nuclei into ^{37}Ar, which could then be detected by its radioactive decay. The number of argon nuclei tracked the number of neutrino interactions during the running period. This instrument uncovered a surprising deficit of solar neutrinos relative to theoretical expectation, providing evidence that neutrinos have mass. Two other detectors of this type, Gallex at Gran Sasso, Italy, and the Russian-American SAGE experiment, use gallium as the target medium.

Other detectors search for higher-energy neutrinos by detecting the products of CC interactions in the matter near the instruments. The usual signature is the detection of upward-going muons (the flux of downward-going muons is dominated by muons from cosmic-ray induced airshowers). The MACRO detector, in Gran Sasso, is a large-area detector designed to search for magnetic monopoles, but it is also quite effective for searching for high-energy neutrino fluxes using this technique. Instead of using optically opaque rock as the target medium for the neutrino interactions, one can use large, naturally occurring volumes of water or ice. The AMANDA experiment is an array of phototubes placed deep in the Antarctic ice, 1.5 to 2.2 kilometers under the surface. Upward-going high-energy muons from CC neutrino interactions produce Cerenkov light in the ice, which is collected by the phototubes. The arrival times of the light at the spatially distributed phototubes gives a measurement of the muon trajectory and hence the original neutrino direction. A vastly expanded version of AMANDA, called ICECUBE, is currently being planned. The BAIKAL experiment in Siberia, the NESTOR experiment in Greece, and the ANTARES experiment in France use water instead of ice as the medium. The detection principles are the same as for the ice experiments, though the details of the light propagation are different.

Gravity Waves

Gravity waves have not yet been detected, but there is a good chance this situation will change dramatically over the coming decade as new detectors are completed and begin operation. According to theory, as a gravity wave propagates, space is stretched and compressed by tiny amounts. The size of the strain depends on the strength of the disturbance that created the waves. Typical astrophysical sources that are believed to generate gravity waves (supernova core collapse, neutron star and black hole mergers) are expected to cause strains around 10^{-20} or smaller. Thus, a typical gravity wave passing through the solar system would change the distance between the earth and the moon by an amount that is less than the radius of a single proton! Such tiny changes can be detected using laser interferometry. The largest interferometer system under construction is LIGO. There are two LIGO sites, one in Hanford, Washington, and one in Louisiana. Each site has a 4-kilometer, two-arm interferometer with state-of-the-art control and noise reduction systems. LIGO will be able to detect gravity waves in the frequency band between a few Hz and a few thousand Hz, with strain sensitivity as low as 10^{-23}. Other gravity wave interferometers are VIRGO and GEO in Europe,

TAMA in Japan, and AIGO in Australia. By comparing signals from these geographically distributed detectors, the direction of the wave can be inferred. Seismic noise and other terrestrial disturbances, along with the interferometer arm length, limit the ultimate capabilities of these detectors. A space-based interferometer system would not have these limitations. LISA, a space-based gravity wave interferometer, is an international project currently in planning. The LISA interferometer will be formed by three spacecraft separated by 5 million kilometers and will be able to detect gravity waves in the frequency band between 10^{-4} Hz and 1 Hz with strain sensitivity as low as 10^{-23}. Together, ground-based and space-based gravity wave interferometers will open a completely new window through which scientists can observe the universe.

Direct Searches for Dark Matter

If the models are correct, we are immersed in a local density of particle dark matter equivalent to about one proton mass in every 3 cubic centimeters. The possibility that there is a flux of a new form of matter passing through the Earth undetected is extremely compelling to particle physicists. There are a number of creative, but indirect, ways to detect particle dark matter (e.g., searches for high-energy neutrinos from the sun and sharp peaks in the galactic gamma ray spectrum), but there is also the possibility of detecting these interesting particles directly. Although they interact only very weakly with ordinary matter, there is a small chance one of the dark matter particles will collide with a nucleus of ordinary matter, imparting approximately 10 keV of energy to the nucleus and causing it to recoil. The challenge is to detect these small recoil energy deposits, and exclude the backgrounds, in enough target detector material so that the rate of the collisions is measurable in a practical amount of time. Many years of new detector research and development have paid off, and there are now many direct dark matter detection experiments, only a few of which can be mentioned here. All of them are conducted in low-background environments, often deep underground.

CRESST is a cryogenic detector located in the Gran Sasso facility. It consists of sapphire target material and sensitive superconducting thermometers to detect the energy deposited by the dark matter particle interaction. The DAMA experiment, also in Gran Sasso, uses scintillators as target detectors. It is worth noting that, as of 2002, the DAMA group has detected a potential signal for dark matter events by examining the annual variation in the event rate in their 100 kg sodium iodide scintillation detectors; however, this must still be confirmed or refuted by other experiments. In particular, the CDMS experiment, which uses sophisticated silicon and germanium cryogenic detectors that can identify classes of backgrounds on an event-by-event basis, has failed to see events at the rate one would expect if the DAMA results were to be confirmed, but it is still too early to be conclusive. CDMS II, the next phase of the CDMS experiment, will operate in the Soudan mine in Minnesota and will achieve dramatically improved sensitivity. UKDMC is a collection of different detectors operating in the Boulby mine in the United Kingdom. These include sodium iodide scintillator detectors, liquid xenon scintillation and drift detectors, and a new kind of detector, called DRIFT, which can provide directional information about the dark matter event. Finally, searches for a different type of dark matter particle, called the axion, are being carried out by the appropriately named AXION experiment in the United States.

Nothing like these detectors is used in 2002 accelerator-based experiments. It is worth noting, however, that the same kinds of particles that could compose the galactic dark matter might also soon be produced and discovered in accelerator-based experiments, using very different measurement techniques, so the searches are complementary.

The adaptation of accelerator-based particle detection techniques for use in astrophysical detectors, along with advances in detector technologies targeted to solve unique experimental problems, has opened up the universe as a laboratory for fundamental physics. These investigations draw together the communities of particle physicists, astrophysicists, and cosmologists.

See also: CASE STUDY: GRAVITATIONAL WAVE DETECTION, LIGO; CASE STUDY: SUPER-KAMIOKANDE AND THE DISCOVERY OF NEUTRINO OSCILLATIONS; DETECTORS; DETECTORS AND SUBSYSTEMS; DETECTORS, COLLIDER; DETECTORS, FIXED-TARGET; DETECTORS, PARTICLE

Bibliography

AMANDA. <http://amanda.uci.edu/>.

CRESST. <http://www.lngs.infn.it/site/>.

GLAST. <http://www-glast.slac.stanford.edu/default.htm>.

HEGRA. <http://wpos6.physik.uni-wuppertal.de:8080/>.

OWL. <http://owl.gsfc.nasa.gov/detector.html>.

Super-Kamiokande. <http://www-sk.icrr.u-tokyo.ac.jp/doc/sk/>.

VIRGO. <http://www.virgo.infn.it/>.

Steven Ritz

DETECTORS, COLLIDER

The experimental study of the basic constituents of matter and their interactions requires detectors that are able to measure the important characteristics of particle interactions, whether they are produced in large accelerators, such as the Fermilab Tevatron, the Stanford Linear Accelerator Center (SLAC) Asymmetric *B* Factory (PEP-II), the Relativistic Heavy Ion Collider (RHIC), or the future CERN Large Hadron Collider (LHC), or whether they are produced in high-energy cosmic ray interactions or in the rare (and as yet unseen) decays of the proton in a large tank of water.

Detectors installed at large high-energy accelerators can be further divided into two categories: fixed-target and colliding beam. In both types of detectors, the aim is the same: to identify as many of the characteristics of the products of a high energy collision as possible. The two types differ mainly in their geometrical layout. In a fixed-target experiment, an accelerator beam of protons, electrons, neutrinos, or pions, which may have energy as high as several hundred GeV, impinges on a target (typically liquid hydrogen, although solid targets such as carbon are also employed) in the laboratory. The detector is arrayed upstream of the target, where it can intercept the products of the beam-target interaction. Because of the relativistic motion of the center of mass of the collision, the products of the collision are thrown forward and can be intercepted with high efficiency in this arrangement.

In a colliding beam detector, two high-energy beams of various combinations of protons, antiprotons, electrons, or positrons are brought into a direct collision. If the beams have equal energy, as in most colliders, the collision center of mass is stationary in the laboratory. In several installations, such as the HERA electron-proton collider or the PEP-II and KEK-B electron-positron colliders, the two beams have unequal energy, and the collision products are boosted in the direction of the motion of the collision center of mass in the laboratory. In both types of colliders, the experimental challenge is to detect and measure the properties of as many of the particles produced in the collision as possible.

The products of a high-energy collision consist of charged particles ($\pi^{\pm}$ mesons, $K^{\pm}$ mesons, protons, electrons, and muons) and neutral particles (primarily photons from π^0 decay, but also including neutrinos, neutrons, and K_L^0 mesons). The aim of detector design is to produce an instrument capable of measuring, with the highest possible efficiency and precision, the direction and momentum (or energy) of each collision product and identifying the particle species. These functions are performed by a variety of devices. Some typical approaches are discussed below.

In many cases, the particles actually detected were not produced directly in the collision but resulted from the decay of unstable particles produced in the primary interaction. These unstable particles typically fall into two classes: those that decay in a very short time ($\sim 10^{-21}$ to 10^{-15} second) via strong or electromagnetic interactions, and those that decay more slowly ($\sim 10^{-13}$ to 10^{-12} second) via weak interactions and can travel a measurable distance (typically 100 μm to 1 cm) within the detector before decay.

Short-lived particles are identified by constructing a quantity called the invariant mass, which combines the measured momenta and directional information of putative decay products in such a way as to isolate individual parent particles, such as π^0 or ρ mesons. Longer-lived particles are isolated by reconstructing their decay vertex, which is displaced from the primary interaction point. Long-lived charged particles ($D^{\pm}$, $B^{\pm}$, . . .) produce detached vertices with an odd number of prongs, whereas neutrals (D^0, B^0, . . .) produce an even number of prongs. These prongs may be the tracks of pions, kaons, protons, electrons, or muons. In order to ascertain that this decay vertex is detached from the interaction

point, it is necessary to reconstruct the origin and direction of each track of the vertex with sufficient spatial resolution to distinguish its origin from the interaction point. These measurements are often made in a silicon vertex detector, which typically consists of three to five planes of thin ($\sim$300 μm thick) high-resistivity silicon on which a series of fine lines (typically of 25 μm pitch and several centimeters long) form a series of diodes. When the charged particles pass through the junction of the diode, energy deposited by ionization is collected and amplified to produce a signal that can be used to locate the position at which the particle passed through the silicon plane to a precision of the order of 5 to 10 μm. Often, the two sides of the silicon wafer have orthogonal diode structures, allowing the simultaneous measurement of two coordinates in the plane for each particle. A computer is used for pattern recognition, that is, to associate the numerous measurements indicating the passage of particles though the series of precisely positioned silicon planes into a series of tracks and then to associate the tracks into a vertex. Pixilated silicon planes are now coming into use. In these devices, a single electronics channel measures both coordinates of the track, eliminating ambiguities that can arise in high multiplicity situations when each coordinate is measured separately.

With knowledge of the initial direction of each of the charged particles produced in the collision, the momentum of each particle can be measured. This is done by surrounding the interaction point with a strong magnetic field of 1 to 3 Tesla, causing each charged particle to bend in the field, with a radius of curvature in the plane perpendicular to the field that is proportional to its momentum. It is then necessary to determine all particle trajectories and to measure their radii of curvature and thus their momenta. This is often done in a drift chamber. A drift chamber can be built in planar (for fixed-target experiments) or cylindrical (for colliding beam experiments) geometry. Each plane or layer of the chamber is comprised of a set of individual cells a few centimeters in diameter. The cell perimeter may be defined by an array of fine wires (composed of, e.g., 80 μm gold-plated aluminum) or by a very thin mylar tube with a layer of aluminum deposited on its inner surface. In the center of each cell is a very fine (25 μm) gold-plated tungsten wire kept at a positive potential with respect to the perimeter. This array is placed in a volume of gas (80% helium + 20% isobutane in a typical gas mixture). The charged particles produced in the collision pass through the silicon vertex detector and then into the drift chamber, ionizing the gas along their trajectory. Electrons thus liberated in the gas then drift onto the fine central wire of each cell under the influence of the carefully calibrated electric field in each cell; the drift time is proportional to the drift distance within the cell. When the drifting ionization reaches the region immediately surrounding the high electric field near the central wire, an avalanche is created, producing a sufficient number of secondary electrons on the central wire to allow the recording of an electrical signal with well-defined amplitude and time characteristics. This allows the reconstruction of the particles' position with respect to the central wire to a precision of ~150 μm in each cell. A typical drift chamber has fifty layers, allowing the reconstruction of the individual particle trajectories in the magnetic field to high precision and the measurement of particle momenta to within a few percent accuracy.

With particle momenta measured, identifying the particle species remains. Since a particle's momentum is the product of its mass and velocity, the particle mass and thus its species may be identified if the particle's velocity can be independently determined. Several approaches are in common use. The first uses the details of the ionization left by the particle trajectory in a drift chamber. The ionization energy loss per unit length dE/dx in a gas is an essentially universal function of the particle velocity, independent of mass. Thus, the sum of all the ionization left by a track in the drift chamber, normalized to the length of the trajectory, provides a measure of a particle's velocity. If plotted against the momentum of a given particle, each species produces a characteristic energy loss in the gas at a given momentum. As a result of the shape of the universal dE/dx curve, however, there are regions of ambiguity that effectively limit this method of particle identification to momenta below 0.5 GeV/c.

Another method of particle identification is to measure the time of flight of the particle over a known distance with high resolution. A typical time of flight from the point of creation to a point at which a time can be recorded is 5 nsec, which can be measured to a precision of ~100 psec. This precision suffices to distinguish pions from kaons up to momenta of 0.6 GeV/c.

A third class of techniques makes use of the phenomenon of Cherenkov radiation, a shock wave emitted when a particle's velocity exceeds the speed of light in a medium. Cherenkov radiation devices come in several forms. There are threshold counters, in which the index of refraction of a high-pressure gas is chosen so that, for example, electrons emit Cherenkov radiation while pions do not, and there are devices that measure the angle of the Cherenkov shock cone with respect to the particle direction, since the cosine of this angle is proportional to the reciprocal of the particle's velocity. The latter device, known as a DIRC, can, using quartz as a Cherenkov medium, distinguish pions from kaons up to 4 GeV/c.

The identification of muons makes use of the fact that these particles do not have strong interactions and are thus able to more readily penetrate material, whereas pions, although close in mass, interact strongly and are absorbed. Detectors with solenoidal magnetic fields require a steel flux return of the order of a meter thick to control the field distribution. Muon detectors are thus typically made by distributing the flux return into several layers and inserting devices that can record the passage of a particle between the layers. These devices may be large arrays resembling crude drift chambers, large planar chambers with a well-defined gap maintained at high voltage and filled with gas, called resistive plate chambers (RPCs), or they may be made of plastic scintillators. In such arrays, a muon is identified as a particle penetrating many layers, while a pion typically does not survive all the way through the flux return.

The detection of high-energy photons presents unique challenges. Photons do not ionize the media through which they pass. Photons in the energy range of interest in elementary particle physics, from tens of MeV to tens of GeV, lose energy in matter through pair production, the photoelectric effect, and Compton scattering. They are detected in devices called calorimeters, which come in many varieties, but which have certain common characteristics. When a high-energy photon enters matter it creates an electromagnetic shower, a cascade of electrons, positrons, and photons. The electrons and positrons are ionizing particles. If the material in which the shower is initiated is large enough to contain the shower products, then the charged particle component of the shower deposits an amount of energy in ionization that is closely proportional to the energy of the initiating high-energy photon. This ionization energy can be detected in several ways. The highest-quality devices are arrays of large high-atomic-number crystals, such as CsI, PbWO$_4$, or LSO that emit scintillation light, which is detected by photomultiplier tubes or silicon photodiodes. Calorimeters that collect the ionization energy deposited by electromagnetic showers in noble liquids such as krypton or xenon have also been employed. Other devices consist of alternating layers of high-atomic-number material, such as lead, with a layer of material, such as plastic scintillator or liquid argon, that is sensitive to the ionization produced in the showers initiated in the high Z material.

See also: DETECTORS; DETECTORS AND SUBSYSTEMS; DETECTORS, ASTROPHYSICAL; DETECTORS, FIXED-TARGET; DETECTORS, PARTICLE

Bibliography

Barger, V., and Phillips, R. *Collider Physics* (Addison-Wesley, Redwood City, CA, 1987).

Pellegrini, C., and Sessler, A. M. *The Development of Colliders (Key Papers in Physics)* (Springer-Verlag, New York, 1995).

David Hitlin

DETECTORS, FIXED-TARGET

Fixed-target experiments are those that study the collisions of a highly relativistic particle beam with a target that is stationary in the laboratory. This technique is complementary to collider experiments that study the collisions of particles from two opposed beams.

Typically in a fixed-target experiment, the total momentum in the laboratory rest frame is large compared with the energy available to produce new particles in the final state, for example, the total energy of the reaction in the center of momentum frame. Therefore, high-energy particles produced in interactions at fixed-target experiments will be observed to be close in direction to the incoming relativistic particle beam, as shown in Figure 1.

Quantitatively, the γ of the Lorentz boost from the center of momentum frame into the lab frame is given by

$$\gamma = \frac{E_{\text{beam}} + m_{\text{targ}}}{\sqrt{m_{\text{beam}}^2 + m_{\text{targ}}^2 + 2E_{\text{beam}}m_{\text{targ}}}}$$

$$\approx \sqrt{\frac{E_{\text{beam}}}{2m_{\text{targ}}}}$$

where m_{targ} is the mass of the target particle, and E_{beam} and m_{beam} are the energy and mass of the particles in the incoming beam. The approximation is valid for the case where $E_{\text{beam}} \gg m_{\text{beam}}, m_{\text{targ}}$. The Lorentz factor γ gives the ratio between the momentum in the laboratory along the beam direction to the center of momentum energy for the case of particles produced at right angles to the incoming beam in the center of momentum frame. $1/\gamma$ is a typical angle

FIGURE 1

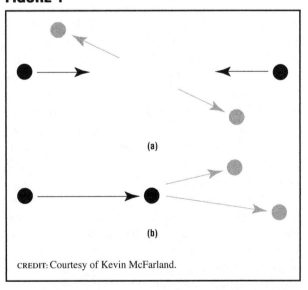

(a)

(b)

CREDIT: Courtesy of Kevin McFarland.

Schematic illustrations of the same particle reaction in (a) a collider experiment and (b) a fixed-target experiment. Note that in the fixed-target experiment, the produced particles both travel in nearly the same direction as the incoming particle beam.

with respect to the incoming particle beam for an observed particle produced in a fixed-target interaction.

The high ratio of incoming beam momentum to total center of momentum frame energy,

$$\sqrt{s} = \sqrt{m_{\text{beam}}^2 + m_{\text{targ}}^2 + 2E_{\text{beam}}m_{\text{targ}}},$$

in typical fixed-target experiments has two important consequences. First, the total energy available for creation of new particles is significantly lower then in colliding beam experiments. This limits the production of massive final states to masses typically well below E_{beam}. Second, the produced final states are usually highly relativistic. Because of time dilation, these particles thus have a long observed lifetime in the lab, and therefore a long flight path. The latter feature makes fixed-target experiments particularly well suited for measuring time evolution of particle states, such as in meson lifetime or mixing measurements.

Fixed-target experiments are also particularly useful when the desired reaction has a very low cross-section, such as in a neutrino scattering experiment, or when the goal of an experiment is to observe the decay of a long-lived particle, such as a weakly decaying meson. In the former case, the fixed target can be very massive in order to increase the interaction probability. In the latter case, the near collinearity of the outgoing particles in a fixed-target reaction means that particle detection apparatus need only subtend a limited solid angle from the point of particle production. This affords the opportunity to build small sophisticated detectors, for example, high resolution vertex detectors for observation of charmed particles, or allows for the decay detectors to be located a great distance from the target, as in a kaon decay experiment.

Geometry of a Fixed-Target Detector

As with collider detectors, a typical detector configuration for a fixed-target experiment is layered, with produced particles passing through a sequence of detectors. After the target, a typical detector configuration is illustrated in Figure 2. Nondestructive tracking detectors, such as segmented scintillator or proportional counters, wire chambers, ring-imaging Cerenkov

FIGURE 2

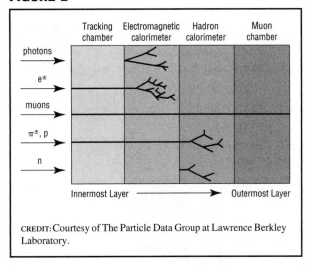

CREDIT: Courtesy of The Particle Data Group at Lawrence Berkley Laboratory.

The layered geometry of subdetectors comprising a typical fixed-target detector.

detectors, or transition radiation detectors, provide information on charged-particle direction and velocity. The addition of a magnetic field in the tracking volume allows the tracker to measure the product of particle charge and momentum *via* the deflection in the field. Subsequent calorimetric, or energy-measuring detectors, destructively measure the total energy carried by electrons, photons and strongly interacting metastable hadrons, such as charged pions. Finally, muon detectors identify and track particles, notably muons, that pass through the calorimeters.

A representative implementation of this geometry in a recent fixed-target experiment is the NA48 Experiment at the European Laboratory for Particle Physics (CERN) whose detector is shown in Figure 3. NA48 seeks to identify decays of neutral K mesons into $\pi^+\pi^-$ and $\pi^0\pi^0 \to 4\gamma$ final states. Highly relativistic K mesons with kinetic energy of approximately 100 GeV are produced in two targets, 220 m and 100 m in front of the decay. The weakly decaying kaons produce decay products nearly collinear to the beam direction. The momenta of charged pions are measured in the tracking volume, and the photons from π^0 decays are measured in the liquid krypton calorimeter. Anticounters surrounding the tracking chambers register the escape of wide-angle particles from the detector volume.

FIGURE 3

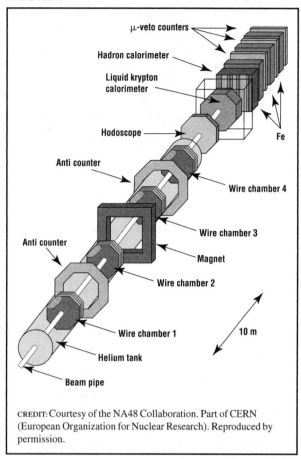

μ-veto counters

Hadron calorimeter

Liquid krypton calorimeter

Hodoscope

Anti counter

Fe

Wire chamber 4

Wire chamber 3

Anti counter

Magnet

Wire chamber 2

Wire chamber 1

10 m

Helium tank

Beam pipe

CREDIT: Courtesy of the NA48 Collaboration. Part of CERN (European Organization for Nuclear Research). Reproduced by permission.

The Detector of the NA48 Fixed-Target Experiment at CERN. Relativistic decay products of neutral K mesons travel from lower left to upper right through the detector in this diagram.

Active Target Material

A unique feature of fixed-target detectors is the capability for target material to serve as an active detector. In the case where interactions are rare, for example, neutrinos, this is essential since interactions can occur anywhere in the target. This feature may also be desirable in identifying extremely short-lived particles, such as τ leptons or hadrons containing heavy quarks.

Active target material may either be fully active or part of a "sampling" detector, one that observes particle interactions in only a fraction of its total material. The FOCUS experiment at the Fermi National Accelerator Laboratory (Fermilab) in Batavia,

Illinois, is an example of the latter type of active target applied to the detection of short-lived particles. FOCUS produces charmed hadrons, with a typical lifetime of 0.5×10^{-12} s, from selected interactions of a wide-band photon beam with an average energy of 180 GeV. Produced charm hadrons have a mean flight path before decay of ~2 cm. As shown in Figure 4, the segmentation of the target, along with the active silicon planes inside the target region itself, allows for identification of multiparticle vertexes from decays of charm in the empty regions of the target.

A classic fully active detector is a bubble chamber or emulsion detector. In such detectors, the target and detection media are the same material that fills the volume of the detector. Modern realizations of this technology include large emulsion experiments (CHORUS at CERN), cryogenic time projection chambers (ICARUS at Gran Sasso), solid-state detectors (CDMS at Soudan, CRESST at Gran Sasso), and water Cerenkov detectors (Super-Kamiokande at Kamioka, SNO at Sudbury). The experiments noted all search for very weakly interacting particles, such as neutrinos or dark matter candidates, which can interact anywhere in the volume of the detector. Fully active detectors typically utilize sensor technology that is not local in space-time to the observed interaction. For example, when produced particles travel through water Cerenkov detectors, such as the SNO detector shown in Figure 5, they create cones of visible light (Cerenkov radiation) that are observed with photomultiplier tubes located at the outer perimeter of the detector. These Cerenkov cones appear as rings when projected onto the perimeter and allow for measurement of the velocity and the identification of the produced particles.

Physics Goals of Planned Fixed-Target Experiments

Planned fixed-target experiments exist that take advantage of all the unique aspects available to fixed-target experiments. High statistics K meson decay experiments, which study rare decays of long-lived kaons and take advantage of high-rate kaon production on fixed targets, are planned to study the rare semileptonic decays $K \rightarrow \pi \nu \bar{\nu}$ in charged kaons

FIGURE 4

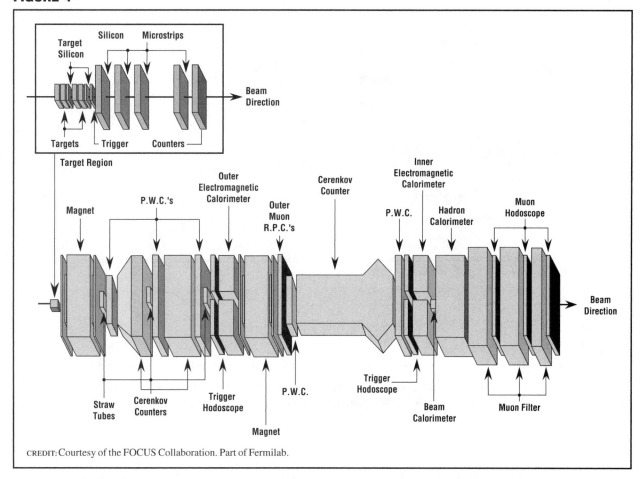

CREDIT: Courtesy of the FOCUS Collaboration. Part of Fermilab.

The Detector of the FOCUS Fixed-Target Experiment at Fermilab. Charmed hadrons are produced from interactions of a photon beam in the target material, including active silicon strip detectors, at the upstream end of the detector.

(CKM at Fermilab) and neutral kaons (KOPIO at Brookhaven National Laboratory and E391A at the Japanese High-Energy Accelerator Research Organization [KEK]). Experiments studying high statistics lepton interactions on polarized electron (E158 at Stanford Linear Accelerator Center) and polarized nuclear targets (COMPASS at CERN and a number of experiments at Jefferson National Accelerator Laboratory) will investigate electroweak interference and nuclear structure.

A large number of planned neutrino fixed-target experiments with very large active targets are currently under construction. Large detectors devoted to the study of neutrino oscillation phenomena in long-baseline accelerator neutrino beams include K2K at KEK/Kamoika (50 kilotons of water), Mini-BooNE at Fermilab (100 tons of mineral oil), MINOS at Fermilab/Soudan (10 kilotons of steel-scintillator sampling calorimeter), and OPERA and ICARUS at CERN/Gran Sasso (2 kilotons of lead-emulsion and up to 5 kilotons of liquid argon, respectively). In addition, a variety of planned experiments searching for very rare ultrahigh energy neutrinos in cosmic rays employ targets of unprecedented mass to search for these rare events. As an example, the IceCube experiment will search for these high-energy interactions in a cubic kilometer of Antarctic ice.

See also: DETECTORS; DETECTORS AND SUBSYSTEMS; DETECTORS, PARTICLE

FIGURE 5

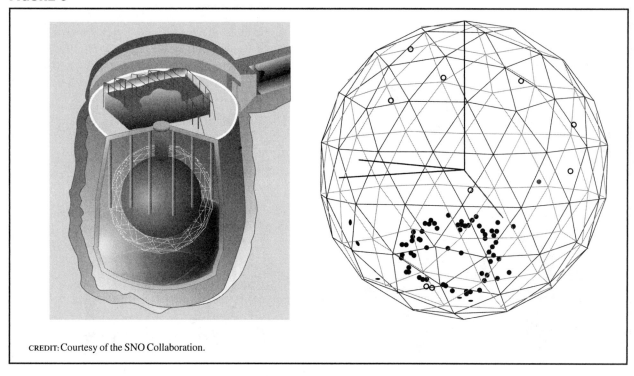

A schematic view of the SNO Detector and a candidate interaction of a solar neutrino. Neutrinos interact in the fully active volume of the detector, and particles produced in the interactions are observed via Cerenkov radiation that travels to light sensitive detectors on the perimeter of the water volume.

Bibliography

Barker, A. R., and Kettell, S. H. "Developments in Rare Kaon Decay Physics." *Annual Review of Nuclear & Particle Science* **50**, 249–297 (2000).

Boger, J., et al. "The Sudbury Neutrino Observatory: The SNO Collaboration." *Nuclear Instruments and Methods in Physics Research A* **449**, 172–207 (2000).

Hughes, E. W., and Voss, R. "Spin Structure Functions." *Annual Review of Nuclear & Particle Science* **49**, 303–339 (1999).

Primack, J. R.; Seckel, D.; and Sadoulet, B. "Detection of Cosmic Dark Matter." *Annual Review of Nuclear & Particle Science* **38**, 751–807 (1988).

Kevin McFarland

DETECTORS, PARTICLE

When a particle, such as a proton or an electron, goes through a gas, liquid, or solid, it can interact in various ways and leave evidence that it passed through or stopped. The interactions that take place can be recorded by building a particle detector that is designed to observe a specific process. The most common method involves using ionizations caused by a charged particle going through a material.

Of the many types of particle detectors, seven common detectors will be described. Some are in large-scale use. Some, such as the cloud chamber, were very important in the earlier history of the study of particle physics. An important distinction between them is whether they can be triggered to record an event by the particles passing through them or can only take a snapshot of what is in the detector at a specific time.

Cloud Chamber

The cloud chamber, invented by Charles T. R. Wilson in 1912, played a vital role in the early development of particle physics. This detector produces visible trajectories (tracks) when charged particles traverse the apparatus. It was used to discover the positron in 1933 and was used extensively in the early studies of muons, which are particles produced by cosmic rays interacting high in the Earth's atmosphere. A cloud chamber works by suddenly decompressing a vapor-filled container, cooling the vapor rapidly so that it becomes supersaturated. When a gas is supersaturated, any areas of nonuniformity can form regions of condensation that can be observed or photographed. These areas are referred to as nucleation centers.

Ionized gas molecules along the path of a charged particle will be the nucleation centers on which droplets will form, giving rise to the visible track seen in the cloud chamber. Since the ionization is velocity-dependent, a measurement of the amount of ionization can determine the velocity of the particle, allowing one to calculate its mass if the momentum of the particle is also known. The momenta of the particles can be measured by placing the chamber in a magnetic field and measuring the curvature of the photographed tracks. It is important that the vapor in the cloud chamber be dust-free, or the vapor will condense around the dust particles rather than the ionized gas regions.

A variation on the cloud chamber is known as the diffusion cloud chamber, which was originally suggested by Alexander Langsdorf Jr. in 1939. This type of detector remains sensitive continuously by establishing a temperature gradient so that vapor will diffuse from the heated top portion of the chamber to a cooled bottom portion.

The cloud chamber was replaced by the bubble chamber in particle physics because of the bubble chamber's higher density liquid media and its higher repetition rates. Although the diffusion chamber has reasonable cycle times, they have only a relatively thin sensitive area, and the interaction rate is significantly lower in a gas as compared to a liquid. The one advantage of the cloud chamber over the bubble chamber is the fact that the cloud chamber could be triggered by the particle that traverses the detector because of its long sensitivity time. Thus, while the cloud chamber is rarely used today, it was an important device in the early development of the field of particle physics and is still one of the simplest ways to visually verify the presence of cosmic rays.

Bubble Chamber

The bubble chamber played an important role in the development and understanding of particle physics in what is sometimes termed the "golden age" of the field, from the 1950s to the 1970s. During this time large accelerators were developed to produce beams of protons and electrons at ever-increasing energies and intensities. The invention of the bubble chamber is attributed to Donald A. Glaser, who in 1950 began work on a detector that would record rare events, referred to as pothooks, V particles, and strange particles, better than a cloud chamber. His concept was based on the behavior of a liquid heated past its boiling point. Glaser recognized that the passage of a charged particle through a superheated liquid would produce a trail of bubbles that could be photographed.

Glaser's first success was with diethyl ether in glass containers holding a few cubic centimeters of liquid. The expansion and recompression was done manually using a crank-and-piston. After the expansion cycle, bubbles formed in the superheated liquid along the path of ionization left by a charged particle. Following Glaser's initial work in the years of 1950 to 1952, other liquids, such as liquid hydrogen, were used.

After expansion, the bubbles expand to 0.1 millimeter in about 1 millisecond. They then can be photographed. By placing the chamber in a magnetic field, the momenta of the charged particles can be found by measuring the curvature of the particle trajectories. The particle interactions can be easily visualized and thus verify the existence of a new particle. The discovery of the omega particle was an example of a single photograph producing one of the most famous bubble chamber results.

The beauty of the hydrogen bubble chamber is that the liquid serves as both the interaction target and the detector. Thus, one can record a particle interacting with a proton as long as charged particles are eventually produced somewhere in the interaction. Rare interactions require searching through thousands of photographs. This required automatic scanners, which were developed from 1960 to 1980. Since the bubbles must be produced within nanoseconds of decompressing the liquid to make it superheated, a bubble chamber cannot be triggered by the interacting particle itself, which is a major disadvantage.

Drift Chamber

A drift chamber (DC) has many similarities to the Multiwire Proportional Counter (MWPC) that was invented by Georges Charpack in the 1960s. Both consist of thin wires strung through a gas volume and will produce an electrical pulse indicating that a particle passed through the detector, ionizing gas molecules. The primary difference is that the time the particles are detected is recorded in a DC, allowing a more precise determination of the location of the particle track.

The electrons produced will be attracted to positively charged anode wires, and the positive ions to the cathodes. The cathodes may be wires, conducting planes, or most commonly, a combination of both. The ionization electrons are used to produce the detector signal because they drift much faster in a gas than the positive ions. A DC must have an arrangement of electric conductors to produce a uniform electric field. The uniformity is important so that the electrons will have a constant drift time. The anode wires have a very small diameter, typically 10 to 20 μm, so that the electric field within 50 to 150 μm of the wire is very large, accelerating the electrons. This produces large secondary ionization near the anode wire, referred to as an avalanche.

The avalanche electrons are collected in less than a nanosecond and produce a very small electrical pulse. The main signal seen in a DC is an induced electrical pulse, via the Faraday effect, from the drifting of the avalanche ions toward the cathodes. The electrical pulse produced is amplified and sent to a discriminator. This simple electric circuit produces a digital output pulse only when the analog electronic pulse exceeds a set voltage.

The discriminator sends the digital pulse to a Time-to-Digital Converter (TDC). The TDC measures when the pulse arrives relative to some time signal generated by fast response detectors, such as scintillators. This provides a fast and accurate measurement of the drift time, giving how far from the anode wire the particle passed. Combining signals from several anode wires, it is possible to determine which side of the wire that the particle passed. With the ability to measure accurately times of less than a nanosecond, accuracies of less than 0.15 mm are routine for a DC.

Geiger-Mueller Tube

The Geiger-Mueller tube probably is the particle detector most familiar to the public. These small hand-held devices, developed in the 1920s, are commonly used to check for radiation. They are referred to as Geiger counters and use a Geiger-Mueller (G-M) tube to detect particles produced by radioactive materials. This gaseous detector depends on a particle passing through a gas and causing ionization. In order to detect the electrons produced, the G-M tube has a central wire, referred to as the anode, that is kept at a high positive voltage and is surrounded by a conductive cylinder kept at ground (zero voltage). Since opposite charges attract, the electrons produced will move toward the anode wire.

The anode wire is very small, which produces a high electric field that will accelerate the electrons where they can produce ionization. This process builds on itself to produce a large number of electrons and positive ions near the wire and is referred to as an avalanche. A G-M tube is designed for this avalanche to be large. The discharge will usually involve the entire length of the wire, limiting how fast a G-M tube can respond. As many as 10^{10} electrons, and the corresponding number of ions, can be produced. The electrical pulse produced is amplified and is used to count the number of ionizing particles passing through the Geiger counter.

Scintillation Counter

In addition to ionization processes that take place when a particle passes through a material, a particle can produce ultraviolet and visible light that is easily detected. One such process is referred to as luminescence or scintillation. The first scintillator was built by Sir William Crookes in 1903. It was made out of a ZnS screen, and a microscope was used to view it. The most common materials used for scintillation counters are clear plastics, organic liquids, and inorganic crystals. Plastics that normally do not scintillate can be made to do so by adding materials such as anthracene or tolulene. Examples of inorganic crystals are sodium iodide doped with thallium [NaI(Tl)], and pure crystals such as CsI. Each of these has different response times, light output, and other properties (and cost!)

that help the experimenter decide which material should be used.

Once the light is emitted, it must be detected. This is most commonly done by photomultiplier tubes (PMTs). These devices have a photosensitive surface that emits an electron when hit by light. This electron will be collected by the next stage of the PMT. Each PMT stage has an electrode with an applied voltage that will attract the electrons from the previous stage. Each electron absorbed will cause multiple electrons to be emitted. This can result in amplifications as high as 10^{12}, depending on the voltage and number of stages inside the PMT (usually referred to as anodes or dynodes, depending on the PMT type).

Since the 1990s, more choices have emerged as alternatives to the PMT. These include such devices as PIN diodes, avalanche diodes, and Visible Light Photon Counters (VLPCs). Each of these has different response times, amplification, timing accuracy, maximum rate, and cost. Scintillators see their greatest use in generating fast signals, or triggers, to indicate that a charged particle has passed through the detector and that the other systems should be recorded.

Materials that are excited so they emit light exhibit a wide range of response times. Plastic scintillators typically can respond in a few nanoseconds, and the resulting emission of light goes away in nanoseconds. Materials such as NaI(Tl) respond on the order of 100 nanoseconds and have secondary emission modes that cause the pulse to last on the order of a microsecond.

Silicon Vertex Detector

At high energies, interacting particles can produce large numbers of charged particles originating from the interaction point, that is, the vertex. Near the vertex, the particles are packed tightly together and require a high degree of precision to resolve them into their individual tracks. The silicon vertex detector is able to do this and is based on the ability to produce very narrow strips of silicon used in common devices such as computer processors. The silicon vertex detector is also useful when the experiment being done requires a very precise position

measurement of a particle's passage. Silicon detectors can be used inside a magnetic field, where many other types of detectors will not work because the ions and electrons are deflected or curled up into a small spiral path, leaving them undetected.

Semiconductors are made by taking silicon and adding very small numbers of impurities that either have an extra electron in their outer shell, producing an n-type material, or have only three electrons in their outer shell, resulting in a hole or a p-type material. When a charged particle passes through the semiconductor, the charges produced can be recorded. In addition to strips of silicon, it is possible to make pads that can be read out like a CCD camera. This has the advantage of being able to record at very high intensities but the disadvantage of having much more data and electronic readouts.

Spark Chamber

The spark chamber, like most of the other detectors discussed, makes use of the trail of ionized particles left behind by the passage of a charged particle through a gas. A charged particle typically will produce thirty to fifty ions per centimeter as it goes through a gas. The specific number depends on the energy of the incoming particle and on the gas used. With a path of ions produced in its wake, the passage of the charged particle is seen by suddenly applying a high voltage between metal electrodes and observing the resulting spark that follows the path of ions. Since the spark can be seen, the event easily can be recorded with a camera.

Usually, a spark chamber will have a stack of metal plates insulated from each other, and be placed inside a sealed container filled with a gas. An inert gas such as neon or argon is used to maximize the lifetime of the detector. As in the case of a cloud chamber, it is possible to trigger a spark chamber by a charged particle that goes through it, as opposed to the bubble chamber that cannot be triggered by the particles entering it. In the case of the spark chamber, the ions produced by the passing particle will remain along the path of the particle long enough for a high voltage to be applied. With the use of modern electronics, the trajectories of particles can be photographed, digitized, and analyzed on a computer rather than using a film camera. This allows many more events to be studied than visually inspecting every photograph.

See also: DETECTORS; DETECTORS, ASTROPHYSICAL; DETECTORS, COLLIDER; DETECTORS, FIXED-TARGET

Bibliography

Glaser, D. "The Bubble Chamber." *Scientific American* **192** (2), 46–50 (1955).

Leo, W. R. *Techniques for Nuclear and Particle Physics Experiments, A How-to Approach* (Springer-Verlag, Berlin, 1987).

Particle Data Group (PDG). <http://pdg.lbl.gov>.

Thornton, S., and Rex, A. *Modern Physics for Scientists and Engineers* (Saunders, Philadelphia, PA, 1993).

Tipler, P. A. *Modern Physics* (Worth Publishers, New York, 1969).

L. Donald Isenhower

DEVICES, ACCELERATING

The devices used to bring charged particles from rest to a desired kinetic energy are known as particle accelerators. All particle accelerators depend on a single physical principle, namely, the force exerted by an electric field on a charged particle. The simplest form of accelerator, represented electrically in Figure 1, consists of a high-voltage generator connected to two high-voltage plates (terminals) enclosed in a vacuum chamber. A positively charged particle at the positive high-voltage terminal will feel an electrical force proportional to the product of the

FIGURE 1

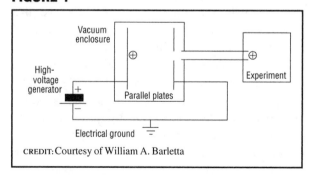

CREDIT: Courtesy of William A. Barletta

The DC Accelerator.

voltage V and the particle charge q. The force will accelerate the particle toward the grounded terminal, and the particle will emerge through the hole in the plate with a kinetic energy qV.

This simplest accelerating structure, the DC accelerator, is the basis for Van de Graaf and Cockroft-Walton accelerators, in which a source of positive ions sits at the high-voltage terminal. Such devices that can accelerate singly charged ions to energies of ~10 mega electron volts (MeV) and have seen widespread use in university nuclear physics laboratories. In large, modern accelerator complexes the DC accelerator (or some variant of it) often serves as the injector or source of particles feeding a much-higher-energy accelerating device.

The energy attainable with a DC accelerator is limited by the electrical breakdown of insulators at very high DC voltages. Furthermore, as the external components of the DC accelerator may also be at high voltage, stringent access controls are required for safety. To a limited degree, both difficulties can be postponed if the vacuum enclosure surrounds the entire accelerating structure, generator, and experiment. Even in such an evacuated enclosure, DC electric fields exceeding ~12 MV/m cannot be sustained without electrical breakdowns flashing across the surface of the structure.

For atoms capable of attaching an extra electron to form a negative ion, the designer can partially circumvent this limitation in ultimate energy. The negative ions will be attracted to the positive, high-voltage electrode where they pass through a very thin foil. The foil strips the excess electron plus an additional electron from the ion, thereby making a positive ion. The positive ion can now continue on its way toward the ground terminal. Using this tandem acceleration (see Figure 2), one obtains particles with energies of $2qV$.

Although the change of sign of the charge of the particle can be performed only once, the tandem "trick" could be applied many times if the sign of the voltage instead of the charge of the particle is changed. The accelerator designer therefore turns to time-varying electric fields.

The most common way to obtain energies higher than those obtainable with DC devices is to rely

FIGURE 2

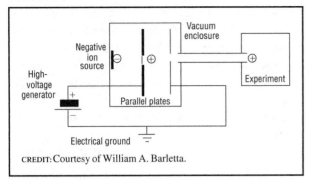

CREDIT: Courtesy of William A. Barletta.

The Tandem Accelerator.

on radio frequency (rf) voltage sources. Consider an arrangement of plates attached to an rf generator as shown in Figure 3. When the voltage at plate 1 is near its maximum value $+V$, positive ions are injected as shown. The particles will accelerate toward plate 2; if the distance between the plates is chosen appropriately, the particles will pass through the hole in the plate just as the voltage falls to zero. As the voltage decreases toward $-V$, the particles will continue to accelerate toward plate 3. If the distance between plates 2 and 3 is just right, the particle will reach the opening in plate 3 as the voltage once again passes through zero. As long as all the spacings are just right (depending on both the radio frequency and the peak voltage), the acceleration process can now continue indefinitely to ever higher energies, without having the peak electric field at any component exceed the limiting values set by electrical breakdown.

The key challenge in designing rf accelerators is to prevent the particles from "seeing" the electrical fields when they have the "wrong" decelerating values. Accelerator designers have used various techniques (and combinations of them) to accomplish this task. Techniques include varying the spacing of accelerating electrodes, varying the frequency of the accelerating fields as the particles gain energy, and providing metal structures to shield the particles from fields of the "wrong" sign. For example, in drift-tube accelerators, these structures (drift tubes) provide field-free regions through which the particles travel at constant velocity (drift) during the periods of decelerating voltage.

At radio frequencies exceeding tens of megahertz, the accelerating structure can exist in the form of resonant rf cavities (see Figure 4). Electromagnetic waves are launched into the structure from a high-quality, high-power source such as a klystron. Once the energy fills the entire structure, particles can be injected for acceleration. Many such structures arranged end to end form a linear rf accelerator (linac). Linacs for protons, which operate in the UHF broadcast band, can provide 3 to 10 MeV per meter of accelerating structure. Because electrons travel at a nearly constant velocity slightly less than the speed of light at energies >1 MeV (as compared to >1 GeV for protons), electron linacs usually employ higher frequencies than the broadcast band. The 2-mile-long Stanford linear accelerator (operating at ~3 GHz) now provides electrons at energies of 50 GeV at a rate (accelerating gradient) of 20 MeV/m. Even larger linacs may achieve energies of 1,000 GeV or more with gradients of 50 to 100 MeV/m.

The accelerating gradients achievable in linacs increase as the square root of the amount of rf power

FIGURE 3

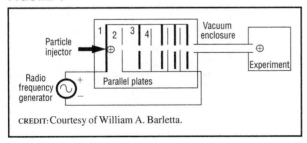

CREDIT: Courtesy of William A. Barletta.

A multicavity radio frequency accelerator.

FIGURE 4

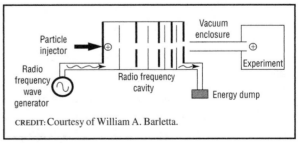

CREDIT: Courtesy of William A. Barletta.

Accelerator with radio frequency cavities.

FIGURE 5

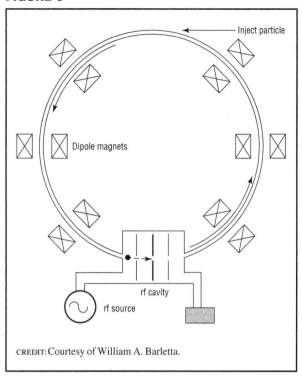

CREDIT: Courtesy of William A. Barletta.

Circular Accelerator.

fed into the cavities. But even with sufficient rf power available, the gradient is eventually limited by electrical breakdowns in the rf cavities. Although these breakdown limits may exceed the DC values by more than an order of magnitude, the desire of scientists for ever-higher-energy beams of particles (particularly ions) from accelerators of manageable sizes using the minimum electrical power necessary leads to a different tactic. If the particles are carried on circular orbits using strong magnets to bend their trajectories, they can pass through the same rf cavity many times. As long as they always enter the cavities when the fields have the correct sign, the particles will continue to accelerate to higher and higher energies, limited only by the strength of the magnets that bend them around through the evacuated beam tube. In circular accelerators such as the synchrotron, the field of the magnets is increased (ramped) so that the particles remain on the same circular orbit and stay in synchronism with the rf field during the acceleration process. See Figure 5 for a diagram of a circular accelerator.

By reusing the same rf cavities many times, the ion synchrotron attains an "effective" gradient that may be increased to hundreds of MeV/m. Electrons in circular accelerators are also limited because

FIGURE 6

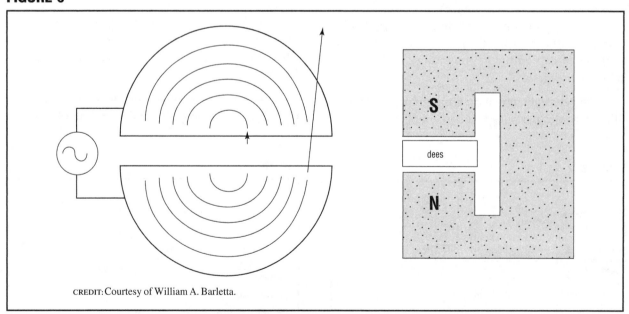

CREDIT: Courtesy of William A. Barletta.

The cyclotron: top view (left) and side view (right).

they emit copious amounts of electromagnetic radiation (synchrotron radiation). Thus, a power balance between the rf source feeding energy to the electrons and the energy lost through synchrotron radiation also limits the ultimate energy of electron synchrotrons.

The first circular accelerator was the cyclotron invented in the 1930s by Earnest Orlando Lawrence. It is a convenient way to realize the scheme of Figure 3 with a compact geometry. In the cyclotron (Figure 6), ions are injected at the center of two hollow, D-shaped electrodes (dees). The dees are connected to an rf generator and placed between the poles of a large electromagnet. Each time the particles cross the gap between the dees, they experience an electric field that can either accelerate or decelerate the particles. Inside the dees, particles are shielded from the electric fields and experience only the static magnetic field, which is perpendicular to their direction of motion and bends them along a circular orbit until they again reach the gap between the dees. The revolution frequency of the particles depends on the ratio of the magnetic field strength and the particle mass. Neglecting the increase of mass with energy due to relativity, this frequency is independent of the particle energy. If the rf frequency is chosen to be in resonance with the particle's revolution frequency, particles injected when the gap voltage is accelerating will continue to be accelerated as they spiral outward on ever-larger orbits.

Accelerator designers continue to seek new schemes, structures, and power sources to achieve higher rates of acceleration. In the twenty-first century, lasers may replace rf sources, and plasmas may supersede metal structures in the drive to achieve gradients of 10 GeV/m or more. The era of innovation in the design of accelerators is far from over.

See also: ACCELERATOR; ACCELERATORS, COLLIDING BEAMS: ELECTRON-POSITRON; ACCELERATORS, COLLIDING BEAMS: ELECTRON-PROTON; ACCELERATORS, COLLIDING BEAMS: HADRON; ACCELERATORS, EARLY; ACCELERATORS, FIXED-TARGET: ELECTRON; ACCELERATORS, FIXED-TARGET: PROTON

Bibliography

Livingston, M. L., and Blewett, J. P. *Particle Accelerators* (McGraw Hill, New York, 1962).

Wiedemann, H. *Particle Accelerator Physics* (Springer-Verlag, New York, 1995).

Wilson, R. R., and Littauer, R. *Accelerators: Machines of Nuclear Physics* (Anchor Books, Garden City, NY, 1960).

William A. Barletta

DILATION/MODULUS

See STRING THEORY

DIRAC, PAUL

Paul Adrien Maurice Dirac (1902–1984) became a member of the Royal Society in 1930, professor at Cambridge University in 1932, and the following year he shared with Erwin Schrödinger the Nobel Prize in Physics for his contributions to the theory of quantum mechanics. Throughout his life, he was occupied with fundamental questions of physics, which he approached in an original and often unorthodox way. Although not himself a particle physicist, his contributions to theoretical physics were of crucial importance to the science of elementary particles that emerged in the 1930s.

Pioneer of Quantum Mechanics

In 1921 Dirac graduated in electrical engineering at Bristol University, England, but was unable to find a job as an engineer. Two years later, he entered Cambridge University as a research student under Ralph Fowler. At that time he knew very little about the quantum theory of atoms, but under Fowler's guidance he quickly mastered the subject. After having become acquainted with Werner Heisenberg's new quantum mechanics, in 1925 he developed his own algebraic formulation of quantum mechanics.

During 1925–1927, Dirac further developed and refined the theory of quantum mechanics, which he presented in a general and logical way. In one of his important papers from 1926, he examined the quantum properties of two or more particles of the same kind, and thereby introduced the distinction between Fermi-Dirac statistics and Bose-Einstein statistics.

English physicist Paul Dirac (1902–1984) shared the 1933 Nobel Prize in Physics with Erwin Schrödinger for contributing to the theory of quantum mechanics. He developed a wave equation for the electron that introduced special relativity into Schrödinger's wave equation, satisfying the principle of relativity for quantum mechanics. CREDIT: COURTESY OF BETTMANN/CORBIS. REPRODUCED BY CORBIS CORPORATION.

(Unknown to Dirac, Fermi-Dirac statistics had been considered by Enrico Fermi half a year earlier.) Many years later, in 1945, Dirac invented the names fermion and boson.

Among the most important of Dirac's early works were his transformation theory and his quantum theory of electromagnetic radiation, both of which were published in 1927. The transformation theory was a general theory of quantum mechanics that comprised both particle and wave aspects, including Max Born's probabilistic interpretation of wave mechanics. His radiation theory treated the emission and absorption of electromagnetic radiation and introduced the notion of second quantization that became of fundamental significance in quantum field theory. Dirac's theory served as the foundation of quantum electro-

dynamics. It initiated a new field of research that would soon occupy center stage in theoretical physics, to which Dirac made important contributions.

The original quantum mechanics of Heisenberg, Schrödinger, and Dirac did not satisfy the principle of relativity and, for this reason, was not considered completely satisfactory. In 1926–1927, several physicists sought in vain to develop a relativistic wave equation that agreed with experiments. The problem was solved in January 1928, when Dirac found a relativistically invariant equation that had the same formal structure as the Schrödinger equation. Moreover, he found that the equation led to the correct spin magnetic moment of the electron. Dirac's equation of the relativistic and spinning electron took the physics community by surprise. Whereas the ordinary Schrödinger equation is of the second order in the space derivatives, the Dirac equation is of the first order. And the wave function does not include two components (one for each spin value), but four components. Although the equation made sense mathematically, it was harder to understand it physically.

Antiparticles

Dirac realized that, in a formal sense, his linear wave equation included solutions that corresponded to particles with negative energy. However, real particles must have positive energy, and he was therefore led to search for a physically valid interpretation of the solutions. In 1929 he came up with a remarkable solution to the puzzle, namely, that the proton is an electron in disguise. Dirac assumed an infinite, unobservable "sea" of negative-energy electrons and suggested that protons were vacancies or "holes" in the sea, hence they had positive energy. He wrote to Niels Bohr: "I think one can understand in this way why all things one actually observes in nature have a positive energy. One might also hope to be able to account for the dissymmetry between electrons and protons; one could regard the protons as the real particles and electrons as the holes in the distribution of protons of [negative] energy" (Kragh 1990, p. 91). He predicted that protons might annihilate with electrons and turn into gamma rays. In spite of Bohr's and most other physicists' rejection of the electron-proton theory, Dirac kept to it for more than a year. By 1930 matter was thought to consist of electrons and protons only, and Dirac felt greatly attracted to the idea—"the dream of philosophers," as he called it—because it promised a reduction to just one fundamental entity.

However, the dream of philosophers remained a dream. It could not account for the difference in mass between the two particles, and Dirac was forced to abandon what he considered a most beautiful idea. Yet he kept to his general picture and deftly turned the defeat into a victory by postulating in 1931 that the hole was an antielectron, "a new kind of particle, unknown to experimental physics, having the same mass and opposite charge to an electron" (Kragh 1990, p. 103). In this remarkable paper, the notion of antiparticles was introduced in quantum physics. However, although the hypothesis agreed with the principles of relativity and quantum theory, in 1931 it had no experimental support. Dirac further suggested that the proton would have its own antiparticle, a negatively charged antiproton, and a few years later he speculated that the symmetry between particles and antiparticles would probably imply the existence of antimatter made up purely by antielectrons and antiprotons.

Initially, the theory of antiparticles was met with skepticism. After all, the only elementary particles known in 1931 were the negative electron and the positive proton. It was only in 1932–1933, when the positron was discovered in the cosmic radiation (by Carl D. Anderson) and identified with Dirac's antielectron (by Patrick Blackett and Guiseppe Occhialini), that the status of the theory changed. In 1933, Dirac subjected the theory of the positron to a detailed mathematical analysis in which he introduced ideas (such as vacuum polarization) that were to become important in the later development of quantum field theory. The theory of antiparticles and the discovery of the positron were among the most important events in the creation of elementary particle physics.

Whereas Dirac's theory of the antielectron was vindicated by Anderson's discovery, it took longer to confirm his prediction of the antiproton. This particle played only a marginal role in the 1930s, and it was detected only in 1955, when it was produced in accelerator experiments. Forty-one years later, in 1996, the first detection of antihydrogen was reported.

Magnetic Monopoles

In his 1931 paper, Dirac predicted yet another elementary particle, the magnetic analogue of an electron, that is, a monopole or single magnetic pole. Such particles cannot exist according to classical electrodynamics, but Dirac showed that they were allowed by the laws of quantum mechanics. He believed that since there were no theoretical reasons barring the existence of monopoles, they would exist somewhere in nature.

The suggestion did not attract much attention, and, contrary to the antielectron, the particle remained elusive. Yet Dirac found the theory compelling, and in 1948 he developed it further. The theory of magnetic monopoles became widely known only in the 1970s, in particular after Paul B. Price in 1975 claimed to have detected a monopole in the cosmic radiation. Neither this claim nor a couple of later claims have been confirmed, and so it is believed that monopoles do not exist or are exceedingly rare. Toward the end of his life, Dirac was "inclined to believe . . . that monopoles do not exist" (Kragh 1990, p. 221). Yet, for him, the actual existence of magnetic monopoles was not what counted most. He considered it more important that the particle, real or not, could be described within the framework of quantum theory.

Attitude to Particle Physics

Although Dirac's 1931 paper was of crucial importance to elementary particle physics, he preferred to occupy himself with fundamental quantum theory rather than follow up the development that the new particles generated. About 1936, he came to doubt quantum field theory, including Fermi's theory of beta decay. Consequently he rejected the neutrino and sought to build up an alternative theory without strict energy conservation. (He soon abandoned the idea.) In 1936 Dirac generalized his wave equation to also cover particles with spin different from that of the electron; for, as he wrote, "It is desirable to have the equations ready for a possible future discovery of an elementary particle with a spin greater than a half" (Kragh 1990, p. 169).

Dirac continued to think of the electron as more elementary than other particles and was reluctant to engage in the physics of other elementary particles.

He thought that the electron first had to be understood on a classical basis, and the theory subsequently turned over into quantum theory. He developed this idea in the 1950s in the hope of reconstructing quantum electrodynamics. However, at that time he was estranged from the fast-growing development in particle physics, and his work was considered unorthodox.

See also: ANDERSON, CARL D.; FERMI, ENRICO; POSITRON, DISCOVERY OF; QUANTUM ELECTRODYNAMICS; QUANTUM FIELD THEORY; QUANTUM STATISTICS

Bibliography

Brown, L. M., and Hoddeson, L., eds. *The Birth of Particle Physics* (Cambridge University Press, Cambridge, England, 1990).

Dirac, P. A. M. *Directions in Physics* (Wiley, New York , 1978).

Hovis, R. C., and Kragh, H. "P. A. M. Dirac and the Beauty of Physics." *Scientific American* **268**, 104–12 (1993).

Kragh, H. *Dirac: A Scientific Biography* (Cambridge University Press, Cambridge, England, 1990).

Monti, D. "Dirac's Hole Model: From Proton to Positron." *Nuncius* **10**, 99–130 (1995).

Pais, A. *Inward Bound: Of Matter and Forces in the Physical World* (Clarendon Press, Oxford, 1986).

Pais, A., et al. *Paul Dirac: The Man and his Work* (Cambridge University Press, Cambridge, England, 1998).

Taylor, J. G., ed. *Tributes to Paul Dirac* (Adam Hilger, Bristol, England, 1987).

Helge Kragh

DOWN

See QUARKS

EIGHTFOLD WAY

The eightfold way is the term coined by Murray Gell-Mann in 1961 to describe a classification scheme for elementary particles that he and Yuval Ne'eman had devised. The name, adopted from the Eightfold Path of Buddhism, refers to the eight-member families to which many sets of particles belong.

In the 1950s Gell-Mann and Kazuo Nishijima invented a scheme to explain a "strange" feature of certain particles; they appeared to be easily produced in cosmic-ray and accelerator reactions but decayed slowly, as if something were hindering their decays. These particles were assumed to carry a property known as strangeness that would be preserved in production but could be changed in decays. Two examples of plots of electric charge Q (in units of the fundamental charge $|e|$) versus strangeness for certain particles, most of which were known in the late 1950s, are given in Figure 1.

Mesons include the π particles, known as pions, whose existence was proposed by Hideki Yukawa in 1935 to explain the strong nuclear force, and the K particles (also known as kaons), discovered in cosmic radiation in the 1940s. Pions and kaons weigh about one-seventh and one-half as much as protons, respectively. Baryons (the prefix *bary-* is Greek for *heavy*)

FIGURE 1

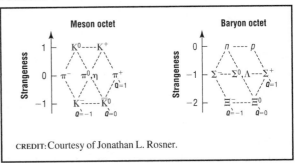

CREDIT: Courtesy of Jonathan L. Rosner.

Plots of electric charge Q versus strangeness for commonly known particles.

include the proton p, the neutron n, and heavier relatives Λ (lambda), Σ (sigma), and Ξ (xi), collectively known as hyperons and discovered in the 1940s and 1950s. The rationale for these families was sought through symmetries of the strong interactions.

According to the Gell-Mann–Nishijima scheme, reactions in which these particles are produced must have equal total strangeness on each side. For example, K^0 and Λ can be produced by the reaction

$$\pi^-(S = 0) + p(S = 0) \to K^0(S = 1) + \Lambda(S = -1).$$

This scheme thus explained another curious feature of the "strange particles": they never appeared to be produced singly in reactions caused by protons, neutrons, and π mesons.

In the 1930s Werner Heisenberg and others had recognized that the similarities of the proton and neutron with respect to their nuclear interactions and masses could be described by a quantity known as isotopic spin. This quantity, called isospin for short, is analogous to ordinary spin with the proton's isospin pointing "up" and the neutron's pointing "down." Mathematically, isospin is described by a symmetry group that is a set of transformations that leaves interactions unchanged, known as SU(2). The 2 refers to the proton and neutron.

Families whose members are related to one another by SU(2) transformations can have any number of members, including the two-member family to which the proton and neutron belong. Collectively, p and n are known as nucleons and denoted by the symbol N. Isospin predicts that certain sets of particles with different charges (e.g., K or Σ) should have similar masses and strong interactions, as is observed.

In 1956 Shoichi Sakata proposed that mesons were composed of the proton p, the neutron n, the lambda Λ, and corresponding antiparticles, with binding forces so large as to overcome most of their masses. Thus, for example, the K^+ would be $p\bar{\Lambda}$. (The bar over a symbol denotes its antiparticle; the electric charges and strangeness of antiparticles are opposite to those of the corresponding particles.) The remaining known baryons (the Σ and Ξ) had to be accounted for in more complicated ways. The Sakata model had the symmetry known as SU(3), where 3 referred to p, n, and Λ.

Gell-Mann and Ne'eman recognized that if electric charge were to be part of the SU(3) description, particles whose electric charges were integer multiples of $|e|$ could belong only to certain families. The simplest of these contained one, eight, and ten members. Other families, such as those containing three and six members, would have fractionally charged members, and fractional charges had never been seen in nature. Both the mesons and baryons mentioned above would then have to belong to eight-member families. The baryons fit such a family exactly, leading Gell-Mann to call his scheme the eightfold way. In addition to the known K and π mesons

shown in the meson octet of Figure 1, there would have to be an eighth meson, which was neutral and had zero strangeness. This particle, now called the η (eta), was discovered in 1961 by Pevsner et al.

A consequence of the eightfold way for describing mesons and baryons was that their masses M could be related to one another by formulae proposed by Gell-Mann and Susumu Okubo in 1962:

$$\text{Mesons: } 4M(K) = M(\pi) + 3M(\eta)$$
$$\text{Baryons: } 2[M(N) + M(\Xi)] = M(\Sigma) + 3M(\Lambda).$$

These formulae, particularly the one for baryons, were obeyed quite well. More evidence for SU(3) soon materialized as the result of another experimental discovery.

Certain baryons known as Δ (delta), Σ^* (sigma-star), and Ξ^* (xi-star) appeared to fit into a ten-member family (a decuplet, which would be completed by a not yet observed particle known as the Ω^- [omega-minus]). The mass of the Ω^- could be anticipated within a few percent because the Gell-Mann–Okubo mass formula for these particles predicted

$$M(\Omega^-) - M(\Xi^*) = M(\Xi^*) - M(\Sigma^*)$$
$$= M(\Sigma^*) - M(\Delta).$$

An experiment at Brookhaven National Laboratory (BNL), discussed by Barnes et al., detected this particle with the predicted mass through a decay that left no doubt as to its nature. Figure 2 shows its place in the baryon decuplet.

FIGURE 2

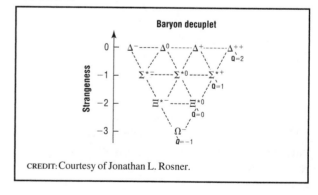

CREDIT: Courtesy of Jonathan L. Rosner.

Ω^- particle's place in the baryon decuplet.

An early application of the eightfold way, building on suggestions by Gell-Mann and Maurice Levy in 1960 and by Gell-Mann again in 1962, was made by Nicola Cabibbo in 1963. Applying the concept to certain decays of baryons, Cabibbo showed that SU(3) symmetry could be used to describe not only the existence and masses of particles but also their interactions.

Underlying the success of the eightfold way and the symmetry group SU(3) is the existence of fundamental subunits of matter, called quarks by Gell-Mann in 1964 and aces by their coinventor George Zweig in 1964. These objects can belong to a family of three fractionally charged members u (up), d (down), and s (strange), as shown in Figure 3.

The fact that fractionally charged objects have not been seen in nature requires quarks to combine with one another in such a way as to produced only integrally charged particles. This is one successful prediction of the theory of the strong interactions, quantum chromodynamics (QCD). Baryons are made of three quarks, whereas mesons are made of a quark and an antiquark (with reversed charge and strangeness). For example, the Δ^{++} is made of uuu; the Δ^- is made of ddd; the Ω^- is made of sss; and the K^+ is made of $u\bar{s}$.

The SU(3) symmetry described here refers to the flavor of quarks (u, d, s). A separate SU(3), associated with quantum chromodynamic degrees of freedom, describes the *colors* of quarks. Each flavor of quark can exist in three colors. Other flavors of quarks—charm (c), bottom (b), and top (t)—were discovered subsequently. They are much heavier than u, d, and s and so do not fit easily into a generalization SU(n) of SU(3) with $n > 3$. The approximate SU(3) symmetry of particles containing u, d, and s quarks remains a useful guide to properties of the strong interactions.

See also: FLAVOR SYMMETRY; PARTICLE PHYSICS, ELEMENTARY; QUARKS, DISCOVERY OF; SU(3)

Bibliography

Barnes, V. E. "Observation of a Hyperon with Strangeness Minus Three." *Physical Review Letters* **12**, 204–206 (1964).

Cabibbo, N. "Unitary Symmetry and Leptonic Decays." *Physical Review Letters* **10**, 531–533 (1963).

Gell-Mann, M. "The Eightfold Way: A Theory of Strong Interaction Symmetry." California Institute of Technology Report. CTSL-20 (1961).

Gell-Mann, M. "Symmetries of Baryons and Mesons." *Physical Review* **125**, 1067–1084 (1962).

Gell-Mann, M. "A Schematic Model of Baryons and Mesons." *Physical Review Letters* **8**, 214–215 (1964).

Gell-Mann, M., and Levy, M. "The Axial Vector Current in Beta Decay." *Nuovo Cimento* **16**, 705 (1960).

Gell-Mann, M., and Ne'eman, Y. *The Eightfold Way* (W. A. Benjamin, New York, 1964).

Ne'eman, Y. "Derivation of Strong Interactions from a Gauge Invariance." *Nuclear Physics* **26**, 222–229 (1961).

Okubo, S. "Note on Unitary Symmetry in Strong Interactions." *Progress of Theoretical Physics* (Kyoto) **27**, 949–966 (1962).

Pevsner, A., et al. "Evidence for a Three Pion Resonance Near 550 MeV." *Physical Review Letters* **7**, 421–423 (1961).

Sakata, S. "On a Composite Model for the New Particles." *Progress of Theoretical Physics* (Kyoto) **16**, 686–688 (1956).

Zweig, G. "An SU(3) Model for Strong Interaction Symmetry and its Breaking" in *Developments in the Quark Theory of Hadrons*, Vol. 1, edited by D. B. Lichtenberg and S. P. Rosen (Hadronic Press, Nonantum, MA, 1980), pp. 22–101.

Jonathan L. Rosner

FIGURE 3

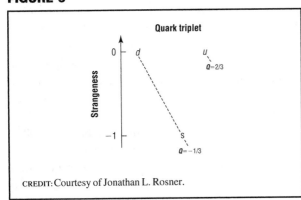

CREDIT: Courtesy of Jonathan L. Rosner.

Quark family of three fractionally charged members.

EINSTEIN, ALBERT

Albert Einstein was born on March 14, 1879, into a Jewish family in the Swabian town of Ulm. He only started to talk when he was three years old, but it is a myth that he was a poor student. What did evidence itself from the start was the single-mindedness that

German physicist Albert Einstein (1879–1955) received the Nobel Prize in Physics in 1921 for his discovery of the law of the photoelectric effect. CREDIT: COURTESY OF UNDERWOOD & UNDERWOOD/CORBIS. REPRODUCED BY CORBIS CORPORATION.

science. In high school, the Gymnasium, he did extremely well in physics and mathematics but was undistinguished in subjects that were of no interest to him.

In 1894 his father's business in Munich failed, and the family went to Italy, leaving Einstein behind to complete his high-school education. Einstein, however, who had little tolerance for the rigid discipline of the Gymnasium, soon dropped out of school and joined his family in Milan. This way, he also avoided being drafted into the German military. After completing his secondary education in Switzerland, he was eventually admitted to the Federal Polytechnic, now the ETH, in Zurich. There he met his first wife Mileva Maric (1875–1948). Einstein frequently skipped class, relying instead on the notes of his classmate Marcel Grossmann (1878–1936), and spent most of his time studying by himself more recent physics than was covered in the university curriculum. He thereby alienated some of his teachers, which was a factor in his failure to find an academic position upon graduation in 1900. In 1902 he finally got a job as a patent examiner third class in Bern. He had become a Swiss national the year before. He and Mileva married in 1903, over the strong objections of his parents. Before they were married, Albert and Mileva had a daughter, Lieserl, who was given up for adoption. No trace of her remains. They had two more children: Hans Albert (1904–1973) and Eduard (1910–1965).

After establishing himself as a serious scholar with several papers on statistical mechanics in *Annalen der Physik,* the leading physics journal of the day, the young patent clerk submitted four ground-breaking papers to this same journal in 1905: one proposing the light-quantum hypothesis, one on Brownian motion that provided crucial evidence for the reality of atoms, one on electrodynamics in moving bodies introducing the special theory of relativity, and a final one on an important consequence of this theory, the inertia of energy or $E = mc^2$. Einstein's approach in these papers was to work with what he later called "theories of principle." He started from generalizations supported by a wealth of empirical evidence, even if such generalizations appeared to be contradictory. With uncompromising logic, he then derived the consequences of these generalizations, in the

became an important characteristic of his later scientific work. He only applied himself when the subject held a strong interest for him. Science was such a preoccupation from early on in his life. When he was only one year old, the Einstein family relocated to Munich, where his father and an uncle went into business together. In his father's factory, the young Einstein marveled at dynamos and other machinery. Two other events appear to have been crucial in awakening his interest in science. At the age of five, he was deeply impressed when his father showed him a compass. At the age of eleven, he discovered what he later called his "holy geometry book." Popular science books showed him that the Bible could not literally be true, and his early religious fervor—which he had developed in spite of his parents who were not practicing Jews—gave way to an enthusiasm for process exposing various preconceived notions as prejudices that had to be cast aside (such as common-sense ideas about simultaneity). Proceeding in a similar vein, Einstein established in 1909 that any satisfactory theory of light must combine aspects of both a wave and a particle theory. This was the very first statement of wave-particle duality.

Einstein presented this result in his first invited lecture as a regular member of the academic community. Earlier in 1909 he had become an associate professor at the University of Zurich. In 1911 he continued his ascent up the academic ladder, becoming a full professor in Prague. A year later he was back in Zurich, this time as a full professor at his alma mater, the ETH. Another year later he was recruited by Max Planck (1858–1947) and Walther Nernst (1864–1941) to come to Berlin where in early 1914 he became a salaried fellow of the Prussian Academy, a position he would hold until 1933, when the Nazi rise to power forced him to leave Germany permanently. The move to Berlin was the final straw that broke his marriage. Mileva and the couple's sons returned to Zurich shortly afterward, and Einstein resumed an affair begun in 1912 with his cousin Elsa Einstein-Löwenthal (1876–1936), with whom he would enter into a marriage of convenience in 1919 shortly after his divorce from Mileva was finalized.

Although Planck and others had recognized the importance of special relativity early on, the growing recognition of Einstein's work by the physics community came mainly from his work on the quantum theory of matter. His quantum theory of light met only with skepticism and strong opposition until the discovery of the Compton effect in 1923. Even the verification by Robert A. Millikan in 1915 of the formula for the photoelectric effect did nothing to change this. When Einstein received the Nobel Prize in Physics in 1921, it was for the formula, not for the light-quantum hypothesis from which the formula had been derived. Einstein's related work on the specific heat of solids at low temperatures was received much better, especially by Walther Nernst who made the fledgling quantum theory the topic of the first Solvay Congress in 1911. At this meeting, Einstein established himself as the leading thinker in this field.

Meanwhile, Einstein also had taken the first steps toward a generalization of special relativity that would at the same time be a new theory of gravity, the general theory of relativity. While special relativity was the work of many, general relativity was essentially the work of Einstein alone. In 1907, still working at the patent office, Einstein had what he later described as "the happiest thought of my life." He realized that the equality of inertial and gravitational mass indicated that there had to be an intimate connection between inertia and gravity. The equivalence principle, as this connection came to be called, was of great heuristic value in finding the new theory of gravity, which was completed in late November 1915, after a final month of intense work on the problem in war-torn Berlin. Within a year, however, Einstein realized that the theory as it stood still contained remnants of absolute space and absolute motion, two notions he had hoped to banish from physics altogether. The problem was that the theory still needed boundary conditions. In 1917, during the course of a lengthy correspondence with the Dutch astronomer Willem de Sitter, Einstein introduced a static spatially closed model of the universe, thereby obviating the need for boundary conditions. The model was static, and this required the addition of the so-called cosmological constant to his theory. Einstein now believed that this theory satisfied what he dubbed "Mach's principle": the geometrical structure of space-time is fully determined by its matter content. De Sitter soon showed that this is not true.

Einstein thereupon lost his enthusiasm for Mach's principle, a position reinforced by the discovery first of expanding models of the universe and then of empirical evidence that the universe is, in fact, expanding. Einstein's paper nonetheless launched the field of relativistic cosmology. In these early years of general relativity, he likewise did pioneering work on gravitational waves, gravitational lensing, and singularities.

By 1920 Einstein had redirected his effort to finding a classical field theory along the lines of general relativity unifying the fabric of space-time (responsible for the effects of gravity) and the electromagnetic field. Rather than reducing the structure of space-time to matter, Einstein now hoped to show how matter emerges from this unified field. He would pursue this new line of research until his death in 1955. His approach in this later period is markedly different from the approach he took in his early years. Rather than building on secure empirical foundation, he came to rely more and more on purely mathematical speculation.

Einstein's hope was that a unified field theory would bring the answer to all the riddles of quantum theory. Before the advent of quantum mechanics in the mid-1920s, Einstein made at least two more fundamental contributions to quantum theory: his radiation theory of 1917, which played an important role in the genesis of quantum mechanics and forms the basis for the laser, and his 1925 work on Bose-Einstein statistics. After that, Einstein's role became more and more that of a critic of the emerging Copenhagen interpretation of quantum mechanics. His most famous contribution to the discussion of the foundations of quantum mechanics is the 1935 paper coauthored with Boris Podolsky and Nathan Rosen known as the EPR paper.

By this time, Einstein had lost touch with the mainstream of physics and shifted his attention more and more to a mathematical audience. He did not contribute, for instance, to the exciting developments in the 1930s in the area of nuclear physics. It is very telling in this connection that Einstein always stuck to unifying gravity and electromagnetism and never included the nuclear forces that ever more clearly emerged as new types of interactions. For that reason alone, Einstein could never have played a substantial role in the development of nuclear weapons. Much has been made of his letter to President Franklin D. Roosevelt of 1939 warning of the possibility of a German atomic bomb. Work on an American bomb, however, was not begun in earnest until the attack on Pearl Harbor over two years later.

The letter to Roosevelt does provide a good illustration of Einstein's standing in the scientific community and in society at large. When it was announced in London in 1919 that measurements of the bending of starlight grazing the Sun during a solar eclipse confirmed the predictions of general relativity, Einstein had become an overnight sensation, the world's first and greatest scientific superstar, whose opinions were sought on all sorts of scientific, political, and moral issues. Einstein, whose political involvement had been tentative up to that point, used his celebrity over the years to support various causes dear to his heart, such as pacifism, Zionism, and disarmament.

The downside of Einstein's fame was that he became a natural target for anti-Semitic forces in German society in the early 1920s. His theories were denounced as "Jewish physics," and there were even rumors that his name appeared on lists of people to be assassinated by ultraright-wing elements. Einstein nonetheless stayed in Berlin, mainly out of loyalty to Planck and others in the Berlin physics community. Einstein also felt solidarity with the German people in the face of the harsh conditions in the aftermath of World War I, further exacerbated by the terms of the Versailles Treaty. Einstein used his position to help Germany regain access to the international scientific community after World War I. After the Nazis came to power in 1933, however, he was forced to leave Germany. He accepted a full-time position at the newly established Institute for Advanced Study in Princeton. He became an American citizen in 1940, although retaining his Swiss citizenship. He never set foot on German soil again.

In 1948 Einstein, who had abdominal problems ever since his period of intensive work on general relativity in the mid-1910s, was diagnosed with an aneurysm of the major abdominal aorta. In April 1955 the aneurysm ruptured and Einstein died. The body was cremated but not before both his brain and his eyes were removed during an unauthorized autopsy. On the day that he died, Einstein had asked his secretary for his latest notes on an unfinished project, finding a classical unified field theory for gravity and electromagnetism.

See also: COSMOLOGICAL CONSTANT AND DARK ENERGY; RELATIVITY

Bibliography

Fölsing, A. *Albert Einstein. A Biography* (Viking, New York, 1997).

Overbye, D. *Einstein in Love. A Scientific Romance* (Viking, New York, 2000).

Pais, A. *Subtle Is the Lord: The Science and the Life of Albert Einstein* (Oxford University Press, New York, 1982).

Schilpp, P. A., ed. *Albert Einstein: Philosopher-Scientist* (Library of Living Philosophers, Evanston, IL, 1949).

Stachel, J., et al., eds. *The Collected Papers of Albert Einstein,* Vols. 1–8 (Princeton University Press, Princeton, NJ, 1987–2002).

Michel Janssen

ELECTROMAGNETIC INTERACTION

See BASIC INTERACTIONS AND FUNDAMENTAL FORCES

ELECTRON

See LEPTON

ELECTRON, DISCOVERY OF

By 1890 many chemists and physicists believed that atoms must have some sort of structure, a belief stemming from a conviction that nature was essentially simple, and that atoms of the many different elements might represent arrangements of a more fundamental unit. This was often taken to be the hydrogen atom, an idea first suggested by William Prout in 1816. Chemists, however, based their atomic ideas on evidence from spectroscopy, while physicists were trying to understand kinetic theory.

In 1860 Gustov Kirchoff and Robert Bunsen showed that each element emitted its own characteristic line spectrum if heated in a flame. Spectral analysis led to a spate of discoveries of new elements, making the variety even more bewildering. The observation that hydrogen and helium were the only two elements present in the Sun prompted both Norman Lockyer and William Crookes to suggest that atoms somehow evolved from these two elements.

Meanwhile physicists were following up the success of the kinetic theory of gases which proposed that macroscopic effects such as heat could be explained by the motion of atoms and molecules. Some idea of atomic structure was needed to establish the types of vibration possible. However, models devised for the kinetic theory disagreed with the results of spectral analysis.

James Clerk Maxwell's electromagnetic theory, published in 1862–1873, and Michael Faraday's work on electrolysis pointed toward a possible solution.

Maxwell had suggested that the vibrations of light were not mechanical, as previously thought, but electromagnetic, while Faraday's laws of electrolysis implied that electricity existed in discrete units with a charge equal to that on the hydrogen ion. Several physicists, notably F. Richarz and H. Ebert, suggested that if the atom contained discrete charge units, then their oscillations might explain the emission of line spectra. In 1891 George Johnstone Stoney named these units of charge electrons and attempted to find out how big they were by reconciling the spectroscopic and kinetic data.

Simultaneously, the theorists Henrik Anton Lorentz and Joseph Larmor were independently trying to accommodate discrete charges within Maxwell's theory that were expressed in terms of a continuous and all-pervading electromagnetic medium known as the ether. They suggested that charges could be modeled by vortices or strain centers in the ether, and Larmor adopted the term electron to describe his charge. In 1896, Pieter Zeeman's discovery of magnetic splitting of spectral lines upheld these ideas for it could be predicted from Lorentz's theory and allowed, for the first time, the size of the vibrating spectroscopic charges to be calculated: they proved to have a mass to charge ratio about 2,000 times smaller than the hydrogen atom, suggesting that the vibration was that of the electron itself. However, there was still no way of manipulating the electrons or measuring them directly.

For Lorentz and Larmor the electron was embedded within the atom but played no role in determining its chemical nature. This view was to change in the years 1895–1905 following the discovery of X rays, radioactivity, and Joseph John Thomson's investigations of cathode rays.

Cathode rays were discovered by Julius Plücker in 1858. They are found when an electric potential is applied across a gas at low pressure and are detected by a fluorescent glow where they hit the glass at the end of the discharge tube. They were known to travel in straight lines and to be deflected by a magnetic field, but by about 1880 the initial interest had died down and most physicists did not consider cathode rays very important. They seemed peripheral to major theoretical concerns and were difficult to experiment on, requiring at least half a day of hand pumping to evacuate the tube, which then frequently broke due to the poor composition of glass available.

However, in 1895, Wilhelm Roentgen discovered X rays, which are emitted when cathode rays hit a target. The discovery caused a furor and the understanding of gaseous discharge and of the behavior of discharge tubes advanced rapidly; it also revived interest in the nature of the cathode rays that caused X rays. Speedy recognition that cathode rays were negatively charged particles about 2,000 times smaller than atoms depended primarily on the work of four men: Philip Lenard, who had followed up Heinrich Hertz's discovery of 1892 that cathode rays could pass out of the discharge tube through a thin foil of metal and showed that cathode rays traveled much further than expected through gases and that their absorption depended on the molecular weight of the gas; Emil Wiechert; Walter Kaufmann; and J. J. Thomson, who measured the charge to mass ratio (e/m) of the rays by various means. While Wiechert's experiments predated Thomson's by a few months and Kaufmann's were often regarded as more reliable, Thomson went the furthest theoretically.

Thomson was unique among British physicists in his concern to explain the chemical properties of atoms. In 1882 he had shown how the then-popular theory that atoms consisted of vortex rings in the ether could account for the periodic table. He spent thirteen years experimenting on gaseous discharge (but not specifically on cathode rays) guided by his own concept of a discrete electric charge modeled by the end of a vortex tube in the ether. In 1896 he established his theory of discharge by ionization on a firm mathematical footing. He then demonstrated that the magnetic deflection of the cathode rays was the same, regardless both of the cathode material and of the gas in the discharge tube, and refined Jean Perrin's experiment of 1895 that showed that the rays carried with them a negative electric charge. Then, guided by the rays' uniform magnetic deflection and by Lenard's absorption data, which he could explain on the assumption that the cathode rays were interacting with the individual components of atoms, and prompted by his previous speculations about discrete charges and structured atoms, Thomson proposed that cathode rays were subatomic, negatively charged particles from which all atoms were built up. He announced his ideas at the Royal Institution on April 30, 1897.

Thomson's proposals suggested something radically new. His particles were not just charges embedded within atoms. They provided the essential mass of the atom and its chemical constitution. To mark the distinction from electrons he called the particles corpuscles. He supported his hypothesis by measuring the charge to mass ratio of the corpuscles. Initially, he did this by comparing the magnetic deflection of the rays with their heating effect when they hit a target. Later in 1897, after he had succeeded in deflecting the rays electrically, he devised his classic e/m experiment that compared the electric and magnetic deflections of the rays. Both sets of results suggested that the nature of the corpuscles was independent of the material in the discharge tube and that they were about 1,000 times smaller than the hydrogen atom. Thomson confirmed the small size of corpuscles in 1899 when he succeeded in measuring their charge independently of their mass using the particles released in the photoelectric effect.

Thomson's suggestion was difficult to accept: to his contemporaries it sounded like alchemy because an atom emitting corpuscles seemed as though it should change its chemical nature. They preferred George Fitzgerald's alternative proposal that cathode rays were free Larmor-type electrons. Thus the name electron became firmly attached to the particles several years before the realization that the particles were indeed essential constituents of the chemical atom. This realization did not come until work on radioactivity demonstrated the identity of beta and cathode ray particles and showed that atoms could and did split up and change their chemical nature.

Fitzgerald's suggestion ensured the early importance of Thomson's cathode ray particles by tying them into the attempt by Lorentz, Henri Poincaré, Kaufmann, and others to formulate an entirely electromagnetic theory of matter; an attempt which fostered, but eventually seemed incompatible with, Einstein's relativity theory.

Meanwhile, Thomson was incorporating corpuscles into his widely applicable discharge theory and his atomic theory. The mass of Thomson's atom was entirely due to the corpuscles (hence there must be thousands of them). He arranged them in rings within a uniform sphere of positive electrification and set them spinning to ensure stability. Radioactive decay became an inevitable consequence, for rotating corpuscles emit energy and slow down; the rate of emission is less if there are lots of corpuscles in the ring, and Thomson's model demanded thousands; yet, the time still comes when the rings have slowed down, become unstable, and the atom flies apart. Prior to decay, the arrangement of corpuscles in rings provided an explanation for the periodic table, valence, and ionic bonding.

Thomson's atom theory proved untenable in 1906 when he showed how to use scattering data to calculate the number of corpuscles in the atom and found that there could be only hundreds, not the thousands, necessary to ensure reasonable stability. But, it was his concept, as interpreted by Ernest Rutherford and Niels Bohr, that ensured that the electron became fundamental to new theories of atoms, chemical bonding, and materials and, together with relativity and quantum mechanics, to ideas of the nature of matter.

See also: CULTURE AND PARTICLE PHYSICS; INFLUENCE ON SCIENCE; INTERNATIONAL NATURE OF PARTICLE PHYSICS; PHILOSOPHY AND PARTICLE PHYSICS

Bibliography

Buchwald, J. *From Maxwell to Microphysics* (Chicago University Press, Chicago, 1985).

Dahl, P. F. *Flash of the Cathode Rays* (Institute of Physics, Bristol, 1997).

Davies, E., and Falconer, I. *J.J. Thomson and the Discovery of the Electron* (Taylor and Francis, London, 1997).

Falconer, I. "From Corpuscles to Electrons" in *Histories of the Electron*, edited by J. Buchwald and A. Warwick (MIT Press, Cambridge, MA, 2001).

Isobel Falconer

ELECTROWEAK INTERACTION

See BASIC INTERACTIONS AND FUNDAMENTAL FORCES

ELECTROWEAK PHASE TRANSITION

Under normal conditions, the electromagnetic force and the weak nuclear force have very different ranges. The electromagnetic force can be important over large distances, whereas the weak nuclear force is only effective on scales smaller than roughly 10^{-18} meters. This vast quantitative difference, between the two forces, which are fundamentally related, is due to electroweak symmetry breaking.

A standard physical example of symmetry breaking is a single magnetic domain of an isolated, idealized ferromagnet. The magnetic moments of the atoms in a ferromagnetic prefer to align with each other, producing a net magnetization of the domain. Nothing about the interactions between the magnetic moments prefers one direction of magnetization over another; yet, the atoms in a piece of ferromagnet randomly choose a direction in which to align. A person living inside the ferromagnet would think that one direction is different from the others; the alignment of the atoms has broken the symmetry between the directions. If you heat a ferromagnet, however, the magnetic moments of the hot atoms start to jiggle. The individual moments are no longer perfectly aligned, and so the net magnetization of the ferromagnet is less. As one increases the temperature, the disorder increases and the net magnetization drops further. There is a temperature (the Curie temperature) at which the individual moments jiggle so much that they become completely disordered and the net magnetization vanishes. The symmetry of directions is then restored—there is no longer a magnetization to pick out a direction. The cold, symmetry-broken (magnetized) behavior of the system, and the hot, symmetry-restored (unmagnetized) behavior of the system, are referred to as phases. The transition between them is an example of a phase transition.

Something analogous is predicted to occur in electroweak theory. If the temperature is high enough, the underlying symmetries relating the electromagnetic and weak forces are predicted to be restored. All the consequences of electroweak symmetry breaking, including the vast difference in ranges, disappear. The minimum temperature required for this restoration, and even whether there is a precise dividing temperature between the phases at all, depends on details of electroweak symmetry breaking (also known as the Higgs sector), not yet measured experimentally in 2002. However, the temperature is expected to be on the order 10^{15} K. Our universe was at that temperature approximately one billionth of a second after the beginning of the Big Bang.

One reason for interest in the electroweak phase transition is that it plays a pivotal role in one set of possible scenarios, known as electroweak baryogenesis, for explaining why there is vastly more matter than antimatter in the universe today (specifically, more baryons than antibaryons or, more fundamentally, more quarks than antiquarks). By theoretically tracing backward the history of the universe, one finds that, a millionth of a second after the Big Bang, there must have been almost as many antiquarks as quarks. But there was a slight imbalance: roughly, for every 30 million antiquarks, there were 30 million and one quarks. As the universe cooled, the quarks and antiquarks paired up and annihilated, leaving just that one part in 30 million excess of quarks to make up (with electrons) us, the Earth, and the stars. The goal for scenarios of baryogenesis is to find an explanation of the tiny early asymmetry between the numbers of quarks and antiquarks. Inspired by early investigations of the Soviet physicist Andrei Sakharov (1921–1989), three requirements have been distilled for any such explanation. First, it must be possible for the difference between the number of quarks and antiquarks to change with time, since otherwise the size of that difference is just an initial condition on the universe—an input into our theories of nature instead of an output. Such a change in the difference has never been observed experimentally, but is theoretically predicted to occur at temperatures above the electroweak phase transition. Second, the universe must be significantly out of equilibrium when the processes which effect such changes slow down and stop. (If allowed to change and equilibrate, the difference relaxes to zero.) Depending on the details of electroweak symmetry breaking, the electroweak phase transition may have been violent, providing this nonequilibrium. The hot symmetry-restored phase of electroweak theory might have experienced significant supercooling,

ending with the formation of symmetry-broken-phase bubbles that violently expanded outward to fill the universe. Finally, there must be some difference between the behavior of matter and antimatter (known as violation of C and CP symmetry), since otherwise the rate for making a few extra quarks would equal the rate for making a few extra antiquarks, and no net asymmetry would develop. Electroweak interactions possess this difference, though whether strongly enough depends on details of electroweak symmetry breaking.

See also: COSMOLOGY; ELECTROWEAK SYMMETRY BREAKING; PHASE TRANSITIONS

Bibliography

Cohen, A.; Kaplan, D.; and Nelson, A. "Progress in Electroweak Baryogenesis." *Annual Reviews of Nuclear and Particle Science* **43**, 27–70 (1993).

Kolb, E. W., and Turner, M. S. *The Early Universe* (Addison-Wesley, Reading, MA, 1994).

Trodden, M. "Electroweak Baryogenesis." *Reviews of Modern Physics* **71**, 1463–1500 (1999).

Peter Arnold

ELECTROWEAK SYMMETRY BREAKING

The electroweak theory proposed by Sheldon Glashow, Steven Weinberg, and Abdus Salam provides a unified description of the electromagnetic and weak forces. At first glance, such a unification hardly seems possible since these two forces mediate very different phenomena. Electromagnetism, for example, is responsible for binding electrons to nuclei in atoms and for binding atoms into molecules. The weak interactions, on the other hand, mediate the transmutation of neutrons into protons via reactions involving electrons and neutrinos, are responsible for the reactions that produce energy in stars, and cause the radioactive decay of unstable nuclei. As will be discussed, the differences between the everyday manifestations of electromagnetism and the weak force arise because the unified electroweak force exists in a broken phase.

Key differences between electromagnetism and the weak interactions are manifest in the properties of the force-carrying particles associated with the interactions. Electromagnetism results from the exchange of photons, quanta of light, whereas the weak interactions result from the exchange of three particles: the W^+, the W^-, and an electrically neutral Z. All four force carriers have an intrinsic spin angular momentum of $\hbar$ and are therefore referred to as spin-1 particles. The photon is a massless particle (which can never be at rest and travels at the "speed of light"), whereas the W and Z particles have masses of roughly eighty-five and a hundred times the mass of a proton ($938 \text{ GeV}/c^2$), respectively. The mass of a force carrier restricts the range over which the corresponding force can act. The massless photon gives rise to the long-range Coulomb potential, whereas the masses of the W and Z particles restrict the distance scale over which the weak interaction can be felt to of order 10^{-16} cm (one one-thousandth the diameter of a proton).

Another, more subtle, distinction between electromagnetism and the weak force is that the weak interactions do not respect the symmetry of parity, whereas electromagnetic interactions do. Parity violation in the weak interactions was initially proposed by Tsung-Dao Lee and Chen Ning Yang in 1956 to explain some properties of the decay of particles that contain strange quarks. Murray Gell-Mann and Richard Feynman showed that parity violation arises because the charged weak force only affects particles whose spin points antiparallel to their direction of motion. In modern terms, this would be stated as follows: only quarks and leptons that are left-handed react via the exchange of W particles. The symmetry of parity involves changing the sign of all the spatial coordinates, from (x, y, z) to $(-x, -y, -z)$, and under parity a left-handed particle would be exchanged for its right-handed complement. The mystery of parity violation in the weak interactions is compounded by the fact that the ordinary quarks and (charged) leptons have mass and may therefore be brought to rest—in which case, the distinction between left-handed and right-handed is meaningless!

Despite these physical differences between the two forces, Glashow, as well as Salam and John Ward, proposed that the electromagnetic and weak interactions were governed by one underlying gauge interaction. Electromagnetism was already known to

display the property called gauge symmetry, arising from the fact that the photon is a spin-1 particle. In fact, gauge symmetry ensures the mathematical consistency of the quantum theory of electromagnetism, quantum electrodynamics. Such consistency was, apparently, only possible for massless spin-1 force-carrying particles like the photon. This was a potential stumbling block to electroweak unification: observed phenomena associated with the weak interactions implied that the W and Z particles must be massive. Moreover, as mentioned above, since the W particles only couple to left-handed matter particles, the mathematical consistency of the theory seemed to imply that all quarks and leptons must be massless, again in contradiction with observation.

Salam and Weinberg realized that an additional theoretical ingredient would enable the fundamental interactions to respect a combined electroweak gauge symmetry, while giving rise to massive W, Z, and matter particles. The key was gauge symmetry breaking, as described by Peter Higgs (and by P. Anderson in the context of superconductivity, and independently discovered by F. Englert and R. Brout, as well as G. Guralnik, C. Hagan, and T. Kibble). Building on the ideas of Jeffrey Goldstone and Yoichiro Nambu, Higgs realized that it is possible for the interactions to respect a gauge symmetry even though the ground state of the system governed by the interactions does not. Consider an ant moving on a surface shaped like a sombrero or the bottom of a wine bottle: perched on the hill at the center, the ant appreciates the rotational symmetry; standing at any point in the circular trough, the symmetry though present is not manifest. Salam and Weinberg realized that the Higgs mechanism of gauge symmetry breaking could be used to break an electroweak gauge symmetry. The fundamental forces would respect the unified symmetry, but this would not be manifest from the vantage point of the ground state in which one lives.

In particular, Salam and Weinberg proposed adding additional scalar fields (fields with spin 0 and therefore no spin angular momentum) to the gauge theory proposed by Glashow, in such a way that the lowest-energy state(s) would not manifest the electroweak gauge symmetry. The theory maintains that the direction in gauge field space picked out by the scalar field leaves one manifest symmetry, that associated with electromagnetism. Some of the additional states of the scalar field can be interpreted as the extra longitudinal polarization state that the massive W and Z particles possess, but the massless photon does not. Salam and Weinberg also showed that the scalar fields could couple to matter particles in such a way as to provide mass to the quarks and leptons as well.

An important prediction of the model proposed by Salam and Weinberg was that one of the additional scalar states should be a new, physical Higgs particle. The Higgs particle has the following property: it couples to pairs of W or Z bosons, quarks, or leptons with a coupling proportional to the masses of these particles. As shown by Gerard 't Hooft, the existence of this Higgs particle enables the electroweak theory to be renormalizable. In other words, the theory is mathematically consistent and calculations of physically observable quantities yield finite answers. Dozens of the predictions of the renormalizable electroweak theory about the properties of the W and Z bosons have been experimentally tested to a fraction of a percent, and thus far, all measurements are in accord with the theory.

The existence of the Higgs particle itself, however, has not yet been directly experimentally confirmed. Unfortunately, although its couplings to other particles are completely determined by the measured masses of those states, the mass of the Higgs particle is related to an arbitrary self-coupling constant and is essentially undetermined. Experimental searches currently (2002) restrict the mass of the Higgs to be higher than approximately 100 GeV/c^2 (slightly higher than the mass of the Z particle). The theory becomes inconsistent if the coupling constant determining the Higgs mass is too large, leading to an upper bound of order 800 GeV/c^2 on its mass.

Although a standard Higgs theory can accommodate electroweak symmetry breaking in a manner consistent with experimental data, such a theory does not explain the necessity for electroweak symmetry breaking or why it should occur at an energy scale that produces masses of order 100 GeV/c^2 for the W and Z bosons. Moreover, if one extends the theory to encompass additional physical dynamics at higher-energy scales that does address this issue, one finds that quantum mechanical correc-

tions in the theory tend to force electroweak symmetry breaking to occur at the highest-energy scale pertinent to the theory. This instability of the symmetry-breaking scale is known as the gauge hierarchy problem. Two theoretical approaches have been taken in constructing a theory of electroweak symmetry breaking that does not suffer from the hierarchy problem: supersymmetry and strongly interacting symmetry breaking.

In a supersymmetric theory, one introduces a symmetry that relates matter (particles with spin $\hbar/2$, known as fermions) and forces (mediated by spin-0 or spin-1 particles, known collectively as bosons). For every fermion, quark, or lepton, in the Standard Model one introduces a boson, scalar squark, or slepton. For every gauge boson, one introduces a fermionic gauge particle, called a gaugino. For the scalar bosons associated with electroweak symmetry breaking (twice as many are required as in the model of Salam and Weinberg), one must add fermionic partners. The couplings of these new particles to each other and to the Standard Model particles are fixed by supersymmetry.

Once the superpartners are included in the calculation of the quantum corrections to the Higgs boson mass, it is found that contributions from bosonic and fermionic species cancel. More precisely, the gauge hierarchy is stabilized so long as the mass splittings between the ordinary particles and their superpartners are of order of the electroweak scale. Consequently, if supersymmetry is relevant to the hierarchy problem, the masses of the supersymmetric partners must lie below about 1,000 GeV/c^2. Moreover, in supersymmetric theories, the masses of the Higgs bosons are related to the gauge couplings (and possibly to other couplings as well). In the simplest models, the lightest Higgs boson must have a mass of less than approximately 130 GeV/c^2.

In a strongly interacting symmetry-breaking theory one introduces a new gauge interaction that becomes strong at an energy scale of order 1,000 GeV, as well as new, massless, weakly charged fermions that experience this new force. The simplest forms of this new interaction are modeled after quantum chromodynamics, the modern theory of the strong nuclear force. When the new force becomes sufficiently strongly interacting, it binds the new fermions together in a way that spontaneously breaks the electroweak symmetry. As a side effect, the binding can also produce additional particles (called technipions, by analogy with the pions of quantum chromodynamics) with masses in the range of a few hundred GeV/c^2. More generally, the new strong dynamics predicts an enhancement over the Standard Model value of the scattering cross section for W and Z bosons, and also the presence of numerous resonances in W and Z boson scattering with masses in the range of a few hundred to a few thousand GeV/c^2. The gauge hierarchy is rendered natural in this kind of theory by the fact that the new gauge coupling must evolve over many decades of energy scale before it becomes strong enough to break the electroweak symmetry.

Experiments planned for the Large Hadron Collider (LHC) currently under construction at the European Center for Particle Physics (CERN) are designed to shed light on the mechanism of electroweak symmetry breaking. If a Higgs boson exists, it will be found; if supersymmetry or new strong dynamics is related to electroweak symmetry breaking, the signatures of this new physics will be discovered. Most likely, uncovering the agent responsible for electroweak symmetry breaking will generate a new set of questions about the fundamental properties of nature.

See also: BASIC INTERACTIONS AND FUNDAMENTAL FORCES; BOSON, HIGGS; HIGGS PHENOMENON; STANDARD MODEL; TECHNICOLOR

Bibliography

CERN. <http://www.cern.ch>.

Coleman, S. "Secret Symmetry" in *Aspects of Symmetry* (Cambridge University Press, Cambridge, UK, 1985).

Fermi National Accelerator Laboratory. <http://www.fnal.gov>.

The Particle Adventure. <http://particleadventure.org>.

Stanford Linear Accelerator Laboratory. <http://www.slac.stanford.edu>.

R. Sekhar Chivukula
Elizabeth H. Simmons

ELEMENTARY PARTICLE PHYSICS

See PARTICLE PHYSICS, ELEMENTARY

ENERGY

In classical physics, energy is defined as the amount of work a body or system is capable of doing against a force. In elementary particle physics, the domain of the smallest known objects, quantum mechanics and special relativity govern physical behavior. Here energy is a more fundamental concept than force, and energy is a measure of the ability of a particle or system to change the state of another particle or system of particles. Energy can also be determined relative to the instruments with which elementary particles are manipulated: energy imparted to a charged particle by an accelerator is equal to the work done on the particle by the electric fields of the accelerator.

Types of Energy

The two most important forms of energy in elementary particle physics are kinetic energy (energy of motion) and rest energy. Rest energy is the energy associated with the mass of an elementary particle. Potential energy is associated with external fields, generally electric and magnetic. If a particle is composite, that is, composed of more fundamental constituents, such as the proton, which is composed of quarks, then internal potential energy can also be considered, but this potential energy, in combination with the internal kinetic and rest energy of the constituents, may also be considered as the rest energy of the composite particle. The difference between the rest energy of the composite particle and the combined rest energies of the constituents is called the binding energy of the composite particle. Albert Einstein showed with his special theory of relativity that this internal energy appears in the mass.

Kinetic energy must be defined relativistically for elementary particle physics because, as noted below, the energies of interest are typically large compared to the rest energies of at least some of the particles being studied. Using special relativity, we find the following relations between total energy E, rest energy E_0, mass m, kinetic energy T, and momentum $\mathbf{p}$, where c is the speed of light:

$$E_0 = mc^2 \tag{1}$$
$$E^2 = \mathbf{p}^2 c^2 + m^2 c^4 \tag{2}$$
$$T = E - E_0 \tag{3}$$

The relativistic momentum of a particle is defined as $\mathbf{p} = \gamma m\mathbf{v}$, where $\mathbf{v}$ is the velocity and γ is the Lorentz factor, $\gamma = 1/\sqrt{1 - \mathbf{v}^2/c^2}$. For speeds much less than c, the kinetic energy from the above equations reduces to the classical value, $\frac{1}{2}m\mathbf{v}^2$.

Conservation of Energy and Virtual Processes

Energy is conserved in physical processes, that is, the total energy is unchanged, although it may be transformed, as in the transformation of kinetic energy into rest energy of new particles as a result of a collision of particles.

Conservation of energy (along with other conservation rules) constrains the possible energy transfers and particle transformations in collisions. The conservation laws are closely related to symmetry principles, that is, symmetries associated with changes that preserve the laws of physics. These symmetries may then tell us what processes are possible. Conservation of energy is related to time translation symmetry, that is, to the fact that the laws of physics are the same at different times.

An example of the constraint introduced by conservation of energy can be seen in the process by which the antiproton was discovered. This discovery was made in 1955 via the reaction $p + p \rightarrow p + p + p + \bar{p}$, by colliding a high-energy proton with a stationary proton, where p refers to the proton, $\bar{p}$ to the antiproton. Conservation of energy requires that the kinetic energy of the incoming proton be sufficient to supply the additional rest energies of the newly created p and $\bar{p}$ as well as the kinetic energies of all the outgoing particles. For this process, we can calculate that the minimum kinetic energy of the incoming proton must be 5.6 GeV, while the proton beam energy in the actual experiment was 6.2 GeV, leaving additional kinetic energy to be distributed among the reaction products.

The Heisenberg uncertainty principle of quantum mechanics permits momentary violations of the conservation of energy. For example, the B meson decays into lighter particles by transforming to a W particle that is nearly 20 times more massive than the B. The W then decays very rapidly into the lighter decay products. These processes that violate strict conservation of energy for very short times cannot be observed directly and are called virtual processes. Nonetheless, events that would not be expected to be observed at all at a particular energy sometimes occur and can be explained by virtual processes. The uncertainty principle places strong constraints on the duration and probability of virtual processes, so these do not violate conservation of energy in directly observed events.

Energy and the Creation of New Particles

Because of the relation between energy and mass discovered in special relativity, it is possible to "trade" kinetic energy for mass in high-energy collisions. Particles that are much more massive than the initial particles can be created if sufficient energy is supplied by an accelerator. Heavy particles have been discovered in this way. For example, two protons, each having an energy of 500 GeV, can be brought to collide head on with the possibility, in principle, that two new particles are produced at rest with rest energies about 500 GeV each, or 500 times as much as the initial protons.

Momentum conservation requires that any momentum carried by the center of mass of a system of particles before collision continue with the center of mass after collision. This momentum associates energy with center-of-mass motion that is not available to create new particles or to explore internal structure. In fact, the only energy available for the purpose of investigating elementary particles is the energy of one particle relative to another. Therefore, when it is possible to make beams of both kinds of particles collide, it is most effective to collide them in such a way that the center of mass of the system of particles is at rest. This approach, used in colliding beam accelerators, makes all of the accelerator energy available to rest energies of the collision products, with any excess energy going to kinetic energies while maintaining the center of mass at rest.

Typical Energies of Interest in Elementary Particle Physics

Elementary particle physics is sometimes called high-energy physics. Energies of interest in elementary particle physics are large compared to the rest energies of at least some of the particles being studied for two reasons: (1) new particles of higher mass are created in collisions of sufficiently high energy;

(2) particles have wavelike properties, with wavelength given by the de Broglie relation, $\lambda = h/p$, where h is Planck's constant, and p is the magnitude of the momentum. This means that particles will have a very short de Broglie wavelength only if they have very large momentum (and hence high kinetic energy). The particle wavelength determines how small a region of space can be probed experimentally, so to study internal structure of very small particles like protons, we need to do scattering experiments at very high momentum.

Typical units of energy in elementary particle physics are multiples of electron volts (eV). An electron volt is the amount of energy acquired by an electron as it falls through a potential difference of 1 V. Since electrons have exceedingly small charge, the eV is a small unit of energy (1 eV = 1.6×10^{-19} J). Rest energies of elementary particles range from zero for photons, to 0.511 MeV (10^6 eV) for electrons, nearly 1 GeV (10^9 eV) for protons and neutrons, to 175 GeV for t quarks, the heaviest quarks.

Energy and Machines: Accelerators and Detectors

Accelerators are designed to produce directed particle beams with energies up to about 1 TeV, as of 2001. Beams of charged particles are accelerated by electric fields and are kept in a prescribed path by magnetic fields, sometimes in combination with electric fields. There are nine major operating accelerator facilities used in elementary particle physics. These include the Stanford Linear Accelerator, which accelerates electrons up to 50 GeV, and large circular synchrotrons at Fermilab in Illinois; the European Laboratory for Particle Physics (CERN) in Geneva, Switzerland; the Deutsches Elekronen-Synchrotron Laboratory (DESY) in Hamburg, Germany; the Japanese High-energy Accelerator Research Organization (KEK) in Japan; the Institute for High-energy Physics (IHEP) in Beijing, China; the Cornell Electron-Positron Collider (CESR); the Budker Institute of Nuclear Physics (BINP) in Russia; and at the Brookhaven National Laboratory in New York. The Fermilab machine accelerates protons to 900 GeV, with center-of-mass energies possible up to 1,800 GeV. A planned new facility, the LHC scheduled for 2007, will increase the energy of proton-pro-

ton collisions possible at CERN to 14 TeV in the center-of-mass reference frame.

Elementary particle detectors depend on the particles of interest interacting with some detector material and exchanging energy that can be measured. Detectors are designed to use some very well understood physical process, such as the interaction of charged particles with matter. Then the presence, properties, and sometimes tracks of the particles can be inferred. The details of these interactions with detectors have been analyzed for the various kinds of detectors in use in elementary particle studies, such as bubble and spark chambers, proportional counters, silicon strip detectors, magnetic spectrometers, and particle calorimeters.

See also: Conservation Laws; Energy, Center-of-Mass; Energy, Rest, Relativity; Symmetry Principles; Virtual Processes

Bibliography

Feynman, R. P.; Leighton, R. B.; and Sands, M. *The Feynman Lectures on Physics,* Vol. 1 (Addison-Wesley, Reading, MA, 1963).

Halliday, D.; Resnick, R.; and Walker, J. *Fundamentals of Physics,* 6th ed. (Wiley, New York, 2000).

National Research Council. *Elementary Particle Physics: Revealing the Secrets of Energy and Matter* (National Academy Press, Washington, DC, 1998).

William E. Evenson

ENERGY, CENTER-OF-MASS

The center-of-mass energy of a system of particles is the energy measured in the center-of-mass reference frame (see below). This energy constitutes all the energy that is available to create new particles or to explore the internal structure of particles, since the energy of the motion of the center of mass itself stays with the center of mass and cannot change the internal properties of the system.

Center-of-Mass Reference Frame

The center of mass is the weighted average position of all the mass in the system and is the point that moves with the total momentum of the system, as if the total mass of the system were concentrated there. The center-of-mass reference frame is then a set of coordinates centered on and moving with the center of mass of the system being studied. In the center-of-mass frame, by definition, the center of mass of the system is at rest, and the total momentum is zero.

In a laboratory reference frame, on the other hand, in which a particle approaches an identical particle at rest, the center of mass is half-way between the two particles. In this case, it is not at rest but moves toward the target particle as the incoming particle approaches, always staying half-way between the particles. The moving center of mass carries its own momentum and energy, in addition to the momentum and energy of the relative or internal motions of the system of particles.

For example, consider the process in which the antiproton was discovered in 1955: a high-energy proton was fired at a stationary proton target, producing a proton-antiproton pair in addition to the two original protons. With the notation p for protons and $\bar{p}$ for antiprotons, this reaction is written as $p + p \rightarrow p + p + p + \bar{p}$. In the laboratory frame, where the experiment was carried out, the center of mass of the two initial protons lies midway between them and approaches the target as the moving proton come closer (Figure 1(a)). After collision, the center of mass continues to move with the same momentum as before, as a consequence of the conservation of momentum (Figure 1(b)).

If this same event is viewed in the center-of-mass reference frame, the two initial protons are seen to approach each other, while the center of mass remains at rest (Figure 2(a)). After the collision, the four particles scatter, and the center of mass is still at rest (Figure 2(b)).

Relation of Center-of-Mass Energy to Laboratory Energy

Conservation of energy means that the energy before and after a collision will be the same. It does not mean that the energy in the center-of-mass reference frame is the same as the energy in the laboratory frame. The laboratory frame includes the energy of center-of-mass motion that does not appear in the center-of-mass frame.

The most direct way to relate laboratory energy and center-of-mass energy is to use a relativistic invariant. A relativistic invariant is a quantity that has the same value in all reference frames related by a constant relative velocity, that is, the reference frames related by special relativity. The quantity $E^2 - \mathbf{p}^2 c^2$ is a relativistic invariant, where E is the total energy of the system of particles, $\mathbf{p}$ is the total vector momentum, and c is the speed of light. Using conservation of momentum in the collision along with this invariant and considering the case of two

FIGURE 1

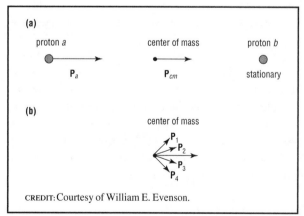

CREDIT: Courtesy of William E. Evenson.

Two-particle collision: stationary target laboratory frame. (a) before collision. (b) after collision.

FIGURE 2

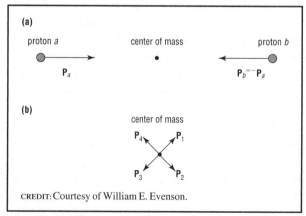

CREDIT: Courtesy of William E. Evenson.

Two-particle collision: center-of-mass frame. (a) before collision. (b) after collision.

FIGURE 3

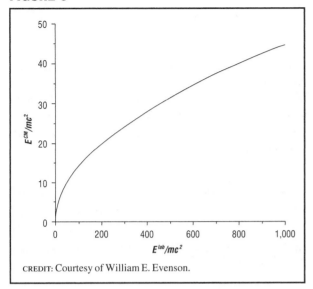

CREDIT: Courtesy of William E. Evenson.

Center-of-mass energy vs. stationary target laboratory energy for an equal mass collision.

initial particles, a and b, with b initially at rest in the laboratory frame, one finds

$$E^{CM^2} = 2E_a^{\text{lab}}m_bc^2 + m_a^2c^4 + m_b^2c^4. \quad (1)$$

Thus, the center-of-mass energy depends on the square root of the laboratory energy for high enough kinetic energies. This means that at higher energies it is less and less effective to increase the laboratory energy in stationary target experiments, as shown in Figure 3 for the case where $m_a = m_b = m$. Notice that the larger E^{lab}, the more the curve showing E^{CM} bends over and falls behind E^{lab}.

Another way to express the relationship between stationary target laboratory energy and center-of-mass energy is through the minimum kinetic energy T^{lab} needed by particle a to produce particles of total mass M by colliding with stationary particle b. Since the rest energies of the new particles produced in the collision come from the center-of-mass energy, $E^{CM} = Mc^2$, and using Equation (1), one finds

$$T^{\text{lab}} = \frac{M^2 - (m_a + m_b)^2}{2m_b}. \quad (2)$$

The antiproton experiment, $p + p \rightarrow p + p + p + \bar{p}$, carried out in the laboratory frame, requires

at minimum $E^{CM} = 4m_pc^2$ ($4 \times 0.938 = 3.75$ GeV) to supply the rest energies of the three protons and the antiproton, with the assumption that they are formed with no kinetic energy in the center-of-mass frame. Equivalently, the total mass of the products of the collision is observed to be $M = 4m_p$. Using Equation (1), one finds for the incoming proton $E^{\text{lab}} = 7m_pc^2$, which means that the accelerator must supply $6m_pc^2$ of kinetic energy ($6 \times 0.938 = 5.6$ GeV), since $E^{\text{lab}} = T^{\text{lab}} + m_pc^2$. Using Equation (2), one also find $T^{\text{lab}} = 6m_pc^2$.

To illustrate the effect of the square root energy dependence, consider the laboratory energy required to provide $E^{CM} = 100m_pc^2$ (94 GeV) in a proton-proton collision. Then Equation (1) yields $E^{\text{lab}} = 4,999m_pc^2$, requiring the accelerator to supply $4,998m_pc^2$ of kinetic energy (4.7 TeV). A 25-fold increase in the center-of-mass energy requires an accelerator more than 800 times more powerful.

Colliding Beams

It is advantageous to do high-energy experiments in the center-of-mass frame because of the square root dependence of center-of-mass energy on stationary target laboratory energy illustrated above. This is accomplished with colliding beam configurations. When it is possible to produce beams of both kinds of particles needed for a planned collision, these beams are accelerated and kept in storage rings until they can be brought together with equal and opposite momentum. Then the resulting collision occurs with the system's center of mass at rest.

For example, one could perform the antiproton experiment by accelerating protons to a kinetic energy of m_pc^2 (0.94 GeV) and storing them in two storage rings that will allow the beams to be brought back into head-on collisions. The total energy of each proton in a collision is then $2m_pc^2$, including the rest energy, and the total center-of-mass energy is $4m_pc^2$, the minimum needed to produce antiprotons. This approach produces the reaction at much lower accelerator energy at the price of a more complicated accelerator and beam arrangement.

The approach of colliding beam accelerators makes all of the accelerator energy available, in principle, to the rest energies of the collision products, with any excess energy going to the kinetic energies of the products. Unfortunately, there is sometimes no possibility of carrying out an experiment in collider configuration because of the problems in producing beams of the particles needed for a particular experiment.

The most powerful accelerator operating in 2001 is the Fermilab Tevatron in Batavia, Illinois, where collisions have been produced between 900-GeV protons and 900-GeV antiprotons. Equation (1) or (2) again allows the calculation of the kinetic energy that would have to be provided to produce the same E^{CM} in a stationary target experiment. One finds $T^{\text{lab}} = E^{\text{lab}} - m_p c^2 = 1,730$ TeV, nearly 2,000 times more energy than the 900 GeV supplied in the collider configuration. This would require an accelerator about 2,000 times as large as the Tevatron, that is, an accelerator diameter of about 2,500 miles!

See also: CONSERVATION LAWS; SYMMETRY PRINCIPLES

Bibliography

Frauenfelder, H., and Henley, E. M. *Subatomic Physics,* 2nd ed. (Prentice Hall, Englewood Cliffs, NJ, 1991).

National Research Council. *Elementary Particle Physics: Revealing the Secrets of Energy and Matter* (National Academy Press, Washington, D.C., 1998).

William E. Evenson

ENERGY, REST

Rest energy is the energy associated with a particle's mass. A free particle of mass m has rest energy $E_0 = mc^2$, and its total energy is the sum of its rest energy and its energy of motion (kinetic energy, T): $E = E_0 + T$. If you imagine traveling in a spaceship along with a particle and measuring the total energy of the particle in your spaceship laboratory, then in that laboratory the particle will be at rest, and the energy you measure will be the rest energy. There is no energy associated with motion in that case, only energy associated with the particle's mass.

Rest Energy and Special Relativity

Albert Einstein discovered the relationship between energy and mass when he formulated the special theory of relativity. Two fundamental equations of relativity for a free particle are

$$E^2 = \mathbf{p}^2 c^2 + m^2 c^4 \qquad (1)$$
$$\mathbf{p} = \mathbf{v}E/c^2, \qquad (2)$$

with total energy E, mass m, momentum $\mathbf{p}$, velocity $\mathbf{v}$, and where c is the speed of light. When the particle is at rest, $\mathbf{v} = 0$, so $\mathbf{p} = 0$ and $E = E_0 = mc^2$, the rest energy.

The relation between energy and mass follows from the two postulates of special relativity: (1) the principle of relativity: the laws of physics have the same form in all reference frames moving at constant speed with respect to one another; (2) the speed of light is the same in all such reference frames.

Rest Energy and Elementary Particle Physics

Elementary particle physics seeks to produce and study fundamental particles that make up all the matter in our universe and study their interactions. This is done by producing and observing collisions between particles at high energies. Some of the kinetic energy of the particles in such collisions can be transformed into the rest energy of new particles, subject to the conservation of energy and momentum.

Energy and momentum conservation mean that the total energy, including rest energy, and total momentum are the same after the collision as they were before that event. In the high-energy collisions of elementary particle physics, rest energy is an essential element of the conservation bookkeeping; energy will appear not to be conserved in most cases if rest energy is not explicitly accounted for.

For example, consider the process in which the antiproton was discovered in 1955: a high-energy proton was fired at a stationary proton target producing a proton-antiproton pair in addition to the two original protons. With the notation p for protons, $\bar{p}$ for antiprotons, we write this reaction as $p + p \rightarrow p + p + p + \bar{p}$. There are two additional particles on the right-hand-side, after the collision, than were there before the collision, a proton and an antiproton. These two additional particles each have rest energies equal to the proton rest energy: 938 MeV. Where did the additional $2 \times 938 = 1,876$ MeV of rest energy come from? It came from converting some of the initial kinetic energy of the incoming proton to the rest energy (and kinetic energy) of the two new particles. In this discovery experiment, the proton beam energy was 6.2 GeV (that is, 6,200 MeV), while we can calculate that the minimum kinetic energy of the incoming proton must be 5.6 GeV to supply the rest energies of the outgoing particles and still conserve momentum. The extra kinetic energy of the incoming particle (6.2 − 5.6 GeV = 600 MeV) was distributed in the motion of the outgoing particles as kinetic energy.

In particle decay processes, some of a particle's rest energy can be converted into kinetic energy, the converse of the antiproton experiment referenced above, in which kinetic energy was converted into rest energy. An example of this is the decay of a negative pion at rest: $\pi^- \rightarrow \mu^- + \bar{\nu}_\mu$. The rest energy of the pion is 139.6 MeV, while that of the muon is 105.7 MeV, and the neutrino has very small rest energy, negligible relative to the other two. The difference in rest energies before and after the decay shows up as kinetic energy of the muon and neutrino.

It is possible to convert all of a particle's rest energy to kinetic energy in particle-antiparticle annihilation. An electron and a positron will annihilate if they approach closely enough. If they start at rest so they have only rest energy, but close enough to interact, then in the annihilation, two photons will be produced with no rest energy. All the rest energy in this case will be converted to light energy.

Rest Energy as an Invariant

A relativistic invariant is a quantity that has the same value in all reference frames related by a constant relative velocity. The rest energy, $E_0 = mc^2$, is an invariant quantity, whether for a single particle or a system of particles. For a system of particles, the total rest energy of all particles in the system is an invariant.

An invariant does not necessarily have the same value before and after a collision as observed in one particular reference frame. In the antiproton experiment discussed above, the total rest energy before the collision was 1.876 GeV (two protons), while after the collision it was 3.752 GeV (three protons and one antiproton). The collision transformed some kinetic energy into rest energy. However, at any instant

during this experiment, the rest energy of the system of particles in existence at that instant would be the same whether measured in the laboratory, or in a spaceship moving with the incoming particle, or in a spaceship moving with the center of mass, or in any other reference frame moving at constant velocity.

Examples of Rest Energies of Elementary Particles

The rest energies of elementary particles range from zero (photons, gluons, gravitons according to current theory) to 174 GeV for the t quark. Some important rest energies, as now known, are as follows (given in MeV, except as noted, with ranges shown in cases where current experiments still leave large uncertainties):

photon	0	electron	0.511	u quark	1–5
d quark	3–9	muon	105.7	pion (π^0)	135
pion ($\pi^\pm$)	139.6	s quark	75–170	proton	938.3
neutron	939.6	c quark	1.15–1.35 GeV	tau	1.78 GeV
b quark	4–4.4 GeV	W boson	80 GeV	Z boson	91 GeV
t quark	174 GeV				

See also: Conservation Laws; Energy; Energy, Center-of-Mass; Momentum; Relativity

Bibliography

Groom, D. E., et al. "2000 Review of Particle Physics." *The European Physical Journal* **C15**, 1 (2000).

National Research Council. *Elementary Particle Physics: Revealing the Secrets of Energy and Matter* (National Academy Press, Washington, DC, 1998).

Okun, L. B. "The Concept of Mass." *Physics Today* **42**, 31–36 (1989).

Taylor, E. F., and Wheeler, J. A. *Spacetime Physics,* 2nd ed. (W. H. Freeman, New York, 1992).

Tipler, P. A., and Llewellyn, R. A. *Modern Physics,* 3rd ed. (W. H. Freeman, New York, 1999).

William E. Evenson

EXPERIMENT: DISCOVERY OF THE TAU NEUTRINO

Most scientific theories originate and evolve from experimental observations. The success of a theory is then based on its experimental application and verification. Often many years may pass between the time that a theory is put forth and experimental measurements are made to confirm or refute the theory. Occasionally, theories become accepted as true or complete before all of the experimental evidence to confirm them has been collected. Nevertheless, it is the responsibility of scientists to follow the scientific method strictly so that no part of a theory is accepted without proof. The discovery of the tau neutrino is just this kind of commitment to the scientific method.

The Elusive Neutrino

In the everyday world observed matter is described by the arrangement of only three subatomic particles: protons, neutrons, and electrons. In the early 1930s the existence of a fourth subatomic particle was postulated by physicist Wolfgang Pauli to explain an apparent nonconservation of energy observed in the radioactive beta decay of nuclei. (Beta decay is the term used to describe the process of a neutron constituent of a nucleus transforming into a proton and emitting an electron, formerly called a beta particle.) Though Pauli's particle could be described as having energy and momentum, it had no easily observable characteristics such as charge or mass. Several years later, physicist Enrico Fermi named the ghostly particle neutrino, Italian for "little neutral one." Fermi subsequently developed the theory of weak interactions that explained the interaction, or force, between the observable electron and its unseen neutrino partner. The essential difference between the electron and its neutrino partner is that the electron has an electrical charge of −1, while neutrinos are chargeless (charge = 0).

Although not detected in normal experience, the electron-neutrino is nevertheless a very common particle. (The identification of this neutrino with the prefix "electron" is an important distinction that will soon become apparent.) It is produced from many natural and humanmade sources, including the Sun, nuclear reactors, and particle accelerators. The difficulty in observing neutrinos is that they rarely interact with matter. Each second, both day and night, 60 billion neutrinos from the Sun pass through every square centimeter on Earth. Essentially all solar neutrinos incident on the Earth pass through it entirely.

Thus it should not be surprising that it was not until 1956 that the neutrino partner of the electron was observed experimentally! When an electron-neutrino hits the constituents of a nucleus (as might happen when a solar neutrino is passing through the Earth), the result is that in addition to the nuclear constituents getting slightly rearranged, the neutrino can be "absorbed," and an electron can be produced. It is the observation of the electron that identifies the neutrino interaction.

The Subatomic Zoo

In the course of the twentieth century, many particles in addition to the proton, neutron, electron, and neutrino have been observed and classified by physicists. These particles are created naturally in very energetic reactions of interstellar protons smashing into atoms of the Earth's atmosphere and also in manmade collisions of high-energy particles in particle accelerators. The most observable characteristic of these particles is that they only exist for a fraction of a second before they decay into lower-mass, less energetic particles carrying off some of the parent particle's original energy. In the 1940s scientists studying the decays of these particles discovered one that had characteristics similar to an electron except that it was 200 times as massive. It was dubbed a muon. Like an electron resulting from a beta decay, the muon also appeared to be accompanied by a neutrino. In 1962, it was conclusively demonstrated that the neutrino partner of the muon was unique. It was not the same as the neutrino partner of the electron. This distinction was able to be made by an experimental verification that when a muon-neutrino interacted with a nucleus, a muon, not an electron, was produced. It was concluded that these elusive particles needed to be distinguished from each other and uniquely paired with their partners the electron and the muon. Collectively these four particles were labeled leptons. In the decades between 1940 and 1970, so many new particles were discovered that the term "subatomic zoo" became the best way to describe them.

In 1964 the theoretical physicist Murray Gell-Mann postulated that the baryons and mesons (general classifications of the subatomic particles that were not leptons) were composed of constituents called quarks. It could be shown that all of the observed particles could be constructed out of three different kinds of quarks, which whimsically were named up, down, and strange. Not surprisingly, it was found that everyday protons and neutrons were made out of the up and down quarks while the strange quark was a building block of the exotic short-lived particles. Experiments conducted in the 1970s confirmed the distributions of quarks in the proton and neutron. The proton contained three quarks, two with charge $+2/3$ (up quark) and one with charge $-1/3$ (down quark). The neutron, which had no charge, was composed of two down and one up quark. Amid the chaos of the subatomic zoo, a simple organization of fundamental constituents was emerging. This organization became more important as particles continued to be discovered and needed to be classified. Before the end of the 1960s the discovery of new particles required that the quark model be expanded to include a fourth quark which became a partner of the strange quark, just as the up and down quark were partnered. In the mid-1970s additional new particles required yet another extension to the quark and lepton picture.

The Third Generation and the Standard Model

In 1975, a third-generation charged lepton was produced and detected at the Stanford Linear Accelerator Center (SLAC) near San Francisco. This third lepton had properties that related it closely to the electron and muon with the exception that it was 3,480 times heavier than the electron and it lived for a very short time before decaying to lighter particles. The leader of the team of experimentalists, Martin Perl, named it the tau lepton (tau is a Greek letter, τ, and a convenient symbol for this lepton). As soon as the discovery of the tau lepton was announced, particle physicists extrapolated from experience and assumed that it would have a neutrino partner, the tau neutrino. Confirming the existence of the tau neutrino would be the obvious next step, but as will be discussed, a variety of circumstances would hold this confirmation at bay for more than two decades! Along with the realization that a third-generation lepton existed, it became obvious that the observed

spectrum of baryons and mesons had expanded to need another set of quarks as well. The fourth quark, which had come to be called charm, was joined by a fifth named bottom. And although no particles had been observed needing a sixth quark to be added, the beauty of symmetry lead physicists to postulate the existence of a sixth quark that they named top, since it was the partner of the bottom quark. Like the tau neutrino, confirmation of the top quark was also going to take some time!

Setting experimental difficulties aside, it was clear that the model of fundamental quarks and leptons provided a simple way to organize the subatomic zoo and was worthy of being the cornerstone of a more general theory that might be able to explain the observed diversity in matter with a set of fundamental parameters. Hence the Standard Model of particle physics evolved.

Over the past thirty years, the Standard Model of particle physics has provided a very accurate description of the properties and interactions of the elementary particles called quarks and leptons. In the model there are six quarks and six leptons. Each quark or lepton is a member of a pair, and a pair of quarks plus a pair of leptons make up a generation. In addition, there are the particles that mediate the interactions between quarks or between quarks and leptons: the photon, and the W and Z bosons. With these particles one can "construct" all of the observed constituents of matter and describe the physical transformations that can occur, such as radioactive decay and nuclear fission and fusion processes. The Standard Model provides an explanation for the origin of the mass of the elementary particles as well as a logical framework for their classification. Quarks are fractionally charged particles ($2/3$ or $1/3$) that are never found isolated and free, but are bound-up as three or two quark configurations. The leptons are electrically charged one unit or electrically neutral.

The Standard Model (SM) requires that each lepton or quark be paired with a partner. These partners share quantum mechanical properties that are incorporated into the SM. The up and down quarks form such a SM pair. The electron, the ubiquitous lepton in our world, has as its partner the electron-neutrino. The up-down quark pair, together with the electron and its neutrino partner, make up virtually all of the matter that is experienced in the everyday world.

In 1995 the top quark was discovered at the Fermi National Accelerator Laboratory (Fermilab) in Batavia, Illinois. This was a great achievement in twentieth-century physics. The top is the second of the pair of quarks comprising the third generation. It is not known if the third generation is the last one. The SM requires quarks and leptons to come in pairs, but it says nothing about how many generations there may be. Nevertheless, following the discovery of the top quark, the tau neutrino remained the only Standard Model particle whose existence had not been experimentally confirmed. By the mid-1990s the demonstrated success of the Standard Model had lead particle physicists to assume the existence of the tau neutrino. Experimental physicists realized that demonstrating its existence would be very challenging.

An Experiment Is Proposed

In order to show that a particle is unique, one needs to measure key properties that distinguish it from all others. For example, charge and mass are two key properties that identify an electron from all other leptons. But neutrinos have charge zero, and the masses of the electron and muon neutrinos are known to be very small and cannot be directly measured at all. The tau neutrino was anticipated to follow suit, so another property would be needed to be sure that one has observed a genuine tau neutrino interaction. The telltale signature for the interaction of a tau neutrino is creation of a tau lepton among other, less interesting, particles. This is necessary because both leptons of a given generation possess a quality that is preserved in the interactions of leptons. For the third generation, this characteristic quality is a "tau-ness" of each lepton in the pair, that is, the tau lepton and the tau neutrino. Similarly the second-generation leptons possess a "muon-ness" and the first an electronlike property. Physicists call such qualities additive quantum numbers, which is either possessed by a particle or not, there being no in-between. In fact, this quantum number is the distinguishing feature of each lepton generation and is strongly conserved, or kept, as these leptons interact or decay. The quark pairs have a similar quantum number for each generation, although the quarks are less strict in the conservation of it.

Thus identification of a tau lepton in the debris of an energetic interaction of a neutrino indicated that the neutrino was a third generation, tau neutrino. The problem facing experimenters was how to unambiguously determine that such a rare lepton was part of a reaction. Tau leptons are much heavier than the second generation lepton, the muon, which is much heavier still than the electron of the first generation. This fact, the massiveness of the tau (3,480 times heavier than the electron), leads to decay at a much quicker rate than the muon (207 times heavier than the electron). The muon, at rest, has a lifetime of 2 microseconds (2 millionths of a second), which is considered to be quite a long time in particle physics. The tau lepton exists for only 300 picoseconds (0.3 trillionths of a second)! A tau lepton produced at the energies available to Fermilab has a velocity very near c, the speed of light, and its apparent lifetime is measured to be about 10 times longer than when it is at rest. This is equivalent to the tau lepton traveling about 1 millimeter before it spontaneously decays into, for example, an electron, an antielectron-neutrino, and (another) tau neutrino. Since the tau lepton usually decays to just one other charged particle, a typical tau lepton event will have one track that travels about a millimeter and then will appear to change direction upon decay to another track. A rule of thumb in particle physics is that only charged particles can produce directly observable tracks, as only they can ionize the atoms in the matter through which they pass. Neutral (charge = 0) particles, such as gamma rays and neutrinos, leave no visible tracks in detectors. In summary, the detection of a tau neutrino interaction requires that the experimenter be able to recognize a tau lepton, which appears in the experiment as a track with a bend in it about 1 millimeter from the interaction point of the neutrino.

This experimental method for observing tau neutrino interactions was originally proposed at Fermilab in the early 1980s; however, the experiment was not carried out. In the peer review process of the proposal, it was deemed that the detector technology, at that time, may not have had the precision required to make a definitive observation while at the same time the construction cost of the experiment was more than $15 million. Since an important criterion in judging experimental proposals is the cost-benefit ratio, it was decided that the experiment should not be carried out at that time.

In 1994, twenty years after the discovery of the tau lepton, a group of forty physicists from the United States, Japan, Korea, and Greece proposed an experiment designed to uniquely establish the existence of the tau neutrino. As was true for the search for the top quark, the world's most powerful proton accelerator (at Fermilab) would be used to create the neutrinos. Specially designed detectors would be employed to create neutrino interactions and record their unique signature (production of a tau lepton), which would be evidence of tau neutrino interactions. This experiment was officially recognized as Fermilab Experiment 872, as it was the 872nd proposal received by a review committee. The experimental group replaced this rather dry moniker with the acronym DONUT, the Direct Observation of Nu-Tau ("nu-tau" is a short way of denoting the tau neutrino symbolically: ν_τ).

Using the technology available in the 1990s, recognition of the rare tau signature would indeed be possible, in principle, with several types of particle detectors developed for such precise measurements. The detectors at Fermilab were also almost completely transparent to neutrinos. There were two ways they could increase the number of these rare interactions: (1) Increase the number of neutrinos produced, and (2) increase the number of quarks in the experimental detector, that is, maximize its mass. The number of neutrinos created is limited by the accelerator, while the detector mass is limited by cost. It is very important to note that the detector must be much more than just mass. It also needs to be able to record a neutrino interaction with enough detail so that it may later be reconstructed accurately and so that the kind of neutrino that was captured within the detector may be identified.

The DONUT Detector

In DONUT the heart of the experiment was a detector consisting of 260 kilograms of nuclear emulsion acting essentially like photographic emulsion or film. An image of each charged particle is formed as it traverses the emulsion, which is made by ionizing silver-halide crystals. Just as in the case of photographic film, the development process amplifies the original, or latent image, so that a particle track passing through the emulsion creates a series of tiny black grains, each less than 1 micrometer (0.00004 of an inch) in diameter. This precision is critical in constructing a picture of the neutrino interaction and resolving the tau lepton track from the rest of the tracks created in the interaction. Using emulsion as the primary particle detector was the key element in the success of the experiment. This was far from the first use of emulsion in particle physics. Emulsion was used extensively in the early years, from the 1940s into the 1960s, and many important results came from the analyses of these early experiments. In the 1970s, emulsion was largely replaced as a detector because the advance in low-cost, sensitive electronic detectors that enabled the data to be easily digitized and stored for computer assisted analysis. Emulsion, by contrast, had to be scanned by human operators, which was a slow process. Nuclear emulsion is also very expensive and unforgiving in the sense that it records everything passing through it, both unwanted background particles as well as physically interesting events.

The use of emulsion in DONUT was made practical by the application of computers and fast digital imaging technology. In fact, the emulsion images themselves are rarely seen directly by eye in DONUT. Tracks recorded in the emulsion are stored on computer disk arrays and are analyzed using the same numerical techniques developed for the electronic detectors of the 1970s and 1980s. The whole process is like a delayed electronic detector, and the emulsion serves as the initial storage medium for the track data. The nuclear emulsion serves as both the recording medium for interesting interactions and as the target material for the beam of neutrinos passing through it.

The Making of a Tau Neutrino Beam

The decision to use an emulsion target and detector was a clear one for the DONUT experimental group. Physicists from Nagoya University in Japan were the world leaders in modern emulsion technology, and the target design and construction became

FIGURE 1

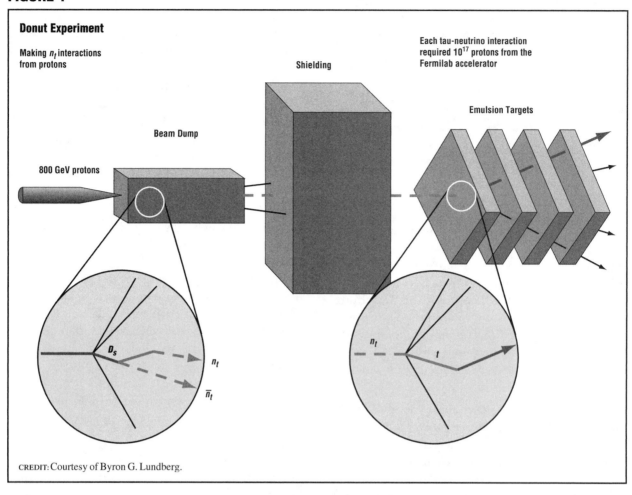

Donut Experiment

Making n_t interactions from protons

Beam Dump

Shielding

Each tau-neutrino interaction required 10^{17} protons from the Fermilab accelerator

Emulsion Targets

800 GeV protons

D_s

n_t

$\bar{n}_t$

n_t

t

CREDIT: Courtesy of Byron G. Lundberg.

The essential pieces of the DONUT experiment are shown schematically. The high-energy proton beam from the Fermilab accelerator interacts with nuclei in the beam dump. Neutrinos produced in these interactions emerge unscathed through the shielding, passing through a series of emulsion targets. About 5 percent of the produced neutrinos are tau-neutrinos, and only a few out of a trillion (10^{12}) actually react in the target.

their responsibility. The design of the neutrino beam rested with the DONUT members from Fermilab. The basic principle in making a neutrino beam is simple: smash as many high-energy protons as possible into a large block of metal. This block is called the "beam dump" and is cooled to prevent melting from the heat generated by the intense proton beam. Neutrinos result from the decays of particles created in the collisions of the protons in the metal block. All of the rest of the particles produced in the dump are absorbed by material following the collision point. Thus, one is left with a beam of neutrinos, since they cannot be significantly absorbed by ordinary matter. In the design of a practical beam

for DONUT, one that would fit within the allocated budget of $1 million, several compromises needed to be made.

First, the distance from the beam dump to the emulsion had to be made as short as possible in order to have most of the neutrinos pass through the emulsion and so have a chance at interaction. Second, because of the first point, many particles created in the beam dump will not be absorbed as there is simply not enough shielding material between the beam dump and emulsion. Most of these unwanted, non-neutrino particles were muons, the charged leptons of the second generation. If nothing special were done to eliminate these muons, there would be so many charged tracks recorded in the emulsion that it would be ruined after only one minute of exposure to protons in the beam dump. To make an analogy to photography, the emulsion detector would be overexposed. The muons were largely swept clear of the emulsion area by a large magnet located immediately after the beam dump. The magnetic field pushed positive muons to one side of the emulsion and negative muons to the opposite side. Since a failure of this magnet would be catastrophic, an electronic fail-safe protection circuit was used for this system. In addition to the magnet, a thousand tons of steel was set between the magnet and emulsion to absorb many products of the interactions in the beam dump. The total system of magnet and steel acted as a shield to protect a small area. The distance from the beam dump to the emulsion was about 36 meters. At this distance the backgrounds of muons and other particles was just tolerable, and about half of all the tau neutrinos that could be seen would be intercepted. This was considered a good compromise. Calculations based on data from other experiments indicated that 5 percent of the neutrinos in the neutrino beam were tau neutrinos, and the remaining 95 percent were the "ordinary" muon-neutrinos and electron-neutrinos, with approximately equal numbers.

The high-energy proton beam was first brought to the beam dump in November 1996, although the first emulsion module was not installed until April 1997. During the first five months of beam (without emulsion), the area around the emulsion location was instrumented to detect muons, neutrons, and gamma rays to test the waters. Several adjustments were made to the shield to correct weak places in the shield. When the level of background flux passing through the emulsion area was acceptable, the emulsion target was installed, the proton beam turned on, and data were recorded. The total number of accelerated protons used in DONUT to make the neutrino beam was 5×10^{17}. About 2×10^{15} neutrinos were manufactured (10^{14} tau neutrinos), of which about 1,000 interacted in the emulsion target (about fifty tau neutrino interactions). The emulsion was not the only detector in DONUT. Electronic charged particle detectors were used to

FIGURE 2

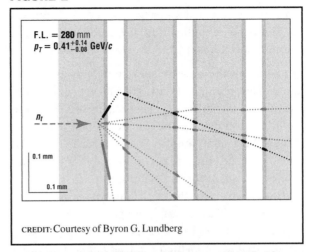

F.L. = **280** mm
$p_T = 0.41^{+0.14}_{-0.08}$ GeV/c

n_t

0.1 mm

0.1 mm

CREDIT: Courtesy of Byron G. Lundberg

One of the four tau-neutrino interactions found in DONUT is shown. The tau neutrino, incident from the left, interacts in a 1-mm steel plate (the lightly shaded thick vertical bands). The emulsion data, recorded digitally, is shown by the thick segments within the emulsion layers (the darker, thinner vertical bands). The tau lepton that was produced (the dark segment traveling northeast along the dotted line) decays into other particles, one of which is directly seen (the more lightly shaded segments). The three-dimensional particle tracks are projected into one plane for viewing.

record tracks after leaving the emulsion. These detectors were essential to pinpointing the position of the interactions inside the emulsion because only 0.01 percent of the total volume can be digitized per year using the best available technology. Only a relatively small volume, about 0.4 cubic centimeters, was digitized and recorded for each interaction. Unfortunately, about 30 percent of the time the position predicted using the electronic detectors was in error so that the interaction point in the emulsion was missed.

There were also several other factors causing interactions to be missed, all of which reduced the overall efficiency for seeing tau neutrino events. The net efficiency for finding a tau neutrino interaction within the emulsion was estimated to be 40 percent. In January 2000 203 interactions had been found in the emulsion (out of 1,000 that existed). Therefore, from the above fraction of expected tau neutrinos times the efficiency, it was expected that about five events should be in this set. Of these five events only four were long enough for both the parent tau and the daughter track to be seen in the detector. What remained in the analysis of the set of 203 events was the decay search: the hunt for a few bent tracks among approximately one thousand tracks.

Four tau events were expected, with an uncertainty of less than one event. This uncertainty was determined from earlier experimental results that were part of the estimate for the number of taus. Four may seem to be a small number, but it is really quite significant in the sense that the probability of seeing zero events when the expectation is four is only 1.8 percent. Conversely, the probability of seeing at least one event is 98.2 percent: very good odds. A more important factor in DONUT was the amount of background that accompanied the true tau events. Great effort was taken to understand all the details and nuances in the analysis so that the number of events that only looked like tau neutrino interactions, but really were not, was small and well understood. In May 2000, the analysis was nearly completed, and four events, each with the telltale bent track, were the result. This was in comfortable agreement with the predictions, but what was the level of background within this signal? A carefully constructed software model was used to determine the background, simulating the neutrino interaction physics as well as the DONUT detector. The number turned out to be 0.36 events. This number is to be viewed as a statistic. Given this average number of background events in a sample of 203, what is the probability that the actual number of background events is four? That is, what is the likelihood all the tau events were actually just background, just a fake? This is easy to calculate, and turns out to be about 4 in 10,000. The ratio of the number of signal events divided by number of background events is an important figure of merit. This ratio is large in DONUT, and the physicists on DONUT unanimously announced the discovery of the tau neutrino in July 2000 in a special seminar at Fermilab.

For twenty-five years particle physicists had assumed that the tau neutrino existed as the neutrino partner of the third-generation lepton, the tau. The tau neutrino was the last of the leptons and quarks to be confirmed, and the Standard Model remained untarnished. There are still pieces of the Standard Model that are not yet in place, and some important questions remain. There are many properties of the ghostly neutrinos waiting to be discovered, one of the most compelling being "Do neutrinos have mass?" It is safe to assume that new results in neutrino physics will continue to be announced, each providing one more clue about an elusive, but important, part of our universe.

See also: FERMILAB; LEPTON; NEUTRINO, DISCOVERY OF; STANDARD MODEL

Bibliography

Cahn, R. N., and Goldhaber, G. *The Experimental Foundations of Particle Physics* (Cambridge University Press, Cambridge, UK, 1991).

Close, F.; Martin, M.; and Sutton, C. *The Particle Explosion* (Oxford University Press, New York, 1987).

Solomey, N. *The Elusive Neutrino* (Scientific American Library No. 65, New York, 1997).

Byron Lundberg
Regina Rameika

EXPERIMENT: DISCOVERY OF THE TOP QUARK

The world around us is made of two types of particles: matter particles and force particles. The former include leptons, such as the electron and the electron-neutrino, and quarks. The two lightest quarks are called up and down, and they make up the proton and the neutron. The matter particles interact by exchanging force particles. These include the photon, mediator of the electromagnetic interaction, the gluon, mediator of the strong nuclear force that holds the nucleus together, and the W and Z bosons, which mediate the weak nuclear force, responsible for nuclear beta decay. Starting in the 1960s, physicists developed the Standard Model, which describes these particles and their interactions.

The electron, electron-neutrino, up quark, and down quark form a fermion generation. This fermion generation contains all the constituents of ordinary matter. However, this is not the only generation. The muon, muon-neutrino, charm quark, and strange quark have the same properties as the particles of the first generation except they are much heavier, and most of them are not stable. They decay to the lighter particles of the first generation and therefore are not usually encountered.

In 1975, a third—yet heavier—lepton, called the tau lepton (τ lepton), was discovered. The Standard Model requires complete generations, consisting of two leptons and two quarks. This implied the existence of a third neutrino and a third generation of quarks. Experiments set out to search for a third-generation quark culminating in the discovery of the bottom or b quark in 1977. At that time, this was the heaviest known fundamental particle with a mass of about 5 GeV/c^2, more than five proton masses. Now there was only one quark missing to complete the third quark generation. It was called the top quark.

The Race for the Top Quark

The race for the top quark was on. It had to be heavier than the b quark. More energetic accelerators were required to produce it. In the early 1980s, experiments in Germany and Japan, using colliding electron and positron beams, ruled out the existence of a top quark with masses below 30 GeV/c^2. In 1989 to 1990 experiments at the Stanford Linear Accelerator Center (SLAC) at Stanford University and at the European Laboratory for Particle Physics (CERN) in Geneva, Switzerland, increased the limit to 46 GeV/c^2. To go to even higher energies experimenters switched to colliding protons and antiprotons. At CERN, proton-antiproton collisions at energies of 630 GeV again failed to see the top quark, setting the limit at 69 GeV/c^2.

In 1985, the Tevatron at Fermilab near Chicago collided proton and antiproton beams for the first time at an energy of 1.6 TeV. The newly commissioned CDF detector pushed the lower limit on the top quark mass to 77 GeV/c^2. In 1992, the energy of the Tevatron was increased to 1.8 TeV, and a second detector, D0, commissioned.

Fermilab is located 30 miles west of Chicago in Batavia, Illinois. The Tevatron is an accelerator ring with a radius of 1 km. Its entire length is enclosed in about 1,000 superconducting magnets that keep the protons on their path. It is the last in a series of ac-

celerators that accelerate protons and antiprotons to 900 GeV each. These were the particle beams with the highest energy in the world. Groups of about 10^{11} protons and 10^{10} antiprotons circle in opposite directions around the Tevatron ring and collide in the center of the two detectors every 3.5 microseconds.

Top Production and Decay

How did CDF and D0 propose to find the top quark? When the protons and antiprotons collide, their kinetic energy can be converted into mass. If enough energy is available, top quarks can be created. Most of the time, top quarks and their antiparticles are created in pairs in proton-antiproton collisions. However, the creation of top quarks is a very rare process. Only once in every 10^{10} proton-antiproton interaction are top quarks produced. This seems much like the proverbial search for a needle in a haystack.

Moreover, the top quark does not live long enough to be observed directly. On average it exists only for 10^{-24} second before decaying into a b quark

The CDF detector. CREDIT: COURTESY OF FERMILAB PHOTO. REPRODUCED BY PERMISSION.

and a *W* boson. An antitop quark decays into an anti-*b* quark and an anti-*W* boson. The *W* bosons are also very short-lived and decay into electron + neutrino, muon + neutrino, or tau + neutrino with a probability of one ninth each. The remaining two thirds of their decays are into quark-antiquark pairs.

Quarks interact so strongly with each other that they cannot exist in isolation but only as constituents of other particles, such as protons. Particles that consist of quarks are called hadrons. The quarks created in the decay of the top quark immediately turn into collimated showers of hadrons, called jets.

Electrons and muons are electrically charged and live long enough to be observed directly in the detector. *τ* leptons decay in a variety of ways and are very difficult to identify.

Neutrinos are electrically neutral and have a fair chance of making it all the way through the earth without interacting. Therefore the chance of catching them in a detector is nil. However, since the initial protons and antiprotons carry only momentum along the beam direction, the total transverse momentum (with respect to the beam direction) of all particles produced in a proton-antiproton collision must be zero. If a neutrino carries away a lot of momentum, that momentum will be missing from the visible particles, and their transverse momenta will not add up to zero. Thus large missing transverse momentum implies that one or more neutrinos were produced that carried away the missing momentum.

In collisions in which a top quark and an antitop quark are created and decay, a *b* quark and an anti-*b* quark are always produced. In addition, other particles appear, depending on the decay channel of the *W* bosons:

- dilepton channel: two electrons or muons, and two neutrinos (5 percent of all top-antitop decays)

- lepton + jets channel: one electron or muon, one neutrino, and two additional quarks (30 percent of all top-antitop decays)

- all-jets channel: four quarks (44 percent of all top-antitop decays)

- decays that include tau leptons (21 percent of all top-antitop decays).

When protons and antiprotons collide, they typically break up into quarks and gluons, which turn into jets. At the Tevatron, this happens several thousand times more often than top quark production. Of the top-antitop signatures above, the ones that contain electrons or muons are easiest to pick out of this background. In the dilepton channels one expects two electrons or muons, missing transverse momentum, and two jets. In the lepton + jets channel one expects one electron or muon, missing transverse momentum, and four jets. Although the other channels make up about two thirds of all top-antitop decays, the experiments did not even try to look for them in the beginning because they are so hard to separate from the background.

The Experiments

Although the D0 and CDF experiments look quite different, they have the same basic detector components.

As the particles emerge from the collision, the experimenters first measure their direction. Thus innermost in both experiments are tracking detectors, which are built very light and with as little material as possible, since they should not disturb the particles very much. They are mostly gas chambers with thin wires strung across. Particles that carry electrical charge ionize the gas, which generates an electrical pulse on the wire closest to the particle path. Many such pulses can be used to reconstruct the trajectory of the particle to about 100-micron precision. CDF also had a detector made of thin slices of silicon. This allows even more precise measurements of the particle path near the collision point. The CDF tracking chambers are placed in a magnetic field such that the particle trajectories curve. The curvature determines the momentum of the particle.

Next the experimenters measure the energy of the particles by stopping them and measuring the energy transferred to the material of the detector, which equals the kinetic energy of the particles. The detector that does this is called a calorimeter, and, in contrast to the tracking detectors, it is built of heavy materials. D0 has a calorimeter made of uranium and liquid argon. CDF's calorimeter consists of lead and scintillator wedges in the center and lead and gas chambers at the ends of the detector. The calorimeter also allows the distinction of electrons and hadrons by their different energy loss patterns. Hadrons are much more penetrating than electrons.

However, not all particles can be stopped. Electrons and hadrons interact strongly with matter and are easily stopped. To contain muons requires tens of meters of steel, not very practical. The experimenters turn this to their advantage. Since muons are electrically charged, they can be detected in tracking chambers, and both experiments have large tracking chambers outside their calorimeters. With high probability, any charged particles that penetrate the calorimeter and are detected in these chambers are muons. Therefore these chambers are called muon detectors. D0 also has a magnet that deflects the muons so that their momenta can be measured.

The Discovery

How do we know when we have discovered the top quark? CDF and D0 were looking for dilepton and lepton + jets events. Seeing such events, however, is not equivalent to discovering the top quark. Other processes can also produce these signatures. For example, in proton-antiproton collisions a W boson and four jets can be produced. Then the W boson decays to an electron and a neutrino, which is exactly the signature of a lepton + jets event. This background is so overwhelming that seeing the actual top signal requires more sophisticated analysis techniques. The top-antitop decays differ in two ways from this W + jets background:

- the top-antitop decays have two jets from b or anti-b quarks

- the jets in top-antitop decays have much higher momentum transverse to the beam.

Hadrons that contain b quarks often live much longer (10^{-12} second) before they decay than other hadrons (10^{-20} second). In this time, they can move several millimeters away from the point at which the proton and antiproton collided. If the detector can measure the particles from their decay very precisely, this displaced secondary decay point can be reconstructed. If such a displaced decay point is found in a jet, it is likely that the jet originated from a b quark. This is called vertex tagging. CDF was able to tag jets that originated from b quarks in this way using the

The D0 detector. Credit: Courtesy of Fermilab Photo. Reproduced by permission.

silicon tracking detector. Both experiments could tag b quark jets using muons. Approximately one in ten hadrons that contains b quarks decays into a muon and other particles. Other hadrons are not likely to decay in this way. The muon can be detected, and a jet in which a muon is found likely originates from a b quark. This is called lepton tagging. It was used by D0 and CDF.

When jets are produced in proton-antiproton collisions, their direction is often close to that of the original proton or antiproton. Such a jet may have high momentum, but the component of its momentum transverse to the beam direction is not large. When a heavy object such as the top quark is produced and decays to jets, however, the jets are emitted in all directions with about the same probability. Thus, in this case, the transverse components of the jet momenta could be quite large. D0 found that the sum of the transverse momentum components of all jets, called H_T, and the angular distribution of the jets provide good discrimination between the top quark signal and the background.

Using all these tricks, the experimenters were able to reduce the background in their data samples so that signal and background were comparable. Since the background originates from known processes, the number of events expected from these processes can be calculated. If the top quark is produced, the observed number of events should be larger than that calculated under the background-only hypothesis. However, even if there is no top quark, the experimenters could observe more events than calculated because of statistical fluctuations. For example, if 3.8 events are expected, most probably 3 or 4 will be observed, but there is a 33 percent probability to observe more than 4, and a 2 percent probability to observe more than 10. The standard for a discovery is much higher. For a discovery, the probability of the observed number assuming there is only background must be around one in a million or less!

D0 observed three dilepton events and fourteen lepton + jets events for a total of seventeen events when a background of 3.8 ± 0.6 was expected. The probability to see seventeen or more, assuming the background-only hypothesis, was 2×10^{-6}. CDF observed six dilepton events with 1.3 ± 0.3 expected. In the lepton+jets channel, CDF did not count events but b-tagged jets. They saw twenty-seven vertex tags (6.7 ± 2.1 expected) and twenty-three muon tags (15.4 ± 2.0 expected). The combined probability to see this many events and b tags without a top signal was 10^{-6}.

The People

The papers published by the D0 and CDF collaborations announcing the discovery of the top quark have 414 authors for the D0 paper and 539 for the CDF paper. This large number indicates the complexity of the effort behind large high-energy physics experiments.

The time scale of such a project from conception to completion is typically a decade or more. For example, the D0 experiment received preliminary approval in 1983. Construction was completed in 1990. The experiment took its first data in 1991. In 1995 the discovery of the top quark was published. To date, the D0 collaboration has published over one hundred research papers.

The D0 detector is about three stories high and weighs about 5,000 tons. It provides about 120,000 electronic signals for every proton-antiproton collision. There are 5,500 km of cables to distribute these signals, to supply power, and to control operation of the detector. All detector components and most of the electronics were developed and built by physicists. This required thousands of person-years of effort. This task was therefore distributed over research groups from many collaborating institutions. For example, parts of the tracking detectors were built at Saclay in France, Lawrence Berkeley Lab in California, State University of New York in Stony Brook, and Northwestern University in Evanston, Illinois. The components were brought to Fermilab and assembled. Every channel had to be tested and problems fixed before data could be taken. This was often the task of Ph.D. candidates and postdoctoral students, who resided at Fermilab. After everything was working and the accelerator started to provide colliding beams, data taking began. The accelerator ran seven days a week, twenty-four hours a day for almost two years, and a crew of six physicists manned the control room of the detector at all times. All collaborators contributed to this, taking eight-hour shifts to acquire the data. When they were not taking shifts, they spent their time analyzing the data to determine hundreds of thousands of calibration constants that were stored in databases, or they developed the millions of lines of computer code necessary to operate the detector and acquire, store, and analyze the data. This comprised a huge body of work that was required before the search for the top quark in the data could even begin. The search for the top quark was of course not the only goal of the experiment. Many other important results were also based on this work.

Over four hundred physicists and uncounted engineers and technicians from forty-two institutions in the United States, Brazil, Colombia, France, India, Korea, Mexico, and Russia contributed to the construction and operation of the experiment. Professor Paul Grannis from the State University of New York at Stony Brook led this operation, from conception to completion.

Once the detector was calibrated and the computer code debugged that found electrons, muons, and jets in the data from the detector, smaller groups of physicists began to analyze the data for signals of different processes of interest. The group that performed the search for the top quark was one of five physics groups. The top physics group consisted of about eighty members, of which twenty were Ph.D. students, and was led by Dr. Boaz Klima from Fermilab and Professor Nick Hadley from the University of Maryland. Within this group, smaller working groups were formed, each concentrating on one channel. These groups reported back to the top physics group in weekly meetings. For example, Meenakshi Narain of Fermilab worked on the dielectron channel with students V. Balamurali and Bob Kehoe from the University of Notre Dame. Kehoe was one of the students whose theses composed the top quark discovery. He worked tirelessly many nights trying to find a candidate event in "his" channel. To his great disappointment the dielectron channel ended up being the only channel in which no candidates were observed.

At the time of the annual winter conference in Aspen in January 1995, no excess close to meeting the standards for a discovery had been seen. The mass limit had increased to 131 GeV/c^2, and it had become clear that the top quark had to be much heavier than originally expected. If it were that heavy, only very few top quarks would be produced, and they would be swamped by background. A concentrated effort was started to develop a strategy to isolate the signal from a very heavy top quark. Studying a computer simulation of the top quark signal, the H_T cut described above was developed, which would have cut out the signal of a lighter top quark but accepted much of the signal from a heavy top quark. The new analysis strategy soon showed hints of a signal. Soon also rumors circulated that CDF was planning to come out with a discovery publication.

The experimenter had agreed with Dr. John Peoples, the director of Fermilab, to give each other notice before they published a discovery. The other experimental group would then have one week to prepare its own paper. This mechanism was designed to prevent one of the groups from jumping the gun with a premature publication that the other experiment would contradict. In February 1995, the CDF collaboration announced that it saw a significant excess and was preparing a publication.

Once the top physics group had concluded that they saw a signal, the next step was to convince the rest of the collaboration. The usual process for publishing results was to set up an editorial board. This was comprised of several physicists from the collaoration who were not involved in the analysis as well as one or two proponents of the analysis. This board would meet and examine the analysis in detail to check for any weak points. It also would review the proposed paper. Once the board was satisfied, the collaboration had a few weeks to comment, and if no objections were voiced, the paper was submitted for publication. In case of the top discovery, there was very tight time pressure, because nobody wanted to let CDF publish first. The editorial board worked tightly with the analyzers, and the collaboration review was accomplished in days. On Friday, February 24, 1995, at 11 a.m. Central Standard Time, Herb Greenlee hit the computer key submitting the paper to *Physical Review Letters* at exactly the same time as his counterpart, Mel Shochet, at CDF. The announcement for the usual weekly top meeting that week read: "Since the top quark has been discovered, there will not be a top group meeting this week."

See also: FERMILAB; QUARKS; QUARKS, DISCOVERY OF; STANDARD MODEL

Bibliography

CDF Collaboration. "The CDF Detector: An Overview. " *Nuclear Instruments and Methods A* **271**, 387–403 (1988).

CDF Collaboration. "Observation of Top Quark Production in Anti-p p Collisions." *Physical Review Letters* **74**, 2626–2631 (1995).

D0 Collaboration. "The D0 Detector." *Nuclear Instruments and Methods in Physics Research A* **338**, 185–253 (1994).

D0 Collaboration. "Observation of the Top Quark." *Physical Review Letters* **74**, 2632–2637 (1995).

Fermilab. <http://fnal.gov>.

Liss, T. M., and Tipton, P. L. "The Discovery of the Top Quark." *Scientific American* **277**, 54–59 (1997).

Meenakshi Narain

EXPERIMENT: G−2 MEASUREMENT OF THE MUON

The electron, muon, tauon, and their neutrinos are fermions with spin $\hbar/2$. These particles (and their antiparticles) form the subclass of fermions known as leptons, which comes from the Greek word λεπτον meaning thin, or light in weight. The muon and tauon are like heavy electrons, except that the electron is stable, having no lighter particles into which it can decay, whereas the muon and tauon undergo radioactive decay through the weak force. Physicists have looked to see if the leptons have any structure, and at the smallest resolution they have been able to achieve, no indication of structure has been seen. It is believed that leptons occupy a physical point rather than filling some volume of space like the proton does (its size is about a femtometer).

The muon has a very long lifetime by subatomic standards, a trait which permits detailed studies of its properties. Precision measurements have been made of the muon's lifetime, mass, and magnetic moment, and at the time of this writing (2002), efforts are under way to improve the precision of the lifetime and anomalous magnetic moment measurements. In its rest frame, the muon exists for 2.19703(4) microseconds (μs), where this value represents a mean or average value of several experiments. The number in parentheses quantifies the uncertainty on the measured value (often referred to as the error), which by convention is one standard deviation, often represented by the Greek letter σ.

Many physical constants have been measured and are tabulated for easy reference (for example, see the Particle Data Group and National Institute of Standards and Technology Web sites). In the tables in these two sources is a listing in the form M ± σ, or M(σ), where M is the "best value," an average computed by the people who compiled the table, and σ is a combined standard deviation which is obtained from all the separate experiments which went into the average value. The number presented in the tables as the uncertainty on a quantity is usually the quadrature of the statistical error σ_E, and the systematic error σ_S; $\sigma = \sqrt{\sigma_E^2 + \sigma_S^2}$. If these two errors are not independent quantities, then they must be added linearly, $\sigma = \sigma_E + \sigma_S$, which gives a larger total uncertainty. In precision measurements (defined below) such as the muon g−2 value, the systematic errors are as important as the statistical errors. The problem faced by the scientist is how to reasonably estimate the systematic errors while not overestimat-

ing them, which would artificially reduce the precision of the measurement.

The size of the uncertainty relative to the value of a quantity is often referred to as the precision of the measurement. A common description of the precision of a measurement is its relative error, $\sigma/$M, which for "precision measurements" can be on the order of 10^{-6}, parts per million (ppm), or 10^{-9}, parts per billion (ppb). While this level of precision is often reached in atomic physics, only a few quantities in particle physics have been measured to this level of precision. One example is the mass of the Z^0 boson, which was measured to a precision of 20 ppm at European Laboratory for Particle Physics (CERN), using the large electron-positron collider (LEP). The lifetime of the muon given above has a precision of 18.2 ppm, and the anomalous magnetic moment (the topic of this article) is currently known to a precision of 1.3 ppm.

Magnetic Moments

The magnetism associated with an elementary particle provides us with information about its structure and also about the forces that can affect it. Because of their spin, the muon, the electron, and the proton behave like tiny magnets, with the direction of the magnetic field pointing along the spin angular momentum of the particle. The classical analog of a spinning particle is sketched in Figure 1a, where an electrically charged sphere (with positive charge) is spinning about its center, just as the Earth spins about its axis. Also sketched are the magnetic field lines that emerge near the north pole and re-enter near the south pole of the magnet. For a spinning object, it is useful to define a spin angular momentum vector $\vec{S}$. To find the direction of the spin vector, curl your right hand fingers in the direction of rotation, and your thumb will point in the direction of the spin vector.

This spinning charge distribution will create a magnetic field, which is much like the magnetic field created by a bar magnet. Such a magnetic field is called a "dipole magnetic field," since it is set up by two magnetic poles. A useful measure of the strength of this magnetic field is a vector called the "magnetic dipole moment," or simply "magnetic moment" or "dipole moment," which is a measure of the strength

FIGURE 1

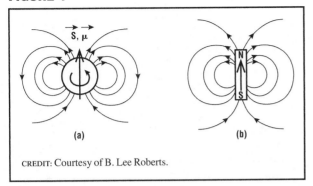

CREDIT: Courtesy of B. Lee Roberts.

(a) The magnetic field created by a sphere of positive charge spinning about its diameter. $\vec{S}$ is the spin, which is in the same direction as the magnetic dipole moment $\vec{\mu}$. The direction in which the sphere is turning is indicated by the curved arrow. (b) The magnetic field lines of a bar magnet.

of the magnetic field. It is traditional to represent this dipole moment by the vector symbol $\vec{\mu}$. (Unfortunately, physicists use the Greek letter μ to represent the magnetic dipole moment and also to represent the muon. It should be clear which is meant from the context.) For a positive charge, the magnetic dipole moment points in the same direction as the spin angular momentum vector.

Also shown in Figure 1b is a bar magnet and its magnetic field lines. For a bar magnet, the dipole moment points from the south to the north pole as indicated by the arrow in the center of the bar. A compass needle is a familiar example of a magnetic dipole. When a dipole is placed in a magnetic field, it experiences a torque that will make it align itself with the magnetic field just as a compass aligns itself with the Earth's magnetic field, indicating which direction is north.

Any magnetic dipole will experience a torque if it is placed in a magnetic field. However, if the dipole is caused by a spin, the dipole cannot align itself with the magnetic field. The situation is like a toy top spinning about its axis in the Earth's gravitational field. Instead of falling over, the top precesses with the tip of the angular momentum vector following a circle as shown in Figure 2. A torque is produced about the contact point by gravity pulling down on the top at its center of mass. This torque causes the angular momentum to precess, since the

rotational form of Newton's Second Law tells us that the angular momentum will change in the direction of the torque labeled N, which points into the page in Figure 2. The precession frequency of a top, ω_p, is proportional to the angular momentum, $\vec{L} = I\vec{S}$, and is given by $\omega_p = mgl/IS$. The symbol ω (the lowercase Greek letter omega) is the angular frequency in radians per second. S is the spin of the top (the angular frequency of rotation about the symmetry axis), and I is the moment of inertia which depends on the distribution of mass about the line through the axis of symmetry of the top.

The magnetic moment of the muon is related to its spin through the relationship: $\vec{\mu} = g \dfrac{e\hbar}{m} \vec{S}$, where g is the constant of proportionality between the spin and the magnetic moment, which is called the g-factor or g-value. The symbol e is the charge of the muon, and m is its mass. For a particle with spin $\hbar/2$, no internal structure, and with no radiative corrections (see below), the relativistic quantum mechanics developed by Paul A. M. Dirac tells us that g is *exactly* two. However, if the particle has internal structure, then g is not equal to two. For example, the g-factor of the proton, which is made up of quarks and gluons, is 5.58, quite different from two. Because of its magnetic moment, a muon which is placed in a magnetic field will precess about the field, just as a toy top precesses about the (vertical) gravitational field of the Earth.

FIGURE 2

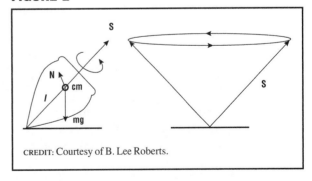

CREDIT: Courtesy of B. Lee Roberts.

A toy top spinning about its axis. The spin angular momentum vector S does not align itself with the gravitational field but proceeds in a circle as shown on the right, sweeping out a cone as it goes around. The direction of the torque (labeled N) which is caused by gravity acting at the center of mass (labeled cm) is into the page. The distance from the contact point to the cm is labeled *l*, and mg designates the top's weight.

Background and History of $g-2$

The development of understanding the electron's *g*-value was at the center of the path to understanding the subatomic world. In 1921 Otto Stern proposed the famous experiment, which was later called the Stern-Gerlach experiment. In their 1924 review paper, Stern and Walther Gerlach concluded that to within 10 percent, the silver atom had a magnetic moment of one Bohr magneton. However, their interpretation was incomplete, and only with the postulate in 1925 by Samuel Goudsmit and George Uhlenbeck that the electron had an intrinsic angular momentum called spin did the full picture emerge. In 1927, motivated by the work of Stern and Gerlach and the proposal of Goudsmit and Uhlenbeck, T. E. Phipps and J. B. Taylor showed that the magnetic moment of the hydrogen atom (and thus the electron) was one Bohr magneton, in agreement with the spin hypothesis of Goudsmit and Uhlenbeck. In modern terminology, these developments told us that the *g*-value of the electron was two.

There were indications in several earlier published results that g might not be exactly two, but the definitive evidence came in 1947 when Polykarp Kusch and H. M. Foley obtained their results on the difference of *g* from two. Julian Schwinger explained this difference with one of the pioneering calculations of what is now called quantum electrodynamics (QED).

In modern language, one measures the anomalous magnetic dipole moment, sometimes called the anomaly *a*, where

$$a = \frac{(g-2)}{2}, \quad \text{where} \quad \mu = (1+a)\frac{e\hbar}{m}$$

is the magnitude of the magnetic moment. It is this latter quantity which is tabulated in the tables of particle properties found at the Particle Data Group's Web site.

The *g*'s difference from two can be understood by examining the Feynman diagrams shown in Figure 3. The Heisenberg uncertainty principle $\Delta E \Delta t \geq \hbar$ permits virtual processes that violate energy conservation to occur, as long as they happen quickly. The particles which appear out of nowhere

FIGURE 3

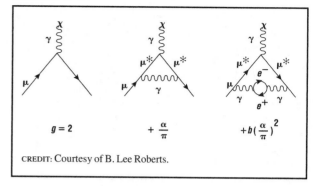

CREDIT: Courtesy of B. Lee Roberts.

Several Feynman diagrams that contribute to the muon *g*-value. The ones at left and at center are the simplest and most important diagrams. The arrows on the muon line show its direction of motion. The symbol γ indicates a photon that is absorbed or emitted by the muon. The constant α is called the fine-structure constant, which is approximately equal to 1/137. The diagram on the right is the simplest "vacuum polarization diagram," where an electron positron pair appear out of the vacuum. The quantity *b* is a constant.

(physicists say the particles appear from the vacuum) are called virtual particles. In the left-hand picture of Figure 3, the muon interacts with the magnetic field by absorbing a photon (labeled γ) from the magnetic field. In the middle picture, the muon emits a virtual photon, then absorbs a photon from the magnetic field, and then reabsorbs the virtual photon. After the virtual photon is emitted, the symbol μ^* is used to remind us that the muon is "off-shell" (meaning that energy conservation must satisfy the uncertainty principle, and the muon is not a free particle). The *g*-value is sensitive to the ratio of charge over mass, and this virtual process changes the mass. The virtual process on the right has the effect of changing the charge distribution around the muon. The muon's magnetic dipole moment (*g*-value) is changed by virtual processes which change the mass, or the charge distribution surrounding the muon. The anomaly caused by the process in the middle is 0.00116140981(5). In nature, there are an infinite number of these virtual processes (called radiative corrections) involving photons and electrons, but the largest effect comes from this one process. Since the anomaly is about one thousandth of the total magnetic moment, experimentally one uses a measurement technique directly sensitive to the anomaly ($g-2$) rather than the total moment (*g*).

The beauty of studying a process that is sensitive to virtual particles is that all particles present in nature can participate, including particles never seen before. The only requirement is that they are able to interact directly or indirectly with (physicists say "couple to") the muon. Angular momentum, charge, and other fundamental quantities (except energy for a brief time) must be conserved. The Standard Model particles that can contribute (virtually) to the anomaly at a measurable level are electrons, muons, tauons, photons, pi and K mesons (quarks and gluons), and the electroweak gauge bosons, the $W^\pm$ and the Z^0.

One of the main motivations for measuring g precisely is its sensitivity to a wide range of new physics beyond the standard model. The g-value of the muon has traditionally served as a calibration point for new theories, since most predict a contribution to the anomaly from the constituents of the new theory. If a new theory predicts an effect on the anomaly which is ruled out by the measured value, then this theory cannot be valid without revision. Over the past decade, great interest has developed in various extensions to the standard model such as supersymmetry . Under some scenarios of this theory, the supersymmetric partners of the $W^\pm$ and the Z^0 could contribute to the anomaly at a measurable level.

In a series of three beautiful experiments at CERN, the muon anomaly was measured to an increasing precision, reaching 7.3 ppm by 1979. The final result confirmed that the muon was a lepton which obeyed the rules of quantum electrodynamics, and it also confirmed the presence of the predicted contribution from virtual pi and K mesons. At this level of precision, the predicted 1.3 ppm contribution of the electroweak gauge bosons was not observable. It was the desire to observe this electroweak contribution to the anomaly, along with the desire to look for effects of new physics with increased sensitivity, that motivated a new proposal to achieve an accuracy for the muon anomaly twenty times more precise than the CERN result.

The technique used in the third CERN experiment was the basis of the new Brookhaven experiment. A beam of muons with their spin pointing in the same horizontal direction is stored in a magnetic ring which forces them to go in a circle of 14 meters in diameter. The muons make up to 4,000 trips around the ring before they decay. Because their g-factor is not exactly two, the muons' spin turns faster than their momentum as they go around the ring (see Figure 4). Every 29.3 turns around the ring, the spin makes one complete revolution relative to the direction the muon is moving (its momentum). Because of relativistic time dilation, the average lifetime of the muon is 64.4 microseconds, and in ten lifetimes essentially all of the muons have been lost to radioactive decay.

The frequency with which the spin turns relative to the momentum (ω_a) is written inside the circle, and this formula is the basis for the measurement. By measuring the magnetic field and the frequency with which the spin turns relative to the momentum, one can determine the anomaly. Since this frequency depends on the anomaly directly, rather than the

FIGURE 4

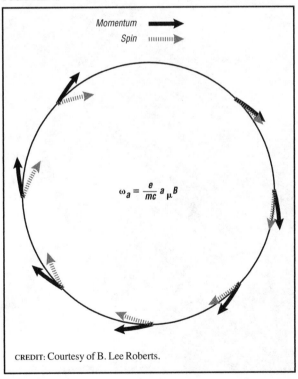

$$\omega_a = \frac{e}{mc} a_\mu B$$

CREDIT: Courtesy of B. Lee Roberts.

A diagram that illustrates the principle of the g−2 experiments. Initially the momentum and spin are parallel. As the muons go around the ring, the spin gets ahead of the momentum, a process which is greatly exaggerated in the diagram.

magnetic moment, it is called a $g-2$ measurement, rather than a g measurement.

The $g-2$ Experiment at Brookhaven National Laboratory

Based on the experience of the third CERN experiment, it was clear that one could improve the precision by a factor of twenty, provided a new, improved apparatus was built. This improvement required extending the current state-of-the art in a number of areas, if the experiment were to be successful. The physics goals were to confirm the predicted Standard Model contribution of the gauge bosons, and/or to search for contributions to the anomaly beyond the standard model. By the early 1980s the Standard Model was well established. It was becoming clear that there were deficiencies in the Standard Model (even

though not all the predicted standard model particles had been discovered), and new ideas such as supersymmetry were being developed.

The new experiment had a goal of 0.35 ppm relative error, which would be adequate to observe the effect of virtual W and Z^0 gauge bosons. This design goal meant that the new experiment might be sensitive to new particles such as the supersymmetric particles (if they exist), and if no effect were found, it would at least restrict what their properties might be. The new experiment would also serve to further restrict muon or W boson substructure. Evidence for any of these effects outside of the standard model would represent a major new discovery.

The proposal was submitted to Brookhaven National Laboratory and was approved by the Laboratory in 1986. A major new feature of the Brookhaven

The Brookhaven muon $g-2$ storage ring. The wide, dark ring is the steel yoke of the magnet. The light-color rings inside of the dark one are the cryostats, which contain two of the three superconducting coils. The magnet is over 14 m in diameter. Several of the 24 detectors are in place. They are the objects inside the ring running radically inward (one of them is shown just to the right of the technician who is working inside the ring). This ring is the world's largest diameter superconducting magnet. CREDIT: COURTESY OF BROOKHAVEN NATIONAL LABORATORY. REPRODUCED BY PERMISSION.

experiment was to form a beam of muons external to the experiment, bring this beam into a storage ring, and then give it a kick to move the beam into a stable orbit within the ring. The heart of the new experiment is a unique storage ring magnet, 14 meters in diameter and weighing 700 tons, which is the world's largest superconducting magnet. This magnet has been shimmed to an average uniformity of 1 ppm over the region where the muon beam is stored. It now appears that the systematic errors associated with the magnetic field will be about 0.3 ppm, which should permit the collaboration to reach close to their design sensitivity.

In February 2001 a new result was reported by the collaboration, $a_\mu = 0.0011659202(14)(6)$, where the first error in parentheses is statistical and the second is systematic. This result, when averaged with previous measurements, was two and one-half standard deviations larger than the value expected from the Standard Model, and statistically there was less than a 2 percent chance that this result was compatible with the Standard Model prediction. This result generated a great deal of excitement among those who believe that supersymmetry is the correct theory to extend the Standard Model, since the magnitude of the discrepancy is easily accommodated by this theory.

Unfortunately, a small piece of the Standard Model theoretical value called the hadronic light-by-light contribution that had been assigned a negative sign turned out to be positive. This mistake was discovered by Marc Knecht and Andreas Nyffeler at the University of Marseille and was soon confirmed by the authors who had originally obtained the negative sign. This incident demonstrates the critical interplay between theory and experiment in the progress of physics. When the correct sign is used, the discrepancy with the theory is reduced to 1.6 standard deviations, which implies an 11 percent chance of agreeing with the Standard Model.

The collaboration has collected seven times as much data as were reported on in 2001, which should reduce the statistical error by the square root of seven, and the systematic error will also be reduced. When these additional data are analyzed, they will determine if there is a meaningful discrepancy with the Standard Model or not.

Organization of the Experiment

With less than 100 collaborators worldwide, the $g−2$ collaboration is a modest-sized collaboration by the standards of particle physics. The collaborators come from Boston University; Brookhaven National Laboratory; Budker Institute of Nuclear Physics, Novosibirsk, Russia; Cornell University; Fairfield University; Rijksuniversiteit Groningen, the Netherlands; University of Heidelberg, Germany; University of Illinois; KEK Japan; University of Minnesota; Tokyo Institute of Technology; and Yale University. The experiment is organized along the lines of almost all large particle experiments, with cospokespersons and a formal management structure. The word spokesperson has different meanings. In high-energy physics, the spokespersons are the leaders of the collaboration, but the duties vary greatly from collaboration to collaboration. In $g−2$, the cospokespersons are Vernon W. Hughes (Yale) and B. Lee Roberts (Boston). Gerry Bunce (Brookhaven) is the project manager, and William M. Morse (Brookhaven) is resident spokesperson.

The principal governing body of the experiment is an executive committee (EC) that consists of the cospokespersons plus representatives from each of the institutions. The chairpersons of the EC have been drawn from the senior members of the collaboration. Over the past twelve years there have been five different chairpersons (B. Lee Roberts, David Hertzog–Illinois, Priscilla Cushman–Minnesota, Klaus Jungmann–Heidelberg), with the 2002 chairperson being James Miller (Boston).

The first paper was signed by over 100 authors from fourteen institutions. The current collaboration list, links to papers, and pictures of the experiment are available on the Muon ($g−2$) Collaboration's Web site. Since the scale of the project was large, teams were set up to carry out the many tasks, and their responsibilities and the names of the team leaders are also given on the Web site.

Many graduate students and postdoctoral research associates have made major contributions to the experiment, and they are also listed on the Web site along with their institutions. The g-2 experiment has served as an important training ground for graduate and undergraduate students.

See also: LEPTON; MUON, DISCOVERY OF

Bibliography

Brown, H. N., et al. "Precise Measurement of the Positive Muon Anomalous Magnetic Moment." *Physical Review Letters* **86** (11), 2227–2231 (2001).

Muon (g-2) Collaboration. <http://phyppro1.phy.bnl.gov/g2muon/index.shtml>.

National Institute of Standards and Technology. <http://physics.nist.gov/cuu/Constants/index.html>.

Particle Data Group. <http://pdg.lbl.gov>.

Rohlf, J. W. *Modern Physics from α to Z⁰* (Wiley, New York, 1994).

B. Lee Roberts

EXPERIMENT: SEARCH FOR THE HIGGS BOSON

The search for the Higgs boson has been the premier high-energy physics goal of the late twentieth century and continues into the twenty-first century. The Standard Model of particle physics has been incredibly successful at describing the electromagnetic and strong and weak nuclear forces. Yet, at the same time, it lacks confirmation of some key elements, most importantly direct evidence for the Higgs boson.

Simply put, the Standard Model is a nonsensical theory without the Higgs boson, or something very much like it. It solves several theoretical problems, providing mathematical consistency and experimental predictability to the theory and giving elementary particles their mass. Even though the Standard Model encompasses all known particles (quarks and leptons, which make up matter, and bosons, which transmit force between matter particles), without a Higgs boson it gives different answers to physical problems; for example, the rate of nuclear fusion reactions in the Sun, depends on how one calculates them from the starting equations. Another problem, which would show up in experiment, is that the probability of the weak bosons W and Z interacting with each other in high-energy collisions is greater than 100 percent, a clear impossibility.

The Higgs boson is actually a scalar set of fields consisting of four components, which are present everywhere in the universe. Scalar means the four field components have numerical values representing their strength for all points in space, much like a frying pan over a stove has a certain temperature at every point inside the pan. The constant interaction of Standard Model particles with these Higgs fields slows them down, making it appear that they have a mass (massless particles must travel at the speed of light). Three components are absorbed by the weak bosons, giving them mass, while the fourth component becomes a physical particle with a mass of its own. Furthermore, the weak force becomes weaker than electromagnetism (hence the name) as well as acting over only very short distances, the size of atomic nuclei, which is an experimental fact. Thus mass, as perceived by us, is not an inherent property, but a result of interactions between matter and the Higgs field.

This has several consequences. First, calculations show that for the Higgs boson to solve the high-energy weak boson scattering probability problem, the Higgs particle cannot have a mass greater than a certain value, called the unitarity constraint, about two million times the electron mass. This value is low enough that physicists are almost certain to see it in the next generation of experiments, starting in 2007 (if not sooner), making the theory experimentally relevant. Second, the Higgs boson interaction strength with particles is proportional to their mass. Third, the Higgs boson is spinless since it is a scalar field and does not have electric charge. These and other quantum properties are all precisely defined and can be tested. To fully understand the Standard Model, scientists must find the Higgs boson, measure its properties, and compare them to theory. The motivation is thus to test current scientific knowledge and learn more about the structure of the universe at a fundamental level.

The LEP Search for the Higgs Boson

To produce a Higgs particle in an experiment, matter must be collided at energies above the mass of the Higgs boson (mass and energy are equivalent). As collider experiments have become more energetic, lack of a Higgs boson observation has pushed the allowable mass to higher values. The limit is still far below the unitarity limit, but the gap is closing fast and will be completely accessible by about 2010.

The most recent collider engaged in this quest was the Large Electron-Positron (LEP) collider at the European Laboratory for Particle Physics (CERN) near Geneva, Switzerland. LEP was a circular collider in a 27-kilometer tunnel 100 meters underground. Small bunches of electrons and positrons (antielectrons) were accelerated around the ring in opposite directions, passing through each other at four interaction regions, where a detector belonging to one of the four experimental collaborations, Aleph, Delphi, Opal, and L3, observed the collisions.

High-energy physics experimental collaborations are an interesting study in their own right. Each LEP group consisted of a few hundred physicists, from all over the world, mostly employed by universities or institutions other than CERN. There is really no way to organize from the top such a diverse group of people, but high-energy physicists are adept at forming self-organizing structures.

The four collaborations took different approaches toward building their detectors, so each had a different performance level of particle identification and measurement, and had to be carefully calibrated, a tedious, months-long process frequently relegated to graduate students. The detectors themselves were enormously large, complicated machines. L3, for example, was four stories tall and had almost as much iron in it as the Eiffel Tower. Supplied continuously by about the same amount of power as a lightning strike, it had millions of electronic components measuring each collision, which occurred 44,000 times every second. Only about three of those collisions every second were potentially interesting, and each event recorded required about one hundred kilobytes of data storage. Computers running the detector had approximately one hundred thousandth of a second to initially judge the value of each potential collision. The other detectors had a similar task and contained about the same number of electronic instruments. Figure 1 shows an illustration of the Aleph detector. Combining the results of the four experiments was a challenge even beyond that of organizing the individual collaborations. For the Higgs boson search this effort was led by CERN physicist Patrick Janot and in the end turned out to be successful. The key to this success is the scientific process itself, the ability to cross-check results with those previously established.

Particles and antiparticles annihilate when they collide, generating a slew of other particles that stream out in all directions from the collision point, or vertex. Many of these outgoing particles are unstable and decay successively in a chain until stable final states such as electrons, protons, and photons are reached. Physicists study collisions for deviations from the expected Standard Model behavior of these collisions, which signals new phenomena. This may consist of anomalous events from production of new particles or may show up as subtly different emission patterns of known particles.

The vast majority of collisions are uninteresting, involving thoroughly studied phenomena. The uninteresting stuff is called background, while new physics is called signal. The purpose of the detector is to identify and measure all of the outgoing particles in a collision, while data analysis separates signal from background; in this case, sifting through data to find the handful of expected Higgs particles amongst the millions of background events.

For the Higgs particle, the experiments at LEP looked for events where the electron and positron annihilated to form a virtual Z boson, which could then radiate a Higgs boson to become a real Z boson. (Virtual means not real and, in this case, too heavy. Real photons, for example, are perceived as light, while virtual photons comprise electric and magnetic fields, which transmit force.) This would happen very rarely, perhaps one in every thousand Z events. Then experimentalists had to consider how the Z and Higgs bosons might be expected to decay. Only some of the decays would be observable, with some decays having larger background than others, mostly from Standard Model processes faking a Higgs boson event by producing the same final state particles in a similar pattern.

Since the Higgs boson mass previously had been excluded to be larger than twice the bottom quark mass, and the Higgs couples preferentially to heavy particles, any Higgs boson LEP could create was expected to decay most of the time to a pair of bottom quarks. This would be fortuitous because bottom quarks are relatively long-lived and will decay a small distance away from the vertex, a fraction of a millimeter to a few millimeters. The experiments could see this separation with devices called micro-vertex detectors. Such an event is often easily recognizable (see Figure 2). The Z boson may decay into any pair

FIGURE 1

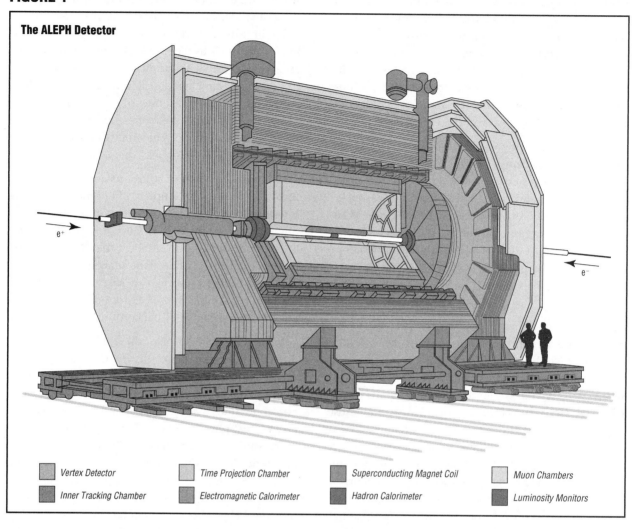

The ALEPH Detector

■ Vertex Detector	■ Time Projection Chamber	■ Superconducting Magnet Coil	■ Muon Chambers
■ Inner Tracking Chamber	■ Electromagnetic Calorimeter	■ Hadron Calorimeter	■ Luminosity Monitors

Cutaway illustration of the Aleph detector. Particle collisions occurred near the center of the vertex detector. Inner detector components measured particle direction while outer components measured energy. The magnet produced a field that caused charged particles to follow a curved path; the direction revealed whether the charge was positive or negative. The detectors of the other LEP experiments were similar.

of matter particles, and all of these channels were considered separately. Neutrinos are nearly impossible to detect, so they leave the detector invisibly, but since the beam energy is known, a deficit of total outgoing energy can easily tag the event as an invisibly decaying Z boson. Thus, experiments looked for Z boson events accompanied by a pair of bottom quarks that together had enough energy to come from a very heavy object, the Higgs boson candidates. The total number of events in all decay channels was compared to the total number expected from all Standard Model processes using a complex statistical procedure to determine if there was a significant excess of events.

Data analysis is a long and painstaking procedure. Experiments must compare measured particle energies against machine performance at the time of the event, since the accelerator energy and collision rate change slightly over time. These changes affect the interpretation of individual events and take additional computer time to analyze. Data must also be corrected for known problems in the detector, mapped out during detector calibration. Extraneous photon radiation from the colliding beams can alter

FIGURE 2

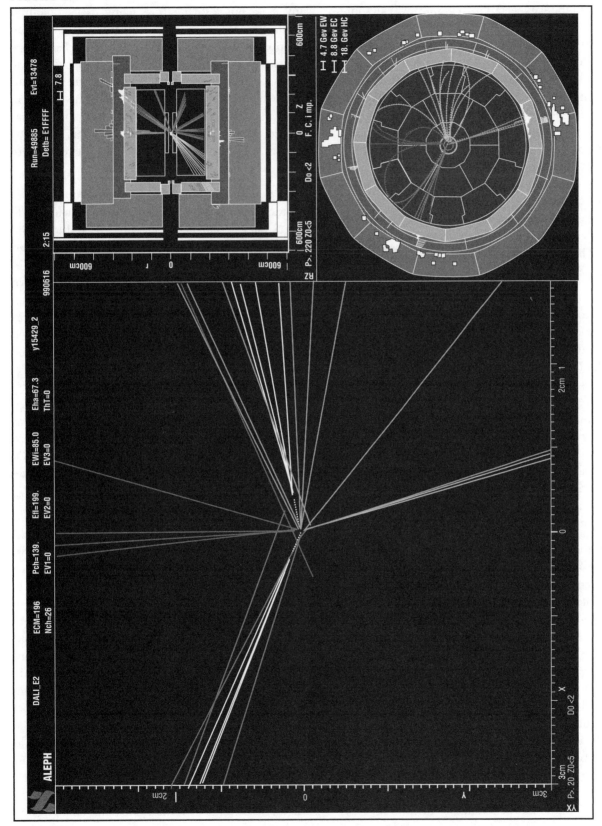

Computer generated display of a ZZ event. One Z decayed to a pair of light quarks, which subsequently decayed a short distance away from the primary vertex, shown by the white dashed line. And the other decayed to a pair of bottom quarks.

energy measurement, throwing off comparison to Standard Model expectations. Cosmic rays can sometimes penetrate the earth even to the experimental halls 100 meters underground and register in the detectors. The complicated electronics will even occasionally "hiccup," producing the exceedingly rare event with strange characteristics that just happens to look like new physics. Data analysis software, written by the physicists, can have hidden bugs that might not ever affect most data but can affect the search for new physics. Many of these problems can be dealt with by careful calibration and comparison with the other running experiments. But in these types of experiments, one cannot take a single event that looks like a Higgs boson to be proof of its existence. This turned out to be exactly the problem with which LEP was faced.

Before LEP, the Higgs boson mass had been excluded to be so high that if LEP could produce it at all, it would be only barely. As the LEP energy was increased, lack of observation meant that the new mass limit was essentially the machine energy, minus the Z boson mass, minus a little bit more—LEP physicists were always hunting right at the edge of accelerator output. The number of Higgs bosons produced at the limit of machine energy would be very small, so the experiments had to rely on analysis of low-statistics events. This is a dangerous situation for an experiment, as the probability for background to give just a handful more events that look signal-like turns out to be not so small.

By the fall of 1999, LEP had been pushed nearly to its design limit of about 200 GeV. (1 GeV = 1 billion electron volts, the energy an electron gains when accelerated through a potential of 1 volt. The mass of an electron is 0.00051 GeV.) Still the Higgs boson had not been found. Since the Z boson has a mass of about 91 GeV, this put the Higgs boson limit at about 105 GeV. However, other indirect but very precise data suggested that for all the parameters of the Standard Model to fit together properly, the Higgs boson must be very light, probably near the energy where LEP was hunting. However, LEP was scheduled to be shut down to make way for the construction of the Large Hadron Collider (LHC), a much more powerful proton-proton collider to be built in the same tunnel and scheduled to turn on in 2007.

The LHC could not only discover a Higgs boson of any mass but also measure many of its quantum properties. However, in the meantime the proton-antiproton accelerator Tevatron at the Fermi National Accelerator Laboratory (Fermilab) outside Chicago, Illinois, was being upgraded for its second run, which would start in March 2001. This was a sensitive issue to LEP, because if the Higgs boson was indeed only slightly heavier than the energy LEP was able to access at that time, as data suggested, discovery might go to another laboratory.

The LEP collaborations petitioned for a delayed shutdown and presented a scheme to squeeze additional energy out of the machine by reducing the number of electrons and positrons being accelerated. Energy was more important than number of collisions for the Higgs search. An ingenious pattern of running was devised that boosted the energy when electrons and positrons were lost as the beams circulated. However, this made data analysis much more difficult as the exact number of particles in the beam and beam energy had to be tracked very precisely as a function of time. Finally, the acceleration cavities would be pushed well beyond their design limits, possibly to failure. The rationale for this was that LEP was going to be dismantled anyway, so there was nothing to lose—"go for broke," literally. CERN management approved the plan and granted a one-year extension.

By October 2000 the mass limit had been pushed several GeV higher, and the collaborations still had not discovered the Higgs boson. But in the last month of extension they announced tantalizing hints of a possible signal. Not all of the collaborations' data agreed, but this was reasonable given the small number of candidate events. L3 had one promising candidate event, while Opal and Delphi saw none, but Aleph had about three candidates, at a mass of about 115 GeV. The Aleph group's Higgs boson search subgroup was led by Professor Sau-Lan Wu of the University of Wisconsin at Madison, who had years before been involved in the discovery of the gluon (the carrier of the strong nuclear force). Her postdoctoral assistant Stephen Armstrong and graduate student Jason Nielsen had been among the first LEP experimentalists to devise a real-time analysis program, to search for candidate events as the data came in, eliminating some but not all of the postrun data analysis that normally took so long.

The collaborations petitioned for and received another month of running but produced no additional candidates. Despite vigorous protest from some members of the collaborations, CERN Director General Luciano Maiani made the decision not to delay LHC construction any longer, and LEP was permanently shut down. This caused a great deal of acrimony among some members, some going so far as to ridicule the director publicly—this was completely in tune with the history of CERN as the experiments there attracted some very colorful personalities with oversized egos. Often these experiment-management clashes became publicly visible. However, the director had made the decision based on the lack of confidence in the result among members of the LEP committee, the scientific advisory board. This perception was mirrored by physicists outside of CERN. When Chris Tully, an assistant professor of physics at Princeton University and member of the L3 collaboration, presented LEP summary results for the Higgs boson search at Fermilab in December 2000, the general reception was skepticism that LEP had seen any signal at all. But in a final rebuke to the shutdown, Aleph made holiday greeting cards from one of their Higgs candidate event displays and sent them to colleagues around the world.

In July 2001 LEP presented an update that lessened confidence in the candidate signal. More thorough analysis had reduced the statistical significance of the data almost to nothing. L3 was decidedly less confident about its event, as the decay particles went into a region of the detector known to have measurement problems.

Tevatron and the LHC

The Fermilab Tevatron began its second run in March 2001, with an upgraded machine that promised to deliver 20 times more collision and 10 percent greater beam energy than its first run. LEP's supreme final performance, however, set a new mass limit that was considerably higher than anticipated. This will make it difficult, but not impossible, for the Tevatron to find the Higgs boson. If the Tevatron performs well before the LHC can analyze its first data, perhaps in 2007, then its two detectors CDF (see Figure 4) and D0 have good potential to observe a Higgs boson up to twice the Z boson mass. However, many

aspects of the new machine and detectors' performance are not yet known well enough to determine the Tevatron's true potential.

Higgs boson search channels at a proton collider are very different, primarily in the production mode. The largest rate would come from a gluon pair fusing to form a Higgs boson. (Massless gluons do not couple directly to the Higgs boson but can produce a virtual top quark pair, which do couple. This is known as a loop-induced process, a feature of quantum mechanics.) Another method is an incoming pair of lighter quarks annihilating to form a real top quark pair, one of which may radiate a Higgs boson. While this rate is quite small, top quarks are very distinctive and have a much smaller background. This has been called the "Cinderella discovery mode" for a Higgs boson by Fermilab theorist Stephen Parke because at first glance the low rate is uninteresting, but the process appears beautiful when one considers it more carefully.

Proton-antiproton collisions are much messier than electron-positron collisions: there are more

Photograph of the upgraded CDF detector before installation. The general configuration is similar to the Aleph detector. CREDIT: COURTESY OF FERMILAB PHOTO. REPRODUCED BY PERMISSION.

Standard Model backgrounds, and they are more complicated to calculate. The tradeoff is the ability to reach much higher energies. The Tevatron has the energy advantage over LEP but must deliver enough collisions to avoid the situation of too few candidates that LEP experienced. Three or four candidates in the Tevatron environment may be less clean than those at LEP and must be considered more carefully.

While it might take six years to confirm the LEP candidate, the Tevatron can rule it out in about two years. The lack of candidate events is a more powerful statistical signal of no Higgs particle than the existence of a few events is of a possible Higgs boson.

The LHC is expected to turn on in 2007. Its role is not just discovery, however, but also to measure a candidate Higgs boson's properties for comparison against theory. This is possible because of the LHC's significantly higher beam energy, seven times that of Tevatron, and enormously greater data taking capability.

While the Higgs boson as discussed in the first section is the most anticipated signal, there are several variant theories for the Higgs mechanism. Some models have two sets of Higgs fields, resulting in additional particles that would be produced, with slightly different properties. Other theories incorporate supersymmetry as well. The LHC has great potential to distinguish these different types of Higgs bosons or even more complicated scenarios. It is also possible that the LHC won't find a Higgs boson but instead will discovers a different mechanism for giving mass.

See also: BASIC INTERACTIONS AND FUNDAMENTAL FORCES; BOSONS, GAUGE; BOSON, HIGGS; CASE STUDY: LHC COLLIDER DETECTORS, ATLAS AND CMS; HIGGS PHENOMENON; STANDARD MODEL

Bibliography

Appenzeller, T. "A Search for the Weight of the Matter." *U.S. News and World Report* **129**, 58 (October 16, 2000).

Henderson, M. "Atom Smasher Shut as 'Holy Grail' Glimpsed." *London Times* (October 10, 2000), p 8.

Kotulak, R. "Chasing the Key to the Cosmos." *Chicago Tribune* (December 3, 2000), p 1.

Monastersky, R. "A Crash Heard Around the World." *The Chronicle of Higher Education* **47**, A13 (January 19, 2001).

Seife, C. "CERN's Gamble Shows Perils, Rewards of Playing the Odds." *Science* **289** (5488), 2260 (2000).

Taubes, G. *Nobel Dreams* (Random House, New York, 1986).

Weiss, P. "Jiggling the Cosmic Ooze." *Science News* **159** (10), 152 (2001).

David Rainwater

EXTRA DIMENSIONS

See STRING THEORY

EXTRACTION SYSTEMS

Beam injection and extraction are critical techniques in circular accelerators. Special conditions must be created to inject and extract the beam. An extracted beam may be transferred to a subsequent accelerator, or it may be directed toward a target.

Some experiments do not require beam extraction, most notably when internal targets are used. An internal target may be a puff of gas or a solid material that is moved into the path of the beam. Perhaps the most important type of internal target, however, is a beam traveling in the opposite direction. This configuration is known as colliding beams and is used extensively for high-energy particle experiments.

Single-Turn Extraction

For single-turn extraction, a special magnet, known as a kicker magnet, is switched on to steer the beam away from its normal orbit. The switching must be almost instantaneous; any beam that passes through the kicker before the full field is obtained tends to be lost because it is not bent far enough. The fast-rise time requirement limits the bending that can be obtained, and kickers are frequently paired with a special type of magnet known as the Lambertson magnet (named for the inventor Glen Lambertson). This magnet does not need to be switched. It has a high magnetic field in one region, which is separated by a thin wall of iron (known as a septum) from a field-free region. The beam usually circulates in the field-free region, and the kicker displaces the beam across the septum to be bent further by the strong magnetic field. Single-turn extraction is used almost universally for transferring beams between machines.

Slow Extraction

Single-turn extraction can be used for directing the beam toward an external target, but the duration of the beam pulse is equal to the time that it takes the beam to make a single turn around the accelerator, typically 1 to 10 μsec for high-energy accelerators. The resulting rate of interactions would generally be too high for effective experimentation. An alternative extraction technique, known as slow extraction, can extend the duration of the extracted beam pulse from milliseconds to many seconds.

During normal accelerator operation, particles oscillate around a nominal closed orbit. When the number of oscillations per turn (the tune) is such that the particle motion repeats after a small number of turns, certain small deviations of the magnetic fields from their nominal values have large cumulative effects after many turns. This condition is known as a "resonance," and particle oscillations may not be stable, that is, they may grow in time. Resonances are usually undesirable but can be controlled and utilized to extract the beam slowly.

Extraction systems have been built using both the 1/2 and 1/3 resonances. The particle motion repeats every two or three turns, respectively, for these resonances. For the 1/2 resonance, quadrupole magnets are introduced deliberately to create deviations from the nominal magnetic field; the 1/3 resonance requires sextupole magnets. A beam may be ex-

tracted by changing the accelerator tune toward the resonance. Typically particles with large oscillation amplitudes become unstable first (when the accelerator tune is relatively far from the resonance); particles with smaller oscillation amplitudes become unstable as the tune approaches the resonant value. The rate of change of tune controls the rate at which particle oscillations become unstable and therefore controls the rate of extraction.

If no special measures were taken, the particles with growing amplitudes would eventually run into the accelerator walls and be lost. They are extracted in a controlled way by placing two parallel plates with an electric field between them in the accelerator beam chamber. The growing particle amplitude eventually crosses into the region between the plates and is deflected by the electric field. The plate nearest the circulating beam is known as a septum and is invariably made of a grid of wires to minimize the mass and hence the number of beam particles that interact in the septum. The wire septum performs for slow extraction the same function that the kicker performs for single-turn extraction. A Lambertson magnet is normally used as a second bending device in slow extraction systems.

See also: ACCELERATOR; ACCELERATORS, COLLIDING BEAM: ELECTRON-POSITRON; ACCELERATORS, COLLIDING BEAM: ELECTRON-PROTON; ACCELERATORS, COLLIDING BEAM: HADRON; ACCELERATORS, EARLY; ACCELERATORS, FIXED-TARGET: ELECTRON; ACCELERATORS: FIXED-TARGET: PROTON; BEAM TRANSPORT; INJECTOR SYSTEM

Bibliography

Edwards, D. A., and Syphers, M. J. *An Introduction to the Physics of High Energy Accelerators* (Wiley, New York, 1992).

John Marriner

F

FAMILY

Investigations into the structure of matter at short distances have revealed three families (sometimes called generations) of elementary particles and antiparticles, identical in all respects except for the values of their masses. Stable matter consists only of particles in the first family. The members of the other two families are heavier unstable particles that existed at the earliest moments of the universe and are routinely produced in particle accelerators.

All family members share some attributes, such as mass, momentum, and spin-$\frac{1}{2}$, but their interactions differentiate them into two classes, quarks and leptons. Quarks are subject to strong, weak, and electromagnetic interactions, whereas leptons have no strong interactions. Each family consists of one negatively charged lepton, its associated neutral lepton, and two quark flavors, of fractional electric charges $\frac{2}{3}$ and $-\frac{1}{3}$, each coming in three colors. The first family that makes up all matter contains the electron e with charge -1 and its associated neutrino ν_e, the charge $\frac{2}{3}$ up quark (u), and the down quark (d) of charge $-\frac{1}{3}$. The second-family members are the negatively charged muon μ and its neutrino ν_μ, the charm quark (c), and the strange quark (s). The third family is made up of the tau τ lepton and its neutrino ν_τ, the top quark (t), and the bottom quark (b). The third-family members are the heaviest.

This classification is remarkably simple by historical standards. By the end of the nineteenth century, chemists and physicists had determined all matter to be arrangements of a finite number of electrically neutral atoms, each with definite chemical properties but an unknown structure. Their findings were summarized in Dmitry Mendeleyev's periodic table of elements. A century later, the nature of atoms, the forces that hold them together, and their interactions had all been understood. This knowledge was embodied in the Standard Model of elementary particle physics.

An earlier portrait of elementary particles, circa 1935, shows two nuclear particles, the neutron and proton, and two leptons, the electron and its associated neutrino. This picture of the first family stayed essentially unchanged for thirty-five years. In 1970 a Rutherford-type experiment was performed by a SLAC-MIT group at such high energy as to probe directly atomic nuclei. This produced an amazing result: neutrons and protons are not elementary but rather an assembly of fractionally charged particles, quarks, as earlier hypothesized by Murray Gell-Mann and George Zweig. Protons and neutrons each contain three quarks, *uud* and *udd*, respectively.

TABLE 1

The Three Families of Elementary Particles		
Family	**Quarks**	**Leptons**
First	*u, d*	e, ν_e
Second	*c, s*	μ, ν_μ
Third	*t, b*	τ, ν_τ

CREDIT: Courtesy of Pierre Ramond.

Quarks do not exist as free particles but are found only in tightly bound triplets, or quark-antiquark pairs. Quantum chromodynamics (QCD) explains this peculiar behavior in terms of the interactions of quarks and massless spin-1 gluons, which couple to the quark colors. This yields a new portrait of the first family: three colors of up and down quarks, the electron, and its neutrino. These are subject to strong interactions, described by eight massless gluons, and electromagnetic interactions, due to exchanges of massless stable photons γ. Weak interactions result from exchanges of two massive unstable particles, the charged W bosons and the neutral Z bosons, both predicted by the Standard Model and found experimentally at the European Laboratory for Particle Physics (CERN) in the early 1980s.

Over a seventy year period, two other families emerged, starting with the discovery in the 1930s of the muon in the by-products of cosmic-ray collisions with the atmosphere. I. I. Rabi's famous quip "who ordered that?" remains unanswered today. This was followed by the discovery of unstable nuclear particles, interpreted as containing a new quark flavor, the strange quark. The muon was found to have its own neutrino, but it was not until 1974 that evidence for the charm quark was produced at the Brookhaven National Laboratory (BNL) and at the Stanford Linear Accelerator Center (SLAC), in the form of a long-lived charm-anticharm bound state, called J/ψ, completing the second family. In 1977 the first sign of a third family, the heavy tau lepton, was also found at SLAC. The rest of the third family was discovered at the Fermi National Laboratory: the bottom quark in 1979, the top quark in 1995, and the tau neutrino in 2000.

There is no experimental sign of a fourth family of elementary particles, and there is a strong argument against its existence. The Z boson lifetime was measured with great precision at SLAC and CERN in the early 1990s, and all its decay channels have been identified with particle-antiparticle pairs of the first three families. A carbon-copy fourth family with a light neutrino would yield a new way for Z bosons to decay into the fourth neutrino-antineutrino pair, in disagreement with experiment: the Z boson counts the number of weakly interacting neutrinos and finds only three. There is no fourth family with a light neutrino!

The wide disparity between elementary particle masses remains a mystery and the subject of much theoretical research; for example, the recently discovered top quark is about 175 times heavier than hydrogen, whereas the up quark mass is two-hundredths that of hydrogen. The situation is similar for the charged leptons: the electron mass is 1/200 of the muon, which is ten times lighter than the tau. Until 1998, when the Super-Kamiokande (SuperK) underground detector in Japan presented convincing evidence to the contrary, the three neutrinos were believed to have escaped this mass hierarchy as each appeared to have zero mass. In 2001 the Sudbury Neutrino Observatory (SNO) in Ontario, Canada, presented further evidence, corroborating a long string of earlier experiments that had found a deficit of solar neutrinos. According to SuperK, ν_μ's produced in the collision of cosmic rays with the atmosphere oscillate into ν_μ's, whereas SNO determined that electron neutrinos produced in the solar core oscillate on their way to Earth into other neutrino flavors, ν_μ's and/or ν_τ's, so that an electron-neutrino detector will see a deficit. Oscillations require neutrino flavors of different masses. In addition, neutrino massess are extremely tiny compared to that of their charged partners; for instance, ν_e is at least 1 million times lighter than an electron.

Much current research centers around the following questions: why do three families exist? Nature appears to operate like a bureaucracy, requiring triplicate copies, without any apparent reason. Is this a sign that further unification is needed? However, this tripling also brings about important new phenomena such as CP violation and neutrino oscillations. What is the origin of the quark and charged lepton mass hierarchies? Why are neutrino masses so tiny? What type of new matter awaits discovery?

As the investigation of the fundamental constituents and their interactions at higher energies continues at Fermilab's Tevatron and CERN's Large Hadronic Collider (LHC), physicists expect to find new types of matter. One noteworthy speculation, supersymmetry, suggests that the new matter is also organized in three families of elementary particles, with exactly the same attributes, except for spin! All elementary spin-$\frac{1}{2}$ particles of the three families have heavy counterparts with no spin. The spinless partners of the light spin-$\frac{1}{2}$ neutrinos, called sneutrinos, are naturally heavy and do not affect the Z boson lifetime. Such speculations await the verdict of experiments.

See also: EIGHTFOLD WAY; FLAVOR SYMMETRY; LEPTON; QUARKS; STANDARD MODEL

Bibliography

Rolnick, W. B. *The Fundamental Particles and Their Interactions* (Addison-Wesley, Reading, MA, 1994).

Pierre Ramond

Italian physicist Enrico Fermi (1901–1954) won the Nobel Prize in Physics in 1938 for identifying new radioactive elements produced by neutron irradiation and for building and controlling the first nuclear chain reaction. CREDIT: COURTESY OF BETTMANN/CORBIS. REPRODUCED BY CORBIS CORPORATION.

FERMI, ENRICO

Enrico Fermi was born on September 29, 1901, and grew up in Rome; his father, Alberto Fermi, was a chief inspector in the Railway Ministry; his mother, Ida de Gattis, was an elementary school teacher. Enrico, his sister, and his brother went to secular schools and were raised as agnostics. A friend of his father lent the young Fermi mathematics books and influenced him to attend the university in Pisa. While an undergraduate, he became Italy's foremost expert on Einstein's theory of general relativity. After Pisa, Fermi received a fellowship for a year in Göttingen. A special professorship in theoretical physics was created for Fermi at the University of Rome in 1927, and at age twenty-six he became the youngest professor in Italy since Galileo. In 1934, he married Laura Capon, who was Jewish. The couple had two children, Nella and Giulio.

Fermi wrote more than 270 articles during his life and made outstanding contributions to most areas of twentieth century physics. At age 24, in January 1926, he wrote the Fermi statistics, one of his most significant and lasting theoretical contributions. It set forth a method for calculating the behavior and properties of systems that obeyed quantum mechanical rules for two or more electrons. Seven months later, Paul Dirac independently derived the same statistical mechanics. Almost all known elementary particles are called fermions because they obey the rules of Fermi-Dirac statistical mechanics.

In 1933 Fermi successfully constructed a formal theory of the beta-decay of radioactive nuclei. In the radioactive process a neutron changes to a proton by creating an electron and a neutrino. Fermi's theory of beta decay introduced a fourth fundamental force, called the weak interaction, to be added to the previously known three: the gravitational force between masses, the electromagnetic force between electric charges, and the strong force between the particles in nuclei. A new fundamental constant, called the

Fermi constant, determines the strength of the weak interaction. In the 1960s the electromagnetic and the weak interactions were combined in a unified theory, and Fermi's weak interaction and well-measured constant have continued to play a major role in particle physics more than half a century later.

After the discovery of the neutron by Chadwick in 1932, Fermi's group in Rome started an experimental program bombarding all available elements with neutrons. Fermi was awarded the Nobel Prize in Physics for the neutron work. He used the trip to Sweden to escape with his family from Italy and Mussolini's anti-Semitic laws. Many people know Fermi as the architect of the atomic age—the scientist who built and controlled the first nuclear chain reaction, the basis for the peaceful uses of nuclear energy. From 1939 until the end of World War II, he played a major role in the U.S. war project that led to the atomic bomb.

After the war, his major interest was in elementary particle physics which includes the properties of the particles, their decays, their production, and their interactions. Measurements are made of what happens in collisions between particles or when particles decay.

Fermi's most extensive particle experiments were studying the interaction of pi mesons and protons. In 1935 Yukawa postulated that pi mesons account for the strong nuclear force that binds nucleons together in short distances. The pi meson is pictured as being emitted by one nucleon and absorbed by another nucleon that binds the two nucleons together. Pi mesons were first observed in cosmic rays. Later, they were found to be produced in the accelerators built in the late 1940s. Fermi and his associates hit protons with beams of negatively or positively charged pi mesons (pi$^-$ or pi$^+$) using the synchrocyclotron at the University of Chicago. They measured the angles and energies of particles emitted from the collisions. They found that in some of the collisions between a pi$^-$ and a proton the particles turned into a neutron and an uncharged pion. The combination of a neutron and an uncharged pion constituted the discovery of short-lived baryons; baryons are particles like the neutron and proton with similar or heavier masses, and the masses are conserved in reactions.

During the 1940s and 1950s, Fermi also was active as a theoretical physicist in a variety of fields. In particle physics, Fermi became concerned that some of the twenty-one particles known in 1950 might not really be "elementary particles," that is, structureless particles. He wrote a provocative paper with C. N. Yang entitled "Are Mesons Elementary Particles?"

Fermi was a fabulous teacher and had an enormous influence on several generations of experimental physicists. He tried to impress on students the importance of understanding theory so that, when planning experiments, they could estimate in advance possible results. He enjoyed interacting with students informally and working with them in the laboratory. He put a great deal of effort into preparing classroom and formal lectures. His lecture notes have been widely circulated and some are published.

Fermi's lectures frequently started with simple explanations of an abstract theory, an example of which can be found in his booklet "Elementary Particles" (based on Fermi's 1950 lectures at Yale). Fermi begins his discussion with quantum field theory as a theoretical framework for the interaction of almost all particles. In later chapters, he shows how to obtain quantitative results when calculating a broad variety of phenomena including the range of nuclear forces, the products of the annihilation of an antinucleon with a nucleon, and the decay of weakly interacting particles. In Fermi's "Lectures, on Pions and Nucleons," edited by B. T. Feld and published in *Nuovo Cimento* in 1955, he explains simple group theory. Group theory provided the classification schemes that became so important in the next decade. Feld's 1969 book *Models of Elementary Particles* exemplifies Fermi's style of combining theory and quantitative experimental results.

Three of Fermi's graduate students (O. Chamberlain [1959], T. D. Lee [1957], and C. N. Yang [1957]) won Nobel Prizes for their work in elementary particle physics. Fermi died of intestinal cancer in Chicago on November 28, 1954.

See also: DIRAC, PAUL; EINSTEIN, ALBERT; NEUTRINO, DISCOVERY OF; PAULI, WOLFGANG; QUANTUM FIELD THEORY; QUANTUM STATISTICS

Bibliography

Amaldi, E.; Anderson, H. L.; Persico, E.; Rasetti, F. Segrè, E.; Smith, C. S.; and Wattenberg, A., eds. *Collected Papers of*

Enrico Fermi, Vols. I and II (Chicago University Press, Chicago, IL, 1962).

Fermi, E. *Elementary Particles* (Yale University Press, New Haven, CT, 1951).

Fermi, E. "Lectures on Pions and Nucleons." *Nuovo Cimento Supplemento* **2** (1955).

Segrè, E. *Enrico Fermi Physicist* (University of Chicago Press, Chicago, IL, 1940).

Wattenberg, A. "The Fermi School in the United States." *European Journal of Physics* **9**, 88–93 (1988).

Albert Wattenberg

FERMILAB

In December 1966 the 6,800-acre site called "Weston," in the Chicago suburbs of northern Illinois, won the national competition for the U.S. Atomic Energy Commission's newest facility, a 200-GeV (billion electron volt) particle accelerator that would become the National Accelerator Laboratory. This frontier research center built in America's heartland was to become, for the last quarter of the twentieth Century and into the new millennium, the highest energy hadron accelerator in the world.

The site competition had been a heated one attempting to reconcile physics interests on both coasts and in the Midwest while navigating the problems of planning to build a physics facility for the future in a difficult period of American history characterized by an unpopular war with its related budget problems and social unrest. In addition to difficult but necessary political cooperation, the project would be a challenge with its unprecedented scale and demands on technology. A new organization, Universities Research Association, Inc. (URA), was formed to manage the new facility according to a new approach, as a "truly national laboratory." Physicist Norman Ramsey of Harvard University, skilled in diplomacy and astute in the political sphere, presided over URA in its early years.

A strong leader was needed to create the Laboratory, and in early 1967 URA offered Robert Rathbun Wilson the position of director. Wilson was an experimental physicist and a specialist in accelerator design and construction. He had a distinguished career at Berkeley, Princeton, Los Alamos, Harvard, and Cornell and was respected by the scientists of the Atomic Energy Commission and the academic leaders of URA.

Wilson selected Edwin L. Goldwasser, an experienced experimental physicist from the University of Illinois, to be his deputy director. The two worked effectively with Ramsey, guiding all aspects of federal funding, local support, development, design, construction, and management of the laboratory, as well as shaping its research program. The first offices were established in rented space in Oak Brook, Illinois, on June 15, 1967, while the land to be donated by the state of Illinois was obtained from its former owners, some fifty farming families and the residents of Weston.

Wilson recruited a corps of explorers for his frontier laboratory and launched an aggressive construction program to deliver the machine that would lead American high-energy physics into a new realm of human comprehension of nature. Having moved from Oak Brook to the Weston site in late 1968, Wilson imparted his aesthetic sense of design, feeling "a laboratory needn't be ugly to be inexpensive." Implementing design innovations and applying the latest understanding of accelerator physics to the four miles of magnets in the underground tunnel called the Main Ring, Wilson drove the machine to completion ahead of schedule, beyond its original design energy of 200 to 400 GeV, with more experimental areas in which to conduct research, and the project still came in under the authorized budget of $250 million. The United States Atomic Energy Commission was pleased and appreciative to have the highest-energy accelerator in the world when it was officially completed and successfully operated at design energy on March 1, 1972.

Wilson had been given assurance from President Lyndon B. Johnson and the Atomic Energy Commission that the Laboratory would be considered "*the* National Accelerator Laboratory," not only in name but also in reality, and it was given priority in funding and development. Wilson therefore expected first-rank recognition for the Laboratory and stressed an enlightened vision in all areas, among them physics research, respect for human rights, preservation of the environment, ethical conduct, an aesthetic sense of the whole of the Laboratory, and an idealistic approach to all aspects of research and life at his "Science City" in order to "produce a small acceleration to society." (Wilson 1968).

In 1970, physicists from around the world submitted eighty-two proposals to conduct particle physics research at the new facility. Each physicist would bring support and students from his or her university if granted the new accelerator's beamtime for proposed experiments. A carefully coordinated schedule was developed to allow maximum use of the accelerator and its beamlines by its many users. Supremely complex plans of construction, utilities, materials, and access were needed to maintain the operation of the physics research program. By 1972, the Main Ring was ready.

In May 1974, NAL was dedicated and renamed the Fermi National Accelerator Laboratory in dedication to Enrico Fermi, the Italian winner of the 1938 Nobel Prize in Physics. Fermi's legacy in experimental and theoretical physics, which extended from Pisa and Rome to Columbia University and the University of Chicago to Los Alamos, was bestowed upon the Laboratory. Fermi's widow, Laura, proudly participated in the dedication ceremony. The identity of Fermilab derives from this historic moment.

Research of a comprehensive scope commenced in the fixed-target experimental areas with ambitious forays into understanding, among other topics, exotic new particle interactions, total scattering cross-section measurements, neutral currents, lepton production, and into searching for quarks. Wilson considered the research areas temporary and deliberately left them unfinished and adaptable for each experiment installed. He designed attractive architectural features for all areas, but the interiors were Spartan; some were without sufficient heating and bathrooms. Conditions were frontierlike: cold, damp, and unpleasant, but the research was exciting. A 15-foot bubble chamber was installed in the Neutrino Area to reveal and detect neutral currents. The Meson Area included several experiments surveying particle production. In the Proton Area one lepton production experiment led to a major discovery: in 1970, Columbia University's Leon M. Lederman started an experiment that evolved over the next seven years into a better equipped, more reliably performing one, which in the summer of 1977 yielded the discovery of the bottom quark.

Work on extending the frontier reach of the accelerator continued under Wilson in hopes of developing an Energy Doubler, a machine implementing the untapped technology of superconductivity. This advance would enable Fermilab's accelerators to achieve one trillion electron volts (TeV). Research for this future plan was not officially authorized by the Department of Energy; nevertheless it proceeded under Wilson. But in 1976, under pressure, Wilson returned to the federal treasury the surplus of original construction money from the Main Ring instead of being allowed to use it to exploit Fermilab's capability with the Energy Doubler.

Fermilab's funding had deteriorated by 1978. Expressing dismay that original promises of priority had not been kept and that Fermilab had not received sufficient recognition for its achievements for the DOE and the U.S. taxpayer, Wilson resigned. The Department of Energy felt it had to maintain a balance of support for its facilities and therefore had not approved Wilson's urgent plea to support the Energy Doubler.

Wilson's second Deputy Director, Philip V. Livdahl, served as Acting Director of Fermilab in mid-1978, and URA announced the selection of Lederman as Director Designate in the fall. A decision was made in November 1978, at the "Armistice Day Shootout," to pursue authorized research and development funding for and construction of the Energy Doubler. Lederman arrived in June 1979, and funding was promised in July. Lederman dispatched a group of physicists from the former Doubler Division to work with the Accelerator Division to build the Doubler. Success was essential, not only to save the Lab from its foundering status, but also to strengthen its position as a viable competitor on the international particle physics frontier.

Lederman's decision launched the Doubler era at Fermilab, marked by years of difficulties with ever-changing designs of magnets and cryostats, new systems, frequent tests, multiple reviews, very hard work, sleepless nights, and dead ends, but finally it produced results. By March 1983 the last superconducting magnet was installed into the Energy Doubler and by February 16, 1984, the 800 GeV experimental program operated successfully. With its higher performance in the TeV range, the Doubler became a critical component in the new Tevatron.

In 1983, the Tevatron's fixed-target experiments were upgraded. After the antiproton source was completed in 1985, the colliding proton and antiproton beams program demonstrated its potential for producing millions of collisions at unprecedented energies that could be observed by huge, complex detectors. These collisions produced many events for analysis by the teams of experimenters from two very large competing collaborations at Fermilab: CDF and DZero. Their search for the top quark began as the Tevatron achieved higher energies and improved luminosity. Computing power was recognized as crucial, and the Advanced Computer Project was developed to coordinate experimental data with its analysis.

An effort to enrich math and science education was launched by Fermilab in the early 1980s. Initially seen as a way to bring the physics of Fermilab to the broader population, including students and teachers from Northern Illinois, its programs have become international successes. Fermilab is acknowledged as a model of laboratory outreach for improving science literacy around the world.

A dazzling distraction captured the attention of physicists around the world in 1982: the Higgs boson. What was it, and where was it? A machine capable of exploiting still-higher energy domains was thought necessary to search for the Higgs, the mechanism responsible for the mass of elementary particles. Lederman was involved with international physics facility planners who spoke of a Very Big Accelerator (VBA) with high enough energy to search for the Higgs. He, like Wilson, thought of Fermilab as the natural site for such a forefront machine. The Tevatron's infrastructure was there, and Fermilab's credibility was now sound. Plans developed in the Department of Energy between 1983 and 1988 for a new machine to probe the frontiers of 20 TeV, called the Superconducting Super Collider. Fermilab scientists involved with magnet and accelerator technology hoped to win the next-generation machine.

In October 1988 Lederman received the Nobel Prize in Physics for his 1962 Brookhaven experiment that distinguished two different types of neutrinos.

One month later Waxahachie, Texas, was named by the Department of Energy as the site for the Superconducting Super Collider, suggesting the possible end of Fermilab.

John Peoples Jr. became the third director of Fermilab in 1989. Peoples's work on the Tevatron's Antiproton Source had led to the successful colliding beams program. His support of the computing project that contributed to the early growth of the World Wide Web was critical for communication and collaboration in Fermilab's expanding international experiments. Transfer of this information technology from basic research to the global marketplace has been rapid and revolutionary. Peoples streamlined Fermilab's experimental program while supporting innovative experimental physics ideas, such as the Pierre Auger Project, the Cold Dark Matter Search (CDMS), KTeV, the Sloan Digital Sky Survey, and Neutrinos at the Main Injector (NuMI). He endorsed further theoretical work on the early universe, supersymmetry, and superstrings.

In 1993, Congress canceled the Super Collider's funding. Peoples was asked to direct its shutdown. Fermilab would remain the highest-energy accelerator for another generation.

In 1995 nearly 1,000 physicists from around the world working on CDF and DZero announced the discovery of the top quark at Fermilab. News of the discovery went out over the World Wide Web at the same time as to the traditional media. The discovery of the top quark strengthened physicists' confidence in the Standard Model, the descriptive means of explaining the interactions of the elementary particles in terms of the fundamental forces of nature. This discovery was possible only at Fermilab because of its state-of-the-art technology and resources assembled for the search. The Tevatron, upgraded between 1993 and 1999 and enhanced with the Main Injector, remains the highest-energy accelerator in the world. Peoples stepped down as director in 1999.

At the start of the new millennium, Michael S. Witherell became Fermilab's fourth director. Still managed by URA for the U.S. Department of Energy, Fermilab employs over 2,000 people from northern Illinois and provides research facilities for thousands of physicists from around the world. The Laboratory has an annual budget of $300 million. Confidently pursuing discoveries in inner and outer space with care for its people and its environment, Fermilab is strategically positioned at the frontier of science and technology.

See also: BENEFITS OF PARTICLE PHYSICS TO SOCIETY; FUNDING OF PARTICLE PHYSICS; WILSON, ROBERT R.

Bibliography

Fermi National Accelerator Laboratory. <http://www.fnal.gov/projects/history/index.html>.

Giacomelli, G.; Greene, A. F.; and Sanford, J. R. "A Survey of the Fermilab Research Program." *Physics Reports* **19C** (4), 1–20 (1975).

Lederman, L. M. "The Upsilon Particle." *Scientific American* **239** (4), 72–80 (1978).

Wilson, R. R. "Particles, Accelerators and Society." *American Journal of Physics* **36**, 490–95 (1968).

Adrienne W. Kolb

FEYNMAN DIAGRAMS

Feynman diagrams are a bit like pictures of processes involving elementary particles. The diagram in Figure 1, for example, represents electron-electron scattering. Time runs upward in this diagram, so it describes two electrons *e* (straight lines) entering at the bottom of the picture, then interacting by exchanging a photon γ (wiggly line), and finally leaving at the top of the picture. This description is, however, a simplified one. To understand Feynman diagrams more fully, one has to consider the role of field theory (both classical and quantum) and the notions of real and virtual particles.

Field Theory: Classical and Quantum

How do two electrons interact with each other? It was discovered by Charles Augustin Coulomb in 1785 that charges of the same sign repel each other and those of opposite sign attract, in each case with a force inversely proportional to the square of the distance between the charges. Taken at face value, this interaction could be described as action at a distance—the presence of a charge at one point causes a force and a resultant acceleration to be felt by another charge at another point, *with nothing happening in between.* During the nineteenth century, however, this reliance on the idea of action at a distance came to be regarded as unsatisfactory, and an alternative description was introduced, relying on the notion of a field. According to this description, one electron, by virtue of its electric charge, produces an electric field that fills all the space around it. Another electron (or any other charged particle), when placed in the field, feels a force due to the presence of the field at the location of the second charge. If this field is real, the idea of action at a distance has been refuted, but is the field real? The answer is that, remarkably, it is! In radio transmission, for example, the oscillating current in a wire causes energy to be transferred from the electrons to the field, which then radiates it away. This was Heinrich Hertz's great discovery in 1888. The electric (in general, electromagnetic) field is real because it can carry energy and momentum; therefore it is perfectly satisfactory to describe this interaction by means of a field.

According to quantum theory, the energy in this electromagnetic field is quantized. The field is not continuous but consists of almost countless photons, each carrying a definite, indivisible amount of energy. It therefore follows that the simplest, most primitive way in which two electrons can interact is to exchange one quantum of the field, one photon. This is what

FIGURE 1

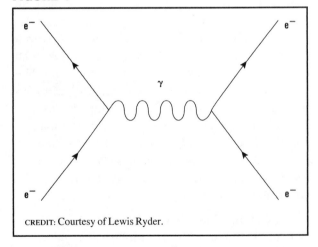

CREDIT: Courtesy of Lewis Ryder.

An electron-electron scattering.

is happening in Figure 1, and this type of diagram is sometimes called a one-photon-exchange diagram.

The Standard Model (SM) of particle physics is simply a generalization of electrodynamics; that is, particles of matter carry a conserved quantity, such as electric charge, and interact with each other through the exchange of a quantum of the corresponding field. For example, inside a hadron, quarks interact through the exchange of gluons, as in Figure 2. Quarks carry a chargelike label called color, which is the source of a field; the quanta of this field are gluons. Figure 2 shows the interaction between a red up quark u_R and a green up quark u_G. The interaction is carried by a gluon (curly line). As a further example, Figure 3 shows the quark decay $d \rightarrow u + e^- + \bar{\nu}_e$. The down quark decays into an up quark with the emission of an electron and an antineutrino. (This is actually the process involved in neutron decay; a neutron is a udd bound state and a proton a uud bound state, so a d decaying into a u results in a neutron decaying into a proton.) In the Standard Model this process is mediated by the exchange of a W particle, the quantum of the weak field.

Feynman Rules

Feynman diagrams are more than pretty pictures. Each diagram represents an amplitude for a process to happen. This amplitude is a complex number, which, for a given diagram, is calculated by us-

FIGURE 2

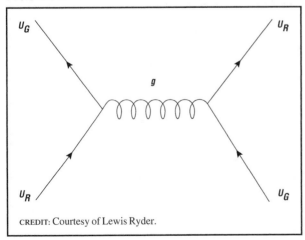

CREDIT: Courtesy of Lewis Ryder.

Quarks interact through the exchange of gluons.

FIGURE 3

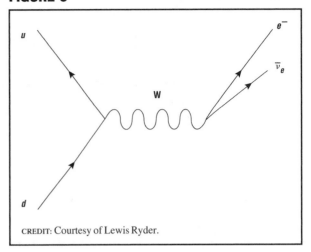

CREDIT: Courtesy of Lewis Ryder.

A down quark decays into an up quark with the emission of an electron and an antineutrino.

ing the so-called Feynman rules. The rate at which a process happens is proportional to the square modulus of the amplitude. Feynman diagrams are thus essential calculation tools in high-energy physics.

Lines and Vertices; Real and Virtual Particles

Feynman diagrams are composed of lines. The convention is that lines representing fermions (e.g., electrons, protons, quarks) are continuous straight lines, and those representing field quanta (photons, gluons, or W or Z bosons) are wavy or curly. (Sometimes dashed lines also appear; these correspond to bosons with spin 0—the field quanta mentioned above all have spin 1.)

Feynman diagrams may be constructed by combining more primitive structures. Figure 4 contains three examples of vertices, that is, a junction of three lines at a point. The first vertex is an electron-photon vertex, the second a duW vertex, the third a down quark–gluon vertex. It should be clear that the diagrams of Figures 1 through 3 can be constructed by putting together combinations of appropriate vertex diagrams. When this is done, however, one gets two types of lines. For example, when two vertices of Figure 4(a) are combined to make the diagram of Figure 1, the photon line both begins and ends at a vertex. It is called an internal line. By contrast, the electron lines are external lines—they either begin or end

FIGURE 4

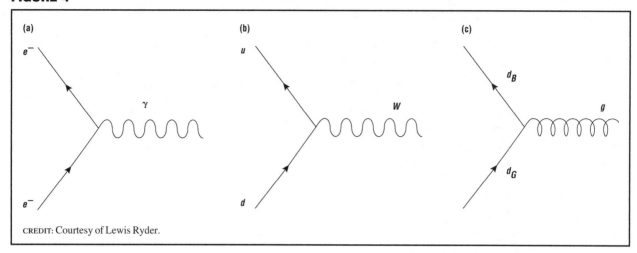

CREDIT: Courtesy of Lewis Ryder.

An electron-photon vertex.

at a vertex, but not both. External lines generally describe real particles, and internal lines virtual particles; however, fermions can also propagate on internal lines. In Figure 1 the photon is a virtual photon. This means that its existence is governed by the uncertainty relation $\Delta E \Delta t \approx h/2\pi$; it lives for such a short time that it is, in principle, impossible to detect it; its existence is only allowed because of quantum theory. It follows that one must be careful about the interpretation of Feynman diagrams. Time is represented by the vertical direction, but only in a loose sense can the horizontal direction be taken to represent space, since the lines do not simply represent paths or trajectories.

A Note on Perturbation Theory

Figures 1 through 4 are the simplest ones possible for the processes they describe, but more elaborate diagrams may be constructed by adding additional internal lines to a given diagram. Figure 5 shows two dressed up versions of Figure 1. It is easy, and rather good fun, to create many more such diagrams, with more and more vertices. In reality *all* of these diagrams will contribute to a given process, but it turns out that diagrams with more vertices are much less important than those with fewer as their amplitudes are much smaller. Hence, to a good approximation, they may be neglected.

FIGURE 5

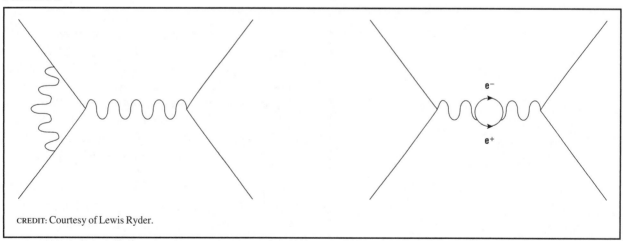

CREDIT: Courtesy of Lewis Ryder.

Examples of more elaborate Feynman diagrams.

See also: ANNIHILATION AND CREATION; QUANTUM FIELD THEORY; QUANTUM MECHANICS; QUANTUM STATISTICS; RELATIVITY; RESONANCES; SCATTERING; VIRTUAL PROCESSES

Bibliography

Feynman, R. P. *QED: The Strange Theory of Light and Matter* (Princeton University Press, Princeton, NJ, 1988).

Taylor, J. C. *Hidden Unity in Nature's Law* (Cambridge University Press, Cambridge, UK, 2001).

Lewis Ryder

FEYNMAN, RICHARD

Richard Phillips Feynman (1918–1988) was one of the most original physicists of the second half of the twentieth century. He was born on May 11, 1918, in Far Rockaway, New York. His father came to the United States from Russia when he was very young and grew up in Patchogue, Long Island. He obtained a degree in homeopathic medicine after graduating from high school but never practiced medicine. Feynman's mother was born into a well-to-do New York family and attended the Ethical Culture School but did not go to college thereafter. Feynman had a younger brother, born when Feynman was 3, who died shortly after birth. He also had a sister, Joan, who was nine years younger than Feynman.

Feynman attended both junior and senior high school in Far Rockaway and had some very competent and talented teachers for his chemistry and mathematics courses who nurtured his interest in the sciences. He entered MIT in the fall of 1935 and was immediately recognized as an unusually gifted student by all his teachers. In 1939 he went to Princeton University as a graduate student in physics and served as John Archibald Wheeler's assistant. Wheeler, who had just arrived at Princeton as a twenty-six year old assistant professor in the fall of 1938, proved to be an ideal mentor for the even younger Feynman. Full of bold and original ideas, a man who had the courage to explore any problem, Wheeler gave Feynman viewpoints and insights into physics that would prove decisive later on.

In the spring of 1942 Feynman obtained his Ph.D. and immediately thereafter started working on prob-

American physicist Richard P. Feynman (1918–1988) shared the 1965 Nobel Prize in Physics with Sin-itiro Tomonaga and Julian Schwinger for their work in quantum electrodynamics. CREDIT: COURTESY OF BETTMANN/ CORBIS. REPRODUCED BY PERMISSION.

lems related to the development of an atomic bomb. In 1943 he was one of the first physicists to go to Los Alamos. He was quickly identified by Hans Bethe, the head of the theoretical division, and by Robert Oppenheimer, the director of the laboratory, as one of the most valuable members of the theoretical division. He was also acknowledged by everyone to be perhaps the most versatile and imaginative member of that community of outstanding scientists. In 1944 he was made a group leader in charge of computations for the theoretical division. Feynman introduced punch-card computers to Los Alamos, and he there developed his life-long interest in computing and computers.

While at Los Alamos, Feynman accepted an appointment at Cornell University as an assistant professor and joined its department of physics in the fall of 1945. In 1951 he left Cornell to become a member of the faculty of the California Institute of Technology, and he remained there until his death from stomach cancer on February 15, 1988.

One aspect of Feynman's genius was that he could make precise what was unclear and obscure to most of his contemporaries. His doctoral dissertation and well-known 1948 *Reviews of Modern Physics* article that presented the path integral formulation of nonrelativistic quantum mechanics helped clarify and make explicit the assumptions underlying the usual quantum mechanical description of the dynamics of microscopic entities. Moreover, he did this in the very act of extending the usual formulation with a startling innovation. His reformulation of quantum mechanics and his integral over paths may well turn out to be his most profound and enduring contribution. They have deepened scientists' understanding of quantum mechanics and have significantly enlarged the number and kinds of systems that can be quantized. His path integral enriched mathematics and has provided new insights into spaces of infinite dimensions.

Feynman was awarded the Nobel Prize in Physics in 1965 for his work on quantum electrodynamics (QED). In 1948, simultaneously with Julian Schwinger and Sin-itiro Tomonaga, he showed that the divergences plaguing QED could be consistently identified and removed by a redefinition of the parameters that describe the mass and charge of the electron in the theory, a process that is called renormalization. Schwinger and Tomonaga had done this by building on the existing formulation of the theory. Feynman, on the other hand, invented a completely new diagrammatic approach that allowed the visualization of space-time processes, which in turn simplified concepts and calculations enormously and also made possible the exploration of the properties of QED to all orders of perturbation theory. Using Feynman's methods, it became possible to calculate quantum electrodynamic processes to amazing precision. Thus, the magnetic moment of the electron has been calculated to an accuracy of one part in 10^9 and found to be in agreement with an experimental value measured to a similar accuracy.

In 1953 Feynman developed a quantum mechanical explanation of liquid helium that justified the earlier phenomenological theories of Lev Landau and Laslo Tisza. Because a ^{4}He atom has zero total spin angular momentum, it behaves as a Bose particle: the wave function describing a system of N helium atoms is therefore symmetrical under the exchange of any two helium atoms. The ground-state wave function of such a system is nondegenerate and everywhere positive. When in this state, the system—even when N is of the order 10^{23}, and the system is macroscopic—behaves as one unit. This is why helium near 0K is a superfluid, acting as if it has no viscosity. Near 0K, pressure waves are the only excitations possible in the liquid. At somewhat higher temperatures, around 0.5K, it becomes possible to form small rings of atoms that can circulate without perturbing other atoms; these are the rotons of Landau's theory. With increasing temperature, the number of rotons increases, and their interaction with one another gives rise to viscosity. An assembly of rotons behaves like a normal liquid, and this liquid moves independently of the superfluid. At a certain point, when the concentration of normal liquid becomes too large, a phase transition occurs, and the whole liquid turns normal. This was Feynman's quantum mechanical explanation of why at any given temperature helium could be regarded as a mixture of superfluid and normal liquid.

In 1956 Tsung Dao Lee and Chen Ning Yang analyzed the extensive extant data on nuclear beta decay and concluded that parity symmetry is not conserved in the weak interactions. This was soon confirmed experimentally by Chien-Shiung Wu, Ernest Ambler, Raymond W. Hayward, Dale D. Hoppes, Ralph P. Hudson, and others. Subsequent experiments further indicated that the violation of parity is the maximum possible. On the basis of these findings, Robert Marshak and George Sudarshan, and somewhat later and independently Richard Feynman and Murray Gell-Mann, postulated that only the "left-handed" part of the wave functions of the particles involved in the reaction enter in the weak interactions. Feynman and Gell-Mann further hypothesized that the weak interaction is universal, that is, that all the weak particle interactions have the same strength. This hypothesis was later corroborated by experiments.

In the late 1960s experiments at the Stanford Linear Accelerator on the scattering of high-energy electrons by protons indicated that the cross section for inelastic scattering was very large. Feynman found that he could explain the data if he assumed that the proton was made up of small, pointlike entities, which interacted elastically with electrons. He called these subnuclear entities partons. The partons were soon identified with the quarks of Gell-Mann and George Zweig. The study of quarks and their interactions, and in particular, an explanation of their confinement inside nucleons and mesons, was an important component of Feynman's research during the 1980s.

Feynman disliked pomposity and frequently made fun of pretentious and self-important people. He was always direct, forthright, and skeptical. These traits have been beautifully captured in the volume of reminiscences that Laurie Brown and John Rigden have edited and in the stories that Feynman told Ralph Leighton. His uncanny ability to get to the heart of a problem—whether in physics, applied physics, mathematics, or biology—was demonstrated repeatedly. As a member of the presidential commission that investigated the Challenger disaster, he was able to simply convey the central problem by dropping a rubber O-ring into a glass of ice water and demonstrating its shriveling. In his physics Feynman always stayed close to experiments and showed little interest in theories that could not be experimentally tested. He imparted these views to undergraduate students in his justly famous *Feynman Lectures on Physics* and to graduate students through his widely disseminated lecture notes for the graduate courses that he taught. His writings on physics for the interested general public, *The Character of Physical Laws* and *QED,* convey the same message.

See also: QUANTUM ELECTRODYNAMICS; QUANTUM FIELD THEORY; VIRTUAL PROCESSES

Bibliography

Brown, L. M., ed. *Selected Papers of Richard Feynman: With Commentary* (World Scientific, River Edge, NJ, 2000).

Brown, L. M., and Rigden, J. S., eds. *Most of the Good Stuff: Memories of Richard Feynman* (American Institute of Physics, New York, 1993).

Feynman, R. P. (with R. B. Leighton and M. Sands). *The Feynman Lectures on Physics,* 3 vols. (Addison Wesley, Reading, MA, 1963).

Feynman, R. P. *The Character of Physical Laws* (MIT Press, Cambridge, MA, 1965).

Feynman, R. P. *QED: The Strange Theory of Light and Matter* (Princeton University Press, Princeton, NJ, 1985).

Feynman, R. P. (as told to Ralph Leighton), edited by E. Hutchings. *"Surely You're Joking, Mr. Feynman!": Adventures of a Curious Character* (Norton, New York, 1985).

Feynman, R. P. (as told to Ralph Leighton). *What Do YOU Care What Other People Think?: Further Adventures of a Curious Character* (Norton, New York, 1988).

Gleick, J. *Genius. The Life and Science of Richard Feynman* (Pantheon Books, New York, 1992).

Mehra, J. *The Beat of a Different Drum: the Life and Science of Richard Feynman* (Oxford University Press, New York, 1994).

Robbins, J., ed. *The Pleasure of Finding Things Out: The Best Short Works of Richard P. Feynman* (Perseus Books, Cambridge, MA, 2000).

Schweber, S. S. *QED and the Men Who Made It* (Princeton University Press, Princeton, NJ, 1994).

Silvan S. Schweber

FLAVOR

See QUARKS

FLAVOR SYMMETRY

The label that distinguishes different types of quarks, *u* for up, *d* for down, *s* for strange, *c* for charm, *b* for bottom, and *t* for top, is called the flavor of the quark. In this context, flavor is a technical term that bears no relation to the experience associated with the sense of taste. The term flavor symmetry refers to relationships between hadrons composed of different flavor quarks. These relationships exist because the strong force, responsible for binding quarks into hadrons, acts with identical strength on all quarks, regardless of their flavor. The relationships are, however, only approximate since the much feebler electroweak interactions do distinguish between flavors, and, in addition, quarks of different flavors have different masses.

Quarks are classified as light or heavy according to whether their masses are small or large compared to the mass of a proton. The up, down, and strange quarks are light, whereas the charm, bottom, and top quarks are heavy, and, moreover, their masses differ from each other by a large multiple of the proton mass. Flavor symmetry is a good approximation for hadrons composed of the light quarks because the differences between light quark masses are small when compared to the proton mass. Conversely, flavor symmetry does not hold at all for heavy quarks.

In units of the proton mass, the masses of the up, down, and strange quarks are approximately 0.005, 0.010, and 0.100, respectively. Hence, the flavor symmetry relating up and down quarks holds to excellent accuracy, whereas the flavor symmetry relating all three light quarks holds somewhat less accurately. The former is known both as isospin or SU(2) symmetry, while the latter is commonly known as SU(3) symmetry. The theory of the SU(3) symmetry of hadrons was first proposed in 1961 by American physicist Murray Gell-Mann and, independently, by Israeli physicist Yuval Ne'eman as a scheme for classifying and relating properties of a multitude of observed particles. It was not until 1964 that Gell-Mann and American physicist George Zweig advanced the quark hypothesis to explain the observed SU(3) symmetry.

The dynamics of subatomic particles is best accounted for by quantum mechanics. Particles are described by state vectors. Much like the position of an object in space is specified by three real numbers *x, y, z*, for example, latitude, longitude, and altitude, that form a vector, the state of a light quark can be described by two complex numbers that can be thought of as the degree to which the quark is a *u* or a *d* quark, or by three complex numbers that can be thought of as the degree to which the particle is a *u, d,* or *s* quark. And just like the laws of physics are invariant under transformations that rotate vectors, the strong interactions are approximately symmetric under transformations that rotate the quark state vectors. SU(2) refers to the group of transformations that rotate state vectors with two components, whereas SU(3) refers to transformations of vectors with three components.

Because hadrons are composed of quarks, their state vectors also transform under SU(2) or SU(3) rotations in specific ways but not necessarily the same way as quarks. Figure 1 shows eight states that comprise a state vector called an octet with components that transform among themselves under SU(3) rotations. Similarly, the ten states in the decouplet of Figure 2 rotate into themselves only. The quark content of the proton and neutron is *uud* and *udd*, respectively. Since they differ in their quark content by one light quark, they are described by a two-component state vector, just like the lightest quarks

FIGURE 1

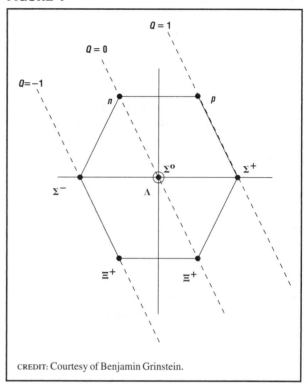

CREDIT: Courtesy of Benjamin Grinstein.

Eight elementary particles that comprise the baryon octet. Their properties are related by flavor symmetry, formally as a transformation in the group SU(3). In the figure, particles on the same horizontal line have the same strangeness, and they form an isospin multiplet. Strangeness advances by one unit from one horizontal line to the next one up. From the top, the isospin multiplets are the nucleon doublet containing the proton and the neutron, the Σ triplet and Λ singlet, both with strangeness −1, and the Ξ doublet.

are. These two states, collectively known as the nucleon *N*, are said to form a doublet of SU(2). Replacing one light quark by a strange quark gives three particles collectively called Σ, to wit the Σ^- (*dds*), the Σ^0 (*uds*), and the Σ^+ (*uus*), and a fourth particle, the Λ (*uds*). The Σ is said to form a triplet of SU(2), whereas the Λ is a singlet of SU(2). Replacing one more light quark by a strange quark gives another doublet of SU(2), known as the Ξ with components Ξ^- (*dss*) and Ξ^0 (*uss*). The properties of the two components of the *N* are related by isospin, as are those of the three components of the Σ and the two components of the Ξ. Thus, for example, the masses of the proton and neutron are 938.3 and 939.6 MeV/c^2, respectively; those of the Σ^-, Σ^0, and Σ^+ are 1,197.4, 1,192.6, and 1,189.3 MeV/c^2, respectively; and those

FIGURE 2

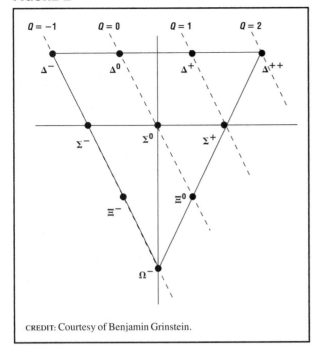

Ten elementary particles that comprise the baryon decouplet. Their properties are related by flavor symmetry. Strangeness advances by one unit from one horizontal line to the next one up, while isospin relates particles on any single horizontal line. From the top, the isospin multiplets are the Δ quadruplet, the Σ triplet, the Ξ doublet, and a singlet, the Ω. The Σ and Ξ here share their name and some properties with the Σ and Ξ in the octet, but they are distinct particles.

of the Ξ^- and Ξ^0 are 1,321.3 and 1,314.8 MeV/c^2, respectively. The Λ, with a mass of 1,115.7 MeV/c^2, remains unchanged under the action of SU(2) symmetry transformations.

The slightly less accurate SU(3) symmetry relates the properties of the N, Σ, Λ, and Ξ. These eight spin-$\frac{1}{2}$ baryons form an octet, a mathematical object that, like a vector, has specific SU(3) transformation properties. Similarly, there exists a spin-$\frac{3}{2}$ baryon decouplet with ten states: Δ^{++} (uuu), Δ^+ (uud), Δ^0 (udd), Δ^- (ddd), Σ^+ (uus), Σ^0 (uds), Σ^- (dds), Ξ^0 (uss), Ξ^- (dss), and Ω^- (sss). The existence and mass of the Ω^- were predicted by SU(3) symmetry three years before its discovery in 1964. Had SU(3) been an exact symmetry, the masses of all states in the decouplet would be the same. Making the assumption that SU(3) fails to be an exact symmetry only because the strange quark is heavier than the up and down

quarks, SU(3) symmetry predicts the Ω to be heavier than the Ξ by the same amount that the Ξ is heavier than the Σ and that this must be the same amount by which the Σ is heavier than the Δ. The three mass differences are experimentally determined to be 139, 149, and 152 MeV/c^2, respectively. Similarly, for the baryon octet the approximate SU(3) symmetry implies that the Λ is heavier than the N by the same amount that the Ξ is heavier than the Λ and that the Λ and Σ have equal masses. The observed mass differences are 177 and 203 MeV/c^2, respectively. The magnitude of these mass differences in units of the proton mass, about 20 percent, is a measure of how accurate SU(3) symmetry is.

The particle content of the octet of spin-$\frac{1}{2}$ baryons and the decouplet of spin-$\frac{3}{2}$ baryons is summarized in Figures 1 and 2. The vertical axis represents the number of strange quarks in a particle, and the oblique axis represents its charge. SU(2) relates particles on a horizontal line, whereas SU(3) transformations relate all particles in a multiplet.

The branch of mathematics known as group theory gives the number of states that must be grouped into an object which has specific SU(2) or SU(3) transformation properties. Since baryons contain three quarks, there are $2 \times 2 \times 2 = 8$ combinations of u and d flavors for a baryon, $2 \times 2 \times 2 = 2 + 2 + 4$. Group theory instructs that these are to be grouped into one object with four components and two objects with two components each (two doublets). The Δ and N are examples of four and two component objects, respectively. Incorporating the s quark, group theory determines that $3 \times 3 \times 3 = 27 = 1 + 8 + 8 + 10$. Examples of decouplet and octet baryons are the spin-$\frac{3}{2}$ and spin-$\frac{1}{2}$ multiplets given above. The lightest singlet baryon is the Λ_1, a spin-$\frac{1}{2}$ particle of mass 1,406 MeV/c^2.

Mesons are hadrons composed of a quark and an antiquark. Group theory also determines the size of a meson multiplet. If made out of u and d quarks and antiquarks, the $2 \times 2 = 4$ combinations are $2 \times 2 = 1 + 3$, a singlet and a triplet. The π^+, π^0, and π^- spin-0 mesons form a triplet of SU(2), whereas the η meson is a triplet. Including the s quark, the $3 \times 3 = 9$ combinations are grouped into an octet and a singlet of SU(3). With the π and η mesons, the K^+, K^0, $\bar{K}^0$, and K^- mesons complete the octet,

263

whereas the η' meson is an example of a singlet. There are similar examples for spin-1 mesons: The ρ^+, ρ^0, and ρ^- mesons form a triplet of SU(2), and these with the four K^* mesons and the two ω mesons complete the octet and a singlet.

As opposed to the strong force, which preserves flavor, the weak force can change the flavor of a quark. Nuclear beta decay is an example of a process in which a weak force induces flavor change. For example, the neutron can decay into a proton, an electron, and an antineutrino, $n \rightarrow pe\bar{\nu}$. In this process, one of the d quarks in the neutron is transformed into a u quark, so the transformation $d \rightarrow u$ gives $(udd) \rightarrow (uud)$, that is, $n \rightarrow p$. Electromagnetic forces do not change flavor but act differently on the charge $+2/3$ u quark than on the charge $-1/3$ d and s quarks. Thus, for example, the electromagnetic force is responsible for the difference in mass between the $\pi^\pm$ mesons and the π^0 meson.

Flavor symmetry can also be used in the context of hadrons that contain heavy quarks in addition to light quarks. For example, the B^+ ($u\bar{b}$) and $\bar{B}^0$ ($d\bar{b}$) mesons form a doublet of SU(2). Together with the B_s ($s\bar{b}$) meson, they form a triplet of SU(3). As such, their properties are related. The B^0 and B^+, of almost equal mass, 5,279 MeV/c^2, are 90 MeV/c^2 lighter than the B_s. The mass difference is of about the size expected given the approximate nature of SU(3) symmetry.

See also: BROKEN SYMMETRY; EIGHTFOLD WAY; FAMILY; LEPTON; QUARK; STANDARD MODEL; SU(3)

Bibliography

Commins, E. D., and Bucksbaum, P. H. *Weak Interactions of Leptons and Quarks* (Cambridge University Press, Cambridge, UK, 1983).

Groom, D. E., et al. "Review of Particle Physics." *European Physics Journal* **C15**, 1 (2000).

Perkins, D. H. *Introduction to High Energy Physics*, 4th ed. (Cambridge University Press, Cambridge, UK, 2000).

Benjamin Grinstein

FUNDING OF PARTICLE PHYSICS

Elementary particle physics is the study of those particles that are considered not to have measurable spatial dimensions or further constituents. The designation of particles as "elementary" has changed in time as substructures of what were previously thought to be elementary particles were found. During the last century, particle physics evolved from atomic physics, to nuclear physics, to what is now called elementary particle physics. It was recognized that atoms were constituted of nuclei surrounded by electrons and then that nuclei were composed of neutrons and protons. Information was developed about the forces acting between neutrons and protons, and this led to some understanding of nuclear structure. Beta decay was discovered which converted neutrons into protons and vice versa with the emission of an electron and a neutrino. During the last quarter of the century, it was found that neutrons and protons are composed of quarks of six "flavors," and the electron was found to have two "brothers," the muon and the tau, constituting the lepton family. This evolution in knowledge was furthered by three branches of particle physics: theoretical physics, accelerator physics, and experimental physics. In turn, experimental physics uses accelerators, radioactive sources, or cosmic rays.

Before World War II, this work was supported by largely private sources—either industry or foundations. Some of the biggest accelerators operated in industrial laboratories such as the Westinghouse research laboratories that housed large electrostatic accelerators. Facilities at universities were generally sponsored by foundations. The Radiation Laboratory at Berkeley founded by Ernest O. Lawrence was unusually successful in obtaining private funding and supported the construction of families of cyclotrons, some of which were contributed to other laboratories.

During World War II, physicists demonstrated that if adequately supported, they could create and organize effective laboratories and produce spectacular results. Based on this wartime experience government became interested in supporting particle physics research on a large scale. Part of the support came from unobligated funds of the military agencies; the Office of Naval Research and also the Office of Scientific Research of the Air Force supported fundamental research at universities. Separately the Atomic Energy Commission, the follow-on agency to the wartime Manhattan District, supported fundamental research in addition to its applied missions. In fact some of the senior physicists, such as Enrico Fermi, returning from wartime to academic research were urged by the government to accept grants for the construction of particle accelerators.

This postwar expansion of government support of particle physics was partially accidental due to funds remaining in government coffers but was also a deliberate effort to encourage physicists to do in peacetime what they had so ably demonstrated in war—to organize large successful laboratory efforts. Thus World War II led to a shift from private to government support of elementary particle physics. With very few exceptions this shift proved irreversible, and elementary particle physics has become the ward of the federal government in the United States and of governments abroad. In the United States the Department of Energy as successor to the Atomic Energy Commission has remained the "custodian" of the particle physics program with the National Science Foundation supporting university based activities and one accelerator center.

Elementary particle physics attacks some of the most fundamental questions of inanimate nature—that is, the search for the fundamental building blocks of the universe and the forces between them. It thus attracts extremely capable people, and these tend to be intolerant of the limitations imposed by available tools. While practical applications of elementary particle research rarely result from the discoveries from that research, the invention and development of the tools that elementary physicists devise to further their work has extensive economic consequences. The electromagnetic cavity invented by W. W. Hansen, which is now an essential component of most microwave devices, was originally devised to provide high voltages for particle research with only moderate amounts of radio frequency power. Microwave linear accelerators that were developed for elementary particle physics have become a near-billion-dollar industry supplying radiation sources for cancer therapy. The large variety of radiation detectors developed for elementary particle physics have been of enormous value for monitoring devices for the reactor industry and in medical practice. Particle physics has led to the world's most intense X-ray sources—synchroton radiation—applied to many industrial uses.

A similar pattern prevails in the field of data handling and communications. The World Wide Web was initiated first at the European Laboratory for Elementary Particle Physics (CERN). The first communication link between the United States and China was established in connection with the need to transmit vast quantities of data in the collaborative effort between those two countries. Many of the algorithms for identifying very rare events among a large class of phenomena, for recognizing specific patterns, and for modeling complex sequential phenomena were developed in elementary particle physics but then widely applied.

While research in elementary particle physics has produced many dramatic economic consequences, private industry has been reluctant to support fundamental research. Practical results from elementary particle physics are delayed, and financial returns can rarely be recovered by the particular entity that supports the work.

Governmental agencies recognized that support of elementary particle physics is difficult to justify economically by its direct results, but the history cited above has amply demonstrated that such research has provided a dramatic return on the public investment. While it is extremely difficult to develop an "audit trail" between the funds invested in elementary particle physics and the returns to society, many economic analyses have been made to estimate the rate of return of investment in fundamental research. The results of such analyses vary widely; calculated rates of return range from 20 to 50 percent—very large figures, but difficult to pin down precisely.

However, by the beginning of the twenty-first century, government began frequently to forget these facts and would like to see a definite demonstration of a direct causal relationship between investment and returns. Should governmental funding be invested in directed research to answer specific questions of an applied nature rather than rely on "spinoff" from fundamental work? History should be persuasive: the most fundamental questions of nature attract highly capable people, and they in turn provide solutions that then result in practical applications.

FIGURE 1

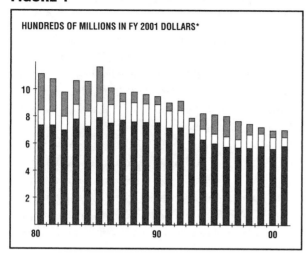

High-energy physics (HEP) funding, excluding SSC, in $100M, corrected at CPI +2%

As the energy of accelerators has grown by seven orders of magnitude during the last century, the cost per unit of energy has shrunk by about a factor of 10,000. Thus the construction cost of a single new machine continues to increase, and therefore worldwide the number of "accelerator centers" has shrunk. These laboratories are operated as facilities for a large community of "users," generally faculty and students at universities. By this method the educational role of particle physics through graduate education is maintained even when the actual collection of data is concentrated at a decreasing number of centers.

Notwithstanding the growth in energy at the particle physics frontier, the large community of particle physicists, the history of production of profound basic revelations, as well as the generation of practical technologies, U.S. funding for particle physics has shrunk by about 20 percent in real terms over the past twenty years as shown in Figure 1. That chart shows the funding history in constant 2001 dollars adjusted by the Consumer Price Index (CPI) + 2 percent per year. The latter correction is necessary since a large fraction of the cost of particle physics research is salaries which in technical fields have grown faster that the CPI.

Particle physics is an international enterprise. 2002's annual budgets for the field are distributed approximately as follows: Europe $1.0 Billion; U.S. $0.7 billion; Japan $0.3 billion; Other ~$0.1 billion. The budget of CERN is almost as large as the sum of the large centers of other countries combined.

The future economic needs of particle physics is difficult to predict because a number of factors are expected to affect the pattern of work within the field:

- Nonaccelerator physics, such as large cosmic ray and neutrino experiments, is becoming of increasing interest.

- There is an increasing overlap of subject matter in studies of particle physics and cosmology, that is, the study of the very small and the very large.

- New accelerator technologies are under intensive study.

- To continue the remarkable history of contributions to knowledge, future particle physics facilities operating at the frontier must demonstrate large increases in three respects: particle collision energy, rate of collision, and data analysis capability.

The general consensus remains that future accelerator centers at the frontier will be one-of-a-kind laboratories constructed and operated with international support. Today (2002) large detectors at regional facilities are internationally financed, constructed, and operated, and that practice may be extended to the accelerators and their infrastructure also. This does not imply that a single nation or region could not "afford" greatly increased support for basic science; the budgetary level of funding for fields such as particle physics is a matter of policy, not fiscal necessity.

See also: BENEFITS OF PARTICLE PHYSICS TO SOCIETY; INTERNATIONAL NATURE OF PARTICLE PHYSICS

Bibliography

American Institute of Physics. "FYI: the AIP Bulletin of Science Policy News." <http://www.aip.org/enews/fyi/>.

Wolfgang K. H. Panofsky

G

GAUGE THEORY

Gauge invariance is the pillar of modern theories of particle physics. All the known fundamental force fields of particle physics, namely, electromagnetic, weak, and strong (the "gluon field"), are gauge fields. Gauge field theories, even when spontaneously broken or hidden by confinement, are renormalizable, which means that an infinite amount of potentially observable quantities are calculable in terms of a finite number of parameters (masses and couplings). (There is one more known force field of nature, which so far has played no role in particle physics, namely, gravity; it too is a gauge field, but in a somewhat different sense from the others, and most of the present discussion does not apply to it.)

A gauge field is a 4-vector, its components transforming as a space-time vector under space-time rotations, and so the quantum of the field is a spin-1 boson. Unless the gauge symmetry is broken by the Higgs mechanism as in the Standard Model, like photons, these quanta have mass zero and helicity ± 1 only, not 0. (The helicity of a particle is the component of its spin in the direction of its momentum; the absence of photons of zero helicity is equivalent to the polarization of electromagnetic waves being transverse only.) Like any boson field, a gauge field produces a force by the exchange of quanta of the field between two objects (emission of a virtual quantum from one and absorption by the other), thus transferring momentum between the two. The resulting force is long-range, depending on distance as an inverse square just as in electromagnetism, if the corresponding gauge invariance is neither broken nor hidden. Of course, any other quantities carried by a quantum of the field are also transferred by its exchange; for example, charge is transferred in nuclear beta decay (a "charge-current weak interaction"), where the exchanged gauge quantum is a charged weak boson, W^+ or W^-.

Each gauge field corresponds to a gauge invariance, that is, an internal symmetry transformation varying arbitrarily from point to point. The concept of gauge invariance arose in electromagnetism as follows: The sourceless Maxwell equations (in natural units) $\nabla \times \mathbf{E} = -\partial_t \mathbf{B}$ and $\nabla \cdot \mathbf{B} = 0$ ($\partial_t \mathbf{B}$ is short for $\partial \mathbf{B}/\partial t$) are equivalent to the existence of scalar and vector potentials ϕ and $\mathbf{A}$, in terms of which the electric and magnetic fields $\mathbf{E}$ and $\mathbf{B}$ are given as $\mathbf{E} = -\nabla \phi - \partial_t \mathbf{A}$, $\mathbf{B} = \nabla \times \mathbf{A}$. (Relativistically, ϕ and $\mathbf{A}$ are the space and time components, respectively, of the 4-vector gauge field of electromagnetism.) These potentials are not completely determined by $\mathbf{E}$ and $\mathbf{B}$: $\mathbf{E}$ and $\mathbf{B}$ are left unchanged by the changes $\phi \to \phi - \partial_t \chi$ and $\mathbf{A} \to \mathbf{A} + \nabla \chi$, where χ is an arbitrary function of space and time. This is known as a gauge transformation of the potentials, and the property that **E** and **B** are unchanged is known as gauge invariance.

Quantum mechanics gives rise to a radical revision of the concept of gauge invariance: In the Schrödinger equation for a particle of charge Q, the effect of the **E** and **B** fields enters through the replacement of the momentum operator $-i\hbar\nabla$ by $-i\hbar\nabla - Q\mathbf{A}$, and the energy operator $i\hbar\partial_t$ by $i\hbar\partial_t - Q\phi$. The Schrödinger equation then remains unchanged by a gauge transformation of the potentials if also the phase change $\psi \rightarrow \exp[i(Q/\hbar)\chi]\psi$ is made to the particle's wave function ψ. (This change is best thought of as a change of reference frame, i.e., as a rotation of the complex plane.) This change of ψ is called an internal symmetry transformation, where "internal" means a transformation (here of ψ) that does not involve a transformation of the space or time coordinates. In the absence of the potentials, the phase change of ψ is a symmetry of the Schrödinger equation (i.e., leaves it unchanged) only if its χ is constant; it is then called global symmetry. (This symmetry yields the conservation of probability.) However, as just seen, the presence of the potentials allows the internal symmetry transformation of ψ to vary arbitrarily from point to point. It is then referred to as a gauged internal symmetry transformation, or gauge transformation for short, and the resulting invariance is called a gauge symmetry or gauge invariance.

This scheme carries over into quantum field theory where the role of ψ is played by a quantum field operator Ψ that absorbs a particle (quantum) carrying charge Q and emits an antiparticle carrying charge $-Q$. Abstractly speaking, the internal symmetry group that is "gauged" (made a gauge symmetry) by the electromagnetic field is U(1), the group of multiplication of complex numbers of unit magnitude. Each field Ψ transforms as did ψ, that is, as a representation of U(1) characterized by the charge Q. The field Lagrangian has global charge conservation symmetry if each of its terms is the product of fields whose Q add to zero, since then the product of the transformation phase factors $\exp[i(Q/\hbar)\chi]$ equals 1. This symmetry becomes gauged if all space-time gradients of Ψ in the Lagrangian are replaced by covariant gradients, components $\partial_t + i(Q/\hbar)\phi$ and $\nabla - i(Q/\hbar)\mathbf{A}$; otherwise, the electromagnetic gauge field (4-potential) only occurs in the Lagrangian as the gauge invariant fields **E** and **B**.

In principle, any continuous internal symmetry group (abstractly speaking, a compact Lie group) can be gauged; there is then a gauge field corresponding to each generator of the Lie group. Most Lie groups are non-Abelian, meaning that some pairs of generators do not commute with one another. A non-Abelian group has nontrivial matrix representations (dimension >1), so a field Ψ is generally many-component and transforms $\Psi \rightarrow \mathbf{M}\cdot\Psi$, where **M** is a representation matrix. The field Ψ is then said to belong to this representation. If the group is gauged, the form of the covariant gradient applied to the field Ψ, and hence the probability amplitude for emission or absorption of a gauge quantum by a quantum of the field Ψ, depends only on the representation to which the field Ψ belongs but not to any of its other properties. For instance, all quarks (u, d, s, . . .) are SU(3)^{color} triplets, and therefore all couple to (absorb or emit) gluons the same way; this is called the universality of the strong interaction. Similarly, the universality of the charge-current weak interaction results from all weak-ispin doublets (both left-handed quarks and left-handed leptons) coupling the same way to the $W^\pm$. The collection of gauge fields transforms the same way as the generators, that is, as the adjoint representation of the group, and so the gauge quanta of a non-Abelian part of a gauge symmetry group carry adjoint charge and couple to one another. In a nonbroken non-Abelian gauge theory such as quantum chromodynamics (QCD) an important consequence of this "self-coupling" of the gauge quanta is asymptotic freedom and confinement (see below).

There exist in nature twelve known gauge fields, the eight color (glue) fields of QCD that gauge the SU(3)^{color} group and whose quanta (gluons) bind quarks together to form hadrons, the three fields whose quanta are the massive weak bosons W^+, W^-, and Z^0, and the electromagnetic field that gauges the U(1)^{charge} group and whose quantum is the photon. Only the last produces a long-range (inverse square) force. Nature has made the others less obvious by either spontaneous symmetry breaking (the Higgs mechanism) or confinement.

Spontaneous symmetry breaking of a gauge theory makes some of its gauge bosons massive, with the resulting effect that the forces from their exchange are short-range. (Just as in Yukawa's nucleon-nucleon force theory, the range of the force is of the order of the Compton wavelength of the exchanged boson, $\hbar/m_{\text{boson}}c$.) In the Standard Model of the electroweak interactions, the gauged group $SU(2)^{\text{weak ispin}} \times U(1)^{\text{hypercharge}}$ is spontaneously broken by the Higgs mechanism in order to make three of the four gauge bosons massive, namely W^+, W^-, and Z^0, whose exchanges produce weak interactions. The remaining gauge boson is the massless photon. Grand Unified Theories (GUTs) conjecture that there are at least twenty-four gauge fields, of which a Higgs mechanism has given all but the known twelve such large masses that reactions due to their exchange (e.g., proton decay) are so weak as to be undetectable thus far.

QCD is hidden by the confinement of colored particles into colorless compounds (hadrons). Roughly speaking, this is the consequence of vacuum polarization (radiative corrections to the exchange of gauge quanta). Since gluons are colored, a gluon can emit or absorb a gluon; hence, a gluon can virtually turn into two gluons, just as a photon can virtually turn into an electron-positron pair with ordinary vacuum polarization as the result. Nonetheless, there are two important differences: first, since gluons are massless, the least mass of the two-particle virtual state is zero, and so the resulting modification to the inverse square force is not short-range. Second, the sign of the effect is opposite, which makes the gluon-exchange force weaker than inverse-square at short range (this is asymptotic freedom, making high-energy processes amenable to perturbative calculations in QCD) and stronger at long range. The large strength of the color force at long range leads, rather paradoxically, to no observable long-range color force at all (except that the masses of high-spin hadrons indicate that the long-range binding force between their quarks is roughly constant). The large potential energy of a system in which oppositely colored bodies are far from one another leads to the rapid decay of any such state of quarks and gluons (the process often involving the creation of quark-antiquark pairs) into another state in which all colored particles are bound together to make hadrons, which are colorless systems and therefore have no long-range strong interaction. Thus, QCD may be considered a hidden gauge theory: it gives no long-range force, nor are its gauge quanta observable as freely propagating particles.

See also: BOSON, GAUGE; QUANTUM FIELD THEORY; RENORMALIZATION; SALAM, ABDUS; STANDARD MODEL

Bibliography

Barger, V. D, and Phillips, R. J. N. *Collider Physics* (Addison Wesley, Redwood City, CA, 1997).

Halzen, F., and Martin, A. D. *Quarks and Leptons* (Wiley, New York, 1984).

t'Hooft, G. "Gauge Theories of the Forces between Elementary Particles." *Scientific American* **242**, 104–138 (June 1980).

Charles Goebel

GLUON

See BOSON, GAUGE

GRAND UNIFICATION

According to grand unification, all nongravitational forces are manifestations of one single fundamental force. In everyday life the influence of two forces is readily apparent: gravity holds one in a chair and electromagnetism, with the help of quantum mechanics, prevents the chair from collapsing under one's weight. As matter is probed in increasingly smaller bits, the influence of gravity is surpassed by the much stronger electromagnetic force. In a hydrogen atom, for example, the electromagnetic force between the proton and its orbiting electron is tremendously stronger than the gravitational force. At even smaller distances, at or below the size of a proton, two new forces appear. One is the strong (or color) force, which binds quarks together in a proton. The other is the weak force that (among other phenomena) gives rise to particle decays, such as the decay of a neutron into a proton, electron, and antineutrino. It is the strong, weak, and electromagnetic forces that are united in a Grand Unified Theory (GUT).

The bits of matter that feel these forces are subatomic particles called quarks and leptons. Quarks are spin-$\frac{1}{2}$ fermions that interact through the strong, weak, and electromagnetic forces. The proton, for example, is made up of two up quarks and one down quark. Leptons are spin-$\frac{1}{2}$ fermions that do not interact through the strong force. The electron is a lepton that interacts through both the weak and electromagnetic forces, whereas neutrinos are leptons that experience only the weak force. The forces themselves arise through the virtual exchange of spin-1 gauge bosons: eight gluons for the strong force, the W^+, W^-, and Z for the weak force, and the photon for electromagnetism.

The world of particle physics does not end with merely up quarks, down quarks, electrons, and electron neutrinos. These comprise only the first of three "generations"; see Table 1. These three generations of quarks and leptons have identical couplings to the strong, weak, and electromagnetic forces. The generations differ from one another only in their masses and lifetimes. The reason for three generations and the pattern of masses remains a mystery.

The most elegant proposal to unify both the force structure as well as the quark and lepton fermion matter fields is SU(5) grand unification. SU(5) takes its name from the mathematical notation for the special unitary group of symmetry transformations with a five-component fundamental representation. It is the smallest group that incorporates the known forces SU(3) (strong), SU(2) (weak), and U(1) (hypercharge) as part of its symmetry transformations. At long distances the electroweak (weak

and hypercharge) forces dissolve into the electromagnetic force. Just as there are three colors of each quark in the fundamental representation of the SU(3) strong force, there are five varieties of fermions that comprise one matter multiplet in the fundamental representation:

$$\overline{5} = (d^*_R; d^*_R; d^*_R; e_L; \nu_e) \tag{1}$$

where the three d_R's correspond to the three colors of the right-handed down quark and (e_L, ν_e) is the electron and electron neutrino that make up the two components of the SU(2) left-handed lepton doublet. Here $n(\overline{n})$ denotes an n-dimensional (conjugate) representation of the grand unified group. It is literally true that $5 = 3 + 2$, which means that the five-dimensional representation of SU(5) incorporates both a three-dimensional (triplet) representation of SU(3) plus a two-dimensional (doublet) representation of SU(2).

This $\overline{5}$ comprises only part of one generation of matter. The other particles of a given generation are accommodated in the next larger representation of SU(5):

$$10 = \begin{pmatrix} 0 & u^*_R & -u^*_R & -u_L & -d_L \\ -u^*_R & 0 & u^*_R & -u_L & -d_L \\ u^*_R & -u^*_R & 0 & -u_L & -d_L \\ u_L & u_L & u_L & 0 & -e^*_R \\ d_L & d_L & d_L & e^*_R & 0 \end{pmatrix} \tag{2}$$

called the antisymmetric tensor representation. Here u_L and d_L are the left-handed up and down quarks, u^*_R is the right-handed up quark, and e^*_R is the right-handed electron. Each generation of quarks and leptons is reduced to just a $10 + \overline{5}$ combination of SU(5) representations. The fact that the Standard Model is chiral (only left-handed fields feel the weak force) is embedded in the simple fact that each generation is a $10 + \overline{5}$ and *not*, say, a $5 + \overline{5}$ or $10 + \overline{10}$. One of the fascinating consequences of embedding matter into a GUT like SU(5) is the prediction of electric charge quantization. This follows from the embedding of quarks and leptons into SU(5) representations and the requirement that the strengths of the forces are equal at some high energy scale (see below). Specifically, in terms of the proton charge e, SU(5) predicts that the electron has charge $-e$, the

TABLE 1

Matter Content of the Standard Model		
Generation	**Particles**	**Symbols**
First	up quark, down quark, electron, electron neutrino	u, d, e, ν_e
Second	charm quark, strang quark, muon, muon neutrino	c, s, μ, ν_μ
Third	top quark, bottom quark, tau, tau neutrino	t, b, τ, τ_μ

CREDIT: Courtesy of Vernon Barger.

up-type quarks have charge $+2e/3$, the down-type quarks have charge $-e/3$, and the neutrinos are neutral. This is a major piece of circumstantial evidence in favor of grand unification.

The full SU(5) force comprises twenty-four gauge bosons. Only those associated with strong, weak, and electromagnetic forces have been experimentally observed. This means SU(5) cannot be an exact symmetry. Particle physicists are very much accustomed to "broken" symmetries in nature. For example, the weak force is spontaneously broken at a characteristic energy of about 100 GeV (which corresponds to a distance of about 10^{-19} meters), called the weak scale. Spontaneous symmetry breaking leaves the interactions unchanged but makes the force carriers massive. The result of SU(5) breaking is that twelve of its twenty-four force carriers acquire a very large mass; the remaining twelve stay massless, namely, the photon, the three weak gauge bosons, and the eight gluons. (The three weak gauge bosons acquire a mass of about 100 GeV after the electroweak symmetry is broken.) The twelve heavy gauge bosons, given the names X and Y, have masses comparable to the characteristic energy scale of the spontaneous SU(5) breaking.

If the Standard Model forces are embedded into an SU(5) GUT, then there is one parameter that characterizes the strength of the SU(5) force. At low energies, however, the strong, weak, and electromagnetic forces have quite different strengths. How is this reconciled? A consequence of the fully quantum mechanical nature of the particles and forces of the Standard Model is that the strengths of forces depend in a calculable way on the distance scale (or energy scale) at which one is probing them. The change in the strengths of the forces from experimentally measured energies (approximately 100 GeV) up to the Planck scale (2×10^{18} GeV) is shown in Figure 1. This graph assumes that there are no additional kinds of matter beyond the known quarks, leptons, and gauge bosons, except for the Higgs fields needed to break the weak SU(2) gauge symmetry. This absence of new matter (called a particle desert) is assumed to persist up to the unification scale.

Intriguingly, the gauge couplings nearly intersect around 2×10^{14} GeV. The largest uncertainty is associated with the strength of the strong force, il-

FIGURE 1

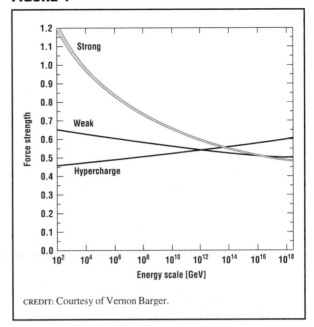

CREDIT: Courtesy of Vernon Barger.

Strength of the forces as a function of the energy scale at which one is probing them. Electromagnetism results after the weak and hypercharge forces "break" near 250 GeV.

lustrated by the width of the shaded band in the figure. The following general picture emerges: starting from a very large energy scale, the grand unified symmetry breaks, leaving the strong, weak, and electromagnetic forces as the unbroken remnants. No relics of the unified theory are to be found because the unification scale is so high. The huge disparity between the unification scale and the weak scale allows the strengths of the forces to deviate quite significantly from their near unified value. The deviations are determined by the type and amount of matter at low energies, and remarkably the predicted deviations from a unified SU(5) lead to strong, weak, and electromagnetic force strengths not too far from their experimentally observed low-energy values.

Since both quarks and leptons are unified into GUT representations, there is no fundamental distinction between them. This means, for example, that quarks can transmute into leptons and vice versa. Such processes are mediated by the twelve heavy X and Y GUT gauge bosons. This is entirely analogous to the transmutation of an electron into an electron neutrino upon emitting a weak gauge

FIGURE 2

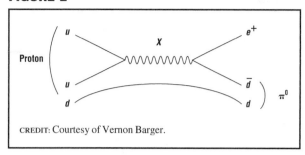

CREDIT: Courtesy of Vernon Barger.

One of several Feynman diagrams that represent the microscopic process leading to proton decay in an SU(5) grand unified theory.

boson. The most striking consequence of the transmutation of quarks into leptons is that, in principle, the proton can decay!

One of the Feynman diagrams representing a proton decay process is shown in Figure 2. The general idea is that two quarks within the proton fuse into a virtual X or Y gauge boson that then promptly decomposes into an antiquark and a lepton. The third quark of the proton does not participate in the proton decay process, but it does combine with the antiquark emitted from X or Y decay to form a meson (a quark-pair bound state). In the SU(5) unified theory the dominant experimental signal for proton decay is a positron and a neutral pion. The lifetime of a proton is estimated to be

$$\text{proton lifetime} \simeq 10^{30} \text{ years} \left(\frac{\text{mass of } X, Y \text{ gauge bosons}}{10^{15} \text{ GeV}} \right)^4. \quad (3)$$

Thus, one would have to wait about 10^{20} times the age of the universe to see a single proton decay. Fortunately, this tiny decay rate can be overcome by looking at more than 10^{30} protons and simply waiting a few years.

Heroic experiments conducted in underground laboratories in the United States, Japan, and elsewhere have done just this. The general principle is to assemble a huge quantity of protons in the form of ultrapure water in a large underground tank that is lined with photomultiplier tubes. These detectors are installed 1 to 2 km deep underground to minimize the possibility of mistaking the collision of a

high-energy cosmic ray for the decay of a proton. Radioactive impurities in the water may also lead to signals resembling proton decay, hence the use of ultrapure water. The positron from the decay of a proton is emitted at such high speed that it emits Cherenkov radiation, resulting in a characteristic cone of light that can be detected by the surrounding photomultiplier tubes.

Using this method, the tightest limits on proton decay have been set by the Super-Kamiokande experiment in Japan. They find

$$\text{proton lifetime} > 2.6 \times 10^{33} \text{ years} \quad (4)$$

to a 95 percent confidence level. This lower limit is several orders of magnitude beyond the lifetime predicted by the SU(5) GUT! Combining this null result with the mismatch of the intersection of the gauge couplings (Figure 1) leads to the conclusion that embedding the Standard Model into SU(5) with no other matter or gauge fields is strongly disfavored by experiment.

Extrapolating the Standard Model to very high energies is also problematic for theoretical reasons. The basic difficulty is that the mass of the Higgs scalar particle needed to break the weak symmetry is extremely sensitive to and dependent on the GUT physics (this is often called the gauge hierarchy problem). The Higgs particles must also be embedded into a unified representation, which requires enlarging the number of scalar particles to include a color triplet that interacts through the strong force. These color triplet scalar particles can also lead to proton decay, and so they must have a large GUT scale mass. How this happens such that the uncolored Higgs scalars stay light remains a mystery.

The preferred solution to the gauge hierarchy problem is supersymmetry. Supersymmetry is a symmetry that relates fermions to bosons. If nature is supersymmetric, there is a supersymmetric particle ("superpartner") for every Standard Model particle, differing by $\frac{1}{2}$ unit of spin. The addition of supersymmetry to the Standard Model removes the extreme sensitivity of the Higgs mass to GUT scale physics. However, like SU(5), supersymmetry cannot be an exact symmetry of nature since no su-

FIGURE 3

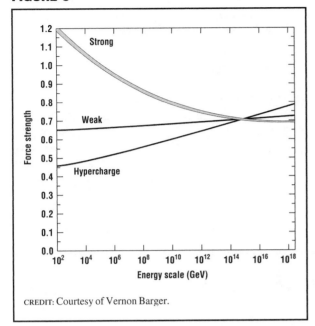

CREDIT: Courtesy of Vernon Barger.

As in Figure 1, except that supersymmetric particles are introduced near the weak scale.

perpartners have been found. The result of breaking supersymmetry is that the masses of the supersymmetric particles are lifted above the masses of the Standard Model particles. To ensure the insensitivity to GUT scale physics is preserved, the superpartners cannot have masses too far above the weak scale.

The novel features of supersymmetry in the context of grand unification are threefold: First, the unification of the gauge couplings is much more accurate than in the Standard Model, as illustrated in Figure 3. Second, the unification scale is higher, near 2×10^{16} GeV. A direct consequence of the higher unification scale is that the rate of proton decay through X and Y GUT gauge boson exchange is about 10^{35} years and is thus not inconsistent with current experimental bounds. Finally, there are new contributions to proton decay that lead to completely different signals in the underground proton decay detectors. In fact, the dominant mode for proton decay in a world that is both supersymmetric as well as SU(5) grand unified is $p \rightarrow \bar{\nu}_\mu K^+$ (antimuon neutrino plus a charged kaon). Unification may also relate the masses of fermions within a generation.

Finally, for the sake of brevity, only the SU(5) unification proposal has been focused on, but there are certainly other possibilities. One particularly interesting alternative is unification into SO(10), the mathematical group of orthogonal matrices corresponding to rotations in a ten-dimensional space. Interestingly, one **16** representation of SO(10) includes an entire generation of matter fermions *plus* an additional particle that interacts through none of the Standard Model forces. This additional field can be naturally incorporated into an extension of the Standard Model that includes both left-handed and right-handed neutrinos. This extension is well motivated by the recent evidence for neutrino masses. SO(10) contains SU(5) as a subgroup along with an additional U(1) symmetry that may or may not survive to the weak scale.

Grand unification remains an extremely active area of frontier research in particle theory nearly thirty years after SU(5) grand unification was suggested. Some of the most recent ideas propose supersymmetric unification in extra physical or deconstructed dimensions.

See also: FAMILY; GAUGE THEORY; PLANCK SCALE; STRING THEORY; UNIFIED THEORIES

Bibliography

Barger, V., Berger, M. S., and Ohmann, P. "Supersymmetric Grand Unified Theories: Two Loop Evolution of Gauge and Yukawa Couplings." *Physical Review* D **47**, 1093–1113 (1993).

Csáki, C., Kribs, G. D., and Terning, J. "4D Models of Scherk-Schwarz GUT Breaking via Deconstruction." *Physical Review* D **65**, 015004, 1–10 (2002).

Dimopoulos, S., and Georgi, H. "Softly Broken Supersymmetry and SU(5)." *Nuclear Physics* **B193**, 150 (1981).

Dimopoulos, S., Hall, L. J., and Raby, S. "A Predictive Framework for Fermion Masses in Supersymmetric Theories." *Physical Review Letters* **68**, 1984–1987 (1992).

Georgi, H. "A Unified Theory of Elementary Particles and Forces." *Scientific American* **244**, 40–55 (1981).

Georgi, H., and Glashow, S. L. "Unity of All Elementary Particles and Forces." *Physical Review Letters* **32**, 438–441 (1974).

Georgi, H., Quinn, H. R., and Weinberg, S. "Hierarchy of Interactions in Unified Gauge Theories." *Physical Review Letters* **33**, 451–454 (1974).

Haber, H. E., and Kane, G. L. "Is Nature Supersymmetric?" *Scientific American* **254**, 42–50 (1986).

Hall, L. J., and Nomura, Y. "Gauge Unification in Higher Dimensions." *Physical Review* D **64**, 055003, 1–10 (2001).

Mohapatra, R. N. *Unification and Supersymmetry: The Frontiers of Quark-Lepton Physics* (Springer-Verlag, Berlin, 1992).

Shiozawa, M., et al. (Super-Kamiokande Collaboration) "Search for Proton Decay via $p \rightarrow e^+\pi^0$ in a Large Water Cherenkov Detector." *Physical Review Letters* **81**, 3319–3323 (1998).

Weinberg, S. "The Decay of the Proton." *Scientific American* **244**, 52–63 (1981).

Vernon Barger
Graham D. Kribs

GRAVITATIONAL INTERACTION

See BASIC INTERACTIONS AND FUNDAMENTAL FORCES

GRAVITON

See BOSON, GAUGE

H

HADRON, HEAVY

The term "hadron" refers to a bound state of quarks, and a heavy hadron is a bound state that contains at least one heavy quark. Of the six quarks that are known to exist, three (up, down, strange) are considered to be light because their masses are much smaller than the mass of the proton. The other three (charm, bottom, top) are the heavy quarks. The lifetime of the top quark is too short for it to have time to form a bound state with other quarks, so it is charm and bottom quarks which are found in heavy hadrons. Because both of these quarks live for only about 10^{-12} seconds before they decay, heavy hadrons are found in nature only when they are produced in high-energy collisions.

There are two primary types of hadrons: mesons, which contain a quark and an antiquark, and baryons, which contain three quarks. For example, there is the Λ_c baryon, made of a charm, an up and a down quark; the B^+ meson, made of a bottom antiquark and an up quark; and the charmonium state J/ψ, made of a charm quark and a charm antiquark. The most exotic of the heavy hadrons, discovered by the CDF Collaboration at Fermilab in 1998, is the B_c, made of a bottom antiquark and a charm quark.

Heavy Quark Decays

Heavy quarks are interesting primarily because the pattern of their decays to lighter quarks can help physicists understand the mechanism responsible for the masses of the fundamental particles. The vacuum through which particles move is not empty; rather it is filled with a background energy density known as the Higgs condensate. Some of the properties of this condensate are known, although the reason for its existence is not yet understood. In a true vacuum quarks would move at the speed of light and therefore would be massless. They appear to have masses only because interacting with the Higgs condensate slows them down. The more strongly a quark interacts with the condensate, the larger its mass.

Quark decays also are governed by their interactions with the Higgs background. For example, 99 percent of the time a bottom quark decays to a charm quark (plus other particles), while only 1 percent of the time does it decay to an up quark. Understanding the origin of this ratio would provide clues to the physics of the bottom quark mass because both properties depend on the Higgs condensate.

An important feature of quarks is that they are never observed in isolation. They are always found in hadrons, which are complex bound states of quarks, antiquarks, and gluons (the force carrier which holds the quarks together). This complicates the problem of studying quarks, since what physicists observe in experiments are not transitions of quarks but transitions of hadrons. For example, the decay of a bottom quark into a charm quark manifests itself as the decay of bottom meson (B) into a charm meson (D). Because the mathematics is so difficult, it is still not known how to compute the structure of hadrons from first principles, and one must use tools more sophisticated than brute force to disentangle information about quarks from experimental data on hadrons.

Heavy Quark Symmetry

An important example of such a tool is heavy quark symmetry. This is an analog of the isotopic symmetry famous in chemistry, namely, that the chemical properties of an element depend only on the charge but not on the mass of the nucleus. The reason is that the nuclear charge determines the number of electrons, and the electrons in turn are so light that the nucleus appears infinitely heavy by comparison, whatever its precise mass may be. For example, the chemistry of deuterium is essentially identical to that of hydrogen, even though the nucleus of deuterium (a deuteron) is twice as heavy as that of hydrogen (a proton). What matters is that a deuteron and a proton have the same electric charge. Similarly, a heavy hadron consists of (1) a bottom or charm quark and (2) light quarks, antiquarks, and gluons, collectively known as the "brown muck" (a term coined by Nathan Isgur). Since the brown muck is much lighter than the heavy quark, it is insensitive to whether it is bound to a charm or a bottom. Therefore there is a symmetry, which is that every charm hadron has a bottom hadron analog for which the brown muck is exactly the same. (There is an equally useful symmetry among the three light quarks, known as SU(3) flavor, which originates in their masses being much less than the proton's, rather than much greater.)

The most important application of heavy quark symmetry is to measure the fundamental quantity V_{cb}. This parameter, which gives the probability for a bottom quark to decay to a charm quark, is one of a collection of nine parameters (the CKM matrix) which determine the transition rates between each of the down-type quarks (down, strange, bottom) and each of the up-type quarks (up, charm, top). Along with the masses of the quarks, the CKM matrix contains all the information about the interaction of the quarks with the Higgs condensate, so its elements must be measured as accurately as possible. The best process for measuring V_{cb} is the semileptonic bottom quark decay $b^- \rightarrow ce\nu$, which produces a charm quark, an electron, and a neutrino. Unfortunately the physical hadronic transition $B \rightarrow De\nu$ depends both on V_{cb} and on the unknown probability for the brown muck initially in the B meson to reassemble itself around the recoiling charm quark to make a D meson. The problem would be intractable, except that there is a special configuration: occasionally the lepton and the neutrino are emitted with equal and opposite momenta, with the charm quark left at rest (Figure 1). For the brown muck, all that happens then is that the motionless bottom quark is replaced by a motionless charm quark. By heavy quark symmetry, the two situations are indistinguishable. Since the brown muck does not have to rearrange itself at all, a D meson will always be produced, and the observed probability for $B \rightarrow De\nu$ is exactly the same as for the quark decay $b \rightarrow ce\nu$. The power of heavy

FIGURE 1

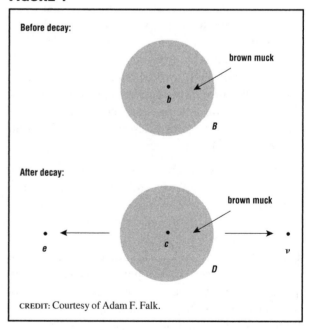

CREDIT: Courtesy of Adam F. Falk.

Semileptonic B meson decays to a D meson at rest.

quark symmetry is that *even though the properties of the brown muck are almost entirely unknown,* physicists can measure V_{cb} to a precision of better than 5 percent.

CP Violation

Heavy quarks, especially bottom quarks, are also of interest because their transitions can manifest CP asymmetry, which is a difference in behavior between a particle and its antiparticle. For example, both a B^0 meson (made of a bottom antiquark and a down quark), and the anti-B^0 (made of a bottom quark and a down antiquark) can decay to the final state $J/\psi + K_S$, where a K_S is a combination of strange and down quarks and antiquarks. If CP symmetry were respected in nature, the probability for a B^0 to decay in this way would be exactly the same as the probability for an anti-B^0 to do so. In 2001, experiments at the *B* Factories at the Stanford Linear Accelerator Center (SLAC) in California and the Japanese High-Energy Research Organization (KEK) showed conclusively that this equality does not hold and therefore that CP is violated strongly in bottom quark transitions. On the other hand, CP violation has never been observed in charm quark transitions. These properties are another important clue to the incompletely understood nature of heavy quark interactions with the Higgs condensate.

See also: B FACTORY; CP SYMMETRY VIOLATION; FAMILY; PARTICLE; PARTICLE PHYSICS, ELEMENTARY; QUANTUM CHROMODYNAMICS; QUARKS

Bibliography

Bigi, I. I., and Sanda, A. I. CP *Violation* (Cambridge University Press, Cambridge, UK, 2000).

Brando, Gustavo C.; Lavoura, L.; and Silva, J. P. *CP Violation* (Oxford University Press, Oxford, UK, 2000).

Manohar, A. V., and Wise, M. B. *Heavy Quark Physics* (Cambridge University Press, Cambridge, UK, 2000).

Adam F. Falk

HIGGS PHENOMENON

In 1934 Enrico Fermi published a descriptive theory of the weak interactions. At the time these were the feeble forces seen at work in nuclear processes. An example is beta decay, in which the neutron decays into a proton, and an electron and a neutrino. This process is slow, and for neutrons trapped in atomic nuclei, it can range from much less than one decay per second to much greater than one decay per many millions of years. Fermi had to introduce a new fundamental constant into physics, later called G_F (the *F* stands for Fermi), that sets the scale of this process and that controls the overall decay rates in beta decay. This fundamental constant can be mathematically converted into a fundamental unit of mass, which sets the scale of the weak forces, and is approximately 175 GeV. (This equals about 175 times the proton mass; 1 GeV = 1 giga electron volt; energy is used to describe mass because $E = mc^2$; the proton has a mass of approximately 1 GeV). This is called the mass scale of the weak interactions.

In the intervening years physicists have come to understand a great deal about the weak forces. In the early 1970s the greatest stride along this path occurred when the Standard Model was theoretically and experimentally established. This is a true unified theory of weak, electromagnetic, strong, and gravitational forces under one fundamental symmetry principle, called the gauge principle. Like the discovery of DNA as the basic information carrier of all living things, the gauge principle is the basic underlying defining concept of all known forces in nature. Yet, despite this triumph, the origin of the scale of weak forces as embodied in Fermi's original theory, the 175 GeV, remains a subtle mystery.

The vacuum state in any quantum theory is complicated. Although it is the state of lowest energy, it is not empty and contains vibrational motion of all fundamental particles, known as quantum zero point motion. It is known that the vacuum itself can have bizarre physical properties leading to very dramatic consequences for the observed excited states, which are the particles found in nature. Indeed, nature is mostly controlled by the laws of physics together with the properties of the vacuum.

A superconductor is a block of metal, usually a relatively poor conductor of electricity at room temperatures (such as lead or nickel) that becomes a perfect conductor of electricity when it is cooled to within a few degrees above absolute zero. Superconductors can be readily made in the laboratory (they are used in many commercial devices, such as medical magnetic imaging systems, sensitive magnetometers, etc.). The phenomenon of superconductivity is a quantum effect. At very low temperatures the ground state (vacuum) of the superconductor is rearranged. Electrons become bound together into pairs, known as Cooper pairs, held together by quantum vibrations of the crystal lattice of the material (phonons). Each Cooper pair has an electric charge of -2, and the Cooper pairs act as though they were bosons, particles that can readily occupy the same quantum state (while free electrons are fermions, and no two fermions can occupy the same quantum state). The Cooper pairs form a kind of densely packed "quantum soup" in which every Cooper pair has exactly the same motion as every other. When a low-energy photon, the particle of light, enters the superconducting material, it blends together with the Cooper pair soup and becomes effectively a massive particle. Outside of a superconductor, in free space, the photon is perfectly massless. Hence, it always travels at the speed of light. However, in a superconductor a photon acts as if it were heavy, with a mass of about 1 electron volt, and, in principle, it can be brought to rest. This quantum condensation of the electrically charged Cooper pairs, and the concomitant mass generation for the photon, gives rise to the peculiar features of superconductors, for example, they have absolutely zero electrical resistance to current flow.

In the Standard Model of the electromagnetic and weak interactions (the electroweak theory) there are four gauge particles, including the γ or photon. If the symmetry of the electroweak theory were exact, these four particles would be identically massless. There are symmetry operations that are abstract mathematical "rotations" that allow one to rotate one particle into another in the electroweak theory. These rotations do not occur in ordinary space and time but rather in an abstract mathematical world known as the internal symmetry group of the electroweak theory. The dynamics of these particles, for example, their interactions, masses, etc., is unaffected by these symmetry rotations, just as the color or shape of a chess piece is unchanged when it is rotated in space.

However, at low energies these four particles all behave very differently. The γ, or photon, remains massless in free space in the Standard Model and can be described at low energies by itself in the context of quantum electrodynamics (QED). However, the three other particles, closely related to the photon, are the W^+, W^-, and Z^0. These particles are very heavy in free space: the masses of the W^+ and W^- (representing the particle and antiparticle and therefore having the same mass) are approximately 80.419 ± 0.056 GeV/c^2, while the Z^0 is heavier still with a mass of 91.1881 ± 0.0022 GeV/c^2. The forces that are mediated by the quantum exchange of W's between other particles are exactly the weak forces that Fermi's early theory of beta decay described. Indeed, the weak forces are weak because the $W^\pm$ (and Z) are very heavy, and the quantum exchange of heavy particles is a very short-range interaction. The differences between the four particles γ, $W^\pm$, and Z^0 mean that the abstract symmetry interrelating them is broken. The symmetry becomes apparent only at very high energies, energies much higher than the masses of $W^\pm$ and Z^0.

What physical mechanism breaks the symmetry of the electroweak theory at low energies and gives rise to the masses of $W^\pm$ and Z^0? Indeed, it is natural to take a cue from the phenomenon of the superconductor. One conceives of some kind of quantum effect, analogous to what occurs in a superconductor causing the photon to become heavy, but now acting in the vacuum of free space and acting everywhere throughout the universe. This phenomenon must give the W's and Z their masses but unlike the superconductor must leave the photon massless. Therefore, whatever undergoes "condensation" in the vacuum must be an electrically neutral particle (unlike the Cooper pairs of the superconductor, which had a net electric charge of -2 and thus affect the photon).

Hence, the question becomes "What condenses in the vacuum to give rise to mass in the Standard Model?" Physicists often build "toy mathematical models" to explain a phenomenon, awaiting additional experimental or theoretical information that will lead to an exact theory of the phenomenon. The toy model is usually incomplete but contains the essence of the gross features of the phenomenon. Indeed, before the correct theory of superconductors (resulting from the work of John Bardeen, Leon

Cooper, and John Schreiffer) was constructed, a "toy" model that explained superconductivity was proposed by Vitalii Ginzburg and Lev Landau , building on the earlier ideas of Fritz London. This idea was adapted to particle physics by many authors to give mass to particles such as the W and Z and has come to become known as the Higgs mechanismm, named after Peter Higgs of the University of Edinburgh, one of its early proponents. Steven Weinberg incorporated the Higgs mechanism into his famous paper "A Model of Leptons," which was one of the earliest works to construct the electroweak Standard Model.

In the Standard Model, to explain the symmetry breaking and masses of $W^\pm$ and Z^0, one introduces a Higgs field. The Higgs field forms what is called a complex doublet and has an electrically neutral component that develops a condensate in the vacuum. The dynamics of the formation of the condensate is largely put in the model "by hand," awaiting a detailed explanation from future experiments. The condensate may be viewed as a nonzero value of the field filling all of space throughout the universe, analogous to an electric or magnetic field filling all of space. The strength of the Higgs field in the vacuum is measured as an energy, and it is postulated to be exactly the Fermi scale, 175 GeV.

The complex doublet Higgs field has four dynamical components (two complex numbers), and three of these components become blended to form the massive $W^\pm$ and Z^0. One remaining component corresponds to small local changes in the vacuum field strength of the condensate. This remaining part of the doublet can show up in the laboratory as a heavy, electrically neutral, spin-0 particle. This particle is often referred to as the Higgs boson although it is really a part of the original Higgs field.

The vacuum condensate is felt by the various particles as they propagate through the vacuum, by their coupling strengths to the Higgs field. This gives rise to their masses. For example, the electron has a coupling strength g_e. The electron mass is then determined to be $m_e = g_e \times (175 \text{ GeV})$. Since $m_e = 0.0005$ GeV, $g_e = 0.0005/175 = 0.0000029$. This is a very feeble coupling strength, so the electron is a very low mass particle. Other particles, like the top quark that has a mass $m_t \approx 175$ GeV have a coupling strength to the Higgs field that is almost identically equal to 1.

Still other particles, like neutrinos, have nearly zero masses and therefore nearly zero coupling strengths.

The Standard Model does not predict in any fundamental way the values of the coupling strengths of quarks and leptons to the Higgs field, that is, these numbers are also put into the theory "by hand." The Standard Model does, however, predict the coupling strength of the $W^\pm$ and Z^0 particles to the Higgs, so their masses M_W and M_Z are predicted (correctly) by the theory. The couplings of the particles $W^\pm$ and Z^0, called gauge particles, have coupling strengths that are related to known quantities, such as the electric charge e and the weak mixing angle θ_W, which are directly measured in various experiments. Thus, if one measures e, and θ_W and G_F, the Standard Model correctly and precisely predicts M_W and M_Z. Indeed, apart from the properties of the Higgs field and coupling strengths of quarks and leptons to the Higgs field, the Standard Model explains correctly and precisely all the phenomena seen in weak and electromagnetic interactions.

The Standard Model as a quantum theory has been subject to precise tests by experiments at LEP at the European Laboratory for Particle Physics (CERN), Tevatron at Fermilab, and SLC at the Stanford Linear Accelerator Center (SLAC). The Higgs boson, if it exists according to the simple mathematical model, has not yet been seen and is therefore heavier than an experimental lower limit from LEP-II of 115 GeV/c^2. One can infer an approximate bound on the allowed mass of the Higgs boson from indirect precision measurements of M_Z, M_W, and m_{top}, and one finds that the Higgs boson should not be heavier than approximately ~200 GeV. This assumes that the Higgs boson is a weakly coupled fundamental particle and that no additional physics is involved in the symmetry-breaking mechanism of the electroweak theory.

So what is the Higgs field in reality? Beyond the simple mathematical model, nothing is certain. However, it is clear there must be something that either really is the Higgs field or that imitates one in a very faithful way. Physicists do know that, with a sufficiently high-energy particle accelerator, they can produce a Higgs-boson-like particle or dynamics in the laboratory.

Two theoretical possibilites have been advanced for the true dynamical origin of the Higgs field. One possibility is that there exists a larger symmetry than the Standard Model structure, known as supersymmetry. Supersymmetry is a very compelling idea for a large number of reasons beyond the scope of this discussion. Supersymmetry is intimately connected with theories of quantum gravity, called superstring theories. In supersymmetry spin-$\frac{1}{2}$ particles must be related to spin-0 particles, and hence the Higgs field must be associated with additional, as yet undiscovered, spin-$\frac{1}{2}$ particles that would appear as heavy leptons, like the electron and neutrino. Supersymmetry has many desirable theoretical properties, and since it can readily accommodate the Higgs fields, it is perhaps the most popular theory involving the Higgs boson. In supersymmetry there are several Higgs fields, each one of which is a truly fundamental pointlike elementary particle. The lowest-mass Higgs boson could appear in experiments fairly soon with a mass less than of order 140 GeV/c^2 (some evidence for a low-mass Higgs boson may have been observed in 2000 at the end of LEP-II at CERN at a mass scale of 115 GeV/c^2).

Another possibility is that the Higgs field is composite, associated with new strong dynamics, that is, it is a bound state of other elementary particles, held together by new forces. This idea is closer to the dynamical phenomena that occur within superconductors and has recently been seen to work well with the idea of extra unseen compact dimensions of space at or near the electroweak scale, of order 1 TeV. One possibility is that the strong interactions, described by the theory of gluons and quarks, when extended to extra dimensions can naturally form a bound state of top and bottom quarks and antiquarks (and possibly their excitations in the extra dimensions, known as Kaluza-Klein modes) that has exactly the correct properties to be the Higgs boson. This theory, known as the top quark seesaw model, may ultimately explain why the top quark is much heavier than other quarks and leptons. Although the Higgs boson is characteristically heavy in these schemes, of order 1 TeV, the theories remain consistent with the precision limits because they are nonminimal and contain many additional particles and interactions. In some versions of new strong dynamics, new low-mass spin-0 particles can occur, known as pseudo-Nambu–Goldstone bosons, and these might be confused early on for Higgs bosons.

The Higgs boson, if it is of low mass and in accord with supersymmetry, may be discovered at the current Run-II of the Tevatron (Fermilab). In 2007 the Large Hadron Collider (LHC) at the European Laboratory for Particle Physics (CERN) will begin operations at seven times the energy of the Tevatron, will explore a larger range of Higgs boson masses, and can detect evidence of a new strong dynamics as well as supersymmetry. Beyond these explorations, higher-energy accelerators, such as a Very Large Hadron Collider (VLHC) or an e^+e^- Linear Collider, will be required to unravel the details of the true mass-generation mechanism of the Standard Model.

See also: BOSON, HIGGS; ELECTROWEAK SYMMETRY BREAKING; EXPERIMENT: SEARCH FOR THE HIGGS BOSON; PARTICLE PHYSICS, ELEMENTARY; STANDARD MODEL; SUPERSYMMETRY; TECHNICOLOR

Bibliography

Bardeen, J.; Cooper, L. N.; and Schrieffer, J. R. "Theory of Superconductivity." *Physical Review* **108**, 1175 (1957).

Ellis, J. R. "Supersymmetry for Alp Hikers." <http://arxiv.org/PS_cache/hep-ph/pdf/0203/0203114.pdf>.

Fermi, E. "An Attempt of a Theory of Beta Radiation." *Zeitschrift für Physik* **88**, 161 (1934).

Ginzburg, V. L., and Landau, L. D. "On the Theory of Superconductivity." *Journal of Experimental and Theoretical Physics Letters* **20**, 1064 (1950).

Hill, C. T., and Simmons, E. H. "Strong Dynamics and Electroweak Symmetry Breaking." <http://arxiv.org/PS_cache/hep-ph/pdf/0203/0203079.pdf>.

Weinberg, S. "A Model of Leptons." *Physical Review Letters* **19**, 1264 (1967).

Christopher T. Hill

HIGH-ENERGY PHYSICS

See PARTICLE PHYSICS, ELEMENTARY

HUBBLE CONSTANT

The Hubble constant (H_0) is a measure of the rate at which the universe is currently expanding. Together with the total energy and matter content of the universe, it sets the size of the observable universe and its age. The Hubble constant is one of the most important parameters in Big Bang cosmology: the square of the Hubble constant relates the total energy plus the matter density of the universe to its overall geometry. In addition, a comparison of the age derived from the Hubble constant and the age of the oldest stars in our galaxy provides constraints on the cosmological model that describes the dynamics of the expansion of the universe. The density of light elements (hydrogen, deuterium, helium, and lithum) synthesized after the Big Bang also depends on the expansion rate. Finally, the determination of numerous physical properties of all the galaxies and quasars (mass, luminosity, and energy density) requires knowledge of the Hubble constant.

The expansion of the universe was first established by the Carnegie Institute's astronomer, Edwin Hubble, in 1929. Determination of the Hubble constant requires the measurement of distances to galaxies d as well as their velocities of recession v: $H_0 = v/d$. Velocities are simply measured from the observed shift of lines in the spectra of galaxies. (For sound, a similar phenomenon is the Doppler effect, in which, for example, the pitch of an oncoming police siren changes as the police car first passes and then recedes.) In the case of galaxies that are moving away from Earth, their light is shifted (and stretched) to redder wavelengths, a phenomenon referred to as redshift. The shift in wavelength is proportional to velocity.

Measuring distances presents a greater challenge. Distances to the nearest stars can be measured using a method called parallax, which uses the Earth's orbit as a basis for triangulation, permitting the distance to be calculated using simple, high-school geometry. Moving out to the nearest galaxies is accomplished using a type of star known as a Cepheid. There is a well-established relationship between these stars' luminosities and period of variation, discovered by astronomer Henrietta Leavitt in 1908. This unique property allows the distance to be obtained using the inverse square law of radiation. This law states that the brightness of an object decreases in proportion to the square of its distance from the Earth. (One also experiences this effect in everyday life. This is the reason, for example, that car headlights in the distance appear fainter than those nearby.)

Using the Hubble Space Telescope, distances to galaxies with Cepheids can be measured out to the nearest massive cluster of galaxies—the Virgo cluster, located about 50 million light years away. Beyond this distance, other methods—for example, bright supernovae—are used to extend the extragalactic distance scale and measure the Hubble constant. These supernovae are believed to result from the explosion of a star near the end of its lifetime. The brightnesses of these objects are so great that for brief periods, they may be as luminous as an entire galaxy. Hence, they may be seen to enormous distances, about half the radius of the observable universe. A key project of the Hubble Space Telescope was the measurement of the Hubble constant to an accuracy of 10 percent. A number of different groups and methods have converged on a value of the Hubble constant in the range of about 60 to 70 km/sec/Mpc.

Implications for Cosmology

The dynamics of the evolution of the universe are described within Einstein's general theory of relativity by what is referred to as the Friedmann equation. The Friedmann equation relates the Hubble parameter (H, where H_0 is the value of this parameter at the current epoch), the average density of matter, the curvature of the universe, and the amount of energy associated with the vacuum of space (or dark energy). Einstein's original equation contained a term that he called the cosmological constant, a term that forced the universe to be static. When Edwin Hubble discovered the expansion of the universe, Einstein later referred to the cosmological constant as his greatest blunder. However, a discovery of a component of dark energy in the universe, based on observations of very distant supernovae, suggests that Einstein may have been correct after all.

One of the classical tests of cosmology is the comparison of timescales as given by the age of the oldest stars and the amount of time the universe has been expanding. The best estimates of the oldest stars in the universe are obtained from systems of stars within our galaxy known as globular clusters. Stars spend most of their lifetime undergoing nuclear burning of hydrogen into helium in their central cores. Detailed computer models of the evolution compared to observations of globular cluster stars yield ages of about 12 or 13 billion years. Integration of the Friedmann equation yields the expansion age of the universe. An accurate determination of the expansion age requires knowledge of the Hubble constant, as well as the average density of matter and the contribution of dark energy. Calculating the expansion age of the universe for a Hubble constant of 70, for a flat universe with no dark energy, yields an expansion age of only 9 billion years, younger than the oldest observed stars in the galaxy. This led to an earlier paradox with a universe that appeared to be younger than its oldest stars.

Much progress has been made toward measuring these individual cosmological parameters, yielding a Standard Model with a Hubble constant of 70, with matter contributing one-third and dark energy approximately two-thirds of the overall mass-energy density. The resulting age for the universe is then calculated to be 13 billion years, in very good agreement with the ages of the oldest stars. Taken together, the results from globular cluster ages and a value for a Hubble constant of 70 favor a model for the universe dominated by dark energy, consistent with the results from distant supernovae.

See also: ASTROPHYSICS; BIG BANG; COSMOLOGICAL CONSTANT AND DARK ENERGY; COSMOLOGY

Bibliography

Croswell, K. *The Universe at Midnight: Observations Illuminating the Cosmos* (Free Press, New York, 2001).

Ferguson, K. *Measuring the Universe* (Walker & Co., New York, 1999).

Freedman, W. "The Expansion Rate of the Universe." *Scientific American Quarterly* **1**, 92–97 (1988).

Wendy L. Freedman

I

INFLATION

The standard hot Big Bang cosmology has developed into a remarkably precise, well-tested theory of the evolution of the universe from its primordial state into the complex cosmos of today. Measurements of high precision have been performed over the last decade, confirming the basic predictions of the theory including Hubble's expansion law, the thermal spectrum of the cosmic background radiation, and the primordial abundances of the elements. A concordance model has emerged that with only a few parameters succeeds in fitting a great array of astronomical data ranging from the abundance and distribution of galaxies to the temperature pattern of the radiation left over from the Big Bang.

Despite its extraordinary success, the standard cosmology is clearly incomplete. It fails to answer many basic questions: Why is the universe so large, so smooth, and so geometrically flat on large scales? What was the original driving force behind the expansion of the universe? What was the origin of the density inhomogeneities that eventually grew to form galaxies, stars, and planets?

In the standard theory, it is assumed that the universe began in a hot, dense state that was almost uniform over macroscopic scales and undergoing very rapid expansion. Such a state is very special. Gravity tends to amplify density variations because denser regions undergo gravitational collapse and less dense regions fly apart. The beginning of the Big Bang had to be carefully adjusted to be very uniform and flat on large scales, with small variations at a level of about one part in a hundred thousand, just sufficient to lead to the formation of the observed structures.

The near uniformity of the temperature of the cosmic microwave sky provides a dramatic illustration of how special the initial conditions for the hot Big Bang had to be. Radiation received from opposite points on the sky is found to be at very nearly the same temperature. However, at least naively, the two emitting regions could not have been in causal contact because light from both is only now reaching the Earth. The puzzle therefore is the apparently strong correlation in the state of the universe on scales so large that no communication has been possible over the entire age of the universe.

Cosmic inflation was proposed by Alan Guth in 1979, although it was prefigured in the prior work of Alexei Starobinsky and others. It provides a beguilingly simple resolution of these basic puzzles. The basic idea is that an exotic form of matter (the simplest being scalar field matter) can have repulsive gravitational fields. If such matter was the dominant component of the early universe, it would have caused the universe to expand exponentially, making it very large, homogeneous, and flat. The most remarkable side-effect is that microscopic short-wavelength quantum mechanical variations in the scalar field matter are exponentially stretched and amplified during inflation, leading to very large-scale density variations. This provides a beautifully economical mechanism for the formation of structure in the universe, within a smooth background. So compelling, in fact, that for two decades inflation has dominated theoretical cosmology and to a large extent even set the agenda for observations.

The basic idea for inflation dates back to the 1930s when cosmological solutions to the equations of Einstein's theory of general relativity were still being developed. It was realized by Willem de Sitter, among others, that these equations allowed one to introduce an arbitrary constant, which became known as the cosmological constant. If this constant is positive, it causes the scale factor of the universe to grow exponentially and without bound. The cosmological constant was later interpreted as describing the energy density of the vacuum, that is, empty space. This vacuum has to be invariant under the Lorentz transformations of special relativity, and it follows that it possesses a *negative* pressure P equal in magnitude to its density ρ times the speed of light c squared. In equations, the stress energy tensor

$$T_{\mu\nu} = \rho g_{\mu\nu} \Rightarrow P = -\rho c^2 \qquad (1)$$

where $g_{\mu\nu}$ is the space-time metric.

In general relativity, one equates $8\pi G T_{\mu\nu}$ to the Einstein tensor $G_{\mu\nu}$, measuring the curvature of space-time. A homogeneous, isotropic universe only evolves via an overall scale factor R, which is a function of time. Einstein's equations read:

$$\ddot{R} = -\frac{GM(R)}{R^2}, \quad M(R) \equiv \frac{4\pi G}{3}(\rho + 3P)R^2 \qquad (2)$$

written in "Newtonian" form: G is Newton's gravitational constant, and $M(R)$ is the mass in a sphere of radius R. Units have been chosen so that the speed of light c is unity. If the pressure is small, $P \ll \rho$, then one sees that for ordinary matter with positive energy, gravity is attractive. A universe that starts out static will tend to collapse. However, for matter in

the form of a cosmological constant, equation (1), $P = -\rho$, and instead one finds gravity to be repulsive. The two solutions to (2) are easily obtained:

$$R(t) = e^{(\pm H_I t)}, \quad H_I \equiv \sqrt{\frac{8\pi G \rho}{3}} \qquad (3)$$

as long as ρ is constant. The physical consequence is that a universe dominated by positive vacuum energy density will generally expand exponentially, since the exponentially growing solution quickly dominates over the contracting one. Homogeneity is assumed here, but a nearly homogeneous region will also start to grow exponentially. And once exponential expansion begins, it will dilute the density of matter (which scales as R^{-3}) or radiation (which scales as R^{-4}), whereas the cosmological constant, or vacuum energy, remains constant. One finds in consequence that a positive cosmological constant causes expansion, which dilutes any matter or radiation initially present. It may also be shown that any curvature of space is likewise expanded away. Literally, the expansion of the universe blows the universe up, and if one assumes that the universe prior to inflation was smooth and flat on small scales, inflation caused it to become smooth and flat on large scales.

All of this is interesting but not useful. After all, the universe has large amounts of matter and radiation today, and these must have dominated in the early universe if the theory of the abundances of the elements is to work. What Guth realized is that a certain form of matter postulated in unified theories of high-energy physics could provide a *temporary* cosmological constant in the early universe. The density ρ can be nearly constant for some time but then decay into other forms of matter. When it does so, the exponential expansion ceases, and the standard hot Big Bang evolution can begin.

The form of matter in question is known as scalar field matter. This was developed in the early days of quantum field theory as a description of elementary particles (pions) before it was realized that pions are actually made out of quarks. Later, Peter Higgs realized that scalar fields could be used to give other fields a mass in particle physics, and Steven Weinberg and Abdus Salam built a detailed model of the weak interactions incorporating this mechanism. Particle physics experiments continue to decisively probe for the existence of the Higgs particle. If it is found, it will be the first proof that scalar field matter of the type needed for inflation actually exists. (The Higgs scalar field does not drive inflation because its potential $V(\Phi)$ [see below] is not sufficiently flat).

For simplicity, consider a spatially uniform scalar field ϕ. In an expanding universe, it evolves in time according to the equation

$$\ddot{\phi} + 3 \frac{\dot{R}}{R} \dot{\phi} = - \frac{dV}{d\phi} \qquad (4)$$

where $V(\phi)$ is an arbitrary function of ϕ, known as its potential. This equation is just for a ball moving in a one-dimensional potential, along with a damping term that depends on the rate of expansion of the universe. To describe cosmology with scalar field matter, the density and pressure must be determined:

$$\rho = \frac{1}{2} \dot{\phi}^2 + V(\phi), \quad P = \frac{1}{2} \dot{\phi}^2 - V(\phi),$$

$$(\rho + 3P) \equiv 2(\dot{\phi}^2 - V(\phi)). \qquad (5)$$

From the last equation here and equation (2), it follows that if the kinetic energy of the field is small, the scalar field causes R to accelerate, leading to exponential expansion.

A potential of the form shown in Figure 1 (e.g., a simple quadratic, $V\alpha\phi^2$) allows one to start the universe with a temporary cosmological constant as follows. The slope of $V(\phi)$ is chosen to be shallow, so that it is possible for ϕ to roll slowly downhill. Then, one chooses the initial conditions for the universe so that ϕ starts out uphill, moving slowly. Finally, the initial conditions for the scale factor of the universe R are chosen so that it is expanding. If these conditions are fulfilled, then ϕ will gradually roll downhill, with its motion damped by the second term in (4). As it rolls slowly downhill, the scale factor R will drive an epoch of exponential expansion, making the universe very large, very smooth, and very flat. When the scalar field nears the minimum of the potential $V(\phi)$, then its kinetic energy $\dot{\phi}^2$ begins to overwhelm $V(\phi)$, and it is seen from (5) that

FIGURE 1

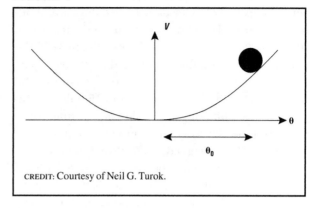

CREDIT: Courtesy of Neil G. Turok.

Potential energy PLC.

$\rho + 3P$ becomes positive so the acceleration of the scale factor ceases.

When the field ϕ is incorporated into a unified theory of particle physics and forces, it automatically couples to all the particles of the Standard Model. This coupling provides a natural mechanism through which the energy stored in ϕ during inflation is eventually released into the radiation and matter needed to start the hot Big Bang. When ϕ reaches its potential minimum, it starts to oscillate, and these oscillations cause the creation of particles to which ϕ couples, particles that comprise the radiation and matter required for the hot Big Bang. The energy initially stored in ϕ is thus redistributed amongst all the particle species present. Generally, it is easy to arrange that sufficient energy is transferred to heat the universe to temperatures well above those needed for the production of the primordial elements in the Big Bang.

Precisely what was gained from this early epoch of exponential expansion? The puzzle is to explain why the universe is so smooth today, where the natural expectation from "random" initial conditions for the universe would be a lumpy universe becoming ever more lumpy under the influence of gravitational collapse. Exponential expansion improves this situation. A characteristic length scale in gravitational physics is the Planck scale

$$L_{\text{Planck}} \equiv \left(\hbar \, \frac{G}{c^3} \right)^{\frac{1}{2}} \approx 10^{-33} \text{ cm.} \qquad (6)$$

By random initial conditions, one might mean a universe composed of many Planck-sized domains of differing energy densities and expansion rates. Compare this picture with the picture obtained if the current observed universe is traced all the way back to the Planck time, L_{Planck}/c or 10^{-43} seconds. The observed universe is now some 13 to 15 billion light-years across. Following this scale back in time using the known densities of matter and radiation, the size of the region it occupied at the Planck time can be calculated. It turns out to be just a millimeter, which, even though small, is a *huge* scale in Planck units. The problem of the initial conditions needed for the hot Big Bang cosmology is that the universe had to have started in a state almost perfectly uniform on a scale of 10^{32} times the natural length scale.

Inflation greatly improves this situation. Take instead one Planck-sized region, 10^{-33} centimeters across, within which the conditions required for inflation happen to be satisfied, with the scalar field ϕ large enough to ensure that inflation lasted for 100 expansion times within it. The size of such a region would have grown by $e^{100} \approx 10^{43}$ during inflation, so that by the end of inflation it would be approximately a kilometer across. Following inflation, this region would then undergo standard hot Big Bang expansion, during which a millimeter-sized subregion would grow to the size of the visible cosmos today. Thus, given just one Planck-sized region with the right inflationary initial conditions, one can account for the present vast cosmos of great uniformity and flatness (see Figure 2).

This argument is plausible but not necessarily compelling. After all, the field needed to drive inflation has been added by hand, and the discussion of the likelihood of obtaining inflation is hardly rigorous. What happened to all those regions that did not inflate? Is inflation still occurring somewhere else in the universe? Scientists have struggled with these questions, but no convincing answer has yet emerged.

What made inflationary theory much more convincing was the fact that it was able to explain the density inhomogeneities needed for cosmic structure formation in a quite magical way. This became apparent soon after inflation was invented and was a surprising, and for some a compelling, success.

FIGURE 2

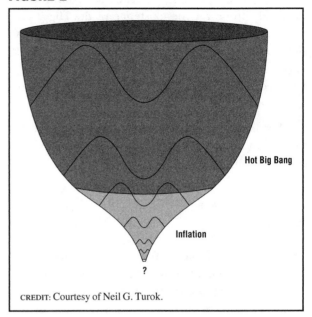

CREDIT: Courtesy of Neil G. Turok.

Schematic of inflationary evolution of the universe.

Relativity and quantum mechanics lie at the heart of the inflationary mechanism for generating inhomogeneities. If ϕ is a field, then according to relativity, if it can vary in time, it must also be allowed to vary in space. But according to quantum mechanics (the Heisenberg uncertainty principle), there must be a minimal level of fluctuations in such a field so that even in empty space, it is constantly fluctuating about its minimal value.

In the context of inflation, the quantum "jitter" in the inflation field ϕ becomes stretched and amplified into large-scale density homogeneities. First, as the universe is blown up by inflation, the scale of any fluctuation in ϕ grows exponentially in time. When the scale of the fluctuation is short, it oscillates like a sound or light wave.

But when the scale of the fluctuation is stretched beyond a certain point, different regions of the fluctuation no longer communicate, and the fluctuation becomes "frozen," thereafter simply undergoing a continuous stretching until the end of inflation. The remarkable consequence of this mechanism is that the final spectrum of fluctuations is "scale-free." That is to say, over an exponentially large range of scales, at the end of inflation one finds that the density of the universe fluctuates with the same amplitude on all length scales.

It was realized long before inflation that scale-free primordial fluctuations could plausibly explain the observed distribution of galaxies in the universe. This spectrum became known as the Harrison-Z'eldovich spectrum, named after the physicists who first postulated it. Therefore, the realization that inflation, invented for very different purposes, automatically produced the Harrison-Z'eldovich spectrum was quite spectacular and convinced many physicists that inflation must have actually occurred.

The Harrison-Z'eldovich hypothesis can be understood as follows. The early universe is filled with many different particles: nucleons such as protons and neutrons, electrons, photons, and neutrinos, and some form of dark matter needed to explain the structure of galaxies as observed today. The simplest possibility is that the overall density of the universe varied from place to place, but the relative abundances of the different particle species were the same everywhere. This possibility is realized in the simplest inflation models. Again, the simplest possibility is that the density variations take the form of a Gaussian random field. That is to say, the amplitudes of plane waves of each wavelength are chosen at random from a Gaussian probability distribution, and there is no correlation between modes of different wavelengths. More prosaically, the density variations are like small ripples on the surface of the sea, with random locations and no special features.

Recent observations of the cosmic background radiation have provided spectacular confirmation of the Harrison-Z'eldovich spectrum, combined with the simplest form of primordial density variations. The Boomerang and Maxima experiments used balloon-borne telescopes to map the cosmic background radiation over hundreds of square degrees on the sky, to a level of tens of microKelvin. They measured the amplitude of the temperature fluctuations as a function of angular scale. What was found was that as one goes from large to smaller scales, the amplitude grows and then oscillates. So far there is evidence for three peaks, on scales of a degree, half a degree, and one-third of a degree. This is in astonishingly close accord with the expectations from theory, under the assumption of the simplest form

of perturbations. These measurements have also allowed a measurement of the spatial geometry of the universe on the largest visible scales. Again, the measurements are in accord with the simplest models of inflation, according to which there was a lot of inflation and the universe became spatially flat with exponential accuracy.

These measurements are a considerable success for inflation. However, one should not infer that inflation has been proven. The observations really confirm something much simpler: scale invariance and the "simplest" type of fluctuations. There is no direct evidence for the existence of the inflation field ϕ. Also, apart from the qualitative successes of inflation—the observed flatness of the universe and the scale invariance of the inhomogeneities—one does not have a quantitative prediction which provides specific evidence that inflation occurred. On the contrary, inflationary theory can only be reconciled with the observed amplitude of the primordial fluctuations if certain parameters in the theory are adjusted to very small values, to fit the data. It is possible that another physical mechanism could make the universe large, flat, and smooth and produce density variations of the observed form.

What would convince skeptics that inflation did, in fact, occur? One of the most distinctive inflationary predictions is that during the period of exponential expansion, gravitational waves would have been amplified and stretched to large scales just as the fluctuations in ϕ were. This leads again to a scale-invariant spectrum of gravitational waves, which would in principle be detectable in today's universe. The most powerful way to search for these waves employs the polarization of the cosmic microwave sky. If gravitational waves are present, they lead to a pattern of polarization that is impossible to obtain from ordinary density inhomogeneities. Observation of this pattern would be a much more direct confirmation of the inflationary mechanism. The Planck satellite, scheduled to be flown by the European Space Agency in 2007, should be able to see this signal, at least for the simplest inflation models.

Even if inflation did provide the mechanism for the Big Bang and the density variations in the universe, many theoretical questions are unresolved. The biggest problem in physics is that of quantizing the gravitational field. However, inflationary fluctuations involve both quantum mechanics and the effects of gravity. How can they be treated consistently, when there is no consistent theory? Current calculations of inflationary fluctuations are performed in an approximation (linear theory) in which the theoretical inconsistencies do not yet appear. Calculations to the next level of accuracy produce meaningless infinities and ambiguities that are still not resolved. Thus within its current framework, inflation can only be viewed as a provisional theory, recognized to be inconsistent at a deep level. More consistent theories of quantum gravity, including string theory and supergravity, so far do not seem to produce inflationary models of the form needed to match observation.

It is also important to emphasize that inflation does not solve many of the most fundamental puzzles in cosmology: Did the universe begin? And if so, how? Recent observations indicate the presence of a positive vacuum energy (or cosmological constant). This discovery was entirely unexpected within inflationary cosmology.

Why is there a cosmological constant today? The current cosmological constant is smaller than that needed for inflation by at least 100 orders of magnitude. So, inflationary cosmology requires two cosmological constants, differing by 10^{100} in magnitude. Since the goal of inflation was to avoid "fine tuning," the actual need for it is quite disturbing. If the expansion of the universe is accelerating today, as observations indicate, where will that expansion lead? What is the future of the universe?

These questions illustrate that inflation is only a theory of the *early* universe, one that neither addresses how the universe began nor its current state nor its future direction. The successes of inflation are considerable, but they may yet be reproduced by other theories that are more complete.

See also: ASTROPHYSICS; BIG BANG; BIG BANG NUCLEOSYNTHESIS; COSMOLOGY; COSMOLOGICAL CONSTANT AND DARK ENERGY; HUBBLE CONSTANT

Bibliography

Guth, A. H. *The Inflationary Universe* (Addison-Wesley, Reading, MA, 1998).

Guth, A. H., and Steinhardt, P. J. *Scientific American* **250** (5), 90–128 (1984).

Linde, A. D. *Particle Physics and Inflationary Cosmology* (Harwood, New York, 1990).

Neil G. Turok

INFLUENCE ON SCIENCE

Particle physics is sometimes criticized as expensive and irrelevant, though the same criticism is rarely made of astronomy and space exploration, which are more costly. What then is expected of a branch of science in terms of impact and cost? What is the impact of particle physics on philosophical, astronomy, cosmology, scientific practice, technology, business, and daily life?

Philosophical Impact

Particle physics challenges intuition. The experimental observations of atomic physics force the acceptance of the intellectual framework of quantum mechanics. As energy is raised into the realm of particle physics, relativity must also be accepted as the norm and not as a difficulty to be hedged about and avoided. A range of new phenomena, such as the creation of new matter out of energy, the existence of antimatter (predicted by Dirac from the combination of relativity and quantum mechanics), time dilation for fast-moving objects, and the speed of light as a limit, plus a wealth of more technical detail, attest to the accuracy of the theory with extraordinary precision. These phenomena, many predicted by theory before they were observed, force a full adoption of special relativity as a sound working basis on which to proceed. Matter-antimatter oscillations provoke consideration of the reality of quantum mechanical amplitudes as more than merely a convenient mathematical construction behind a theory governed by nonnegative probabilities.

This willingness to stretch the imagination using mathematical rigor as a touchstone is one of the stimuli given by particle physics. Three-dimensional space becomes four-dimensional space-time. Mathematical difficulties still to be overcome create interest in strings in ten dimensions or surfaces in eleven. By working in this space, and "compactifying" the unobserved dimensions, theorists hope to unify the theory of the smallest objects (quarks and leptons) with the gravity that dominates the universe at large distances. Is this approach correct? Only future experimental data can verify or disprove that.

If one makes such a unified Theory of Everything (TOE) is it worthy of the name? This is controversial territory. This extension of the reductionist approach must be approached with care. Could it explain everything including life, love, music, and free will? It is difficult to tell. One aspect of the question that has received attention recently is the Anthropic Principle. In its weak form this states that the laws of nature are such as to permit human existence. This appears to need very careful tuning of some of the numbers found in nature. For example, if the charge on the electron differed by more than a percent or two from its actual value, stars could not produce both carbon and oxygen, so human life could not exist. Is this evidence of divine design, or is it solved by assuming endless repetitions of universes with randomly different laws, most of which have no one to observe them? Isaac Newton referred to physics as "experimental philosophy." This is a very appropriate name.

Impact on Cosmology and Astronomy

In cosmology, the greatest impact of particle physics is found. Particle physics provides the rules that governed the crucial first seconds after the Big Bang. Distances were then tiny and energies gigantic, so that full play was given to the realm of high-energy physics. What exists now is a result of what happened then. Astronomers are convinced of the historical reality of the Big Bang because its echo is seen in the cosmic microwave background in the blackness between the stars. The Big Bang started with radiation ("Let there be light"), which created matter and antimatter equally. Yet today no antimatter can be found in nature, even after exhaustive searches. The solution to this puzzle lies in particle physics theory. Particle physics methods have been the driving force in the question of inflation (of why the universe is so uniform) and of its large-scale structure.

Astronomy poses questions such as "What makes stars shine?" "What is a supernova?" and "Is there dark invisible matter all around us?" The first two questions are answered, and the last addressed, by particle physics.

Impact on Science

The treaty which set up the European Laboratory for Particle Physics (CERN) in 1954 is one of the earliest examples of European cooperation. Europe needed this scale and structure of operation if it were to compete with America where scientific activity had been given a tremendous boost by the atomic bomb project. Only by combining its nations' strength could Europe hope to stem the "brain drain." Yet curiously, out of this competition has come wider cooperation. Even during the Cold War, large-scale collaborations existed with the Soviet Union based on a shared enthusiasm for science. The HERA electron-proton collider in the Deutsches Elektronen-Synchotron Laboratory (DESY) in Hamburg, Germany, was built by voluntary international agreement with Germany as the lead partner. By 2000, countries from all around the world, including the United States and Russia, were joining the now twenty CERN member states to build a common project (the Large Hadron Collider). Financial and operational "Memoranda of Understanding" provide the formal structure within which autonomous national funding agencies manage their own institutes and obligations. Big commonly owned and operated facilities at the world's best research sites are now a feature of many fields of science, such as telescopes (for example, the European Southern Observatory) and the European Synchrotron Radiation Source.

Doing science within a large international collaboration as a research student is training for leadership in the "real world." To find one place among say 400 co-workers, to collaborate usefully, to question the work of others, to offer one's own work for criticism and suggestion to a group from several nations, and finally to lead an international activity to successful completion and publication gives the training needed for being effective in large national and multinational commercial organisations.

Computer data handling is an area in which for forty years particle physics has been pushing the limits of what the computer scientists can deliver. The process continues through the World Wide Web and the computer grid. The approach to managing immense data sets with complex calibration methods, and the international access to them, makes particle physics the ideal testing ground for developing the analytic and computational techniques needed for, say, elucidating protein structure at new synchrotron x-ray sources.

Impact of Particle Physics Technology

The World Wide Web is the greatest gift of particle physics to humankind. It was invented by Tim Berners-Lee at CERN to provide easy document exchange around a widely dispersed international group of collaborators and transformed the Internet from an academic tool into a telecommunications revolution. The Web is free to all users, in contrast to some proprietary software used for, say, word processing. This libertarian approach stemmed deliberately from the open collaborative approach pioneered by particle physicists. Berners-Lee's view is that the Web would never have taken off if CERN had tried to exploit it. e-business conducted over the web amounted to $657 billion in 2000. The optoelectronics industry is a major supplier to the Internet and in 2000 was worth $140 billion and growing at 25 percent per year.

Superconducting magnet technology was pushed by and for particle physics. The electric current producing the magnetic field in superconducting magnets will circulate forever without any power loss or need for external supply. This is accomplished by cooling certain materials to within a few degrees of the absolute zero of temperature. Rutherford cable is the key to stably operating magnets, now in use in nuclear magnetic resonance imaging machines at hospitals worldwide.

The 1992 Nobel Prize in Physics was awarded to Georges Charpak of CERN for his invention of detector techniques for particles that he had made and then adapted for medical imaging purposes. Positron-emission tomography is one such noninvasive technique. It brings antimatter out of the research laboratory and into hospitals as a diagnostic tool. To make it affordable requires the accelerator techniques enabled by superconducting magnets, detector instrumentation such as that developed by Charpak, and the fast data handling power pioneered by particle physics. Radiation therapy for cancer treatment was

an early spin-off from accelerator technology: new techniques are still being developed.

Theoretical particle physics has an interesting spin-off application in high finance. The same techniques used in solving abstruse problems in particle theory have shown ability to predict the movements of financial markets. Merchant banks like to hire particle theorists. Another application of particle theory computational methods is in warship design.

The list of spin-off applications is indeed diverse. It provides a classic illustration of the need to give rein to curiosity-driven science. Highly focused application-driven research of course is vital for future prosperity, but real innovation can often come from research planned for a different reason. As with lasers, (theorized in 1905 by Einstein, invented over 50 years later but with no obvious use, and now used in every CD player as well as in carrying web messages around the world), one can have surprises.

See also: BENEFITS OF PARTICLE PHYSICS TO SOCIETY; CULTURE AND PARTICLE PHYSICS; PHILOSOPHY AND PARTICLE PHYSICS; UNIVERSE

Bibliography

Greene, B. *The Elegant Universe: Superstrings, Hidden Dimensions and the Quest for the Ultimate Theory* (Jonathan Cape, London, 1999).

Barrow, J. D., and Tipler, F. *The Anthropic Cosmological Principle* (Oxford University Press, New York, 1986).

Fraser, G. *The Quark Machines: How Europe Fought the Particle Physics War* (Institute of Physics Publishing, Bristol, UK, 1997).

Fraser, G., ed. *The Particle Century* (Institute of Physics Publishing, Bristol, UK 1998).

Naughton, J. *A Brief History of the Future: The Origins of the Internet* (Weidenfield and Nicolson, London, 1999).

David H. Saxon

INJECTOR SYSTEM

The injector system for typical particle accelerators consists of a source of charged particles, a DC electric field to give the charged particles an initial kinetic energy, followed by a radio frequency (rf) acceleration stage that prepares the charged particles for injection into the main accelerator system. The main accelerator system accelerates the charged particles to their final energy. These particles are then used either for injection into a storage-ring collider or to provide a beam to produce secondary particles for a variety of scattering or particle production experiments.

Particle Sources

Charged particle sources vary from simple thermionic emission from hot tungsten filaments for low-current electron sources to carefully engineered solid-state photo cathodes for high-pulsed-current polarized electron beams. Low-energy beams of protons are normally produced using an rf plasma discharge with either magnetic confinement of the plasma (magnetron) or electric field confinement (penning trap) of the discharge. Several methods are used to convert these beams to H^- beams via a charge exchange reaction in a low-pressure gas with suitable characteristics.

Initial Acceleration

In the case of electrons, the initial acceleration mechanism can either be a DC electrostatic field or a very-high-gradient rf field, used in conjunction with a pulsed laser illuminating a photocathode timed so that the accelerating rf electric field is at a maximum. With a DC electrostatic field as the initial accelerator, a combination of rf fields is then used to bunch the beam so that it can be captured and accelerated by the rf system. In modern electron accelerators, the rf accelerator usually consists of a disk-loaded circular waveguide, with the disks providing a propagating rf wave with a phase velocity matched to the velocity of the electrons. These accelerators typically operate with an rf frequency of 2,856 MHz, and the rf power is provided by pulsed klystron amplifiers capable of an output of up to 100 MW peak with pulse lengths of a few microseconds. Typical electron energies at the end of this initial rf acceleration section are a few hundred MeV.

Because protons or H^- ions are much heavier, more elaborate initial acceleration systems are required to increase their velocity to the point where they can be accelerated by an rf disk-loaded waveguide structure. In the first method, they are accel-erated by a DC electric field of several hundred kilovolts. This is then followed by a linear accelerator consisting of an rf tank, in which the beam passes through a series of drift tubes of increasing length to shield the particles from the rf field when it is of the wrong phase to accelerate the particles. Usually, these drift tubes also contain DC magnetic fields to provide focusing for the beam. At the end of this structure, the particles have an energy of approximately 200 MeV and a velocity of 0.2 c (c = speed of light). This velocity is sufficient so that the beam can be captured in a disk-loaded waveguide structure. This system typically provides an additional 200 to 400 MeV of energy to the particles. With a final energy ranging from 400 to 600 MeV, the ions are sufficiently relativistic so that they can be injected into a rapid cycling synchrotron and accelerated to their final energy for injection into the main accelerator. With the invention of the rf quadrupole accelerator in the early 1980s, a lower DC accelerating field can be used, and the initial part of the drift-tube linac can be replaced by a compact and efficient acceleration system. The rf quadrupole accelerator consists of a tank with four precisely machined vanes orientated at 90° to one another, extending toward the center of the tank. The inner edges of the vanes are machined with a wave shape that increases in wavelength as the particles gain energy. The rf field inside the tank induces an accelerating gradient via the vanes along the axis of the cylinder that then accelerates the particles. An rf quadrupole can provide energies of up to 2 MeV. Most modern proton accelerators incorporate one of these devices as part of the initial acceleration chain.

Main Acceleration

For electrons, the main acceleration stage is either a synchrotron (described below) or more of the same disk-loaded waveguide to accelerate the electrons to their final energy. The nominal accelerating gradient in these disk-loaded structures is 15 MV/m. The Stanford Linear Accelerator is the highest-energy accelerator of this type, achieving 50 GeV with a length of 3,200 meters.

For protons or H^- ions, the next element following the linear accelerator in the injection system is a circular accelerator. It accelerates the particles by confining them to a circular orbit using electromagnets that provide both bending and focusing and then passing the particles through an rf-accelerating cavity system many times. These machines are called synchrotrons since the magnetic fields have to increase in synchronism with the energy increase of the particles. A system of pulsed electric and magnetic fields deflects the incoming protons onto the stable orbit. In the case of H^- ions, injection into the circular accelerator is accomplished by passing the ions through a thin foil that strips the two electrons from each of the ions, leaving protons that are then guided by the magnetic field of the synchrotron. Modern proton accelerators use H^- injection since much higher beam intensities can be achieved using this technique. Electron synchrotrons use a system of pulsed elements to inject onto the stable orbit.

Synchrotrons are necessarily cyclic machines. The ratio of peak energy of the particles to the injection energy of the particles is typically between 10 and 20. Remnant field effects in the iron-based electromagnets limit the dynamic range of synchrotrons to this range. The power supplies that provide the current for the guide field magnets can be either programmable supplies or be configured as a resonant circuit with a DC bias. Usually, lower-energy synchrotron magnet systems are configured as resonant circuits and high-energy synchrotrons use programmable power supply systems. An example of the former is the 8-GeV booster at Fermilab, whereas the new main injector at Fermilab uses a programmable power supply system for its magnets. The 8-GeV booster beam is injected into the main injector, which then accelerates the protons to 120 GeV. These 120-GeV protons are subsequently injected into the Tevatron that then accelerates them to 980 GeV. The cycle time for synchrotrons varies from 1/60th of a second to minutes, depending on available power for the magnets and the rf acceleration system. When the particles reach their full energy at the end of the chain of accelerators, they are either stored for use in colliding-beam experiments, or they are extracted from the accelerator by the extraction system. The extracted beam is then used to produce secondary particle beams to carry out scattering and particle production experiments.

See also: ACCELERATOR

Bibliography

Chao, A., and Tigner, M. *Handbook of Accelerator Physics and Engineering* (World Scientific, Singapore, 1998).

Edwards, D. A., and Syphers, M. J. *An Introduction to the Physics of High Energy Accelerators* (Wiley, New York, 1993).

Donald Hartill

INTERNATIONAL NATURE OF PARTICLE PHYSICS

Nature's laws are universal, and logic and experiment prevail in their formulation. Their validity ignores national boundaries and cultural differences. Research in physics has therefore always demonstrated an international character, and peer recognition at the international level has been eagerly sought. However, whereas the dissemination of new results and ideas through international journals and international conferences has always taken place, original work long originated from individuals or small research groups with their distinct national styles.

By the beginning of the twenty-first century, particle physics experiments in all the major laboratories of the world had become truly international ventures, drawing researchers from around the world to participate. Particle physics has played a leading role in enhancing a more pronounced internationalization of science, with the pooling of resources from individual nations and its requirement of large international teams of scientists working together to obtain new results. The European Laboratory for Particle Physics (CERN) in Geneva, Switzerland, provides an ideal example of this internationalization not only because it draws researchers from around the world but also because it was created by a collaboration of nations.

The Creation of CERN

CERN indeed illustrates well the success of international scientific collaboration. This first looked possible in those academic domains with no immediate applications but where several neighboring nations were still interested in developing their research in an important way. This was the case for particle physics. The need for this new way to do research was first strongly realized in Europe after World War II, when it became clear that needed instruments were financially out of the reach of individual nations. In 1949 French Nobel laureate Louis de Broglie made a vibrant appeal for an international laboratory where physicists from different European nations could work together. Encouragement from prominent American physicists culminated with the address of Nobel laureate Isidor Rabi to the United Nations Educational, Scientific and Cultural Organization (UNESCO) conference of 1950 in Florence. The response took only a few years to materialize; in 1954 CERN was created.

International collaboration in Europe first appeared to be most needed for the construction and operation of large accelerators, and the purpose of CERN was to build a laboratory for the study of elementary particles with high-energy accelerators in Western Europe. The facilities offered to European scientists had to compare favorably with those available in the United States (e.g., the Brookhaven National Laboratory) in order to attract physicists. In a period of time, a "pyramidal approach" was considered where the base was the universities, the middle level the national laboratories with their medium-size machines, and the top CERN with the largest machines. However, natural evolution soon eliminated most of the nation-based machines in Europe, leaving the national laboratories dependent on CERN for research in particle physics. The Deutsches Elektronen-Synchrotron Laboratory (DESY), in Germany, has remained a national laboratory with top-class machines, but it is used on an international basis, and several countries contributed to the construction of its newest machine, HERA. Very soon also after its creation, all experimental groups at CERN involved physicists from different institutes, in different countries, working together, building detectors and exploiting them. Full international collaboration thus became a new style in research.

Back in the early 1950s it was difficult for the promoters of CERN, such as Pierre Auger and Eduoardo Amaldi, to convince many of their colleagues that pooling resources (financial and also human resources) in a common endeavor was the only way to progress.

Some indeed argued that any extra international funding should be distributed among the existing national structures. However, eventually, everyone was convinced that full European collaboration centered on an international laboratory was the proper choice.

Scientists from different countries and backgrounds worked together. They decided together what to build, how to construct accelerators and detectors, and how to exploit them best. In that way, they found they had to learn much from each other, breaking national and cultural boundaries. This was not always easy, but the benefits of pooling resources and ideas were soon realized. Nobody would any longer say that the national resources that are transferred to CERN by its member states should rather stay in the home nation, and nobody would any longer consider a purely national experiment in particle physics. In Europe, the example set by CERN was eventually followed in other areas of research: ESO (the European Southern Observatory for astronomy, ESA (the European Space Agency) for space research, the ILL (the Langevein-von Laue Institute) and the ESRF (European Synchrotron Radiation Facility) in condensed matter physics, and the EMBL (European Molecular Biology Laboratory in biology.

Working at the International Level

Proper collaboration at the European level called for the implementation of new and flexible decision-making structures. They have been working well. Physicists have the great advantage of sharing the same passion and speaking a common language. Yet, what was achieved at CERN can also serve as an important, fruitful example for other professions.

For instance, any decision on the construction of new machines has not been left to CERN alone but has involved international discussions and reviews, and an international body was created representing the users of the laboratory in its member states. It was called the ECFA (the European Committee for Future Accelerators). One may summarize by saying that European physicists gradually learned how to fully collaborate on their research in particle physics, and, in that respect, Europe has paradoxically shown far more unity than the United States, where resources were distributed over several large laboratories. The concentration of resources at the CERN site has allowed the construction of many machines of a new type, unique in the world, such as the Intersecting Storage Rings in the 1970s, the proton-antiproton collider in the 1980s, the LEP machine in the 1990s, and most recently the LHC, now under construction and scheduled to be completed in 2007. These unique machines have attracted many scientific users from all over the world and, in particular, from the United States, bringing an international world dimension to an initially European endeavor.

Since it quickly grew from small to big in both size and budget, CERN was granted a very large amount of autonomy from the outset, control from the member states being essentially present at the global budget level and for the approval of new major projects, but not in the ongoing operation and research programs of the laboratory. This contributed much to its success as an international organization. Yet, in the beginning of the twenty-first century, one now sees nations less willing to grant the same liberty to newly created agencies, instead preferring to maintain a stronger control. Sometimes national officials have to be reminded that CERN is their country's own laboratory for particle physics and not a foreign institute drawing on their financial resources. Thus, international collaboration, which European physicists have learned to appreciate so much, still calls for lasting effort and cannot simply be taken for granted.

Opening to the East and to the World

The collaborative spirit of CERN was quickly extended to the East. Anything that could be done to bring together physicists from the West and from the East during the tense Cold War years was extremely beneficial in paving the way, in a modest but tangible manner, to the eventual thaw. Soon after the creation of CERN, the countries of the eastern block established an international laboratory, the Joint Institute for Nuclear Research, in Dubna, near Moscow. Some collaboration between CERN and Dubna started in a modest way in the late 1950s; however, it quickly developed. For example, through over 20 years of East-West confrontation, CERN and Dubna formed a joint school that every other year brought together for two weeks fifty young physicists from the East and as many from the West. More im-

portantly, European groups worked at the large Soviet laboratory Serpukhov in the early 1970s, and many Soviet groups later worked at CERN as part of large international collaborations. It took courage and good will to exploit any crack in the Iron Curtain, but the effort paid off. CERN and Dubna were both nominated for the Nobel Peace Prize in 1997. Physicists were talking physics, but they were also talking about Andrei Sakharov and Yuri Orlov! They were first of all getting to know one another.

At the beginning of the twenty-first century, some cooperation in particle physics extends to the whole world. It is partly monitored by ICFA (the International Committee for Future Accelerators), and worldwide research collaborations are often at work for research. It was strongly the case for the LEP experiments and is extending further for the LHC experiments.

International collaboration between western Europe and the United States has always been strong. It first included European and American physicists as individuals, mainly many Europeans learning particle physics in the United States, and some Americans coming to Europe on sabbatical leave, but, by the 1970s, this connection had already turned into full university groups with professors, postdoctoral physicists, and students working for several years on the other side of the Atlantic within the context of international collaborations. By the 1980s, with the unique machines at CERN, the number of American particle physicists in Europe eventually became much larger than that of their European counterpart particle physicists in the United States. The United States, together with Japan and other countries, has agreed to contribute to the construction of the new LHC at CERN, extending scientific collaboration beyond the European nations. Additionally, starting in the 1970s, Fermilab, in Batavia, Illinois, made a big effort to associate to its research not only Europeans but also scientists from the Soviet block, from Asia, and from South America.

An Example of International Experimental Collaboration

A good example (among many) of international collaboration is offered by the L3 detector at LEP, built and operated under the leadership of Samuel

Ting from MIT. It shows how collaboration in particle physics can shatter barriers for the benefit of a scientific endeavor. Crystals of bismuth and germanium oxide (BGO), for detecting gamma rays and electrons, were a key part of this big LEP detector that was built by a collaboration of physicists from seventeen countries. The team consisted of several hundreds physicists from western and eastern Europe, the United States, the Soviet Union, and Asia, with researchers from China and also from Taiwan. The construction of the detector required 12 tons of BGO of high purity, something never before realized at that level. This called for the joint effort of 100 physicists and engineers from China, France, Germany, the Nertherlands, Italy, the Soviet Union, Switzerland, and the United States. Any reluctance generated by national sensibilities were overcome, and the Soviet Union agreed to provide 5 tons of germanium oxide, a product deemed "strategic" at that time. China brought in the necessary quantity of bismuth oxide, and the ceramic institute of Shanghai produced in two years the required number of crystals, over 10,000 altogether. Machines for cutting and polishing the crystals and also those to control the achieved quality were built in France and shipped to China in the early 1980s. Production commenced in 1985, and the crystals were delivered to CERN in 1987. Everything was ready by 1989 when LEP started. The sophisticated electronics coupled to the crystals came from the United States.

This key piece of the overall detector was an important element in the success of the L3 experiment. The BGO crystals and their associated electronics were also quickly found to be useful in increasing, in an important way, the power of the positron emission tomography (PET) machines (an earlier spin-off of CERN), which play a great role worldwide in cardiology, neurology, and oncology. None of this could have been achieved without much good will, but also the motivation, generated by the challenges of physics.

Physics as a Link among People

Since that the East-West cleavage has happily largely disappeared, it is natural to find particle physicists at the origin of new endeavors, in particular, attempting to draw together scientists from the countries of the Middle East to an international laboratory built for a synchrotron radiation source. This would be a highly upgraded machine made from one recently decommissioned in Germany and offered for that purpose. This budding project, named SESAME, is trying to take shape with the help of UNESCO. It is no longer particle physics, but particle physicists led the way. Indeed, all this started with the efforts of Sergio Fubini of CERN and Turin, who organized a first, and very fruitful, meeting in the Sinai, in 1995. This one week physics conference brought together Egyptian, Israeli, Jordanian, and Palestinian physicists together with their American and European colleagues. It was followed two years later by another one in Turin where the SESAME was first discussed. It is hoped that participation in this project, a major common endeavor, will result in other collaborations among these nations. They may find with physics the possibility to get to know one another, and this may lead—let us hope—to a better understanding.

See also: CERN (EUROPEAN LABORATORY FOR PARTICLE PHYSICS); CULTURE AND PARTICLE PHYSICS; DESY (DEUTSCHES ELEKTRONEN-SYNCHROTRON LABORATORY); FERMILAB; JAPANESE HIGH-ENERGY ACCELERATOR RESEARCH ORGANIZATION, KEK

Bibliography

CERN. "Science Bringing Nations Together." <http://outreach.web.cern.ch/outreach/public/cern/Brochures/science-together.pdf>.

Jacob, M., and Schopper, H., eds. *Large Facilities in Physics* (World Scientific, Singapore, 1995).

Krige, J., ed. *History of CERN,* 3 vols. (North Holland, Amsterdam, New York, 1987,1990, and 1996).

Weisskopf, V. *The Joy of Insight* (Basic Books, New York, 1991).

Maurice Jacob

J

J/Ψ

The particle known as the J/ψ has a double name because of the history of its discovery. Particles are traditionally named by their discoverers. What happens when two separate experimenters announce their discovery on the same day? The solution was to keep both names.

The J/ψ particle is a meson made from a charmed quark together with an anticharmed quark. Its discovery was a turning point in the development of the Standard Model of particle physics. All previously known particles could be explained in terms of just three quark types or flavors: up, down, and strange. The down and strange quarks have the same charge. Observations put very stringent limits on transitions between these two.

A proposed theory of the weak interactions, involving the W and Z bosons, needed an additional type of quark to avoid a wrong prediction. With only three quarks, the theory predicted quark flavor-changing, Z-mediated processes at rates inconsistent with observation. Sheldon Glashow, John Illiopoulos, and Luciano Maiani showed that adding a fourth quark type, which Glashow had earlier dubbed charm, provided an additional contribution canceling the wrongly predicted rate. But in mid-1974 only

a small part of the physics community took these ideas seriously. The discovery of the J/ψ particle and subsequent related measurements proved that the fourth quark existed, paving the way to the current Standard Model.

The group that named their discovery J worked at Brookhaven National Laboratory in New York, studying the production of electron-positron pairs in the collision of protons with nuclei. They plotted the rate of pair production as a function of the mass of the combined system. Any produced particle that can decay into an electron and a positron shows up as a peak in this plot, centered at the mass of the particle and with a width inversely related to its lifetime.

Starting in mid-September 1974, the group began to see that a new type of particle was being produced at a mass of about 3.1 GeV/c^2. The data showed a clear and very narrow peak, indicating a particle with an anomalously long lifetime for a meson of this mass. The experiment was led by MIT professor Samuel Ting, known for his cautious attention to detail. He would not let his group announce a new and somewhat anomalous effect without first making cross-checks. Over the next month and a half these checks steadily confirmed the effect. Ting revealed the result to very few people outside his group, and those he swore to secrecy. He continued to require further checks.

Meanwhile a second group, at Stanford Linear Accelerator Center (SLAC), was hot on the trail of the same particle. The SLAC experiment is essentially the reverse of the Brookhaven one. Starting with colliding electron and positron beams with equal and opposite momentum, it measured the rate of events that produce hadrons at each energy setting. The narrow peak found at Brookhaven translates into an increased rate for the SLAC experiment only if the accelerator is tuned to precisely the right energy.

The SLAC experiment had earlier scanned the rate as a function of energy in steps of about 0.1 GeV. Dr. Roy Schwitters, one of the physicists working on the experiment, noticed that in two measurements when the machine energy was nominally set at the same energy, 3.1 GeV, this rate was about 30 percent higher than in the other data at similar energies. This warranted checking. The SLAC group decided to study the energy region around 3.1 GeV in smaller energy increments. Perhaps the anomalous rate had occurred when the energy setting was slightly different from the intended one.

Immediately this approach yielded dramatic results. At 3.12 GeV they found that hadrons were produced at three times the normal rate, at 3.11 GeV it was almost a factor of seven. With great excitement collaborators came rushing to the site as they heard this news. They mapped out an extremely prominent and narrow peak in the rate, centered at 3.105 GeV, which was an indisputable indication of a new particle, which they chose to name ψ. The group leader Burton Richter began to draft a paper describing the results. Word of the discovery spread around the world within a day.

Credit for a scientific discovery is based not on when the measurement is made, but on when it is formally announced in a paper submitted to a journal or conference. Ting was on his way to SLAC to attend an advisory group meeting at the time the SLAC group was making their discovery. That night he heard from his collaborators about the SLAC discovery. The time for caution was over. Overnight the Brookhaven group sent data plots to Ting. The next day, at SLAC, Schwitters presented the SLAC results and Ting the Brookhaven results in a joint public seminar. Both groups immediately submitted their results

for publication. The papers appeared in the same issue of *Physical Review Letters*. Ting and Richter shared the 1976 Nobel Prize in Physics for this discovery.

See also: CHARMONIUM; QUARKS

Bibliography

Aubert, J., et al. "Experimental Observation of a Heavy Particle." *Physical Review Letters* **33**, 1404–1406 (1974).

Augustin, J. E., et al. "Discovery of a Narrow Resonance in E+ E− Annihilation." *Physical Review Letters* **33**, 1406–1408 (1974).

Glashow, S.; Illiopoulos, J; and Maiani, L. "Weak Interactions with Lepton-Hadron Symmetry." *Physical Review* **D2**, 1285–1292 (1970).

Richter, B. "From the Psi to Charm: The Experiments of 1975 and 1976." *Reviews of Modern Physics* **49**, 251–266 (1977).

Riordan, M. *The Hunting of the Quark: A True Story of Modern Physics* (Simon and Schuster, New York, 1987).

Ting, S. C. C. "The Discovery of the J Particle: A Personal Recollection." *Reviews of Modern Physics* **49**, 235–249 (1977).

Helen R. Quinn

JAPANESE HIGH-ENERGY ACCELERATOR RESEARCH ORGANIZATION, KEK

The Japanese High-Energy Accelerator Research Organization (KEK) was established in 1971 by the Japanese government for the purpose of promoting experimental research in elementary particle physics. Although Japanese physicists were actively contributing to the theoretical developments in this field at that time, their experimental research activities were limited to cosmic ray observations even though high-energy particle accelerators had been the standard research tools since the 1950s.

The new laboratory's first mission was to build a proton synchrotron capable of accelerating protons up to an energy of 12 billion electron volts. One billion electron volts approximately corresponds to the energy needed to create one proton out of a vacuum. This accelerator began operating in 1975 and provided high-energy beams consisting of π mesons, K mesons, antiprotons, and protons for a wide range of particle physics experiments conducted for the first time in Japan. It was an important milestone for the development of particle physics in Japan.

Two more accelerators have been added to KEK's research facilities since then: the Photon Factory in 1982 and TRISTAN in 1986; the latter was then converted to a *B* Factory in 1999. The Photon Factory produces intense beams of light (or equivalently photons, thus the name Photon Factory) in the wavelength range stretching from ultravioletlight to X rays and has been used in research in material and biological sciences as well as for industrial applications. Accelerated electrons are stored in a circular orbit in this accelerator, and intense beams of light are generated from the circulating electrons.

In the 1980s, one of the most urgent issues in elementary particle physics was to find the top quark, the heaviest and only missing member among the theoretically proposed six-quark family. KEK joined this search by building a high-energy electron-positron colliding accelerator (called TRISTAN). In 1987, TRISTAN reached a collision energy of 64 billion electron volts, the highest electron and positron collision energy in the world at that time. The top quark was out of reach with the available energy, and the experimenters could only conclude that the top quark must be more massive than 32 billion electron-volts. When it was finally discovered at Fermi National Laboratory in 1995, the top quark turned out to have a mass of 174 billion electron volts.

In 1999, TRISTAN was converted to a new type of accelerator which generates particle-antiparticle pairs called *B* mesons and anti-*B* mesons. The *B* meson is an unstable particle, about five times heavier than a proton, and decays into several more stable particles immediately after being created. The anti-*B* meson is its antiparticle. Such a system can provide a laboratory for observing differences between particles and antiparticles, provided they can be generated in the millions.

Some important new findings were made as a result of KEK experiments. In 1989, a team from Japan, America, Korea, and China, working at a TRISTAN experiment, observed for the first time that gluon particles do interact among themselves. The gluon is a particle that carries the strong force between the quarks. Unlike the electromagnetic forces, where photons only

FIGURE 1

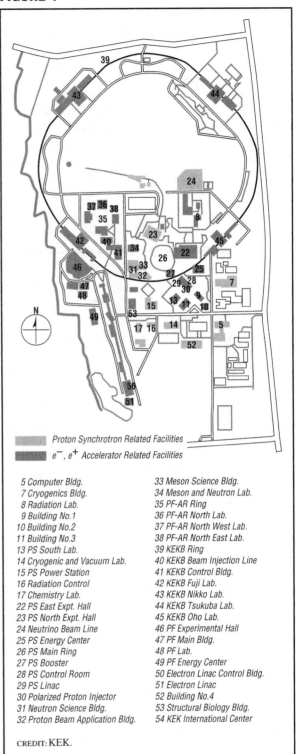

Proton Synchrotron Related Facilities

e^-, e^+ Accelerator Related Facilities

5 Computer Bldg.	33 Meson Science Bldg.
7 Cryogenics Bldg.	34 Meson and Neutron Lab.
8 Radiation Lab.	35 PF-AR Ring
9 Building No.1	36 PF-AR North Lab.
10 Building No.2	37 PF-AR North West Lab.
11 Building No.3	38 PF-AR North East Lab.
13 PS South Lab.	39 KEKB Ring
14 Cryogenic and Vacuum Lab.	40 KEKB Beam Injection Line
15 PS Power Station	41 KEKB Control Bldg.
16 Radiation Control	42 KEKB Fuji Lab.
17 Chemistry Lab.	43 KEKB Nikko Lab.
22 PS East Expt. Hall	44 KEKB Tsukuba Lab.
23 PS North Expt. Hall	45 KEKB Oho Lab.
24 Neutrino Beam Line	46 PF Experimental Hall
25 PS Energy Center	47 PF Main Bldg.
26 PS Main Ring	48 PF Lab.
27 PS Booster	49 PF Energy Center
28 PS Control Room	50 Electron Linac Control Bldg.
29 PS Linac	51 Electron Linac
30 Polarized Proton Injector	52 Building No.4
31 Neutron Science Bldg.	53 Structural Biology Bldg.
32 Proton Beam Application Bldg.	54 KEK International Center

CREDIT: KEK.

Site layout of KEK laboratory.

carry the force between electrons and positrons but do not interact among themselves, gluons have been predicted to interact among themselves in addition to working as the carrier of strong force. Observation of this peculiar property of gluons was a welcomed experimental verification for the theory of strong force.

KEK developed a method of using the proton synchrotron as an intense source of neutrino particles, which are aimed at a large underground neutrino detector (called Super Kamiokande) located approximately 250 km away. Neutrinos, known to exist in three types, interact with other particles only very weakly. This elusiveness has prevented any experimental measurement of their masses in spite of the recognition that their tiny, if not zero, masses, can play a vital role in the evolving process of the universe. In 1999, a team of Japanese, Korean, and American scientists, counting the neutrinos entering their detectors at both the Super-Kamiokande and KEK sites, succeeded in detecting the neutrinos that traveled from KEK to Super-Kamiokande. This was the first time such measurements had been performed, and it opened the possibility of determining the neutrino masses using a phenomenon called "neutrino oscillation." Neutrinos are believed to go back and forth between different types, oscillating back and forth with a characteristic time frequency depending on their masses. Counting neutrinos that travel a long distance opens the possibility of measuring their oscillation frequencies.

In 2000, an international team of more than 200 scientists from eleven nations, working at the *B* Factory site, found convincing evidence that the *B* meson behaves differently from the anti-*B* meson, as seen in certain decay patterns. This result is an important step toward the comprehensive understanding of tiny and subtle differences between particles and antiparticles. In spite of many years of study, it has been difficult to pin down the origin of particle-antiparticle differences because they appear in only very limited processes. Whatever is causing these differences, a similar mechanism is believed to be responsible for the creation of our universe in its present form. In spite of a widely accepted belief that the Big Bang originally created equal amounts of particles and antiparticles, our universe is now completely dominated by matter (particles) with no trace of antimatter (antiparticles).

Located in Tsukuba Science City, 60 km north of Tokyo, KEK has evolved into a major research laboratory for elementary particle physics and other fields of science that use the particle accelerators as research tools. All the accelerator facilities are open to the international scientific community. Anyone or any team can use the KEK facilities as long as their research proposals are approved by a scientific committee of the laboratory. Quite often international collaborations are formed to face the scientific challenges more effectively. Besides providing a variety of high-energy particle beams to the experimenters, KEK leads advanced research and development toward building more powerful, next-generation particle accelerators.

See also: BENEFITS OF PARTICLE PHYSICS TO SOCIETY; FUNDING OF PARTICLE PHYSICS; INTERNATIONAL NATURE OF PARTICLE PHYSICS; UNIVERSE

Bibliography

KEK. <http://www.kek.jp/>.

Kazuo Abe

JETS AND FRAGMENTATION

The quantum field theory quantum chromodynamics (QCD) is the theory of quarks and gluons, which are collectively referred to as partons. The most direct evidence for the existence and properties of partons is found in jets.

What are jets? When particles collide in very-high-energy accelerators, they usually produce many secondary particles, some of which travel at wide angles from the initial directions. Particle detectors are designed to identify and measure the directions and energies of these particles. When particles emerge from a collision at high energies, they often appear grouped into a few, highly collimated sprays. This happens with a frequency much greater than can be accounted for by chance. Such groups of nearly parallel-moving particles are called jets. In QCD jets are understood as the visible manifestations of partons. These partons may have collided and been scattered, created in a collision, or emitted in the decay of short-lived particles, such as electroweak bosons. Each such parton evolves into a collection of hadrons, a process known as fragmentation. This term is a little misleading because in QCD the partons are elementary, without substructure, and the hadrons are composite, made up of partons. A special case of interest is the very heavy top quark, which itself decays into lighter quarks, which then produce jets.

The situation in electron-positron (e^+e^-) collisions is particularly well suited to illustrate these processes. An electron and a positron collide head on and combine (annihilate) to form a photon of considerable energy but with little momentum. This is impossible for a photon in classical physics, whose momentum and energy must be related by $p = E/c$, with c being the speed of light. In quantum field theory, however, the uncertainty principle allows such a photon to exist for a short period of time. This virtual photon quickly transforms itself (decays) into a pair of electrically charged particles: particle plus antiparticle. This process is illustrated in Figure 1. In quantum field theory, it is not possible to say beforehand which species of charged particles will appear. In fact, every kind is possible, as long as the total energy E of the colliding electron and positron is large enough to create the pair. Sometimes, the charged pair will be a pair of quarks.

FIGURE 1

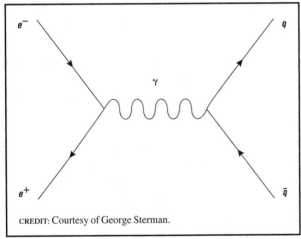

CREDIT: Courtesy of George Sterman.

Electron-positron annihilation to a quark q and antiquark $\bar{q}$ pair through a virtual photon γ. The arrows for particles (electron, quark) point to the right, those for antiparticles (positron, antiquark) point to the left.

FIGURE 2

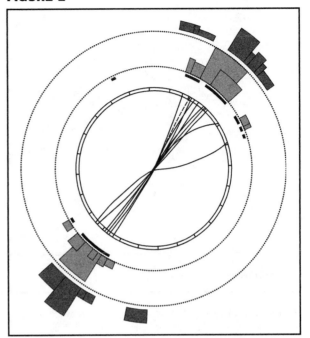

Two-jet event from the Opal detector at CERN. The curved tracks carry relatively low energy.

If one takes a closer look at energies that are much higher than the masses of the quarks, the majority of e^+e^- annihilation events appear like the one in Figure 2, which shows particles produced at the Opal detector at the European Laboratory for Particle Physics (CERN) in Geneva, Switzerland. This event has two, highly collimated, and nearly back-to-back jets of particles. The tracks in the middle of the picture are made by charged particles, primarily hadrons, and the histograms show how much energy was deposited in the outer shells of the detector. The probabilities and angles at which these jets appear are given to good approximation by the process of Figure 1. It is as if each quark simply becomes a jet.

The fragmentation process, through which a quark turns into a jet, is illustrated in Figure 3. A newly produced quark is not yet surrounded by a color field, and it begins to emit the quanta of that field, the gluons, in much the same way that an electron emits photons when it is accelerated, from radio waves at low energies to gamma rays under extreme circumstances. Unlike the photons of quantum electrodynamics

FIGURE 3

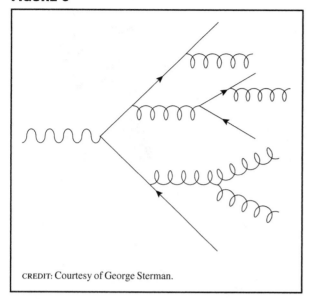

CREDIT: Courtesy of George Sterman.

Beginning of fragmentation as a partonic cascade. The wavy line represents the initial virtual photon, the curly lines gluons, straight lines with arrows to the right quarks, and arrows to the left antiquarks.

(QED), however, gluons themselves carry color charge, and when they are created, they also radiate. This process of rapid particle creation is known as a partonic cascade. In a short time, the original quark is surrounded by a cloud of partons: quarks, antiquarks, and gluons.

QCD predicts how the cascade develops over time. With fairly good accuracy, it reduces to individual steps of the sort shown in the Figure 3. For example, let $P_{q \to qg}(E, \theta)$ be the probability per unit energy and unit angle for a quark to emit a gluon of energy E at angle θ to its own momentum. Then $P_{q \to qg}(E, \theta)$ is given approximately by

$$P_{q \to qg}(E, \theta) = \frac{\alpha_s}{E \sin \theta} \qquad (1)$$

where α_s is the QCD coupling, the analog of the fine-structure constant of quantum electrodynamics. Equation 1 shows that the probability of emitting gluons increases as the angle between the gluon's and quark's momenta decreases. One can understand this effect in the following manner.

If one could move alongside the quark just as it came into being, it would appear to establish its color field by sending out waves of the strong force more or less equally in all directions. On the other hand, as seen "in the laboratory," the quark is moving rapidly, and the waves it emits, made up of gluons, are extremely Doppler-shifted.

They appear to have much higher energy and are primarily moving in the same direction as the quark. This collection of Doppler-shifted radiation is the jet.

As the high-energy partons produced in the cascade travel further and further outward, they separate, penetrating into the surrounding vacuum, which actually repels their color fields, by raising the energies of gluons with wavelengths larger than about 1 fermi (10^{-13} cm). The wavelength associated with a gluon of energy E is $\lambda(E) = hc/E$, with h being Planck's constant and c the speed of light. For the first few gluon emissions at an event such as the one shown in Figure 2, E can easily be of order 10^{10} electron volts (10 GeV), corresponding to wavelengths much less than a fermi. Thus, in the first few steps of these cascades, the unfriendly vacuum is not too important, and Equation 1 can be used. Eventually, however, as the energies of the emitted gluons begin to fall, and their wavelengths grow, an extra energy is required for them to penetrate the vacuum, compared to $E = hc/\lambda$. For such long wavelengths and low energies, the approximations implicit in Equation 1 fail. Nevertheless, as the color charges of the remaining energetic partons grow farther and farther apart, they must be connected by lines of the color field, just as electric charges are connected by lines of the electric field.

Although it is not yet possible to describe this process quantitatively, it is certain that the energy of the color field between a quark and antiquark grows without bound as they separate, unless their color charges are neutralized. For this reason, it is always energetically favorable to create pairs of the lightest quarks and antiquarks, until all partons are grouped into color-neutral hadrons. This stage of jet formation is known as hadronization. The process of hadronization, although inexorable, hardly ever uses up more than a small fraction of the total energy. After hadronization the particles can separate freely, and it is the hadrons that create the tracks that are seen in detectors. The numbers and energies of hadrons within a given jet depend only on whether the jet started out as a quark or a gluon. This property of jets is known as factorization. Thus, quark jets in electron-proton collisions are indistinguishable from quark jets of the same energy in proton-antiproton collisions.

Despite the complexity of the cascade and of hadronization, it is possible to compute the probability of finding a jet with a specified total energy and direction because when gluons are emitted at small energies or angles, where Equation 1 is not useful, a jet's total energy and overall direction are unchanged. Properties like the total energy, which are insensitive to low-angle and low-energy gluons, are sometimes said to be infrared safe, a term chosen to emphasize their independence of long wavelength gluon emission. Infrared safe quantities can be calculated with analogs of Equation 1.

Equation 1 implies that the likelihood for extra jets to appear from gluon radiation is proportional to α_s. Because QCD is asymptotically free, the coupling α_s in Equation 1 decreases as $E \sin \theta$ increases. At the highest-energy accelerators, jet energies can exceed 100 GeV, and at this scale, α_s is approximately

equal to 0.1. This means that about one out of every ten events with a quark whose energy is more than 100 GeV includes a gluon with a comparable energy, separated from the quark by a substantial angle. Compelling evidence for the gluon was first provided in "three-jet" events in $e^+ e^-$ annihilation, originating from a quark, an antiquark, and an energetic gluon emitted at wide angles. An example of such an event is shown in Figure 4.

Figure 5 shows the relative numbers of events for jets observed at the Tevatron accelerator at Fermilab in Batavia, Illinois, as a function of their energy. Over many orders of magnitude, calculations based on more elaborate versions of Equation 1 track the data. When, as in this case, the collisions are of protons and antiprotons, the observed jets come mainly from the scattering of partons already present in these particles. The data then reveal how energy is shared among the partons inside a proton.

Jets are at the center of QCD studies at high energy, as well as in the search for new particles created at high energy. In the formation of jets also lies an essential challenge for the theory of QCD: to

FIGURE 4

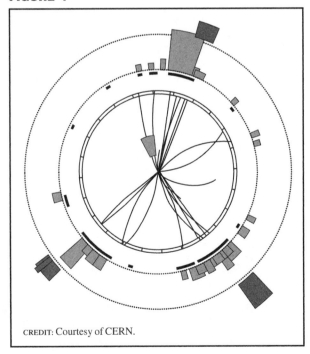

CREDIT: Courtesy of CERN.

Three-jet event from the Opal detector at CERN.

FIGURE 5

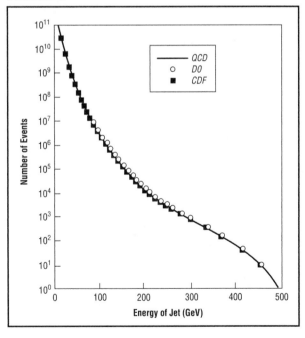

Relative numbers of jets as a function of jet energy, as observed by the D0 and CDF detectors at Fermilab.

create a quantitative description of how quarks and gluons evolve into hadrons.

See also: QUANTUM CHROMODYNAMICS; QUARKS

Bibliography

Hey, T., and Walker, P. *The Quantum Universe* (Cambridge University Press, Cambridge, 1987).

Kane, G. *Modern Elementary Particle Physics* (Perseus, Cambridge, MA, 1993).

Particle Data Group. "The Particle Adventure." <http://particle adventure.org>.

George Sterman

K

KENDALL, HENRY

Henry Way Kendall shared the 1990 Nobel Prize in Physics with Jerome Friedman and Richard Taylor for their pioneering studies of the scattering of electrons from protons, studies that produced the first solid evidence that quarks exist and are the basic constituents of neutrons and protons.

Kendall was born on December 9, 1926, in Boston, Massachusetts, the oldest son of one of the wealthiest families in New England. As he wrote in his autobiography, written upon receiving the Nobel Prize, until he went to college, he considered himself a poor student "more interested in non-academics matters and bored with school work." Encouraged and supported by his father, Kendall devoted most of his time to exploring "things mechanical, chemical and electrical." Near the end of the World War II, he entered the U.S. Merchant Marine Academy, spending a winter on a troop transport in the North Atlantic. Older and more experienced, he was a serious student interested in many disciplines when he enrolled at Amherst College. Kendall majored in mathematics but was so interested in other fields that he could also have majored in English, history, biology, or physics. He spent the summers learning to be an expert underwater diver and photographer. These efforts resulted in two successful books, written with a schoolboy friend, on shallow-water diving and underwater photography, both of which became Kendall's life-long hobbies.

By the end of Kendall's undergraduate years, physics began to dominate his studies. He chose to do his senior thesis in physics, and, with the support of his father who understood that his son wanted a life in science rather than business, he chose to study physics in graduate school. At that time, he also made the decision to become self-supporting, without monetary assistance from his family.

Kendall described the years from 1950 to 1954 that he spent as a graduate student in the physics department of MIT as "a continuing delight—the first sustained immersion in science at a full professional level." His mentor was Martin Deutsch, the discoverer of positronium, the bound state of a positron and an electron, the simplest possible atom. Kendall's attempt to measure the Lamb shift in positronium was unsuccessful, but his interest in electromagnetic interactions continued throughout his professional career.

After a two-year National Science Foundation Postdoctoral Fellowship at MIT, Kendall accepted an appointment at Stanford University in 1956. He joined the group of Robert Hofstadter who was deep

American physicist Henry Kendall (1926–1999) shared the 1990 Nobel Prize in Physics with Jerome I. Friedman and Richard E. Taylor for their studies that provided evidence that quarks exist and are the basic constituents of neutrons and protons. CREDIT: COURTESY OF AIP EMILIO SEGRE VISUAL ARCHIVES, W. F. MEGGERS GALLERY OF NOBEL LAUREATES. REPRODUCED BY PERMISSION.

into his Nobel Prize–winning studies of the scattering of electrons from protons. These experiments showed that the proton was not a point but had an extended structure, with no hint, however, that the structure was anything but uniform.

Kendall remained at Stanford for five years. It was there that his fascination with mountain climbing and mountain photography began. Over the years he indulged this passion with climbs all over the world, most notably the Andés and Himalayas. It was also at Stanford that he commenced his lifelong collaboration with Jerome Friedman.

Kendall left Stanford's faculty in 1961 to take an assistant professorship at MIT. In doing so, he again joined Friedman, who had joined the faculty a year earlier. They formed a research group to continue experiments at Stanford, most notably, the construction of the world's most powerful electron accelerator, a 2-mile linear machine that produced intense beams of 20-GeV electrons. This was to be a national facility for use by physicists of every nationality. It was called the Stanford Linear Accelerator Center (SLAC).

The electrons in Hofstadter's experiments had energies of a few hundred MeV. When considered as wave packets, the electrons had wavelengths of about 10^{-13} cm, sufficient to determine the size of the proton but insufficient to probe its internal structure. The new SLAC accelerator was expected to probe more than an order of magnitude more deeply.

In concept, the deep-inelastic experiment was simple: use an electron spectrometer to precisely measure the energy and angular distributions of the electrons scattered from a hydrogen target and make deductions from the analysis of the electrons alone. The experiment was fundamental; it needed to be done. But a meaningful deep-inelastic collision scattering result was definitely a long shot. Physicists were not eager to tackle it for two reasons. It was generally accepted that the results would be difficult, perhaps impossible, to interpret, and even if the roadblocks to interpretation could be overcome, the results were likely to be not very interesting since it was widely assumed that there was no structure to be found inside the proton. The MIT group under the leadership of Kendall and Friedman, in collaboration with a group led by Richard Taylor of SLAC, took on the challenge.

The daunting problem in interpreting the results was how to account for the electromagnetic radiation that would inevitably be produced in the scattering of electrons, obscuring the effects due to nuclear structure. The experiments would have to be carried out at unexplored energies. The radiative corrections would be large and increasingly important the deeper the electron probed the proton. Kendall and Friedman spent several years studying this problem until they had confidence that the uncertainties in the radiative corrections would be no greater than approximately 10 percent.

The experiments began in the fall of 1967. Results from the very first runs are shown in Figure 1, taken from Kendall's Nobel lecture of 1990. The measured cross section was expected to drop precipi-

FIGURE 1

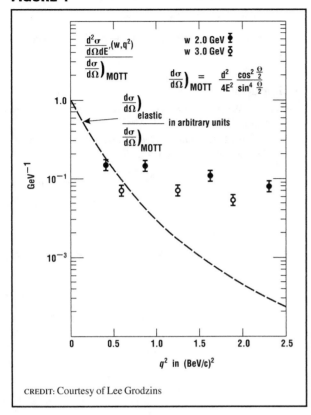

$$\frac{\frac{d^2\sigma}{d\Omega dE'}(w,q^2)}{\frac{d\sigma}{d\Omega}\Big)_{\text{MOTT}}}$$

w 2.0 GeV
w 3.0 GeV

$$\frac{d\sigma}{d\Omega}\Big)_{\text{MOTT}} = \frac{d^2}{4E^2}\frac{\cos^2\frac{\Theta}{2}}{\sin^4\frac{\Theta}{2}}$$

$$\frac{\frac{d\sigma}{d\Omega}\Big)_{\text{elastic}}}{\frac{d\sigma}{d\Omega}\Big)_{\text{MOTT}}} \quad \text{in arbitrary units}$$

CREDIT: Courtesy of Lee Grodzins

Inelastic data for $W = 2$ and 3 GeV as a function of q^2. This was one of the earliest examples of the relatively large cross sections and weak q^2 dependence that were later found to characterize the deep inelastic scattering and which suggested point-like nucleon constituents. The q^2 dependence of elastic scattering is shown also; these cross sections have been divided by σ_M.

tously, following the downward slope of the inelastic cross section, shown as a dashed line. Instead, the cross sections stayed unexpectedly high. (The cross sections, in units of inverse energy, are presented as a ratio to the idealized cross section expected if the entire charge were concentrated in a point. They are plotted as a function of the square of the momentum transferred to the proton, a convenient measure of the ability of the electron to probe the structure.) A rapidly falling cross section was expected if the proton charge were uniform throughout the proton's volume since, as the electron traveled deeper inside the proton, there would be less and less proton material to scatter from. Clearly, this was not the case. The cross sections were 10 to 100 times greater than could be accounted for by a uniform structure. Elec-

trons, as they probe to a tenth of the proton size, were still scattering copiously. The radiative issues that Kendall and Friedman had tried so hard to understand had turned out to be negligible compared to the observed enhancements.

The phenomenon was reminiscent of Ernest Rutherford's revelation in 1911 that the backward scattering of alpha particles from gold nuclei meant that the atomic mass was concentrated in a central "point" and could not be spread uniformly over the atomic volume. The results showing deep-inelastic scattering clearly indicated that there were hard, pointlike entities inside the proton. James Bjørken, whose theoretical guidance was important to the SLAC experiments, made correct predictions of the deep-inelastic results from a model based on the possible particlelike constituents in the proton. Richard Feynman, on seeing the early SLAC data, identified the entities with his pointlike partons that he conjectured were the building blocks of nucleons. The specifics of the structure of the proton could not be resolved by deep-inelastic studies alone, but these experiments were the foundation for the next wave of new discoveries and theoretical insights that culminated in quantum chromodynamics (QCD), one of the pillars of the Standard Model.

It is worth remarking on another parallel between the deep-inelastic scattering experiments and the alpha-particle-scattering experiments a half-century earlier. The Rutherford atom could not be reconciled with electrodynamics, which demanded that orbiting electrons radiate energy, leading to the collapse of the atom. It was not until Niels Bohr's introduction of the quantum concepts into atomic physics that the Rutherford atom was accepted. The quark model, which could explain the deep-inelastic results and so much more, faced the dilemma that experimenters could find no evidence of particles with fractional charge, despite diligent and ingenious efforts. The reality of quarks did not become generally accepted until the believable theory of asymptotic freedom convinced the physics community that quarks would not be found individually but would remain bound in hadron structures.

In 1969, even as Kendall continued his central role in the studies of deep-inelastic scattering, he entered a phase of his career that would propel him into the public arena. In that year the Union of Concerned Scientists (UCS) was founded by faculty members of MIT, Kendall among them, to educate the public about the science and technology issues that impacted society. Kendall became the chairman of UCS in 1974 and proceeded to transform it into one of the nation's most effective venues for public awareness of science. Educating the public and policy makers on science issues, especially those that threatened the environment, became an increasingly important focus for Kendall. His work, which contin-

ued until his untimely death in 1999, was recognized with several international awards.

See also: QUARKS; SLAC (STANFORD LINEAR ACCELERATOR CENTER)

Bibliography

Kendall, H. W. *A Distant Light: Scientists and Public Policy* (Springer-Verlag, New York, 1999).

Riordan, Michael. *The Hunting of the Quark: A True Story of Modern Physics* (Simon & Schuster, New York, 1987).

Lee Grodzins

L

LATTICE GAUGE THEORY

Asymptotic freedom has defined the history of quantum chromodynamics (QCD), the gauge theory of quarks and gluons that describes subnuclear physics. Asymptotic freedom means that the interaction energies between quarks weaken relative to their kinetic energies at short distances, less than about 1/3 fm. This weakening allows us to analyze short-distance interactions using an expansion in powers of the interaction energy divided by the kinetic energy. This expansion is called perturbation theory, and perturbation theory was well developed when QCD's asymptotic freedom was discovered in 1974. Consequently most tests of QCD during the next two decades focused exclusively on its short-distance behavior.

Unfortunately hadrons, such as the proton and the neutron, are several times larger than 1/3 fm. Thus the physics of hadronic structure is highly nonperturbative and, in 1974, was impossible to compute. Within months of the discovery of asymptotic freedom, however, Kenneth Wilson introduced a new formulation of QCD, called lattice QCD, that facilitated nonperturbative, numerical simulations of QCD. An early triumph of lattice QCD was Wilson's demonstration that quarks are confined within hadrons in the strong-coupling limit of the theory, but further progress was very slow until the 1990s. Today Wilson's theory—the first lattice gauge theory studied by particle physicists—provides the only rigorous approach for computing long-distance properties of QCD, including such things as the masses and structure of hadrons.

The Lattice Approximation

Wilson's innovation was to replace continuous space and time by a rectangular lattice of discrete points or sites in space and time, separated from each other by a fixed lattice spacing in each direction. In the lattice approximation, the fields that describe quarks and gluons are specified only at the lattice sites. Thus the fields within a hadron can be specified by a finite number of numbers—the field values at each lattice site inside the hadron—and the problem becomes tractable on a computer.

The QCD path integral, which defines the quantum theory, becomes an ordinary multidimensional integral in the lattice approximation. In principle any property of the theory can be computed using this integral. The integration variables are the values of the eight gauge fields at each of the lattice sites. Since a typical lattice today has a lattice spacing between 0.1 and 0.2 fm, and covers a volume of about 3 fm, lattices with 20^4 sites or more are common.

Consequently the path integral involves the evaluation of millions of nested integrals. Special numerical techniques, called Monte Carlo simulations, are used to evaluate these integrals. The simplest simulations, using very coarse lattices and severe approximations, can be completed on a laptop within an hour. High-quality simulations, however, require months of running on clusters composed of hundreds of PCs.

Quantum fields, unlike most classical fields, have important structure at all length scales. This suggests that a grid approximation of the sort used in lattice QCD should fail because it omits all structure at distances smaller than the lattice spacing. In fact, the effects of this missing structure can be mimicked by modifying the integrand of the path integral. The modifications are computed using perturbation theory because they come from short distances, where asymptotic freedom renders QCD perturbative. Thus lattice QCD is actually a hybrid of perturbative techniques for physics at scales smaller than the lattice spacing and with numerical, nonperturbative techniques for physics at scales of order the lattice spacing or larger.

Application in QCD and Beyond

Lattice QCD simulations are commonly used to compute the properties of single hadrons. They are particularly effective for hadrons that are stable or nearly stable with respect to strong interactions. Particle masses, radii, magnetic moments, and other aspects of a particle's structure are all readily computed. In addition, QCD simulations are used to calculate electroweak form factors, structure functions, and decay amplitudes that couple photons, W bosons, or Z bosons to a hadron. This last application is particularly important for heavy-quark physics, with its focus on heavy-quark decays mediated by weak interactions.

Lattice QCD simulations are also used to study the behavior of QCD at high temperatures. Such simulations provide insights into the behavior of matter in extreme conditions, such as might be found in stellar interiors or in the very early universe.

After slow progress in the 1970s and 1980s, the 1990s saw rapid improvements in lattice QCD tech-

niques and in the computer hardware needed for the simulations. As a result, simulation errors were reduced from 100 percent to between 10 and 20 percent for a wide variety of nonperturbative quantities. The first decade of the twenty-first century will see these errors fall by another order of magnitude, and lattice QCD will play an increasingly important role in high-precision studies of the weak interactions of heavy quarks.

Lattice methods are applicable to other field theories as well. They have been used to explore the Higgs sector of the Standard Model in the limit of large Higgs mass, where the interactions become strong. These techniques could well be important for studying physics beyond the Standard Model. Most realistic quantum field theories have strong interactions either at low energies (e.g., QCD) or at high energies (e.g., gravity). The only exceptions are theories in which symmetries are spontaneously broken (e.g., electroweak interactions). But even in these theories, the most natural mechanism for spontaneous symmetry breaking is dynamic in origin and again involves strong coupling. Lattice methods, which must be extended to cover such models, offer the best hope of dealing with all such strong-interaction phenomena.

See also: ASYMPTOTIC FREEDOM; QUANTUM CHROMODYNAMICS

Bibliography

Weingarten, D. H. "QCD by Monte Carlo." *Scientific American* **274**, 119 (1996).

G. Peter Lepage

LAWRENCE, ERNEST ORLANDO

Ernest Orlando Lawrence was born August 8, 1901. He was one of two children of Carl and Gunda Lawrence. His family was well-educated; his father was a superintendent of schools in Canton, South Dakota, and his grandfather also taught. His ancestors originally came to the United States from Norway.

Young Lawrence came of age just as radio was under development, and the new radio technology brought out his interest in science. He and his boyhood friend Merle Tuve, who also became a physicist, became ham radio operators during high school. The young Lawrence spent his summers working at money-making schemes that would allow him to pay his way through school.

After graduating from high school, Lawrence attended St. Olaf College for a year intending to study medicine (his brother John did become a doctor). Lawrence was not vitally interested in medicine, and his grades at St. Olaf's showed it. He then attended the University of South Dakota, where he came under the tutelage of Lewis Ackley, who influenced him in his choice of science as a course of study. Lawrence was able to build a ham radio station at the University of South Dakota. Dean Ackley was impressed enough to allow Lawrence to teach a physics course his senior year. Lawrence graduated with a degree in chemistry in 1922.

After graduation from the University of South Dakota, Lawrence went to graduate school to study physics at Minnesota with Tuve, where he met his advisor, William F. G. Swann. In Swann's laboratory, Lawrence exhibited his knack for making machines and other physics apparatus work. Swann went to Chicago the succeeding year bringing Lawrence along with him, and Lawrence again followed Swann the next year when Swann took a professorship at Yale. Lawrence received his Ph.D. from Yale under the Swann's direction in 1925 with a thesis on the photoelectric effect in potassium vapor.

Lawrence received a National Research Fellowship and remained at Yale for the next two years and then became an assistant professor there. During his time at Yale, he worked to determine the time between emission of a photon and the change in the state of the electron in the photoelectric effect, finding that it was beyond the ability of his apparatus to discern (this was in the era contemporaneous to the discovery of quantum mechanics; it is believed the transition occurs instantaneously).

In 1928, the University of California, Berkeley hired Lawrence as an associate professor. Two years later, they made him the youngest person ever promoted to professor of physics at Berkeley. During this interval, he had found a paper on the acceleration

American physicist Ernest Orlando Lawrence (1901–1958) received the Nobel Prize in Physics in 1939 for his invention of the cyclotron. CREDIT: COURTESY OF CORBIS. REPRODUCED BY PERMISSION.

of ions by Rolf Wideröe (in German, which Lawrence couldn't read) and invented the principle of the cyclotron after examining one of the article's diagrams. His first cyclotron was just 10 cm across and accelerated ions to 80 kilo electron volts (keV). This invention and its consequences influenced Lawrence for the rest of his life.

By the late 1930s cyclotrons had exploded in size and Lawrence's entrepreneurial spirit had led to the foundation and directorship of the Radiation Laboratory (then called the Rad Lab by its denizens, later known as Lawrence Berkeley National Laboratory). The Rad Lab brought scientists from around the world to work at Berkeley with Lawrence, and Lawrence appeared on the cover of *Time* magazine in November 1937. Lawrence organized his postdoctoral and graduate students into groups who worked together on machine and physics problems. Well-known alumni of the Lab included Philip Abelson and Robert R. Wilson as well as Nobel Prize winners Edwin McMillan, Glenn Seaborg, Emilio Segrè, and Luis Alvarez.

J. Robert Oppenheimer, the brilliant theoretical physicist, had extensive conversations about physics with Lawrence at Berkeley. Their work together in the early 1930s led to advances in understanding of nuclear processes and improvements of the Lawrence group's experimental apparatus. This group work at the Rad Lab was the harbinger of the development of large group collaborations in modern experimental nuclear and particle physics.

In the 1930s, Lawrence's Rad Lab was a hotbed of cutting edge physics work; new particles were discovered, and isotopes were categorized and used in medicine. However, the boosterism inherent in Lawrence's character locked him into building "bigger and better." His focus blinded him to the possibility of investigating neutrons, artificial radioactivity, and fission, which were discovered first in other laboratories.

During this time, Lawrence's brother John joined him in Berkeley's Medical Physics Laboratory to work on the medical applications of radioisotopes. Many modern medical techniques were first conceived and tested at Berkeley.

In 1939 Lawrence, still in his thirties, was awarded the Nobel Prize in Physics for his invention of the cyclotron. He is one of only a handful of physicists who have ever won the Nobel Prize for building a piece of apparatus. He was also the first physicist working at a state university in America to win a Nobel Prize.

Ever the patriot, Lawrence was deeply involved in top secret work during World War II, serving as one of three civilian chiefs of the Manhattan Engineering District (better known now as the Manhattan Project) and building a machine known as "the racetrack" at Oak Ridge to try to produce enriched uranium for a bomb. In this endeavor he was not ultimately very successful, and it was the gas centrifuges operating at Oak Ridge that produced the uranium used in the "atomic" (nuclear) bombs tested in the New Mexico desert and dropped on Hiroshima and Nagasaki in August 1945, near the end of the war. Lawrence was also involved in creating the MIT radiation laboratory, which developed radar, and in other war-related endeavors.

As a chief of the Manhattan Project, Lawrence recommended Robert Oppenheimer as the director of the Project at Los Alamos, a suggestion that was accepted by General Leslie Groves, the Manhattan Project military commander. This choice put Oppenheimer and Lawrence in frequent contact throughout the war. While their friendship continued, it was strained by disagreements.

Lawrence hoped that the detonation of nuclear weapons was the last that would ever be needed and that the world would remain at peace. He was always concerned about people, and his noble humanitarian impulses led him to hope that his dream would be realized.

After the war, Lawrence returned to trying to build "bigger and better" at the Rad Lab. He also established a University of California laboratory at Livermore, California, which was unsuccessful in developing a machine to produce enriched uranium for weapons. During this time, his friendship with Robert Oppenheimer suffered because Lawrence felt that Oppenheimer was responsible for the lack of support for his machine.

International politics intervened in 1949 as the Soviet Union exploded its first nuclear weapon. In response, Lawrence gathered a new group of physicists and changed the focus of the Livermore Laboratory to meet the Soviet threat. The thermonuclear, or hydrogen, bomb was first developed there by a group led by Edward Teller. Lawrence's laboratory eventually became the Lawrence Livermore Laboratory, where weapons work has proceeded ever since. Oppenheimer, who had uttered prophetic words about physicists knowing sin after the explosion of the first nuclear device in New Mexico, opposed the development of the hydrogen bomb, straining the Oppenheimer-Lawrence friendship still further.

After the beginning of the McCarthy era, accusations of subversive activities by Communists were flying everywhere. Oppenheimer was tarred as a closet Communist sympathizer. Lawrence interpreted Oppenheimer's silence in response to accusations as a personal betrayal. The Atomic Energy Commission held hearings on Oppenheimer's security clearance. Teller testified in favor of the removal of the security clearance. Lawrence intended to testify against Oppenheimer but became ill and could not make the trip. He asked Luis Alvarez not to testify, but Alvarez ultimately did testify against Oppenheimer. The removal of security clearance would preclude Oppenheimer from continuing to give advice on nuclear matters, and the Atomic Energy Commission did lift Oppenheimer's security clearance. The Oppenheimer affair became a cause célèbre, and many physicists chose sides. The hard feelings engendered have died only as the principals in the affair themselves have died.

In 1932, Lawrence married his Yale sweetheart, Mary Kimberly "Molly" Blumer. Molly Blumer Lawrence was the daughter of a Dean of the Yale Medical School. The Lawrence family ultimately included six children: John (1934), Margaret (1936), Mary (1939), Robert (1941), Barbara (1947), and Susan (1949).

Lawrence remained an inveterate tinkerer his whole life, and, in response to a challenge from his children to build a color television developed improvements in television tubes in his garage that led to several patents. It is believed that the cumulative effects of stress from his many responsibilities during the war years as well as the heavy schedule of his postwar years led to a condition of progressive ulcerative colitis complicated by atherosclerosis. He acted unaware of the danger, playing energetic tennis matches shortly before his death even as his body failed. The disease eventually caused his death on August 27, 1958, in Palo Alto, California, shortly after his 57th birthday.

See also: ACCELERATORS, EARLY; CYCLOTRON

Bibliography

Childs, H. *An American Genius: The Life of Ernest Orlando Lawrence* (E. P. Dutton, New York, 1968).

Davis, N. P. *Lawrence and Oppenheimer* (Simon and Schuster, New York, 1968).

Heilbron, J., and Seidel, R. W. *Lawrence and his Laboratory*, Vol. 1 (University of California Press, Berkeley, 1989).

Kevles, D. *The Physicists* (Harvard University Press, Cambridge, MA, 1987).

Weiner, C., and Hart, E. *Exploring the History of Nuclear Physics* (American Institute of Physics, New York, 1972).

Gordon J. Aubrecht II

LEPTON

The name lepton derives from the Greek word *leptos*, meaning thin or light. This name is appropriate because leptons are a set of particles with no measurable dimensions, and hence they are elementary. One of the members of this family, the electron, was the first elementary particle to be discovered. Two other family members that carry electric charge are the muon and the tau. For each of these three charged leptons, there is an uncharged partner particle, a neutrino. There is an electron neutrino, a muon neutrino, and a tau neutrino.

Leptons were discovered much earlier than the other set of elementary fermions, the quarks, because they appear individually in nature rather than as composite particles. The defining feature of a lepton is that it does not participate in the strong interaction, allowing it to exist for substantial periods of time as an independent particle.

The set of leptons can be arranged into three generations, as shown in Table 1. There is an electron, muon, and tau lepton family. Each generation has two particles and two antiparticles, where the antiparticles have the same mass as the particle but opposite quantum numbers.

Each force has an associated charge. By historical convention, the electrically charged leptons are assigned one unit of negative, rather than positive, electric charge. Leptons do not participate in the strong interaction, so it is said that they carry zero strong (color) charge. All fermions participate in the weak interaction and carry weak charge. Through the weak interaction, the more massive charged leptons may decay into their less massive counterparts.

It has been experimentally observed that the net difference in the number of leptons compared to antileptons before and after an interaction is unchanged. This is known as lepton conservation, which has an associated quantum number of lepton L. Leptons have $L = +1$ and antileptons have $L = -1$, whereas quarks have $L = 0$. As an example of L conservation, consider the case where an electron and positron annihilate and create a muon and an antimuon ($e^- e^+ \rightarrow \mu^+ \mu^-$). Prior to the interaction $L = (+1) + (-1) = 0$, and after the interaction $L = (+1) + (-1) = 0$.

TABLE 1

Characteristics and Quantum Numbers Associated with Leptons

Generation (family name)	Name	Symbol	Particle or Antiparticle	Mass (MeV)	Spin	Charge (e)	L_e	Lepton Number L_μ	L_τ	L
First (electron)	Electron	e^-	Particle	0.511	±1/2	−1	+1	0	0	+1
	Electron Neutrino	ν_e	Particle	<0.000003		0	+1	0	0	+1
	Positron	e^-	Antiparticle	0.511		+1	−1	0	0	−1
	Electron Antineutrino	$\bar{\nu}_e$	Antiparticle	<0.000003		0	−1	0	0	−1
Second (muon)	Muon	μ^-	Particle	106	±1/2	−1	0	+1	0	+1
	Muon Neutrino	ν_μ	Particle	<0.19		0	0	+1	0	+1
	Antimuon	μ^-	Antiparticle	106		+1	0	−1	0	−1
	Muon Antineutrino	$\bar{\nu}_\mu$	Antiparticle	<0.19		0	0	−1	0	−1
Third (tau)	Tau	τ^-	Particle	1777	±1/2	−1	0	0	+1	+1
	Tau Neutrino	ν_τ	Particle	<18.2		0	0	0	+1	+1
	Antitau	τ^-	Antiparticle	1777		+1	0	0	−1	−1
	Tau Antineutrino	$\bar{\nu}_t$	Antiparticle	<18.2		0	0	0	−1	−1

CREDIT: Courtesy of Janet Conrad.

It is also observed that the net number of leptons and antileptons within each generation is conserved in each interaction. Therefore, a quantum number is introduced for each family: the lepton family number. The reaction $e^- e^+ \rightarrow \mu^+ \mu^-$ has $L_e = (+1) + (-1) = 0$, $L_\mu = 0$, $L_\tau = 0$ prior to the interaction, and $L_e = 0$, $L_\mu = (+1) + (-1) = 0$, $L_\tau = 0$ after the interaction and so conserves lepton family number. The only case where lepton family number is violated occurs in the quantum mechanical effect called neutrino oscillations, and in this case the total lepton number L is still conserved.

The charged lepton masses are similar in magnitude to the quark masses. There is no direct evidence that neutrinos have mass. Experiments have only placed upper bounds on the neutrino masses. Neutrino masses are so tiny that direct measurement in the near future will be very difficult. However, it may be possible to infer that neutrinos have mass through the observation of neutrino oscillations, a quantum mechanical effect that can be observed only if each neutrino species has a different mass. In the Lagrangian that describes the fermions, the masses of the charged leptons are arbitrary parameters. The neutrinos are explicitly assumed to be massless.

Assuming that neutrinos are massless provides an explanation for neutrino *handedness,* a property observed in the weak charged-current interaction. To understand handedness, it is simplest to begin by discussing helicity, since for massless particles helicity and handedness are identical. For a spin-$\frac{1}{2}$ particle, helicity is the projection of a particle's spin along its direction of motion. Helicity has two possible states: spin aligned opposite the direction of motion (negative or left helicity) and spin aligned along the direction of motion (positive or right helicity). If a particle is massive, then the sign of the helicity of the particle is frame-dependent. For example, in a frame where one is moving faster than the particle, the sign of the momentum changes but the spin does not, and therefore the helicity flips. However, for massless particles, traveling at the speed of light, one cannot boost to a frame where helicity changes sign so helicity is conserved.

Handedness (or chirality) is the Lorentz invariant (i.e., frame-independent) analogue of helicity for both massless and massive particles. There are two states: left-handed (LH) and right-handed (RH). For the case of massless particles, including Standard Model neutrinos, helicity and handedness are identical. A massless fermion is either purely LH or RH and, in principle, can appear in one or the other state. Massive particles have both RH and LH components. It is only in the high-energy limit, where particles are effectively massless, that handedness and helicity coincide.

Unlike the electromagnetic and strong interactions, the weak interaction has a definite preferred handedness. In the late 1950s, in Madam Wu's famous parity violation experiment, it was shown that neutrinos are LH and antineutrinos are RH. No RH neutrino interactions or LH antineutrino interactions have ever been observed.

RH neutrinos (and LH antineutrinos) could in principle exist but be undetected because they do not interact. Neutrinos do not interact via the electromagnetic interaction because they are neutral or via the strong interaction because they are leptons. In addition, the RH neutrinos do not participate in the left-handed weak interaction. Because RH neutrinos are noninteracting, these hypothetical leptons (not a part of the Standard Model) are called sterile neutrinos.

They raise obvious theoretical and experimental questions. From a theoretical viewpoint: how do sterile neutrinos come into existence since they cannot interact? This is solved relatively easily if the Standard Model is extended to include, at energy scales well beyond the range of present accelerators, a right-handed W interaction that could produce the RH neutrino. From a experimental viewpoint: if there are sterile neutrinos out there, how can they be observed if they do not interact? The quantum mechanical effect called neutrino oscillations provides one method—if neutrinos have mass.

See also: CASE STUDY: SUPER-KAMIOKANDE AND THE DISCOVERY OF NEUTRINO OSCILLATIONS; EXPERIMENT: DISCOVERY OF THE TAU NEUTRINO; EXPERIMENT: G-2 MEASUREMENT OF THE MUON; NEUTRINO; NEUTRINO, DISCOVERY OF; NEUTRINO OSCILLATIONS; PARTICLE; QUARKS; STANDARD MODEL

Bibliography

Kane, G. *The Particle Garden* (Perseus Publishing, Cambridge, MA, 1995).

Ne'eman, Y., and Kirsh, Y. *The Particle Hunters* (Cambridge University Press, Cambridge, UK, 1996).

Sutton, C. *Spaceship Neutrino* (Cambridge University Press, Cambridge, UK, 1992).

Janet Conrad

LEPTON NUMBER

See CONSERVATION LAWS

M

MACHO

See DARK MATTER

METAPHYSICS

Metaphysics is concerned with human thinking about the nature of reality in the widest and most general sense. A general worldview of this kind will be influenced by what physics has to say about the character of the universe, but it will not be determined by this alone. Other forms of knowledge must also be brought into play. Physics constrains metaphysics, but it does not entail it, as the foundations of a house constrain what can be erected upon them, but they do not fix the form of the edifice. The proper relationship between physics and metaphysics is that of consonance and not one of deductive necessity.

In this kind of intellectual exchange, particle physics has been one of the most metaphysically influential branches of physics because of its ability to discuss the basic constituents out of which the variety of the physical world appears to be constructed. Its influence on general human thinking has been varied and contentious, however, just because of the ambi-

guity inherent in attempting to move from the particularities of physics to the generalities of metaphysics.

History

The pre-Socratic philosophers, such as Thales and Anaximander, considered that the variety of the world resulted from different states of a single kind of basic stuff. Thales assigned water to this fundamental role, while Anaximander favored air. The intuition that simplicity underlay apparent complexity was brilliant (one could think of the pre-Socratics as proto-particle physicists), but these early thinkers were unable to develop their ideas in any plausible detail. It was Leucippus and Democritus in the fifth century B.C.E. who took the further important step of proposing the existence of irreducible atoms (the Greek word means "uncut"), whose motions in the void constituted the nature of the world. The idea was developed more than a century later by Epicurus to form the basis of his atheistic philosophy. This way of thinking was fluently and fervently expounded by the Latin poet Lucretius in his influential poem *De Rerum Natura* (On the nature of things, 58 B.C.E.).

Atomism became of renewed interest in the seventeenth century with the rise of modern science. On this occasion, however, its principal proponents, such as Robert Boyle and Isaac Newton, were theists.

It was only later, principally in eighteenth-century France, that atomism again began to be associated with atheism. Atomism in a recognizably modern form dates from the beginning of the nineteenth century, starting with John Dalton's development of his atomic theory of chemistry. This enabled Dalton to bring impressive order into a complex collection of data relating to chemical interactions. He himself was a Quaker.

Modern Developments

For the contemporary particle physicist, atoms are very large-scale systems, and they are certainly not uncuttable. Current candidates for basic constituents are at least as small as quarks and gluons and perhaps as infinitesimal as superstrings. Two aspects of the chain of discovery linking Dalton to the Standard Model of contemporary particle physics have particularly impressed themselves on the thinking of a wider intellectual public. One is simply the success of the program in accounting for many of the properties of the physical world in terms of the behavior of a small set of basic constituents. The other striking feature has been that at each level in the exploration of the structure of matter the resulting theories have been characterized by elegant economy and simplicity. Just as Dalton's atomic theory succeeded in bringing order into a bewildering welter of chemical facts, so contemporary particle physics uses the search for underlying simplicity as a successful guide to the discovery of yet more basic theories.

Such success raises the question of whether these elemental entities are not entitled to the metaphysically more profound epithet "fundamental." Are they, in fact, the stuff of reality, so that a particle physics perspective is the right way to approach metaphysics, with the other levels of human experience of the world simply being the complex corollaries of the aggregation of elementary entities? To ask the question raises the issue of reductionism (thinking in terms of constituents) versus holism (thinking in terms of totalities). If the former stance were the totally correct one, particle physics would not only influence human thought—it would be the proper basis for all fundamental human thinking. But this would be a metaphysical, rather than a scientific, conclusion.

Reductionism

Reductionism is the way of thinking that concentrates on an understanding founded in terms of the properties of constituents, rather than in terms of entities considered as a whole. The atheistic thinking of Epicurus was strongly reductionist—atoms and the void constituted the actual nature of reality. The examples of Boyle and Newton show, however, that atomistic thinking need not lead either to atheism or to the view that atomism is the all-sufficient account of reality. In the introduction the ambiguity inherent in the relation between physics and metaphysics is noted. The issues at stake for human thought can be clarified by recognizing that there are a variety of distinct forms of reductionist thinking.

First, there is a strategy of enquiry that can be called *methodological reductionism*. The complex character of totalities means that if they are decomposed into their component parts, these constituents will often be much easier to understand. Much can be learned by this technique of intellectual "divide and rule," and particle physics is the ultimate reductionist subject in this sense. Its successes indicate that this is indeed a very fruitful procedure to follow. It by no means follows, however, that one can learn all that is knowable and worth knowing by following this reductive tactic alone. The success of regarding subatomic matter as composed of quarks and gluons does not at all imply that this is all that needs to be said about physical reality.

A second form of reductionism corresponds to what may be called *ontological reductionism*. This asserts that when the decomposition of systems into their smallest parts is actually carried out, the entities that will be found among the fragments are indeed those that particle physics describes. Adopting this point of view implies, for example, that there is no "extra ingredient" necessary for living entities of the kind that vitalism supposed to be required. This form of reductionism is very widely accepted, and it is reinforced by the successes of molecular biology in giving an account of the processes taking place within living cells. However, it does not follow from this acceptance that higher-level entities are "nothing but" collections of quarks and gluons. That this caveat is necessary can be seen by considering the truth or falsehood of a third form of reductionism.

This is *conceptual reductionism.* This would claim that the concepts used in physics are all reducible to being expressed in terms of elementary particle physics; that all of chemistry is reducible to physics; and so on. Many thinkers reject this view. Its implausibility seems clear enough within physics itself. Subjects such as condensed matter physics and fluid mechanics make use of concepts that cannot be expressed just in terms of elementary entities. Single atoms have neither temperature nor viscosity. Although there has been a tendency among some particle physicists to call the currently unknown Grand Unified Theory a "Theory of Everything," that claim is overblown. More persuasive is the remark made by the condensed matter theorist, Philip Anderson, that "more is different." New properties emerge at higher levels of complexity, and they have to be treated on their own terms.

Some emergent properties are intricate but, in principle, unproblematic. An example is wetness. H_2O molecules individually do not exhibit this property, but it is scarcely surprising that when large numbers of them are brought together, the resulting redistribution of energy due to intermolecular interactions results in the collective effect that we call surface tension. This insight is an example of *causal reductionism.* The phenomenon of wetness is capable of being understood as due to effects operating at the atomic, and ultimately at the particle, level of causation. It is not at all clear, however, that all emergences are capable of being satisfactorily understood in this causally reductive way. There are emergences that appear to differ qualitatively from the properties of the substrate that sustains them. The most striking example is consciousness. Many see a yawning gap between neuroscience's talk of neural interactions, however interesting and sophisticated such talk may be, and the simplest mental experience, such as that of seeing pink. The execution of a willed intention obviously involves lower-order processes, such as muscle contractions, but it is not at all obvious that there is not also a new kind of mental causality that has also been brought into play. The issue here is essentially metaphysical, and it certainly cannot be settled by particle physics.

The most significant issue concerning the relationship of particle physics to human thought concerns the status of the entities that it describes. Are

they simply an interesting and significant level in the structure of reality, or are they to be considered as "more fundamental" than other entities, such as cells or human beings? The question is a metaphysical one. Strong opinions can be expressed on either side of the issue, but it is not one that particle physics itself can settle.

See also: BENEFITS OF PARTICLE PHYSICS TO SOCIETY; CULTURE AND PARTICLE PHYSICS; INFLUENCE ON SCIENCE; PHILOSOPHY AND PARTICLE PHYSICS; UNIVERSE

Bibliography

Barrow, J. D. *Theories of Everything: The Quest for Ultimate Explanation* (Oxford University Press, Oxford, UK, 1991).

Cushing, J. T. *Philosophical Concepts in Physics: The Historical Relation between Philosophy and Scientific Theories* (Cambridge University Press, Cambridge, UK, 1998).

Penrose, R. *The Large, the Small and the Human Mind* (Cambridge University Press, Cambridge, UK, 1997).

Polkinghorne, J. C. *Beyond Science: The Wider Human Context* (Cambridge University Press, Cambridge, UK, 1996).

Weinberg, S. *Dreams of a Final Theory: The Search for the Fundamental Laws of Nature* (Pantheon Books, New York, 1992).

John Polkinghorne

MOMENTUM

Momentum and energy are among the most important quantities in physics. Their importance arises from the fact that they are conserved, which means that energy is never created or destroyed, although it can be transformed from one form to another. There is always an exact accounting so that in the end the books balance to exactly zero. For example, the Earth absorbs solar energy, but the energy is transformed into thermal energy of the Earth. Most of this energy is radiated back into space, but, if the Earth is warming, some remains as thermal energy. When all such energy changes are added up, the result is always zero net energy change. This is conservation of energy. Momentum obeys a similar conservation law.

Definition of Momentum

While energy is a scalar (a number), momentum is a vector. A moving particle has a momentum,

p, equal to its mass, *m*, times its velocity, **v**—**p** = *m***v**. The velocity vector **v** can be represented by an arrow. The length of the arrow equals the speed of the particle—how fast it is moving and the direction of the arrow indicates the direction of the motion. When **v** is multiplied by the mass *m*, the result is the momentum vector, which points in the same direction as the velocity vector. Momentum is defined this way because, with this definition, momentum is conserved.

The Law of Conservation of Momentum

Conservation of momentum is illustrated in Figure 1. A moving particle of mass, *m*, strikes a stationary particle of mass, *M*. Initially, the total momentum of the system is simply the momentum of the moving particle, *m***v**. After the collision, the initial momentum is shared between the two particles as shown. Although they may be moving at wildly different speeds and directions, by the law of momentum conservation the sum of the two final momenta equals that of the original incoming particle. To add the two vectors, one places the vectors head to tail (Figure 1b): the sum is the vector drawn from the tail of the first to the head of the second. The figure shows that the sum of the final momenta equals the original momentum.

FIGURE 1

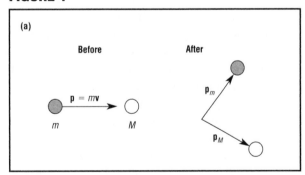

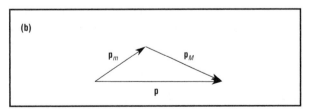

(a) Momentum vectors before and after collision. (b) The sum of final momentum equals the initial momentum vector.

According to relativity any ordinary vector is always paired with a scalar quantity to form a four-vector. The most familiar case is the space-time four-vector consisting of position (described by a vector) and time. Momentum (a vector) and energy (a scalar) also form a four-vector. Relativity requires that all components of a four-vector be conserved if any of them are. Thus, conservation of energy and momentum are really the same conservation law, the conservation of the energy-momentum four-vector.

Symmetries and Conservation Laws

All conservation laws are believed to come from symmetries (Noether's Theorem). Energy conservation comes from the symmetry that the laws of nature are the same at all times. This is a symmetry in the same sense that a circle is symmetric because it is the same no matter how you rotate it. Changing the angle of rotation changes nothing; for the time symmetry we change time and nothing changes, that is, the laws of physics stay the same.

Relativity suggests that position (the three-vector part of the space-time four-vector) should show a similar symmetry and that this symmetry should give rise to the conservation of momentum. Indeed, this is the case: the laws of physics are unchanged when we move from one place to another. Otherwise, physicists in Hong Kong would have to use different laws and theories than physicists in Canada. But they do not, and conservation of momentum results.

Colliding beam accelerators

Particle accelerators are designed to produce new particles in order to test predictions of theories of elementary particles. For example, the Higgs boson is predicted by current models, and its discovery would be a major confirmation of those theories. The primary reason for wanting to build the Superconducting Super Collider, which was canceled in 1993, was to look for the Higgs boson. Europe's Large Hadron Collider is designed to carry out the same search.

The demands of conservation of momentum are a significant obstacle for particle production and have required a major redesign of particle accelerators. New particles are produced by accelerating familiar particles to large energies and aiming the beam at a stationary target. When a beam particle strikes a target particle, new particles can be formed if there is enough energy available to create the new particle, that is, to create its rest energy. Ideally, we would like all of the kinetic energy of the incoming particle plus the rest energies of the two initial particles to go into creating the new particle.

The problem is that momentum conservation requires that there be some net forward momentum of the new particle equal to the momentum of the incoming particle (Figure 2). Thus, some of the energy must go into this motion and is not available for creating the new particle. For new particles with large masses like the Higgs boson (about 1,000 times the mass of a proton), only a tiny fraction of the incoming particle's energy is available for particle creation; the vast majority of the energy is used up in satisfying conservation of momentum. For example, in producing a particle with a mass of 50 times that of a proton, only 4 percent of the incoming particle's energy is available, meaning that the beam particle must have a kinetic energy of 1,250 times the rest energy of a proton, or about 1,200 GeV. This exceeds the capabilities of the most energetic accelerators, and creating the heavier Higgs boson is even further out of reach.

The solution is straightforward, in principle. Instead of a beam of particles colliding with a stationary target, let two beams of identical particles collide head-on (Figure 3). The net initial momentum is zero so the final momentum is also zero. No energy

FIGURE 3

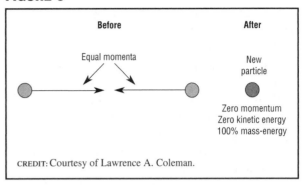

CREDIT: Courtesy of Lawrence A. Coleman.

Conservation of momentum in the creation of a new particle: zero net momentum system.

has to go into post-collision kinetic energy so all of the kinetic energy of both beam particles plus their rest energies are available to create the new particle.

Still, there is a downside. Since the spacing between particles in a beam is much greater than in a material target, the rate of collisions is correspondingly less. Thus, one has to run the experiment for a long time in order to produce and detect the desired particle. Nonetheless, this is far better than not being able to produce the particles at all.

See also: CONSERVATION LAWS; SYMMETRY PRINCIPLES

Bibliography

Lederman, L. "Accelerators: They Smash Atoms, Don't They?" in *The God Particle* (Houghton Mifflin, Boston, MA, 1993).

Smith, C. L. "The Large Hadron Collider." *Scientific American* **283**, 70–77 (2000).

Lawrence A. Coleman

FIGURE 2

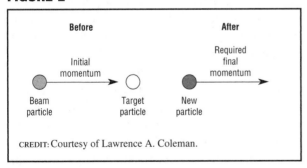

CREDIT: Courtesy of Lawrence A. Coleman.

Conservation of momentum in the creation of a new particle and stationary target particle.

MUON, DISCOVERY OF

The discovery of the muon, the first and lightest unstable subatomic particle, involved many experiments over a period of fourteen years. The effort to understand the nature of this particle did much to create the framework in which the science of particle physics developed.

The discovery grew out of studies of "cosmic rays" in the 1930s by Carl D. Anderson of the California Institute of Technology. Our planet is continually bombarded by high-energy radiation that originates outside the solar system, most of which consists of atomic nuclei. These nuclei collide with the nuclei of atoms in the upper atmosphere. By conversion of kinetic energy to particle mass, a cascade of particles is produced. Most of these are highly unstable and do not survive to reach the surface of the Earth, where one finds instead the products of their disintegration.

Anderson used a device called a "cloud chamber" to identify the particles that did reach the surface. In a cloud chamber, a moving electrically charged particle leaves a trail of water droplets along its path, which can be recorded in a photograph. If the chamber is in a magnetic field, the tracks will be curved, and the curvature can be used to measure the momentum of the particles. The spacing between droplets on the track gives a rough measure of the velocity of the particle. If one knows momentum and velocity, the mass of a particle can be estimated.

By 1934 Anderson had determined that a major fraction of the particles he was observing had a mass much more than that of an electron but less than that of a proton. A more exact measure of the mass was not yet possible but Anderson could clearly rule out any known particle. These particles could carry either positive or negative electric charge.

In 1935, the Japanese theorist Hideki Yukawa proposed that a particle two hundred to three hundred times heavier than an electron could transmit a new force that acts only on the scale of the nucleus, explaining why a nucleus holds together despite the mutual repulsion of its protons. By 1938, most physicists in the field believed that Anderson's mysterious particle was the one proposed by Yukawa. The name "mesotron" was given to this particle, later shortened to "meson."

In 1940 it was discovered that the meson was unstable, breaking up into an electron and some unseen neutral particles in an average lifetime of two microseconds. As short as this time may seem, it is long enough to allow some of the particles to reach the Earth's surface. This is possible because of an effect from the theory of relativity; the "internal clock" of a particle slows down when it is moving close to the speed of light.

In a series of experiments that began in 1944 under difficult wartime conditions three Italian physicists at the University of Rome, Marcello Conversi, Ettore Pancini, and Oreste Piccioni, allowed mesons to stop in a variety of materials after separating the positive ones from the negative by means of a magnet. It was expected that positive mesons would survive for their normal lifetime, while negative ones would be attracted to nuclei where they would be quickly absorbed. But in 1946 they found that in light elements the negatives managed to survive. This showed that the mesons were not affected by the nuclear force and thus could not be Yukawa's particle.

A Mystery Resolved

The puzzle was resolved in 1947 by Cesar Lattes, Giuseppe Occhialini, and Cecil Frank Powell, working at Britain's Bristol University. They studied cosmic rays at mountaintop altitudes, where it is possible to observe some of the particles directly produced in cosmic ray collisions. They used detectors called "nuclear emulsions," sheets of light-sensitive material similar to that used in ordinary photographic film. While the thickness of the sensitive layer in camera film is usually no more than a few hundredths of a millimeter, in nuclear emulsions it can be as much as 2 millimeters. When the emulsion is developed, the path of a particle appears as a trail of tiny grains of metallic silver, which can be followed in three dimensions with the aid of a medium-power microscope.

Lattes, Occhialini, and Powell studied what happened when particles came to a stop in their emulsions. They discovered that in some cases a particle that appeared to be a meson would stop and then emit another particle of somewhat lower mass. The second particle would also stop within a fraction of a millimeter and decay into an electron and neutral particles, just like a meson. They called the first particle a "pi meson" because it was the primary particle in the two-step process, and pi is the Greek equivalent of the letter p. They hoped that the pi meson would prove to be Yukawa's particle. They assumed that the second particle was the familiar meson, which they called a "mu meson" (the Greek letter mu had by then become the accepted symbol for a meson). This was later shortened to "muon" when physicists decided to reserve the name "meson" for the members of a large family of particles that interact strongly with nuclei.

After these discoveries, the origin of cosmic ray muons became clear. The pi meson has a lifetime eighty times shorter than that of the muon. Thus a major share of them decay into muons near the collision that produced them, high in the atmosphere. The muons, having longer lifetimes, can reach the surface.

In the following year, Lattes took a stack of emulsions to the University of California in Berkeley, where he observed pi mesons produced in the collision of helium nuclei with larger nuclei in a cyclotron. This showed that the pi meson interacted strongly with nuclei, finally establishing it as Yukawa's particle. In the same year, Edward Hincks and Bruno Pontecorvo, at Canada's Chalk River Laboratory, demonstrated that the breakup of muons released two neutral particles in addition to the electron. Neither of the neutrals was a gamma ray, and they interacted weakly with matter. This made it likely that they were neutrinos, neutral counterparts of the electron.

Who Ordered That?

Like the electron, neutrinos spin with an angular momentum one-half of the fundamental quantum unit. Because the muon breaks up into an odd number of particles with half-integer spin, it too must be a half-integer spin particle, called "fermions" after the Italian-American physicist Enrico Fermi. Thus the muon is simply a heavier version of the electron. In response to this discovery the celebrated American physicist Isidor Rabi is reported to have remarked: "Who ordered *that?*"

Today the muon is classified as one of six members of a family of particles called "leptons." Three of these, the electron, muon, and tau, are electrically charged and come in both positive and negative forms. The other three are neutrinos, designated the electron, muon, and tau neutrino because each is associated with one of the charged leptons. When a neutrino collides with a nucleus, it can be transformed into its companion charged lepton. All leptons interact weakly with nuclei, so such collisions are very rare.

See also: ANDERSON, CARL D.; YUKAWA, HIDEKI

Bibliography

Piccioni, O. "The Discovery of the Muon" in *History of Original Ideas and Basic Discoveries in Particle Physics,* edited by Harvey B. Newman and Thomas Ypsilantis (Plenum Press, New York, 1996).

Galison, P. "The Discovery of the Muon and the Failed Revolution Against Quantum Electrodynamics." *Centaurus* **26,** 262–316 (1983).

Robert H. March

N

NEUTRINO

In spite of its history of more than seven decades, Wolfgang Pauli's mystery particle, the neutrino, still remains the least known particle. The mystery is attributed to its feeble interaction with others, and recently, Leon Lederman called it a barely existing particle.

In an open letter to a 1930 Tübingen conference, which starts with "Dear Radioactive Ladies and Gentlemen," Pauli introduced the concept of a neutral fermion (a spin-$\frac{1}{2}$ particle) and called it a neutron. This was his desperate attempt to rescue the laws of energy and momentum conservation that seemed to be violated by beta-decay processes. However, after James Chadwick's discovery of a heavy particle with no charge in the nucleus, Pauli's neutral fermion, which was actually created in beta decay, was renamed neutrino by Enrico Fermi in 1933. Chadwick's neutral particle kept the name neutron. In the same year, at the Solvay conference, Pauli also speculated that neutrinos may be massless. Fermi recognized the importance of the neutrino and developed his famous beta-decay theory in 1934. Pauli's neutrino is now represented by ν_e since it is accompanied by an electron in most interactions.

In 1934 Hans Bethe and Rudolf Peierls expressed a pessimistic view on the direct detection of neutrinos because their interaction strength was so weak. In the early 1950s Frederick Reines and Clyde Cowan looked for a way to detect neutrinos by observing the inverse beta decay (a process by which ν_e turns into a positron). Since this required a large target and an enormous flux source of neutrinos, Reines and Cowan even considered a nuclear explosion; however, they chose a nuclear reactor instead. In June 1956 they sent a telegram to Pauli informing him that neutrinos had been detected.

Second and Third Neutrinos

A large discrepancy between the predicted rate of muon decay $\mu \rightarrow e + \gamma$ and the assumption of only one kind of neutrino resulted in the speculation about the existence of a second neutrino. A new neutrino with a different lepton number or flavor, the muon neutrino ν_μ, was necessary. This would not permit the process $\nu_\mu + p \rightarrow e^+ + n$. Indeed, the absence of this process was confirmed in 1962 at the Brookhaven National Laboratory, establishing the existence of ν_μ. After the dramatic discovery of charm quarks in 1974, the presence of two families of quarks and leptons was established.

Soon after, the third charged lepton τ was discovered by a collaboration led by Martin Perl in 1975 at the Stanford linear electron-positron collider. This immediately suggested the possibility of a third family of quarks and leptons. A doublet of bottom (*b*) and top (*t*) quarks was subsequently discovered, and the existence of the third neutrino, the tau neutrino ν_τ, was generally accepted. This is because, in each family, quarks and leptons must appear as a doublet so that no anomalies (unwanted infinities) appear in the Standard Model. However, a direct observation of the process $\nu_\tau \rightarrow \tau$ has remained elusive because the accelerator-produced ν_τ did not have enough energy to produce a clear signal of the production of τ by ν_τ. Finally in 2000 the signal was observed at Fermilab, confirming the existence of the tau neutrino and completing the three families of quarks and leptons.

Neutrino Mass

The mass of neutrinos is complicated. Originally, Wolfgang Pauli proposed that the neutrino mass was approximately that of an electron. Later, at the 1933 Solvay conference, he speculated that it could be massless. For some period, the only way to measure the neutrino mass was to see possible deviations of the Kurie plot of the electron energy spectra in beta decays. At the highest electron energy, the neutrino carries the least energy so that the shape of the spectrum is sensitive to the mass of the neutrino. A large number of such experiments have been carried out. The most popular process used is $^3\mathrm{H} \rightarrow {}^3\mathrm{He} + e^+ + \nu_e$, which has the maximal kinetic energy of the electron. So far, all the results based on various beta decays have been negative, yielding the following limits: $m(\nu_e) < 3$ eV, $m(\nu_\mu) < 0.19$ eV, and $m(\nu_\tau) < 1.92$ MeV. A substantial improvement in these results is not feasible in the near future.

Neutrino Oscillations

An entirely new approach to probe neutrino mass began with Bruno Pontecorvo's proposal of neutrino oscillations in a series of papers in the late 1950s. In 1957 he proposed the conversion of a neutrino into antineutrino, which would be possible for neutrinos with mass. Recall that only one neutrino was known at that time. When the second neutrino ν_μ was discovered in 1962, Bruno Pontecorvo, and Ziro Maki, Masami Nakagawa, and Shoichi Sakata independently entertained the possibility of $\nu_\mu \leftrightarrow \nu_e$ oscillations. The physics of neutrino oscillations can be understood as follows.

When those particles produced and detected in an experiment differ from those that govern the propagation, oscillation phenomena can occur. For example, when ν_μ is produced from the π^- decay, this process produces a neutrino with a definite flavor (in this case, ν_μ); however, when ν_μ travels, its motion is governed by an equation of motion for a particle with a definite mass. If neutrinos have mass, the ν_μ state produced is not a state with a definite mass (a pure mass eigenstate). Instead, it is a linear combination of three mass eigenstate neutrinos. That is, ν_μ does not have a definite mass! Moreover, when it propagates, it cannot be described by a single equation of motion because all three components have different values of mass.

Suppose one tries to detect ν_μ with a detector a distance L from ν_μ source. Since all three components (mass eigenstates) have different values of mass, they propagate in different manners, creating phase differences. This means the neutrino that reaches the detector is no longer the original ν_μ because the detected neutrino contains, in addition to the original ν_μ, additional ν_e and ν_τ components. Hence, the detector can detect all three neutrinos of different flavors. The necessary and sufficient conditions for the oscillation are as follows: (1) The neutrinos have mass, and (2) neutrinos with flavor are linear combinations of mass eigenstate neutrinos. Oscillations are characterized by the quantity $\Delta m^2 = m_2^2 - m_1^2$, where m_2 and m_1 are the mass of ν_2 and ν_1, respectively. Thus, although the observation of neutrino oscillations cannot pin down the values of mass, it shows that some neutrinos have mass. This is the reason why the search for neutrino oscillations has gained a great deal of attention. Since the Standard Model of particle physics has been constructed with the assumption of massless neutrinos, the discovery of neutrinos with mass signals new physics beyond the Standard Model.

Numerous attempts to observe neutrino oscillations, using ν_e from nuclear reactors and ν_μ from accelerators, all failed. The problem was the lack of information of the values of Δm^2. In order to observe oscillations, one has to design the experiment so that $\Delta m^2 L/4E$ (the combination appearing in the oscillation formula) approximately equals 1. In reality, the energy of neutrinos E is fixed by an accelerator. One then has to determine the location L of a detector. This cannot be done unless one knows the value of Δm^2. With past experimental setups with L ranging from about 10 m to 1 km, no oscillations have been seen.

In 1998 oscillations of atmospheric neutrinos were confirmed for the first time by the detector at Kamioka in Japan. The detector was originally designed as a nucleon decay experiment and was thus named Kamiokande. The larger version at Kamioka is called the Super-Kamiokande (SK). The atmospheric neutrinos, produced by cosmic rays, consist of ν_μ and ν_e whose ratio on the surface of the Earth is roughly two to one. Both Kamiokande and SK observed a smaller number of ν_μ than expected whereas the number of ν_e was consistent with theory. The latest experimental results have confirmed that ν_μ's from cosmic rays oscillate into ν_τ's that escape detection by the current experimental setups, whereas ν_e's do not oscillate into other neutrino flavors. The latest quantitative results are expressed by the finding that $\Delta m^2 = m_3^2 - m_2^2 = 3 \times 10^{-3}$ eV^2 and that the mixing is almost maximal. This has been supported independently by other groups. It is generally believed that these findings are very convincing evidence for neutrinos with mass.

A more significant experiment in neutrino physics is the measurement of the solar neutrinos (located in Homestake, Idaho), which was originally designed to probe the solar core. The observed rate of solar neutrinos is roughly one-half the value expected with the standard solar model developed by John Bahcall and others. Recently, it has been concluded that solar ν_e's are depleted because their oscillations, most likely into ν_μ's that cannot be detected since the energy of the original ν_e is too low for a converted ν_μ to produce μ. The results of all the solar neutrino experiments suggest that $\Delta m^2 = m_2^2 - m_1^2 = 10^{-5} \sim 10^{-4}$ eV^2. Although two different values of Δm^2 appear to be determined, thus confirming massive neutrinos, the determination of individual neutrino mass value still remains a major task for future experiments.

The recent K2K long-base-line experiment, so named because ν_μ's from the accelerator at KEK in Tsukuba, Japan, are sent to Kamioka located 250 km away, is exclusively designed to confirm the SK re-

sults, $\Delta m^2 = m_3^2 - m_2^2 = 3 \times 10^{-3}$ eV^2, without relying on the atmospheric neutrinos. The distance of 250 km is long enough to see the oscillations suggested by the atmospheric neutrino experiment. Since 1999 the SK detector has observed ν_μ-induced events (forty four) that are about one-third less than the expected number (sixty-six) without oscillations, indicating that about one-third of ν_μ's have most likely transformed into ν_τ's, consistent with the atmospheric neutrino results. This may be, when confirmed independently, the first positive result of neutrino oscillations with human-made neutrinos. Similar long baseline experiments, for example, with a distance of 730 km, are under construction including MINOS (Fermilab to Soudan mine in Minnesota) and OPERA (CERN to Gran Sasso National Laboratory in Italy).

See also: CASE STUDY: SUPER-KAMIOKANDE AND THE DISCOVERY OF NEUTRINO OSCILLATIONS; CONSERVATION LAWS; LEPTON; PARTICLE; STANDARD MODEL

Bibliography

Boehm, F., and Vogel, P. *Physics of Massive Neutrinos* (Cambridge University Press, Cambridge, UK, 1987).

Kim, C. W., and Pevsner, A. *Neutrinos in Physics and Astrophysics,* Vol. 8: *Contemporary Concepts in Physics* (Harwood, Langhorne, PA, 1993).

Chung W. Kim

NEUTRINO OSCILLATIONS

Neutrino oscillations are a phenomenon of quantum mechanical nature, intrinsically connected to the question of neutrino mass. The neutrino was originally theorized by Wolfgang Pauli in 1931 to reconcile data on radioactive decay of neutrons with energy conservation. While the postulated neutrino had no mass, no electric charge, and essentially did not react with matter, its inclusion as an emitted particle balanced energy conservation—the observed energy range for the electron corresponded to the many ways in which the emitted proton, electron, and neutrino can share energy.

Scientists now know that there are three "flavors" of neutrinos (labeled after their leptonic companions): the electron neutrino, ν_e, the muon neutrino, ν_μ, and tau neutrino, ν_τ. Similarly, up quarks and down quarks, which make up neutrons and protons, each have two siblings. No one yet knows why two apparently useless copies of the "useful" particles exist.

Pauli calculated that for the neutrino to function as theorized, the upper limit on its mass must be less than 1 percent of the proton mass. Subsequent experiments derived an upper limit of 10^{-8} times the proton mass—thus, the neutrino would be virtually massless. This is somewhat puzzling because no basic principle such as gauge invariance prevents neutrino mass, as it does for photons. However, modern theories have ways to accommodate small but nonvanishing neutrino masses. Flavor oscillations of neutrinos traveling from a source to a detector provide a powerful experimental signature of such small masses.

The concept of oscillating neutrinos originated in the late 1960s with Bruno Pontecorvo and requires the introduction of quantum states, in particular those describing quarks and leptons participating in weak interactions. It is an empirical fact that the states describing down (d) and its heavier sibling strange (s) quarks of definite mass are not the states of definite quark flavor that participate in weak interactions. Mixtures of the d and s flavor states, quantified by the Cabibbo angle, are the relevant quark quantum states with definite mass. So, it is also expected that the neutrino quantum states of definite flavor that participate in the weak interactions by which neutrinos are created and detected are also admixtures of neutrino quantum states of definite mass. Conversely, a neutrino of definite mass does not have a definite flavor—it carries an admixture of flavors. For the simplified case of two neutrinos, the ν_μ and ν_τ flavor states are related to states ν_1 and ν_2 of mass m_1 and m_2 by

$$\nu_\mu = \nu_1 \cos\theta - \nu_2 \sin\theta,$$
$$\nu_\tau = \nu_1 \sin\theta + \nu_2 \cos\theta$$

where θ is the Cabibbo-like "mixing angle," a constant of nature whose value will hopefully be understood some day in the context of a theory beyond the Standard Model in which neutrinos have non-

vanishing masses. A neutrino created, for instance in the weak decay of a charged pion

$$\text{pion} \rightarrow \text{muon} + \text{neutrino},$$

is born in the pure flavor state ν_μ. This is the meaning of the statement that "the flavor states participate in the weak interaction." This state is a superposition of the ν_1 and ν_2 mass states and their admixture is described by the above equation. The propagation of these mixed states is described by the Schrödinger equation; their interference causes the probability of detecting a particular flavor to change with the distance traveled by the neutrino. In other words, because the ν_1 and ν_2 states slightly differ in mass, the flavor admixture of the states fluctuates back and forth as the neutrino propagates through space. For instance, for a neutrino born as a ν_μ, the probability P that it will be observed as a ν_τ after traveling a distance L has a sinusoidal (i.e., oscillating) dependence on $\Delta m^2\, L/E$, where $\Delta m^2 = m_2^2 - m_1^2$ is the mass-squared difference and E is the neutrino energy:

$$P(\nu_\mu \rightarrow \nu_\tau) = \sin^2 2\theta \sin^2 (\Delta m^2\, L/4E).$$

The probability that a ν_μ is observed at L is

$$P(\nu_\mu \rightarrow \nu_\mu) = 1 - P(\nu_\mu \rightarrow \nu_\tau).$$

This is illustrated in Figure 1. The appearance of tau-neutrinos in a beam of muon-neutrinos, or the disappearance of muon-neutrinos, are thus signatures that neutrinos have different masses. Conversely, massless neutrinos do not oscillate because all Δm^2 values vanish.

Examples of experimental evidence for neutrino oscillations include

- A deficit of electron neutrinos born in nuclear processes that make the sun shine has been observed in several deep-underground detectors. Recently, by combining data from experiments in Japan and Canada, the first evidence has been produced that the missing electron neutrinos have indeed transformed into muon neutrinos and tau neutrinos.

- Cosmic rays interacting with the nitrogen and oxygen in the Earth's atmosphere at an average height of 20 kilometers produce pions that de-

FIGURE 1

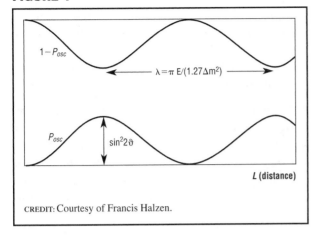

CREDIT: Courtesy of Francis Halzen.

The probability P_{osc} (or oscillation probability) that a neutrino of a particular flavor will, after traveling a distance L, be observed as a neutrino of a different flavor (bottom curve) or the same flavor (top curve).

cay into muon neutrinos. The observed neutrino flux agrees with relatively straightforward computations. However, for neutrinos produced by exactly the same mechanism on the other side of the Earth, a deficit of neutrinos of muon flavor relative to expectations is observed after they travel roughly 10,000 meters through the Earth. (Because they only participate in the weak interaction, atmospheric neutrinos penetrate the Earth with no attenuation.) There is mounting evidence that they have oscillated into tau neutrinos. The accumulated evidence for neutrino oscillations is summarized in Figure 2.

The quest for precise information on neutrino masses and mixings has only just begun. There are plans on three continents to shoot accelerator beams of neutrinos to underground detectors over baselines of hundreds or even thousands of kilometers. One such experiment has already produced supporting evidence for oscillations of atmospheric neutrinos by observing a beam produced at a laboratory in Tsukuba, Japan at the SuperKamiokande detector, 250 kilometers away.

See also: CASE STUDY: SUPER-KAMIOKANDE AND THE DISCOVERY OF NEUTRINO OSCILLATIONS; NEUTRINO; NEUTRINO, SOLAR

Bibliography

Bahcall, J. *Neutrino Astrophysics* (Cambridge University Press, Cambridge, UK, 1989).

FIGURE 2

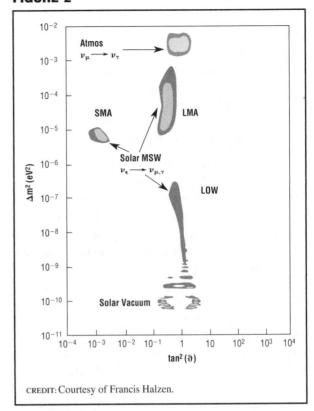

CREDIT: Courtesy of Francis Halzen.

The shaded areas delineate values of neutrino mass-squared difference Δm^2 and mixing angle θ that can accommodate the observed oscillations of atmospheric and solar neutrino beams. S(L)MA stands for oscillations with small (large) mixing angles, or small Δm^2 (LOW). Vacuum corresponds to oscillations of the flux between Sun and Earth.

Caldwell, D. O., ed. *Current Aspects of Neutrino Physics* (Springer-Verlag, Heidelberg, Germany, 2001).

Kim, C. W., and Pevsner, A. *Neutrinos in Physics and Astrophysics* (Harwood Academic Press, Langhorne, PA, 1993).

Sutton, C. *Spaceship Neutrino* (Cambridge University Press, Cambridge, UK, 1992).

M.C. Gonzalez-Garcia
Francis Halzen

NEUTRINO, DISCOVERY OF

The particle called the neutrino was conceived in 1930 by the Austrian-Swiss theoretical physicist Wolfgang Pauli (1900–1958) as a possible solution to two vexing problems confronting a widely accepted model of the structure of the atomic nucleus, which used the two elementary constituents of matter then known: the electron and the proton. The neutral atom of mass number A and atomic number Z was supposed to contain in its nucleus A protons and $A - Z$ electrons; Z electrons made up the shells of the atom. This picture seemed reasonable because protons were knocked out of light nuclei by alpha particles from radioactive decay, and, while in the beta-decay form of radioactivity, electrons emerged from the nucleus.

However, there were puzzles concerning the nuclear electrons. Beta decay causes the positive nuclear charge Z to increase by one unit and decreases the energy of the nucleus by a definite amount; but, the electron emerges with varying (lesser) amounts of energy, so that a part of the energy loss of the nucleus is unaccounted for. Another puzzle was related to a property known as spin angular momentum (by analogy with a spinning top). Electrons and protons each have spin of $\frac{1}{2}$ (in units $h/2\pi$, where h is Planck's quantum of action). In those nuclei where the total number of nuclear particles is odd, such as the common element nitrogen ($_7N^{14}$), the total nuclear angular momentum should be half an odd integer. However, in the case of $_7N^{14}$, it was shown to be the integer one. Nevertheless, most physicists accepted the electron-proton model, even though it contradicted the well-known laws of conservation of energy and angular momentum, believing that different physical laws might hold within the tiny space of the nucleus. Indeed, physicists had recently learned that the new puzzling laws of quantum mechanics ruled within the atom. The influential atomic physicist Niels Bohr believed that the law of conservation of energy held only in a statistical sense, like the law of increase of entropy in statistical mechanics.

The Idea of the Neutrino

Pauli, who was unwilling to give up the conservation laws, conjectured the existence of a new particle in order to solve the two difficulties mentioned. This was a neutral particle of spin $\frac{1}{2}$ with a mass "not larger than 0.01 proton mass," as Pauli suggested in a famous letter sent on December 4, 1930, to nuclear physicists who were holding a meeting in Tübingen, Germany. He proposed that each electron in the nucleus was accompanied by one of the new particles,

which he provisionally named neutrons. This solved the problem of $_7N^{14}$ and analogous cases. When a nucleus underwent beta decay, a neutron would emerge with each electron, carrying away the energy that appeared to be lost. Pauli's particle would have been almost undetectable.

There Pauli let the matter rest, presenting his idea publicly in October 1933 at an international conference held in Brussels. He renamed his particle the neutrino, following a suggestion by the Italian Enrico Fermi. The nuclear particle that we now call the neutron was discovered in 1932 by the Englishman James Chadwick.

Fermi's Theory of Beta Decay

A Russian, Dmitri Iwanenko, suggested earlier that Chadwick's neutron was a kind of neutral proton—that is, a massive elementary particle of spin $\frac{1}{2}$—and that the nucleus contains no electrons, only neutrons and protons. He also proposed that an electron and a neutrino are created together in the process of beta decay, much as a photon is created in an ordinary atomic transition.

Shortly after the Brussels conference, in 1933, Fermi put forth a quantum field theory of beta decay. It used the relativistic theory of Paul Dirac, which provides the possibility of creation and annihilation of particles in pairs. In Dirac's theory, the electron is accompanied by a matching positive particle of the same mass and spin, called the positron. In Fermi's theory, the spin-half neutrino also has an analogous partner called the antineutrino. The beta decay of neutrino-rich nuclei produces an electron-antineutrino pair, while the beta decay of proton-rich nuclei produces a positron-neutrino pair. With some important later modifications, Fermi's theory (when generalized) forms a part of the modern electroweak theory that unifies electromagnetism with the weak nuclear interaction, of which beta decay is one example.

As a result of Chadwick's discovery of the neutron and the success of Fermi's theory of beta decay, nuclear electrons were soon rejected and other nuclear models took their place. By 1936, Bohr also agreed that the conservation laws were valid in each individual nuclear event.

Detection of the Neutrino

The neutrino was theoretically indispensable, but it was necessary to detect it directly. This formidable task took two decades to accomplish, since the neutrino can pass through light-years of matter without interacting. It was first observed in 1956 by a group led by Clyde L. Cowan and Frederick Reines of Los Alamos National Laboratory, who used the enormous flux of antineutrinos from a nuclear reactor at the Savannah River Plant in South Carolina, using a "target" consisting of cadmium chloride dissolved in water, surrounded by large detectors filled with a liquid scintillator. They detected the nuclear reaction known as inverse beta decay, in which a proton captures an antineutrino. In this process, a neutron and a positron result; the capture of these particles produces characteristic flashes of light in the scintillator. In 1995, Frederick Reines was awarded the Nobel Prize in Physics for the discovery of the neutrino. (Clyde Cowan died in 1974.)

Additional Neutrinos

In 1962 at Brookhaven National Laboratory on Long Island, New York, Leon M. Lederman, Melvin Schwartz, and Jack Steinberger used a very large apparatus, consisting of spark chambers and scintillators, to detect a second type of neutrino, whose existence theory had suggested. This new neutrino is produced together with another elementary particle called the muon in the decay of an elementary particle, the pion. The decay process, first observed in the cosmic rays in 1947, takes one-hundred millionth of a second. The pion belongs to the class of particles with strong nuclear interaction called hadrons, and they were produced copiously at Brookhaven in a proton accelerator called the Cosmotron. The muon is a lepton, the general name for particles whose nuclear interaction is weak (such as the electron and the neutrino). The muon also decays, in about a microsecond, into an electron and two different neutrinos, the electron neutrino (Pauli's original neutrino) and the muon neutrino. A third kind, the tau neutrino, corresponds to a third charged lepton, the tau, discovered in 1975 by a group at the Stanford Linear Accelerator Center led by Martin Perl.

The mass of the electron neutrino is very small and was for a long time believed to be zero. It is now thought that neutrinos have mass, but the masses are not well determined. Upper limits are listed as about 0.01, 0.5, and 40 electron masses, respectively, for the electron, muon, and tau types of neutrinos.

Neutrinos in High-Energy Physics and Astrophysics

Beta decay and its inverse play an essential role in the nuclear reaction cycles that produce energy in stars; and neutrinos, with their high degree of penetration, carry information to scientists on earth about processes occurring deep in stellar interiors. Neutrinos from the sun have been monitored by Raymond Davis Jr. of Brookhaven National Laboratory and his collaborators for more than a quarter-century. They used a large tank of cleaning fluid (C_2Cl_4) located 1,500 meters underground to reduce charged cosmic ray background. Neutrinos absorbed by chlorine nuclei convert chlorine to argon at a measurable rate (about one atom per day). The number of solar neutrinos detected this way has been smaller than theoretically expected; the reason for this is an open question. Other large underground installations are used for detecting very high-energy neutrinos from outer space (neutrino astronomy). In 1987, two of these detectors observed neutrinos produced by a supernova in the skies of the Southern Hemisphere.

The number of solar neutrinos observed in underground detectors is about one-third to one-half the number expected on theoretical grounds for massless neutrinos. However, if neutrinos have nonzero mass, a phenomenon known as neutrino oscillation is expected, in which one type of neutrino transforms into another. This process could account for the missing electron neutrinos from the sun. Some experiments performed at the Japanese underground neutrino detector Super-Kamiokonde look at muon neutrinos produced in the earth's atmosphere by cosmic rays. These experiments suggest that oscillation occurs and thus that neutrinos have mass.

After it was shown in the Brookhaven experiment that found the second neutrino that high-energy neutrinos from large particle accelerators can be detected, neutrino beams were used as effective probes of the proton and neutron, supplementing the use of high-energy electron beams. Both types of beams, produce simpler, more easily interpretable interactions than beams of hadrons. Neutrino experiments carried out at the European Laboratory for Particle Physics (CERN) in Geneva, Switzerland, and at Fermilab in Batavia, Illinois, beginning in 1972, formed the basis for the electroweak theory. Other neutrino experiments have helped to establish the color quark model, the other sector of the Standard Model of elementary particle interactions which has dominated the theory for the past two decades. Accelerator experiments are in preparation for testing neutrino oscillation.

See also: FERMI, ENRICO; PAULI, WOLFGANG; REINES, FREDERICK

Bibliography

Boehm, F., and Vogel, P. *Physics of Massive Neutrinos,* 2nd ed. (Cambridge University Press, Cambridge, England, 1992).

Brown, L. M. "The Idea of the Neutrino." *Physics Today* **31**, 23–28 (1958).

Brown, L. M., and Rechenberg, H. *The Origin of the Concept of Nuclear Forces* (IOP Publishing, Bristol, England, 1996).

Fitch, V. L., and Rosner, J. R. "Elementary Particle Physics in the Second Half of the Twentieth Century," in *Twentieth Century Physics,* Vol. II, edited by L. M. Brown, A. Pais, and B. Pippard (IOP Publishing, Bristol, England, 1995).

Franklin, A. *Are There Really Neutrinos?* (Perseus Books, Cambridge, MA, 2001).

Lederman, L. M. "Resource Letter Neu-1 History of the Neutrino." *American Journal of Physics* **38**, 129–136 (1970).

Pais, A. *Inward Bound* (Oxford University Press, New York, 1986).

Reines, F. "50 Years of Neutrino Physics," in *Neutrino Physics and Astrophysics,* edited by E. Fiorini (Plenum Press, New York, 1982).

Reines, F., and Cowan, C. L. "Neutron Physics." *Physics Today* **10**, 12–18 (1957).

Schwartz, M. "The Early History of High-Energy Neutrino Physics," in *The Rise of the Standard Model,* edited by L. Hoddeson, L. M. Brown, M. Riordan, and M. Dresden (Cambridge University Press, New York, 1997).

Sutton, C. *Spaceship Neutrino* (Cambridge University Press, Cambridge, England, 1992).

Wu, C. S. "The Neutrino," in *Theoretical Physics in the Twentieth Century,* edited by M. Fierz and V. F. Weisskopf (Interscience, New York, 1960).

Laurie M. Brown

NEUTRINO, SOLAR

In the summer of 1965 workers deep within the Homestake Gold Mine, Lead, South Dakota, completed the excavation of a $30 \times 60 \times 32$ ft^3 cavern. This excavation, nearly a mile underground, was the first step in an experiment proposed by Ray Davis Jr. and his Brookhaven National Laboratory (BNL) collaborators. The cavern was soon filled by a large tank containing 610 tons—equivalent in volume to ten railway tankers—of the chlorine-bearing cleaning fluid perchloroethylene. The purpose of this detector was to record, for the first time, the neutrinos produced as a by-product of the thermonuclear reactions occurring in the Sun's core. The results of the Davis experiment presented the physics and astrophysics communities with a puzzle that is only now being resolved.

The Standard Solar Model

It is known that the Sun has burned for about 4.6 billion years, sustaining itself against the crushing effects of its own gravity. The energy required to maintain the pressure of solar gases is produced by thermonuclear reactions. Four protons are converted into a helium nucleus (which contains two protons and two neutrons) plus two electrons and two electron neutrinos:

$$4p \rightarrow {}^4\text{He} + 2e^- + 2\nu_e \qquad (1)$$

with a net release of energy. The series of reactions by which almost all solar helium synthesis occurs is called the *pp* chain (see Figure 1). Roughly half the hydrogen fuel that was initially in the Sun's core has been converted into helium over the past 4.6 billion years.

The photons that make up ordinary solar radiation scatter repeatedly within the Sun, and take millions of years to diffuse outward from the solar core. Thus, the sunlight that arrives on Earth last scattered near the Sun's outer surface, called the photosphere. In contrast, neutrinos pass through matter almost unaffected—they lack the electromagnetic interactions by which photons scatter, instead having only "weak" interactions. Consequently, the Sun is transparent to neutrinos, which arrive at Earth directly from the core. As they carry, in their number and energy distribution, detailed information about the nuclear reactions by which they were produced, neutrinos allow one to "see" directly into the Sun's center.

Originally, the motivation for solar neutrino measurements was to test the standard theory of

FIGURE 1

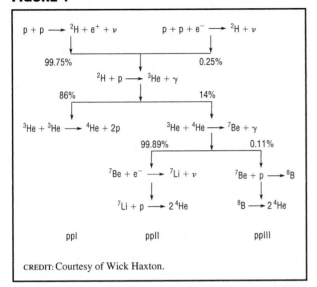

CREDIT: Courtesy of Wick Haxton.

The *pp* chain by which the Sun and similar stars synthesize ^{4}He from protons. Note that the three cycles composing the chain are "tagged" by the neutrinos from $p + p$ beta decay (I + II + III), from electron capture on ^{7}Be (II), and from the beta decay of ^{8}B (III).

stellar energy generation and evolution, as applied to the Sun. The Standard Solar Model (SSM) postulates that the Sun burns in hydrostatic equilibrium, with the gravitational force balanced at each point within the star by the gas-pressure gradient. Energy is generated by hydrogen burning and transported by radiation (in the Sun's interior) and by convection (outer envelope). The initial composition of the Sun, by mass roughly 75 percent hydrogen and 25 percent helium, with traces ($\sim$2%) of heavier elements, is chosen so that today's luminosity is reproduced after 4.6 billion years of evolution.

Three cycles (I, II, and III) make up the *pp* chain, with each producing a distinctive neutrino spectrum (see Figure 1). The relative importance of these three cycles depends critically on the Sun's central temperature. The Davis experiment was sensitive primarily to the high-energy ^{8}B neutrinos produced in the *pp*III cycle. As the flux of these neutrinos varies as T_c^{22}, where T_c is the core temperature, Davis hoped to measure that temperature with an accuracy of a few percent.

The flux of neutrinos at Earth is enormous, with about 65 billion neutrinos passing through each square centimeter each second. Nevertheless, be-

cause matter is so transparent to neutrinos, detecting them requires heroic efforts. The clever idea behind the Davis detector was to exploit the reaction

$$^{37}\text{Cl} + \nu_e \rightarrow {}^{37}\text{Ar} + e^- \qquad (2)$$

to measure the ^{8}B solar neutrinos that, though only 0.01 percent of the total flux, interact more readily because of their higher average energy. Because argon is a noble gas, the few atoms of radioactive ^{37}Ar produced in the Davis detector (about one every 2 days) could be flushed from the large volume of perchloroethylene and counted by observing their subsequent decays. As the half life of ^{37}Ar is 35 days, the Davis tank was flushed about once every 2 months. The exotic site, a mile underground, provided a thick rock shield to screen out cosmic rays, which also trigger production of ^{37}Ar. Davis found about one-third the predicted number of ^{8}B neutrinos, a result that many scientists initially attributed to the Sun's core being somewhat cooler than expected.

Neutrino Oscillations

The mystery deepened some years later with data from new experiments, sensitive to different combinations of the neutrinos from the three *pp* cycles. An experiment to measure solar neutrino reactions event by event was performed in a detector mounted in the Kamioka mine in Japan. This detector, originally constructed to search for proton decay, consisted of 4,500 tons of ultrapure water, surrounded by a large array of phototubes. Solar neutrinos scatter off electrons in the water, which then emit a cone of Cerenkov radiation that the phototubes record. Two experiments similar to that done by Davis, but using gallium instead of chlorine, were performed in Russia and Italy. Gallium was chosen because the lowest-energy solar neutrinos, produced in the initial $p + p$ reaction of Figure 1, can change ^{71}Ga into ^{71}Ge.

Together, the chlorine, gallium, and Kamioka experiments determined the principal neutrino fluxes produced by the *pp* chain. Remarkably, the pattern was not compatible with simple adjustments of the SSM, such as a cool solar core: something more interesting was happening.

It had been recognized for many years that the lack of solar neutrinos might have nothing to do with deficiencies in the SSM but might instead reflect a lack of understanding of the properties of neutrinos. In particular, if neutrinos have a small mass—a possibility not envisioned in the current Standard Model of particle physics—a natural explanation could be offered for the observations. Electron neutrinos produced by the nuclear reactions in stars can then transform (or oscillate) into neutrinos of a different flavor, thereby escaping detection on Earth. (The other flavors, muon and tauon neutrinos, are not recorded by the chlorine and gallium detectors and have a probability for interacting in water that is only 15% that of electron neutrinos.) Because solar neutrinos are low in energy and travel a great distance before they are detected on Earth, solar neutrino oscillations can arise for neutrino masses much smaller (e.g., 10^{-6} eV) than those detectable by any other means. Furthermore, it was shown in 1985 that as neutrinos make their way from the Sun's core to its surface, the probability of oscillation can be greatly enhanced. This phenomenon, known as the MSW or Mikheyev-Smirnov-Wolfenstein effect, can distort the spectrum of solar neutrinos in distinctive ways.

Super-Kamiokande and SNO

The possibility of discovering massive neutrinos stimulated new efforts to measure solar neutrinos. In Japan a much more massive successor to the Kamioka experiment, the 50,000- ton water Cerenkov detector Super-Kamiokande, was built by a collaboration of Japanese and American physicists. This experiment not only sharpened the case for solar neutrino oscillations but also provided direct evidence that oscillations alter another source of neutrinos, those produced by cosmic ray interactions in the atmosphere.

A second detector, the Sudbury Neutrino Observatory (SNO), is similar in design, except that the water in SNO's central vessel is "heavy," with the hydrogen replaced by deuterium. The SNO detector, which was built by physicists from the United States, the United Kingdom, and Canada, is located two kilometers underground, within the Creighton nickel mine in Sudbury, Ontario, Canada. The acrylic vessel containing 1,000 tons of heavy water is surrounded by a shield of 7,000 tons of ordinary water, with events in the entire volume viewed by 9,500 photomultiplier tubes. The great depth all but eliminates

cosmic ray backgrounds. In addition, the detector was constructed with extraordinarily pure materials to reduce backgrounds from natural radioactivity: "clean room" conditions had to be maintained in the mine, as the introduction of even a thimbleful of dust in the 10-story-high detector cavity would cause the experiment to fail.

The heavy water allows the experimentalists to measure, in addition to the elastic scattering (ES) of neutrinos off electrons, the following charge-current (CC) and neutral-current (NC) reactions off deuterium:

$$\begin{aligned} \nu_e + d &\rightarrow p + p + e^- \\ \nu_x + d &\rightarrow n + p + \nu_x. \end{aligned} \tag{3}$$

Only electron neutrinos can induce the first (CC) reaction, whereas neutrinos of any type can stimulate the second (NC). The SNO results, announced in April 2002, are shown in Figure 2. Indeed, two-thirds of the solar neutrinos arrive at Earth as muon or tauon neutrinos, not the expected electron neutrinos. Solar neutrinos do oscillate, and neutrinos do have mass.

FIGURE 2

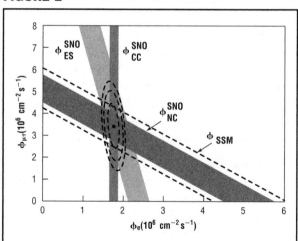

CREDIT: SNO Collaration. "Direct Evidence for Neutrino Flavor Transformation from Neutral-Current Interactions in the Sudbury Sudbury Neurino Observatory," *Physical Review Letters* **89**, 011301 (2002).

Results from the three SNO detector measurements, indicating that approximately one-third of the solar neutrinos are of the electron type and two-thirds are muon or tauon neutrinos.

The implications of the Super-Kamiokande and SNO experiments are startling. The Big Bang filled the universe with a sea of neutrinos: it is now known that the total mass in these neutrinos is at least equal to that in all the visible stars. There is great hope that the pattern of neutrino masses and oscillations emerging from experiments will provide theorists with the clues they need to construct a new Standard Model of particle physics to replace one that is now known to be incomplete.

See also: ASTROPHYSICS; CASE STUDY: SUPER-KAMIOKANDE AND THE DISCOVERY OF NEUTRINO OSCILLATIONS; NEUTRINO; NEUTRINO, DISCOVERY OF; NEUTRINO OSCILLATIONS

Bibliography

Bahcall, J. N. *Neutrino Astrophysics* (Cambridge University Press, Cambridge, UK, 1989).

Super-Kamiokande Collaboration. "Measurements of the Solar Neutrino Flux from Super-Kamiokande's First 300 Days." *Physical Review Letters* **81**, 1158–61 (1998).

SNO Collaboration. "Direct Evidence for Neutrino Flavor Transformation from Neutral-Current Interactions in the Sudbury Neutrino Observatory." *Physical Review Letters* **89** (1), 011301-1-6 (2002).

Wick C. Haxton

NEUTRON, DISCOVERY OF

The discovery of the neutron by James Chadwick in 1932 was the central discovery that opened up the field of nuclear physics in succeeding years. Earlier, physicists believed that the nucleus of every atom was composed of only two elementary particles, the positively charged proton (the nucleus of the hydrogen atom) and the much lighter negatively charged electron; now that no longer could be maintained, although the question of whether or not the neutron was a new elementary particle remained open for more than two years after its discovery.

The first suggestion that a neutron, a particle with no electric charge but with a mass comparable to that of a proton, might exist in the nucleus was made by Ernest Rutherford in a Bakerian Lecture before the Royal Society in London on June 3, 1920, a year after he had succeeded J. J. Thomson as Cavendish Professor of Experimental Physics in Cambridge. Rutherford believed that the alpha particle, the doubly charged, mass-4 nucleus of the helium atom, consisted of four protons and two electrons, and he also believed that he had just found evidence for a new doubly charged, mass-3 nuclear particle consisting of three protons and one electron. Thus, he argued, since two electrons could bind four protons, and one electron three protons, one electron should be able to bind two protons, which would be a new mass-2 isotope of hydrogen, and one electron should be able to combine with one proton, which would be a mass-1 neutron. To Rutherford the neutron also was needed to explain how the nuclei of heavy elements could be built up. Convinced therefore that the neutron should exist, he set some of his research students in search for it experimentally. Chadwick, his right-hand man in the Cavendish Laboratory, also joined that search at odd times throughout the 1920s, to no avail.

As it turned out, the discovery of the neutron was a Tale of Three Cities, with Walther Bothe and his assistant Herbert Becker working in the Physikalisch-Technische Reichsanstalt (Imperial Physical-Technical Institute) in Charlottenburg, a suburb of Berlin; Irène Curie and her husband Frédéric Joliot working in the Institut du Radium in Paris; and James Chadwick working in the Cavendish Laboratory in Cambridge. The story reached its crescendo between June 1930 and February 1932.

By June 1930, when Bothe and Becker published a preliminary report of their experiments, Bothe had worked in the field of nuclear physics for three years, bombarding various light elements with the alpha particles emitted by polonium. He had become convinced that the incident alpha particles excited the nuclei of these elements to higher energy levels, and when they dropped back down to lower energy levels, they emitted high-energy gamma rays. By October he and Becker had found experimentally that such gamma rays were emitted by six light nuclei, including beryllium and boron, whose energies were as high as those of the most energetic gamma rays being emitted spontaneously from heavy radioactive elements.

Bothe and Becker's experiments exerted a powerful influence on Irène Curie and Frédéric Joliot, who had begun to collaborate scientifically in 1928, two years after their marriage. They were drawn to Bothe and Becker's work following an international conference on nuclear physics—the first major one of its kind—that Enrico Fermi organized in Rome in October 1931, which Irène's mother Marie Curie attended and where she heard Bothe give a lecture on his experiments. She also heard Niels Bohr from Copenhagen question whether the laws of conservation of energy and momentum remained valid in the nucleus, and she heard Robert A. Millikan from Pasadena argue strenuously that cosmic rays consist of photons of energy even higher than that of gamma rays. On returning to Paris, she reported these ideas to her daughter and son-in-law, and they all exerted perceptible influences on them.

Curie and Joliot first repeated and then extended Bothe and Becker's experiments, bombarding lithium, beryllium, and boron with polonium alpha particles, and finding that the energy of the gamma rays emitted by beryllium, for example, was much higher than Bothe and Becker had reported; indeed, it lay somewhere between the energies of the gamma rays emitted by radioactive elements and the energies of Millikan's cosmic rays. To investigate the gamma rays emitted by beryllium further, Curie and Joliot inserted sheets of lead and other substances in their path and in front of an ionization chamber. Nothing surprising happened—until they inserted thin sheets of paraffin and other hydrogenous substances, which, in the case of paraffin, caused the ionization current to suddenly double. They reasoned, following a suggestion of Marie Curie, that their high-energy gamma rays were striking and dislodging protons in these hydrogenous (proton-rich) substances which then entered their ionization chamber, greatly increasing its current. They measured the energies of the protons and calculated that to produce them the energies of the incident gamma rays from beryllium and boron had to be 50 and 35 million electron volts, respectively—enormous energies comparable to those of Millikan's cosmic rays. The real problem was that they exceeded the energies that were available from the nuclear reactions that presumably had produced them in the first place. Conservation of energy thus was violated—but that, according to Niels Bohr, was possible in the nuclear realm.

Curie and Joliot reported their experimental findings and conclusions on January 18, 1932, and before the end of the month the journal in which they appeared arrived in Cambridge—where James Chadwick was astonished by them. He then showed their paper to Rutherford, who burst out, "I don't believe it"—a reaction, Chadwick recalled, that he never heard before or since. They agreed that Chadwick should repeat Curie and Joliot's experiments immediately. Chadwick did and convinced himself that their observations were correct but that their interpretation of them was not. The radiations from beryllium and boron did not consist of highly energetic gamma rays but of neutrons.

The neutrons, Chadwick reasoned, were being produced by the nuclear reactions

$$_4Be^9 + {}_2He^4 \rightarrow {}_6C^{12} + {}_0n^1 \text{ and}$$
$$_5B^{11} + {}_2He^4 \rightarrow {}_7N^{14} + {}_0n^1,$$

in other words, the alpha particle ($_2He^4$) was striking either a beryllium or boron nucleus ($_4Be^9$ or $_5B^{11}$) and producing either a carbon or nitrogen nucleus ($_6C^{12}$ or $_7N^{14}$) and a neutron ($_0n^1$), where the subscripts denote atomic numbers and the superscripts atomic masses. These neutrons then were striking and dislodging protons from the paraffin and other hydrogenous substances. From the second reaction, knowing the kinetic energy of the incident alpha particle and the masses of the boron, helium, and nitrogen nuclei, and measuring the kinetic energies of the nitrogen nucleus and neutron, Chadwick assumed that energy was conserved and calculated the mass of the neutron from the mass-energy balance of the reaction, finding it to be 1.0067 atomic mass units (amu). He sent off a preliminary note on his discovery on February 17 and a full report around May 10, 1932.

The fundamental question remaining was whether the neutron was a new elementary particle or a stable proton-electron compound particle. Chadwick's discovery had been conditioned psychologically and institutionally by his long and close association with Rutherford while working in the Cavendish Laboratory. Chadwick's answer to the above question also was conditioned by his knowledge that Rutherford had envisioned the neutron as a stable proton-electron compound in his Bakerian Lecture of 1920—and that was precisely what Chadwick took it to be in 1932. Moreover, he had quantitative support for this view because the sum of the masses of the proton and electron was 1.0078 amu, or 0.0011 amu larger than the mass of the neutron at 1.0067 amu, which translated into a proton-electron binding energy of 1 to 2 million electron volts, taking into account the experimental uncertainties involved in his calculation. To Chadwick that was convincing evidence that the neutron was a stable proton-electron compound and not a new elementary particle.

This question remained in dispute for over two years as Curie and Joliot and also Ernest O. Lawrence in Berkeley weighed in on it as well. Thus, at the seventh Solvay conference in Brussels in October 1933, Chadwick again argued for his value of 1.0067 amu for the mass of the neutron, while Curie and Joliot presented evidence supporting a much higher value of 1.012 amu, and Lawrence a much lower one of 1.0006 amu. Five months later, in March 1934, Lawrence was forced to withdraw his low value, admitting that he had misinterpreted his experiments. The issue was finally settled after Maurice Goldhaber, who had found refuge from Nazi Germany in Cambridge, pointed out to Chadwick in April 1934 that the energetic gamma rays emitted by a certain radioactive nucleus probably could be used to disintegrate the nucleus of heavy hydrogen (which Harold C. Urey had discovered in December 1931) into a proton and a neutron and therefore that the mass of the neutron could be calculated from the mass-energy balance of this reaction. Chadwick tested Goldhaber's idea in a preliminary way some weeks later, found that it worked, and invited Goldhaber to join him in pursuing it further. They reported their finding in August 1934: the mass of the neutron was 1.0080 amu, not quite as high as Curie and Joliot's value, but definitely higher than Chadwick's earlier value of 1.0067 amu. The mass of the neutron was unquestionably greater than the sum of the masses of the proton and electron at 1.0072 amu. The neutron therefore was not a stable proton-electron compound but a new elementary particle. In fact, it was a new *un*stable elementary particle

that would decay spontaneously into a proton, electron, and neutrino.

See also: CHADWICK, JAMES

Bibliography

Brown, A. *The Neutron and the Bomb: A Biography of Sir James Chadwick* (Oxford University Press, Oxford, 1997).

Chadwick, J. "Some Personal Notes on the Discovery of the Neutron" in *Cambridge Physics in the Thirties,* edited by J. Hendry (Adam Hilger, Bristol, 1984).

Feather, N. "The Experimental Discovery of the Neutron" in *Cambridge Physics in the Thirties,* edited by J. Hendry (Adam Hilger, Bristol, 1984).

Stuewer, R. H. "Mass-Energy and the Neutron in the Early Thirties." *Science in Context* **6**, 195–238 (1993).

Stuewer, R. H. "The Seventh Solvay Conference: Nuclear Physics at the Crossroads" in *No Truth Except in the Details: Essays in Honor of Martin J. Klein,* edited by A. J. Kox and D. M. Siegel (Kluwer Academic, Dordrecht, 1995).

Roger H. Stuewer

NOETHER, EMMY

Emmy Noether was a very important mathematician whose work profoundly influenced twentieth-century physics. Her 1918 paper "Invariante Variationsprobleme" contains theorems and their converses that reveal deep, fundamental connections between symmetries and conservation laws. Although she knew that these theorems were of great importance for physics, she regarded the work as something of a departure from the main line of her research. The work was done just after the completion of the general theory of relativity, when Albert Einstein, her Göttingen colleague David Hilbert, and others were seeking a principle of conservation of energy in the general theory of relativity. Her work clarified the issue of conservation of energy and solved the problem of the apparent absence of the law in Einstein's theory. After completing this work, she returned to her main line of research, which was development of abstract algebra. In the publication of her *Collected Works*, editor Nathan Jacobson writes, "Abstract algebra is one of the most distinctive innovations of twentieth century mathematics, and it is largely due to her" (Jacobson, 1983). Concepts and methods of modern algebra are now widely used in all areas of physics. Noether is distinguished as a contributor to physics not only for her work on symmetries and conservation laws but also for her contributions to modern abstract algebra whose importance for twentieth-century physics cannot be overstated.

Emmy Noether was registered as a student in the University of Erlangen in 1904, the first year women were admitted. She is an outstanding example of the fact that, after women had been excluded for centuries, female scientific genius emerged as soon as women were allowed to study in institutions of higher learning. Noether was born March 23, 1882, in Erlangen, Germany. Her father, Max Noether, was Professor of Mathematics at the University of Erlangen. His father was the first in the family to take up academic studies, but he became a merchant, and the family assumed that Max's mathematical talent came from his mother's side. Max Noether was a distinguished and highly respected mathematician who became even more distinguished in later years as the father of Emmy. Emmy attended the Municipal School for Higher Education of Daughters in Erlangen from age seven to age fifteen, when her formal schooling ended, as was normal for girls at that time. She then studied French and English and was certified to teach in girls' schools. To further her education, she sought permission to audit lectures at the University of Erlangen. This was granted, but at the discretion of the lecturer; some professors refused to lecture when a woman was present. In the winter of 1903–1904, she went to Göttingen to attend the lectures of the great mathematicians there. She returned to Erlangen in 1904 when, after years of debate, the university finally admitted women. She wrote a doctoral thesis under the direction of Paul Gordan and was awarded Ph.D. summa cum laude in 1908.

For more than a decade after receiving her Ph.D., Emmy Noether worked unpaid at the University of Erlangen, and later at Göttingen, teaching and doing mathematical research. During this period, she published fifteen papers in important mathematical journals, became a member of the prestigious Cicolo Mathematico di Palermo and the Deutsche Mathematiker Vereinigung (German Association of Mathematics [DMV]), and gave two lectures to the DMV. David Hilbert and Felix Klein in-

German mathematician Emmy Noether (1882–1935) solved the problem of the apparent absence of the law of conservation of energy in Einstein's theory of general relativity. CREDIT: COURTESY OF THE BRYN MAWR COLLEGE LIBRARY.

vited her to join their group at the University of Göttingen, then a world center for mathematics and physics. In 1915, she went to Göttingen after her father died. Prior to his death, his health had not been good, and she had been filling in for him as lecturer at the University of Erlangen. She was refused appointment as lecturer (Privatdocent) by the University of Göttingen in spite of the very strong recommendations of mathematicians such as Hilbert and Klein who were two of the university's most distinguished scholars. The reason the university refused to appoint her was because she was a woman. This so enraged Hilbert that he stormed out of a faculty meeting saying, "I do not see that the sex of a candidate is an argument against her admission as *Privatdocent*. After all, we are a university, not a bathing establishment."

On July 16, 1918, Noether's paper "Invariante Variationsprobleme" was read to the Königliche

Gesellschaft der Wissenschaften zu Göttingen (Royal Society of Sciences of Göttingen) by Felix Klein. Presumably Klein presented it because Noether was not a member of the Society; it seems likely she wasn't even there when the paper was read. Records of the Society were lost in World War II, and it is not known when women were first admitted; counterpart societies in London and Paris did not admit women until after World War II. For example, the Royal Society (London) elected its first female member in 1945, and the Académie des Sciences of Paris did so in 1962.

This paper was immediately seen to be of enormous importance because it shows with great generality logical connections between symmetries and dynamical properties of the fundamental forces of nature. The results led to a deeper understanding of the principle of conservation of energy and of a vast variety of other conservation laws as well. Since the paper gives proofs of theorems and their converses, the insight it provides has led to discoveries of new symmetries of nature following empirical discoveries of new conservation laws. Examples include gauge field symmetries in the Standard Model of particle physics. In 1919 Noether was given *Habilitation* and was finally able to lecture as Privatdocent and be paid. In 1922 she became *Lehrauftrag für Algebra und nicht-beamteter ausserordentlicher Professor* and was paid a salary, although this was not an ordinary faculty appointment.

Historians of mathematics date the creation of modern abstract algebra to the work of Emmy Noether and Emil Artin and their school during 1921 to 1933. Prominent mathematicians came from all over Germany and abroad to consult with Noether and attend her lectures. It was in 1921 that she published "Ideal Theorie in Ringbereichen," which is regarded as a truly monumental work, the first paper published on this vast subject. A pillar of this subject is her 1929 paper "Hyperkomplexe Grössen und Darstellungstheorie," which gave a general representation theory of groups and algebras, valid for arbitrary ground fields.

In 1933, when Adolf Hitler came to power, all Jews were dismissed from the University of Göttingen, and she could no longer teach there. A world-renowned mathematician, she had only two job offers. They were from two leading women's colleges, Bryn Mawr in Pennsylvania and Somerville College in Oxford, UK. Lacking financial resources, they could not offer her a secure position. Somerville offered her room and board and a small stipend. The Bryn Mawr job was subsidized in part by the Rockefeller Foundation. She went to Pennsylvania and was also invited to lecture at the Institute for Advanced Study in Princeton, New Jersey. Weekly, she took the train from Bryn Mawr to Princeton to lecture there. As in Göttingen, her lectures were very well attended. Presumably, if she had not been female, she would have been invited to be a member of the institute.

Following what was expected to be a minor surgery, Emmy Noether died at Bryn Mawr in April 1935. In memoriam Albert Einstein wrote, "In the realm of algebra, in which the most gifted mathematicians have been busy for centuries, she discovered methods which have proved of enormous importance . . . Pure mathematics is, in its way, the poetry of logical ideas. . . . In this effort toward logical beauty, spiritual formulas are discovered necessary for deeper understanding of laws of nature" (Einstein, 1935). Einstein may have been referring, in part, to formulas Noether gave in her 1918 paper. They do indeed yield deeper understanding of laws of nature. What must have been most impressive to Einstein shortly after he completed the general theory of relativity was the deeper understanding of the law of conservation of energy her work offered. Conservation of energy is one of the most important conservation laws in physics. In the early days of the general theory, Einstein, Hilbert, and others were perplexed by the apparent absence of a law of conservation of energy in Einstein's theory. Noether's theories solved this problem. Her Theorem I shows that space-time symmetry implies local energy conservation in classical field theories that are nonrelativistic or governed by the special theory of relativity—a familiar result. On the other hand, in the presence of gravitational fields, the general theory applies, and there is no such local energy conservation law. Instead, her Theorem II shows that space-time symmetry implies a different formula. This is necessary to accommodate gravitational radiation and the principle of equivalence. These results give deeper meaning to the principle of conservation of energy. What is truly extraordinary about her theorems is that they are very general and apply to a vast variety of symmetries and conservation laws.

See also: CONSERVATION LAWS; SYMMETRY PRINCIPLES

Bibliography

Bourbaki, N. *Elements of Historical Mathematics,* translated by J. Meldrum (Springer-Verlag, Berlin, 1994).

Byers, N. "The Life and Times of Emmy Noether: Contributions of Emmy Noether to Particle Physics" in *History of Original Idea and Basic Discoveries in Particle Physics, II,* edited by B. Newman and T. Ypsilantis (Plenum, New York, 1996).

Byers, N. "E. Noether's Discovery of the Deep Connection between Symmetries and Conservation Laws." *Israel Mathematical Conference Proceedings* **12**, 215–230 (1999).

Dick, A. *Emmy Noether (1882–1935),* translated by H. I. Blocher (Birkhauser, Boston, 1981).

Einstein, A. "Letter to the Editor." *New York Times* (May 5, 1935).

Jacobson, N. Introduction to *Emmy Noether, Collected Papers,* edited by N. Jacobson (Springer-Verlag, Berlin, 1983).

UCLA. "Contributions of 20th Century Women to Physics." <http://www.physics.ucla.edu/~ewp/>.

Nina Byers

O

OUTLOOK

Elementary particle physics is a highly developed field. Over 2,000 technically sophisticated Ph.D. scientists in the United States alone work in elementary particle physics. Many features characterize the status, activities, plans, and constraints of the field as it exists.

Scientists have an elaborate and detailed understanding of the constituents and the relationships between the strong, electromagnetic, and weak forces of nature: quarks and leptons interacting through gauge forces. This is embodied in the Standard Model of particle physics, a structure that stands as a major achievement of twentieth-century science. In addition, it is known that new phenomena beyond the Standard Model must exist, most likely including a massive boson (the Higgs particle) that gives rise to the different masses of the particles.

The best approach to this new physics is at the energy frontier, which means experimenting at the highest-energy colliding beam facilities that can be constructed. Experiments and accelerator technology are highly sophisticated. These technologies include very high-speed electronics, real-time processing, and data transmission at very high bandwidth, as well as accelerator beams of unprecedented intensity and stability. New accelerators are in the multibillion-dollar category and take well beyond a decade to develop. Observational cosmology has also revealed phenomena beyond the Standard Model, such as dark matter, dark energy, and inflation.

The benefits of discoveries in particle physics are influential beyond contributions to the field. Many other fields of physics use discoveries made by particle physics. Atomic physics and certain areas of astrophysics are among those fields. In addition, many by-products of work in elementary particle physics find their way into the public sector, such as medical imaging and the World Wide Web.

Many theorists in particle physics work on string theory (where particles are viewed as incredibly tiny vibrating loops), which holds the promise of incorporating gravity among the other forces. However, the fear is that string theory may be untestable because, for consistency, it requires several extra spatial dimensions that might be so small that laboratory-based studies will never be sensitive to them.

Particle physicists have traditionally attempted to isolate the fundamental constituents in nature and study their interactions. Indeed, it remains a major thrust of the field. It is most enlightening to know what particles exist, what symmetry patterns they obey, and how to accelerate them to very high energies and perform controlled experiments with them. This field has continually revealed important new phenomena and states of nature, and this knowledge provides richer and more complete picture of the world.

Over the past forty years, many of the most important advances resulted from accelerator-based experiments. During the 1960s, the violation of CP symmetry was found: for the first time, the asymmetric decay of the neutral *K* meson was observed, and it distinguished matter from antimatter. Immediately this led to the question of whether the violation was connected to the matter-antimatter asymmetry in the universe. This is still a mystery. Through the scattering of electrons from protons, pointlike constituents within the proton, later identified as the quarks, were discovered. Both of these discoveries were unexpected, and they had a very significant impact on the thinking and on the development of the field.

In the 1970s came the discovery of two new quarks, charm and bottom, and a new lepton, the tau. These were striking and also largely unexpected discoveries. In addition, during this same timeframe the gluon, the force carrier of the strong interaction, was isolated. In the 1980s the force carriers of the weak interaction, the *W* and *Z*, were finally discovered; they had been sought for decades, but by the 1980s theory was able to predict their masses, and experiments found them in just the right place. Much was also learned about particles containing the bottom quark, and this allowed the determination of quark couplings with even more precision, firming up the Standard Model of particle physics.

The 1990s saw the observation of the top quark (the last one in the Standard Model) with a mass much larger than that of any of the other quarks. This brings home the mystery of the vastly different masses of the fundamental constituents of matter; however, the mass was in the range allowed by consistency with all other measurements and the Standard Model. Physicists were able to copiously produce and accurately study the decays of the *Z* boson, pinning down the interactions of the quarks and leptons with remarkable precision. In quark decay, a second manifestation of CP violation was also clearly observed. This came from very difficult measurements, and it helped verify the Standard Model means of accommodating CP nonconservation.

At the onset of the twenty-first century, CP violation with *B* mesons was observed; these measurements required the construction of what are commonly called *B* factories to make *B*'s in sufficient quantities. The results were expected in the Standard Model.

Particle Physics is at a crossroad. Tremendous progress has been made over the past few decades: the questions now being asking could not have been conceived of earlier. Since so much is known about the constituents of matter and their patterns, the focus has shifted to explaining their origin. The forces and their relative strengths are known at low energies, and the possibility exists that at very high energies, there is only a single force. However, one must be open to surprises. The major activities today in accelerator-based particle physics are the search for the Higgs boson, the search for supersymmetric particles, the study of CP violation, and the study of neutrino oscillations.

Particle physics is strongly linked to cosmology. From observations of the cosmos, the fact that our universe has been expanding and cooling since the Big Bang has been learned. At accelerators, particle interactions are studied that occurred in the universe when it was less than a microsecond old. And, the understanding of particle physics has had great impact on cosmology. For example, from the knowledge of how forces change with particle energies, scientists are boldly able to extrapolate the forces studied at scales of a few hundred giga electron volts to much earlier epochs in the Big Bang. Such an approach is valid but must be taken with caution. Particle physics thus enables cosmology, and cosmology demonstrates the promise of new particle physics.

There are three major puzzles in cosmology that have a direct bearing on particle physics. The first is the existence of dark matter. Dark matter was discovered by simply using Newtonian gravity to analyze the motion of stars in galaxies and galaxies in clusters of galaxies. This analysis makes it clear that there is more than meets the eye, that is, the dominant matter that holds galaxies and galaxy clusters together is not luminous. It is also believed that this nonluminous matter is not the stuff that humans are made of or that has been studied in the laboratory. This belief is supported by the successful theory of how the light elements were created during the first few minutes of the universe. The prediction for the relative fractions of H, He, Li, and D holds together only if the density of such ordinary matter is much smaller than the density of all the matter, which can be determined by a number of independent techniques.

Particle physics provides good candidates for particles that could comprise this dark matter, particles that were created very early in the history of the universe and have survived to the present day. The extension of the Standard Model known as supersymmetry posits a new set of particles that mirrors known particles but with different quantum numbers. The lightest of these is likely stable, that is, it has a lifetime exceeding that of the universe, and if it has a mass of approximately 100 GeV, it could comprise the dark matter and be detectible by a number of means. The search for such particles is a major objective of the field.

Another puzzle from cosmology that directly impacts particle physics is that of dark energy, a dominant component of the universe whose constitution is completely unknown and which is apparently causing the universe's expansion to accelerate. This could result from some residual energy in the vacuum, or it might be some new dynamical field. There are new observations of the cosmos that may help unravel this phenomenon, but how to address this puzzle within the context of particle physics is not yet known. The third deep puzzle beyond the Standard Model is the apparent acausal nature of the universe. If one looks at the sky with detectors sensitive to microwave radiation, one sees the radiation left over from the birth of the universe, the so-called cosmic microwave background radiation. This radiation split off from the hot soup of photons, electrons, and protons less than a million years after the birth of the universe and has been traveling unperturbed, as it cools, to the present day. The surprise is that everywhere one looks one sees, to very high precision, the *same* temperature for this radiation, about 3° absolute. The puzzle is that regions far from each other (just a few degrees on the sky) have never been in causal contact with each other since the birth of the universe. Accelerated expansion very early in the history of the universe has been postulated to explain why the universe is so extremely smooth: this theory says that everything was initially in causal contact, but by expansion faster than the speed of light, these regions that had come into thermal equilibrium with each other disappeared from each other's causally connected regions. Further tests of cosmic microwave background lend more support to this notion, called inflation, and definitive tests are in the planning stage.

There is another connection: it is now known, as a result of some very incisive experiments studying neutrinos emitted from the Sun and as by-products of the interactions of cosmic rays in the atmosphere, that neutrinos oscillate. They have a rich structure not unlike what has been known for a long time about the weak interactions of the quarks. Experiments under construction at accelerators will further characterize this new system.

To get to every smaller distance scale requires, through a basic relation in quantum mechanics, ever higher energies. This means larger, more costly devices and inevitably international machines. The Large Hadron Collider (LHC) in Geneva, Switzerland, will be the next major instrument in the field. It has an energy about seven times greater than the Tevatron currently operating at Fermilab in Batavia, Illinois, and can make collisions more than ten times more frequently. A wealth of new phenomena will very likely be discovered and explored at that facility, which is scheduled to begin operations around the year 2007. The Higgs boson and several of the supersymmetric partners are prime candidates for discovery.

Beyond the LHC, a worldwide consensus is developing for a facility that collides electrons and positrons at an energy of about 1,000 billion electron volts. This machine could perform precision studies of the Higgs boson and of some supersymmetric states within its energy range. It is technically very challenging and will require the intellectual and financial support of the United States, Europe, and Japan. It is being planned from the start as an international facility.

There is another class of experiments that deals with focused in-depth studies. These experiments tend to be smaller and more limited in scope than those searches discussed above. These include studies of particle decays in beams. Here one can create a well-defined beam of a known particle, such as a *K* meson, and build a detector to study a particular decay that is expected to occur rarely. These experiments could not be performed in the collider environment: to reach needed sensitivity, it requires 10,000 or more times the number of protons that can be obatined in collider experiments. Another example of a special purpose experiment is building a storage ring to capture muons and make a precison measurement of the structure of the muon. Recently such an experiment was performed at the Brookhaven National Laboratory (BNL) with intriguing results. Dedicated experiments are mounted to study and search for the violation of certain fundamental symmetries: How do matter and antimatter behave differently? How does nature distinguish left from right?

As the scale of experiments grew ever larger and as there were fewer focused experiments, the field experienced some changes. The physics that the largest facilities reveal cannot be studied any other way: it simply takes large accelerators and large collaborations to explore that kind of science. However, as the domain of particle physics is expanding, there are other opportunities. For example, smaller efforts in observational cosmology address fundamental science, as do experiments performed deep underground to look for proton decay or to detect dark matter particles. There are many new initiatives to explore this science, and the prospects for new discovery are bright.

See also: DARK MATTER; FUNDING OF PARTICLE PHYSICS; INTERNATIONAL NATURE OF PARTICLE PHYSICS; STRING THEORY; SUPERSYMMETRY; UNIFIED THEORIES

Bibliography

DOE/NSF High Energy Physics Advisory Panel Report (January 2002). <http://doe-hep.hep.net/HEPAP/lrp_report0102.pdf>.

Kane, G. *Modern Elementary Particle Physics: The Fundamental Particles and Forces* (Addison Wesley, Reading, MA, 1995).

Weinberg, S. *The Discovery of Subatomic Particles* (Scientific American Press, New York, 1983).

Winstein, B., et al. *Elementary Particle Physics: Revealing the Secrets of Energy and Matter* (National Academy Press, Washington, DC, 1998).

Bruce Winstein

P

PARITY, NONCONSERVATION OF

The 1957 discovery that parity was not conserved in weak interactions hit the world of particle physics like a minibombshell. Before then it had been assumed that parity was conserved in all interactions. However, on close scrutiny this assumption turned out to have no firm foundation, and nature, as it were, took advantage of the loophole. Even now some of the repercussions of parity violation are not understood at a profound level, but nevertheless parity violation has been incorporated successfully into the Standard Model.

Definition of Parity

Parity is concerned with the inversion of the space coordinates:

$$x \rightarrow -x, \qquad y \rightarrow -y, \qquad z \rightarrow -z.$$

The question is, "Are the laws of nature invariant under this operation?" In a world that differs from our world in this way, are the laws of nature the same as those we know, or not? It is important to remark that other questions of a very similar type have very clear answers. For example, one knows that the laws of nature are the same when a rotation is performed in space. Invariance under rotations is inti-

mately related to the conservation of angular momentum, and there is abundant evidence that this holds. But it turns out, surprisingly, that parity is different: the laws of nature are not quite invariant under parity.

The Puzzle of K Decays

The first hint of trouble came with the observation of the decay of neutral kaon particles. These are spinless particles that are generally expected to have a parity quantum number $\eta = \pm 1$; in fact, the kaons, and also the pions into which they decay, actually have negative parity, $\eta = -1$. The problem began when it was found that kaons can decay in two distinct ways—either into two pions or into three:

$$K^0 \rightarrow \pi^+ + \pi^-; \, K^0 \rightarrow \pi^+ + \pi^- + \pi^0.$$

In these decays the parity of the decay products in each case is simply the product of the parities of the pions, that is, $(-1)^2 = +1$ in the first decay and $(-1)^3 = -1$ in the second. If parity were conserved, since the kaon has negative parity, only the second decay would be allowed (and, of course, if it had positive parity, then only the first decay). The fact that K^0 decays into two final states with opposite parity posed a real puzzle in 1956.

Parity Violation in Beta Decay

The solution to this puzzle was suggested by two Chinese-American physicists, Tsung Dao Lee and Chen Ning Yang, who made the highly interesting suggestion that weak interactions, as a class, do not in general conserve parity. The previously mentioned decays of the kaon result from weak interactions. This proposal clearly solves the problem of K decays (the solution being that the first decay above violates parity, while the second conserves it), but it also implies that parity is violated in nuclear beta decay, since this is also due to the weak interaction. The most compelling feature of this prediction is that it should be possible to see directly—with the naked eye, as it were—whether or not parity is conserved in beta decay. Nuclear beta decay is, in essence, neutron decay (the beta particles being the electrons e^-):

$$n \rightarrow p + e^- + \bar{\nu}_e.$$

Lee and Yang proposed an experiment that was carried out by Chien-Shiung Wu and collaborators in 1957. They investigated the decay of ^{60}Co nuclei, in which a neutron decays into a proton, emitting an electron, as described above. The decaying nuclei are polarized by placing them in a strong magnetic field. This means that the neutron spin is aligned in space. What Wu and her collaborators found was that the electrons are emitted in an *opposite direction* to the nuclear spin. This is direct proof of parity violation. Perhaps the easiest way to see this is to note that the parity operation just defined may be expressed as a combination of two operations: (1) $x \rightarrow -x$, $y \rightarrow -y$, z unchanged, and (2) x and y unchanged, $z \rightarrow -z$.

The first operation is simply a rotation about the z-axis through 180^0—and it is already known that the laws of nature are invariant under rotations. The second operation is a mirror reflection, the mirror being in the xy-plane. A test of invariance under parity is therefore a test of invariance under mirror reflection. Figure 1 shows ^{60}Co decay, with the nuclear spin aligned in a magnetic field generated by a solenoid. The experimental result (left half of the diagram) is that the electron momentum is *antiparallel* to the nuclear spin. In the mirror (right half of diagram), however, the magnetic field is *reversed* because the solenoid windings are reversed, but the electron momentum is *not* reversed, so in the "mirror experiment" the elec-

FIGURE 1

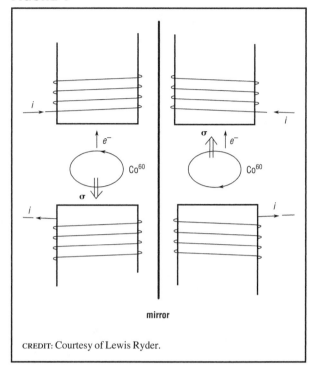

CREDIT: Courtesy of Lewis Ryder.

Nuclear beta decay and its mirror image.

tron momentum and nuclear spin would be parallel. The experiment and its mirror image are thus different: beta decay violates parity. It distinguishes between left and right and is the only fundamental interaction to do so. (If it were the case that weak interactions, and therefore beta decay, conserved parity, then in the Wu experiment it would have been observed that electrons were emitted with no preferential direction, that is, they would travel equally in all directions. This would clearly look the same in a mirror.)

Left-handed Neutrinos

Why does beta decay violate parity? It turns out that the blame can be laid on the neutrino. In 1957 Lee and Yang, and Lev Landau in the Soviet Union and Abdus Salam in England, made the suggestion that the neutrino was a purely left-handed particle. That is, the projection of its spin in the direction of motion is always negative. This immediately has the consequence that any experiment involving neutrinos is bound to violate parity. The experiment observed in a mirror is bound to look different since a (left-handed) neutrino, looked at in the mirror, will be a right-handed neutrino, which does not exist. Now this suggestion is not a trivial one for it can only hold if the neutrino has no rest mass. To see this, consider a neutrino observed in the lab traveling at a speed of v ($<c$). It is left-handed, so its spin is in the opposite direction of its momentum. Now consider the situation in a moving frame of reference. If one "overtakes" the neutrino, its velocity will be reversed, but not its spin, so in this frame it would then appear right-handed. If the neutrino is only ever to be left-handed, this observation must be forbidden. It must be impossible to overtake the neutrino, which means that it must travel at the speed of light, and this, in turn, means it must be massless (like a photon). Traditionally, neutrinos have been considered to be massless, but recently doubt has been cast on this assumption, particularly in the theory of neutrino oscillations. Interestingly, this idea does not deal a deathblow to parity violation. For, even if neutrinos do have a mass (and oscillations are therefore possible), it may be so small (in comparison with their kinetic energy) that they behave as if they were massless.

The way in which parity violation is built into the Standard Model is actually quite straightforward. It is simply stated, as an axiom, that the fundamental leptons in electroweak theory are the (weak isospin) doublet of left-handed particles ($\bar{v}_e, e_L$) and the right-handed singlet e_R. The neutrino is purely left-handed, and the left-handed and right-handed parts of the electron enter the theory on a different footing. This automatically yields parity violation in beta decay, and in the weak interactions in general, and almost miraculously nowhere else, just as desired!

See also: BOSON, HIGGS; ELECTROWEAK SYMMETRY BREAKING; HIGGS PHENOMENON; QUANTUM CHROMODYNAMICS; SUPER-SYMMETRY; SYMMETRY PRINCIPLES

Bibliography

Chen, N. Y. "The Law of Parity Conservation and other Symmetry Laws of Physics" in *Nobel Lectures: Physics 1942–1962* (Elsevier, Amsterdam, 1964).

Okun, L. B. α, β, γ . . . Z: A Primer in Particle Physics (Harwood Academic Publishers, Chur, Switzerland, 1987).

Tsung, D. L. "Weak interactions and Nonconservation of Parity" in *Nobel Lectures: Physics 1942–1962* (Elsevier, Amsterdam, 1964).

Lewis Ryder

PARTICLE

In day-to-day usage, the term particle is used to describe very small objects. Physicists use the term particle in a more precise way: they use it to describe the behavior of an object without reference to any internal structure. Thus astronomers might refer to the Earth or the Sun as a particle if they are studying its motion in a crude enough way such that, say, tidal forces are not important.

Originally, scientists thought that atoms were elementary objects with no structure. As such, they would have been the ultimate particles. Indeed, the word "atom" means uncuttable. However, shortly after the existence of atoms was firmly established (perhaps most convincingly by Einstein's explanation of the Brownian motion in 1905), Ernest Rutherford, bombarding gold atoms with alpha particles, discovered that the atom is largely empty space, with electrons revolving around a tiny nucleus. Over the next decades, the proton and neutron were discovered and established as the basic entities making up the nucleus.

With the development of quantum mechanics, the notion of particle took on a new aspect. Particles such as the electron were seen to exhibit characteristics of waves. Electrons exhibited interference phenomena, much like light, and could undergo diffraction. On the other hand, as first postulated by Albert Einstein, light often exhibited particle characteristics. For example, light comes in discrete packets of energy and momentum. Electromagnetic radiation can be thought of as consisting of large numbers of particles, called "photons." This discrete character of light currently underlies much of electronics technology.

Traditionally, the objects that make up the atom, the electron, proton and neutron, as well as the photon, are referred to as "elementary particles." The electron and photon are, as far as we know, without structure. They are completely described by their mass, charge, and spin angular momentum. The electron has nonzero mass and charge, as well as $\frac{1}{2}$ unit of spin angular momentum; the photon has neither mass nor charge and carries one unit of spin angular momentum. The proton has mass nearly 2,000 times larger than that of the electron, the same spin, and opposite electric charge. To ask whether particles such as electrons or protons have structure requires microscopes capable of resolving extremely short distances. Particle accelerators are such microscopes.

The most energetic accelerators today can resolve structures as small as 10^{-17} cm. On this scale, neither the electron nor the photon exhibits structure. The proton and neutron, however, have a size of about 10^{-13} cm. Experiments in the late 1960s, similar to Rutherford's in spirit, but involving high-energy electrons scattered off nuclei, demonstrated that nuclei are made of smaller entities called quarks. While there have occasionally been suggestions that the quarks themselves might have structure, there is so far no evidence for this, and some theoretical arguments have been put forward that they do not.

Many other particles have been discovered in cosmic rays and accelerators. Most of these are known as "hadrons" and are strongly interacting like the proton and neutron. They are composed of quarks and can be put together in tables similar to the periodic table. Six others, known as "leptons," are more similar to the electron. These include the muon and tau particles. These have the same electric charge as the electron, and, like the electron, they experience the electromagnetic and weak force (responsible for beta decay) but not the strong force. The muon is about 200 times as massive as the electron; the tau particle about 3,000 times. The other three are the neutrinos. The neutrinos are electrically neutral and extremely light (recent experiments show that the neutrinos have mass less than one millionth that of the electron). They experience only the weak force and thus don't readily stop in matter.

For every known particle, there is also an antiparticle. This is a particle of the same mass but opposite charge. For example, the antiparticle of the electron is the positron, which has been well studied experimentally. The antiparticle of the proton, the antiproton, was discovered in the early 1950s (when the proton was still widely believed to be an elementary particle). More recently, antihydrogen has been created in the laboratory. Some particles, like the photon, are their own antiparticles (neutrinos are not).

The fact that light has both particle and wave aspects emerges immediately if one applies the rules of quantum mechanics to Maxwell's theory of electricity and magnetism. The resulting theory is known as quantum electrodynamics. Just as the photon is described in terms of electric and magnetic fields, the electron is also described by a field. In this theory, Einstein's principle of relativity, and knowledge of the charge and spin of the electron fully determine its other properties. For example, the fact that electrons obey the Pauli exclusion principle is automatic. In this theory it is possible to calculate the properties of the electron and photon, as well as of simple atoms, to extraordinary precision. The magnetic moment of the electron can be calculated to twelve significant figures and measured with comparable accuracy. Beginning in the 1970s, quantum electrodynamics was generalized to a theory that includes the weak and strong interactions. The equations of this theory are similar to Maxwell's equations (Maxwell's equations are a special case). The quantization of this theory leads to a theory called the Standard Model. In addition to the quarks and leptons and the photon, the Standard Model predicts three other particles analogous to the photon, called the $W^{\pm}$ and the Z. Like the photon, these particles carry spin one, but they are massive, and the W bosons are charged. Just as the photon is responsible for the electromagnetic force, the W and Z particles are responsible for the weak force. The model predicts the masses of these particles to be approximately eighty and ninety times the mass of the proton, respectively. It also predicts their half-lives (these particles are very radioactive, with half-lives of order 10^{-23} seconds). Both the masses and lifetimes of these particles have been measured in particle accelerators (LEP at the European Laboratory for Particle Physics [CERN] in Geneva and the SLC at the Stanford Linear Accelerator Center) to the level of parts per thousand and agree closely with the theory.

There is one other particle predicted by the Standard Model, known as the Higgs boson, which has yet to be observed. Within the theory, this particle is responsible for the masses of the other elementary particles. From the precision measurements of the properties of the W and Z bosons, the mass of this particle is predicted to lie between approximately 80 and 230 times the mass of the proton. Searches for this particle have so far excluded a Higgs particle with masses about 115 times that of the proton. Accelerators at Fermilab and CERN will search for this crucial particle over the next few years.

Einstein's theory of general relativity predicts another particle, the graviton. In much the same way that the photon is responsible for the electromagnetic force, the graviton is responsible for the gravitational force. This particle interacts so weakly with matter that it is not possible to count gravitons one at a time, as one can photons. However, with very sensitive instruments it should be possible to detect gravitational waves emitted by violent events in the universe. This is the goal of the LIGO project.

Is this the sum total of possible particles? Probably not. Experiments are searching for a variety of particles. Among these are a particle called the axion and one called the neutralino. These have been proposed as possible solutions to puzzles in the Standard Model. Either of these could well comprise the dark matter which astronomers believe constitutes most of the mass of the universe. Other particles, predicted by supersymmetry, are subjects of search at large particle accelerators.

See also: ANTIPROTON, DISCOVERY OF; AXION; BOSON, GAUGE; BOSON, HIGGS; LEPTON; MUON, DISCOVERY OF; NEUTRINO; NEUTRINO, DISCOVERY OF; NEUTRINO, SOLAR; NEUTRON, DISCOVERY OF; POSITRON, DISCOVERY OF; QUARKS

Bibliography

Kane, G. L. *The Particle Garden: Our Universe as Understood by Particle Physicists* (Addison-Wesley, Reading, MA, 1995).

Michael Dine

PARTICLE IDENTIFICATION

Elementary particles are studied by looking at the production and decay of particles in high-energy collisions, where the initial state energy is converted (via $E = mc^2$) into the mass of new particles. These collisions and decays obey the laws of probability so that to build up a picture of reality, a library of the events must be accumulated that are the observed outcomes of collisions. A collision will typically produce a number of particles whose identities must be unraveled to label a particular event correctly.

Elementary particles have a wide range of masses, interactions, and average lifetimes against decay. Table 1 gives a representative selection of particles—for each charged particle there is an antiparticle with the opposite electric charge. The average distance to decay is calculated allowing for relativistic time dilation at a momentum equal to 10 times the particle mass.

Ionization of the Detector Medium

Electrically charged particles ionize (remove electrons from) matter as they pass through it. This disturbance can be used to detect the path they follow. If the particle detector, typically a large volume of a suitable gas such as argon, is placed in a magnetic field, the trajectory is deflected into a circular path (by the same law that makes electric motors turn when an electric current (moving charge) passes

TABLE 1

Masses, Interactions, and Average Lifetimes against Decay for Selected Particles					
Particle	Charge (proton charge = 1)	Mass (proton mass = 1)	Absorption length in iron (cm)	Mean lifetime (sec)	Average distance to decay (cm)
Proton (p)	+1	1	20	Stable	Infinite
Electron(e^-)	−1	0.0005	2	Stable	Infinite
Muon (μ^-)	−1	0.113	Very long	2×10^{-6}	600,000
Pion (π^+)	+1	0.149	30	3×10^{-8}	8,000
Kaon (K^+)	+1	0.527	30	1×10^{-8}	4,000
Photon (γ)	0	0	2	Stable	Infinite
K^0_s	0	0.531	—	9×10^{-11}	30
ϕ^0	0	1.086	—	1×10^{-22}	4×10^{-11}
B_s	0	5.724	—	1×10^{-12}	0.4

CREDIT: Courtesy of David H. Saxon.

through a magnetic field). The radius of curvature depends on the particle momentum (mass times velocity with relativistic corrections). Thus the momenta and directions of charged particles can be measured.

The density of ionization along the track depends on the particle velocity. (Slow particles have a longer time available to disturb each atom as they pass and so are more efficient at ionizing gases.) So, for a known measured momentum the particle velocity will depend on the mass. Thus, if the momentum is measured by the curvature of a track, and the ionization density provides the information about the velocity, the ionization can be compared to that expected for e^+, μ^+, π^+, K^+, and p, and the particle's identity can be inferred. This measurement is rather delicate as the differences in ionization are not large compared to the sample-by-sample fluctuations obtained from measurements taken while moving down the track. So, other methods are preferred.

Cherenkov Radiation

A direct method of inferring the particle velocity, and hence the mass, once the momentum is known, is to look for Cherenkov radiation. It is not possible for a particle to travel faster than light in a vacuum, but in a material medium, light has a reduced velocity, and a particle may exceed that (reduced) light velocity in the medium. In doing so, it gives out a flash of blue/ultraviolet light, analogous to the sonic boom of a plane traveling faster than sound. This has been used to infer velocities, and hence particle identities, most notably in the case of electrons, which have a dramatically lower mass than any other charged particles and so travel much faster for the same momentum.

Electrons and Muons

Electrons and muons are produced only rarely in particle collisions and often arise from the decay of heavy quarks. It is therefore important that they are identified efficiently and unambiguously. We take advantage of their very different interaction lengths in solid material (Table 1). Compared to more commonly produced particles such as pions and kaons, electrons are absorbed after only a short

distance in material, and muons pass through great thicknesses without being affected.

Figure 1 shows an event produced in an e^+e^- collision. (The incident particles enter at right angles to the plane displayed and annihilate at the center of the detector to produce new matter.) A back-to-back quark-antiquark pair is produced. Each one materializes as a jet of charged and neutral particles. The circular tracks of charged particles passing through a gas are seen (some of them identified by ionization measurements as kaons), followed by signals from a detector made of successive layers of passive metal absorber interleaved with active detector layers that show whether interactions have occurred. (Such a detector is known as a calorimeter because one can measure the total energy of the incident particle by adding up the signals from all the layers.) The electron has interactions in the inner detector layer only, the pions and kaons give signals also in the outer detector layer.

Vertex Detectors

Neutral particles leave no tracks. One can detect them only by absorption in a calorimeter (which treats photons just like electrons) or by their decay in flight to charged particles. The blow-up in Figure 1 shows a complex chain of events. By placing several layers of extremely precise position measurements close to the production vertex, one can reconstruct the outgoing particle trajectories and show whether they originated from the primary interaction point or from a separate vertex arising from the decay in flight of a heavier particle. Such vertex detectors are readily made using silicon-microchip technology and can measure to a precision of a few millionths of a meter.

In Figure 1a, $\bar{B}_s$ is produced at the interaction point (IP) and decays after a few millimeters to a D_s^+ plus a π^-. The D_s^+ travels a further 0.4 mm and decays to $K^+K^-\pi^+$. (Note that the π^+ track, for example, does not point at the B_s decay and so could not have been produced there.)

Mass of Decaying Particle

From the measured momenta and directions of the outgoing kaons one can reconstruct the mass a

FIGURE 1

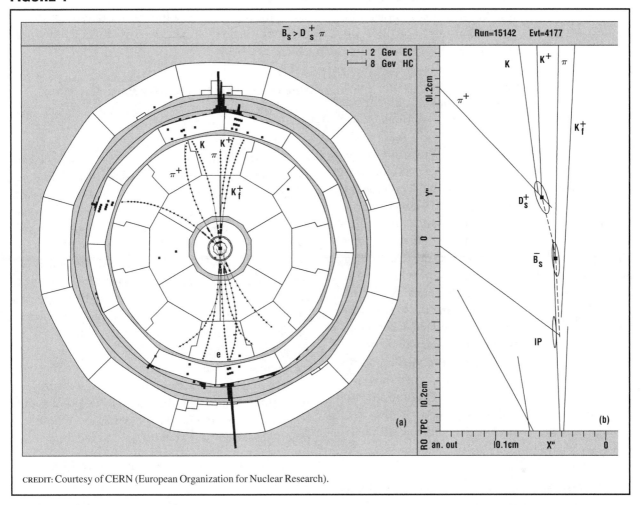

CREDIT: Courtesy of CERN (European Organization for Nuclear Research).

View of an e^+e^- annihilation event. The detector layers, starting at the center, are (a) vertex detector, (b) large gaseous detector with tracks curved by magnetic field, (c) inner calorimeter layer with indicated signals, (d) coil to produce magnetic field, (e) iron layers of magnet structure with indicated calorimeter signals. Blow up on right shows details of vertex region—note scales.

possible parent particle that produced them. Finding a mass consistent (within the accuracy of the measurement) with the expected ϕ^0 mass, one identifies the K^+K^- pair as originating from a ϕ^0, itself produced in the D_s^+ decay. The ϕ^0 lives only fleetingly and so travels a negligible distance before itself decaying to the K^+K^- pair.

See also: ANTIPROTON, DISCOVERY OF; ELECTRON, DISCOVERY OF; NEUTRINO, DISCOVERY OF; QUARKS, DISCOVERY OF

Bibliography

Fernow, R. *Introduction to Experimental Particle Physics* (Cambridge University Press, Cambridge, UK, 1986).

Fraser, G. *The Particle Century* (Institute of Physics, Bristol, UK, 1998).

Kleinknecht, K. *Detectors for Particle Radiation* (Cambridge University Press, Cambridge, UK, 1998).

David H. Saxon

PARTICLE PHYSICS, ELEMENTARY

Elementary particle physics is the investigation of nature at a level below current understanding. Its driving questions are as follows: What are the basic constituents needed to build everything that is observable? How do these constituents interact with each other? What is the relationship among the constituents and the interactions? This quest has progressed from everyday objects to molecules, molecules to atoms, atoms to electrons and nuclei, nuclei to protons and neutrons, and protons and neutrons to quarks. Does this progression to smaller and smaller components go on forever, or is there in the end a single fundamental particle? Current understanding of elementary particle physics is expressed in what is called the Standard Model. The universe consists of quarks and particles related to electrons called leptons. These basic constituents interact under the influence of only four forces: the strong force binding quarks to make protons, the electromagnetic force holding electrons to protons to give atoms, the weak force responsible for radioactive decay, and the gravitational force tying the Earth to the Sun. Why do these particular forces and constituents exist? Are these constituents fundamental, or are they made of more basic objects? Are the four forces fundamental, or are they different attributes of a more basic force? Are the constituents and interactions observed only different patterns in the basic geometry of space and time? The ultimate goal of elementary particle physics is to have one single explanation for everything.

The process of asking and answering questions is the most basic human endeavor. People design experiments and decide which measurements to make. People determine the implications of those measurements. People imagine theories that they connect to reality through their interpretation of these measurements. In the ensuing competition among ideas, the theory accepted as being closest to underlying reality is the simplest theory that explains existing measurements, predicts the results of new measurements, and suggests experiments to test those predictions. Since elementary particle physics is an expression of the human desire to understand and control nature, its record is as old as recorded human history. In the past, elementary particle physicists have been called philosophers, natural philosophers, chemists, or physicists. These scientists became elementary particle physicists by searching for a new level of reality underlying the complex behavior of matter at the frontiers of their knowledge.

Because the fundamental questions remain the same, history provides a framework that helps to understand the current perspective of elementary particle physics.

Illustrations of Elementary Particle Physics from the Far Past

Ancient Greece: Seeking Unification

The desire to understand the universe in terms of its constituent particles, its fundamental interactions, or the geometry of space and time is illustrated by three rival elementary particle theories from ancient Greece. One focused on the constituents and held that everything could be explained by atoms (the fundamental constituents) and the space between them, the void. The interactions of objects could be explained in terms of the innate properties of the atoms such as their shape or smoothness. Another theory made four basic interactions of nature fundamental. Those interactions had the properties exemplified by Earth, Air, Fire, and Water. The objects were different because of their combination of these interactions. The third theory held that geometry determined the fundamental nature of both the objects and their interactions.

From the Renaissance to the Nineteenth Century: The Basics

Progress in mathematics, technology, observational techniques, and intellectual rigor over the next 2000 years led to the work of physicists such as Nicolaus Copernicus, René Descartes, Galileo Galilei, Johannes Kepler, and Isaac Newton in the sixteenth and seventeenth centuries. Their investigations resulted in a general theory of force as a description of the interaction between any objects, together with a specific mathematical description of the gravitational force. The concept of mass was invented to characterize both a property of objects and the strength of the gravitational interaction. The search for the fundamental constituents of objects, called atoms or elements, was begun by the next generation of chemists exemplified by John Dalton, Joseph Priestly, and Antoine-Laurent Lavoisier. It took another 200 years for physicists such as Benjamin Franklin, Charles-Augustin Coulomb, Michael Faraday, and James Clerk Maxwell to characterize the electric and magnetic forces well enough to combine them into a unified theory of electromagnetism. Along the way, the concept of charge was invented to characterize both a property of objects and the strength of the electric interaction.

Nineteenth Century: Domination and Puzzles

By the end of the nineteenth century, physicists had developed a powerful theory of the universe. All of the esoteric experiments and theoretical work had paid off handsomely for society. The successful synthesis of classical mechanics had given rise to the first Industrial Revolution that was still in full swing. Civil engineers were building larger and more useful structures, railroads and steamships made the large-scale movement of goods and people possible, personal transportation by bicycle and automobile contributed to a growing sense of freedom, and soon people would be able to fly. Atoms were a theoretical construct of debatable reality, but the atomic theory of elements and their classification in the periodic table gave rise to a thriving chemical industry. Synthetic substances were being constructed and manufactured. The profound effects of the equally successful synthesis of electromagnetism had just initiated the second Industrial Revolution. Messages could be sent across long distances first by telegraph and then by telephone. Cities and even individual homes could be illuminated by electric lighting. Large electrical systems turned mechanical energy into electrical energy to run machines. Even the very abstract concept of electromagnetic waves would find practical application in radio.

There were some anomalies that worried the elementary particle physicists. The theory of classical mechanics was fundamentally inconsistent with electromagnetism. Both theories were very successful, but both could not be correct. It was difficult to completely understand the nature of light. It must be a wave as predicted by electromagnetic theory. However, its behavior did not correspond to either a wave or particle. Furthermore, when an element was heated, it emitted light of only certain colors. Another element would emit different colors of light. This was useful for identifying elements, but no one had shown that the theory explained this behavior. It was clear that matter contained two types of electric charges, but no stable configuration of positive

and negative charges could be constructed that allowed this to happen. At an even more fundamental level, what were charge and mass? Were they related? Could a particle have any amount, or was there a smallest possible unit? It was also known that not all matter was stable. What caused radioactive decay? The Sun was a puzzle. Geological time was very long, and there was no known energy source that could keep the Sun burning. The structure of the universe was also a mystery. What keeps the stars distributed in the sky when gravitational force should be pulling them together? The big question was whether all these anomalies could be explained with a better understanding of mechanics and electromagnetism, or whether they were the result of other interactions in nature that required a new theory. The most radical possibility was that both mechanics and electromagnetism were not correct, but only approximations of nature. If that were the case, the observed anomalies could never be explained without building a new theory from the ground up.

First Third of the Twentieth Century: A New Framework

The resolution of the inconsistency of mechanics and electromagnetism required a redefinition of the concepts of space and time—the special theory of relativity. Special relativity also allowed the conversion of matter into energy, which explained how the Sun could keep shining for billions of years. Going further, Albert Einstein and his colleagues were able to determine the nature of the gravitational force from the geometry of space-time—the general theory of relativity. Geometry seemed to be the key to understanding everything. Meanwhile the experiments of J. J. Thomson and Ernest Rutherford showed that the many different atoms that made up the chemical elements were not fundamental particles. Atoms consisted of a small, dense, positively charged nucleus surrounded by very light negatively charged electrons. The existing framework of physics based on classical mechanics, electromagnetism, and special relativity predicted that an atomic structure of this type could not be stable. The negative electrons would quickly spiral into the positive protons. A new theory, quantum mechanics (invented by Niels Bohr, Max Born, Werner Heisenberg, Erwin Schrödinger, and others), avoided this catastrophe. Even the weirder predictions of the new theories were confirmed by experiments. Rapidly moving particles did live longer than those at rest, light was bent by gravity, and electrons did exhibit interference patterns. Soon experiments showed that the nucleus of an atom was itself made of positively charged protons and electrically neutral neutrons.

During the first third of the twentieth century, elementary particle physics was radically different than it had been just 30 years before. The universe could be understood in terms of a simple and satisfying model. There were two fundamental constituents, the electron and proton. There were two fundamental interactions, the gravitational and electromagnetic. The constituents of this theory were characterized by their mass and their charge. There was one particle of each kind of charge: positive and negative. The neutron was thought to be made of a positive proton and a negative electron because it decayed into a proton and electron with a lifetime of about 15 minutes. The elements were the atoms of all possible configurations of electrons, protons, and neutrons held together by electromagnetic forces obeying quantum mechanics and relativity. This theory gave stable atoms that could emit only certain colors of light when heated. Light was neither a classical wave nor a classical particle. Light was a particle, called a photon, that behaved as predicted by quantum mechanics. All particles really behaved this way, but with light the behavior was obvious because it was massless. The new formulation of elementary particle physics not only explained many of the old anomalies, it predicted new phenomena. The antielectron (positron) was discovered in cosmic ray interactions as predicted from the symmetry of relativistic quantum mechanics. The discovery of the antiproton was just around the corner. The spectra from star light were shifted to the red, as predicted by relativity, if those stars were moving away from the Earth and each other. The universe was not collapsing due to gravitation, it was expanding. A fundamental theory was not yet constructed, but its formulation was surely based on quantum mechanics and relativity.

There were details that needed to be explained. For example, the proton and electron had the same magnitude of charge, but the proton was about 2,000 times more massive then the electron. Why should the masses of the fundamental constituents be so different and their charges be exactly the same? If a nucleus consisted of protons and neutrons, the positive protons should repel and tear it apart. How does nature prevent this nuclear catastrophe? Some nuclei were observed to decay with lifetimes ranging from seconds to thousands of years, yet others seemed absolutely stable. This range of lifetimes is allowed by quantum mechanics, but careful measurements of the products of those nuclear decays could not account for all the energy or momentum. Some was missing. Could the very successful principles of conservation of energy, conservation of momentum, and special relativity be incorrect? One way out was to invent a new invisible particle, the neutrino, that carried off the missing energy and momentum. Meanwhile another hypothetical particle, the meson, was invented to hold nuclei together. Unlike the neutrino, the meson was charged and interacted strongly with matter. It might be detected in cosmic rays or even produced in the more powerful versions of the exciting new invention, particle accelerators. It was obvious that the rate of expansion of the universe had to be slowing down due to gravitational attraction, but was the mass of the universe enough to pull it back together, or would it keep expanding? Was there a Big Bang that started the expansion? If so, how did the energy released in the Big Bang result in the galaxies, stars, planets, and particles that make up everything?

The mystery of cosmic rays needed to be investigated. These very-high-energy particles interacted in the Earth's atmosphere. What were they, where did they come from, and how did they acquire such large energies? One component of the cosmic rays was unusual, a charged particle that did not interact as strongly as either a proton or an electron. It was not the expected hypothetical meson because its interaction was not strong enough to hold nuclei together. Further investigation showed that this new particle, called a muon, was just like an electron except it was about 200 times more massive. How did it fit into a framework of elementary particle physics?

Despite this long list of questions, elementary particle physicists were optimistic. To tie together the loose ends, all that was needed was a single theory that unified the forces of electromagnetism and gravity within a framework that encompassed quantum mechanics and general relativity. This would be a unified field theory of everything. Within this theory, it was hoped that the symmetry of space and time would explain how the elementary particles fit together. How could a family that included protons, neutrons, electrons, neutrinos, and mesons also include the particles of light, photons, and this new particle that no one wanted, the muon?

Second Third of the Twentieth Century: Satisfaction Then Confusion

The second third of the twentieth century started well. Experiments had found the predicted hypothetical particles: the neutrino, meson, and antiparticles. Electromagnetism, special relativity, and quantum mechanics were unified in the theory of quantum electrodynamics (QED). This theory explained electromagnetic interactions as the exchange of photons between charged particles. In this theory, elementary particles determined the behavior of the universe. Properties of space were determined by pairs of particles and antiparticles that were everywhere. Called virtual particles, they could not be directly observed, but they did affect the behavior of observable particles in a way that could be predicted and measured. Using QED, physicists calculated precisely the properties of the electric and magnetic interactions of particles. There were certain terms that gave infinity, but they just had to be ignored. Over the next half-century it was learned that those terms cancelled out in the theory, so they did not really exist in nature.

Following the lead of QED, proton and neutron interactions, called the strong interaction, were formulated in terms of the exchange of mesons. If fundamental particles determined everything, a quantum theory of gravity would require the exchange of a new hypothetical particle, the graviton. Although theorists struggled without much success to put these parts together, experimenters were using the new particle accelerators to make additional elementary particles whose existence was not predicted. Other experiments showed that fundamental interactions were not as symmetric as expected. As the second third of the twentieth century progressed, elementary particle physics was beginning to look very complicated indeed.

New Particles: Who Needs Them?

Newer and larger particle accelerators gave protons and electrons ever higher energies. When they smashed into the stationary nuclei of ordinary atoms, new particles emerged. These new particles were not just pieces of the original projectiles or target particles because they were often heavier than either. It was as if a bullet was shot into a wall and created a car. This conversion from energy to mass is exactly what special relativity predicts in its famous equation, $E = mc^2$. These new particles could be distinguished primarily by their different masses. There were soon too many of these elementary particles to remember, but they could be classified into groups. Some of these new particles were related to the proton and neutron and were called baryons (heavy particles). They were more massive than the proton and survived less than a nanosecond. When they finished decaying, either a proton or neutron was left.

Some of these new baryons were electrically positive, some were negative, and some were neutral. They all had an obscure property called spin. The concept of spin was invented in the first third of the twentieth century to explain the behavior of electrons in atoms. It was called spin because its behavior is like that of a spinning top when described by quantum mechanics. Baryon spin always came in half-integer units such as $\frac{1}{2}$ or $\frac{3}{2}$, never 0 or 1. Careful measurement of the lifetimes of these new baryons revealed that some decayed more than a billion times more rapidly than others.

The other type of new particle that was produced by accelerators was related to the first meson, now called a pi meson. These new mesons were more massive than the pi meson. Their spin was in integer units such as 0 or 1, never $\frac{1}{2}$ or $\frac{3}{2}$. Again, careful measurement revealed that some of these mesons decayed more than a billion times more rapidly than others.

For every particle that was discovered, its antiparticle was also found. The particle and antiparticle had the same mass but the opposite charges. Neutrino interactions had finally been detected using the intense neutrino fluxes generated by nuclear reactors. These interactions were so rare that most of the neutrinos would pass through the Earth without being stopped. The behavior of neutrinos required an interaction that was much weaker than electromagnetism but much stronger than gravitation. This interaction was called the weak interaction. It was seen to be the mechanism of the slower decays of baryons and mesons. To make the situation even more complicated, experiments determined that there were two distinct types of neutrinos. They had the same properties such as charge, mass, and spin. One produced electrons, but never muons, when it interacted and was called an electron neutrino. The other produced muons, but never electrons, and was called a muon neutrino.

Since there were hundreds of particles, it was hard to believe that they were all fundamental. They could be classified into three major groups: baryons, mesons, and leptons. The leptons included the electron, the muon, the electron neutrino, and the muon neutrino. The photon did not belong to any of these categories. Although the photon and the neutrino were both shown to be massless, electrically neutral, and stable, they did not seem to be related. The neutrino had spin $\frac{1}{2}$, while the photon had spin 1. The photon easily interacted with matter, while the neutrino rarely did. Even though there were hundreds of elementary particles, there were only four fundamental interactions. Perhaps it was the four interactions that were fundamental, and the particles were all possible results of those interactions. On the other hand, it was possible that there were so many particles because they were made of more basic constituents that were the real elementary particles.

Symmetry Shattered

It had been assumed that a theory of fundamental interactions would be symmetric. For example, if space were uniform, it should make no difference if an interaction occurred in San Francisco or Minneapolis. Indeed, that type of symmetry agrees with experimental results. It also seemed reasonable that a reaction would give the same results if all particles were swapped for antiparticles, the directions of all particles were reversed, or the reaction run forward or backward in time. Contrary to common sense, experiments showed that nature does have

preferences in each of those cases. For example, as a neutrino raced away from the interaction that created it, its spin was always in the opposite direction to its velocity. Antineutrino spin, on the other hand, was always in the same direction as its motion. Reversing the direction of the neutrino would mean that its spin would now be in the same direction as its velocity, something that does not occur. Three ways of changing interactions were special because a combination of all of them—switching all particles for antiparticles (called C symmetry), then switching the directions of all particles (called P symmetry), and running the reaction in the opposite direction (called T symmetry)—was mathematically shown to give the same result for all interactions that satisfied a few reasonable criteria. One of these criteria was that the theory included special relativity. Another was that particles could only be affected by interactions at their location. Measurements showed that C and P symmetries were violated only for the weak interaction. One mystery was that a combination of C and P symmetry for an interaction (i.e., swapping particles for antiparticles as well as swapping the direction of particles) gave almost the same results even in the weak interaction. The CP symmetry violation existed but was very small. It seemed reasonable that nature might respect a symmetry or not. What could cause an interaction to violate a symmetry only a little?

Bigger Is Better

Particle accelerators were the primary tool used to produce these new particles and study their interactions. It was hoped that the extensive investigation of particle properties would uncover an underlying simplicity in this complex situation. More powerful accelerators were built to obtain higher-energy proton or electron projectiles that could produce more and heavier particles. These accelerators grew from the size of a machine that could be built on a table by a few people to machines that would fill a large aircraft hanger requiring dozens of people to build and operate. Experiments to analyze the properties of particles also got larger. Soon the apparatus would fill several rooms and require ten or twenty people to operate and determine the results. Elementary particle physics had become too large and complex for a single scientist and a few students.

It now required a team of scientists and students working together in a collaboration that spanned several different universities.

The Last Third of the Twentieth Century: Consolidation and Puzzles

A More Fundamental Constituent: Quarks

As the second third of the twentieth century drew to a close, it was proposed that all baryons and mesons were made of more basic particles called quarks. The hundreds of known baryons and mesons could be reproduced with only three quarks, called up (u), down (d), and strange (s). A baryon was a combination of three quarks, and a meson was a combination of a quark and an antiquark. The spin of a quark was $\frac{1}{2}$ unit. The charge of an up quark was $+\frac{2}{3}$ that of a proton, the down quark $-\frac{1}{3}$, and the strange quark $-\frac{1}{3}$. Thus a proton was made of two up quarks and a down quark (uud), a neutron of two down quarks and an up quark (ddu), and a positive pi meson consisted of an up and an antidown quark.

The quark theory also explained a puzzling set of measurements. When a particle has a spin and a charge, it can be affected by a magnetic field. The strength of this interaction is called its magnetic moment. The magnetic moment of an elementary particle depends on its electric charge, spin, and mass. Measuring the magnetic moment of nuclei in the body is the principle behind the medical diagnostic tool of magnetic resonance imaging (MRI). When the proton's magnetic moment was measured, it was too large for an elementary particle. Since the neutron is electrically neutral, it should have no magnetic moment. However, measurements showed that it also had a large magnetic moment with the opposite sign of the proton. If protons and neutrons were made up of quarks, their magnetic moments would be the sum of the magnetic moments of their quarks. Thus, the neutron would have a magnetic moment. When baryon magnetic moments were measured, they all agreed reasonably well, but not perfectly, with the quark model.

The quarks were bound together to form either baryons or mesons via a force strong enough to overcome the electric repulsion between like charges. This was the real strong force. The force binding protons and neutrons in the nucleus was a remnant of

the force between quarks. The force between neutrons and protons was similar to the force between two magnets. In a single magnet, there is a magnetic force between the north and south poles. Since the magnet has both a north and a south pole, it is magnetically neutral. Nevertheless, because the poles are separated, another magnet experiences a reduced amount of that force. Experiments were launched to find free quarks, but none were found. The first concrete manifestation of quarks appeared when high-energy electrons were used to probe inside a proton. The experiment revealed that there were smaller particles inside. It would take some time to develop a theory that explained how the strong force prevented the existence of free quarks.

Tools, Technology, and Discovery

The final third of the twentieth century saw a synthesis of elementary particle physics into what is called the Standard Model. Developing and testing this theory required still larger accelerators producing higher-energy particles and probing smaller distances. These particle accelerators would no longer fit into a building. Their sizes were measured in miles. National laboratories with staffs of hundreds were required to build and operate these machines. Experimental teams grew to include hundreds of physicists from around the world. New and faster communication was needed to exchange data and other information between the accelerator site and the scientists at their home institutions. This need pushed the development of the Internet and motivated researchers at the European Laboratory for Particle Physics (CERN) to invent an easy way for elementary particle physicists to use it. This invention, known as the World Wide Web, is the most recent example of the huge effect that fundamental scientific investigations can have on everyday life. Even particle accelerators became everyday tools used in treating disease and designing integrated circuits. Toward the end of the century, elementary particle experiments were limited more by economics than by desire, ideas, and technology. Building an experiment now costs about the same as a military aircraft. For the cost of a new accelerator, a medium-size city could operate its school system for a few years.

Soon particle accelerators reached higher energies by colliding beams of particles instead of smashing them onto a fixed target of ordinary material. Beams of antiparticles were created and stored so that they could be collided with beams of particles inside the same accelerator. An unexpected new particle was discovered almost simultaneously by teams at two different accelerators: a new electron-positron collider at Stanford Linear Accelerator Center (SLAC), and an older proton accelerator using a fixed target at Brookhaven National Laboratory (BNL). The properties of the new meson could only be explained by the existence of a fourth quark, called the charmed quark (c). Interestingly, the existence of this quark had been postulated by some theorists to explain why the heavier strange quark did not easily decay to the lighter down quark. As is often true, what does *not* happen is at least as important as what does.

Consolidation: Particles Are Basic

Quarks were finally fit into a complete theory modeled after quantum electrodynamics. In electrodynamics the strength of the interaction is determined by the charge of the object. Similarly, the strong force would need a property of the quark that acted like a charge and determined the strength of the interaction. The electric interaction could be described with two charges, positive and negative. The strong interaction, on the other hand, needed three "charges." In electrodynamics, combining all the different kinds of charges gives a neutral charge (a plus charge and a minus charge gives a neutral charge). Similarly, combining a charge with its anticharge also gives a neutral charge since the anticharge of plus is minus. With the strong force one would have to combine its three charges to get neutral. In an analogy with mixing colored light, where combining the three primary colors gives white or neutral, the term "color" was used for the strong charge. This theory was called quantum chromodynamics (QCD).

Now it was clear why there were baryons and mesons. A system of three quarks, each with a different color, would be neutral and thus not attract any more quarks. These were the baryons. A quark and an antiquark would also be neutral as a combination of a color and an anticolor. These were the mesons. The particles that were exchanged between the quarks to give the strong force were called gluons, since they glued together the quarks. Like the

photon, the gluon had no mass or electric charge. However, they did have color, which meant that the gluons themselves would interact strongly. This theory predicted that the strong force holding quarks together in a proton would behave differently than the electromagnetic force holding an electron and a proton in an atom. Unlike the electric force that becomes weaker with increasing distance, the strong force did not diminish as the distance between quarks became greater. It was not possible to isolate a single quark.

Another theoretical breakthrough unified quantum electrodynamics (QED) and the weak interactions into an electroweak theory. This theory confidently predicted the mass of the particles exchanged to give the weak force, called the W^+, W^-, Z^0, at about eighty times the mass of a proton. A new type of accelerator colliding protons and antiprotons was built at CERN to reach this energy, and two experiments were built to surround the collision regions. The particles appeared as predicted, and the electroweak theory appeared to be on firm ground. One consequence of this theory was that it predicted the existence of a new hypothetical particle, the Higgs. The Higgs was responsible for masses of the W^+, W^-, and Z^0 particles exchanged in weak interactions.

By the end of the twentieth century, the new accelerators in the United States had unearthed two more new quarks, the bottom (b) and the top (t) at Fermi National Accelerator Laboratory (Fermilab), and two new leptons, the tau (τ) at SLAC and the tau neutrino at Fermilab. The Standard Model was almost complete. There were six quarks and six leptons that each came in three families of two. Each quark family consisted of a quark with $+\frac{2}{3}$ electric charge and $-\frac{1}{3}$ charge. (u, d), (c, s), (t, b). Each lepton family consisted of a lepton with a charge of -1 and 0, (e, ν_e), (μ, ν_μ), (τ, ν_τ). Even the $\frac{1}{3}$ charges of the quarks no longer appeared so odd. If the proton charge were redefined as 3 units of charge instead of 1, the electric charges represented by fundamental particles would be 0, 1, 2, and 3.

In the Standard Model the photon was responsible for the electromagnetic interaction, the W^+, W^-, Z^0 for the weak interaction, and the gluons for the strong interaction. The W^+, W^-, and Z^0 had been produced by particle accelerators. Their properties

FIGURE 1

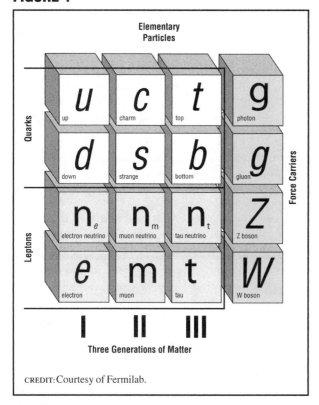

Elementary
Particles

Quarks	u up	c charm	t top	g photon
	d down	s strange	b bottom	g gluon
Leptons	n_e electron neutrino	n_m muon neutrino	n_t tau neutrino	Z Z boson
	e electron	m muon	t tau	W W boson

Force Carriers

I   II   III

Three Generations of Matter

CREDIT: Courtesy of Fermilab.

The Standard Model of particle physics.

were measured to be in agreement with electroweak theory. Precision measurements of Z^0 decay showed that there were only three massless (or even low mass) neutrinos. This implied that all the families of Standard Model particles had been found. According to quantum chromodynamics (QCD), gluons, like quarks, could not be produced in an isolated state, but their effects in high-energy interactions were distinctive, and those patterns were observed. There were only two particles in the Standard Model that had, as yet, not been discovered. These were the Higgs required by electroweak unification and the graviton needed for the still unformulated quantum theory of gravity. This was a great simplification over the hundreds of baryons and mesons that were previously called elementary particles. However, many elementary particle physicists thought that there were still too many particles and too many interactions for either of them to be truly fundamental.

The obvious similarity of having three quark families and three lepton families made it seem natural to place quarks and leptons into a single theory that would unify electroweak and strong interactions. These grand unified theories (GUT) were constructed, and they predicted that quarks would decay into leptons. This meant that protons could not be stable but must decay into leptons and mesons. The lifetime of such decays would have to be long since protons last long enough to form stars and populate a universe about 10 billion (10^{10}) years old. Unification of the strong and the electroweak interaction was supported by the results from the particle accelerators that were reaching higher and higher energies. The higher the energy of the interaction, the closer the interacting particles became. Experiments showed that the strength of the strong and the electroweak forces changed as the particles became closer. The strengths of the interactions were becoming similar with increasing energy. At the energy at which the strengths of the interactions were the same, they merged into a single interaction. Using these data, GUT theories predicted that the lifetime of the proton was about 10^{30} years. This is very stable on the scale of the universe but within reach of experiment. Very sensitive experiments were built kilometers underground to shield them from cosmic rays. In these experiments detectors watched thousands of tons of material for several years. No proton decays were found, and it was determined that the lifetime of the proton was at least 100 times longer than the most straightforward grand unified theories predicted.

The Beginning of the Twenty-first Century: Puzzle and Promise

The end of the twentieth century found elementary particle physics in a similar situation as at the end of the nineteenth century. There was a very successful framework of basic constituents and their interactions, the Standard Model. Unfortunately, the framework was incomplete. Just as there was no known way to unify classical mechanics and electromagnetism at the end of the nineteenth century, there was no known way to unify electroweak, strong, and gravitational theories within the Standard Model. At the end of the nineteenth century, the structure of the periodic table was known but could not be explained. At the end of the twentieth century, the organization of quarks and leptons into three families was not explained. There are also many questions about the nature of the elementary particles that remained unanswered. What is mass? What causes the quarks and leptons to have such a wide range of masses? Of all the quarks and leptons, what causes only the neutrinos to have zero mass? Why are there three families and not just one? Why not four? Why do quarks and leptons have separate families? What is charge? Why do the quarks and leptons have electric charges of 0, 1, 2, and 3? Why not charge 4? Why don't they all have the same charge? What is spin? Why do the particles that are the constituents of matter, the quarks and leptons, have $\frac{1}{2}$ unit of spin, whereas the particles responsible for the forces have integer spin? Why don't all elementary particles have the same spin? Why do interactions obey some symmetries and not others? Why is CP violation so small? If a symmetry is violated, why not maximum violation? How does nature determine a property that is between zero and maximum?

From the Smallest to the Largest

Elementary particle physics was getting closer to explaining the origin of the universe. It was clear that the universe is expanding. All known space was once very small but, for some unknown reason, exploded. The energy released in that Big Bang eventually became the galaxies, stars, and planets. Photons, the Big Flash from the Big Bang, filled space having cooled to an energy within 3 degrees of absolute zero. Measurements determined that mass makes up only one billionth of the energy of the universe. Elementary particle physics had satisfied the ancient human desire to describe the birth of everything. The universe started with the Big Bang. Space was very small but contained a lot of energy. Space expanded, and the energy created matter (quarks and leptons) and antimatter (antiquarks and antileptons) that annihilated and turned back into energy. As space continued to expand, the surviving quarks cooled and combined into baryons. This process continued, finally leaving protons and electrons along with a few simple nuclei. These nuclei were gravitationally attracted to make large objects. When the objects were large enough, the gravitational force squeezed together the nuclei igniting thermonuclear reactions that created stars. The light elements such as hydrogen, deuterium, and helium were made in the early universe and became part of stars. Measurements of the ratios of these light elements agreed with the Standard Model predictions. Nuclear reactions inside of stars made more complex nuclei, and eventually the star exploded as a supernova spreading the nuclei through space. The gravitational force pulled together the complex nuclei making new stars and planets. Meanwhile some of the stars had enough mass to collapse into black holes. These black holes have been detected and seem to become the centers of galaxies.

This is a coherent picture, but there are some obvious questions to be asked. When a particle accelerator turns energy into matter, equal amounts of matter and antimatter are made. That should have happened in the Big Bang. Why did the antimatter not annihilate the matter in the very early universe, leaving nothing but energy? Since our existence shows that at least some quarks were left over to form protons and neutrons, what happened to the equal amount of antiquarks that would make antiprotons and antineutrons? After decades of searching, there is no evidence of an equivalent amount of antimatter in the universe. To have only matter left over from the furnace of creation that was the early universe, quark and antiquark formation and annihilation reactions must have occurred at different rates. Within the framework of the Standard Model, the existence of such a rate difference requires both proton decay and that a CP symmetry violation occur. CP symmetry violation has been observed in the decay of some mesons. However, proton decay is a phenomenon that has not been observed. Either the proton lifetime is so long that it will take much larger experiments to detect it, or something is wrong with the Standard Model.

Other measurements of the properties of the universe also provide a challenge for elementary particle physics. The amount of matter in space can be determined by measuring the strength of the gravitational force on stars. Since stars are held in galaxies by gravity, measurements of their orbital velocities determine how the matter is distributed. The surprising result of these measurements is that about 90 percent of the mass of the universe is not accounted for by visible objects. Just what is this dark matter? Does this indicate the existence of new fundamental particles that are too massive to be made in current particle accelerators? At least some of it could be due to neutrinos if they had even a very small mass. A more shocking result comes from measurements of the velocities of supernovas at different distances, and thus different times, in the history of the universe. These measurements indicate that the expansion of the universe is not slowing down but is speeding up. This is possible if space has a stored up energy like a compressed spring. What fundamental interaction allows the fabric of space to store this energy?

Neutrinos Give Hints

Closer to home, the Sun is the star with the most direct impact on humans. To probe its inner workings, a method for looking deep within its core was needed. Neutrinos could provide the means. Because the neutrino interaction is weak, these neutrinos can escape the Sun's core and reach the Earth. Measuring the properties of these neutrinos was thought to be a good way to probe the inner workings of the Sun. However, when the rate of electron neutrinos coming from the Sun was measured, it was much too low. The differences between the predictions from nuclear reactions and the measurements were too big to be explained by uncertainties of the details of the Sun's structure. This very sensitive measurement required detectors operated kilometers underground, often in mines or tunnels through mountains, to shield them from the cosmic rays that bombard the surface of the Earth. As technology improved, larger and even more sensitive underground experiments determined that, although there were too few electron-type neutrinos, other neutrinos were coming from the Sun. The total number of neutrinos of all types from the Sun was equal to the number of predicted electron neutrinos.

The solar neutrino puzzle that spanned almost half a century now had an explanation. Electron neutrinos originating from the Sun's nuclear reactions change into a different type of neutrino before they reach Earth. It is as if a dog walking across the yard became a cat when it reached the other side. This identity confusion is actually possible in quantum mechanics because an object's behavior is determined by probability. In quantum mechanics, a particle can be created with a probability of being each of two or more types. Only one of the types will be detected but which type it is will depend on how the mixture evolves with time. Identity changing requires that the two identities have a small mass difference and the same charge. This weird behavior was originally observed in the decays of mesons that contained strange quarks and later for those that contain bottom quarks.

Meanwhile another neutrino anomaly surfaced. When the protons and nuclei that make up the cosmic rays strike the Earth's atmosphere, they make mesons, baryons, and leptons in a manner that can be replicated at particle accelerators for all but the highest energy cosmic rays. All the decays of the produced particles have been measured and some of the decay products are neutrinos. The resulting ratio of electron neutrinos to muon neutrinos can be calculated. This ratio is not sensitive to the details of the cosmic ray interactions. When large underground detectors measure this ratio, they obtain a number approximately a factor of 2 different than the prediction. Careful measurements, primarily from a large underground water detector in Japan, show that there are the predicted number of electron neutrinos but too few muon neutrinos. This result has been verified by a large underground iron detector in the United States. The cosmic ray neutrino anomaly can also be explained if some of the muon neutrinos produced in the atmosphere change identity to tau neutrinos as they travel toward the detector. Changing identity is only possible if the neutrinos have mass. An accelerator in Japan has verified this result by producing a muon neutrino beam and shooting it toward the same large underground water detector.

What Next?

It is now clear that the Standard Model is not the fundamental theory of elementary particles. New theories are waiting in the wings. A theory of supersymmetry would unify the spin-$\frac{1}{2}$ constituent particles, quarks and leptons, and the spin-1 interaction particles. This theory predicts that another set of particles mirrors those of the Standard Model but at a mass high enough that existing particle accelerators cannot produce them. Each spin-$\frac{1}{2}$ quark and lepton would have a hypothetical spin-0 partner.

Each spin-1 interaction particle would also have a hypothetical spin-$\frac{1}{2}$ partner. Supersymmetry has the added advantage of reducing the rate of proton decay to a size that experiments would not yet have tested. If supersymmetry is correct, there are many elementary particles waiting to be discovered.

Geometry Again

Another alternative to the Standard Model goes back to the fundamental importance of the geometry of space and time. Perhaps the many particles of the Standard Model are not fundamental after all. The particles could be minute but regular vibrations of the fabric of space itself. In this type of theory, the fundamental structure of space and time only allows a certain set of vibrations. This is like the musical notes from the string of a violin that are determined by where the ends of the string are held, the density of the string, and how tight the string is stretched. In this theory the string would be the underlying structure of space and the notes the elementary particles. Such a theory, called superstring theory, has been formulated. The mathematical structure of this theory allows the possibility of unifying the strong, electroweak, and gravitational interactions into a single framework for the first time. However, superstring theory predicts that space has at least ten dimensions. It is hard to visualize more than the usual four dimensions of height, width, depth, and time. The six extra dimensions do provide the flexibility of having neutrinos with mass, supersymmetric partners, and an energy density for space. If these dimensions exist at every point in space why haven't they been noticed?

The theory is still in its early stages and so far has not made any definitive predictions that could be tested with an experiment. It is possible that the extra dimensions are curled up so tightly that they have little effect on everyday life. Perhaps only very precise experiments will reveal the influence of extra dimensions. For example, experiments are underway to determine whether the gravitational force changes behavior at distances of less than 1 mm. It is even possible that the extra dimensions have large effects that were mistakenly thought to be connected to a different cause or were not noticed because no one looked for them. Perhaps there might be a violation of CPT symmetry because interactions in the four dimensional world are influenced by the other dimensions. Examples of such symmetry violations would be different masses or lifetimes of particles and antiparticles. Of course, superstring theory may just be an interesting mathematical diversion with nothing to do with reality. Maybe history will repeat itself, and the Standard Model particles will be found to be made of a smaller number of more fundamental particles that have not yet been found.

On the Horizon

On the experimental front, a new and more powerful colliding beam accelerator is being constructed at CERN to deliver a high enough energy to produce the hypothetical Higgs particle needed for electroweak unification. This will be a crucial test of that theory. Experiments at the lower-energy colliding beam accelerator at Fermilab have already begun to determine if the Higgs is at the lower end of its possible range of masses. These experiments may find the Higgs and begin the task of understanding it by measuring its properties. On the other hand, they may not find it but instead open the door to a new and more fundamental level of matter by finding unexpected particles or interactions. Meanwhile the United States, Europe, and Japan are all building powerful neutrino beams to be shot at huge detectors hundreds of miles away. These experiments will probe the nature of mass by investigating the identity-changing properties of the neutrino and making precision measurements of the mass differences of the neutrino types.

As the twenty-first century begins, elementary particle physics is on the brink of a new understanding of the fundamental workings of nature. Will investigations of CP violation together with new and larger proton-decay experiments finally show why the universe is made of matter? Not all the hidden dark matter can be neutrinos. What is the rest of it? It is clear that the Standard Model is at best incomplete. Will superstring theory emerge to explain the masses, families, and charges of the quarks and leptons? Perhaps results from larger and more sensitive underground experiments or larger and more powerful accelerators will send theory in completely different directions. Will there be a new level of particles below those known? Is humanity about to learn it lives in a universe with more dimensions than imagined? Will the interactions of the Standard Model be unified or, are they only approximations of a different set of fundamental interactions. Elementary particle physics continues its quest to explain everything. In the coming decades which view of the universe will take center stage: that particles are fundamental, interactions are fundamental, or geometry of space is fundamental? It is even possible that a new and original paradigm will emerge. How will these new ways of viewing the universe affect everyday life? One lesson learned from elementary particle physics is that the universe is both stranger and simpler than imagined. Even its most abstract discoveries tend to find practical application. The search for the fundamental components of the universe is certainly never boring.

See also: BIG BANG; EIGHTFOLD WAY; HADRON, HEAVY; HIGGS PHENOMENON; LEPTON; QUANTUM CHROMODYNAMICS; QUANTUM FIELD THEORY; QUARKS; STRING THEORY; STANDARD MODEL; SYMMETRY PRINCIPLES; UNIFIED THEORIES

Bibliography

Asimov, I. *Atom: Journey Across the Subatomic Cosmos* (Plum, New York, 1992).

Einstein, A., and Infeld, L. *The Evolution of Physics* (Simon & Schuster, New York, 1967).

Feynman, R. *QED* (Princeton University Press, Princeton, NJ, 1988).

Greene, B. *The Elegant Universe: Superstrings, Hidden Dimensions, and the Quest for the Ultimate Theory* (W. W. Norton, New York, 1999).

Hawking, S. *The Universe in a Nutshell* (Bantam, New York, 2001).

Hoffmann, B. *The Strange Story of the Quantum* (Dover, New York, 1959).

Particle Data Group. "The Particle Adventure." <http://particleadventure.org>.

Toulmin, S., and Goodfield, J. *The Architecture of Matter* (University of Chicago Press, Chicago, 1982).

Weinberg, S. *The First Three Minutes* (Basic Books, New York, 1993).

Kenneth J. Heller

PAULI, WOLFGANG

The Austrian-Swiss physicist Wolfgang Ernst Pauli was born in Vienna on April 25, 1900, the son of Bertha (Schütz) and Wolfgang Joseph Pauli. His father, originally from Prague, became a professor of chemistry at the University of Vienna in 1922 and was one of the founders of the science of colloid chemistry. Bertha Pauli was a writer, as was her daughter Hertha who was also an actress, and they belonged to the cultural elite of Vienna. The family was originally of Jewish origin, but Wolfgang Sr. became a Catholic, and his son was baptized—his godfather being the famous physicist and philosopher Ernst Mach.

In high school, Pauli was an outstanding student, with a strong interest in mathematics and astronomy. In 1918, he enrolled at the University of Munich, Germany, to study with Arnold Sommerfeld, a famous expert on relativity and atomic physics and the teacher of future Nobel Prize winners, including Werner Heisenberg and Hans Bethe. Pauli completed his Ph.D. in only three years, writing a dissertation on the quantum theory of the hydrogen molecule ion. In 1920, while still a student, Pauli wrote a 250-page article on relativity for the 1921 *Encyclopedia of Mathematical Physics* at Sommerfeld's request. It was highly praised by Einstein and is still regarded as a major treatise on the subject.

The Exclusion Principle

After receiving his Ph.D., Pauli spent a year at the University of Göttingen, with James Franck and Max Born. He then worked for a year at Copenhagen with Niels Bohr, who had originated the quantum theory of the atom. It was at this time that he first took up the problem of the Zeeman effect, the splitting of spectral lines in the presence of a magnetic field, which was a subject of major interest because it seemed to be an insoluble problem in the Bohr-Sommerfeld quantum theory that dominated atomic physics. This theory, an extensive elaboration of Bohr's 1913 hydrogen model, placed the atomic electrons in classical orbits that were restricted by a general set of quantum conditions. For example, angular momenta and their vector components were restricted to being integer multiples of $\hbar = h/2\pi$, where $h = 6.63 \times 10^{-34}$ J/s is Planck's constant.

However, the number of atomic states in a magnetic field was double the number that the theory predicted. In the simple case of sodium, for example,

Austrian-Swiss physicist Wolfgang Pauli (1900–1958) received the Nobel Prize in Physics in 1945 for his discovery of the Exclusion Principle, later known as the Pauli Exclusion Principle. CREDIT: COURTESY OF CERN (EUROPEAN ORGANIZATION FOR NUCLEAR RESEARCH). REPRODUCED BY PERMISSION.

there is one electron outside of a closed shell of electrons (the core) that Bohr-Sommerfeld theory predicts should have zero angular momentum. To account for the extra atomic states, it was proposed that the core should instead have an angular momentum of $\frac{1}{2}\hbar$, an idea that Pauli rejected. His solution was to say that the electron itself has a "non-classically describable two-valuedness," so that it was not the core but the external valence electron that was responsible for the doubling of the number of atomic states

This was a suggestion of the greatest importance, for it played an essential role in explaining the periods in Mendeleev's table of chemical elements. Bohr had already made a start in this direction with his building-up principle, which asserted that in passing from one atom in the table (characterized by the

atomic number Z, or number of electrons) to the next one, the inner electrons kept the same quantum numbers. However, to complete this picture, Pauli's new two-valuedness was needed: each set of the old quantum numbers labeled not one, but two states. The new form of the principle became known as the Pauli Exclusion Principle, and it was for this discovery that Pauli was awarded the Nobel Prize in Physics in 1945.

Quantum Mechanics and Quantum Electrodynamics

In 1925 Heisenberg, Pauli's close friend and collaborator, replaced the Bohr-Sommerfeld orbit theory with a new quantum mechanics from which the modern theory of physics and chemistry has originated. Pauli, however, was the first to apply Heisenberg's theory to a real physical problem, namely, the hydrogen atom, which he solved completely. Meanwhile, also in 1925, two young Dutch physicists, Samuel Goudsmit and George Uhlenbeck, identified Pauli's quantum number as belonging to electron spin. That is, every electron has an intrinsic spin angular momentum of $\frac{1}{2}\hbar$, and an associated intrinsic magnetic moment $e\hbar/mc$, which can take up one of two orientations in a magnetic field. Pauli resisted this rotating electron interpretation for almost a year, but in March 1926 he wrote to Bohr that he would "capitulate completely" (Mehra and Rechenberg 1982, p. 709). He then applied the spin and the exclusion principle to explain the magnetic properties of normal metals (paramagnetism) and thus initiated in 1927 a new research field, the quantum electron theory of metals.

In that same year, the English quantum theoretician Paul Dirac made a quantum theory of the electromagnetic field and also a relativistic generalization of the wave function, introduced by the Austrian physicist Erwin Schrödinger in his version of quantum mechanics. Dirac's theory predicted the existence of a positive electron (positron) that could be produced together with an ordinary negative electron, providing enough energy was available (at least $2mc^2$, with m being the electron mass and c the velocity of light). Pauli and Heisenberg then wrote two important papers providing a relativistic treatment of the interaction between radiation and matter. They discovered important difficulties in their theory, which had to wait until the late 1940s for a satisfactory resolution. Problems of quantum field theory, as it came to be called, occupied Pauli for the rest of his life, especially the relation between spin and quantum statistics, which is crucial for the collective behavior of identical particles (whether they form shells, for example, or collapsed states, as in a laser).

As professor at the Swiss Federal Institute of Technology (ETH) from 1928, after serving at the University of Hamburg, Pauli continued his research on wave mechanics. This led in 1933 to another remarkable treatise, published as an encyclopedia article.

Nuclear Beta Decay and the Neutrino

In December 1930, convinced that a puzzling situation in nuclear beta decay required a "desperate solution," Pauli suggested that a new extremely penetrating neutral particle of very small (perhaps zero) mass accompanied each electron emitted in beta decay. Pauli took this step to account for what appeared to be energy "missing" from the process. Now called the neutrino, Pauli's particle became an ingredient of a new and successful quantum field theory of beta decay worked out by the Italian physicist Enrico Fermi at the end of 1933. (A generalized version of Fermi's theory forms part of the so-called Standard Model of elementary particle interactions developed in the 1970s.)

In 1940 Pauli, fearing a possible German invasion of Switzerland, moved to the Institute for Advanced Study in Princeton, New Jersey, where Einstein was also in residence. He returned in 1945 to the ETH in Zurich, where he remained until his death on December 14, 1958. Pauli had many important accomplishments in physics, and he was also a philosopher. In studying the psychology of creativity, he collaborated with the Swiss psychoanalyst Carl Gustav Jung. Because of the profoundly high standards that he brought to his work, Pauli is sometimes referred to as the "conscience of physics."

See also: NEUTRINO, DISCOVERY OF

Bibliography

Enz, C. P. "W. Pauli's Scientific Work" in *The Physicist's Conception of Nature,* edited by J. Mehra (Reidel, Boston, 1973).

Enz, C. P. *No Time to Be Brief: A Scientific Biography of Wolfgang Pauli* (Oxford University Press, New York, 2000).

Fierz, M., and Weisskopf, V.F. *Theoretical Physics in the Twentieth Century* (memorial volume to Pauli) (Interscience, New York, 1960).

Mehra, J., and Rechenberg, H. *The Historical Development of Quantum Mechanics,* Vol. 1, Part 2 (Springer-Verlag, New York, 1982).

Pauli, W. "Remarks on the History of the Exclusion Principle." *Science* **103**, 213–215 (1946).

Pauli, W. "Relativitätstheorie" in *Encyclopädie der Mathematischen Wissenschaften* 5, Part 2, 539–775 (B.G. Teubner, Leipzig, 1921). English translation: *Theory of Relativity* (Pergamon, New York, 1958).

Pauli, W. *Lectures on Physics,* Vols. 1–6 (MIT Press, Cambridge, MA, 1973).

Pauli, W. "Die allgemeine Prinzipien der Wellenmechanik" in *Handbuch der Physik,* Vol. 24, Part 1, edited by H. Geiger and K. Scheel (Springer-Verlag, Berlin, 1980). English translation: *General Principles of Quantum Mechanics* (Springer-Verlag, New York, 1990).

Peierls, R. E. "Wolfgang Ernst Pauli." *Biographical Memoirs of Fellows of the Royal Society* **5**, 175–192 (1959).

Weisskopf, V. F. "Personal Memories of Pauli." *Physics Today* **38**, 36–41 (1985).

Laurie M. Brown

PHASE TRANSITIONS

Most substances exist in three distinct phases: solid, liquid, and vapor. From everyday experience, in particular with water, it is observed that a substance in a given phase may transition into another phase as a response to some imposed change, for example, of temperature or pressure. These phase transitions occur at very specific temperatures and pressures for different substances. At sea level water transitions from a liquid to a vapor phase—it boils—at 100° Celsius; at an altitude of 3,000 meters, due to the lower atmospheric pressure, water boils at only 90° Celsius.

Historically, the study of phase transitions has been of interest mostly to condensed-matter and statistical physicists. However, since the early 1960s, it has become clear that several parallels can be drawn between the physics of phase transitions and the changes in the properties of elementary particles of matter and their interactions at different energies.

The theories describing elementary particles and their interactions can also be understood as having different phases, which reveal themselves at different energies or temperatures. The physics of these phase transitions can be probed in two different arenas: high-energy collisions between subatomic particles in particle accelerators, and during the early stages of the universe's history when, according to the prevailing Big Bang theory of cosmology, the temperatures were high enough to promote these changes.

Phase Transitions and Symmetry

At the microscopic level, phase transitions can be understood as a spatial rearrangement of the molecules of a given substance or a mixture of substances resulting from externally imposed changes in temperature, pressure, and, in some cases, magnetic field. In general, the phase of a substance is determined by a competition between the chemical bonding of its molecules and their thermal agitation. In the vapor phase, which occurs at high enough temperatures, thermal agitation is the dominant factor, causing the molecules to move freely, colliding with one another but never forming large clusters. At lower temperatures, the chemical bonding between the molecules starts to counterbalance their thermal agitation, and a liquid phase sets in, where the molecules are held closely together but still in a disorderly fashion. At sufficiently low temperatures, the further loss of thermal energy facilitates the arrangement of the molecules in rigid clusters characterizing the solid phase. The phase of a given substance is defined by the spatial ordering of its molecules.

A phase change can also result from structural changes within different solid phases of the same substance—known as solid-solid transitions, or, in the case of magnetic materials, from the rearrangement of magnetic fields within a collection of atoms. As with ordinary phase transitions, it is often possible to associate these structural rearrangements with changes in the underlying spatial symmetry of the substance. When water is in its liquid phase, the probability of finding a molecule anywhere within a given volume is approximately the same, as all molecules are on average equally spaced; one can say that liquid water has a large spatial symmetry. However, as water freezes, its molecules rearrange themselves in a crystal lattice, and the probability of finding a molecule is no longer approximately the same everywhere. The spatial symmetry of liquid water is lost when it freezes. In other words, a drop in temperature may decrease the amount of symmetry in a substance.

As a second example, magnetic materials also exhibit a phase transition, related to the spatial ordering of its atoms. Each atom can be considered as a small magnet, which interacts with its neighbor. At high temperatures, the small magnets point at random directions, and the net magnetization of a sample of the material is zero. This is called the paramagnetic phase. However, as the temperature drops, the neighboring magnets tend to align in the same direction in order to minimize their interaction energy. Below a temperature known as the critical temperature the material separates into domains with a net magnetization, as indicated in Figure 1. If an external magnetic field is applied, the magnetization of the separate domains aligns in the same direction of the magnetic field. This is referred to as the ferromagnetic phase. Again, the symmetry that exists at high temperatures, when all directions are equiva-

FIGURE 1

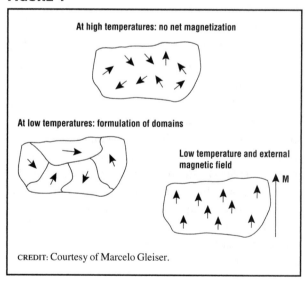

CREDIT: Courtesy of Marcelo Gleiser.

A drop in temperature triggers the formation of magnetic domains in a paramagnetic material. Without an external magnetic field, the domains will point in arbitrary directions and will compete with one another for survival. An applied magnetic field will force the net magnetization in the domains to align with it, until the whole sample has a net magnetization.

lent, is broken below the critical temperature, when only one direction prevails.

Symmetry Breaking in Particle Physics

According to the Standard Model of particle physics, there are four fundamental forces—or interactions—in nature: the long-range gravitational and electromagnetic forces, and the strong and weak nuclear forces. The two latter forces act only within the atomic nucleus; thus, they are very short-range forces. These four forces describe the interactions between the twelve elementary particles of matter, which, according to the Standard Model, can be divided into two groups: six quarks and six leptons. The quarks make up particles that interact via the strong nuclear force, such as protons and neutrons, whereas the leptons participate in the weak interactions, responsible, for example, for radioactive decay. Each of the forces has an associated symmetry, related to quantities that are conserved during the interactions. For example, the associated conserved quantity of electromagnetic interactions is the electric charge.

The crucial link between particle physics and the physics of phase transitions is that the nature of the four fundamental interactions, and thus their symmetries, change with energy. At high enough energies, the interactions start to behave in similar ways: at energies greater than 100 times the mass of the proton (times the square of the speed of light, as required by the $E = mc^2$ relation, where c is the speed of light), the weak interaction becomes long range and is indistinguishable from the electromagnetic interaction. Thus, above these energies, the interactions between matter particles can be described in terms of three fundamental forces and not four: gravity, strong, and electroweak. At much higher energies (on the order of a thousand trillion proton masses times the square of the speed of light), it is expected that the strong interactions join the electroweak force to become the grand unified force. If this proves to be correct, at these enormous energies nature can be described in terms of two forces: gravity and the grand unified force. Some theories presently under investigation, such as superstring theories, attempt to include gravity in the unification scheme. This high-energy unification of the fundamental interactions comes with an increase in the underlying symmetry, very much as the increase in symmetry in a solid-to-liquid transition or when a ferromagnetic material becomes paramagnetic at high temperatures.

To summarize, each of the four fundamental interactions has an associated symmetry. At high energies, the interactions can be described in a unified way, and their underlying symmetry increases. In other words, at extremely high energies nature is highly symmetric. As the interactions between particles are probed at lower energies, this large symmetry is progressively broken, until we reach the current Standard Model description in terms of four fundamental forces.

Phase Transitions in Particle Accelerators and in the Early Universe

There are two ways to test the prediction of increased symmetry at high energies: at particle accelerators and during the early stages of the universe's history. The unification of the electromagnetic and weak interactions was verified at the European Laboratory for Particle Physics (CERN) in 1983. The theory of strong interactions, known as quantum chromodynamics (QCD) predicts that above a certain energy protons, neutrons, and other composite particles known as hadrons break into a plasma of free quarks and gluons, the particles that promote their interactions. This prediction, which is also interpreted as a phase transition, is presently being tested at the Relativist Heavy Ion Collider (RHIC) at the Brookhaven National Laboratory in New York.

The early universe offers the best laboratory to test unification ideas, as temperatures were high enough to probe the larger symmetry regimes predicted by particle physics. As the universe expanded and cooled from its extremely hot and dense initial state, it may have undergone a succession of phase transitions related to the breaking of the initial large unified symmetry into smaller ones associated with the cited four interactions (see Figure 2). Each of these phase transitions may have left imprints and remnants that could be observed today.

As is true with the transition from water to ice or from a paramagnet to a ferromagnet, it is possible that these cosmological phase transitions also

FIGURE 2

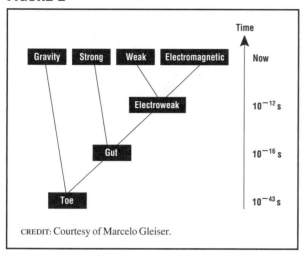

CREDIT: Courtesy of Marcelo Gleiser.

Schematic sequence of symmetry breakings that may have occurred during the universe's history. "TOE" signifies Theory of Everything, the conjectured unification of all four fundamental forces of nature. The time axis gives the approximate times where a given force separates from the others.

developed domains and other imperfections, which carry the particular signature of the symmetry breaking process. One possibility of great interest is that the observed excess of matter over antimatter, a puzzle still unsolved in particle physics, could be explained due to the particular details of the electroweak phase transition, predicted to have occurred when the universe was one trillionth of a second old. This transition, which has some similarities with the vapor-water transition, may have produced the conditions necessary to generate a small excess of particles of matter over particles of antimatter, which eventually led to the existence of complex material structures such as galaxies, stars, and people.

See also: COSMOLOGY; QUANTUM CHROMODYNAMICS; QUARK-GLUON PLASMA

Bibliography

Bak, P. *How Nature Works: The Science of Self-Organized Criticality* (Springer-Verlag, New York, 1996).

Ferris, T. *Coming of Age in the Milky Way* (William Morrow, New York, 1988).

Gleiser, M. *The Dancing Universe: From Creation Myths to the Big Bang* (Dutton, New York, 1997).

Gleiser, M. *The Prophet and the Astronomer: A Scientific Journey to the End of Time* (W. W. Norton, New York, 2002).

Weinberg, S. *Dreams of a Final Theory* (Pantheon, New York, 1992).

Weinberg, S. *The First Three Minutes: A Modern View of the Origin of the Universe* (Basic Books, New York, 1993).

Marcelo Gleiser

PHILOSOPHY AND PARTICLE PHYSICS

There are many claims in the literature about the impact of modern particle physics on the way philosophers conceptualize the world. The issues are complex and the conclusions not as decisive or clear-cut as is sometimes supposed.

Individuality and Quantum Statistics

First, there is the question of what confers individuality on the particles, or indeed whether they are individuals at all in the sense of being particulars that transcend in some sense the properties (universals) they exhibit. Arguments from quantum statistics are often adduced to show how one would simply get the wrong (Boltzmannian) statistics if the particles possessed individuality in the classical sense. The argument here is persuasive but by no means decisive. The assignments of statistical weights may just reflect limitations on the accessibility of certain states rather than nonindividuality. Pursuing the nonindividuality route, however, chimes in with the view that particles are really quantized excitations of a field, and this is of course the point of view taken in quantum field theory (QFT). One of the big advantages of QFT is that the evanescent character of elementary particles, their creation and annihilation in high-energy collision experiments for example, seems a great deal less mysterious than the creation and annihilation of particles conceived of as individual particulars in their own right. The philosophical origins of Greek atomism, the ancient precursor of modern atomic theories, was quite at odds with such possibilities.

Quantum Field Theory

However, the ontological status of a quantum field is somewhat problematic. In classical physics, field theories were sharply distinguished from particle theories.

In the former, the role of the individual was played by the space-time points that were endowed with or in some sense associated with the properties of exhibiting a field amplitude or excitation. Notice that the role of space-time is quite different in field theories as opposed to particle theories, where spatiotemporal location is treated as a property of the particle.

Quantum fields come in two varieties, those like the electron field associated with matter, and those like the electromagnetic field associated with interactions. The interaction fields obey Bose-Einstein statistics, or in the language of QFT the fields satisfy microcausality conditions, expressing the fact that they cannot transmit influences faster than the speed of light. Their interpretation in terms of discrete quantized excitations or "quanta" as surrogate "particles" seems clear enough, setting aside all the interpretational problems of quantum mechanics itself. But the matter fields obey Fermi-Dirac statistics, and in the language of QFT they fail to satisfy the microcausality condition imposed by the special theory of relativity, so the matter fields are not observable; they belong to the so-called surplus mathematical structure. In a sense the matter fields don't "exist"! In particular, the classical limit of such a quantum field theory is not a classical field! Of course one can construct quantities like charge and current densities out of the fields, which do satisfy microcausality, and are observable, but the point is that these constructions are not the fields themselves. The upshot of this discussion is that QFT does not in any simple way resolve the ancient philosophical puzzles of particle versus field, of atom versus plenum.

Relativistic Particle Theories

Returning to the particle option, there is, however, another quite distinct difficulty. The elementary particles are usually thought of as unextended points (this is modified in string theory but will be ignored for present purposes). Considered as point particles, they should have precise spatial locations. But the apparently innocuous condition that if a particle is localized at one spatial point there must be zero probability for finding it at that very moment located at a different point turns out to be inconsistent in relativistic theories with the objectivity of localization in the sense that observers in different states of uniform relative motion will not agree on

whether the particles are in fact localized at all! This is closely related to the fact that relativistic wave packets that are sharply localized in one reference frame disperse superluminally relative to that frame. These unpleasant features of relativistic particle localization have generally militated after all in favor of the quantum field approach.

The Quantum Vacuum

Particular interest, philosophically speaking, attaches to the concept of the vacuum in relativistic QFT. In nonrelativistic theories the global vacuum identified with the absence of unlocalized particles (quanta with a definite momentum) implies a local vacuum in the sense of the absence of localized particles. This is no longer true in the relativistic vacuum, where the global vacuum actually implies the violation of a local vacuum. This violation is often described in terms of the creation and annihilation of so-called virtual particles that indeed violate energy conservation provided their lifetimes are governed by the time-energy uncertainty relations of quantum mechanics. Virtual particles of mass m exist for a time bounded by h/mc^2, where h is Planck's constant and c the velocity of light.

The properties of virtual particles show how far removed the particle concept is from any classical picture. Talking of classical pictures reminds one that the dynamics of elementary particles, whether conceived as field excitation or as "true" particles, is governed by the laws not of classical mechanics but of quantum mechanics.

Quantum Mysteries

Physical magnitudes on the orthodox or so-called Copenhagen interpretation only have sharp definite values in special states called eigenstates. In general, the formulation of quantum mechanics provides rules for calculating the probabilities that possible values will turn up on measurement. In particular, so-called conjugate quantities have reciprocally related spreads of possible values governed by the famous Heisenberg uncertainty relations. As a result the particle theories no longer allow a notion of spatiotemporally continuous trajectories. The notion of causality as mediated by continuous processes has to be significantly revised. Essentially it is reduced to conservation laws for energy and momentum. Determinism survives in the time-development of the quantum-mechanical state in accordance with the time-dependent Schrödinger equation. The failure of determinism comes in with measurement interactions which play a privileged role in the theory. The sorts of questions that can be posed and answered are relativized to specific experimental setups. This leads to a form of perspectivalism, in which perspectives may be incompatible, but all are necessary for a complete conspectus of reality. All of this makes for a heady revision of traditional realist metaphysics.

But again the arguments are not decisive. There are other interpretations of quantum mechanics in which hidden variables are introduced, and it is our ignorance of these variables which allow for epistemic rather than ontic probabilities in the interpretation of the theory. So it is possible to restore determinism at the level of the hidden variables. But all this comes at a severe price. John Bell in 1964 showed that any such theory must exhibit nonlocality in the sense that mysterious changes in possessed values of local observables must be produced by operations carried out even at spacelike separation from the local observables. This causes prima facie problems for a relativistic theory of such hidden variables. But the exact interpretation of this nonlocality is a subtle matter that has been the subject of much philosophical debate. Under some interpretations the nonlocality is better described in terms of nonseparability, that the properties of composite systems in so-called entangled states cannot be analyzed in terms of local attributes of the constituents, thus introducing a holistic aspect to the interpretation of multiparticle states. Indeed this holistic aspect of quantum phenomena is also emphasized in the orthodox Copenhagen interpretation. The upshot of such arguments is that quantum mechanics may ultimately be inimical to the reductionist philosophy of understanding wholes in terms of their parts. It is ironic that particle physics that seems so conducive to reductionism may actually provide a counterexample to it!

The Mathematization of Nature

As previously noted fermionic fields do not have direct physical significance. They belong to the mathematical "surplus structure," as it is often called, that

has become endemic in modern theoretical particle physics. This is particularly true of the popular gauge theories of particle interactions where the physically significant quantities are invariant under transformations of the gauge symmetry, but these transformations can themselves only be specified in terms of quantities which are *not* gauge invariant. Much modern particle theory works with surplus structure which elegantly "controls" the physically significant magnitudes, rather than formulating the theories directly in terms of the physical magnitudes themselves. This situation has led to much philosophical debate ranging from a revival of Pythagorianism (that reality *is* mathematical) to an uneasy reflection that modern particle physics may have entered a decadent phase, losing touch with its empirical roots, and attempting Theories of Everything guided to a large extent by purely mathematical considerations.

See also: INFLUENCE ON SCIENCE; METAPHYSICS; UNIVERSE

Bibliography

Redhead, M. L. G. "A Philosopher Looks at Quantum Field Theory" in *Philosophical Foundation of Quantum Field Theory,* edited by H.R. Brown and R. Harré (Clarendon Press, Oxford UK, 1988).

Redhead, M. L. G. *From Physics to Metaphysics* (Cambridge University Press, Cambridge, UK, 1995).

Sklar, L. *Philosophy of Physics* (Oxford University Press, Oxford, UK, 1992).

Teller, P. *An Interpretive Introduction to Quantum Field Theory* (Princeton University Press, Princeton, NJ, 1995).

Michael L. G. Redhead

PHOTON

See BOSON, GAUGE

PLANCK SCALE

The Planck scale is named in honor of the famous German physicist Max Planck. He was the first to realize that three constants of nature can be combined to give fundamental units of mass

$$M_P = \sqrt{\frac{\hbar c}{G}} = 2.2 \times 10^{-5} \text{ g},$$

length

$$L_P = \sqrt{\frac{G\hbar}{c^3}} = 1.6 \times 10^{-33} \text{ cm}$$

and time

$$T_P = \sqrt{\frac{G\hbar}{c^5}} = 5.4 \times 10^{-44} \text{ s}.$$

The Planck mass, length, and time are equivalent ways to describe the Planck scale. They are constructed from the speed of light c, the gravitational constant G, and the quantum of angular momentum $\hbar$. In particle physics units, M_P works out to be 1.2×10^{19} GeV/c^2, about 10^{19} times larger than the mass of a proton. The question of why the Planck mass is so large—or why the proton mass is so small—is at the heart of modern particle physics.

Why is the Planck scale important? It contains c, G, and $\hbar$, so it connects relativity, gravity, and quantum mechanics. In fact, the Planck scale marks the place where quantum gravity replaces Einstein's relativity. At energies higher than the Planck mass, at distances shorter than the Planck length, or at times shorter than the Planck time, classical notions cease to hold. Planck wrote his expressions in 1899—before relativity and quantum mechanics were discovered—even before he presented his famous formula for blackbody radiation! Planck's remarkable intuition has stood the test of time.

Physics at the Planck scale is very different from the physics of the everyday world. At the Planck scale, gravity is a strong force—so strong that it changes the behavior of subatomic particles. Space-time itself is torn apart by quantum fluctuations. In 1957 American physicist John Wheeler proposed that Planck-scale space-time is a quantum foam, bubbling with virtual processes. Wheeler was the first to recognize the Planck scale's role in quantum gravity.

In quantum electrodynamics, the photon couples to the electron through a gauge coupling denoted by e. The fine-structure constant, α, is measured to be

$$\alpha = \frac{e^2}{4\pi\hbar c} \approx \frac{1}{137}.$$

At short distances, quantum effects renormalize α. In a scattering experiment with particles of energy E, the renormalized coupling increases logarithmically with energy.

In gravity, the analog of the photon is called the graviton. It couples to mass—or energy—through a gravitational coupling g. The strength of this coupling changes *linearly* with energy, $g = E\sqrt{G}$. Therefore in gravity, the analog of the fine-structure constant is not constant, but

$$\alpha_G = \frac{g^2}{4\pi\hbar c} = \frac{E^2}{4\pi M_P^2}.$$

In today's particle physics experiments, in which elementary particles are scattered with energies $E \approx$ 300 GeV, the force of gravity is about a factor of 10^{-32} weaker than electromagnetism. Gravity is weak because the Planck mass is large.

At very short distances, the story is different. In a scattering experiment with Planck-scale energies, the gravitational coupling is of order 1, and quantum effects are critically important. Einstein gravity is nonrenormalizable, which means that it is not a consistent theory of quantum theory. It is a low-energy, long-distance approximation to a deeper, more fundamental theory of quantum gravity. At the Planck scale, Einstein's relativity becomes part of whatever takes its place.

At present, string theory is the best candidate for a fundamental theory of gravity. In string theory, point particles are tiny strings, Planck length in size. Viewed from afar, the strings appear to be points. But with Planck-scale resolution, their stringlike character becomes evident. It is believed that string theory gives a consistent theory of quantum gravity—but only in ten (or eleven) dimensions.

In string theory, all the known particles—quarks, lepton, even gravitons—appear as quantized vibrations of strings. In this sense, string theory unifies all the forces and particles of nature. At present, there is no direct experimental evidence in favor of strings. Nevertheless, there is compelling indirect evidence for new physics near the Planck scale. As in electrodynamics, the couplings of the Standard Model gauge particles change logarithmically with energy. In certain extensions of the Standard Model, they become equal at about 10^{16} GeV, close to the Planck scale, 10^{19} GeV. This suggests that the Planck scale—and string theory—might play an essential role in the ultimate unification of physics.

Why, then, is the Planck mass so much larger than the mass of the proton? There are two ways to approach the question. The first is to suppose that the Plank mass is fundamental. If so, it is reasonable to assume that the forces of nature are unified near M_P. In particular, near M_P, the "strong" coupling of quantum chromodynamics (QCD) is no different than the other gauge couplings. At lower energies, however, the QCD coupling grows stronger (unlike electrodynamics, where it grows weaker). The coupling changes slowly—only logarithmically—so it is not until approximately 1 GeV that the QCD coupling is strong enough to bind quarks and gluons into protons. The mass of the proton is so much smaller than M_P because of the logarithmic evolution of the QCD coupling.

A second point of view is to assume that the proton mass is close to the fundamental scale. In a theory with extra dimensions, for example, it is possible for the *true* Planck scale to be 10^3 GeV and the *apparent* Planck scale to be 10^{19} GeV. To see how this works, suppose that quarks and leptons are restricted to a three-dimensional membrane embedded in six-dimensional space-time, where the extra dimensions are not infinite, but circles of radius R. Also suppose that gravity is not confined to the membrane, but can extend into the two extra dimensions.

At very short distances, shorter than R, the gravitational force law is that of six dimensions. It is not inverse square, but rather

$$F = \frac{G_* M_1 M_2}{r^4}$$

where G_* is Newton's constant in six dimensions.

At large distances, much larger than R, the gravitational lines of force cannot extend into the extra dimensions. Therefore at large distances, the force law is inverse square:

$$F \approx \frac{G_* M_1 M_2}{r^2 R^2}.$$

The effective four-dimensional gravitational constant is

$$G = \frac{G_*}{R^2}.$$

As in four dimensions, a fundamental "Planck mass" can be constructed for the six-dimensional theory. It is

$$M_* = \sqrt[4]{\frac{\hbar^3}{c G_*}}$$

where G_* is the six-dimensional Newton constant. Combining equations, one can write the apparent Planck mass M_P in terms of the fundamental Planck mass and the radius R:

$$M_P = \left(\frac{cR}{\hbar} \right) M_*^2.$$

In this theory, quantum gravity becomes important at M_*, which is assumed to be $M_* \approx 3 \times 10^3$ GeV. The apparent Planck mass is $M_P \approx 10^{19}$ GeV. The two can be reconciled provided $R \approx 0.1$ mm—in other words, provided there are new spatial dimensions of macroscopic size! From this point of view, gravity is weak because the extra dimensions are large.

Exquisitely beautiful (and careful) experiments are being carried out to test the inverse-square nature of gravity at submillimeter scales. A deviation from the inverse square law could be a hint of new macroscopic dimensions. The discovery of new dimensions would spark a revolution in physicists understanding of the universe—and humankind's place within it.

See also: GRAND UNIFICATION; STRING THEORY; UNIFIED THEORIES

Bibliography

Arkany-Hamed, N.; Dimopoulos, S.; and Dvali, G. "Large Extra Dimensions: A New Arena for Particle Physics." *Physics Today* (February 2002).

Greene, B. *The Elegant Universe: Superstrings, Hidden Dimensions, and the Quest for the Ultimate Theory* (W.W. Norton, New York, 1999).

Planck, M. *"Über Irreversible Strahlungsvorgänge."* *Sitzungsberichte der könglich Preussischen Akademie der Wissenshaften zu Berlin* **5**, 440 (1899).

Wheeler, J. "On the Nature of Quantum Geometrodynamics." *Annals of Physics* **2**, 604 (1957).

Wilczek, F. "Scaling Mount Planck I: A View from the Bottom." *Physics Today* (November 1999).

Jonathan Bagger

POSITRON, DISCOVERY OF

The positron, the antiparticle of the electron, was discovered in two steps. The first and crucial one was by Carl D. Anderson, who in 1932 concluded the existence of a positive particle of electronic mass (positive electron) from the tracks left by cosmic rays in a cloud chamber and in 1936 was awarded the Nobel Prize in Physics for it. The second step entailed the production of positive electrons by means of radioactive sources and the identification of Anderson's particle with the antielectron, whose existence had been suggested by Paul A. M. Dirac in 1931. The whole process took some fifteen months, from Anderson's first communication to *Science* in September 1932 to Dirac's Nobel lecture on his "Theory of Electrons and Positrons" in December 1933.

Beginning in 1930, upon completing his Ph.D. at the California Institute of Technology (Pasadena), Anderson joined in Robert A. Millikan's long-lasting research program on cosmic radiation, regarded since the early 1910s as a very energetic gamma radiation of extraterrestrial origin. Millikan expected to provide new evidence that the energy of incoming cosmic photons corresponded to the mass defect of light atoms as built from hydrogen, as he polemically thought.

Rather than measuring the absorption of primary cosmic rays by means of an ionization chamber, as was common practice, Anderson undertook the study of secondary particles by means of a cloud chamber that fitted into the powerful electromagnet of Caltech's Guggenheim Aeronautical Laboratory. In the cloud chamber, the sudden expansion of a vapor-saturated container prompted the formation of droplets on the ions left by an ionizing particle in its path. Particles were thus visualized as cloud tracks that could be photographed. If the chamber is placed in a magnetic field, the curvature, range, and ionization density along the tracks provide information about the particle mass and velocity as long as the tracks were clearly visible and not affected by turbulence.

By November 1931, Anderson had a dozen good pictures that showed as many positive as negative particles and frequent instances of a "simultaneous ejection." Anderson, who like most physicists stood by the two-particle paradigm that held matter to consist of just electrons and protons, attributed the positive tracks to the one positive particle known at the time, the proton, and pointed to nuclear disintegration as their probable origin. The ionization density of positive tracks, however, was consistent with a particle much lighter than the proton, and Anderson first thought they might well be electrons traveling upward.

Anderson next inserted a lead plate across the chamber. A particle crossing the plate would lose energy, and the increase in the track's curvature would reveal the direction of movement. Through August 1932 Anderson took new photographs and analyzed several hundred tracks. Three of them showed events that were either due to a light positive particle going through the plate or the simultaneous ejection of an electron and a small-positive. This was the basis of Anderson's communication to *Science* announcing "the possible existence of a positive electron" (1932).

Anderson did not relate the new particle to Dirac's antielectron, nor did he refer to Dirac's hole theory as a likely mechanism of production. Dirac's ideas were not widely known nor generally accepted at the time, and Anderson referred instead to Millikan's ideas about the cosmic genesis of elements. The discovery of the positive electron was not prompted by the search for antimatter; indeed, the positive electron was originally not an antiparticle at all.

The new particle was met with caution because of its cosmic descent—the nature of cosmic rays was disputed—and the paucity of visual evidence—Anderson had not published any picture. Anderson's note, however, did not go unnoticed. At the Cavendish Laboratory in Cambridge, UK, Patrick M. S. Blackett and Giuseppe P. S. Occhialini had been working since 1931 on a cloud chamber controlled by an electronic coincidence device. The expansion of the chamber was triggered by the simultaneous discharge of two Geiger-Müller counters, one above and one below the chamber. Cosmic ray particles were thus made "to take their own cloud photographs" (Blackett and Occhialini 1932, 363) and 70 percent of their pictures, as compared with Anderson's 2 percent, showed significant events. Blackett and Occhialini attributed 14 tracks "almost certainly" to positive electrons. They presented their case before the Royal Society on February 16, 1933—an event hailed by the press and reported by *Science Service,* after consulting Anderson, as the "New Particle of Matter Christened 'Positron'" (1933). Dirac's hole theory was referred to by the Cavendish experimentalists as a likely mechanism of production. However, the relationship was not spelled out, and all positive electrons observed so far proceeded from cosmic radiation. Influential physicists such as Niels Bohr and Wolfgang Pauli remained skeptical on both counts.

At the main European centers for radioactive research—the Cavendish in Cambridge, the Institut du Radium in Paris, and the Kaiser-Wilhelm-Institut für Chemie in Berlin—physicists set out to produce positrons in the laboratory by means of radioactive sources, which were far better known and more serviceable than cosmic rays. By late March it was clear that the radiation from a beryllium target exposed to a polonium source—the radiation used to produce neutrons—was also able to produce positrons in lead; early in May several laboratories reported that positrons were ejected from a lead target exposed to high-energy gamma rays, such as those from ThC (^{208}Tl). All the while theoretical physicists, including Rudolf Peierls, Max Delbrück, J. Robert Oppenheimer, and Milton S. Plesset, were trying to make sense of the behavior of positrons by means of Dirac's theory of the electron. Experimental and theoretical developments were assembled in a number of scientific meetings in the fall of 1933, including above all the Seventh Solvay Conference, held in Brussels late in October 1933. By the end of the year, Anderson's particle had been unambiguously identified with Dirac's antielectron, and the positron was born.

The positron thus became the first antiparticle to be discovered. The manipulation of positrons provided early evidence that matter could be created and annihilated—direct confirmation of Einstein's mass-energy relationship. The theoretical analysis of electron-positron interactions helped clarify the stance of early quantum electrodynamics—the quantum theory of the electromagnetic field—and played a key role in the formulation of renormalized quantum electrodynamics in the late 1940s, especially in Richard Feynman's version. Together with the neutron, the positron broke the simple dual paradigm of matter and paved the way for elementary particle physics.

See also: ANDERSON, CARL D.; ANTIMATTER; DIRAC, PAUL

Bibliography

Anderson, C. D. "The Apparent Existence of Easily Deflectable Positives." *Science* **76**, 238–239 (1932).

Anderson, C. D. "Unraveling the Particle Content of Cosmic Rays" in *The Birth of Particle Physics,* edited by L. M. Brown and L. Hoddeson (Cambridge University Press, Cambridge, UK, 1983).

Blackkett, P. M. S., and Occhialini, G. P. S. "Photography of Penetrating Corpuscular Radiation." *Nature* **130**, 363 (1932).

Blackkett, P. M. S., and Occhialini, G. P. S. "Some Photographs of the Tracks of Penetrating Radiation." *Proceedings of the Royal Society of London* **A139**, 613–629 (1933).

Davis, W. "New Particle of Matter Christened 'Positron'." *Science Service* (February 18, 1933).

De Maria, M., and Russo, A. "The Discovery of the Positron." *Rivista di Storia della Scienza* **2**, 237–286 (1985).

Galison, P. *How Experiments End* (The University of Chicago Press, Chicago, 1987).

Hanson, P. *The Concept of the Positron. A Philosophical Analysis* (Cambridge University Press, Cambridge, UK, 1963).

Roqué, X. "The Manufacture of the Positron." *Studies in History and Philosophy of Modern Physics* **28**, 73–129 (1997).

Xavier Roqué

Q

QUANTUM CHROMODYNAMICS

Quantum chromodynamics (QCD) is the component of the Standard Model that describes the strong interactions. QCD is the theory of quarks and gluons. Quarks carry a new charge, called color, that enables them to emit and absorb gluons. (This is the origin of the name chromodynamics, although the "color" of QCD should in no way be confused with the colors of light.) The quarks are also electrically charged and like electrons, are fermions, and carry a spin, or intrinsic angular momentum, of one-half in units of Planck's constant. The gluons are electrically neutral, and, like photons, are bosons of spin one. Together, the fields of quarks and gluons make up a nonabelian gauge theory.

In QCD, quarks interact by the exchange of gluons in much the same way that electrons interact by the exchange of the quanta of light, photons. Like photons, the gluons have no mass and travel at the speed of light. Unlike photons, however, the gluons carry the very color charge that produces them, so gluons can emit and absorb more gluons. The resulting strong force is thus more complicated to analyze than the electromagnetic force.

A convenient measure of the strength of the strong force is the QCD coupling $\alpha_s(Q)$ that controls the probability of a quark emitting a gluon, which produces forces between quarks. The QCD coupling depends on the momentum carried by the emitted gluon, denoted by Q. The strong coupling is large for very low-momentum gluons and decreases as the momentum increases, a variation known as asymptotic freedom. For the highest-momentum gluons that can be produced in modern accelerators, $\alpha_s(Q)$ is relatively small, about 0.1, but at momentum transfers characteristic of nuclear interactions, it gets to be quite large. Asymptotic freedom makes it easier to analyze processes over short times, which generally involve a few gluons, than over long times, which generally involve many.

Quarks come in six varieties, known as quark flavors. In the Standard Model, the six flavors of quarks, together with the six leptons (the electron, muon, tau, and their neutrinos), are truly elementary. The different flavors of quarks have different charges. Three quarks have electric charge $+2e/3$: the up (u), charm (c) and top (t) quarks. Three quarks have charge $-e/3$: the down (d), strange (s) and bottom (b) quarks; $-e$ is the charge of an electron. The masses of these quarks vary greatly, and of the six, only the u and d quarks, which are by far the lightest, appear to play a direct role in normal matter.

Hadrons

QCD binds quarks together into states that can be observed directly in the laboratory as hadrons, particles that feel the strong force. The best-known examples of hadrons are the nucleons, the proton and neutron, from which all atomic nuclei are formed. The idea of quarks arose to explain the regularities of hadron states, and their charges and spins could be readily explained (and even predicted) by simply combining the then known u, d, and s quarks. This is the quark model. Perhaps the most extraordinary feature of quarks in QCD is the confinement of quarks in hadrons. A free quark, one that is separated from a nucleon, would be readily detectable because its charge would be $-\frac{2}{3}$ or $\frac{1}{3}$ that of the charge of an electron. No convincing evidence for such a particle has been found, and it is now believed that confinement is an unavoidable consequence of QCD.

There must be three quarks to make a proton or a neutron. The proton and neutron are different because the proton is a combination of two u quarks and one d quark and hence has a total charge of $2(2e/3) + (-e/3) = +e$. The neutron is made up of one u and two d quarks and hence has total charge $(2e/3) + 2(-e/3) = 0$.

In addition to nucleons, other combinations of three quarks have been observed, and all are known collectively as baryons. For example, from the u, d, and s quarks, it is possible to make ten distinct combinations, and all have been seen (notice that they all have charges that are integer multiples of e). In addition, baryons with c and b quarks have also been observed. All the baryons except for the proton and neutron decay quite rapidly because the weak interactions make all the quarks aside from the u and d unstable. The spins of the quarks that combine to form baryons may line up in parallel and antiparallel combinations, as long as the resulting state obeys the Pauli exclusion principle, which states that no two fermions may have the same set of quantum numbers. As a result, some of the baryons have total spin $\frac{1}{2}$, and others spin $\frac{3}{2}$, and all the baryons are fermions with half-integer total spin. With orbital angular momentum taken into account, even higher spins are possible, although these baryons are very unstable.

In addition to baryons, quarks can combine with their antiparticles, the antiquarks, to form mesons. Antiquarks are usually denoted by an overbar, such as $\bar{u}$ for the antiquark of the u. For example, the combinations $\bar{u}d$ and $\bar{d}u$ can form the π^+ and π^-, the pions, of electric charge ^+e and ^-e, respectively. All the mesons are bosons with integer spins. Other exotic bosons, which are bound states of gluons without quarks, called glueballs, appear to be possible, although very unstable.

The Color Charge and Gauge Theory

The concept of color was introduced to solve a problem in the quark model. The lowest-energy states in the quark model appeared to require that the three quarks in the proton be in identical states, which would violate the Pauli principle. With color, this conflict is avoided. For example, two u quarks can have the same energy and spin quantum numbers as long as their colors are different. Consistency with the Paul principle requires that all the quarks in any baryon have different colors: three quarks, three colors.

QCD is an example of a nonabelian gauge theory. In a nonabelian gauge theory, the concept of charge must be generalized from electromagnetism. An electric charge is just a number, such as e or $2e/3$, and the electric charge stays the same when a charged particle emits or absorbs a photon. A quark, however, actually has three separate charges that make up its color, which are sometimes labeled by the names of colors of light: a red (r) charge, a green (g) charge, and a blue (b) one. When a quark emits a gluon, its color charges change, and the kinds of gluons can be identified by combinations of the color "before" and the color "after," for example, $\bar{r}b$, where the color with the overbar is the final color of the quark. Correspondingly, when a quark with only color b absorbs an $\bar{r}b$ gluon, it changes into a quark with only color r. This way, the total of r, g, and b charges are conserved, and nine possibilities exist for the gluons. Of these nine, one combination, equal parts of $\bar{r}r$, $\bar{g}g$, and $\bar{b}b$, leaves all three of the colors the same and is absent, leaving eight gluons. The color charges in QCD have a surprising property: they can never be distinguished experimentally, and yet their number, three, can be measured. All hadrons have zero net *r*, *b*, and *g* colors; they are all colorless.

Evidence for Color, Quarks, and Gluon

Experimental discoveries beginning in the late 1960s and early 1970s established firmly the reality of hadrons made up of quarks and gluons. As seen, quarks carry an electric charge, as do electrons. When an energetic electron collides with a target, the electron, which is blind to the strong force, nevertheless scatters from nucleons within the target by exchanging a photon. When the energy of the electron is high, the momentum p' of the photon can be large, so large that the corresponding wavelength of light is much smaller than a nucleon in the target. This wavelength is given by $\lambda = h/p$, where h is Planck's constant. The rules of quantum mechanics indicate that a photon cannot be absorbed by an object that is larger than its wavelength λ.

Nevertheless, the photons *are* absorbed at a rate almost independent of p, a phenomenon called scaling. When this happens, the nucleon generally breaks apart into high-energy fragments, a process called deep-inelastic scattering. The scaling of deep-inelastic scattering indicates that there are charged particles within nucleons that are much smaller than nucleons. These are the quarks. In addition, the distribution in angles of the scattered electrons depends on the spin of the charged particles. The electron-nucleon experiments show that all the charge in the nucleons is carried by spin-$\frac{1}{2}$ particles, exactly as suggested by the quark model.

In another set of experiments, electrons and their antiparticles, positrons, collide and annihilate into a virtual state that consists of a single photon. According to the rules of quantum mechanics, this photon then transforms itself into any pair of particle and antiparticle with nonzero electric charge, say, q. The probability for each species is proportional to q^2, and the angular distribution at which they emerge depends on their spin. The quarks created this way are not observed directly because of confinement, but they each quickly evolve into jets of hadrons, which preserve their original directions. The total probability to produce hadrons is given simply by the probability to produce a single fermion of electric charge equal to 1, times the sum of the squares of the electric charges of all the quarks light enough to be produced at the available energy, times 3. The 3 stands for the three possible colors of each quark.

Many other high-energy experiments have confirmed the reality of quarks, gluons, and color. For example, a fraction of annihilation events include an extra jet from a gluon in addition to their quark and antiquark jets. In proton-proton scattering, jets also emerge from collisions, whose angular distributions exactly match predictions based on the elastic scattering of quarks through the exchange of gluons. Each of these predictions depends directly on the numbers of quark and gluon colors as well as their spins.

Physicists' understanding of the confinement and other low-energy properties of QCD is somewhat less complete but is still convincing. Qualitatively, it appears that the "empty" space—the vacuum—is not empty of QCD content. Analysis suggests that the vacuum acts as a sort of QCD superconductor, which repels the QCD lines of force between quarks, the QCD analogs of magnetic and electric fields between electrons. Within hadrons, this repulsion is absent. The larger the separation between isolated quarks, however, the larger the energy necessary to overcome the repulsion. Very quickly, it becomes easier to create enough pairs of quarks and antiquarks to hide all the lines of force within colorless hadrons than to lengthen them through the resistant vacuum.

Beginning in the 1970s, it became possible to simulate QCD on computers. Over time, more and more precise calculations have established that the energy of an isolated quark is essentially infinite and that the observed hadrons are, indeed, combinations of confined quarks. Experiments in which entire nuclei collide may create conditions in which quarks and gluons are temporarily freed, or deconfined. This would be a novel state of matter, called the quark-gluon plasma.

It is now understood that the Standard Model actually requires that there be three colors to avoid quantum inconsistencies, called anomalies, which would ruin it as a quantum theory. Nevertheless, physicists are far from a complete understanding of QCD. For example, quantum fluctuations in the QCD vacuum appear to have the ability to act differently on particles and antiparticles in a manner that is not seen in nature, a puzzle known as the strong CP problem. In this, and in how to reconcile fully the quark-gluon and baryon-meson descriptions of the strong interactions, we have much still to learn about quantum chromodynamics.

See also: AXION; ASYMPTOTIC FREEDOM; BOSON, GAUGE; PARTICLE PHYSICS, ELEMENTARY; QUANTUM MECHANICS; QUARKS; STANDARD MODEL

Bibliography

Feynman, R. *QED, The Strange Theory of Light and Matter* (Princeton University Press, Princeton, 1988).

Hey, Y., and Walker, P. *The Quantum of the Universe* (Cambridge University Press, Cambridge, UK, 1987).

Johnson, G. *Strange Beauty* (Vintage, New York, 1999).

Kane, G. *Modern Elementary Particle Physics* (Perseus, Cambridge, MA, 1993).

Zee, A. *A Fearful Symmetry* (Princeton University Press, Princeton, 1999).

George Sterman

QUANTUM ELECTRODYNAMICS

Quantum electrodynamics, also known by its acronym, QED, is a relativistic quantum field theory that describes at a fundamental level the electromagnetic interactions among electrically charged elementary particles such as electrons, positrons, muons, and quarks. Remarkably simple in form, it nevertheless respects the principles of special relativity and quantum mechanics, two of the great scientific revelations of the twentieth century. In addition, QED is essentially a complete theory of the electron's electromagnetic interactions and therefore provides a dynamical basis for atomic physics and all natural phenomena that spring from it, including chemistry, biology, and technology. At the quantitative level, QED predictions, as later briefly surveyed, have been tested to nearly one part in 100 billion, making it the most successful physical theory ever devised.

In its simplest form, which will primarily be discussed here, QED combines James Maxwell's equations for electric and magnetic fields with Paul Dirac's quantum theory of electrons. When intro-

duced, its novelty was to provide a quantization of electromagnetic fields that provided a particle interpretation in terms of massless quanta called photons. (That idea actually had its origin in Max Planck's quantum light theory that was invented to explain blackbody radiation in 1900.) Also, the relativistic quantum description of the electron requires, as shown by Dirac, that it have an antiparticle partner called the positron (given that name because of its opposite sign "positive" electric charge) which can annihilate or be pair produced with electrons. Thus, pure QED can be viewed as a fundamental theory of interacting electrons, positrons, and photons. Easily extended to other heavier charged particles such as muons and quarks, it can also be applied in the non-relativistic (low-velocity) limit that is often more appropriate for many-body condensed-matter or quantum optics systems.

QED is a special kind of quantum field theory referred to as renormalizable. That means short-distance quantum fluctuations (called quantum loop corrections) which are common to all QED processes and are, in fact, infinite (ultraviolet divergent) can be absorbed into the definition of the measured electron mass m_e and electric charge e. This "renormalization" of charge and mass renders the predictions of the theory finite and unique.

Understandably, the occurrence of infinite quantum effects caused much concern in the early days of QED. To some, renormalization appeared to be a way of sweeping a fundamental defect in QED under the rug rather than confronting it. In more recent times, however, QED has been viewed as an effective theory that must break down at very short distances (high energies) and be subsumed by a larger more complete theory. The infinities are expected to be artifacts of not including that still largely unknown, new short-distance physics. Renormalization of QED absorbs all those unknown short-distance physics effects into the measured value of m_e and e. Predictions from QED in terms of those renormalized parameters are then insensitive to such unknowns. The validity and power of this approach are confirmed by the quantitative success of QED as a stand-alone theory.

In contemporary elementary particle physics, QED is actually only part of a more complete theory called the Standard Model that describes strong, weak, and electromagnetic interactions. It, like QED, is based on symmetry considerations and the principle of local gauge invariance that will be illustrated via QED.

To appreciate the origins of gauge symmetries as fundamental descriptions of nature, it is instructive to consider how one introduces electromagnetic interactions into Dirac's theory of electrons. Consider the electron field $\psi(x)$ that depends on the space-time coordinate x. $\psi(x)$ itself is called a four-component spinor field which indicates that under relativistic Lorentz transformations (i.e., space-time translations, rotations, and velocity boosts), it acts like a field with spin $\frac{1}{2}$, where spin is the intrinsic angular momentum in units of $\hbar$, Planck's constant divided by 2π, carried by the electron. With no interactions the four-component field $\psi(x)$ should satisfy Dirac's free field equation (hence, natural units are employed with $\hbar$ and the speed of light c set equal to 1):

$$\left(i \frac{\partial}{\partial x^{\mu}} - m_e^0 \right) \gamma^{\mu} \psi(x) = 0 \qquad (1)$$

where m_e^0 is the bare electron mass (i.e., before interactions are turned on), γ^{μ}, $\mu = 0, 1, 2, 3$ are 4×4 Dirac matrices, and the repeated index μ is summed over. Solutions to that equation represent the evolution of a free (noninteracting) electron field as a function of the space-time coordinate. Similarly, the electromagnetic field potential $A_{\mu}(x)$ has four components labeled by the index μ. Under Lorentz transformations it acts like a spin-1 field. It satisfies the free field Maxwell equations.

To go from a theory of classical fields to a quantum field theory, one second quantizes the theory. That entails replacing $\psi(x)$ and $A_{\mu}(x)$ with anticommuting (for spin $\frac{1}{2}$) and commuting (for spin 1) operators that can create or annihilate electron (positron) and photon states.

So far, the equations given above describe noninteracting particles. An elegant formalism for introducing interactions between electrons and photons is the principle of local gauge invariance. One notes that the free theory is invariant under what are called U(1) global phase rotations:

$$\psi(x) \rightarrow e^{ie_0\theta} \psi(x) \qquad (2)$$

where $0 \leq \theta < 2\pi$, and e_0 is the bare electron's charge. If $\psi(x)$ is a solution to the Dirac equation, then $e^{ie_0\theta}\psi(x)$ is also a solution. Physics is invariant under such phase changes. The set of possible phase transformations form a U(1) group. That symmetry is associated with the conservation of electric charge (carried by the electron). It is a global symmetry because the same phase change is made at all space-time points x, that is, θ is a constant.

The free Dirac equation is not invariant under local transformations of the more general form

$$\psi(x) \rightarrow e^{ie_0\theta(x)} \psi(x) \qquad (3)$$

where $\theta(x)$ now varies with x. To render the theory invariant under local phase transformations, one requires the simultaneous change

$$A_{\mu}(x) \rightarrow A_{\mu}(x) + \frac{\partial}{\partial x^{\mu}} \theta(x) \qquad (4)$$

called a gauge transformation on the electromagnetic potential and replaces the derivative in the Dirac equation by the so-called covariant derivative

$$\frac{\partial}{\partial x_{\mu}} \rightarrow \frac{\partial}{\partial x^{\mu}} - ie_0 A_{\mu}(x). \qquad (5)$$

(Maxwell's free equations are already invariant under gauge transformations.) That change makes the full combined theory gauge invariant, at the expense of introducing a coupling or interaction term between the photon and the electron field.

$$e_0 A_{\mu} \overline{\psi} \gamma^{\mu} \psi \qquad (6)$$

When the fields are quantized, that important term causes an interaction that allows quantum excitations of electrons, positrons, and photons characterized by coupling strength e_0 between those particles. It is responsible for all electromagnetic interactions of electrons. The space-time dependent U(1) symmetry is called a local gauge symmetry, and $A_{\mu}(x)$ is called a gauge field. It can be thought of as a connection field that allows invariance under local electron field rephasing or changing of the gauge.

Treating the interaction in (6) as a perturbation on an otherwise free set of fields, one can compute QED interaction effects as quantum loop expansions in the small fine-structure constant

$$\alpha = \frac{e^2}{4\pi} \cong \frac{1}{137} \qquad (7)$$

where e is the renormalized electric charge in which infinite (and finite) short-distance vacuum polarization quantum corrections from e^+e^- virtual pairs have been absorbed. Similarly, the bare mass m_e^0 is replaced by a renormalized measurable physical mass m_e that eliminates the remaining short-distance infinities from electron self-energy quantum corrections.

Although the roots of QED date back much earlier, a systematic program of renormalization and calculational formalism was developed by Sinitiro Tomonaga, Richard Feynman, and Julian Schwinger in the 1940s. In particular, the Feynman diagram approach gave a simple systematic method of calculation. Quantum loop corrections were then carried out for various atomic physics properties, scattering processes, static electron properties, etc. Those calculations were compared with very precise experimental measurements such as the Lamb shift and hyperfine atomic structure, which confirmed the validity of QED with spectacular success. Such measurements also helped advance the state of the art in atomic physics experimentation with developments in laser and other sophisticated technologies. As an illustration of experimental progress in QED, consider the determination of the fine-structure constant α. In Table 1, four precise α values (actually its inverse) obtained in very different electromagnetic situations and their weighted average are given. Two involve quantized condensed matter effects and two come from atomic measurements.

The good agreement of such different methods to many significant figures confirms the validity and universality of QED. Perhaps even more impressive is the comparison of theory and experiment for the anomalous magnetic moment a_e of the electron and positron. That quantity describes the deviation of the gyromagnetic ratio g_e from its Dirac equation value of 2 (without quantum fluctuations):

$$a_e = \frac{g_e - 2}{2}. \qquad (8)$$

A deviation from zero is predicted as a result of QED quantum loop effects. Starting with the famous leading effect $a_e = \alpha/2\pi$ calculated by Schwinger, high-powered analytic and computer-aided calculations now give the prediction

$$
\begin{aligned}
a_e^{\text{QED}} = {} & \frac{\alpha}{2\pi} - 0.328478444\left(\frac{\alpha}{\pi}\right)^2 \\
& + 1.181234\left(\frac{\alpha}{\pi}\right)^3 - 1.5098\left(\frac{\alpha}{\pi}\right)^4 \\
& + \cdots + 1.66 \times 10^{-12} \\
& \text{(strong and weak effects)} \qquad (9)
\end{aligned}
$$

where the last term corresponds to very small calculable strong and weak interaction effects that lie outside the framework of QED. Comparing that prediction with the experimental average of electron and positron measurements (which are in perfect agreement with each other)

$$a_e^{\text{exp}} = 1{,}159{,}652{,}188 \pm 3 \times 10^{-12} \qquad (10)$$

leads to the current best determination of α:

$$\alpha^{-1} = 137.03599959(40). \qquad (11)$$

Agreement with the average in Table 1 is impressive. It confirms QED to about a few parts in 10^{11}! An alternative method of comparison is to predict a_e using the average α of Table 1 in (9). That procedure gives

TABLE 1

Values of the Fine-structure Constant Extracted from Different Experiments

Numbers in parentheses indicate $\pm$ uncertainties in the last three digits.

α^{-1} Value	Method
137.03600300(270)	Quantum Hall Effect
137.03600840(330)	Rydberg + $h/m_{neutron}$ Measurements
137.03598710(430)	AC Josephson Effect
137.03599520(790)	Muonium (μ^+e^-) Hyperfine Structure
137.03600140(183)	Average

CREDIT: Mohr, P., and Taylor, B. "CODATA Recommended Values of the Fundamental Physical Constants: 1998." *Review of Modern Physics* **72**, 351 (2000).

TABLE 2

Elementary Particles

Particle	Symbol	Spin	Charge	Colors	Mass (GeV)	
Electron Neutrino	ν_e	1/2	0	0	$<2 \times 10^{-9}$	
Electron	e	1/2	−1	0	0.51×10^{-3}	1st
Up Quark	u	1/2	2/3	3	5×10^{-3}	generation
Down Quark	d	1/2	−1/3	3	9×10^{-3}	
Muon neutrino	ν_μ	1/2	0	0	$<2 \times 10^{-9}$	
Muon	μ	1/2	−1	0	0.106	2nd
Charm quark	c	1/2	2/3	3	1.25	generation
Strange quark	s	1/2	−1/3	3	0.175	
Tau neutrino	ν_τ	1/2	0	0	$<2 \times 10^{-9}$	
Tau	τ	1/2	−1	0	1.777	3rd
Top quark	t	1/2	2/3	3	174.3	generation
Bottom quark	b	1/2	−1/3	3	4.5	
Photon	γ	1	0	0	0	
W boson	$W^\pm$	1	±1	0	80.43	Gauge
Z boson	Z	1	0	0	91.187	bosons
Gluon	g	1	0	8	0	
Higgs scalar	H	0	0	0	114~200?	

Credit: Courtest of WIlliam Marciano.

$$a_e^{\text{QED}} = 1,159,652,172 \pm 16 \times 10^{-12} \quad (12)$$

which agrees with experiment. Future improvement in a_e^{exp} by a factor of 10 is expected. To utilize such a result will require a much better independent determination of α and further improvements in the QED perturbative calculation of a_e^{QED}.

QED is now generally recognized to be part of a larger theory called the Standard Model that also describes strong and weak interactions. That theory is a generalization of the local gauge invariance principle used in QED. The Standard Model is based on the enlarged local gauge symmetry group $SU(3)_c \times SU(2)_L \times U(1)_Y$ where c denotes color, and the $SU(3)_c$ part of the theory is called quantum chromodynamics, L indicates that only left-handed chiral components of spin-$\frac{1}{2}$ (fermion) particles exhibit the $SU(2)$ symmetry, and Y stands for weak hypercharge. $SU(N)$ is the symmetry group of $N \times N$ unitary matrices. That local gauge invariance requires the introduction of twelve gauge bosons: eight gluons, $W^\pm W^0$, and B spin-1 fields. Unlike QED, $SU(2)_L \times U(1)_Y$ is not an exact symmetry but is broken down to its $U(1)_{\text{QED}}$ subgroup, which remains intact. The breaking mechanism (called the Higgs mechanism) endows three of the gauge bosons $W^\pm$ and Z (a linear combination of W^0 and

B) with masses, while the photon (the orthogonal combination of W^0 and B) and eight gluons remain massless. The particle content of the Standard Model is illustrated in Table 2. All the particles listed there have been discovered except the elusive spin-zero Higgs scalar left over as a remnant of the Higgs mechanism.

The Standard Model, like QED, is renormalizable. All its masses, couplings, and mixing parameters must undergo infinite renormalizations. Quantum loop effects in that framework have been computed and compared with a large body of precise experimental measurements. Those confrontations have confirmed the validity of the Standard Model and its quantum loop corrections at the 0.1 percent level. Although not as impressive as the 10^{-11} tests of QED, the electroweak tests of the Standard Model are much more powerful probes of "new physics" such as additional heavy mass particles or new (as yet undiscovered) interactions. Thus, for example, precise measurements of $W^\pm$ and Z masses along with decay properties of muons, Z bosons, etc. have been used in conjunction with quantum loop calculations to give the bound

$$m_H < 200 \text{ GeV} \quad (13)$$

whereas direct (negative) searches yield

$$m_H > \approx 114 \text{ GeV} \qquad (14)$$

Therefore, with that predicted mass range rather narrow and relatively close at hand, it seems that the discovery of the Higgs particle should be possible either at Fermilab's existing Tevatron proton-antiproton collider facility or, if not, at the higher energy collider called the Large Hadron Collider (LHC) being constructed in Switzerland at the European Laboratory for Particle Physics (CERN). Discovery of the Higgs will complete the minimal content of the Standard Model but will leave unanswered many of the perplexing questions that plagued QED: What additional physics lies at shorter distances than are currently being explored? Does it finally tame the infinities of QED and the complete Standard Model? How do we unify gravity with local gauge interactions? Why is nature governed by local gauge invariance? Those fundamental questions and others drove scientists to explore beyond QED and establish the Standard Model. They are likely to continue guiding them in the quest to further unveil the simplicity and intricacy of nature.

See also: BOSON, GAUGE; DIRAC, PAUL; FEYNMAN, RICHARD; GAUGE THEORY; PAULI, WOLFGANG; QUANTUM FIELD THEORY; RENORMALIZATION; SCHWINGER, JULIAN; STANDARD MODEL; TOMOGANA, SIN-ITIRO

Bibliography

Abers, E., and Lee, B. W. "Gauge Theories." *Physics Reports* **9**, 1–143 (1973).

Kinoshita, T., ed. *Quantum Electrodynamics* (World Scientific, Singapore, 1990).

Marciano, W., and Pagels, H. "Quantum Chromodynamics." *Physics Reports* **36C**, 137–276 (1978).

Mohr, P., and Taylor, B. "CODATA Recommended Values of the Fundamental Physical Constants: 1998." *Reviews of Modern Physics* **72** (2), 351–495 (2000).

Peskin, M., and Schroeder, D. *An Introduction to Quantum Field Theory* (Addison-Wesley, New York, 1995).

Schwinger, J., ed. *Quantum Electrodynamics* (Dover, New York, 1958).

William J. Marciano

QUANTUM FIELD THEORY

Quantum field theory is the widely accepted tool for describing systems such as the electromagnetic field, in accordance with the laws of quantum mechanics as well as with other restrictions depending on the context. In elementary particle physics, relativistic quantum field theory, whose predictions respect the principle of relativity, is used. Quantum field theory is also used extensively in other areas such as superconductivity, the quantum Hall effect, and statistical mechanics.

Knowledge of some ideas from mechanics and quantum mechanics is necessary in order to understand quantum field theory. Consider a single mass, coupled to an anchored spring and vibrating along the *x*-direction on a frictionless surface. This oscillator is a system with one degree of freedom, since only one coordinate *x* is needed to describe it. A system of *N* masses moving along the *x*-axis would have *N* degrees of freedom. A field is a system with infinite number of degrees of freedom. For example, if one had an infinite array of masses in a line, coupled to their neighbors by springs, the displacements of each mass from equilibrium would constitute a field. Another example is the electromagnetic field whose degrees of freedom are the values of the electric and magnetic field at each point in space. Notice that the latter has an infinite number of degrees of freedom within any finite volume, and this causes many problems. The aim of quantum field theory is to describe these degrees of freedom according to the laws of quantum mechanics.

The field can be arrived at by starting with the simplest system: a single mass *m* attached to a spring of force constant *k*. In classical mechanics this mass will vibrate with angular frequency $\omega = 2\pi f = \sqrt{k/m}$. It can have any finite amplitude x_0, and the corresponding energy will be $E = \frac{1}{2} k x_0^2$.

In the quantum version the oscillator can only have energy $E = \hbar\omega(n + \frac{1}{2})$, where $n = 0, 1, 2, \ldots$, and $\hbar = 1.05 \cdot 10^{-34}$ J/s is Planck's constant. Note that the lowest energy is not zero but $\frac{1}{2}\hbar\omega$, which is referred to as zero-point energy. The uncertainty principle in quantum mechanics, which forbids a state of definite location and momentum, does not allow the classical zero-energy state in which the mass sits at rest in the equilibrium position. Next, the fact that the levels are equally spaced means that instead of saying the oscillator is in a state labeled by the quantum number n, one can say that there are n quanta of energy $\hbar\omega$. This seemingly semantic point proves seminal, as shall be seen.

Now consider two masses m attached to springs of force constant k as in Figure 1. The coordinates x_1 and x_2 are coupled to each other, and the motion is quite complicated. But consider the following combination of coordinates called normal coordinates:

$$X_1 = \frac{x_1 + x_2}{\sqrt{2}}, \quad X_2 = \frac{x_1 - x_2}{\sqrt{2}}.$$

These behave like independent oscillators of frequency $\omega = \sqrt{k/m}$, $\sqrt{3k/m}$, respectively. To see this, imagine starting off the masses with equal displacements, so that X_1 is nonzero and $X_2 = 0$. Since the middle spring is undistorted, the masses will begin moving in response to the end springs. Since these are identical, the condition $x_1 = x_2$ (the same as $X_2 = 0$) will be preserved for all times, and the coordinate X_1 will vibrate at $\omega = \sqrt{k/m}$. On the other hand, if both masses are given equal and *opposite* displacements, then $X_1 = 0$ initially. The middle spring is distorted twice as much as the end springs (so the effective force constant felt by the masses is $3k$), and the masses vibrate with equal and opposite displacements at $\omega = \sqrt{3k/m}$. That is, $X_1 = 0$ for all times, and X_2 vibrates at $\omega = \sqrt{3k/m}$. If some arbitrary initial displacements are given so that both X_1 and X_2 are nonzero, one can compute their future values (which is easy since they behave like independent os-

cillators) and go back to x_1 and x_2 at the end. The same strategy used for the two-mass problem also works for any number of masses.

Thus, consider N such masses coupled to their neighbors by N such springs of equilibrium length a arranged around a circle of circumference $L = Na$. (units $m = k = 1$.) Let $\phi(n)$ be the displacement of mass numbered n, where $1 \leq n \leq N$. The system can once again be reduced to those of decoupled oscillators. The normal coordinates are just the Fourier coefficients:

$$\varphi(K) = \sqrt{\frac{1}{N}} \sum_n \phi(n) e^{jKna}.$$

The requirement that the points numbered n and $n + N$ are one and the same implies

$$K = \frac{2\pi r}{Na}, \quad r = 0, \pm 1, \pm 2, \ldots$$

whereas the fact that K and $K + 2\pi/a$ are indistinguishable in the exponential factor limits r to the range of size N and K to an interval of width $2\pi/a$, which is chosen to be $-\pi/a \leq K \leq \pi/a$. The system is equivalent to N oscillators whose frequencies can be shown to be $\omega(K) = 2[1 - \cos(Ka)]$. If one lets $N \to \infty$, the allowed values of K become continuous in the interval $[-\pi/a, \pi/a]$. The field ϕ is a classical field.

The quantum version will be obtained by treating the decoupled oscillators in terms of quantum mechanics. Each oscillator of frequency ω can only have energy $E = \hbar\omega(n + \frac{1}{2})$, where $n = 0, 1, 2, \ldots$. This means that even in the ground state the field has zero-point energy $\frac{1}{2}\hbar\omega$ per oscillator, which can have observable effects. (The zero-point energy of the electromagnetic field leads to the spontaneous decay of atoms.) Next, the fact that the levels are equally spaced means that instead of saying the oscillator is in a state labeled by n, one can say that there are n quanta of energy $\hbar\omega$ and momentum $\hbar K$, where the identification of $\hbar K$ with momentum comes from examining the interaction of the system, described in terms of the decoupled oscillators, with any external probe. Thus, the quantum state of the field is specified by saying how many quanta there

FIGURE 1

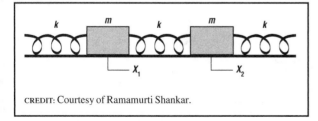

CREDIT: Courtesy of Ramamurti Shankar.

The system of two masses m coupled by springs of force constant k. Their displacements from equilibrium are given by X_1 and X_2.

are at each K. In the present problem of vibrating atoms the phenomenon is just sound, and the quanta are called phonons.

If these methods are applied to the electromagnetic field, which has degrees of freedom at each point in space, that is, $a = 0$, the allowed K values will go from $-\infty$ to ∞, a consequence of which soon follows. The quanta of this field are referred to as photons.

Both quanta above come from bosonic oscillators, for which the quantum number n is not restricted. When a macroscopic number of bosonic quanta, say, photons, are present in a state with some K, it is perceived as a classical (electromagnetic) field at that wave number and the corresponding energy density. To describe fermions like electrons as quanta, one needs to quantize a fermionic oscillator that cannot support more than one quantum. This reflects the Pauli exclusion principle, which says that no two electrons can have the same quantum numbers. There is no classical manifestation of such fermionic fields, which is why they are so unfamiliar.

In the cases considered, the number of quanta in each oscillator stays fixed, a result of the fact that the total energy is a quadratic function in the coordinates and velocities, which, in turn, is why normal coordinates that evolve independently of each other. Upon adding higher-order interaction terms, a change in these numbers may occur. Since the quanta correspond to particles, this means either that particles change their energy or momentum values or that new particles are created. Consider quantum electrodynamics (QED), in which a term quadratic in the electron field and linear in the photon field is added to the total energy with a coefficient e, which is the electron's charge. This cubic term describes a process in which an electron emits or absorbs a photon. This can cause, among other things, two electrons to scatter from their original momentum states to new ones.

The results are computed in a perturbation series in e or, equivalently, $\alpha = e^2/\hbar c \approx 1/137$, called the fine structure constant. Various contributions to the series are represented by Feynman diagrams. For example, Figure 2(a) shows a process second order in e (or first order in α) in which an electron emits a photon and recoils, while another captures the photons and recoils the other way. *Thus, the quanta*

of the field (photons) also mediate interactions between other quanta (electrons). If one goes to higher orders in the expansion, more complicated scattering diagrams may be obtained, say, a diagram where one of the electrons emits two photons, and then either both are absorbed by the other, or one is absorbed by the emitter itself and one is absorbed by the other. The Feynman diagrams go on to infinite order, and one usually stops after a few terms and obtains excellent numbers since α is so small. In a general theory, there may be no such small parameter. But even with the small α, there is a problem extracting predictions in QED due to the following divergence problem.

Consider a lone electron that emits a photon and reabsorbs it, so that for a while one has an electron and a photon, as illustrated in Figure 2(b). This can be shown to change the observed mass of the electron from the value m_0 to

$$m = m_0 + \alpha I$$

where m is the observed mass, calculated to order α in perturbation theory, and I is an integral that sums over the various electron-photon states that can occur between the emission and reabsorption. This sum diverges because of the infinite range of momenta for the quanta. (This does not happen in Fig-

FIGURE 2

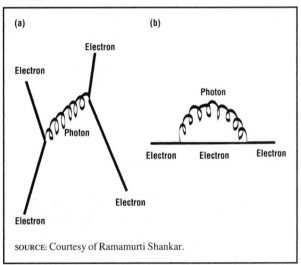

SOURCE: Courtesy of Ramamurti Shankar.

Elementary Feynman diagrams. In (a) two electrons scatter off each other by exchanging a photon. In (b) an electron emits and then reabsorbs a photon.

ure 2(a) since the photon's momentum is fixed by those of the electrons and hence not summed over. In Figure 2(a), the *sum* of the momenta of the electron and photon is fixed by that of the incoming electron but not their individual pieces, one of which can be chosen at will.)

Since the observed mass of the electron is finite, one resorts to renormalization, in which all energy and momentum sums are cut off at some large value Λ. Thus, $m = m_0 + \alpha I(\Lambda)$, but one now requires that m_0 itself be Λ dependent in such a way that m, which is the observed mass, is finite and Λ independent. Next, one finds that at higher orders the scattering rate of two electrons is infinite as well. One now says that the coupling is not given by $\alpha \simeq 1/137$ but by $\alpha_0(\Lambda)$, chosen so that the scattering rate turns out to be finite and cut-off independent and corresponds to the measured value of $\alpha \simeq 1/137$. *Remarkably enough, once these parameters are thus chosen, no new infinities arise,* a feature called renormalizability. Renormalizability has been a guiding principle in arriving at the theory of strong interactions (called quantum chromodynamics or QCD) and in the unified theory of electromagnetic and weak interactions, the *Glashow-Weinberg-Salam (GWS) model.* In recent years, a far deeper understanding of the renormalizability of quantum field theories has emerged by casting them in terms of the mathematically equivalent problem of phase transitions in statistical mechanics.

An important guide in searching for the right field theory, that is, the right set of interaction terms, is symmetry. In particle physics the predictions have to be invariant under space-time symmetries such as translations, rotations, and Lorentz transformations, or internal symmetries, such as isospin symmetry, in which a proton and neutron are exchanged. Invariance under translations or rotations means that an experiment will give the same answer if all the relevant parts of the experiment are translated (shifted) to a new location or rotated. Each symmetry implies a conservation law. For instance, translation symmetry leads to the conservation of momentum, whereas rotational symmetry leads to the conservation of angular momentum. Local gauge symmetry, which is the invariance of the theory to a particular redefinition of the fields that varies from point to point in space-time, is enjoyed by QED, QCD, and the GWS model.

Typically, the lowest-energy state or vacuum state is invariant under all the symmetries of the interactions. For example in a theory invariant under rotations, the lowest-energy state will look the same when rotated. In a magnet, however, the ground state may be magnetized in some direction and will therefore look different when rotated. Now, the direction of magnetization is chosen randomly from all possible directions, none of which is favored intrinsically by the microscopic interactions. This is referred to as a spontaneous breakdown of symmetry and can occur in the field theories of particle physics. According to Goldstone's theorem, for every spontaneously broken symmetry, there is one massless excitation or particle, called a Goldstone boson. The nearly massless pion is the Goldstone boson corresponding to the nearly exact chiral symmetry. There is one way out of Goldstone's theorem. If the variable that breaks the symmetry (like the magnetization) interacts with a massless gauge field, then instead of a massless Goldstone boson, one ends up with a massive gauge field. This Higgs mechanism plays a crucial role in the unified theory of the GWS whereby the massive gauge bosons that mediate weak interactions start out massless and become massive due to the Higgs mechanism.

All quantum field theories are formulated in terms of point particles like electrons and photons. In the last two decades of the twentieth century an alternate view, in which the building blocks are strings, came into prominence. It is being vigorously studied and the correct answer is not yet known.

See also: DIRAC, PAUL; FERMI, ENRICO; GAUGE THEORY; PAULI, WOLFGANG; RENORMALIZATION; SALAM, ABDUS; SCHWINGER, JULIAN; TOMONAGA, SIN-ITIRO

Bibliography

Abrikosov, A. A.; Gorkov, L. P.; and Dzyaloshinski, I. E. *Methods of Quantum Field Theory in Statistical Mechanics* (Dover, New York, 1963).

Green, B. *The Elegant Universe: Superstrings, Hidden Dimensions, and the Quest for the Ultimate Theory* (Vintage Books, New York, 2000).

Itzykson, C., and Zuber, J. B. *Quantum Field Theory* (McGraw Hill, New York, 1980).

Mahan, G. D. *Many Body Physics* (Plenum, New York, 1981).

Peskin, M. E., and Schroeder, D. V. *Introduction to Quantum Field Theory* (Perseus Press, New York, 1995).

Shankar, R. *Principles of Quantum Mechanics* (Plenum, New York, 1994).

Weinberg, S. *The Quantum Theory of Fields* (Cambridge University Press, Cambridge, UK, 1995).

Ramamurti Shankar

QUANTUM MECHANICS

What is quantum mechanics? An answer to this question can be found by contrasting quantum and classical mechanics. Classical mechanics is a framework—a set of rules—used to describe the behavior of ordinary-sized things: footballs, specks of dust, planets. Classical mechanics is familiar to everyone through commonplace activities like tossing balls, driving cars, and chewing food. Physicists have studied classical mechanics for centuries (whence the name "classical") and developed elaborate mathematical tools to make accurate predictions involving complex situations: situations like the motion of satellites, the twist of a spinning top, or the jiggle of jello. Sometimes (as in the spinning top) the results of classical mechanics are unexpected, but always the setting is familiar due to one's daily interaction with ordinary-sized things.

Quantum mechanics is a parallel framework used to describe the behavior of very small things: atoms, electrons, quarks. When physicists began exploring the atomic realm (starting around 1890), the obvious thought was to apply the familiar classical framework to the new atomic situation. This resulted in disaster; the classical mechanics that had worked so well in so many other situations failed spectacularly when applied to atomic-sized situations. The obvious need to find a new framework remained the central problem of physics until 1925, when that new framework—the framework of quantum mechanics—was discovered by Werner Heisenberg. Quantum mechanics does *not* involve familiar things, so it is not surprising that both the results and the setting are often contrary to anything that we would have expected from everyday experience. Quantum mechanics is not merely unfamiliar; it is counterintuitive.

The fact that quantum mechanics is counterintuitive does not mean that it is unsuccessful. On the contrary, quantum mechanics is the most remarkably successful product of the human mind, the brightest jewel in our intellectual crown. To cite just one example, quantum mechanics predicts that an electron behaves in some ways like a tiny bar magnet. The strength of that magnet can be measured with high accuracy and is found to be, in certain units,

$$1.001\ 159\ 652\ 188$$

with a measurement uncertainty of about four in the last digit. The strength of the electron's magnet can also be predicted theoretically through quantum mechanics. The predicted strength is

$$1.001\ 159\ 652\ 153$$

with about seven times as much uncertainty. The agreement between experiment and quantum theory is magnificent: if I could measure the distance from New York to Los Angeles to this accuracy, my measurement would be accurate to within the thickness of a silken strand.

So, what does this unfamiliar quantum mechanical framework look like? Why did it take thirty-five years of intense effort to discover? The framework has four pillars: quantization, probability, interference, and entanglement.

Quantization

A classical marble rolling within a bowl can have any energy at all (as long as it's greater than or equal to the minimum energy of a stationary marble resting at the bottom of the bowl). The faster the marble moves, the more energy it has, and that energy can be increased or decreased by any amount, whether large or small. But a quantal electron moving within a bowl can have only certain specified amounts of energy. The electron's energy can again be increased or decreased, but the electron cannot accept just any arbitrary amount of energy: it can only absorb or emit energy in certain discrete lumps. If an attempt is made to increase its energy by less than the minimum lump, it will not accept any energy at all. If an attempt is made to increase its energy by two and a half lumps, the electron will accept only two of them. If an attempt is made to decrease its energy by four and two-thirds lumps, it will give up only four. This phenomena is an example of quantization, a word derived from the Latin *quantus,* meaning "how much."

Quantization was the first of the four pillars to be uncovered, and it gave its name to the topic, but today quantization is not regarded as the most essential characteristic of quantum mechanics. There are atomic quantities, like momentum, that do not come in lumps, and under certain circumstances even energy doesn't come in lumps.

Furthermore, quantization of a different sort exists even within the classical domain. For example, a single organ pipe cannot produce any tone but only those tones for which it is tuned.

Probability

Suppose a gun is clamped in a certain position with a certain launch angle. A bullet is shot from this gun, and a mark is made where the bullet lands. Then a second, identical, bullet is shot from the gun at the same position and angle. The bullet leaves the muzzle with the same speed. And the bullet lands exactly where the first bullet landed. This unsurprising fact is called determinism: identical initial conditions always lead to identical results, so the results are determined by the initial conditions. Indeed, using the tools of classical mechanics one can, if given sufficient information about the system as it exists, predict exactly how it will behave in the future.

It often happens that this prediction is very hard to execute or that it is very hard to find sufficient information about the system as it currently exists, so that an exact prediction is not always a *practical* possibility—for example, predicting the outcome when one flips a coin or rolls a die. Nevertheless, in principle the prediction can be done even if it's so difficult that no one would ever attempt it.

But it is an experimental fact that if one shoots two electrons in sequence from a gun, each with exactly the same initial condition, those two electrons will probably land at different locations (although there is some small chance that they will go to the same place). The atomic realm is probabilistic, not deterministic. The tools of quantum mechanics can predict probabilities with exquisite accuracy, but it *cannot* predict exactly what will happen because nature itself *doesn't know* exactly what will happen.

The second pillar of quantum mechanics is probability: Even given perfect information about the current state of the system, no one can predict exactly what the future will hold. This is indeed an important hallmark distinguishing quantum and classical mechanics, but even in the classical world probability exists as a practical matter—every casino operator and every politician relies upon it.

Interference

A gun shoots a number of electrons, one at a time, toward a metal plate punched with two holes. On the far side of the plate is a bank of detectors to determine where each electron lands. (Each electron is launched identically, so if one were launching classical bullets instead of quantal electrons, each would take an identical route to an identical place. But in quantum mechanics the several electrons, although identically launched, might end up at *different* places.)

First the experiment is performed with the right hole blocked. Most of the electrons strike the metal plate and never reach the detectors, but those that do make it through the single open hole end up in one of several different detectors—it's more likely that they will hit the detectors toward the left than those toward the right. Similar results hold if the left hole is blocked, except that now the rightward detectors are more likely to be hit.

What if both holes are open? It seems reasonable that an electron passing through the left hole when both holes are open should behave exactly like an electron passing through the left hole when the right hole is blocked. After all, how could such an electron possibly know whether the right hole were open or blocked? The same should be true for an electron passing through the right hole. Thus, the pattern of electron strikes with both holes open would be the sum of the pattern with the right hole blocked plus the pattern with the left hole blocked.

In fact, this is not what happens at all. The distribution of strikes breaks up into an intricate pattern with bands of intense electron bombardment separated by gaps with absolutely no strikes. There are some detectors which are struck by *many electrons* when the right hole is blocked, by *some electrons* when the left hole is blocked, but by *no electrons at all* when neither hole is blocked. And this is true even if at any instant only a single electron is present in the apparatus!

What went wrong with the above reasoning? In fact, the flaw is not in the reasoning but in an unstated premise. The assumption was made that an electron moving from the gun to the detector bank would pass through either the right hole or the left. This simple, common-sense premise is—and must be—wrong. The English language was invented by people who didn't understand quantum mechanics, so there is no concise yet accurate way to describe the situation using everyday language. The closest approximation is "the electron goes through both holes." In technical terms, the electron in transit is a *superposition* of an electron going through the right hole and an electron going through the left hole. It is hard to imagine what such an electron would look like, but the essential point is that the electron *doesn't* look like the classic "particle": a small, hard marble.

Entanglement

The phenomenon of entanglement is difficult to describe succinctly. It always involves two (or more) particles and usually involves the measurement of two (or more) different properties of those particles. There are circumstances in which the measurement results from one particle are correlated with the measurement results from the other particle, even though the particles may be very far away from each other. In some cases, one can prove that these correlations could not occur for any classical system, no matter how elaborate. The best experimental tests of quantum mechanics involve entanglement because it is in this way that the atomic world differs most dramatically from the everyday, classical world.

Mathematical Formalism

Quantum physics is richer and more textured than classical physics: quantal particles can, for example, interfere or become entangled, options that are simply unavailable to classical particles. For this reason the mathematics needed to describe a quan-

tal situation is necessarily more elaborate than the mathematics needed to describe a corresponding classical situation. For example, suppose a single particle moves in three-dimensional space. The classical description of this particle requires six numbers (three for position and three for velocity). But the quantal description requires an infinite number of numbers—two numbers (a "magnitude" and a "phase") at every point in space.

Classical limit

Classical mechanics holds for ordinary-sized objects, while quantum mechanics holds for atomic-sized objects. So at exactly what size must one framework be switched for another? Fortunately, this difficulty doesn't require a resolution. The truth is that quantum mechanics holds for objects of *all* sizes, but that classical mechanics is a good approximation to quantum mechanics when quantum mechanics is applied to ordinary-sized objects. As an analogy, the surface of the Earth is nearly spherical, but sheet maps, not globes, are used for navigation over short distances. This "flat Earth approximation" is highly accurate for journeys of a few hundred miles but quite misleading when applied to journeys of ten thousand miles. Similarly, the "classical approximation" is highly accurate for ordinary-sized objects but not for atomic-sized objects.

The Subatomic Domain

When scientists first investigated the atomic realm, they found that a new physical framework (namely quantum mechanics) was needed. What about the even smaller domain of elementary particle physics? The surprising answer is that, as far as is known, the quantum framework holds in this domain as well. As physicists have explored smaller and smaller objects (first atoms, then nuclei, then neutrons, then quarks), surprises were encountered and new rules were discovered—rules with names like quantum electrodynamics and quantum chromodynamics. But these new rules have always fit comfortably within the framework of quantum mechanics.

See also: QUANTUM CHROMODYNAMICS; QUANTUM ELECTRODYNAMICS; QUANTUM FIELD THEORY; QUANTUM TUNNELING; VIRTUAL PROCESSES

Bibliography:

Feynman, R. *QED: The Strange Theory of Light and Matter* (Princeton University Press, Princeton, New Jersey, 1985).

Milburn, G. J. *Schrödinger's Machines: The Quantum Technology Reshaping Everyday Life* (W.H. Freeman, New York, 1997).

Styer, D. F. *The Strange World of Quantum Mechanics* (Cambridge University Press, Cambridge, UK, 2000).

Treiman, S. *The Odd Quantum* (Princeton University Press, Princeton, New Jersey, 1999).

Daniel F. Styer

QUANTUM STATISTICS

One of the most basic facts about the physical world is that matter is built up from a few fundamental building blocks (e.g., electrons, quarks, photons, gluons), each occurring in vast numbers of identical copies. Were this not true there could be no lawful chemistry, because every atom would have its own quirky properties. But in nature we find accurate uniformity of properties, even across cosmic scales. The patterns of spectral lines emitted by atoms in the atmospheres of stars in distant galaxies match those we observe in terrestrial laboratories.

From the perspective of classical physics, the indistinguishability of electrons (or other elementary building blocks) is both inessential and surprising. If electrons were nearly but not quite precisely identical—say, for example, their masses varied over a range of a few parts per billion—then according to the laws of classical physics different specimens would behave in nearly but not quite the same ways. And since the possible behavior is continuously graded, we could not preclude the possibility that future observations, attaining greater accuracy than is available today, might discover small differences among electrons. Indeed, it would seem reasonable to expect that differences would arise, since over a long lifetime each electron might wear down, or get bent, in a way dependent on its individual history.

The first evidence that the similarity of like particles is quite precise and goes deeper than mere resemblance emerged from a simple but profound reflection by Josiah Willard Gibbs (1839–1903) in his work on the foundations of statistical mechanics. It is known as "Gibbs's paradox," and it goes as follows. Suppose that we have a box separated into two equal compartments A and B, both filled with equal densities of hydrogen gas at the same temperature. Suppose further that there is a shutter separating the compartments, and consider what happens if we open the shutter and allow the gas to settle into equilibrium. The molecules originally confined to A (or B) might then be located anywhere in A + B. Thus, since there appear to be many more distinct possibilities for distributing the molecules, it would seem that the entropy of the gas, which measures the number of possible microstates, will increase. On the other hand, one might have the contrary intuition, based on everyday experience, that the properties of gases in equilibrium are fully characterized by their volume, temperature, and density. If that intuition is correct, then the act of opening the shutter in our thought experiment makes no change in the state of the gas, and so of course it generates no entropy. In fact, this result is what one finds in actual experiments.

The experimental verdict on Gibbs's paradox has profound implications. If we could keep track of every molecule, we would certainly have the extra entropy, the so-called entropy of mixing. Indeed, when gases of *different* types are mixed, say hydrogen and helium, entropy *is* generated. Since entropy of mixing is not observed for (superficially) similar gases, there can be no method, *even in principle,* to tell their molecules apart. Thus we cannot make a rigorous statement of the kind "Molecule 1 is in A, molecule 2 is in A, . . . , molecule n is in A," but only a much weaker statement, of the kind "There are n molecules in A." In this precise sense, hydrogen molecules are not only similar, nor even only identical, but beyond that indistinguishable.

In classical physics, particles have definite trajectories, and there is no limit to the accuracy with which we can follow their paths. Thus, in principle, we could always keep tab on who's who. Thus classical physics is inconsistent with the rigorous concept of indistinguishable particles. It comes out on the wrong side of Gibbs's paradox.

In quantum mechanics the situation is quite different. The possible positions of particles are described by waves (that is, their wave-functions). Waves can overlap and blur. Related to this, there is a limit to the precision with which their trajectories can be followed, according to Heisenberg's uncertainty principle.

When we calculate the quantum-mechanical amplitude for a physical process to take place, we must sum contributions from all ways in which it might have occurred. Thus, specifically, to calculate the amplitude that a state with two indistinguishable particles of a given sort—call them g-ons at positions x_1, x_2 at time t_i will evolve into a state with two q-ons at x_3, x_4 at time t_f, we must sum contributions from all possible trajectories for the q-ons at intermediate times. These trajectories fall into two distinct classes. In one class, the q-on initially at x_1 moves to x_3, and the q-on initially at x_2 moves to x_4. In the other class, the q-on initially at x_1 moves to x_4, and the q-on initially at x_2 moves to x_3. Because (by hypothesis) q-ons are indistinguishable, the final states are the same for both classes of trajectories. Thus, according to the general principles of quantum mechanics, we must add the amplitudes for these two classes. We say there are a "direct" and an "exchange" contribution to the process. Similarly, if we have more than two q-ons, we must add contributions involving arbitrary permutations of the original particles.

It is found by experiment that the particles of nature, from which matter is built, fall into two great classes. For bosons the direct and exchange contributions are simply added. For fermions they are added after supplying a relative sign change—or, to put it more simply, subtracted. We say these two types of particles, bosons and fermions, display different quantum statistics.

Deep understanding of the origin of quantum statistics is obtained in relativistic quantum field theory. Undoubtedly the single most profound fact about nature that quantum field theory uniquely explains is *the existence of different, yet indistinguishable, copies of elementary particles.* Two electrons anywhere in the universe, whatever their origin or history, are observed to have exactly the same properties. We understand this as a consequence of the fact that both are excitations of the same underlying ur-stuff, the electron field. The electron field is thus the primary reality. The existence of classes of indistinguishable particles is the necessary logical prerequisite to a sec-

ond profound insight from quantum field theory: *the assignment of unique quantum statistics,* boson or fermion, to each class. Given the existence of indistinguishability of a class of elementary particles, and complete invariance of their interactions under interchange, the general principles of quantum mechanics require that solutions forming any representation of the permutation symmetry group retain that property in time. But these general principles in themselves do not put any constraint upon the representations that are realized. Quantum field theory not only explains the existence of indistinguishable particles and the invariance of their interactions under interchange but also goes a step further and constrains the symmetry of the solutions. For bosons, only the identity representation is physical (symmetric wave functions); for fermions, only the one-dimensional odd representation is physical (antisymmetric wave functions). Put another way, the wave function for a many-boson system is unchanged when two particles are interchanged, whereas the wave function for a many-fermion system changes sign when two particles are interchanged. This rule is, of course, closely connected with the rule for direct and exchange processes mentioned earlier. Finally, the detailed mathematics of quantum field theory determines the quantum statistics of a particle from what superficially appears to be an entirely unrelated property, namely the magnitude of its spin. The spin-statistics theorem states that objects whose spin is a whole number (measured in units of Planck's constant) are bosons, whereas objects whose spin is half an odd integer spin are fermions.

Among the particles appearing in the Standard Model, quarks and leptons (and their antiparticles) have spin $\frac{1}{2}$ and are fermions; whereas color gluons, photons, W and Z bosons, with spin 1, and the spin-0 Higgs particle, are bosons.

It is straightforward to determine the quantum statistics of composite particles from that of their constituents. The rule, easily derived, is that a composite particle is a fermion if and only if it is built up from an odd number of fermions. Protons and neutrons, according to the naive quark model, are built from three quarks and are therefore predicted to be fermions. In the more sophisticated picture of protons supplied by Quantum Chromodynamics

(QCD), they can contain any number of quark-antiquark pairs and gluons. Nevertheless the "naive" prediction remains valid, since adding any number of quark-antiquark pairs changes the number of fermions by an even number, and adding gluons changes it not at all. A slightly more complicated example, which has dramatic experimental consequences (see immediately below), concerns the isotopes of helium. Atoms based on ^{3}He, the isotope of helium containing two protons and one neutron, are fermions. Indeed, these atoms are built up from two protons, one neutron, and two electrons, for five fermions altogether, an odd number. Atoms based on the more common ^{4}He isotope, on the other hand, are bosons.

At the level of elementary processes, the influence of quantum statistics comes through the different rules for combining direct and exchange processes. Since dominant nonstatistical—that is, electromagnetic—interactions at low energy are essentially the same for both isotopes of helium, differences arising in the results of low-energy scattering experiments involving the three possible combinations ^{3}He-^{3}He, ^{3}He-^{4}He, and ^{4}He-^{4}He must be ascribed to the operation of quantum statistics. For example, the probabilities of scattering through 90° in the three cases are in the ratio 0:1:2, since in the first case the direct and exchange processes cancel, in the second they add as probabilities, and in the third they add as amplitudes. (For simplicity of exposition, I have assumed that the nuclear spins of the ^{3}He are all aligned in the same direction.) More generally, the quantum statistics of highly unstable or even confined particles, such as quarks and gluons, plays an essential role in predicting and interpreting the results of scattering experiments, which are the bread-and-butter of experimental elementary particle physics.

At the level of macroscopic phenomena, the quantum statistics of particles determines major aspects of the behavior of the matter they form. Roughly speaking, identical bosons like to occupy the same quantum-mechanical state. Laser action, wherein many photons of the same kind—same spectral color, same direction, same spatial cross-section, same polarization—are emitted in a correlated beam, is a manifestation of this behavior. The

(closely related) phenomena of superfluidity, superconductivity, and Bose-Einstein condensation are all characteristic of systems of bosons. They tend to occur at low temperature, when the tendency of bosons to occupy a common state overcomes the disordering influence of thermal agitation.

Identical fermions are forbidden to occupy the same quantum-mechanical state. This is the precise formulation of Pauli's exclusion principle. When forced into a small volume, therefore, additional fermions will be forced into ever higher energy states. Thus a system containing many identical fermions resists compression. The fermionic character of electrons, in particular, underlies the stability of matter, the structure of the periodic table, and the properties of metals. White dwarf stars are supported against gravity by the quantum statistical incompressibility of the electrons they contain. Neutron stars are supported by the quantum statistical incompressibility of their neutrons.

Although ^{3}He and ^{4}He are chemically identical, and the gases based on their atoms behave very similarly at ordinary temperatures, their behavior at low temperatures is radically different, reflecting the difference in their quantum statistics. ^{4}He is the original superfluid, with vanishing viscosity below about 4K. ^{3}He, on the other hand, is still a normal liquid down to much lower temperatures, around 10^{-3}K, below which it too becomes superfluid. The difference in temperatures, and in many more subtle properties, reflects the very different mechanisms at work in the two cases. In ^{4}He the bosonic atoms readily organize themselves into a common quantum state. In ^{3}He the atoms are fermions, and only after they form delicately bound quasi-molecular pairs (which are bosons) is superfluidity possible.

In recent years two developments have brought fundamentally new perspectives to the subject of quantum statistics.

Using only the currently established symmetry principles of special relativity and quantum field theory, it is impossible to connect particles of different spin—or therefore, according to the spin-statistics theorem, particles of different quantum statistics. Supersymmetry is a new kind of symmetry that extends the Lorentz symmetry of special relativity.

Supersymmetry postulates the existence of additional purely quantum dimensions. When a particle takes a step into one of the quantum directions, its position in ordinary space-time does not change, but it undergoes changes in its spin and quantum statistics. For example, a spin-0 boson will transform into a spin-$\frac{1}{2}$ fermion. Supersymmetry transformations mix ordinary and quantum dimensions. In order for such transforms to be symmetries of physical law, there must be particles of different spin and statistics with closely related physical properties. At present the evidence is far from conclusive, but there are serious reasons to believe that (spontaneously broken) supersymmetry is a feature of fundamental physical law.

It has been discovered that particlelike excitations arising in condensed matter systems, specifically in the state of matter known as the fractional quantum Hall effect, obey new forms of quantum statistics, intermediate between bosons and fermions. Particles obeying the new forms of quantum statistics are called anyons. The existence of these new possibilities for quantum statistics transcends, but does not contradict, the principles discussed earlier. When a material is in the fractional quantized Hall state, the presence of an anyon within it modifies the wave-functions of all its underlying electrons. Thus anyons do not correspond to a simple (that is, spatially localized) composite of any definite number of electrons, and the usual rule for determining the quantum statistics of composite particles cannot be applied to them.

See also: BOSON, HIGGS; QUANTUM MECHANICS; SYMMETRY PRINCIPLES

Bibliography

Feynman, R.; Leighton, R. B.; and Sands, M. *The Feynman Lectures on Physics,* vol. 3, chapter 4 (Addison-Wesley, Reading, MA, 1964).

Kane G., and Shifman, M., eds. *The Supersymmetric World: The Beginnings of the Theory* (World Scientific, River Edge, NJ, 2001).

Schrödinger, E. *Statistical Thermodynamics* (Cambridge University Press, Cambridge, England, 1946).

Wilczek, F. *Fractional Statistics and Anyon Superconductivity* (World Scientific, Teaneck, NJ, 1990).

Frank Wilczek

QUANTUM TUNNELING

In quantum mechanics particles are treated as waves. Out of that idea come some of the most surprising results in all of science. One of them is the ability of particles to pass through barriers that would be completely impenetrable according to pre-quantum physics. For example, a ball thrown against a brick wall will rebound every time. An electron striking the atomic equivalent of a brick wall can pass through without even slowing down.

An atomic particle approaching a barrier is like a ball rolling toward a hill. If the ball doesn't have enough energy, it will not make it over the hill; the hill is a barrier. In quantum mechanics what makes the difference is that the particle can also be treated as a wave. A wave might be reflected from a barrier, for example, light reflected from a mirror, but waves never stop abruptly when they strike barriers. Light waves penetrate a short distance into the mirror before being totally reflected.

Similarly, particle waves penetrate into barriers. If the barrier is not too thick, some of the wave gets through to the far side even though the wave amplitude decreases rapidly (Figure 1). Since the wave represents the particle, it seems that some part of the particle has gotten through the barrier. But atomic particles such as electrons cannot be partly reflected: either the particle is reflected, or it is not. The proportion of the wave that survives at the far side of the barrier represents the probability that the particle will get through. Thus, some particles are reflected, and the others pass through the barrier, but there is no way to tell what will happen to a particular particle.

Since quantum mechanics applies in principle to everyday objects, there is even a slight possibility that a BB could pass through a steel plate (without damaging the plate or leaving a hole). The BB does not even have to be going very fast. Still, calculations show that the probability that a BB would actually pass through a steel plate is unimaginably small. Thus, we never observe such behavior in everyday processes.

However, the tunneling probability increases as the mass of the particle and the thickness of the barrier decrease. As we approach atomic dimensions

FIGURE 1

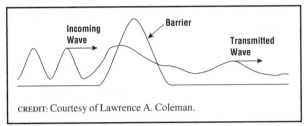

CREDIT: Courtesy of Lawrence A. Coleman.

Quantum tunneling through a barrier.

and masses, the probability can become substantial. It is common to use electron tunneling in semiconductors as part of the design of common electronic devices. It is also used in the scanning tunneling microscope, which is able to image individual atoms.

Alpha Decay

One of the first applications of quantum tunneling was by the physicist George Gamow in 1928, soon after the development of quantum mechanics. Alpha particles, which consist of two protons and two neutrons, are emitted by some nuclei. For example, ordinary uranium, ^{238}U, with a lifetime of 4.5 billion years, decays by emitting an alpha particle.

For decades alpha decay had presented a problem: the emitted alpha particles seemed to have too little energy to get out of the nucleus. The Coulomb barrier arises from the combined effect of the Coulomb repulsion between the alpha particle and the nucleus (both positively charged) and the nuclear force that attracts the two particles. The energy of the emitted alpha particle is less than the top of this barrier. Classically, the particle would be unable to get out of the nucleus, but it obviously does.

Gamow suggested that alpha particles tunnel through the barrier. If so, the half-life of the decay should depend on the width and height of the barrier, and it does: the lower and thinner the barrier, the greater the chance of penetrating it. As the alpha particle's energy increases, the particle sees both a lower and thinner barrier so the probability of getting through increases extremely rapidly. For example, the energies of the alphas emitted by ^{232}Th and ^{212}Po are 4.05 MeV and 8.95 MeV, respectively, while their respective half-lives are 14 billion years and 0.3

millionth of a second. Thus, a factor of about two in energy produces a difference in half-lives of sixteen orders of magnitude (that is, sixteen powers of ten)!

Hydrogen fusion in the Sun

The Sun is mostly hydrogen. Its energy arises from combining hydrogen nuclei (protons) to form helium in a process called hydrogen fusion. In order for the protons to react with each other, they must get close enough for the strong nuclear force to hold them together. But the strong force does not reach very far, which means that the protons must come close enough to touch.

Since protons are positively charged, they repel each other and must approach with a lot of energy in order to get close enough to react. The higher the temperature of a gas, the more energetic the particles. At fifteen million degrees Celsius the center of the Sun is hot enough to give the protons the required energy. The reactions are able to take place, however, because the protons tunnel through the barrier, thereby producing the energy that sustains life on Earth.

See also: QUANTUM MECHANICS

Bibliography

Binnig, G., and Rohrer, H. "The Scanning Tunneling Microscope." *Scientific American* **253**, 50–56 (1985).

Feinberg, G. *What Is the World Made of: Atoms, Leptons, Quarks and Other Tantalizing Particles* (Anchor, Garden City, NY, 1977).

Lawrence A. Coleman

QUARK-GLUON PLASMA

Quark-gluon plasma is a novel form of matter whose existence and properties are predicted by quantum chromodynamics (QCD). QCD is the theory of quarks and gluons and their interactions. QCD describes protons, neutrons, pions, kaons, and many other subatomic particles collectively known as hadrons. Hadrons are quite complicated bound states of many quarks, antiquarks, and gluons that are "color-neutral" and are much heavier than the quarks inside them. Why do the hadrons that QCD describes turn out to be so complicated, relative to the elementary quarks and gluons? The answer to this question relies on the properties of the vacuum in QCD. Furthermore, answering this question reveals that at high enough temperatures, QCD simplifies. At temperatures above 2×10^{12} Kelvin, the complex hadrons fall apart into a plasma of unconfined quarks and gluons. For the first 10 microseconds after the Big Bang, the entire universe was hot enough that it was filled with this quark-gluon plasma. Current experiments at the Brookhaven National Laboratory (BNL) in Brookhaven, New York, and at the European Laboratory for Particle Physics (CERN) in Geneva, Switzerland, seek to recreate these extraordinarily high temperatures last seen during the Big Bang, in order to study QCD by simplifying it.

According to the laws of quantum mechanics, the vacuum is not empty. All states are characterized by quantum mechanical fluctuations, and the vacuum is just the state in which these fluctuations happen to yield the lowest possible energy. In QCD, the vacuum is a sea of quarks, antiquarks, and gluons arranged precisely so as to have the minimum possible energy. QCD describes the excitations of this vacuum, which turn out to be the colorless and heavy hadrons, instead of colorful and light quarks and gluons.

To understand why hadrons are colorless, one must first understand how the QCD vacuum responds to the presence of a single "extra" quark. This quark disturbs (polarizes) the surrounding vacuum, which responds by surrounding it with a cloud of many quark-antiquark pairs and gluons. QCD predicts that the force between quarks is weak when they are close together, much closer than about 1 Fermi (10^{-15} m, about the diameter of a proton). At distances of about 1 Fermi, however, the cloud surrounding a quark acts to ensure that the force between this quark and another quark (surrounded by its own cloud) does not lessen as one tries to separate the quarks. Pulling a single, isolated quark completely out of a colorless hadron requires working against a force that does not weaken with increasing separation—and therefore costs infinite energy. Thus, the energy of a single quark (or of any colored excitation) is infinite, once one includes the energy cost of the resulting disturbance of the vacuum.

Adding a colorless combination of quarks to the vacuum disturbs it much less, creating a finite-energy excitation. This explains why the excitations of the QCD vacuum must be colorless.

Understanding why hadrons are heavy requires a second crucial feature of the QCD vacuum. Part of the description of the vacuum is a specification of what fraction of the quark-antiquark pairs at any location is $\bar{u}u$, $\bar{d}d$, $\bar{u}d$, or $\bar{d}u$. At each point in space, the vacuum is therefore described by a "vector" that can point any direction in an abstract four-dimensional space with axes labeled $\bar{u}u$, $\bar{d}d$, etc. QCD predicts that in order to achieve the lowest energy, all these vectors must be aligned. A sea of quark-antiquark pairs so ordered is called a condensate. The fact that the arrows must pick one among many otherwise equivalent directions is known as symmetry breaking. The condensate that characterizes the QCD vacuum is much like a ferromagnet, within which all the microscopic spin vectors are aligned (see Figure 1). The presence of a hadron disturbs this condensate, and the largest contribution to the mass of the hadron is the energy of this disturbance. In effect, the condensate that fills the vacuum slows down the quarks, and because of its presence, hadrons are much heavier than the quarks of which they are made.

There is one exception to the dictum that hadrons must be heavy. Because QCD does not specify in which direction the arrows point, it should be relatively easy to excite "waves" in which the directions of the arrows ripple as a wave passes by. In quantum mechanics, all such waves are associated with particles, and because these waves are easily excited, the related particles should not have much mass. The requisite particles, called pions, are indeed light, as they have a mass only about one-seventh that of a proton.

Thus, the QCD vacuum is a complex state of matter. The laws describing it are written in terms of colored quarks and gluons, but its natural excitations are colorless hadrons, which are heavy because of their interaction with a symmetry-breaking condensate that pervades all of space. One good way of testing understanding of the QCD vacuum is to create new states of matter that are simpler than the vacuum, although they must, of course, be more energetic. Is there a phase of matter in which quarks can roam free? In which the excitations are individual quarks and gluons rather than complicated hadrons?

QCD provides two methods of deconfining (freeing) the quarks. The first is to squeeze nuclei together until their protons and neutrons overlap. In the resulting dense quark matter the quarks are close together and therefore interact only weakly. The second approach is to take a chunk of matter and heat it. When a magnet is heated, by analogy, the spins in the magnet start to oscillate; eventually, above some critical temperature, they oscillate so wildly that the spins all point in random directions, and the magnet loses its magnetization. Something similar happens in QCD. At low temperatures, the arrows that describe the QCD condensate ripple, yielding a gas of pions. Above a critical temperature, the arrows oscillate so wildly that they point randomly, and the condensate "melts." Above its critical temperature, the matter described by QCD is more disordered but more symmetric (no direction favored in Figure 1) than the QCD vacuum. Theoretical calculations that challenge the world's fastest supercomputers show that at a temperature of about 2×10^{12} K, a phase transition occurs in which the QCD condensate melts and the hadrons "ionize," yielding a quark-gluon plasma in which the quarks are light and free (as shown in Figure 2). At low temperatures, QCD describes a gas of hadrons, mostly pions. Once the condensate melts, the pions ionize, releasing quarks and gluons that are lighter and more numerous (three colors of each of up, down, and strange quarks plus their antiquarks and eight types of gluons) and therefore have a much larger energy density at a given temperature.

The prominent features on the phase diagram of QCD are shown in Figure 3. At low densities and temperatures, QCD describes hadrons. The only known place in which nuclei are squeezed together without being heated is the center of neutron stars. The cores of these extraordinarily dense cinders, with masses about that of the Sun but with radii of only approximately 10 km, may be made of superconducting quark matter. Finding a phase of matter in which QCD simplifies completely requires exploring the high temperature region of the phase diagram. Upon heating any chunk of matter to trillions of degrees, the vacuum condensate melts, the hadrons ionize, and the quarks and gluons are free

FIGURE 1

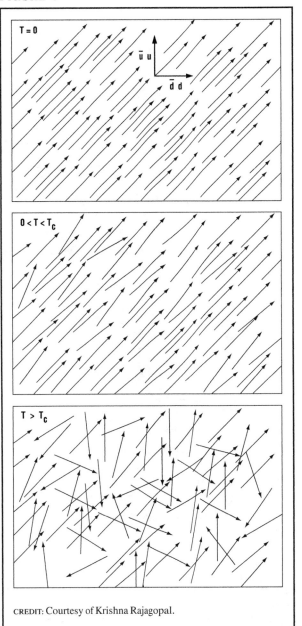

CREDIT: Courtesy of Krishna Rajagopal.

Melting the Vacuum. (Top) The QCD vacuum is a condensate. At each location are quark-antiquark pairs whose type must be specified by an arrow indicating what fraction of the pairs are $u\bar{u}$ versus $d\bar{d}$ versus $u\bar{d}$ versus $d\bar{u}$. Only two of these four directions are shown. The central property of a condensate is that all the arrows are aligned. (Middle) At nonzero temperatures, the arrows describing the condensate begin to undulate. These waves can equally well be described as a gas of particles, called pions. (Bottom) As the temperature increases, the waves on the condensate become more and more violent. Above some critical temperature, the arrows are completely scrambled, and the condensate has melted.

FIGURE 2

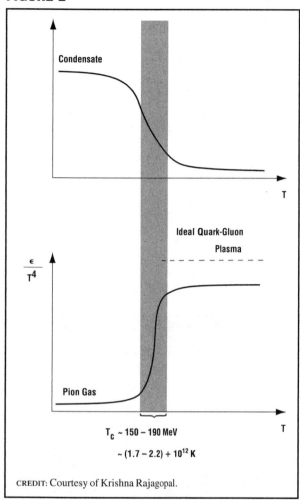

CREDIT: Courtesy of Krishna Rajagopal.

(Top) The strength of the condensate (the "vector average" of all the arrows in Figure 1) decreasing with increasing temperature. The shape of this curve mirrors that of an analogous curve showing how magnetization vanishes when the spins in a magnet get scrambled at a temperature of only a few hundred degrees. (Bottom) The higher the temperature (T), the more energy per unit volume (ε). The ratio ε/T^4 is a measure of how many different types of particles are present. ε/T^4 rises rapidly from the value for a pion gas to close to the value for an ideal quark-gluon plasma in which the interactions between quarks and gluons have become weak. This rise directly reflects the freeing of the quarks.

to move in the resulting quark-gluon plasma. The universe began far up the vertical axis of the diagram: at its earliest moments, shortly after the Big Bang, it was filled with a hot quark-gluon plasma that expanded and cooled, moving down the vertical axis, falling below 2×10^{12} K after about 10 microseconds. Since then, quarks have been confined in hadrons—

with the possible exception of quarks at the centers of neutron stars and those that are briefly liberated in heavy-ion collisions.

In a heavy-ion collision, two nuclei accelerated to enormous energies are collided in an attempt to create a tiny, ultrahot region within which matter enters the quark-gluon plasma phase. As in the Big Bang (but much more quickly), this quark-gluon plasma droplet expands and cools, moving downward on the phase diagram. For a brief instant the quarks are free, but their liberation is short-lived. After about 10^{-22} s they recombine to form an expanding gas of hadrons, which expands for approximately another 10^{-22} s. After that these hadrons are so dilute that they fly outward without further scattering, to be seen in a detector. Detectors record many thousands of hadrons—the end products of a collision in which quark-gluon plasma may have been created. The purpose of these heavy-ion collision experiments is twofold. First, they seek to create a region of quark-gluon plasma—the stuff of the Big Bang—and measure its properties to see whether the complexities of the QCD vacuum have truly melted away. And, second, they seek to study how matter behaves as it undergoes the transition from this plasma back to a mundane hadron gas.

In June 2000, the first collisions occurred at Brookhaven's new Relativistic Heavy Ion Collider (RHIC) whose collisions are about ten times more energetic than those achieved previously at CERN. This increases the initial energy density, and thus the initial temperature, pushing further upward into the expected quark-gluon plasma region of the phase diagram. In addition, these higher-energy collisions produce many more pions, diluting the net quark density. More energetic heavy-ion colliders therefore explore upward and to the left on the phase diagram, more and more closely recreating the conditions of the Big Bang. To date, only the first, simplest analyses of collisions at RHIC have been performed, but it is already clear that these collisions have higher initial energy densities and lower net quark densities than ever before.

Near the vertical axis of the phase diagram (traversed by the Big Bang and by the highest-energy collisions), the phase transition from quarks to hadrons occurs smoothly and continuously. In this way, it is like the ionization of a gas and is quite unlike the boiling of water. The latter phase transition occurs discontinuously at a single sharply defined temperature. Theoretical arguments indicate that at higher net quark density, the phase transition between quark-gluon plasma and hadrons is similarly discontinuous and can be shown in Figure 3 as a sharp line. This line ends at what is called the critical point.

There are phenomena that occur at the critical point and nowhere else on the phase diagram. At this point, the arrows of Figure 1 undulate in a unique, precisely calculable manner. Consequently, distinctive fluctuations occur in the momentum of the pions produced in those heavy-ion collisions that pass near the critical point as they cool. Experiments move leftward in the phase diagram as the collision energy increases and search for the telltale signatures of the critical point.

In addition to studying the transition between quark-gluon plasma and hadrons, the goal of heavy-ion collisions is to measure properties of the quark-gluon plasma itself. This requires observables that reveal something about the earliest, hottest moments of a collision. One method would be to shoot a very fast quark through the plasma and watch how rapidly it loses energy. Estimates suggest that a quark plowing through such a plasma loses much more energy than it would if it encounters only heavy, colorless hadrons. The sign of this rapid energy loss is a paucity of 5 to 10 GeV pions emerging from a heavy-ion collision. Any such energetic pions must have originated as a fast quark. If these quarks have to fight their way through a quark-gluon plasma, they will lose energy and thermalize, and consequently very few pions of 5 to 10 GeV will be seen, relative to what occurs in proton collisions. No evidence of such a deficit has been seen in the lower-energy collisions at CERN. One of the most exciting features of the first, preliminary data from the RHIC experiments is an indication that they yield about five times fewer energetic pions than expected. With time, this measurement should become a quantitative measure that will allow one to test whether high-temperature QCD indeed describes a quark-gluon plasma—and to test predictions of its properties.

FIGURE 3

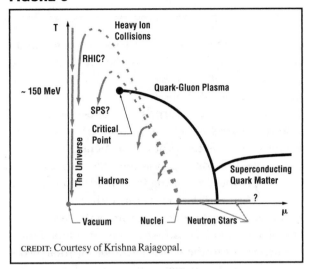

CREDIT: Courtesy of Krishna Rajagopal.

Phases of QCD as a function of temperature and μ, a quantity that is a convenient measure of the net quark density, namely the density of quarks minus that of antiquarks. The vacuum is at the bottom left; nuclei have nonzero net quark density. Squeezing nuclei without heating them pushes matter to the right on the diagram, while heating matter pushes it up. The line separating the quark-gluon plasma from the hadronic phase ends at a critical point.

The study of the quark-gluon plasma has to date been largely accomplished by theoretical methods, working deductively beginning from the laws of QCD. Experimenters hope to soon confirm that they are regularly recreating the material of the Big Bang. As they then begin to measure its properties, scientists shall learn whether QCD behaves as expected. If the vacuum condensate melts and hadrons ionize, freeing the quarks, the simplicity implicit in the laws of QCD will have been realized.

See also: COSMOLOGY; PHASE TRANSITIONS; QUANTUM CHROMODYNAMICS

Bibliography

Brookhaven National Laboratory. "RHIC." <http://www.bnl.gov/RHIC>.

Hallman, T. J.; Kharzeev, D. E.; Mitchell, J. T.; and Ullrich, T., eds. *Quark Matter 2001, Proceedings of the 15th International Conference on Ultra-Relativistic Nucleus-Nucleus Collisions.* (Elsevier, Amsterdam, 2002).

Rajagopal, K., and Wilczek, F. "The Condensed Matter Physics of QCD" in *At the Frontiers of Particle Physics,* edited by M. Shifman (World Scientific, Singapore, 2001).

Krishna Rajagopal

QUARKS

Quarks are subatomic particles that combine in various ways to form all of the known hadrons. In spite of their ubiquity, free quarks are not seen in nature. Besides this anomaly, they have fractional electric charge and a surprising number of nonclassical properties. They only form triplets, called baryons, such as the proton and neutron, or doublets, called mesons, such as the pi meson; they are bound together by a strong force field whose quanta are called gluons. However, this rather simple definition of the quarks does not convey the fascination and challenges of the fifty-year quest that led to their discovery and to the characterization of their properties. Table 1 provides a time line of important events in quark history.

Table 2 lists the quarks, their names (or flavor), and their quantum numbers. Their charge and spin are measured in units relative to those of the electron, and their approximate mass is given in MeV/c^2.

TABLE 1

1947	Discovery of the π and first evidence for strange particles
1951	Discovery that there were both strange mesons and strange baryons
1952	Hypothesis of associated production
1952	First evidence for strangeness 2 baryon (Ξ)
1953	Associated production confirmed at the Cosmotron
1961	Gell-Mann proposed Eight Fold Way for particle classification
1964	Gell-Mann and Zweig propose the quark model of subatomic structure
1964	Charm predicted
1964	Ω^- discovered at Brookhaven
1969	Electron-scattering experiments reveal substructure inside proton
1973	Formulation of Standard Model of forces
1974	Discovery of charmonium J/ ψ at SLAC and BNL
1977	Discovery of bottom quark at FNAL (upsilon)
1979	Experimental observation of gluon jets at DESY
1995	Discovery of top quark at FNAL

CREDIT: Courtesy of Alvin Tollestrop.

This table shows some of the milestones in the discovery of the quark model. The first entry showed that there were two strongly interacting particles that had strikingly different characteristics. This work was done by one or two physicists looking at tracks left in photographic plates by cosmic ray particles. As the quark mass increased, it required increasingly more complicated experimental detectors. The last entry, the top quark, had over 200 experimenters on each of the two experiments and required the full energy of the Fermilab Tevatron Collider.

There are four forces that act between quarks: the strong force that binds together quarks; the electromagnetic force whose quantum is the photon that couples to the quark charge; the weak force that causes beta decay and allows a quark of one type to change into another; and gravity that couples to their mass.

Early Discoveries

The first step in the discovery of quarks occurred in 1947 when a new particle, the K meson, with a mass of about $500\ MeV/c^2$, was discovered in cosmic rays. However, along with the discovery came a mystery. These particles were made in nuclear collisions by very-high-energy cosmic-ray particles via the strong force, with an interaction time of about 10^{-24} s, which is approximated by dividing the radius of a nucleus by the speed of the cosmic-ray particle. The new particles were observed, though, to decay into pions with lifetimes of the order of 10^{-9} s. If the strong force could make these particles so easily, why did it not cause them to decay just as fast? They were named strange particles. Later, in 1953, these particles were made in the Cosmotron at Brookhaven National Laboratory (BNL), and soon whole families of these particles were discovered. By 1964 two octets, one of baryons and one of mesons, and an additional decuplet of baryons had been discovered. Figure 1 lists their modern quark assignments.

In 1964 Murray Gell-Mann and George Zweig independently found the solution for explaining many of the observed properties of these particles. They proposed that all baryons were composed of triplets of quarks (selected from u, d, and s in Table 2) and that mesons were doublets formed by $q\bar{q}$, where the overbar indicates an antiquark.

Conservation Laws

To see how theory and experiment worked together to elucidate the quark model, it is important to understand conservation rules. Classical mechanics revealed the conservation of energy, momentum, and angular momentum. These conservation laws apply to collisions of particles: the total energy and momentum of the incoming particles are equal to those of the outgoing state. This is a powerful constraint on the reactions that can take place. In addition,

TABLE 2

Quarks, Their Names (or Flavors), and Their Quantum Numbers

Quark Property	d	u	s	c	b	t
J Spin	1/2	1/2	1/2	1/2	1/2	1/2
N Baryon number	1/3	1/3	1/3	1/3	1/3	1/3
Q – electric charge	−1/3	+2/3	−1/3	+2/3	−1/3	+2/3
I_z Isospin z component	−1/2	+1/2	0	0	0	0
S – strangeness	0	0	−1	0	0	0
C – charm	0	0	0	+1	0	0
B – bottomness	0	0	0	0	−1	0
T – topness	0	0	0	0	0	+1
Mass in MeV/c²	1–5	3–9	125	1,200	4,200	175,000

CREDIT: Courtesy of Alvin Tollestrup.

electric charge conservation places constraints on reactions: the total charge of the initial state must be equal to that of the final one. How well is it known that charge is conserved? It has been determined that

FIGURE 1

Configuration of Lightest Mesons

K^0, K^+ (495)	S = +1		$d\bar{s}$	$u\bar{s}$
$\pi^{-,0,+}$ (140)	S = 0	$d\bar{u}$	$(u\bar{u} - \bar{d})/\sqrt{2}$	$u\bar{d}$
$\bar{K}^0$, K^- (495)	S = −1	$\bar{u}s$	$\bar{d}s$	

Configuration of $J = \frac{1}{2}$ Baryons

n, p	S = 0		dud	uud
Σ(1,190)	S = −1	sdd	sud	suu
Ξ(1,315)	S = −2	ssd	ssu	

Configuration of $J = \frac{3}{2}$ Baryons

Δ(1,232)	S = 0	ddd	udd	uud	uuu
Σ(1,385)	S = −1	sdd	sud	suu	
Ξ(1,530)	S = −2	ssd	ssu		
Ω⁻	S = −3		sss		

CREDIT: Courtesy of Alvin Tollestrup.

The quark composition of meson and baryon states that played an important role in the discovery of quarks. The Ω⁻ was predicted before the quark model, and its discovery gave strong support to the idea that the new particles were related to each other by an underlying theory. Its composition of three identical quarks in a common state required the invention of a new quantum: color.

the lifetime for the disappearance of an electron into an all-neutral final state (for instance, neutrinos and photons) must be greater than 4.2×10^{24} years. Here, it will be assumed that this time is so long that charge is really conserved, but future experiments may discover otherwise. If the electron does not decay, does the proton? Charge conservation and energy conservation allow many final states, such as $e^+ \pi^0$, into which the proton could decay, but measurements have shown that its lifetime is greater than 1.6×10^{25} years!

Proton stability is incorporated into the quark model by positing the conservation of a new number N. Each quark is assigned an additive baryon quantum number of 1/3 or −1/3 for antiquarks. N is then equal to −1, 0, and 1 for antibaryons, mesons, and baryons. This law not only prohibits the decay of the proton into pions but also requires that any baryon made in a high-energy collision must be accompanied by an antibaryon.

Figure 2 shows a reaction between a 900-GeV/c^2 proton and an equal energy antiproton moving in the opposite direction for a total energy of 1,800 GeV/c^2. The incoming state has charge and baryon numbers zero. The conservation laws discussed assure that in this event the final state conserves energy and momentum, that there are as many positive particles as negative ones, and that an equal number of baryons and antibaryons exists.

We now return to the mystery mentioned above concerning the production of strange particles. It

FIGURE 2

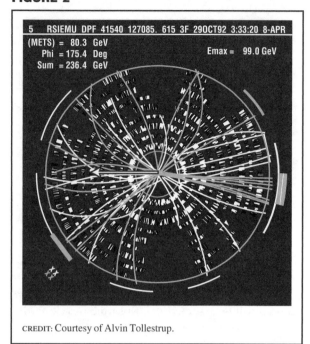

5   RSIEMU DPF 41540 127085. 615 3F 29OCT92 3:33:20 8-APR
(METS) = 80.3 GeV
Phi = 175.4 Deg   Emax =   99.0 GeV
Sum = 236.4 GeV

CREDIT: Courtesy of Alvin Tollestrup.

This is an event involving the collision of a 900 GeV proton with a 900 GeV antiproton as reconstructed by the CDF detector. The view is a projection onto a plane perpendicular to the proton direction. There is a magnetic field perpendicular to the page that causes the particle tracks to bend and allows the particle momentum to be measured. In this collision, there is a preponderance of momentum going to the left. Conservation of momentum is used to infer that neutrinos, which don't leave tracks, carried off the missing momentum in the direction of the arrow.

was discovered that strange particles were made in pairs. The production and decay are given by

$$\pi^- + P \rightarrow \Lambda^0 + K^0$$
$$\Lambda^0 \rightarrow \pi^- + P$$
$$K^0 \rightarrow \pi^+ + \pi^-.$$

The arrow indicates that the particles on the left turned into the particles on the right. The first reaction, which conserves Q and N, results from the strong force. The experiments on the production of strange particles indicated that there is a new additive quantum number called strangness that is conserved in strong interactions. It is zero for the u and d and $(-1, 1)$ for the $(s, \bar{s})$. Figure 1 shows that the Λ^0 has a strangeness of -1 and the K^0 is $+1$, so both sides of the equation balance to zero. In the

two lower equations, which involve decay of the strange particles, S is not conserved and the weak force is responsible for the strangeness-changing decay of the s quark. Note that the decay of the neutron is also due to the weak force changing a d quark into a u quark. The general rule is that the strong force treats all the quarks equally and conserves the quantum numbers shown in Table 2, whereas the weak force allows quarks to change flavor or decay into leptons.

Particles Classified by Their Quark Content

Once the postulate was made that hadrons are composites of quarks, it became possible to explain the properties of the many different hadrons that had been discovered experimentally, in much the same way that the Periodic Table relates the chemical properties of different elements. Figure 1 shows three examples of hadrons and their quark compositions. Since quarks have spin $\frac{1}{2}$, they can have spins parallel for a total spin of 1 or antiparallel for a spin of 0. The baryons, with three quarks, can have a total spin of either $\frac{3}{2}$ or $\frac{1}{2}$. The figure shows examples of spin-0 mesons, spin-$\frac{1}{2}$ baryons, and spin-$\frac{3}{2}$ baryons. Strangeness is plotted on the vertical axis.

Strangeness and spin are used to classify the particles in Figure 1, but there is still another connection between these particles. Isotopic spin I is the quantum number first observed in nuclear reactions, where isotopic states in nuclei with the same number of nucleons but different numbers of protons have very similar properties. Figure 1 shows similar behavior for the u and d quarks, and this feature is embedded in the quark model by assigning isotopic spin $\frac{1}{2}$ to the u and d.

Unlike the additive quantum numbers Q and N, the total isospin for a $q\bar{q}$ state combines the individual isospins in the same mathematical manner as spins are combined, that is, like vectors. Classically, if there are two forces on an object, one downward and an equal sideward force, the total force occurs at 45°. One does not just numerically add the forces, but one takes their vector sum. Since spin is a vector, the same rules apply for combining the spins of two particles. However, quantum mechanics quantizes the rules so that nature only allows integral or half-integral spin. Thus, for example, a neutron and

400

proton, each with a spin of $\frac{1}{2}$, can exist in a state with a total angular momentum of either 1 or 0. The rules for the vector addition of spin were first learned in studying the behavior of electrons around atoms and subsequently were found to be exactly the rules necessary for understanding a new quark quantum number. It was given the name isotopic spin and follows the same rules for addition as spin but has no relation to angular momentum,

Thus, a $q\bar{q}$ state can have either $I = 1$ with $I_3 = (-1, 0, 1)$, $I = 0$ with $I_3 = 0$ for the u plus d combinations, or $I = \frac{1}{2}$ with $I_3 = +\frac{1}{2}$ or $-\frac{1}{2}$ when only a single u or d is present. The π^-, π^0, π^+ in the middle row is an example with $I = 1$, $I_3 = -1, 0, +1$, and the K^+ and K^- doublets show $I = \frac{1}{2}$, with $I_3 = -\frac{1}{2}$ and $+\frac{1}{2}$. The isospin, which is plotted horizontally, is related to the actual charge of the hadron by the general formula $Q = I_3 + N/2 + S/2$ for the strange particles.

In 1962 Murray Gell-Mann and Yuval Ne'emann were the first to recognize the symmetries shown in Figure 1. At the time the Ω^- had not yet been discovered, and they predicted this state with $S = -3$ in the decuplet. There was great excitement in 1964 when experimenters discovered the Ω^- at Brookhaven's Cosmotron. The same year Gell-Mann, and independently Zweig, proposed the first elements of the quark model using u, d, and s quarks. This model naturally predicted the Ω^- but also raised a curious problem that led to an important extension of the theory.

Color

It has been known for a long time that particles come in two types: fermions, whose spins are odd multiples of $\frac{1}{2}$, and bosons, which have zero or integral spin. Fermions obey the Pauli exclusion principle that states that two of them cannot occupy the same physical state simultaneously. On the other hand, one can collect as many bosons as desired into a given state. A simple example is the ground state of the helium atom with two electrons around the nucleus. The spatial state of the two electrons is the same, but because the electrons have spin $\frac{1}{2}$, two electrons, one with spin up and the other with spin down, can occupy this state. The Ω^- was found to have a

spin of $\frac{3}{2}$, and the simplest configuration of the ground state to give this result would require all three quarks to have spin up. This violates the exclusion principle, as the quarks are supposed to be fermions.

This problem led to the surprising conclusion that there must be three different states for each quark. This is referred to as the color charge; each quark comes in one of three different states called red, blue, and green. These are just names for an additional quantum number and do not, of course, refer to an actual color. Thus, the three quarks in the Ω^- have their spins parallel and the same spatial wave function, but are of three different colors. Since all three colors are equally present, it is said that the state is color-neutral, and since this additional quantum number is not seen in real hadrons, one knows that these states must all be colorless. The baryons accomplish colorlessness by being composed of three different-colored quarks; the mesons are color-neutral as they are composed of quarks and antiquarks.

Finally, one comes to the force, carried by the gluons, that binds the quarks together. The development of the theory of quantum chromodynamics (QCD), which describes how these eight gluons interact among themselves and with the quarks, was started by the work of Yoichiro Nambu and by Oscar W. Greenberg in 1966. It was essentially completed by 1973 and remains the underlying theory of strong interactions.

Can the quark hypothesis be tested? The answer is yes. Direct experiments show that baryons are indeed composed of three quarks and that quarks have an additional three-valued quantum number, color. The first experiment that showed that the proton behaved dynamically, as if composed of subatomic particles, was done at the Stanford Linear Accelerator (SLAC) in 1969 using high-energy electrons to scatter off the constituent particles composing the proton. In 1990 a Nobel Prize in Physics was awarded to Henry W. Kendall, Jerome I. Friedman, and Richard E. Taylor for this discovery.

Charm, Bottom, and the Top

The next surprise occurred in 1974. As early as 1964 theorists had speculated on the existence of an additional quark. Ten years later the charm quark appeared simultaneously in experiments at SLAC and BNL, resulting in a Nobel Prizes in Physics for Burton Richter and Samuel C. Ting. They discovered a meson, called the J/ψ, consisting of a charm and anticharm quark. The new quark was more massive than the proton at 1,200 MeV/c^2 and had a charge of $\frac{2}{3}$ like the u. An additional beautiful feature of the J/ψ is that gluons bind the two quarks together in a fashion that leads to a spectrum of many excited states, called charmonium, that has played a role for QCD similar to that played by the hydrogen atom for quantum electrodynamics (QED). Soon experiments found new charmed mesons and baryons in which the charm quark formed states with the u and d; even charmed-strange states with the s quark appeared.

In 1977 two more pieces of the puzzle were discovered. The tau, or τ, meson was discovered at SLAC by a collaboration led by Martin Perl, and the upsilon meson, or Ψ, was discovered at Fermilab by a collaboration led by Leon Lederman. The upsilon had a mass of about 9.5 GeV/c^2 and, in a fashion analogous to the J/ψ, it is composed of a $b\bar{b}$ pair. States have also been discovered in which the bottom quark, whose mass is about 4.5 GeV/c^2, binds via the strong force with the other quarks to form baryons and mesons that are several times as massive as the proton.

The discussion here has centered on the discovery of how quarks bind together to form the particles observed in nature and how they are produced by the strong force. At the same time, a parallel line of experiment and theory clarified how the quarks decay. Since the strong force treats quarks democratically, it is mainly the weak force that exposes their individual properties. It was discovered theoretically that the quarks should be grouped together into three doublets:

$$Q = 2/3 \qquad u \qquad c \qquad t$$
$$Q = -1/3 \qquad d \qquad s \qquad b$$

(The t, or top quark, which was discovered later, is also included.) These doublets are related to the three lepton doublets by the weak force:

$$e \qquad \mu \qquad \tau$$
$$\nu_e \qquad \nu_\mu \qquad \nu_\tau$$

This connection, called the electroweak theory, describes how the weak force mediates the decay of the various quarks. It represents a major triumph of theorists.

After the discovery of the τ lepton and b quark, there was much anticipation, based on the expectations mentioned above, that a sixth quark, the top, would exist. The surprise was that it took nearly twenty years before two groups at Fermilab announced the discovery. The reason for the long delay was that the top has a mass of 175 GeV/c^2. It was not until the Tevatron proton-antiproton collider came on-line that any machine had a high enough energy to make the top. It weighs about as much as a gold atom!

It took fifty years from the discovery of strangeness to the discovery of the top to complete what is called the Standard Model for quarks and leptons and their interactions. The first half of the quest was led by experiments making unexpected discoveries; the last half by theory making predictions of things to be discovered. A very accurate theory now exists for energies up to a few hundred GeV, but questions of why there are only three families of quarks and leptons, and an explanation of their masses, remain to be answered.

See also: ELECTROWEAK SYMMETRY BREAKING; EXPERIMENT: DISCOVERY OF THE TOP QUARK; J/Ψ; LEPTON; PARITY, NONCONSERVATION OF; PARTICLE; PARTICLE PHYSICS, ELEMENTARY; QUANTUM CHROMODYNAMICS; STANDARD MODEL; SUPERSYMMETRY; SYMMETRY PRINCIPLES

Bibliography

Cahn, R. N., and Goldhaber, G. *The Experimental Foundations of Particle Physics* (Cambridge University Press, Cambridge, UK, 1989).

Carrigan, R. A., and Tower, W. P. *Particles and Forces: At the Heart of the Matter* (W. H. Freeman, New York, 1990).

Ezhela, V. V., et al. *Particle Physics: One Hundred Years of Discoveries* (AIP Press, Woodbury, New York, 1996).

Gottfried, K., and Weisskopf, V. *Concepts of Particle Physics*, Vol. 1 (Oxford University Press, New York, 1984).

Lederman, L. M., and Schramm, D. N. *From Quarks to the Cosmos* (W. H. Freeman, New York, 1995).

Liss, T. M., and Tipton, P. L. "The Discovery of the Top Quark." *Scientific American* **277**, 54–59 (1997).

Alvin V. Tollestrup

QUARKS, DISCOVERY OF

The twentieth century began with the confirmation that matter was not continuous but made of tiny atoms and molecules. It ended with the confirmation that matter is made, in part, of even tinier objects called quarks.

Atoms consist of nuclei and electrons, and nuclei consist of neutrons and protons. However, in 1950 the proton and neutron were considered to be the final elementary constituents of matter. The pion was the carrier of the strong force that attracted protons and neutrons to form nuclei, just as the photon was the carrier of the electromagnetic force that bound electrons and nuclei into atoms. But by 1962 many new unexpected particles had been discovered. They were first grouped into families called multiplets and described by the Eightfold Way. By 1966 it became clear that none of the new particles could be really elementary. The neutron, proton, and pion were not qualitatively different like the electron and the photon; they and all the new strongly interacting particles called baryons and mesons were built of the same even smaller building blocks now called quarks.

The Eightfold Way itself had been puzzling because it gave no reason why any particular multiplets should be found. Like the Mendeleev table of the chemical elements, it provided a way to classify the so-called "elementary constituents of matter," but their very number suggested that they could not all be elementary.

In 1963 Hayim Goldberg and Yuval Ne'eman pointed out that all the known particles could be constructed mathematically from the same three building blocks, now called the up (u), down (d), and strange (s) quarks, together with their antiparticles, now called antiquarks.

In 1964 Murray Gell-Mann and George Zweig dared to propose that these were indeed the basic building blocks of matter. But a serious difficulty arose. The electron, neutron, proton, and pion were all discovered experimentally as isolated particles that could be detected and created individually and whose paths through space could be determined. However, with current technology, scientists are still not able to create or study individual quarks. But scientists already believed that matter consisted of atoms and molecules long before anyone had created or detected them individually. Perhaps future discoveries will make the creation and detection of quarks possible.

There are no simple answers to the questions who discovered the atom, who discovered the quark, and how the reality of atoms and quarks was established. One possible answer appears in the book by E. D. Hirsch Jr. *The Schools We Need and Why We Don't Have Them:* "The scientific community reaches conclusions by a pattern of independent convergence (a kind of intellectual triangulation), which is along with accurate prediction, one of the most powerful confidence-building patterns in scientific research. There are few or no examples in the history of science when the same result, reached by three or more truly independent means, has been overturned" (p. 159). Hirsch quotes Abraham Pais's biography of Einstein for an example of this convergence:

> The debate on molecular reality was settled once and for all because of the extraordinary agreement in the values of N [Avogadro's number] obtained by many different methods. Matters were clinched not by a determination of N but by an overdetermination of N. From subjects as diverse as radioactivity, Brownian motion, and the blue in the sky, it was possible to state, by 1909, that a dozen independent ways of measuring N yielded results in remarkable agreement with one another.

In 1966 this kind of circumstantial evidence already convinced Richard Dalitz that matter was made of quarks, when he gave his invited review at the annual International Conference on High Energy Physics in Berkeley, California. This evidence included the existence of experimentally observed regularities in the properties of particles created at high-energy accelerators, the fact that collisions between different kinds of particles were simply related, the fact that the electromagnetic properties of different mesons and baryons were simply related, the observed experimental ratio of the magnetic moments of the neutron and proton, and the fact that the annihilation of a proton and an antiproton at rest nearly always produced three mesons. These were otherwise unexplained and converged on the same conclusion: mesons and baryons were built from the same elementary building blocks. This independent convergence eventually convinced everyone that all of the many particles described by the Eightfold Way were not the basic building blocks of matter, as had been formerly believed, but were themselves built of even smaller building blocks.

Many particle physicists could not understand why quarks were not generally accepted until well into the 1970s. One problem was that the values of the electric charges of the quarks were smaller than the electric charge of the electron. The *u* quark has a positive electric charge two-thirds of the value of the electron's charge, and the *d* and *s* quarks have negative charges one-third of the electron's charge. So far all known particles have values of electric charge that are integral multiples of the charge of the electron and its antiparticle the positron. Neither fractionally charged particles nor isolated quarks have ever been observed.

Yet more and more circumstantial evidence for the existence of quarks as the building blocks from which all matter is constructed has accumulated since 1966. All the particles that are continually being discovered and that fit into the multiplets defined by the Eightfold Way behave as if they are either built from three quarks or from a single quark and a single antiparticle of the quark called an antiquark.

The Search for Evidence for Individual Quarks

Ever since the first quark proposal in 1964, experimenters have searched for particles with electric charges less than the charge of the electron. But none have been found. All the overwhelming evidence for the existence of quarks came from properties of the mesons and baryons that indicated that they were built from quarks.

In the 1970s experiments shooting high-energy electrons at a proton target produced evidence that the electrons were striking and being scattered by single quarks. Here again the evidence was still circumstantial. The quark itself was never observed. But an electron scattered by a pointlike object with an electric charge changes its direction of motion and changes its energy in a well-defined and well-known way. Studying the changes of direction and energy in the electron scattering experiments indicated that the electrons were scattered from pointlike constituents in the proton with the fractional electric charges predicted by the quark model.

These experiments helped to confirm that the peculiar quarks really existed. But they raised two new questions. Although the quarks were hit very hard by the electron, and they absorbed a very high energy and momentum, they were never knocked out of the proton. Isolated free quarks were never observed. This indicated that the quarks were bound by very strong forces inside the proton that kept them confined. But the electron scattering data indicated that the objects scattering the electrons transferred energy and momentum like a free particle, with no evidence of being constrained by any strong forces. These two puzzles have been clarified in the new Standard Model and given the names of confinement and asymptotic freedom.

The forces that bind quarks together into mesons and baryons are so strong at large distances that separating a quark from its neighbors costs a tremendous amount of energy. When a quark in a proton is struck with an energy sufficient to create new particles, a new quark-antiquark pair is created. The created antiquark then combines with the struck quark to create a pion or other meson, and the created quark returns to the other constituents of the original proton. The energy produced by striking a quark in a proton does not drive the quark by itself out of the proton; the quark picks up an antiquark which has been created by the large energy transfer and then goes off as a meson. Thus isolated quarks are never observed as products of high-energy collisions; rather they always find partners created in the collisions and combine with them to form mesons and baryons. They are thus always confined by being bound into mesons or baryons and are never observed as isolated free quarks.

More recent experiments with high-energy collisions show how a struck quark creates quark-antiquark pairs that recombine in different ways to create a chain of mesons and baryons. The struck quark combines with a created antiquark to form a meson, leaving the quark partner of the antiquark to seek a new created antiquark, etc. This appears in the detector of the experiment as a "jet" of particles going out from the initial proton to the struck or leading quark.

An analog to this jet phenomenon from our everyday experience is lightning. When the electric charge on a cloud becomes sufficiently large, the strong force on the air atoms becomes so great that they break up into positively and negatively charged ions. If the cloud is negatively charged, it attracts the positive ions, leaving the negative ions to search for new partners and create a chain or "jet" through the air that one sees as lightning.

The Standard Model now explains how these strong forces do not disturb the electron scattering experiments that give information about the electric charges of the quarks. The field theory called quantum chromodynamics (QCD) states that although the forces between quarks become very strong at long distances, they become so weak at short distances that they are completely negligible in high-energy electron scattering. This difference between short and long distance behavior is called asymptotic freedom.

The Circumstantial Evidence Supporting the Quark Picture

There is much circumstantial evidence supporting the existence of the quark: the agreement with the experimental values of the electric charge, spin, and magnetic moments of particles with quark model predictions have provided striking evidence.

The electric charges of baryons made from three quarks with electric charge values $+2/3$ and $-1/3$ can only be $+2$, $+1$, 0, and -1. The electric charges of mesons made from a quark and its charge-conjugate antiquark can only be 1, 0, and -1. Many hundreds of particles are now known, and so far all have only these values for electric charge.

The spinning motion of the particles and their displaying of behavior similar to tiny magnets provided important clues to their structure. A spinning electrically charged top behaves like a magnet. The strength of the tiny magnet of the electron, called its magnetic moment, was successfully described by Paul Dirac's famous theory and equation.

The magnetic moments of the proton and neutron gave the first indication that they were not elementary but had a more complicated structure. The neutron has no electric charge but behaves like a magnet made of spinning negative charge. This suggests that the neutron is not an elementary object with no electric charge but consists of smaller building blocks having both positive and negative charges spinning in opposite directions. The proton magnetic moment is much larger than that described by Dirac's theory.

One of the first successes of the quark model was showing how the right experimental values of particle spins and magnetic moments were obtained by adding up the contributions of the quark spins and magnetic moments in each. A baryon made of three quarks will have a spin three times the spin of the electron or proton if the spins are parallel and will have a spin equal to the electron spin if the spin of one is opposite to the spin of the other two. A meson made of a quark and an antiquark will have a spin equal to twice the electron spin if the spins are parallel and zero spin if they are opposite and cancel. The spins of all measured particles fit this picture.

To obtain the values of the magnetic moments in the proton and neutron, one must first note that the proton consists of two u quarks with parallel spins and one d quark with opposite spin. The u and d quarks have opposite signs of electric charge, their magnets point in the same direction when they are spinning in opposite directions. Each quark magnetic moment is proportional to its electric charge. Thus the two u quarks in the proton with charge $+2/3$ each contribute $+2/3$ Dirac units of magnetic moment, while the d quark with charge $-1/3$ is spinning in the opposite direction and contributes $+1/3$ Dirac unit. In a crude approximation one adds these to get the proton magnetic moment as $+5/3$ Dirac units. The neutron has two d quarks with charge $-1/3$ units and parallel spins each contributing $-1/3$ units, and one u quark with charge $+2/3$ and opposite spin contributing $-2/3$ units to give a neutron magnetic moment of $-4/3$ Dirac units. This gives $-5/4$ for the ratio of the proton and neutron magnetic moments. A more accurate calculation using the quantum mechanical adding of spins gives

−3/2, which agrees remarkably well with the experimental value of −1.46. The sum of the neutron and proton moments is 1/3 Dirac unit. A reasonable assumption for the value of the quark Dirac unit gives an experimental value of 0.33.

This is typical of the accumulation of circumstantial evidence supporting the belief that quarks are the correct building blocks of matter. First, the electric charges of the neutron and proton and all other particles come out right. Second, the spins and very precise correct values for the magnetic moments of the neutron and proton are explained. All these confirm the picture that particles behave "as if they were made out of quarks." Their electricity, magnetism, and spin would be very hard to understand if they were not built from these building blocks. It would not be clear, for example, why the neutron, which has no electric charge, has a magnetic moment similar to the proton, which has electric charge, or why the neutron also has the opposite sign and

the correct ratio to the proton moment predicted by the quark model.

This is only one example of the circumstantial evidence supporting the conclusion that quarks are the basic building blocks of all matter. The Standard Model that guides all theoretical and experimental investigations in particle physics begins with this knowledge, even though isolated individual quarks have never been observed.

See also: EIGHTFOLD WAY; STANDARD MODEL; SYMMETRY PRINCIPLES

Bibliography

Hirsch, E. D., Jr. *The Schools We Need and Why We Don't Have Them* (Doubleday, New York, 1996).

Pais, A. *Subtle Is the Lord: The Science and the Life of Albert Einstein* (Oxford University Press, New York, 1982).

Lipkin, H. J. "The Structure of Matter." *Nature* **406**, 127 (2002).

Harry J. Lipkin

R

RADIATION, CHERENKOV

When a charged particle travels through a transparent material at a speed greater than that of light in that material, it emits a characteristic blue light, called Cherenkov radiation. Around 100 years ago, Marie and Pierre Curie enjoyed seeing a beautiful, if slightly eerie, blue glow coming from their concentrated radium solutions, but their observations occurred long before the complex light emitting effects in these solutions were understood or, indeed, before the health dangers of the ionizing radiation producing the effects were realized. Inventive experimental investigations to fully explore the phenomena now called Cherenkov radiation, carried out by Pavel Cherenkov between 1934 and 1944, with rudimentary apparatus and under trying physical conditions, were explained theoretically by Ilya Frank and Igor Tamm using classical electromagnetic theory, resulting in the award of the Nobel Prize to these three physicists in 1958.

Cherenkov light is an electromagnetic analog of the more familiar sonic boom, produced by an aircraft moving faster than the speed of sound in air, and is possible only because the phase velocity of light (v_{light}) in transparent materials is slower than the speed of light (c) in a vacuum. Even though Albert Einstein's Special Theory of Relativity prohibits a rapidly moving particle with velocity ($v_{particle}$) from traveling faster than the speed of light in a vacuum, it may still exceed the speed of light in the material that it enters and, therefore, may exceed the Cherenkov threshold velocity ($v_t = v_{light}$) at which an electromagnetic shock wave forms (Figure 1). This shock wave can be observed as a very fast pulse of light emitted uniformly in a cone around the particle direction with a characteristic Cherenkov cone-opening angle θ_c, the cosine ($\cos \theta$) of which is given by

$$\cos (\theta_c) = \frac{v_{light}}{v_{particle}} = \frac{c}{n v_{particle}}.$$

Here, $n = c/v_{light}$ is an optical parameter of a material called the index of refraction. The Cherenkov angle θ_c approaches a maximum value as the particle speed approaches c.

Cherenkov radiation is emitted at all light frequencies for which the particle speed exceeds v_{light}, with most of the light being observed at shorter wavelengths. This leads to the characteristic blue color. Above the threshold, the amount of light emitted increases as $\theta_c(v_{particle})$ increases and is proportional to the length of the particle's path in the material.

Particle detectors that use Cherenkov light are called Cherenkov counters. They are used in a number of scientific fields, such as elementary particle

FIGURE 1

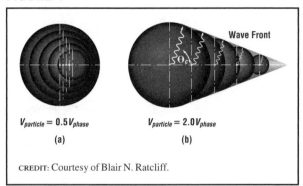

$V_{particle} = 0.5V_{phase}$
(a)

$V_{particle} = 2.0V_{phase}$
(b)

Wave Front

θ_c

CREDIT: Courtesy of Blair N. Ratcliff.

Schematic of Cherenkov light production. (a) A particle with velocity below Cherenkov threshold, and (b) a particle with velocity above Cherenkov threshold showing the formation of the shock wave front.

physics, nuclear physics, studies of cosmic rays, and neutrino astronomy. Typically, they make particular use of one or more of the properties of Cherenkov light: (1) the fast emission, (2) the velocity threshold, and (3) the dependence of the light emission angle and the amount of light emitted on the particle velocity. The latter two features are especially useful in elementary particle physics, as they may be combined with a measurement of the momentum of the particle to determine the particle's mass and, thus, identify the kind of particle that has been observed. Detectors used in this manner are called particle identification detectors (PID).

Cherenkov counters have two main elements: (1) a radiator through which the charged particle passes and (2) a photodetector to interact with the emitted light and produce an electrical signal that can be further processed, counted, and displayed. At the very low light levels of the Cherenkov effect, light is usually detected as a small number of individual particles called photons. The most common kind of photon detector is the photomultiplier tube, which can detect a single, visible light photon with an efficiency of about 20 percent and its time of arrival to better than one nanosecond (10^{-9} second). Other common photon detectors include wire chambers that can determine the positions of photons within the detector.

Radiators can be chosen from a wide variety of transparent materials. The choice between a gas

(e.g., Nitrogen [$v_t = 0.9997c$]), a liquid (e.g., water [$v_t = 0.730c$]), or a solid (e.g., glass or plastic [$v_t = 0.667c$]) radiator is made to best match the velocity range of the particles under study. As the amount of light emitted per unit length is very small for the high velocity threshold materials, the radiator must be much longer for a gas counter than for a solid counter. An extremely light "designer" material called silica aerogel (made mainly from the same molecular material as ordinary glass but with much more space between the molecules) may be used to cover the velocity region between the gases and liquid or solid materials.

Cherenkov counters are classified as either imaging or threshold types, depending on whether they do, or do not, make use of the Cherenkov angle (θ_c) information. Imaging counters are sometimes used to track particles as well as to identify them.

Cherenkov counters may be of almost any size. Some threshold detectors fit in the palm of a hand. A typical high-energy physics detector is much larger. For example, the imaging Cherenkov counter called the DIRC (Detection of Internally Reflected Cherenkov Light) that is part of the *B*-factory detector BaBar at the Stanford Linear Accelerator Center has a radiator of pure fused silica glass that covers about 5 m^2 and a photon detector with about 11,000 3-centimeter diameter photomultiplier tubes. The detectors used for neutrino astronomy, or cosmic ray studies, are even larger. For example, a large imaging neutrino detector in Japan called Super-Kamiokande is around 40 meters high, contains 50,000 tons of pure water as a radiator, and has over 11,000 very large (50 centimeter diameter) photomultipliers, while the AMANDA detector at the South Pole plans eventually to have 5,000 large photomultipliers imbedded in a cubic kilometer of ice.

See also: RADIATION, SYNCHROTRON; RELATIVITY

Bibliography

Amanda. <http://amanda.berkeley.edu/amanda/amanda.html>.

BaBar. <http://www.slac.stanford.edu/BFROOT>.

Kleinknecht, K. *Detectors for Particle Radiation,* 2nd ed. (Cambridge University Press, Cambridge, UK, 1998).

Stanford Linear Accelerator Center. <http://www2.slac.stanford.edu/vvc>.

Super-Kamiokande. <http://www-sk.icrr.u-tokyo.ac.jp/doc/sk/super-kamiokande.html>.

Blair N. Ratcliff

RADIATION, SYNCHROTRON

Electromagnetic radiation can be generated by accelerating electrically charged particles. This phenomenon is well known from the electrons moving up and down in an antenna or from the electrons that are decelerated at the surface of the anode in an X-ray tube, which results in the emission of Bremsstrahlung. The term synchrotron radiation is employed when an accelerated charge moves with a velocity close to the speed of light relative to an observer. This occurs in storage rings where highly relativistic free electrons or positrons, moving in a closed orbit at a constant energy, are deflected by strong magnetic fields. The centripetal acceleration stimulates the emission of synchrotron radiation. Due to the relativistic velocity the radiation is emitted within a small cone tangential to the direction of the particle, which results in a high degree of collimation of synchrotron radiation.

The power emitted by a relativistic particle with rest mass m and energy E per revolution in the storage ring is proportional to $(E/m)^4/R$, with R being the radius of curvature. Since a proton is 1,836 times heavier than an electron, the total radiated power per revolution is much larger for lighter electrons or positrons than for heavier protons. The unique properties of synchrotron radiation, which distinguish it in particular from conventional X-ray tubes, are its continuous spectrum, extending from the far-infrared to the hard X-ray region; its high intensities; its small source size, determined by the electron beam; its high degree of natural collimation; as well as its linear polarization in the orbit plane and its elliptical, nearly circular polarization above and below the orbit plane. Furthermore, synchrotron radiation provides a well-defined time structure, and its intensity distribution can be calculated quantitatively with high accuracy.

The first experimental observation of synchrotron radiation emitted by centripetally accelerated relativistic electrons was made in 1947 at the General Electric 70 MeV synchrotron in Schenectady, New York. Since then, dedicated synchrotron radiation facilities with large user communities have developed from the early synchrotrons with low electron currents to today's third generation storage rings dedicated exclusively to the production of synchrotron radiation.

Several terms characterize the emitted X-ray beam: *flux* describes the number of emitted photons per second, *brightness* refers to how much the beam diverges as it propagates, while *brilliance* includes additionally the source size as defined by the electron beam.

The properties of synchrotron radiation emitted from a bending magnet is fully determined by the energy E of the electrons in the storage ring and the radius R of the orbit of the bending magnet. For an actual storage ring, R is fixed and E can only be varied over a limited range. More flexibility is offered by straight sections into which arrays of magnets can be installed. These insertion devices force the electron beam to wiggle, and by adjusting the magnetic field and thereby the amplitude of the oscillations, one can adjust the X-ray properties to the needs of a particular experiment. Depending on the strength of the electron deflection, one distinguishes wigglers and undulators. In a wiggler the electron beam is deflected by strong magnetic fields in a sinusoidal transverse motion. At each oscillation the electrons emit synchrotron radiation, and the radiation emitted in the different poles is incoherently superimposed. The wiggler radiation is therefore a superposition of the radiation fans from N individual bending magnets, and the intensity is on the order of N times that of a corresponding bending magnet source.

An undulator is a similar arrangement of successive small bending magnets like a wiggler, but the field strength is smaller such that the electron-beam deflection is small compared to the natural opening angle of the emitted synchrotron radiation. Therefore, the properties of undulator radiation are based on a coherent superposition of the X-rays emitted from each individual electron in the poles of the device with itself, and the spectral and spatial distributions are characterized by these interference effects. The superior properties of an undulator are its high brightness proportional to N^2. Undulators with $N = 100$ magnetic poles are common and can theoretically achieve an increase in brightness by a factor of 10^4 over that of a bending magnet.

These extremely bright X rays can be used to investigate objects of atomic and molecular size and a large variety of experimental challenges in physics, chemistry, geology, biology and applied sciences benefit from the possibility of fine-tuning the required properties of synchrotron radiation. With the use of appropriate X-ray monochromators, one can choose a particular photon energy, which is best suited for the experiment. This makes photon and electron based spectroscopic techniques, such as X-ray absorption spectroscopy, photoemission, and X-ray scattering as well as X-ray microscopy and X-ray microtomography, powerful techniques for the investigation of the electronic and geometric structure of materials.

See also: RADIATION, CHERENKOV; RELATIVITY

Bibliography

Koch, E. E., ed. *Handbook on Synchrotron Radiation* (North-Holland Publishing Company, New York, 1983).

Winick, H., and Doniach, S. *Synchrotron Radiation Research* (Plenum, New York, 1980).

Katharina Baur

RADIOACTIVITY

If an atomic nucleus is to be radioactive, two conditions must be satisfied: 1) the nucleus must be unstable; 2) the instability must be rapid enough to be observable in a reasonable amount of time.

Some nuclei are completely stable. For example, if the $^{16}_{8}O$ nucleus is in its ground state, it will remain in that state forever, unless it is perturbed. This is because the $^{16}_{8}O$ ground state is the configuration of sixteen nucleons with the lowest possible energy. Any other configuration has more energy, and thus the ground state $^{16}_{8}O$ will continue to exist unless an outside agent supplies energy to it. The ground state $^{16}_{8}O$ nucleus is said to be energetically stable, or absolutely stable.

Other nuclei are energetically unstable. For example, the $^{238}_{92}U$ nucleus in its ground state is not the lowest energy combination of 238 nucleons. If these nucleons were rearranged to form a $^{234}_{90}Th$ nucleus in its ground state plus an α particle (a $^{4}_{2}He$ nucleus), the total energy would be lower by 4.2 MeV (4.2×10^6 electron volts). Thus an undisturbed $^{238}_{92}U$ nucleus has enough energy to emit an α particle, and every ground state $^{238}_{92}U$ nucleus will eventually disintegrate in this way.

The force that drives the α particle out of the $^{238}_{92}U$ nucleus is the electrical repulsion between the two protons in the α particle and the ninety protons in the $^{234}_{90}Th$ nucleus that is left behind. This repulsion must overcome the attractive nuclear forces that tend to keep the $^{238}_{92}U$ nucleus together. The competition between these attractive and repulsive forces determines how long the $^{238}_{92}U$ nucleus is likely to last before it emits the α particle. In general, nuclei that emit more energetic α particles will emit these α particles more quickly. This property is expressed in terms of the half-life of the nucleus, which is the time it takes for half of the nuclei in a sample to decay. The half-life of $^{238}_{92}U$ is 4.5×10^9 years. But $^{224}_{92}U$, which emits an 8.46 MeV α particle, has a half-life of only a thousandth of a second.

The $^{234}_{90}Th$ nucleus left behind by the α emission from $^{238}_{92}U$ is itself energetically unstable. It disintegrates by a process called β-decay, with a half-life of only 24 days, leading to the $^{234}_{91}Pa$ nucleus, which is also unstable. This succession of decays continues until $^{206}_{82}Pb$ is reached. $^{206}_{82}Pb$ is also unstable (with respect to disintegration into $^{202}_{80}Hg$ plus an α particle), but the excess energy of $^{206}_{82}Pb$ is so small (about 1 MeV), that the half-life of $^{206}_{82}Pb$ is very long—many orders of magnitude longer than the age of the universe. Thus, for all practical purposes, $^{206}_{82}Pb$ can be regarded as a stable nucleus, in the sense that it is highly unlikely that anyone will ever see a $^{206}_{82}Pb$ nucleus decay.

Thus we can distinguish between three types of nuclei: those that are energetically stable, those that are energetically unstable but have half-lives that are so long compared to the age of the universe that they can be regarded as stable, and those that are unstable with half-lives short enough for us to be able to observe their decay. Nuclei in this last class are said to be radioactive.

The sequence of nuclei $^{238}_{92}$U, $^{234}_{90}$Th, $^{234}_{91}$Pa, . . . , $^{206}_{82}$Pb are said to form a radioactive chain. There are three other radioactive chains, beginning, respectively, with $^{232}_{90}$Th, $^{235}_{92}$U, and $^{237}_{93}$Np. All the nuclei in a chain either have the same nucleon number or differ in nucleon number by a multiple of four. This is a consequence of the types of decay processes that occur within a chain (see below).

All the nuclei in our universe that are heavier than iron were either formed in the very energetic processes that accompanied a supernova explosion or are the descendants of nuclei formed in this way. The nuclei formed in this explosion were probably all radioactive, with half-lives that are short compared to the age of the Earth (about 5×10^9 years). Therefore these nuclei have already undergone radioactive decay and are no longer present on the Earth. However, the $^{238}_{92}$U that was formed in a supernova explosion has a half-life of 4.5×10^9 years, which is long enough so that some of this $^{238}_{92}$U still remains. Similarly, each of the other radioactive chains begins with a long-lived relic from a supernova explosion. The radioactive nuclei included in these chains are said to exhibit natural radioactivity. Artificial radioactivity refers to the radioactivity of nuclei that do not occur naturally but can be made when nuclei are bombarded by neutrons produced by a nuclear reactor or by the charged particles in an accelerator beam.

Several types of processes can occur when nuclei undergo radioactive disintegration:

1. Emission of photons. If a nucleus is in an excited state, it can undergo a transition to a lower-energy excited state or to the ground state. The energy difference between the initial and final nuclear states is usually carried away by a photon. This process is precisely analogous to the emission of light by atoms or molecules. The photon energy can vary from a few keV (10^3 electron volts) to several MeV. The lower energy photons (with energy ≤ 50 keV) are usually referred to as X-rays. Higher energy photons are called γ-rays.

2. Emission of electrons or positrons. Due to the so-called weak interaction, it is possible for a neu-

tron to transform into a proton, while emitting an electron and an antineutrino. Alternatively, a proton in the nucleus may transform into a neutron, while emitting a positron and a neutrino. A related process is the capture by a proton in the nucleus of an atomic electron, which causes the nuclear proton to become a neutron. Early in the study of this type of radioactivity, the emitted electrons were called β-rays.

3. Emission of α particles. This process is a special case of the process of fission, the disintegration of a nucleus into two or more parts (fission fragments). As for α emission, the driving force behind fission is the electrical repulsion between the fission fragments. Fission does not occur in the natural radioactive chains.

The activity of a radioactive source is the rate at which disintegrations occur. The units of activity are becquerels (disintegrations/sec) or curies (3.7×10^{10} disintegrations/sec). The unit of 1 curie was defined to be the activity of 1 gram of ^{226}Ra. In general, the activity of a radioactive sample is given by the formula

$$\text{Activity (becquerels)} = \frac{\text{mass of sample (grams)}}{\text{gram atomic weight}}$$
$$\times 6.02 \times 10^{23} \times \frac{0.691}{\text{half-life (seconds)}}.$$

See also: FERMI, ENRICO; NEUTRINO, DISCOVERY OF; PAULI, WOLFGANG; QUANTUM TUNNELING; RADIOACTIVITY, DISCOVERY OF; RUTHERFORD, ERNEST

Bibliography

Krane, K. S. *Introductory Nuclear Physics* (Wiley, New York, 1988).

Benjamin Bayman

RADIOACTIVITY, DISCOVERY OF

Radioactivity was discovered by Henri Becquerel in early 1896. At the time it was considered only a moderately interesting phenomenon. Before long, however, it was recognized as a key to the study of the atom, and it led within two decades to the creation of nuclear physics. After the discovery of nuclear fission in 1938, the subject was regarded as the most significant in all of science, not only for the nuclear reactors and bombs that emerged from World War II, but also for the remarkable new particles and theories of matter that filled the rest of the twentieth century.

Because newly discovered X rays (1895) seemed to stream from a luminescent spot on the cathode ray tube in which they were produced, Henri Poincaré suggested that all glowing bodies, not just those at high voltage, might be sources of these rays. Becquerel, with much experience examining minerals that glowed upon stimulation by light, was well suited to look for the invisible X rays. Among the minerals he tested in his physics laboratory in the Museum of Natural History in Paris was a compound of uranium that responded well. He wrapped a photographic plate in black paper to make it light-tight and placed a piece of the compound on it. Then he set this arrangement on his window sill, where sunlight stimulated the uranium salt for a few hours (had the mineral been moved quickly into a dark closet, it would have glowed a while). When Becquerel developed the plate, he saw a darkened area beneath the rock. Soon, he inserted coins and keys under a crystalline layer of the salt and was rewarded with silhouettes of their patterns. Since one of the striking properties of X rays was the sharp pictures they made, he believed merely that he had confirmed their production from sources other than cathode ray tubes.

Further experiments showed Becquerel that the intensity of the rays was attenuated as they passed through thin sheets of metal and that diffuse, reflected, and refracted light worked equally well in stimulating his uranium source. Other tests at the end of February 1896 were interrupted by almost a week of overcast skies. Since he felt that sunlight was required to excite his crystal, he put away the experimental arrangement in a dark drawer. When cloudy weather persisted, Becquerel developed the plate anyway, so he could give a report at the Monday meeting of the Academy of Sciences. He thought that he could perhaps show a weak exposure from light in his laboratory or no exposure at all—a "control" experiment to confirm his working hypothesis. He was astounded, however, to find his plate blackened.

This called for a reassessment of his ideas. Was it possible that X rays could be emitted without the necessity of first exposing the uranium crystal to sunlight? Thus began a long series of tests in which he kept a crystal in a dark box and periodically checked its activity with photographic plates, finding no detectable diminution. Becquerel also conducted investigations similar to those done with other forms of radiation. When he placed a lump of uranium next to a charged electroscope, he saw the gold leaves fall: the rays emerging from the uranium crystal made air a conductor of electricity. He concluded (incorrectly) that uranium rays were reflected, refracted, and polarized, confirming for him their similarity to X rays and their electromagnetic nature.

Yet other puzzles appeared. When Becquerel tested uranium compounds that did not phosphoresce, and when he destroyed a source's ability to phosphoresce by melting it and allowing it to recrystallize in darkness, he nonetheless got intense images on the photographic plates. By May 1896, Becquerel learned that uranium metal was more active than any compound and concluded that the element itself was the source of activity; it was an atomic phenomenon. Still, he could not abandon his original line of reasoning and proclaimed the discovery of a new property of metals: invisible phosphorescence.

X rays dominated scientific and popular attention, for they yielded sharper pictures more quickly and were useful in diagnostic medicine. Moreover, the equipment for X rays was more common in physics laboratories than uranium crystals. More than 1,000 papers on these penetrating rays were published in 1896 alone, compared with just a handful on uranium rays. Becquerel himself seems to have abandoned his discovery: he wrote seven papers in 1896, two in 1897, and none in 1898. Several other prominent physicists added another dozen papers in the first two years. Becquerel rays, as they were called, were not especially interesting.

The year 1898 saw a resurrection of interest. Gerhard C. Schmidt of the University of Erlangen tested other materials and found thorium compounds emitted somewhat similar rays. Because of Becquerel's errors in determining some of the radiation's properties, Schmidt could not be certain of an exact match.

Soon after, and independently, Marie Curie in Paris also pointed to thorium. It is unclear why she chose to investigate these rays after some very able scientists seemed to have exhausted the subject. Perhaps she sensed it remained important, or possibly she wanted a doctoral dissertation topic without any likely competition.

A careful and thorough investigator, Marie Curie used a sensitive electrometer to measure the intensity of the radiation. Designed by her physicist husband, Pierre, and his brother, Jacques, this instrument provided quantitative data, compared with Becquerel's largely qualitative results, and was in the more modern tradition of seeking numerical results from experiments. Like Becquerel and his invisible phosphorescence of metals, Marie Curie had a guiding idea: space was filled with rays similar to X rays but more penetrating. When they struck elements of high atomic weight, such as uranium and thorium, they caused Becquerel rays to be emitted as secondaries.

She too was faced with a puzzle. Her uranium ore was more active than its uranium content should allow. Faced with the possibility of another active element, Pierre Curie dropped his own research to join that of his wife. In the summer of 1898 they announced discovery of a new element, named polonium in honor of Poland, her native country. Marie also gave the phenomenon the name radioactivity. Before the end of the year, they and a chemist colleague named Gustave Bémont revealed yet another constituent of the ore: radium, named for its outpouring of rays. The quantities of these new substances were so small they were at first invisible, yet the Curies persisted in declaring them to be new elements. Eventually, the spectrum and atomic weight of radium were measured, providing the proof.

Also in 1898, Ernest Rutherford began to investigate radioactivity. A graduate student in J. J. Thomson's Cavendish Laboratory at Cambridge University, he quickly became the central figure in radioactivity and in nuclear physics. He showed, by their different abilities to penetrate thin sheets of foil, that the radiation consisted of two components, which for convenience he named alpha and beta, names that have endured. Paul Villard in 1900 revealed a third component, the electromagnetic gamma rays. A

year earlier, Friedrich Giesel in Braunschweig deflected beta rays in a magnetic field, showing they were charged particles. Becquerel, rejoining a now-exciting field, showed that beta particles were identical to the recently discovered electron. In 1903, Rutherford, now a professor at McGill University, bent alpha rays in a magnetic field, proving they were positively charged particles.

Several new radioelements also were discovered around the turn of the century, some that seemed to maintain a constant level of activity and others that lost activity over time, and there was need of a concept to organize and explain them. With the chemist Frederick Soddy, Rutherford in 1902–1903 advanced the transformation theory of radioactivity. The radioelements were placed in just a few series, with uranium and thorium heading their own. All decayed, with different half-lives, until an as-yet unknown, stable, end product was reached (lead). With this insight, research largely shifted from studies of the radiations to investigations of the bodies that emitted the alphas, betas, and gammas. The goal was to ascertain the identity and position of each radioelement in each decay series.

Rutherford and Soddy's explanation of radioactivity overcame the interpretations of Becquerel and the Curies. By placing the energy source within the atom itself, they opened a four-decade-long debate over whether atomic energy could be harnessed. The nuclear weapons and reactors constructed in World War II answered that question in the affirmative.

See also: RADIOACTIVITY; RUTHERFORD, ERNEST

Bibliography

Badash, L. "Radioactivity Before the Curies." *American Journal of Physics* **33**, 128–135 (1965).

Badash, L. "Becquerel's 'Unexposed' Photographic Plates." *Isis* **57**, 267–269 (1966).

Badash, L. "How the 'Newer Alchemy' was Received." *Scientific American* **215**, 88–95 (1966).

Badash, L. "The Completeness of Nineteenth-Century Science." *Isis* **63**, 48–58 (1972).

Badash, L. "The Discovery of Radioactivity." *Physics Today* **49**, 21–26 (1996).

Curie, M. *Pierre Curie* (Macmillan, New York, 1923).

Romer, A. *The Restless Atom* (Doubleday, Garden City, NY, 1960).

Romer, A. *The Discovery of Radioactivity and Transmutation* (Dover, New York, 1964).

Lawrence Badash

REINES, FREDERICK

Frederick Reins was born in Patterson, New Jersey, on March 16, 1918. His parents, Israel Reines and Gussie (Cohen) Reines, met and married in New York City after emigrating from Russia. Israel worked as a weaver, started a silk mill after World War I, and later ran a country store in Hillburn, New York, where Fred spent much of his childhood. Fred, the youngest of four children, attended Union Hill (New Jersey) High School and the Stevens Institute of Technology (SIT; Hoboken, New Jersey). While at SIT he developed a lifelong love of vocal and dramatic performance and received a B.S. in Engineering in 1939 and an M.S. in Mathematical Physics in 1941. Reines continued his graduate studies at New York University and earned a Ph.D. in theoretical physics in 1944. He married Sylvia Samuels in 1950; they had two children.

Following completion of his Ph.D. program, Reines joined the Manhattan Project in the Theoretical Division of the Los Alamos Scientific Laboratory. He continued in the weapons testing program after the end of World War II. This work took him to Eniwetok Atoll, as director of the Operation Greenhouse experiments and included research on the effects of nuclear blasts at that location and at the Bikini and Nevada test sites.

While on a sabbatical-in-residence at Los Alamos in 1951, Reines pondered the possibility of detecting the neutrino, a particle postulated by Wolfgang Pauli in 1930. Although the neutrino postulate proved to be consistent with beta-decay observations, the neutrino itself, which carries no electrical charge, has zero or tiny mass, and interacts only weakly with matter, had never been detected and seemed to be well beyond the limits of detection. Shortly thereafter Reines and Clyde L. Cowan, Jr., who was also a Los Alamos physicist, joined forces in planning an effort to detect neutrinos following the detonation of a nuclear bomb. They set out to exploit a new technology, a large liquid scintillation detector, to distinguish neutrino-induced inverse-beta-decay events from the large background of other products of nuclear fission. In the fall of 1952 they realized that a nuclear fission reactor would also provide an abundant source of neutrinos. Their first attempt to detect neutrinos, with their detector adjacent to a reactor at the Hanford Engineering Works near Richland, Washington, in 1953, was inconclusive due to factors such as a large cosmic-ray muon background that mimicked inverse beta-decay reactions. They subsequently redesigned their detector and installed it in an underground location adjacent to a reactor at the Savannah River (South Carolina) Plant of the U.S. Atomic Energy Commission. The Reines-Cowen detector was separated by 11 meters of concrete from the reactor core and by 12 meters of overhead shielding from cosmic-ray muons. There, between February and June 1956, Reines and coworkers obtained unequivocal evidence that identified the neutrino as a free particle. Additional measurements at Savannah River later in 1956 and in 1963–1964 provided more evidence and secured a definitive value for the cross section for the inverse-beta-decay reaction.

In 1959, Reines left Los Alamos to become professor and head of the Physics Department at Case Institute of Technology in Cleveland, Ohio, where he undertook new neutrino research while continuing work previously begun at Los Alamos. Fission reactors continued to be an important tool in this work. For example, a twenty-year effort using reactor neutrinos to observe the direct elastic-scattering of neutrinos by electrons finally was brought to culmination in 1976 with the successful measurement of the tiny cross section for this reaction. In other reactor experiments that began at Los Alamos and bridged his time at Case, Reines and his coworkers observed both the "charged current" and the "neutral current" interactions of neutrinos with deuterons, completing these measurements in 1969 and 1979, respectively. The first studies of neutrino stability and of neutrino oscillations also employed reactor neutrinos.

Part of the experimental program begun at Case took Reines to deep underground locations, in the Morton Salt Company mine 600 meters beneath the shores of Lake Erie near Cleveland, and in a 3,200-

American physicist Frederick Reines (1918–1998) shared the 1995 Nobel Prize in Physics with Martin L. Perl. Reines received the Nobel for his detection of the neutrino. CREDIT: COURTESY OF LOS ALAMOS NATIONAL LABORATORY, AIP EMILIO SEGRE VISUAL ARCHIVES. REPRODUCED BY PERMISSION.

meter-deep gold mine near Johannesburg, South Africa, where reduced cosmic ray muon fluxes made it possible to study additional neutrino properties, to search for solar neutrinos, or to probe the limits of fundamental conservation laws such as conservation of charge and of baryon number. In 1964–1965 at the South African site, Reines and his coworkers were the first to observe muons produced in their detector by neutrinos generated in the earth's atmosphere by high-energy cosmic rays.

In 1966, Reines left Case to become founding dean of the School of Physical Sciences at the University of California, Irvine (UCI) campus, a position he held until he returned to full-time teaching and research in 1974. While at Irvine Reines served as spokesman for the Irvine Michigan Brookhaven

(IMB) collaboration which operated a Cerenkov detector containing 8,000 tons of water in the aforementioned salt mine beginning in 1982. The primary motivation for this effort was to extend the measured lower limit on the lifetime of the proton. This experiment ran until 1991 without observing a proton decay. The lower limits determined for lifetimes of several proton decay channels provided a significant test of particle theories. In addition, this detector, along with a similar detector in Kamiokande, Japan, detected a burst of neutrinos from the supernova 1987A, confirming the role of neutrinos in stellar collapse.

Reines was appointed Distinguished Professor of Physics at UCI in 1987 and retired in 1988. He received many honors during his lifetime, among them the 1981 Oppenheimer Memorial Prize, the National Medal of Science, the 1992 Franklin Medal, the 1992 W. K. H. Panofsky Prize of the American Physical Society, and the 1995 Nobel Prize in Physics ("for the detection of the neutrino"), which he shared with Martin Perl, discoverer of the tau lepton. His fame rests not only on the discoveries made by him and his coworkers and the experimental techniques that they pioneered but also on demonstrating that neutrino physics is both doable and immensely fruitful. Reines died in Orange, California, on August 26, 1998.

See also: FERMI, ENRICO; NEUTRINO, DISCOVERY OF; PAULI, WOLFGANG

Bibliography

Kropp, W.; Moe, M.; Price, L.; Schultz, J.; and Sobel, H.; eds. *Neutrinos and Other Matters: Selected Works of Frederick Reines* (World Scientific, Singapore, 1991).

Kropp, W. R.; Schultz, J.; and Sobel, H. W. "Obituary: Frederick Reines." *Physics Today* **52**, 78–80 (1999).

Reines, F. "The Neutrino: From Poltergeist to Particle." *Reviews of Modern Physics* **68**, 317–327 (1996).

Sutton, C. *Spaceship Neutrino* (Cambridge University Press, Cambridge, UK, 1992).

Robert G. Arns

RELATIVITY

Relativity is the physics of the nature of space and time, a branch of physics that plays a minor role in technology but is crucial in the world of elementary particles.

Overview

At the end of the nineteenth century, the cornerstones of physics were Isaac Newton's mechanics (the theory of forces and motion) and James Clerk Maxwell's electrodynamics. This classical physics referred to locations in space and changes in time and therefore embodied assumptions about the structure of space and time. Among these assumptions was the relationship, or relativity, of measurements. When one reference frame moves with respect to another, different values of some quantities are measured in the two frames. As an example, a hawk might be flying at 80 mph east relative to the ground but 300 mph west relative to an eastbound airplane. The classical conceptions of relativity were too obvious to get much attention: space had a three-dimensional Euclidean geometry, and the flow of time was universal, a clock ticking the same way for all reference frames. Most important, classical space and time were separate.

Questions raised by Maxwell's theory changed all this. The equations of his electrodynamics could only work in a single reference frame. Experimental searches were undertaken for this special reference frame, most notably the 1887 experiment of Albert Michelson and Edward Morley. No trace of such a frame was found; experiments showed Maxwellian electrodynamics to work in all reference frames.

In 1905 Albert Einstein gave the explanation: the classical assumptions about space and time are wrong. Einstein's space-time description, special relativity theory (SRT), gave new relationships of quantities in different reference frames. With these new relationships Maxwellian electrodynamics worked in all reference frames. The mathematics of the new relationships was easy, requiring nothing more sophisticated than a square root. The formulas, the Lorentz transformations for the new relationships, had, in fact, been worked out one year earlier by Hendrik Lorentz, who did not understand their real impact. Though SRT is universally accepted in the scientific community today, its violation of intuition made it revolutionary in 1905. Most revolutionary was the SRT view that space and time were not distinct but had some of the character of different directions in a single space-time entity.

The historical association of SRT and electromagnetism can be blamed for a common misconception. In SRT there is a fundamental constant of nature, denoted as c, that has the same units as a velocity and has a numerical value very close to 300,000 km/sec. This constant plays an important role in the fundamental structure of space-time. One of its well-known roles is as a velocity limit; the relative velocity of any two objects must have a value less than c. Because light (and certain other) waves propagate at c, it is often called the speed of light. This name, avoided here, too often gives the false impression that SRT is a consequence of observations based on light signals.

The classical conception of space-time, and the Newtonian mechanics that uses it, lasted so long because they were highly accurate for most laboratory and astronomical science. Einsteinian relativistic effects become noticeable only for relative motions at speeds comparable to c, and such speeds are rarely encountered. Even a speed of 1,000 km/hr is only one millionth of c. The situation is opposite for much of elementary particle physics: Speeds are extremely close to c, and special relativistic effects are crucial.

Special Relativity and Elementary Particles

In Newton's mechanics, particles are characterized by an inherent property called mass m. For a particle moving at velocity $\vec{v}$ two other classical quantities are useful, its energy $E = \frac{1}{2}mv^2$ and its momentum $\vec{p} = m\vec{v}$. The usefulness of E and $\vec{p}$ lies in the fact that they are conserved; they do not change in time. For example, suppose you have a set of particles that undergo complicated interactions. Some of the particles could exert forces on each other; some of the particles could join together to form a new particle; some could break apart, forming new particles in another way. Add up all the particle energies at any time, and at any other time, and one gets the same number. In particular the total energy is the same before and after the particle interaction. The same is true for the total momentum.

Since the classical E and $\vec{p}$ are constructed from the velocity $\vec{v}$, they will have different values in different reference frames. This means, for example, that the total energy in a particular reference frame is different from that in a second frame. This was not a problem in Newton's mechanics. What was important was that in any one reference frame the total E and $\vec{p}$ are conserved.

In SRT, the relativity of $\vec{v}$ is different from that of classical physics. A consequence of this is that if $E = \frac{1}{2}mv^2$ is conserved in one frame, it is not conserved in another; the same frame-dependent conservation applies to $\vec{p} = m\vec{v}$. Clearly it is useless, and even meaningless, to have a conservation law that depends on the reference frame. Einstein found that the conservation laws could be saved by changing the definitions of energy and momentum. The SRT definitions, incorporating the Lorentz factor $\gamma = 1/\sqrt{1 - v^2/c^2}$, are:

$$E = \frac{mc^2}{\sqrt{1 - v^2/c^2}} = \gamma mc^2 \quad \text{and}$$

$$\vec{p} = \frac{m\vec{v}}{\sqrt{1 - v^2/c^2}} = \gamma m\vec{v}.$$

If these energy and momenta are conserved in one frame, they are conserved in any other frame.

At least one feature of these equations is immediately surprising: when v is set to zero, the energy of a particle of mass m is not zero, as in Newtonian mechanics, but rather is given by the famous formula $E = mc^2$. A particle has energy even if it is not moving. There exists specialized vocabulary to help sort out the many possible meanings of energy. The value mc^2 is called the rest energy of a particle, and γmc^2 can be called the mass-energy. Often the explicit expression kinetic energy is used to mean the mass-energy minus the rest energy, $(\gamma - 1)mc^2$. This kinetic energy has the familiar feature that it is zero for a stationary particle.

It is important to understand that it is the total mass-energy E, not the total kinetic energy, that is the same before and after something happens. A dramatic example of this is the decay of a kind of elementary particle called a pion. Sitting still in a laboratory, a pion can transform itself into two pure bursts of electromagnetic energy, two gamma rays. These bursts have no rest mass, so the death of the pion is

a total conversion of rest energy into electromagnetic energy. A more familiar example is nuclear fission. Certain nuclei, such as the isotope uranium-235, spontaneously break into several fragments. The sum of the rest energies of the fragments is less than the rest energy of the original uranium nucleus, so the fragments must have kinetic energy in order for energy to be conserved. This kinetic energy is converted into great amounts of heat in bombs and in reactors.

Two issues of terminology can cause confusion about the meaning of mass. In some older books mass was taken to be what is now denoted as $m\gamma$. (In potentially confusing contexts the term rest mass is used to denote what we call m.) The second source of confusion is the nearly universal practice of referring to the rest energy of an elementary particle as the particle mass. The rest energy, furthermore, is stated in an unusual system of units: electron volts, or eV (1 kilo electron volt [keV] = 10^3 eV, 1 mega electron volt [MeV] = 10^6 eV, 1 giga electron volt [GeV] = 10^9 eV, and 1 tera electron volt [TeV] = 10^{12} eV). As an example, a proton with a mass of $m_p = 1.672610 \times 10^{-24}$ grams has a rest energy of $m_p c^2 = 1.503 \times 10^{-3}$ ergs or 938 MeV. A scientist will speak then of the mass of a proton being 938 MeV and similarly of an electron having a mass of 511 keV.

The practical aspects of other terminology is more obvious. For particles moving at everyday speeds, v is a useful number, but the Lorentz factor γ is not. For a particle at rest γ is exactly unity, and for a particle moving at 1,000 km/sec the Lorentz factor is unity up to the thirteenth decimal place. For elementary particles produced in accelerators and found in cosmic rays, on the other hand, γ and particle energy are useful, but v is not. For example, the Large Hadron Collider (LHC), to be built at the European Laboratory for Particle Physics (CERN) near Geneva, will be capable of creating protons with a $\gamma \approx 3,700$, or equivalently with energy $\gamma \times 938$ MeV = 3.5 TeV. It is inconvenient to talk of the velocity of such a proton; to eight significant figures it is equal to c.

As an example of the crucial, and surprising, relativistic effects for elementary particles we can imagine a fictitious experiment at the LHC designed to create an exotic new high-mass particle by the joining of two protons in a collision. The left side of Figure 1 shows an appropriate "before" and "after" if one of

FIGURE 1

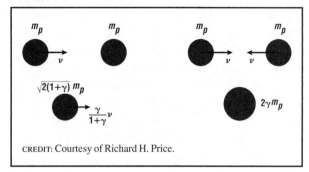

m_p m_p m_p m_p

v v v

$\sqrt{2(1+\gamma)}\,m_p$

$\dfrac{\gamma}{1+\gamma}v$

$2\gamma m_p$

CREDIT: Courtesy of Richard H. Price.

In an accelerator two equal-mass particles collide to form a new, more massive particle. In the collision shown on the left, one of the initial particles is a stationary target. In the collision on the right, the initial particles have oppositely directed velocities, as in the Large Hadron Collider.

the protons is a stationary target that is hit by a proton moving at velocity v. From only the conservation of relativistic energy and momentum it can be shown that the final particle that is formed is moving at a velocity $\gamma v/(1 + \gamma)$ and has a mass of $\sqrt{2(1 + \gamma)}\,m_p$. For everyday speeds, γ is approximately unity, and the final particle mass and velocity have approximately the classically computed values $2m_p$ and $v/2$. For high values of γ, however, the answers are very different. The final particle produced can have a mass much greater than twice m_p and will be moving very nearly at the speed of light.

Though it entails considerable engineering difficulty, the LHC is built to accelerate particles both clockwise and counterclockwise, so that collisions can take place as shown on the right side of the figure, creating an unmoving final particle. The mass of this particle is calculated, from the conservation of relativistic energy and momentum, to be $2\gamma m_p$. In the LHC, with $\gamma = 3{,}700$, this is a mass of 7 TeV. For this value of γ, the particle created for a stationary target proton (the left side of the figure) has a mass of 81 GeV, only around 1 percent of 7 TeV. From this it is clear that particles with much larger masses can be created in colliding beam accelerators.

General Relativity

Einstein's 1905 theory is called special relativity because it applies only in the special circumstance that gravitational forces are absent. Around 1915, Ein-

stein successfully extended his space-time ideas to general relativity, a more general theory that can deal with gravitational interactions. This theory is built on the equivalence principle of gravity, the principle that all particles undergo the same acceleration due to gravity. Einstein saw that what was crucial to gravity was the special space-time curves describing how a particle, any particle, could move. For that reason his theory is a geometric one in which the special curves are features of a curved space-time geometry.

Like SRT, general relativity is not crucially important for much of science. Newton's much simpler seventeenth-century theory of gravitational forces is an excellent approximation except for the description of the universe as a whole or for objects like black holes in which gravitational forces are extreme. Unlike SRT, general relativity requires advanced mathematics. Another difference is that general relativity has not been solidly verified to the same extent as SRT. It has, however, passed all experimental tests yet tried.

Ultimately, general relativity cannot be correct; it does not include the quantum effects that scientists believe must be part of every physical interaction. Such a complete theory, combining curved space-time gravity and quantum effects, has been a foremost goal of theoretical physics for more than a half-century. For the past several years some physicists have felt that a new approach called string theory shows promise in providing a path to the complete theory and at the same time resolving longstanding problems in quantum theory.

See also: EINSTEIN, ALBERT

Bibliography

Mermin, N. D. *Space and Time in Special Relativity* (McGraw-Hill, New York, 1968).

Misner, C. W.; Thorne, K. S.; and Wheeler, J. A. *Gravitation* (W. H. Freeman, San Francisco, 1970).

Price, R. H. "A General Relativity Primer." *American Journal of Physics* **50**, 300 (1982).

Taylor, E. F., and Wheeler, J. A. *Spacetime Physics: Introduction to Special Relativity,* 2nd ed. (W. H. Freeman, San Francisco, 1992).

Thorne, K. S. *Black Holes and Time Warps: Einstein's Outrageous Legacy* (W. W. Norton, New York, 1994).

Richard H. Price

RENORMALIZATION

Renormalization is a technique required for extracting meaningful predictions from a quantum field theory. The description of particles and their interactions involves various physical parameters, such as the masses and charges of particles. These parameters are not known a priori and must be measured. Quantum effects can shift the values of these parameters and the measured physical value will include the quantum modifications. These notions form the basis of the renormalization program. When considered more carefully, these concepts can be used to show that the predictions of many theories are independent of any new physics that might appear at very high energies, such as the issue of whether or not the divergences found in perturbation theory exist. Renormalization theory also leads to the idea of a "running" charge, that is, one that depends on the energy scale at which it is measured.

The Renormalization Program

Briefly stated, renormalization refers to the process by which the sum of all contributions to a physical parameter is adjusted so that the total is equal to the experimentally measured value of this parameter, and all predictions are expressed in terms of the experimental value.

First consider the concept of different contributions to a parameter. As examples, consider two specific parameters: the mass and the charge of an electron. Normally, one just thinks of these as two numbers. However, the values that one sees in nature reflect various physical effects. Since electric fields carry energy, some of the rest energy of the electron is carried by its electric field. If one were able to turn off the electromagnetic interaction, the mass of the electron would be different. One can distinguish the value of the electron mass with the electromagnetic interaction turned off (sometimes referred to as the bare mass) and the physical mass including electromagnetic effects. While the bare mass is independent of the electric charge e, the energy in the electric field would be of order e^2. Similarly, the electron's charge can be modified by electromagnetic interactions. Quantum mechanics allows the temporary production of a virtual electron-positron pair in the electromagnetic field of an electron. This virtual pair can partially screen the charge of the electron, an effect referred to as vacuum polarization. This is shown in the Feynman diagram of Figure 1. In this figure, a photon splits into an electron-positron pair in the neighborhood of a charged particle, and the external photon interacts with the virtual pair rather than the original charged particle. This modifies the interaction of the external photon with the charged particle. Again, one often refers to a bare charge, which can be denoted by e_0, plus quantum corrections. Although normally the electric field is linear in the electric charge e_0, the modification of the field due to vacuum polarization depends on this charge cubed. In perturbation theory, there would also be corrections at higher powers of the charge. The physical value is the sum of these various contributions.

Next consider the measurement process. To accomplish a measurement, the experimenter must set up some interaction with the particles. Typically, this involves scattering experiments at particular energies and angles. Since there are also quantum corrections to each process, one uses the theory to provide the relationship of the experimental measurements to the underlying parameters. Given a specific condition for the measurements, one measures the physical values for the parameters.

One would also need to describe other predictions of the theory in terms of the physically measured values of the parameters. This is the process of renormalization. The original parameters with

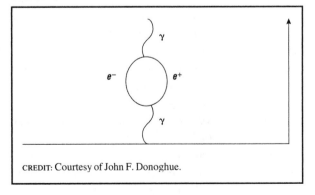

CREDIT: Courtesy of John F. Donoghue.

Feynman diagram of vacuum polarization.

which one starts calculating would generally be analogous to the bare values. Using these original parameters, one calculates the quantum corrections to the parameters and also the quantum corrections to the other processes that one is predicting. One then expresses the new predictions in terms of the physical values of the parameters, referred to as the renormalized parameters. The bare parameters never appear in these results—the true predictions of the theory are expressed in terms of well-defined measured renormalized parameters.

Importance of Renormalization

In the history of physics, the renormalization procedure was first important in the treatment of infinities that one finds in perturbation theory. For a class of theories that are called renormalizable, the predictions of the theory are all finite when expressed in terms of the measured values of a small number of parameters. This was a surprising and important result because at intermediate steps infinities appear in the calculation of the quantum corrections. By expressing the calculation in terms of the physically measured parameters, all the infinities disappear, and one is left with finite predictions. The theory unambiguously connects one physical process (the measurement of the parameters) to another (the process that is being predicted). However, note that the idea of renormalization is independent of the possible existence of infinities. If one had a fully finite theory, or a method of calculation that did not include infinities, one would still need to express the predictions in terms of measured parameters.

It is the success of renormalization that allows physicists to make quantum predictions in the face of incomplete knowledge. One is always in the situation where one does not know the ultimate theory at the highest energies. Physics is an experimental science, and only the nature of the fundamental theories up to a given energy has been explored. There could be new particles and new interactions that will only show up at higher energy, and this may be a problem for quantum predictions. In calculating quantum corrections, one is instructed to sum over all possible intermediate states and integrate over all energies. If the true physics at high energies are not yet known, how can such a calculation be conducted? The an-

swer is definitely not that the effects of high energies are unimportant in general. In some of the intermediate steps there are contributions that are sensitive to the highest energies. However, it is a general result that the major effects of unknown new physics at very-high-energy scales can be absorbed (along with the infinities) into the values of the renormalized parameters. The relationships between physical processes are not influenced by the unknowns when expressed in terms of the measured parameters.

Example

One of the most sensitively measured quantities in physics is the magnetic moment of the muon. This provides a fine example of the success of the renormalization program. The theoretical calculation has proceeded to many orders of perturbation theory. In each order there are divergences that arise in individual diagrams. However, when all diagrams are considered and the result expressed in the measured value of the fine structure constant, one obtains a finite result. It is traditional to express the result using the gyromagnetic ratio g_μ that describes the fractional deviation of the result from the standard Dirac magnetic moment (which corresponds to $g_\mu = 2$). The theoretical result is

$$
\begin{aligned}
\frac{g_\mu - 2}{2} &= \frac{\alpha}{2\pi} + 0.765857388(44) \left(\frac{\alpha}{\pi}\right)^2 \\
&+ 24.050509(2) \left(\frac{\alpha}{\pi}\right)^3 + 126.04(41) \left(\frac{\alpha}{\pi}\right)^4 \\
&+ 930(170) \left(\frac{\alpha}{\pi}\right)^5 + \cdots \\
&= 11659177(7) \times 10^{-10}
\end{aligned}
$$

where α is the fine-structure constant $\alpha = e^2/(4\pi\hbar c)$, and the uncertainties in the last digits are listed in parentheses. In a recent experiment at Brookhaven National Laboratory (BNL), the value of

$$
\frac{g_\mu - 2}{2} = 11659202(14)(6) \times 10^{-10}
$$

was measured. Although physicists are intrigued as to whether the modest 1.6 standard deviation difference ($25 \pm 16 \times 10^{-10}$) might be a harbinger of some new physics, for the purposes of the discussion here, one may marvel that theory and experiment agree so well through many orders of perturbation theory.

Running Charges

There are many different processes or experimental conditions that one could use for the measurement of a physical parameter, and each situation has its own set of quantum corrections. This leads to a potential ambiguity involving which corrections should be included in the definition of the renormalized parameter, and indeed different choices are allowed. The different choices are referred to as different renormalization schemes, and each scheme involves a precise definition of the basic renormalized parameter. It may also be observed that it is useful to change the definition of the basic coupling when working at different energies. For example, the quantum corrections that contribute to Coulomb scattering depend on the momentum transferred to the scattered electron. At large momentum transfer, the Fourier transform of the Coulomb potential becomes

$$V(q^2) = \frac{\alpha}{q^2} \left[1 + \frac{\alpha}{3\pi} \ln\left(\frac{q^2}{m_e^2}\right) + \cdots \right]$$

where $q = p' - p$ is the momentum transfer. This relation can be used to define the coupling at large values of momentum transfer to be

$$\alpha(q^2) = \alpha \left[1 + \frac{\alpha}{3\pi} \ln\left(\frac{q^2}{m_e^2}\right) + \cdots \right].$$

This is referred to as the running coupling. Predictions can be expressed in terms of either the static coupling or the running coupling, and the experimenter could choose to measure either α or $\alpha(q^2)$ at a given value of q^2. However, the definition of the running coupling is found to be very useful. In the case where the logarithm factor is large, the use of the running coupling will absorb the large logarithms in *all* processes with an energy scale $E^2 \sim q^2$. In a theory such as quantum electrodynamics (QED), one can express predictions in terms of the low-energy coupling $\alpha = 1/137$. However, at higher energies one could also calculate using the running coupling in order to avoid large logarithms.

In quantum chromodynamics (QCD), there is a coupling constant that gives the strength of the in-

teractions of quarks and gluons. When formulated as a running coupling constant, it is found to become weaker at high energies (a phenomenon referred to as asymptotic freedom) and grows strong at low energies. Therefore, one can treat the theory perturbatively at high energy where the coupling constant is weak enough. This has been verified experimentally. Since the predictions occur at energies that are large compared to the masses in the theory, the only relevant scale for the prediction is the energy at which the experiment is being conducted. Therefore, the predictions are expressed in terms of the running coupling at that energy scale. This notion can be made quite precise. The results of different experimental measurements at different energies are shown in Figure 2, where the phenomenon of the running coupling is clearly visible.

There are three separate charges in the Standard Model. Listed in order of decreasing size, one has the strong charge of QCD, then the charge associated with the weak gauge bosons, and finally the electric charge. When formulated as running charges,

FIGURE 2

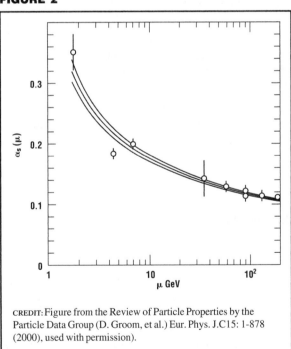

CREDIT: Figure from the Review of Particle Properties by the Particle Data Group (D. Groom, et al.) Eur. Phys. J.C15: 1-878 (2000), used with permission.

The measurement of the strong coupling constant measured at different energy scales μ.

the first two decrease at high energy, with the strong charge changing the most rapidly, while the electric charge grows. This raises the possibility that all three charges might meet at some very high energy. Associated with this possibility is the theory of forces referred to as the Grand Unified Theory (GUT). The idea is that at some very high energy there is only one interaction that encompasses the strong, weak, and electromagnetic forces, tied together in a simple gauge theory. At this scale, only a single coupling constant would exist. This unified theory could undergo spontaneous symmetry breaking, leaving the three separate interactions at lower energy. (This would be much like how the unified theory of the electromagnetic and weak interactions in the Standard Model breaks to leave only electromagnetism at low energies.) The three couplings would then evolve to their observed values at low energy. To determine if the couplings observed are compatible with this hypothesis, one would need to evaluate the running couplings at high energy and see if there is some energy at which they are all the same. The way that the couplings change with energy depends on the particles that are able to be excited at such an energy. If it is assumed that the only particles existing at any energy are those already discovered, then the running couplings come very close to each other, at an energy about thirteen orders of magnitude greater than that of present accelerators. Although it appears that the couplings do not exactly all coincide at a single point, it is possible that some new particles exist in those many orders of magnitude which slightly modify the running of the charges and which allow the couplings to be fully unified.

See also: GAUGE THEORY; QUANTUM FIELD THEORY; SALAM, ABDUS; STANDARD MODEL

Bibliography

Brown, L. M., ed. *Renormalization: From Lorentz to Landau (and beyond)* (Springer-Verlag, New York, c.1993).

Collins, J. C. *Renormalization: An Introduction to Renormalization, the Renormalization Group, and the Operator-product Expansion* (Cambridge University Press, Cambridge, UK, 1984).

Feynman, R. P. *QED: The Strange Theory of Light and Matter* (Princeton University Press, Princeton, NJ, 1985).

Kane, G. L. *Modern Elementary Particle Physics* (Addison Wesley, Redwood City, CA, 1987).

John F. Donoghue

RESONANCES

Most of the several dozen "elementary" particles listed in the tables published annually by the Particle Data Group at the Lawrence Berkeley Laboratory are unstable: if produced, they decay according to an exponential decay law familiar from radioactive decay. Their mean lifetimes vary over a wide range. For instance, the mean lifetime of the first excited state of the proton, the so-called Δ particle, is about 10^{-23} seconds, that of the muon is a few microseconds, and that of the proton (if it is unstable at all) is comparable to the known lifetime of the universe (about 10 billion years.)

Typically, a short-lived unstable particle can be produced if one scatters its decay products on each other. For instance, the excited states of the proton can be produced by scattering a beam of π mesons (pions) or photons on a hydrogen (proton) target. In such an experiment, the incident particle (the pion or photon) spends a time comparable to the mean lifetime of the excited state close to the proton. This leads to an enhancement of the scattering probability, measured by the scattering cross section. The cross section has a peak at a center-of-mass energy of the incoming particles nearly equal to the rest energy of the unstable particle produced. The peak is of a finite width: it is inversely proportional to the mean lifetime of the unstable particle or resonance, corresponding to the uncertainty relation between time and frequency.

Unlike other uncertainty relations known in quantum mechanics, the uncertainty relation between frequency and time exists in classical physics as well. For instance, if one wants to measure the frequency of a pendulum, one has to measure it over several periods. The error of the frequency measurement is inversely proportional to the time spent on the measurement. This purely classical result translates into quantum theory by the use of Planck's relation: $E = h\nu$, where h is Planck's constant and ν is the frequency of oscillation.

In a situation like that just described, one speaks about a resonance in the cross section. The shape of the cross section at energies close to the energy of the unstable particle excited is well described by the

Breit-Wigner formula. This expression for the scattering cross section is identical to the response of a damped, forced harmonic oscillator at forcing frequencies close to the eigenfrequency of the oscillator, hence the name.

In general, the existence of resonances in a scattering cross section indicates that the target particle is a composite one, just as in atomic or nuclear physics. (For instance, the scattering of a beam of photons on an atomic target excites the various electronic levels and it leads to peaks in the scattering cross section.) In the example quoted, the proton is now known to be composed of quarks and gluons. String models may be different in this respect from other physical theories. In such models, resonances result from a novel concept of space-time rather than from the composite nature of elementary particles.

See also: SCATTERING; VIRTUAL PROCESSES

Bibliography

Green, M. B.; Schwarz, J. H.; and Witten, E. *Superstring Theory,* Vol. 1 (Cambridge University Press, Cambridge, UK, 1987).

Griffiths, D. J. *Introduction to Elementary Particles* (Harper and Row Publishers, New York, 1987).

Particle Data Group. <http://pdg.lbl.gov>.

Gabor Domokos

RUTHERFORD, ERNEST

Ernest Rutherford, the central figure in the science of radioactivity and the founder of its extension, nuclear physics, was born in Brightwater, near Nelson, on the southern island of New Zealand, on August 30, 1871. He died in Cambridge, England, on October 19, 1937. Rutherford's parents, seeking economic opportunity, were part of the mid-nineteenth-century migration from the British Isles. Before her marriage, his mother was a teacher, while his father pursued a variety of jobs, from cutting railroad ties to flax farming.

Young Ernest, one of a dozen siblings, won a scholarship to Nelson College, an excellent sec-

ondary school, and a further award in 1889 to Canterbury College in Christchurch, one of the few institutions of higher education in New Zealand. He received his B.A. degree in 1892, won another scholarship for his M.A. (1893), and stayed still another year for his bachelor of science degree (1894). The Canterbury faculty, although quite few in number, offered Rutherford solid grounding in mathematics and the sciences and, better yet, inspired in him a love of physical investigation.

For the B.Sc. degree, he magnetized iron by high-frequency electrical discharges. James Clerk Maxwell had predicted the existence of radio waves, Heinrich Hertz had found them less than a decade before, and Rutherford was one of a number of people to construct detectors of these waves. Seen as a promising scientist, he won a scholarship that required him to

British physicist Ernest Rutherford (1871–1937) received the Nobel Prize in Chemistry in 1908 for his investigations of radioactivity.
CREDIT: COURTESY OF BETTMANN/CORBIS. REPRODUCED BY CORBIS CORPORATION.

attend another institution; he chose Cambridge University because the director of its Cavendish Laboratory, Joseph John Thomson, was the leading authority on electromagnetic phenomena.

When Rutherford arrived in England in September 1895, he became part of the first class of "research students" (today called graduate students, but the doctoral degree would not be awarded there for another quarter-century) admitted by new university regulations. He continued to improve his wireless wave detector, increasing its range. Thomson was so impressed that he invited Rutherford to collaborate with him in studying the recently discovered X rays. Without the foresight that wireless would become big business, and also lacking the entrepreneurial skills of a Guglielmo Marconi, Rutherford accepted his professor's offer.

Thomson had spent years studying the discharge of electricity in gases. X rays provided another means of making a current flow readily. Thomson and Rutherford determined that equal numbers of positively and negatively charged particles were formed and advanced a theory of ionization. In 1897, Thomson asserted that the negative particles were smaller than atoms, and soon they were called electrons.

Since the radiation that Henri Becquerel found issuing from uranium was at first thought to be X rays, it was only natural that Rutherford would test its ability to ionize air. Indeed, this property became a major indication of a source's strength. In 1898, Gerhard C. Schmidt and Marie Curie independently found thorium was similarly "radioactive," and Curie and her associates soon detected other sources, named polonium and radium. Testing uranium radiation's ability to penetrate metal foil, Rutherford found one component that was easily stopped and another that passed through some layers of foil. These he named alpha and beta, respectively, "for convenience."

In 1898, Rutherford was appointed as a full professor of physics at McGill University in Montreal. Blessed with a wealthy patron, the laboratory was probably the best equipped in the Western Hemisphere. Even better, the department chairman, who wanted a research star, willingly took over some of Rutherford's teaching duties. Rutherford now began

a careful study of thorium and soon found that it evolved a radioactive gaseous product, which he called emanation (now called thoron). Significantly, some radioactive bodies maintained a steady level of activity, while others exhibited a rise or fall. These latter each changed over a period unique to it, which was soon called its half-life. This measure served as a means of identifying sources which contained too few atoms to be determined by ordinary chemical tests.

Always adept at drawing others into his investigations, Rutherford and a chemical colleague, Frederick Soddy, in 1902 recognized that freshly prepared thorium increased in activity at the same rate as a constituent found in its ore, thorium X, decreased. They reasoned that a genetic relationship existed, with the parent decaying into a daughter product, which also decayed if it was radioactive. The several active bodies could be arranged in decay series, which ultimately would end in inactive products. Since each product was regarded as an element, what they proposed was atomic transmutation—alchemy—which had been driven out of scientific chemistry centuries before. Yet, few challenged this new idea of unstable atoms, for the evidence fit into the theory exceedingly well. For this explanation of radioactivity, Rutherford received the 1908 Nobel Prize in Chemistry, an indication of this subject's position on the borderline between physics and chemistry.

In 1907, desiring to be closer to the centers of scientific activity, Rutherford accepted the directorship of the physics laboratory at Manchester University, the best in Britain after Cambridge. While in Canada he had shown that the alpha ray was a positively charged particle but could not decide if it was hydrogen or helium. Availing himself of the university's skillful glassblower, who constructed a tube that alphas could enter but not leave, Rutherford's student Thomas Royds in 1908 showed that alphas produced the spectrum of helium.

With his assistant, Hans Geiger, Rutherford developed a means of visually counting the flashes of light made by alpha particles striking a scintillating screen. Geiger later extended this valuable measure of a source's strength to electrical and then electronic counting. At McGill, Rutherford had observed a certain fuzziness when experimenting with alphas, which he supposed was due to slight scattering when they hit an object. To explore this phenomenon further, in 1909 he asked an undergraduate, Ernest Marsden, to allow alphas from a naturally decaying source to strike a foil target and measure the scattering. Most hit the detector without any deflection, many were bent through small angles, but some were, surprisingly, turned more than ninety degrees. Rutherford's (embellished) reaction was that "It was almost as incredible as if you fired a fifteen-inch shell at a piece of tissue paper and it came back and hit you."

Two years later Rutherford could explain what happened. The atom, whose diameter was of the order of 10^{-8} centimeters, was not a solid billiard ball-like object or J. J. Thomson's popular "plum pudding" model of a sphere of positive electrification studded with a geometric array of electrons. Instead, it was mostly empty space, with the atom's mass concentrated in a tiny nucleus measuring 10^{-12} centimeters, with electrons orbiting at a distance. When an alpha came near enough to a charged nucleus for electrostatic forces to act, the alpha could be deflected from its original path in this single encounter.

The discovery of the nuclear atom turned few heads at first. But Niels Bohr, who had visited Rutherford's laboratory in 1911 and absorbed the excitement of the new concept, showed its implications in 1913. He explained radioactivity as a nuclear phenomenon, chemical reactions as belonging to outer electrons, and spectral lines as jumps by electrons from one orbit to another. Still more, he incorporated the new quantum ideas to explain that only certain orbits were permitted, a major consolidation of atomic physics. Another former student, Henry Gwyn-Jeffreys Moseley, at the same time explained that the regular sequence of X-ray spectral lines as one went through the periodic table of elements was due to the regular increase of positive charge on the nucleus. Thus, the periodic table was organized upon atomic number, not atomic weight, as previously believed. And still another Manchester alumnus, Kasimir Fajans, added to the significance of the nucleus when he devised the group displacement laws to show how alpha and beta decay transformed radioelements from one box to another of the periodic table.

With his laboratory largely empty during World War I, Rutherford spent some time on submarine-detection apparatus but had the opportunity to pursue something curious that Marsden had found. When alpha particles traveled in a tube filled with hydrogen gas, scintillations were observed on a screen placed beyond the alphas' range. Obviously, an alpha-hydrogen collision sent the latter particle flying toward the screen. But scintillations that looked like those from hydrogen were also seen when the tube was filled with nitrogen. In 1919, Rutherford explained that this was not an elastic collision but an induced nuclear reaction. The alpha and nitrogen (he concluded) transformed into hydrogen (a proton) and oxygen. Along with the explanation of radioactivity and the discovery of the nucleus, this induced nuclear reaction confirmed the view that Rutherford was the greatest experimental physicist since Michael Faraday.

Also in 1919, Rutherford succeeded Thomson as director of the Cavendish Laboratory, continuing it as the preeminent research center and source of physics professors for the British Commonwealth. With James Chadwick, he showed that other light elements succumbed to alpha-induced transformations. However, heavier elements, with larger positive charges on their nuclei, resisted close encounters with alphas. By the late 1920s, Rutherford encouraged engineers to overcome insulation breakdown and other technical problems to build high-voltage apparatus that could accelerate copious quantities of protons toward a target. If the voltage (or energy) were high enough, the projectile might overcome (or tunnel through) the potential barrier around the nucleus and cause a nuclear reaction. This was accomplished in 1932 by John Douglas Cockcroft and Ernest Thomas Sinton Walton in the Cavendish Laboratory, who also used for the first time Albert Einstein's equation $E = mc^2$ in this interpretation of the nuclear reaction. Whereas in natural radioactivity elements spontaneously transmuted into other elements, and in his 1919 experimental arrangement, Rutherford had produced artificial transmutation by natural means, this accelerator experiment involved artificial transmutation produced artificially.

Rutherford's leadership of the Cavendish Laboratory was burnished still more in 1932, when Chadwick discovered the neutron, an uncharged particle able easily to reach atomic nuclei and cause reactions. While Rutherford believed correctly that "harnessing the energy of the atom" was unlikely, given the known reactions and equipment, the phenomenon of neutron-induced nuclear fission was discovered shortly after his death, and both reactors and bombs followed. Curiously, Rutherford was associated also with the other process for constructing nuclear weapons, fusion, when he and colleagues achieved a fusion reaction in 1934.

As one of the world's leading scientists, Rutherford was awarded many honors, including knighthood, then a peerage, the Order of Merit, and presidency of the Royal Society and of the British Association for the Advancement of Science. Not politically active, he nonetheless was drawn into public positions in the 1930s when one of his colleagues, Peter Kapitza, was prevented from returning to Cambridge after a visit to his home in the Soviet Union, and when he accepted the presidency of the Academic Assistance Council, an organization created to help refugee scientists from Hitler's Germany. Rutherford, thus, was one of the last scientists largely able to devote himself single-mindedly to exploring nature. The next generation could not avoid

questions of science and social responsibility, politics, and national security.

See also: CHADWICK, JAMES; NEUTRON, DISCOVERY OF; RADIOACTIVITY; RADIOACTIVITY, DISCOVERY OF

Bibliography

Andrade, E. N. da C. *Rutherford and the Nature of the Atom* (Doubleday, Garden City, NY, 1964).

Badash, L., ed. *Rutherford and Boltwood: Letters on Radioactivity* (Yale University Press, New Haven, CT, 1969).

Badash, L. "Nuclear Physics in Rutherford's Laboratory before the Discovery of the Neutron." *American Journal of Physics* **51**, 884–889 (1983).

Badash, L. *Kapitza, Rutherford, and the Kremlin* (Yale University Press, New Haven, CT, 1985).

Campbell, J. *Rutherford: Scientist Supreme* (AAS Publications, Christchurch, New Zealand, 1999).

Eve, A. S. *Rutherford* (Cambridge University Press, Cambridge, UK, 1939).

Feather, N. *Lord Rutherford* (Blackie & Son, London, 1940).

Oliphant, M. *Rutherford: Recollections of the Cambridge Days* (Elsevier, Amsterdam, 1972).

Wilson, D. *Rutherford, Simple Genius* (MIT Press, Cambridge, MA, 1983).

Lawrence Badash

S

SALAM, ABDUS

Abdus Salam (he added "Muhammad" much later) was born in Pakistan but spent most of his life as a leading theoretical physicist in England and Italy, making outstanding contributions to quantum field theory and particle symmetries.

Education

Salam was born January 29, 1926, at Santokdas in Western Punjab, which was then part of British India and is now part of Pakistan. He spent his childhood in the small town of Jhang, where his father was a teacher. At fourteen, he won a scholarship to Government College, Lahore, with the highest marks ever recorded. His first paper, on an algebraic problem of Srinivasa Ramanujan, was published when he was just seventeen.

In 1946, the Punjab Government awarded Salam a scholarship to Cambridge, where he studied mathematics and physics, and was then taken on as a research student in theoretical physics by Nicholas Kemmer. Another student of Kemmer's, Paul Matthews, was then finishing his Ph.D., working on extending renormalization theory from quantum electrodynamics to meson theories. Salam began work on one of the outstanding problems and very

rapidly succeeded in showing that the process of removal of infinities works also in the case of so-called overlapping divergences, work that immediately established his reputation. He then went with Matthews to spend a very productive year in Princeton.

Lahore, Cambridge, and Imperial College

In 1952, Salam returned briefly to Pakistan as Professor of Mathematics at Punjab University and Government College, but he found it impossible to continue his research and encountered increasing prejudice against the Ahmadiyya sect to which he belonged; they were regarded by many orthodox Muslims as heretical. He returned to Cambridge as a Lecturer and Fellow of St. John's in 1954. Then in 1956 he was invited to Imperial College, London, where he and Matthews set up a very lively theoretical physics group. He remained a professor there until his retirement. In 1959, he was elected as the youngest member of the Royal Society at the age of thirty-three.

Founding of the ICTP

Salam greatly regretted having to leave his native country to pursue his chosen career and determined to do what he could to help others avoid this dilemma. As Pakistan's delegate, he persuaded the International Atomic Energy Agency to set up an

Pakistani physicist Abdus Salam (1926–1996) shared the 1979 Nobel Prize in Physics with Sheldon Lee Glashow and Steven Weinberg for contributions to the unified gauge theory of weak and electromagnetic interactions. CREDIT: COURTESY OF THE BETTMANN ARCHIVE/NEWSPHOTOS, INC. REPRODUCED BY PERMISSION.

International Centre for Theoretical Physics (ICTP) in Trieste, with support from the Italian Government and the City of Trieste. Salam became its founding Director, and so remained until his retirement in 1994, dividing his time between London and Trieste. The ICTP was later cosponsored by UNESCO and has grown to a very large and internationally respected establishment, which was renamed the Abdus Salam ICTP. Through its associateship program, it has helped many physicists from developing countries to maintain contacts with front-line research.

Research

Much of Salam's research was concerned with symmetries in quantum field theories, beginning with hadronic symmetries and the chiral symmetry implied by vanishing neutrino mass. Unification was a major theme. With John Ward in the sixties he wrote a series of papers struggling with the problems of constructing a unified gauge theory of weak and electromagnetic interactions. The culmination was

the proposal of the electroweak theory, independently proposed by Steven Weinberg. For this, Salam won the 1979 Nobel Prize in Physics, together with Weinberg and Sheldon Glashow.

Later, Salam worked on the possibility of extending the unified theory to include strong interactions and gravity. With Jogesh Pati, he was one of the first to propose a grand unified theory predicting instability of the proton. He made important contributions to the ideas of supersymmetry; with John Strathdee, he developed the superfield formalism and was one of the early proponents of supergravity.

International Development

Salam was a passionate advocate of the importance to developing countries of building a scientific base. From 1961 he was scientific advisor to President Ayub Khan, helping to set up research institutes on subjects from nuclear power to wheat and rice. But he was frustrated by his inability to persuade the government to devote to science the resources he thought essential. In 1974, he was outraged when the Government of Zulfikar Ali Bhutto declared the Ahmadiyya sect to be non-Muslim. To emphasize his Muslim credentials, he grew a beard and later adopted the name Muhammad.

After winning the Nobel Prize, Salam was much in demand as a speaker throughout the developing world and particularly in Islamic countries. Tirelessly he used these occasions to argue the case for science; his dream was to revive the spirit of free inquiry that for several centuries once made the Arab world the standard-bearer for science. He argued strongly for the establishment of centers similar to the ICTP. He also took the lead in establishing the Third World Academy of Sciences and became its first President.

Honors

Salam was a member of twenty-four academies, including the Royal Society, the U.S. National Academy of Sciences, and the Soviet Academy. Among many prizes, in additional to the Nobel, he won the Atoms for Peace Award in 1968, the first Edinburgh Medal and Prize (1988), the Catalunya International Prize (1990), and the Royal Society's premier award, the Copley Medal (1990). He received forty-five honorary doctorates from twenty-eight different countries. In 1989 he was awarded an honorary knighthood (K.B.E.).

Final Years

Salam married twice and had six children. His final years were blighted by a degenerative neurological complaint, progressive supranuclear palsy (PSP). He found it increasingly difficult to talk and to move about, and he became confined to a wheelchair, but he bore his affliction with remarkable stoicism. So long as he possibly could, he continued to work. He still made innovative contributions to research on the origin of biological chirality and on models of high-temperature superconductivity. In 1994 he had to give up his post as Director of the ICTP, becoming instead its first President. He died at his home in Oxford on November 21, 1996, and was buried at Rabwah in Pakistan.

See also: CONSERVATION LAWS; GAUGE THEORY; QUANTUM CHROMODYNAMICS; QUANTUM FIELD THEORY; RENORMALIZATION; SYMMETRY PRINCIPLES

Bibliography

Hamende, A. M., ed. *From a Vision to a System: the International Centre for Theoretical Physics of Trieste (1964–1994)* (Fondazione Internazionale Trieste per il Progresso e la Libert à delle Scienze, Trieste, 1996).

Kibble, T. W. B. "Muhammad Abdus Salam, K.B.E." *Biographical Memoirs of Fellows of the Royal Society* 44, 385–401 (1998).

Lai, C. H., and Kidwai, A., eds. *Ideals and Realities, selected essays of Abdus Salam,* 3rd ed. (World Scientific, Singapore, 1989).

Singh, J. *Abdus Salam, a Biography* (Penguin Books, New Delhi, India,1992).

T. W. B. Kibble

SCATTERING

The properties of elementary particles and their interactions are probed in relativistic scattering experiments. In accelerator-based experiments, either two beams of particles collide or a single beam of particles hits a fixed target that is made up of parti-

cles at rest. In both cases, the particle beams carry well-defined energy and momentum, and these quantities, as well as electric charge, are conserved in the scattering process. The more energy an incoming particle beam carries, the more massive the final state that can be produced. (Recall that in the theory of relativity, energy, momentum, and mass are related.) Energy and distance are inversely related; in optical language, a particle's wavelength λ is inversly proportional to its momentum k, $\lambda = 2\pi/k$. Smaller distances are thus probed in scattering experiments with higher-energy beams.

The probability of any particular final state being produced in a scattering experiment is expressed in terms of a quantity called a cross section, denoted by σ. A cross section has units of area. A pictorial description is given by the likelihood of hitting the side of a barn with a baseball; the bigger the barn, the more likely that the baseball will hit it. In fact, the unit of measure for a cross section is called a barn. One barn is 10^{-24} cm^2. The event rate, the number of events produced per second, is given by the product of the cross section for a reaction with the incident luminosity of the particle beams. The beam luminosity is simply the number of particles per second per unit area traveling in the beam. The total number of events collected in the lifetime of an experiment is given by integrating the event rate over the time that the experiment has been operating. In this analogy, the total number of baseballs that hit the barn is given by the area of the barn times the frequency with which the baseballs were tossed, which is then summed, or integrated, over the length of time during which the baseballs were tossed.

In 2002 high-energy accelerator experiments routinely measured cross sections with a magnitude of pico-barns, or 10^{-36} cm^2. Typical collider luminosities were roughly 10^{31} cm^{-2}s^{-1}. This yields 100 events produced for a pico-barn sized cross section when the experiment operates for 10^7 seconds, which is on the order of a year. Clearly, higher luminosities result in a larger number of events.

The quantum theory of scattering is described by a quantity known as the *S*-matrix. The possible set of incoming particles in a reaction, or initial states, are known as in states and are denoted as $|in>$. Likewise, the set of possible final, or outgoing, states are *out* states, $<out|$. Both the *in* and *out* states are complete sets, meaning that they include all possible initial and final states. The probability of a specific incoming state from the full set of *in* states producing a particular final state from the possible *out* states is encoded in a transition matrix. This matrix is called the *S*-matrix, and the transition is written as $<out|S|in>$. There are two contributions to the *S*-matrix: the cases when the particles do not interact and those when they do. Note that the possibility always exists that the particles do not interact, or simply miss each other. The *S*-matrix is then written as $S = I + iT$, where *I* represents the identity matrix (the diagonal elements are unity and the nondiagonal elements vanish), and all the particle interactions are contained in the quantity *T*. The invariant matrix element for a scattering amplitude is thus written as

$$M = <out|\,iT\,|in>. \qquad (1)$$

It contains all information about the interaction between the incoming and outgoing particles. It is Lorentz invariant, meaning that it has the same value in all reference frames.

The cross section for a reaction is obtained by integrating the square of this invariant matrix element over the available phase space, which is the final state momentum space that is kinematically available in a reaction, and dividing this by the incident flux, which is proportional to the relative velocity between the two incoming particles.

High-energy physicists compute *S*-matrix elements from Feynman diagrams using a set of Feynman rules. The diagrams pictorially represent a reaction, and the rules encode all the information from quantum field theory that is needed for the computation. Typical diagrams for two body $\to$ two body scattering are displayed in Figure 1, where the arrows indicate the direction of momentum flow. These graphs represent the contributions to the process known as Bhabha scattering, $e^+e^- \to e^+e^-$. In the Standard Model this process is mediated by the virtual exchange of $V = \gamma, Z$ bosons. Each diagram represents a different virtual momentum being carried by the exchanged bosons. The graph on the left is known as an *s*-channel diagram, where the total initial momentum, or incoming center-of-mass energy, is transferred to the gauge boson. The graph on the right is known as a *t*-channel diagram, where the exchanged boson carries the difference between the initial and final state electron momenta. These quantities, *s* and *t*, as well as a third, *u*, are the so-called invariant Mandelstam variables. As above, they are invariant as they have the same value in all reference frames. They are specifically defined as

$$\begin{aligned} s &= q_s^2 = (p_1 + p_2)^2, \\ t &= q_t^2 = (p_1 - k_1)^2, \\ u &= q_u^2 = (p_1 - k_2)^2 \end{aligned} \qquad (2)$$

where q_i, p_i, k_i represent the particle's four-momentum as labeled in the figure. They satisfy the relation $s + t + u = m_1 + m_2 + m_3 + m_4$, where m_i represents the mass of the particles in the reaction. The *S*-matrix elements are expressed in terms of these variables as they convey the kinematic information of the process, in particular, the angular dependence of the final states. A matrix element is computed for each channel that contributes to a reaction, and then they are added coherently to obtain the full invariant matrix element.

s-channel scattering possesses special properties. The denominator of an *s*-channel *S*-matrix element is proportional to $s - M^2$, where *M* is the mass of the particle being exchanged. When the value of the center-of-mass energy $\sqrt{s}$ approaches the mass of the

FIGURE 1

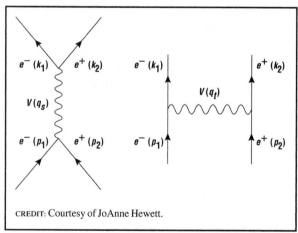

CREDIT: Courtesy of JoAnne Hewett.

Feynman diagrams describing Bhabha scattering, $e^+e^- \to e^+e^-$. S-channel (left) and t-channel (right) contributions.

exchanged particle, a resonance known as a Breit-Wigner resonance appears. This results in a substantial increase, usually a factor of 10 to 1,000, in the production cross section. As the cross section is mapped out as a function of $\sqrt{s}$, one sees that this resonant enhancement has a finite width, that is, it occurs for a finite range of $\sqrt{s}$, and the resonance looks like a bump. The width of this resonance is the total decay width of the exchanged particle. The appearance of resonances is a key tool in the discovery of new particles. Resonance structures do not occur in t- and u-channel processes.

See also: ANNIHILATION AND CREATION; FEYNMAN DIAGRAMS; QUANTUM FIELD THEORY; QUANTUM MECHANICS; RELATIVITY; RESONANCES; QUANTUM STATISTICS; VIRTUAL PROCESSES

Bibliography

Barger, V. D., and Phillips, R. J. N. *Collider Physics* (Addison-Wesley, Redwood City, CA, 1987).

Perkins, D. H. *Introduction to High Energy Physics* (Addison-Wesley, Reading, MA, 1982).

Peskin, M. E., and Schroeder, D. V. *An Introduction to Quantum Field Theory* (Perseus, Cambridge, MA, 1995).

JoAnne Hewett

SCHWINGER, JULIAN

Julian Schwinger was born in New York City on February 12, 1918, to a middle-class Jewish family. His father immigrated to the United States in 1880 when he was in his early teens and later became a successful designer of women's clothing. The family of Schwinger's mother came to the United States when she was very young. Julian was a very precocious and talented child, but it was Harold, Julian's older brother, who was considered the bright child in the family since he won many prizes at school. Like Harold, Schwinger attended Townsend Harris High School, then one of the best secondary schools in the United States, located on the campus of the City College of New York (CCNY). Schwinger was fourteen when he entered the school in 1932. He graduated from Townsend Harris in 1934 and entered CCNY in the fall of that year as a physics major.

Schwinger's precocity and ability in physics had made him a living legend even while in high school. However, he did not do well at CCNY. He spent most of his time in the library reading advanced physics and mathematical texts and rarely went to his classes. His grades reflected his erratic class attendance. The matter became serious enough for Lloyd Motz, one of his physics instructors at CCNY, to bring Schwinger's problems to Isidor Rabi's attention. Motz was aware of Schwinger's talents because Schwinger had given him a copy of a paper he had written as a freshman entitled "On the Interaction of Several Electrons." Additionally, as a sophomore, Schwinger had collaborated with Motz and had calculated the lifetime of the neutron in the Konopinski-Uhlenbeck version of Enrico Fermi's beta-decay theory. Similarly, even as a freshman at CCNY, Schwinger regularly attended the weekly theoretical seminar that Rabi and Gregory Breit ran at Columbia on Wednesday

American physicist Julian Schwinger (1918–1994) shared the 1965 Nobel Prize in Physics with Sin-itiro Tomonaga and Richard P. Feynman for his work on quantum electrodynamics. CREDIT: COURTESY OF CORBIS. REPRODUCED BY CORBIS CORPORATION.

evenings. The 16-year-old undergraduate published a joint paper with Otto Halpern on the problem of the polarization of electrons in double scattering experiments, the young Julian having done extensive and difficult calculations.

Largely through Rabi's efforts, Columbia offered Schwinger a scholarship, and Schwinger became an active participant in Rabi's research activities. During his senior year, Schwinger worked on the problem of the magnetic scattering of slow neutrons by atoms. In early January 1937 he sent a manuscript entitled "The Magnetic Scattering of Neutrons" to the *Physical Review*. The characteristics that distinguished Schwinger's subsequent works are present in this paper: an important physical problem is addressed; the solution is elegant; the methods used are powerful; contact is made with experimental data; and suggestions for empirical tests are given. Edward Teller, who was visiting Columbia in the spring of 1937, suggested that Schwinger's research on the scattering of neutrons be further developed for his Ph.D. thesis. Schwinger worked with Teller and showed that the scattering of neutrons by ortho- and para-hydrogen could yield information about the spin dependence and the range of neutron-proton interaction. That Schwinger had written his Ph.D. dissertation before receiving his bachelor's degree is indicative of his remarkable talents.

After receiving his B.S. from Columbia in 1936, Schwinger continued his graduate studies there. But shortly thereafter, Rabi arranged for Schwinger to receive a traveling fellowship from Columbia for the academic year 1937–8. The plan was for Schwinger to spend six months in Wisconsin studying with Gregory Breit and Eugene Wigner, and then to go on to Berkeley for another six months to work with Robert Oppenheimer. As it turned out, he remained at Wisconsin for the entire year, and there developed his characteristic working habits: staying up at night and sleeping during the day. Thereafter, Schwinger did go to Berkeley for two years: spending the first as a National Research Council (NRC) fellow and the second as a research associate to Oppenheimer. His stay was enormously productive. He collaborated extensively and worked on a wide range of subjects. An analysis of the electromagnetic properties of the deuteron when tensor forces are present led him to predict the existence of the deuteron's quadrupole moment—before it had been measured by Jerome Kellogg, Norman Ramsey, Isidor I. Rabi, and Jerrold Zacharias.

Schwinger left Berkeley in the summer of 1941 to accept a position as instructor at Purdue. An active program in semiconductor research to develop better rectifiers for the detection of radar was being carried out there by Karl Lark-Horovitz for the Radiation Laboratory (Rad Lab). In 1942 Schwinger and several other theorists at Purdue were asked to join a Rad Lab project on the propagation of microwave radiation under Hans Bethe's direction.

When Los Alamos was organized in early 1943 to build an atomic bomb, Robert Oppenheimer invited Schwinger to join the laboratory, but he declined. However, since many leading theorists were leaving their academic posts to go to Los Alamos, Schwinger was offered a full-time position at MIT, which he started in the fall of 1943. In his work at MIT Schwinger indicated how to set up and solve a wide variety of microwave problems. In a memorial lecture for Sin-itiro Tomonaga delivered in 1980, Schwinger commented that his waveguide investigations showed the utility of organizing a theory to isolate those inner structural aspects that are not probed under the given experimental circumstances. That lesson was subsequently applied to the effective-range description of nuclear forces. It was this viewpoint that would lead to the quantum electrodynamic concept of self-consistent subtraction or renormalization. Schwinger also worked on the problem of the radiation emitted by fast electrons traveling in synchrotron orbits. The formulation of this problem taught him the importance of describing relativistic situations covariantly, that is, without specialization to any particular coordinate system.

In 1944 universities began competing with one another for the outstanding talent in physics, and Schwinger was courted by a number of academic institutions and, in particular, by Harvard. In the fall of 1945 Schwinger accepted an appointment there as an associate professor. A year later he was offered a full professorship at Berkeley, and Harvard promptly promoted him. This same year Schwinger married Clarice Carrol of Boston. Harvard provided him with outstanding graduate students, and he became the thesis adviser to many of them. They, together with many of MIT's graduate students and postdoctoral fellows and a fair number of the Harvard and MIT physics faculty, formed the audience for Schwinger's brilliant lectures. It is difficult to exaggerate the impact of these lectures—and of the widely circulated notes based on them—on the generation of physics graduate students in the late 1940s and 1950s. Many of today's texts on nuclear physics, electromagnetic theory, quantum mechanics, quantum field theory, and statistical mechanics have incorporated the approaches, techniques, and examples that Schwinger discussed in his lectures.

During his stay at Harvard, Schwinger's contributions and that of his students to physics were numerous and profound. In the late 1940s he reformulated quantum electrodynamics (QED) in terms of a manifestly covariant formalism which—using the concepts of mass and charge renormalization—allowed him to unambiguously extract the corrections to the magnetic moment of the electron that the theory implied. Similarly, he was able to calculate the level shifts predicted by QED for the energy levels of a hydrogen atom described by the Dirac equation. His covariant formulation, when amended with the notions of renormalization, was the first self-consistent framework in quantum field theory (QFT) from which physical consequences could be extracted and checked with experiments. For this work, Schwinger shared the Nobel Prize in Physics in 1965 with Richard Feynman and Sin-itiro Tomonaga. Schwinger's 1948 formulation of QED could not be easily extended to calculate higher-order effects. He thereafter developed increasingly powerful calculation techniques.

In 1951 in an eight-page-paper published in the *Proceedings of the National Academy of Sciences,* Schwinger gave a concise presentation of his formulation of the equations for Green's functions of quantum fields. He there introduced the use of "sources"—classical sources for Bosonic fields and Grassmann anticommuting sources for Fermionic fields—as functional variables. To this same period belong his formulation of the Schwinger action principle and the use of temperature-dependent many-particle Green's functions for addressing equilibrium and nonequilibrium problems in condensed-matter physics.

In the mid-1960s Schwinger started reformulating the foundations of fundamental physics and expressing these within a new framework: source theory. Source theory represented Schwinger's efforts to replace the prevailing operator field theory by a philosophy and methodology that eliminated all infinite quantities. Schwinger's objections to operator field theory arose at the pragmatic level from the fact that it seemed impossible to incorporate the strong interactions within its framework and at the philosophical level that from the fact it made implicit assumptions about unknown phenomena at inaccessible, very high energies to make predictions at lower energies. Source theory, on the other hand, began with robust knowledge about known phenomena at accessible energies to make predictions of physical phenomena at higher energies. But, Schwinger's insistence on basing his theories on phenomenology led him to reject the quark model of hadrons and quantum chromodynamics. His pursuit of source theory in the early 1970s, at the very time quantum field theory was resurging in the aftermath of the successes of the Glashow-Salam-Weinberg electroweak theory and of the proof by Gerardus 't Hooft that the Yang-Mills theory with a Higgs mechanism for breaking symmetry and giving masses to particles is renormalizable, alienated him from his community and drove him out of the mainstream of modern physics. This alienation was further aggravated when he left Harvard in February 1971 to accept a position at UCLA. He thus had to establish ties to a new community with interests somewhat different from those in Cambridge.

As a young Harvard professor, Schwinger had been the person who set the agenda for the field theory and high-energy community. While at Harvard, he directed some seventy doctoral theses and became an important influence on at least four generations of active, and later influential, theoretical physicists. However, when he left the mainstream of particle physics and challenged the foundations on which numerous theoretical investigations were being carried out, his new endeavors were contemptuously dismissed by the community as mistaken or irrelevant. His research papers, in turn, were rejected in a dismissive manner by *Physical Review Letters* and other leading journals. His response was to resign both as a member and as a fellow of the American Physical

Society. The hostility toward source theory that he had experienced probably contributed to his involvement in such fringe projects as cold fusion.

Starting in the 1980s, after teaching a course in quantum mechanics, Schwinger began writing a series of papers on the Thomas-Fermi model of atoms, and together with Berthold-Georg Englert he elaborated on the approach. These contributions have been deemed extremely important by the atomic physics community. His last scientific endeavor before his death in 1994 was an attempt to explain sonoluminescence.

Schwinger's work extended to almost every frontier of modern theoretical physics. He made far-reaching contributions to nuclear, particle, and atomic physics, to statistical mechanics, to classical electrodynamics, and to general relativity. Many of the mathematical techniques that he developed are to be found in every theorist's toolkit. He was one of the prophets and pioneers in the use of gauge theories. The influence of Julian Schwinger on the physics of his time was profound.

See also: FEYNMAN, RICHARD; QUANTUM ELECTRODYNAMICS; QUANTUM FIELD THEORY; TOMONAGA, SIN-ITIRO

Bibliography

Flato, M.; Fronsdal, C.; and Milton, K. A.; eds. *Selected Papers (1937–1976) of Julian Schwinger* (Reidel, Dordrecht, Netherlands, 1979).

Mehra, J., and Milton, K. A. *Climbing the Mountain: The Scientific Biography of Julian Schwinger* (Oxford University Press, New York, 2000).

Milton, K. A., ed. *A Quantum Legacy: Seminal Papers of Julian Schwinger* (World Scientific, Singapore, 2000).

Schweber, S. S. *QED and the Men Who Made It* (Princeton University Press, Princeton, 1994).

Schwinger, J. *Particles, Sources and Fields,* 3 Vols. (Addison-Wesley, Redwood City, CA, 1989).

Silvan S. Schweber

SLAC (STANFORD LINEAR ACCELERATOR CENTER)

The Stanford Linear Accelerator Center (SLAC) is a National Laboratory funded chiefly by the U.S. Department of Energy. Located on the campus of Stanford University in Menlo Park, California, it began operation in 1966 as a laboratory dedicated to high-energy physics, with a two-mile-long linear electron accelerator as its tool to study matter on very tiny scales. This accelerator followed the design of earlier machines in the Stanford High Energy Physics Laboratory developed by William Hansen. SLAC's early development was led by Edward Ginzton and Wolgang K. H. Panofsky.

The high-energy physics program of physics at SLAC has garnered numerous awards, including three Nobel Prizes for experimental discoveries that are key to the understanding of particles. The program continues to be at the forefront of particle physics, in part because every ten years or so a new addition or upgrade to the facility has been made, opening up new research opportunities. These include the Stanford Positron Electron Asymmetric Ring (SPEAR), a 3.6 GeV electron-positron storage ring (1972); the Positron Electron Project (PEP) a similar but larger facility capable of storing 9 GeV electrons and positrons (1980); the SLAC Linear Collider (SLC) (1989); and most recently PEPII, an upgrade and rebuilding of the PEP ring with the addition of a second lower energy storage ring in the same tunnel to make the SLAC asymmetric B factory (1998).

The first round of SLAC experiments, conducted in the late sixties and early seventies, earned a Nobel Prize in Physics in 1990 for Richard Taylor of SLAC and Jerome Friedman and Henry Kendall of MIT. Collision of electrons from the accelerator with stationary targets (such as a tank filled with hydrogen or deuterium) probed the structure within protons and neutrons and provided evidence that these are made from yet smaller objects known as quarks.

Research done at the SPEAR ring in the mid-seventies garnered two Nobel Prizes. In 1976, the Nobel Prize in Physics went to Burton Richter, who led the project to build the SPEAR facility and its first physics detector. The prize honored the discovery of the particle known as J/ψ, the first particle that indicated the existence of the fourth type of quark: the charm quark. This prize was shared by Samuel Ting, leader of group at Brookhaven National Laboratory,

Aerial view of the Stanford Linear Accelerator Center (SLAC) with the linear accelerator (upper right) and PEP-II collider highlighted. Electrons from the electron gun and positrons from the positron source are accelerated in the linear accelerator. In PEP-II (large ring) the electrons and positrons circulate in opposite directions. These particles then collide in the BABAR detector, creating B mesons and anti-B mesons. Differences in these particles' rates of decay may help explain why there is more matter than antimatter in the Universe. CREDIT: COURTESY OF STANFORD LINEAR ACCELERATOR CENTER/SCIENCE PHOTO LIBRARY/PHOTO RESEARCHERS, INC. REPRODUCED BY PERMISSION.

who announced the same discovery on the same day. The 1995 Nobel Prize in Physics was shared by Martin Perl for the discovery of the tau lepton, which is the third electronlike particle (the second being the muon), and Federick Reines of the University of California, Irvine, for detection of the neutrino. The discovery of the tau lepton suggested an entire third generation of quarks and leptons, all of which have since been found. These two discoveries were key in the development of the theory now known as the Standard Model in particle physics. Particle physicists refer to the discovery of the J/ψ as the November Revolution, so great was its impact on their worldview.

Beginning as a sideshow to the high-energy physics program at SPEAR, a new idea was explored that led to a worldwide program of synchrotron light sources being used to perform a great variety of scientific research. The Stanford Synchrotron Radiation Laboratory (SSRL) at SPEAR was a pioneer in this field. In a synchrotron electron storage ring the particles are made to circulate by bending their path with strong magnetic fields. This causes them to radiate energy. This effect must be compensated for by reaccelerating the particles at intervals around the ring. So, as far as the high-energy physicists were concerned, synchrotron radiation was an annoying but unavoidable side effect of putting electrons into a storage ring. However, this radiation at SPEAR was also found to produce an intense swath of X rays. SLAC and Stanford physicists recognized this as the world's best source of X rays for diffraction scattering and other studies of the atomic-scale structure of materials. The rich program that developed from this recognition continues, with the SPEAR ring now (2002) fed by its own cyclotron accelerator and devoted solely to SSRL use. Researchers from industry and academia worldwide come to SSRL to carry out their research, studying topics as diverse as the structure of an enzyme (knowledge that helped develop the protease inhibitor treatment of AIDS), and the distribution of impurities in a silicon wafer. Worldwide there are now a number of other synchrotron-based light sources built specifically to do this work

The PEP storage ring experiments also made significant contributions to particle physics. The patterns of particles produced in the electron-positron collisions gave evidence for the part of the Standard Model theory that describes strong interaction physics, a theory known as Quantum Chromodynamics (QCD). Particles were produced in groups or jets. The existence and angular distribution of the jets confirmed predictions from the QCD theory. Meanwhile the linear accelerator, working with a fixed target, made another key contribution. Physicists devised a way to create a polarized beam of electrons, that is, a beam in which electron spins were preferentially aligned in a predetermined direction. The dependence of the outcome of collisions on the direction of the polarization of the beam tested details of the emerging Standard Model theory of weak interactions. The results confirmed predictions of this theory.

The SLC facility was built for two reasons. The first was to demonstrate that the principle of a linear collider was a feasible approach for exploring very high-energy electron-positron collisions. In a storage ring the bunches collide many times. With a linear collider, which uses two linear accelerators head to head, one avoids the problem (for high-energy physics) of synchrotron radiation energy losses. The price is that there is only one chance at colliding each bunch of electrons with a bunch of positrons. So to make the payoff in interesting events large enough (to do the experiments in reasonable

time), one must make the bunches much smaller and denser at the collision point than in a storage ring. This required new technology to control and monitor the beams, and, while it took some time to get the facility running well, SLC has shown this can be done. Because SLC has only one two-mile accelerator, a design with two arcs (and some concomitant energy loss) was used. Designs for a higher-energy true linear collider are under development worldwide as a likely next step in the high-energy physics agenda.

The second role for SLC was to produce Z bosons and study their decays. SLC began operation a little earlier and was the first to show that the Z boson decays into only three types of neutrinos, an indication that the three known repeating sets of quarks and leptons may be the complete set. This result was later confirmed with higher precision in the LEP storage ring at the European Laboratory for Particle Physics (CERN) near Geneva, Switzerland. One area where SLC could make measurements that were not feasible at LEP was in using polarized electron beams. Measuring the dependence of the Z production on the beam polarization gave an additional probe of predictions of the Standard Model.

The next new addition at SLAC was not a new higher-energy facility but instead a rebuilding of the PEP ring into an asymmetric B factory. B mesons are mesons containing b quarks. The neutral B mesons, made from a b quark and an anti-d quark, or vice versa, provide a laboratory in which to study the predictions of the Standard Model about the differences in the laws of physics for matter and antimatter. Physicists think that these differences are key to understanding why our universe contains predominantly matter and very little antimatter. When physicists try to understand how this imbalance developed in the history of the universe using the Standard Model theory, they fail to get answers that match the observations for the ratio of matter to radiation in the universe. So it is possible that physics beyond the Standard Model comes into play here. One way to look for such effects is to carefully check the patterns of differences between the decay time distribution of B and anti-B mesons to see whether they match the Standard Model predictions. The first such difference was observed at SLAC, in the decays of B and anti-B mesons to a J/ψ and a K-short meson. Similar observations were made at about the same time at a B factory facility at the KEK laboratory in Tsukuba, Japan. An asymmetry between the B and the anti-B results was established. The magnitude of the effect is consistent with the Standard Model expectations. There are still many other rates to be measured and cross-checks to be made. The facility will continue to operate, possibly with some increases in its rate of B production. It will take all this and more, experiments elsewhere also contributing to the picture, to check whether the full pattern of Standard Model predictions is borne out, or whether some anomalies suggesting new physics are found.

While the primary purpose of the laboratory is basic research in high energy physics and the synchrotron radiation applications to both basic and applied science, there are a number of ways in which this work has developed tools that have much broader application. Electron accelerators of the type developed at Stanford are found in hospitals around the world as the source of X rays for medical treatments. The computer code EGS that models the interactions of electrons and photons with matter, developed at SLAC to allow the design of radiation shielding for the experiments, has provided ways to refine X-ray treatments to give a greater radiation dose to a tumor and a lesser dose to surrounding tissue. SLAC mounted the first U.S. web site, helping to develop this particle-physics-initiated technology that has so changed the world of information technology. Synchrotron radiation studies have provided clues to help develop new medical treatments, new ways to detect small quantities of pollutants and to develop and test pollution remediation approaches, and improvements in the production of silicon wafers, to name but a few developments. Technology developed for particle physics detectors is now being used at SLAC to build a gamma ray observatory (the Gamma Ray Large Area Space Telescope [GLAST]) to be stationed in space. These rich and varied effects, often called spin-offs, are a second payoff for the money invested in such a facility.

See also: INTERNATIONAL NATURE OF PARTICLE PHYSICS

Bibliography

Bienenstock, A., and Winick, H. "Synchrotron Radiation Research—An Overview." *Physics Today* **36** (6), 48–58 (1983).

Feldman, G., and Steinberger, J. "The Number of Families of Matter: How Experiments at CERN and SLAC, Using Electron-Positron Collisions, Showed that There Are Only Three Families of Fundamental Particles in the Universe." *Scientific American* **352** (2), 70–78 (1991).

Kendall, H. W., and Panofsky, W. "The Structure of the Proton and the Neutron." *Scientific American* **224** (6), 60–77 (1971).

Perl, M. L., and Kirk, W. T. "Heavy Leptons." *Scientific American* **238** (3), 50–57 (1978).

Richter, B. "Nobel Lecture." *Reviews of Modern Physics* **49**, 251–266 (1977).

Stanford Linear Accelerator Center. <http://www.slac.stanford.edu>.

Witherell, M., and Quinn, H. "The Asymmetry Between Matter and Antimatter." *Scientific American* **279** (4), 76–81 (1998).

Helen R. Quinn

SSC

The Superconducting Super-Collider Project, SSC, was a hadron colliding-beam accelerator which was first proposed by the United States in 1982. It was named Super-Collider because its beam energy of 20 tera electron volts (TeV) was sixty times the energy of Europe's proton-antiproton collider, then beginning operation at the European Laboratory for Particle Physics (CERN) in Switzerland. It was intended to re-establish U.S. supremacy in the field of high-energy particle physics, but although funding was approved in 1987, and construction commenced soon after in Texas, its cost escalated from an initial estimate of $3 billion to almost $12 billion. The project was terminated by the U.S. Congress in 1994.

The SSC consisted of a pair of synchrotron accelerators installed in a single tunnel, 54 miles in circumference. The synchrotron rings were designed to interlace and cross at a number of collision points where large-particle detectors would record and analyze the products of collisions between the two beams of protons. These beams circulated in opposite directions in a guide field provided by electromagnets—

dipoles and quadrupoles—excited by coils wound from an alloy of niobium titanium that becomes superconducting when cooled to within 4 degrees of absolute zero.

Anticipated Outcomes of SSC

At the time that the SSC was first proposed, the Standard Model was emerging as the underlying explanation of what had seemed a large number of subatomic particles. In this model, three generations of quarks and leptons interact through particles, bosons, which carry the forces of nature. The existence of three of these bosons, the neutral Z boson and the two charged W bosons, was about to be verified and this, together with the earlier discovery of the J/ψ, provided final confirmation of the model. Nevertheless, some very important questions remained unanswered, notably an explanation of the masses of the quarks and leptons and the particles they comprise.

A theoretical concept based upon spontaneous symmetry breaking of the vacuum field—the Higgs phenomenon—predicted the existence of another boson at an energy less than 1 TeV. If discovered, the Higgs boson would confirm the theory, explain the masses, and indicate a threshold in energy beyond which strong and weak interactions would become of comparable strength.

Unfortunately, to produce a particle with a mass of 1 TeV it is not sufficient to collide two protons with an energy of 500 giga electron volts (GeV), otherwise Fermilab's Tevatron collider (completed in 1985) would have been powerful enough. In such a collision only one of the three quarks in each proton interact and together the pair will have less than one sixth of the total beam energy. Ten, or preferably twenty, tera-electron-volt protons are needed to be sure to create the Higgs boson.

Special Aspects of the SSC

The most obvious special feature of the SSC is its size, due simply to the difficulty of deflecting high-energy particles in a circle. The magnetic rigidity of a beam of particles is $B\rho = e/p$, where B is the magnitude of the deflecting field in Tesla, ρ is the radius of curvature of the machine, e is the particle's charge, and p is its momentum. Early synchrotrons and storage rings used conventional magnets whose iron yokes saturated at a field of 2 Tesla, limiting the peak energy of the machine. The quest for higher-energy collisions has led inevitably to larger rings—roughly an extra kilometer of circumference for each 70 GeV of energy. A 20-TeV ring of conventional magnets would be about 300 kilometers in circumference.

It is possible, by passing large currents in the coils, to drive magnets to even higher fields, beyond saturation. In this regime the coils determine the field shape, and the iron yokes are merely there for mechanical stability and to contain stray fields. However, such magnets are only feasible if their windings are superconducting. This reduces the resistive losses to virtually zero. The coils must be very precisely constructed, and moreover, at low temperature the thermodynamic efficiency of refrigerators is so low that 40 megawatts of electrical power is still required to remove the few kilowatts of heat that leak into many kilometers of the SSC magnet.

The SSC was not the first such collider to use superconducting magnets. An earlier 1-TeV superconducting ring (the Tevatron) was completed at the Fermi National Accelerator Laboratory (Fermilab) near Chicago in the early 1980s and was fed with protons and antiprotons. While the guide field of the Tevatron was twice that of a conventional synchrotron magnet, the field of the SSC magnets, 6.6 Tesla, was three times more than a warm magnet.

In spite of the higher field, the SSC circumference was 87,120 meters, almost fourteen times larger than the Tevatron, and three times as large as the largest tunnel available in Europe, which was then under construction for the electron-positron collider LEP at CERN and later destined to house the LHC. Its two semicircular arcs formed a racetrack with two straight sides, one of which accommodated three major experiments. The other side was principally dedicated to a chain of three boosters: injector synchrotrons of 11, 200 and 2,000 GeV, fed by a 600-MeV linac. There were also two more experimental halls. Boosters are used to feed such a large synchrotron because as a proton beam is accelerated, it shrinks and needs a smaller magnet aperture. The chain of injectors exploits this fact so that expensive, wide-aperture magnets are only needed for the smaller, low-energy rings.

FIGURE 1

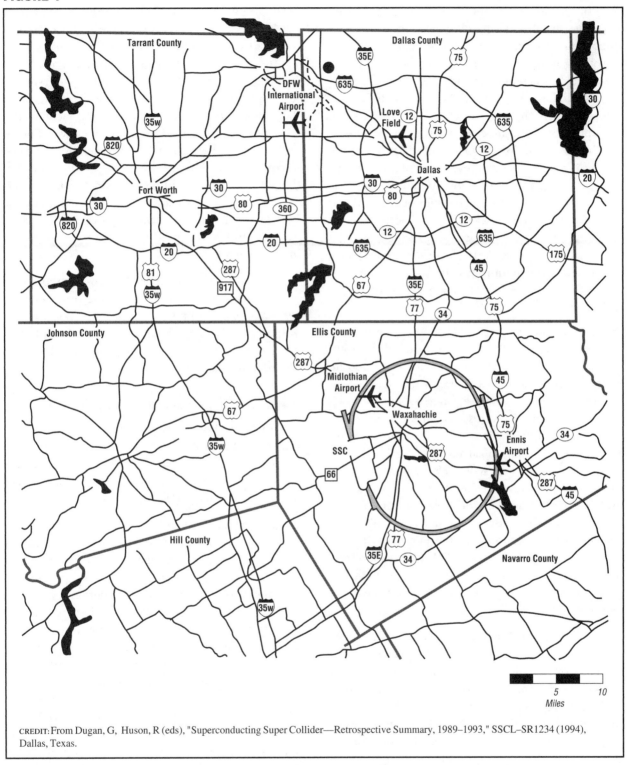

CREDIT: From Dugan, G, Huson, R (eds), "Superconducting Super Collider—Retrospective Summary, 1989–1993," SSCL–SR1234 (1994), Dallas, Texas.

SSC project site region.

High beam intensities were needed. In a collider each particle in one beam passes through the other oncoming beam at each collision point and has an opportunity to interact with all the particles in the oncoming beam. As it does so, a particle presents a certain cross section—depending on the nature of the interaction under study. The probability of a collision between any two particles is small, but when multiplied by the number of particles in each beam and by the revolution frequency of the beam in the machine, many interesting events per second may be expected. A quantity called the luminosity, the measure of the probability of such events per second and per unit interaction cross-section, and is typically in the range 10^{30} to 10^{33}. Processes for which particles present a cross section of 10^{-33} cm^2 will appear once per second if the luminosity is 10^{33}. Cross sections of interesting processes may be many orders of magnitude smaller.

The Tevatron, like CERN's proton-antiproton collider before it, collides protons with antiprotons. Antiprotons, being of the same mass but opposite charge as protons, will circulate in the opposite direction in the same ring of magnets, thus avoiding the construction of two distinct rings. However, antiprotons are difficult to produce, and the Tevatron reached a luminosity of at most 10^{31} cm^2. The de Broglie wavelength of a 20-TeV proton is twenty times smaller than at the Tevatron's 1-TeV hadron, and hence the detail it will reveal in structure is twenty times finer. But such detail is 400 times smaller in cross section, and the luminosity has be over 10^{33} to produce an acceptable observation rate. To reach this luminosity one needs to collide two proton beams of very high density—hence the SSC's twin rings.

To reach the highest luminosity both beams must be focused down until a limit is reached when the electromagnetic field from the oncoming beam becomes large enough to disturb the precise magnetic focusing properties of the ring. Other intensity limits come from fields produced by the particle's neighbors and their images reflected in the walls of the vacuum chamber. Any sudden change in the transverse dimensions of the vacuum chamber will be excited by the electromagnetic wake field of the beam passing through it as if it were a parasitic ac-

celerating cavity. The fields set up in the cavity act back on the beam and like an amplifier with a feedback system can become unstable if the current is too high. Yet another potential limit to the luminosity comes from the fact that protons at 20 TeV are beginning to radiate significant flux of synchrotron light just as electrons at much lower energy. This falls on the inner, cold, surface of the vacuum tube adding to the heat load of the refrigerators. Finally, another difficulty with such large machines is the precision required for the magnetic guide field. The largest computers cannot simulate the beam's behavior in these fields for more than a million or so turns, a small fraction of the required lifetime. Such studies stretch the predictive power of nonlinear mathematics to the limit.

A practical concern is that considerable care must be applied to the design of the protection devices and energy dumping circuits, which must safely dispose of the energy stored in the magnets' field if their superconducting properties are suddenly lost due to a mishap.

Reasons for the Cancellation of the SSC

In 1982, the Snowmass Study, organized by the American Physical Society, first proposed the SSC. Their initial cost estimate was $2.9 to $3.2 billion, a figure that was confirmed in 1983 by the U.S. Department of Energy (DOE).

Subsequently, detailed design issues were studied by a Central Design Group set up under Maury Tigner at Lawrence Berkeley Laboratory—a body of experienced accelerator designers who produced a convincing design, building on experience gained in constructing the successful 1-TeV Tevatron at Fermilab. By 1986 the conceptual design study was complete, and in 1987 President Reagan set in motion the search for a site. In 1988 Waxahatchie, Texas, was announced as the successful candidate. This decision was perhaps influenced by Vice President George Bush of Texas; Jim Wright of Fort Worth, then Speaker of the House of Representatives; and a powerful senator, Lloyd Bentsen, also from Texas.

In the past, the management of large accelerator projects, once approved, had been entrusted to their designers. However, in this case the DOE

FIGURE 2

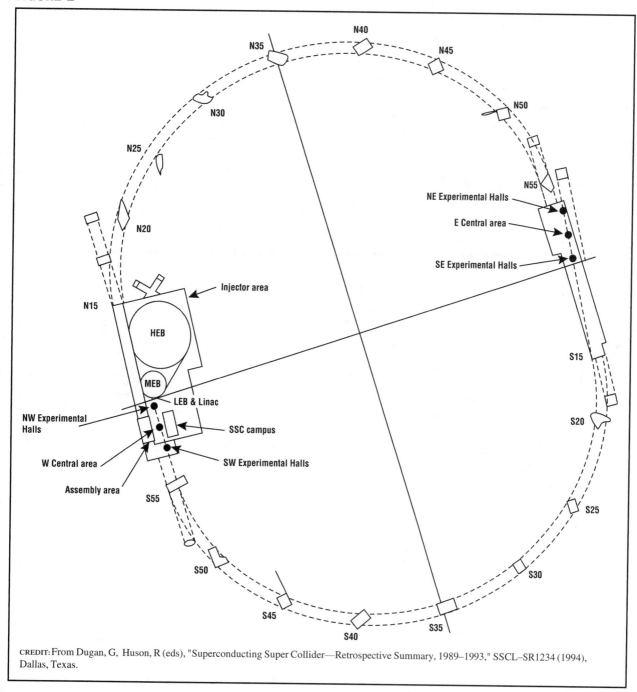

CREDIT: From Dugan, G, Huson, R (eds), "Superconducting Super Collider—Retrospective Summary, 1989–1993," SSCL–SR1234 (1994), Dallas, Texas.

Layout of the SSC Accelerators.

judged the SSC to be too mammoth an undertaking to be constructed without the aid of industrial firms with considerable expertise in the management and operating of large projects—but, it must be said, with very little knowledge of accelerators. The management and operation of the project was contracted to EG&G, Inc., which had managed the Nevada Test Site and the other DOE facilities, and the Sverdrup

Corporation, which was involved in defense-based contracts. The Central Design Group Team leader was replaced by Roy Schwitters as director who reformed a design team in Waxahatchie alongside the contractors. It was said that communication between the two communities had its problems and that this contributed to the escalation of cost.

The designers made a number of costly but necessary design modifications. They increased the magnet aperture to ensure the beam was further away from uneven fields near the coils. The strength of focussing magnets was augmented, the energy of the accelerators in the injection chain was increased, and the experimental areas enlarged. Meanwhile, it seemed to some that contracts were placed, not always to the lowest bidder, but with a view to giving every state in the Union a stake in the project. The cost rose steeply from the Central Design Group's estimate of $3.9 billion in 1986 to $5.3 billion in 1987, which was estimated by the DOE for a construction period that was longer by one year. This estimate became $5.9 billion in 1991, but review teams, taking into account the site-specific costs, adjusted this to $7.2 billion and then to $8.2 billion. The final estimate by an independent cost estimating team of the DOE, which added $2.5 billion for peripheral expenses that would not be incurred if the SSC had not been there, was $11.8 billion.

The result of this cost escalation was to trigger the U.S. House of Representatives to cancel funding in 1993. There was a rival project—the space station—which many in the House preferred. There were also those who believed both projects should be sacrificed in order to balance the budget. The Senate restored funding for one year, but in 1994, after some unseemly maneuvering by both sides, Congress finally canceled the SSC.

Those who regret the demise of the SSC blame the way in which DOE set up the project and particularly its choice of contractors. Its management team was headed by a procession of able project managers who left or were replaced. It was said they were frustrated by the lack of sound DOE leadership and were never in the saddle long enough to restrain the rising costs. Another factor was the choice of the site that discouraged many experienced people from joining the team. For those who opposed the SSC and its funding, it was just too much money for the general public to provide to support a science that they were convinced was irrelevant to their everyday life. At that time it was even becoming intellectually fashionable in some circles to question whether science had enhanced the quality of human life at all. Meanwhile, it was left to Europe to construct a more modest collider, the CERN LHC, with beams of 7 TeV, and to the Tevatron at Fermilab to hunt the for Higgs boson, hoping that it might be found below its rather limited reach in energy.

See also: ACCELERATOR, COLLIDING BEAMS: ELECTRON-POSITRON; ACCELERATORS, COLLIDING BEAMS: ELECTRON-PROTON; ACCELERATORS, COLLIDING BEAMS: HADRON; ACCELERATORS, FIXED-TARGET: ELECTRON; ACCELERATORS, FIXED-TARGET: PROTON

Bibliography

Huson, R., et al. "20 TeV Colliding Beam Facilities: New, Low-Cost Approaches." *Proceedings of the DFP Summer Study on Elementary Particle Physics and Future Facilities,* 315–322 (1982).

Jackson, J. D., ed. *Conceptual Design of the Superconducting Super Collider SSC* (Central Design Group, SSC-SR-2020, 1986).

Report of the DOE review Committee on the Baseline Validation of the SSC (DOE/ER-0594P 1993).

Report of the HEPEP Subpanel on Future Facilities (DOE/ER-0169, July 1983).

Report on the SSC Cost and Schedule Baseline (DOE/ER-0468P, 1991).

Edmund J. N. Wilson

STANDARD MODEL

All matter is believed to be made up of a small number of building blocks that are structureless and fundamental: six quarks, six leptons, their antiparticles, and a set of particles which carry the forces between the quarks and leptons. In addition, there is a postulated spin-0 particle, the Higgs boson, which is the remnant of the mechanism used to give mass to the particles that carry the weak force. The quarks, leptons, and their antiparticles are spin-$\frac{1}{2}$ particles, whereas the force carriers are spin-1 particles. The description of these particles and their interactions with each other is contained in the Standard Model of particle interactions.

The six quarks are called up (u), down (d), charm (c), strange (s), top (t), and bottom (b). The up, charm, and top quarks all have electric charge $\frac{2}{3}$ $|e|$, where $e = -4.803206$ esu is the electric charge of the electron. The up, charm, and top quarks have different masses, but identical electromagnetic interactions, since they have the same electric charges. The down, strange, and bottom quarks have electric charge $-\frac{|e|}{3}$ and again have identical electromagnetic interactions, but different masses. Associated with each quark is an antiquark, which is identical to the corresponding quark except that it has the opposite electric charge. For example, the antiup quark is written as $\bar{u}$ and has electric charge $-\frac{2}{3}|e|$.

The quarks and their approximate masses are shown in Table 1. With the exception of the top quark, the quarks do not exist as free particles so the masses given in Table 1 involve significant theoretical and experimental uncertainties. The underlying reason for the pattern of quark masses is an unsolved problem in particle physics. In particular, there is no understanding of why the top quark is so much heavier than the up and down quarks. Models that attempt to explain the large top quark mass require the introduction of new physics beyond the Standard Model. Physicists using high-energy accelerators continue to search for higher-mass quarks, since the Standard Model also does not predict the total number of quarks.

Most matter is made from up and down quarks and their antiquarks. Baryons consist of three quarks bound together, whereas antibaryons contain three antiquarks. All baryons and antibaryons have half-integer spin. For example, the proton is composed of two up quarks and one down quark. The proton thus has electric charge $\frac{2}{3}$ $|e| + \frac{2}{3}$ $|e| - \frac{|e|}{3} = |e|$. The neutron is made from two down quarks and an up quark and so has electric charge zero.

Mesons are composed of a quark and an antiquark and have integer spin. The lightest mesons are the pions with mass 139.6 GeV/c^2, which have three electric charge states: π^+, π^-, and π^0, and are composed of the following combinations of quarks, $u\bar{d}$, $\bar{u}d$, and $\frac{1}{\sqrt{2}}$ $(\bar{u}u + \bar{d}d)$. No particle with fractional charge has ever been observed, and to the best of current knowledge, quarks are always bound into mesons and baryons with integer charge (termed hadrons). (The top quark decays before it can bind into a hadron).

There are three negatively charged leptons with electric charge e: the electron (e^-), muon (μ^-), and tau (τ^-). The three charged leptons have identical electromagnetic interactions but different masses. Each lepton also has associated with it an antilepton with opposite electric charge, e^+, μ^+, and τ^+. Corresponding to the three charged leptons and their antileptons are three neutrinos (v_e, v_μ, and v_τ) and three antineutrinos ($\bar{v}_e$, $\bar{v}_\mu$, and $\bar{v}_\tau$). Table 2 lists the charged leptons along with their electric charges and masses. Evidence for nonzero neutrino masses is just beginning to be amassed.

The quarks and leptons are grouped together into families or generations. Each generation consists of a charged $\frac{2}{3}$ $|e|$ quark, a charged $-\frac{|e|}{3}$ quark, a charged lepton, its neutrino, and the associated

TABLE 1

Quark Masses	
u	$1 - 5$ MeV/c^2
d	$3 - 9$ MeV/c^2
c	$1.15 - 1.35$ GeV/c^2
s	$75 - 170$ MeV/c^2
t	174 ± 5.1 GeV/c^2
b	$4 - 4.4$ GeV/c^2

CREDIT: Courtesy of Sally Dawson.

TABLE 2

Lepton Masses	
e^-	$.510998902 \pm .000000021$ MeV/c^2
μ^-	$105.658357 \pm .000005$ MeV/c^2
τ^-	$1777.03 + 0.30 - 0.26$ MeV/c^2

CREDIT: Courtesy of Sally Dawson.

antiparticles. The first generation contains the up and down quarks, the electron, the electron neutrino:

$$\begin{pmatrix} u \\ d \end{pmatrix}, \begin{pmatrix} v_e \\ e^- \end{pmatrix}$$

and the corresponding antiparticles. The second generation contains the charm and strange quarks as well as the muon and muon neutrino:

$$\begin{pmatrix} c \\ s \end{pmatrix}, \begin{pmatrix} v_\mu \\ \mu^- \end{pmatrix}$$

whereas the third generation consists of the bottom and top quarks, the tau lepton, and the tau neutrino:

$$\begin{pmatrix} t \\ b \end{pmatrix}, \begin{pmatrix} v_\tau \\ \tau \end{pmatrix}$$

There is no evidence of a fourth generation.

Forces

Besides having different masses, the particles are distinguished by the forces with which they interact. There are four forces that communicate between the quarks and leptons: the electromagnetic force, the weak force, the strong force, and gravity. All massive particles feel gravity. Since it is the weakest of the forces, gravity is usually neglected in discussions of the Standard Model. Furthermore, there is no consistent quantum theory of gravity.

The strong, weak, and electromagnetic forces are carried by particles called gauge bosons. The forces are described by a Yang-Mills (or gauge) theory based on the product of two special unitary groups, SU(3) $\times$ SU(2), and one unitary group, U(1). The gauge structure of the Standard Model is then a product, SU(3) $\times$ SU(2) $\times$ U(1). The SU(3) group corresponds to the strong interactions, whereas the SU(2) $\times$ U(1) product groups describe a unification of the weak and electromagnetic forces, usually called the electroweak force.

Associated with each group are gauge bosons and a single independent coupling constant describing the strength of the interaction between the quarks and leptons and the gauge bosons. An SU(N) group has $N^2 - 1$ gauge bosons, while a U(1) group has a single gauge boson. Therefore, the SU(3)

gauge group has eight gauge bosons that are the carriers of the strong force, while the SU(2) gauge group has three gauge bosons.

The coupling constants of the SU(3) $\times$ SU(2) $\times$ U(1) gauge groups are denoted by g_3, g_2, and g_1, respectively, and represent the relative strengths of the corresponding forces. In a quantum theory, these couplings scale with energy. At an energy scale of 91 GeV, they are in the approximate ratio $g_3{:}g_2{:}g_1$ = 1:0.5:0.3. The SU(3) and SU(2) coupling constants g_3 and g_2 decrease with increasing energy, whereas the U(1) coupling constant g_1 increases with increasing energy. In the Standard Model, all three coupling constants become approximately equal at a scale near 10^{15} GeV, leading to speculation that the strong and electroweak forces are unified at this scale. Conversely, as the energy scale is decreased, the SU(3) and SU(2) coupling constants g_3 and g_2 increase with a decreasing energy scale, and the strong coupling constant g_3 becomes approximately equal to 1 at 1 GeV. At this scale, the strong interactions dominate the other forces and provide the interactions that bind the quarks into hadrons.

Left-Handed Particles

Massive spin-$\frac{1}{2}$ particles such as the quarks and leptons have two possible spin states. One way to describe these spin states is by whether the spin of the particle is parallel or antiparallel to the momentum of the particle. These two possibilities are called helicity states. Particles whose spin and momentum are parallel are said to have positive helicity and are called right-handed, whereas those whose spin and momentum are antiparallel are called left-handed. The electroweak interactions treat right- and left-handed particles very differently, while the strong interactions do not distinguish between helicity states.

A massless spin-$\frac{1}{2}$ particle, such as a neutrino or antineutrino, has only one helicity state. The Standard Model assumes that neutrinos are massless and left-handed. A nonzero neutrino mass therefore requires the introduction of a right-handed neutrino and hence physics beyond the Standard Model. The quarks and charged leptons have both right- and left-handed helicity states.

Electroweak Interactions

The electroweak interactions are a unification of the weak and electromagnetic interactions. The weak interaction describes, for example, the beta decay of the neutron, $n \rightarrow p + e^- + \bar{v}_e$. The strength of the weak interaction in this decay has been measured very precisely and is $G_F/(\hbar c)^3 = 1.16639 \times 10^{-5}$ GeV^{-2}, where G_F is called the Fermi constant. In 1914, James Chadwick observed that the energy spectrum of the electrons emitted in beta decay was inconsistent with a two-body decay. This led Wolfgang Pauli to postulate the existence of the neutrino so that momentum could be conserved in this decay. Subsequently, in 1934, Enrico Fermi advanced his theory of beta decay that included the neutrino.

The modern quark model explains beta decay as the four-fermion interaction of $d \rightarrow ue^- v$. The transition of the down quark to an up quark occurs by the exchange of a charged W^- boson, which then is absorbed to create the negatively charged electron and its neutrino. The W boson is postulated to be the carrier of the weak force. This theory was later incorporated into a gauge theory by Sheldon Lee Glashow, Steven Weinberg, and Abdus Salam.

The electroweak interactions are described by the SU(2) × U(1) gauge groups and represent a unification of the electromagnetic and weak interactions. Associated with the SU(2) group are three gauge bosons, typically denoted W^a, $a = 1, 2, 3$. There is a single gauge boson B corresponding to the U(1) gauge group. Particles are classified according to how they interact with the gauge bosons. The left-handed fermions couple to the SU(2) gauge bosons, while the right-handed fermions do not, and the strength of this coupling is proportional to the SU(2) coupling constant g_2. The couplings of the fermions to the U(1) gauge boson B are proportional to the product of the hypercharge Y of the fermions and the U(1) coupling constant g_1. The left-handed charge $2|e|/3$ quarks have a hypercharge of $Y = \frac{1}{6}$, whereas the right-handed charge $2|e|/3$ quarks have hypercharge $Y = \frac{2}{3}$.

The electroweak gauge symmetry is a broken symmetry. At an energy scale of approximately 80 GeV, the electroweak gauge group SU(2) × U(1) is broken to U(1)$_{em}$, where U(1)$_{em}$ is the electromagnetic gauge group whose gauge boson is the mass-

less photon γ. The electromagnetic force corresponds to the attraction (or repulsion) of particles with unlike (or like) electric charge.

When the gauge groups are unbroken (at energy scales above 80 GeV), the electroweak gauge bosons are massless and correspond to long-range forces. The weak interactions, however, are known from beta decay to be short range, and thus they must be mediated by massive gauge bosons. Masses for the SU(2) × U(1) gauge bosons are generated using a theoretical mechanism proposed by Peter Higgs and others in the 1960s. This mechanism involves introducing a scalar field that is a doublet under SU(2) gauge transformations. The interactions are arranged in such a way as to give masses to linear combinations of three of the four gauge bosons of the SU(2) × U(1) gauge theory, while leaving the photon massless.

The three gauge bosons that receive masses are the W^+, W^-, and Z bosons. These gauge bosons were first observed in 1983 in $p\bar{p}$ collisions at the European Laboratory for Particle Physics (CERN) ISR accelerator in Geneva, Switzerland. The masses are given experimentally by

$$M_W = 80.419 \pm 0.056 \text{ GeV}/c^2$$
$$M_Z = 91.1882 \pm 0.0022 \text{ GeV}/c^2$$

The gauge boson masses are predicted by the electroweak theory in terms of the Fermi constant G_F and the SU(2) × U(1) coupling constants g_2 and g_1:

$$M_W^2 = \frac{g_2^2}{4\sqrt{2}G_F} ,$$
$$M_Z^2 = \frac{g_1^2 + g_2^2}{4\sqrt{2}G_F}$$

The experimentally measured values of the weak gauge boson masses agree well with those predicted from measurements of the gauge coupling constants and the Fermi constant.

An inevitable consequence of the breaking of the SU(2) × U(1) symmetry is the existence of a scalar particle that remains after the symmetry is broken. This particle is called the Higgs boson. The mass of the Higgs boson is a free parameter of the theory, but its couplings to gauge bosons and to quarks and charged leptons are completely fixed. The discovery of the Higgs boson is necessary to confirm the validity of the Standard Model of electroweak interactions. Current experimental searches at the Large Electron Positron (LEP) collider at the CERN laboratory restrict the Higgs boson mass to be greater than 114 GeV/c^2, whereas measurements of the top quark and W boson masses imply that the Higgs boson mass is less than approximately 200 GeV/c^2. It is expected that the next round of collider experiments at the Fermilab Tevatron in Batavia, Illinois, or the CERN Large Hadron Collider will discover the Higgs boson if it exists.

Quantum Chromodynamics

The force that binds quarks together to make hadrons is called the strong force, or quantum chromodynamics (QCD). Quantum chromodynamics is a Yang-Mills gauge theory corresponding to an $SU(3)$ symmetry group. The $SU(3)$ symmetry is believed to be unbroken and thus an exact symmetry at all energy scales. Associated with the $SU(3)$ symmetry are eight massless gauge bosons called gluons, G^a, $a = 1 \dots 8$, which remain massless at all energy scales. The gluons interact with the quarks and antiquarks, but not with the leptons. It is the interactions of the gluons with the quarks and antiquarks that binds the quarks into hadrons. A summary of the interactions of the particles with the strong, weak, and electromagnetic forces is given in Table 3.

Outlook

Many aspects of the Standard Model have been verified experimentally. The next frontier is the discovery of the Higgs boson. A theoretical understanding of the pattern of quark masses and the neutrino masses and the possible unification of all forces are still missing, however, and these questions motivate experiments at still higher-energy scales.

See also: BOSON, HIGGS; ELECTROWEAK SYMMETRY BREAKING; HIGGS PHENOMENON; LEPTON; PARITY, NONCONSERVATION OF; QUANTUM CHROMODYNAMICS; QUARKS; SUPERSYMMETRY; SYMMETRY PRINCIPLES

Bibliography

Glashow, S. L. "Towards a Unified Theory: Threads in a Tapestry." *Reviews of Modern Physics* **52**, 539–543 (1980).

Groom, D. E., et al. "2000 Review of Particle Physics." *The European Physical Journal* **C15**, 1–878 (2000).

Quigg, C. *Gauge Theories of the Strong, Weak, and Electromagnetic Interactions* (Addison-Wesley, Reading, MA, 1985).

Salam, A. "Gauge Unification of Fundamental Forces." *Reviews of Modern Physics* **52**, 525–538 (1980).

Weinberg, S. "Conceptual Foundations of the Unified Theory of Weak and Electromagnetic Interactions." *Reviews of Modern Physics* **52**, 515–523 (1980).

Sally Dawson

TABLE 3

Interactions of Particles with Forces			
Particle	Weak Force	Electromagnetic Force	Strong Force
Quarks, (u, d, s, c, t, b)	Yes	Yes	Yes
Leptons, (e, μ, τ)	Yes	Yes	No
Neutrinos, (υ)	Yes	No	No
Higgs boson	Yes	No	No
CREDIT: Courtesy of Sally Dawson.			

STRING THEORY

String theory is a proposed unified theory of fundamental physics, incorporating both particle physics and gravity. It is based on the idea that the basic building blocks of nature are strings, one-dimensional objects of zero thickness, which form either closed loops or open curves (Figure 1). This theory has not yet been experimentally tested, but it has attracted the attention of theoretical physicists from a wide range of fields because it unifies many of the central concepts of physics and resolves a number of long-standing theoretical problems.

FIGURE 1

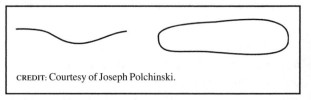

CREDIT: Courtesy of Joseph Polchinski.

Closed and open strings.

A Brief History

Two of the central questions in physics are the nature of matter and the nature of gravity. In the twentieth century, very successful theories of each were discovered: the Standard Model of matter and the general theory of relativity. However, both theories are incomplete. The Standard Model is based on a complicated pattern of particles and forces, similar to the Periodic Table of the elements, and this pattern must be explained. General relativity, when combined with quantum mechanics, suffers from several problems and paradoxes when applied to very short distances or to black holes. Further, ultimately matter and gravity should not be described by two unrelated theories but should be understood in a unified way. String theory is believed to solve all of these problems.

The idea of building blocks that are one-dimensional, rather than zero-dimensional points, is rather novel, and it has had an odd history. It was first developed between 1968 and 1973 as a theory of the strong interaction: mesons such as the pion behave in some respects like open strings. This idea was superceded by the 1973 discovery of the true theory of the strong interaction, quantum chromodynamics, but a small handful of theorists regarded string theory as a compelling idea and continued to develop it. In the following years it was discovered that string theory is actually a theory of gravity, that it implies a symmetry between bosons and fermions (which was named supersymmetry), and that it is free of the unphysical infinities that plagued all previous theories of quantum gravity.

In 1984 a discovery by Michael Green and John Schwarz, known as anomaly cancellation, showed that string theory could also describe quarks, leptons, and gauge interactions. This led to a tremendous wave of research activity, often called the first superstring revolution, as theorists who had been pursuing other approaches to unification began to develop string theory. The discovery of Calabi-Yau compactification by Philip Candelas, Gary Horowitz, Andrew Strominger, and Edward Witten and of heterotic string by David Gross, Jeff Harvey, Lance Dixon, and Ryan Rohm strengthened the evidence for string theory. However, work at this time was limited to a certain approximation, known as perturbation theory, which applies only to small numbers of

strings interacting weakly. In 1995 Witten, extending results of Chris Hull, Paul Townsend, and others, identified the principle of string duality, which governs the behavior of strongly interacting strings. The ensuing period, known as the second superstring revolution, has produced many further discoveries. One has been a new understanding of the quantum mechanics of black holes, resolving some long-standing puzzles. Another has been the understanding that string theory contains extended structures known as branes, which has led to new ideas for realistic models and for experimental and cosmological tests.

String theory is still an incomplete theory. It is widely believed that the description of the theory in terms of one-dimensional building blocks is not ultimately the simplest or most complete but rather is a stepping-stone toward a more fundamental principle, which is being actively sought by string theorists today. String theory is equivalently referred to as superstring theory, reflecting the central role of supersymmetry. Since 1995 the term M theory has also been used, reflecting the fact that the string picture is believed to be just a stepping-stone to a final theory. What the "M" stands for is deliberately left unspecified, reflecting the unknown nature of the final theory; "magic," "mother," "mystery," "membrane," and "matrix" have all been suggested.

String Theory and Particle Physics

Experimentally, the electron, quarks, photon, and other Standard Model particles are all points: no experiment has revealed any substructure, down to the distance scale of 10^{-16} cm. The idea of string theory is that under sufficient magnification the Standard Model particles will be seen to be loops or segments of string. The magnification needed is very large, as the size of the string is expected to be of order 10^{-32} cm: this is the Planck length, the distance scale where gravity and quantum mechanics come together. This scale is far beyond the reach of particle accelerators, so that experimental tests of the theory will have to be indirect. (Recent ideas, to be discussed below, raise the possibility that the strings are larger and so more accessible to experiment.)

The string of string theory is much like a violin string: it vibrates, and this vibration can be decomposed into a sum of notes or harmonics. Depending upon which harmonics are excited, and to what degree, the string will behave like different kinds of particles, and all the particles found in the Standard Model can be obtained as different states of vibration of this one building block. In particular, one of the states of vibration of the closed string is a massless particle of spin two. This was a problem when string theory was supposed to describe the strong interaction, as there is no such hadron, but this particle has precisely the properties of the graviton, the particle associated with the gravitational field, and this is why string theory must incorporate gravity. In addition to vibrating, strings can break in two and join together (Figure 2). All of the basic processes of nature, such as an electron emitting a photon or a graviton, or a Z boson decaying into a quark-antiquark pair, arise from this one basic string process.

Before the first superstring revolution, many other ideas were explored for unifying the Standard Model and explaining its patterns. Three ideas of particular note are grand unification, supersymmetry, and extra dimensions. These can be thought of as new symmetry principles, meaning that Standard Model particles that appear to be different are really the same kind of particle but with, in some sense, a different orientation. Each of these ideas had some successes, and one of the attractive features of string theory is that it automatically incorporates all three of these enlarged symmetries.

Extra Dimensions

The idea that space-time has more than the four visible dimensions is almost as old as general relativity. Gravity and electromagnetism are similar in that both forces fall off as the inverse square of the distance; they differ in that gravity couples to energy and momentum, and electromagnetism

couples to charge. Theodore Kaluza in 1919 and Oscar Klein in 1926 put forward the idea that electromagnetism would actually originate from gravity if space-time were five-dimensional, with the fifth dimension too small to be seen directly (Figure 3). The gravitional field has a polarization (spin). If this polarization is fully aligned along the large dimensions, the five-dimensional graviton behaves like a four-dimensional graviton and produces the gravitational force. If it is partly aligned along the small dimension, then it behaves like a four-dimensional photon and produces the electromagnetic force. What is seen as electric charge is actually momentum that is directed along the small dimension. This elegant unification of the two then-known forces fascinated some of the greatest physicists of the early twentieth century. Albert Einstein and Wolfgang Pauli each spent substantial periods trying to develop it further.

String theory requires that space-time have ten dimensions (nine space and one time). Ultimately this originates from the mathematical structure of the supersymmetry algebra. There are actually five consistent string theories, known as types I, IIA, IIB, heterotic SO(32), and heterotic $E_8 \times E_8$. These differ primarily in the way that the supersymmetry acts on the states of the string; also, type I theory has both open and closed strings, while the other four have only closed strings.

FIGURE 3

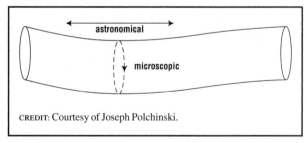

CREDIT: Courtesy of Joseph Polchinski.

Space-time with extra dimensions. The large dimensions are of cosmic size, 10 billion light-years or more. The small dimensions are microscopic, too small to have yet been discovered. In this schematic picture only one large dimension is drawn, representing the three large spatial dimensions plus time. Only one small dimension is drawn, as in the original Kaluza-Klein theory; in string theory there are six or seven extra spatial dimensions.

FIGURE 2

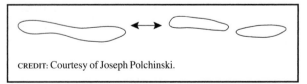

CREDIT: Courtesy of Joseph Polchinski.

Two closed strings joining, or one splitting into two.

In Kaluza-Klein theory, the single extra dimension forms a circle. The six extra dimensions of string theory can have a much more complex topology and geometry. Although we do not see these dimensions directly, their shape determines the physics that we do see—the spectrum of particles, and their masses and couplings. A relatively simple set of spaces known as Calabi-Yau manifolds, when combined with the $E_8 \times E_8$ heterotic string, give a result very much like the grand unified supersymmetric Standard Model. One of the central problems in string theory is that there are many different Calabi-Yau manifolds, as well as other possible spaces, and to account for the precise details of the Standard Model requires knowing the precise shape of the extra dimensions.

The simplest estimate of the size of the extra dimensions in string theory gives the Planck length, which would put them far beyond direct experimental detection. As will be discussed below, more recent ideas raise the possibility that they are much larger and might have a variety of observable effects.

String Theory and Quantum Gravity

General relativity and quantum mechanics are two of the central principles in physics, and each has been verified experimentally in great detail. General relativity is important at astronomical scales, but its effects are negligible in the microscopic regime of atomic and particle physics. Quantum mechanics is essential to microscopic physics, but its effects are negligible at astronomical scales. Thus, in ordinary circumstances one does not encounter general relativistic and quantum effects together.

However, general relativity and quantum mechanics conflict with one another, and this conflict will appear in certain extreme situations. At very short distances one encounters the problem of space-time foam. General relativity states that space-time is curved and that the effect of this curvature is gravity. Quantum mechanics states, roughly speaking, that nothing sits still (the uncertainty principle). Taken together, these imply that space-time does not sit still, its shape is constantly fluctuating. These fluctuations are totally negligible on astronomical scales and even on the scales of particle physics, but as one goes to very short distances they become more evident. At the

Planck length of 10^{-32} cm they become so severe that shorter distances do not make sense at all—space-time, in a sense, tears itself to pieces. In the language of particle physics, this means that quantum gravity is not renormalizable: when the effects of virtual gravitons are included in quantum mechanical amplitudes, the result is infinite (Figure 4).

The problem of renormalization arose for the three particle interactions as well and in each case was an important clue to the correct theory. In the case of gravity, string theory removes the problem by changing the theory at distances below the Planck scale.

Other approaches to this problem are still under study, but to date string theory is the only known finite theory of quantum gravity. The problem of quantum foam means that distances smaller than the Planck scale cannot make sense, and so the historic progression toward ever-smaller constituents must end. In string theory, the size of a string represents a minimum length, the shortest distance that can be probed.

Modern String Theory

String Duality and D-branes

As physics has progressed toward more unified theories based on more fundamental principles, there has been a growing expectation that there is

FIGURE 4

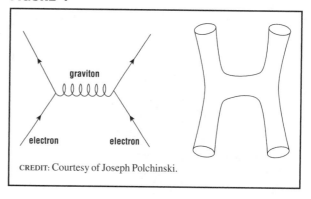

CREDIT: Courtesy of Joseph Polchinski.

Two electrons exchanging a virtual graviton (conventionally drawn as a coiled line), as represented by a Feynman graph in quantum field theory and in string theory. The graph represents the time-history of the process. In the field theory process the graviton is emitted and absorbed at precise space-time points, leading to infinities. In string theory each particle is replaced by a loop and the process is thus smeared out.

a unique theory that incorporates all of the laws of physics. The existence of five different string theories was therefore a puzzle (though the situation is much better than quantum field theory, where there is an infinite number of theories that are characterized by different symmetries, particles, masses, and couplings). In 1995 it was understood, through string duality, that these are all part of a single theory. Essentially, they are different phases (Figure 5), related to each other much like the liquid, solid, and gas phases of water. In the case of water, one varies the pressure and temperature to change one phase into another. In the case of string theory, one varies the shape and size of the extra dimensions, and in certain regimes the theory behaves like one or the other of the string theories. There are also new phases, most notably a phase known as $D = 11$ supergravity where a new space-time dimension, the eleventh, appears.

In addition to the strings themselves, string theory contains a variety of higher-dimensional objects known as branes. These were discussed before 1995, but in the context of string duality it became clear that they play a central role. A particular class known as D-branes (Figure 6) were shown by Joseph Polchinski to have the special property that strings can end on them, and play an important role in understanding many of the phases of the theory.

FIGURE 5

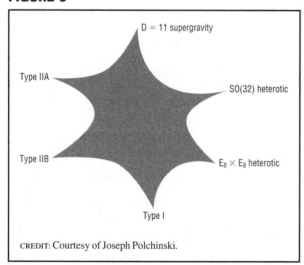

CREDIT: Courtesy of Joseph Polchinski.

Phase diagram of string theories. The shapes and sizes of the extra dimensions vary as one moves around the diagram.

FIGURE 6

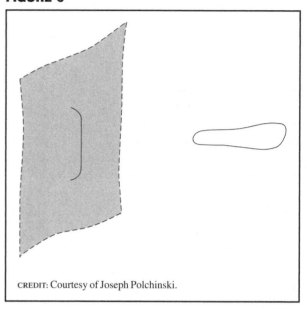

CREDIT: Courtesy of Joseph Polchinski.

D-brane, with attached string. A freely moving closed string is also shown.

Black Hole Entropy and Information

Black holes are among the most extreme objects in physics, and they present another situation where the laws of relativity and quantum physics come together. In the early 1970s it was found that black holes satisfy laws parallel to those of thermodynamics. In particular, Jakob Bekenstein and Stephen Hawking argued that they have an entropy, implying a microscopic structure of states. The nature of these states was mysterious until 1996, when Strominger and Cumrun Vafa showed that they were accounted for by string theory. In particular, for certain charged black holes, D-branes give a precise construction of the microscopic states.

Hawking also discovered that black holes radiate and eventually disappear and that this leads to a paradox. The particles produced in the decay do not depend on what initially forms the black hole, so information is lost in the process of black hole formation and decay; this information loss is inconsistent with the laws of quantum mechanics. Hawking argued the quantum mechanics must therefore be modified. This claim inspired much further work that showed that if quantum mechanics were not modified then the principle that physical processes are local in space-time must break down in a subtle way (there are other alternatives, but these are generally regarded as less likely). This issue is not yet decided, but string duality relates black holes to ordinary systems that do satisfy the laws of quantum mechanics, and this suggests in fact it is the principle of space-time locality, not quantum mechanics, that must be modified.

Braneworlds, Large Extra Dimensions, and Low-energy Strings

The existence of branes of various dimensionalities suggests that we might actually live on a three-brane, a brane with three space dimensions (plus time, of course). String theory still requires nine space dimensions, or ten in the $D = 11$ supergravity phase, so our brane would be embedded in this higher-dimensional space. The open strings that attach to the D-brane can give rise to all the particles of the Standard Model except the graviton. Thus there are string models in which everything seen in nature except for gravity is attached to a brane, while the gravitational field lines spread out in all the dimensions. This is different from the previous extra-dimensional ideas, where there are no branes, and all particles live in the full set of extra dimensions.

For extra dimensions without branes, dimensional analysis indicates that both the size of the dimensions and the string scale are near the Planck length and so remote from experiment. The situation with branes is more complicated. The size of the dimensions is not fixed by theory, and it could be very much larger. It is then important to consider the experimental limits on the size. In 1998, Nima Arkani-Hamed, Savas Dimopoulos, and Gia Dvali argued that, although particle accelerators probe physics down to 10^{-16} cm, the extra dimensions could be much larger than this. If everything but gravity were attached to a brane, the extra dimensions would not be seen readily at accelerators but only in gravitational experiments. They would show up as a change in the gravitational force law, from the inverse square law that comes from the field lines spreading in three space dimensions, to a different behavior. Since the inverse square law is tested only down to 0.1 cm, the extra dimensions could be as large as this.

The large extra dimensions might be seen at a particle accelerator in an indirect way. If two particles moving along the brane collide with enough energy to leave the brane and move into the extra dimensions, they become undetectable, and the result is an event in which energy seems to disappear. This process could also occur in such environments as the core of a supernova, and the observation of neutrinos from supernovae actually gives a more stringent upper limit than the force law experiments, around 10^{-4} cm. Arkani-Hamed, Dimopoulos, and Dvali also showed that in theories with large extra dimensions the string size is larger than the Planck length, so that string physics, and gravitational effects like microscopic black holes, might be seen at particle accelerators. There is no definite prediction yet for the size of the extra dimensions, and many theorists still expect that they are too small to be observed, but it is an exciting new possibility that is under theoretical and experimental study.

The Future of Theory and Experiment

The existence of branes of all dimensions raises the issue of whether the one-dimensional strings are truly fundamental. So also does the string-duality phase diagram, which shows that strings exist only in certain phases. These and other arguments have led string theorists to believe that the defining principle of the theory, the analog of the equivalence principle of general relativity and the uncertainty principle of quantum mechanics, has yet to be found. Many ideas are under investigation. A common theme is that physics is expected to be nonlocal, even before the Planck length. Two concepts being considered are the holographic principle, which is connected with the black hole information problem, and noncommutative geometry, which is an extension of the uncertainty principle involving only lengths and not momenta.

Experimentally, string theory does not yet make firm predictions. The part of the theory that is most likely to be accessible to accelerators is supersymmetry; large extra dimensions are a striking but less probable signature. String theory may eventually make distinctive predictions for cosmology. Finally, experience shows that as the theory is understood better, unexpected new possibilities are found.

See also: PARTICLE PHYSICS, ELEMENTARY; PLANCK SCALE; GRAND UNIFICATION; UNIFIED THEORIES

Bibliography

Arkani-Hamed, N.; Dimopoulos, S.; and Dvali, G. "The Universe's Unseen Dimensions." *Scientific American* **283**(2), 62–70 (2000).

Duff, M. "The Theory Formerly Known as Strings." *Scientific American* **278**(2), 64–69 (1998).

Greene, B. R. *The Elegant Universe: Superstrings, Hidden Dimensions and the Quest of the Ultimate Theory* (Norton, New York, 1999).

Horowitz, G. T., and Teukolsky, S. A. "Black Holes." *Reviews of Modern Physics* **71**(2), S180–S186 (1999).

Mukerjee, M. "Explaining Everything." *Scientific American* **274**(1), 88–94 (1996).

Schwarz, J. H., and Seiberg, N. "String Theory, Supersymmetry, Unification, and All That." *Reviews of Modern Physics* **71**(2), S112–S120 (1999).

Susskind, L. "Black Holes and the Information Paradox." *Scientific American* **276**(4), 52–57 (1997).

Weinberg, S. "A Unified Physics by 2050?" *Scientific American* **281**(6), 68–75 (1999).

Joseph Polchinski

STRONG INTERACTION

See BASIC INTERACTIONS AND FUNDAMENTAL FORCES

SU(3)

The symmetry group SU(3) figures prominently in elementary particle physics. There are two important and distinct SU(3) symmetries that are relevant for the strong interactions: SU(3) color symmetry of the quark and gluon dynamics and SU(3) flavor symmetry of light quarks. Each of these symmetries refers to an underlying threefold symmetry in strong interaction physics.

Mathematically, SU(3) is the group of special unitary 3×3 matrices U. The SU(3) group consists of all symmetry transformations that preserve the unit magnitude of 3 vectors:

$$\psi = \begin{pmatrix} \psi_1 \\ \psi_2 \\ \psi_3 \end{pmatrix},$$

where the ψ_i are complex numbers satisfying

$$|\psi_1|^2 + |\psi_2|^2 + |\psi_3|^2 = 1.$$

In quantum mechanics, the 3-vector ψ is called the probability amplitude or wavefunction for finding a particle in any one of three possible states, whereas the dot product $\psi^* \cdot \psi$ is the probability for measuring the particle in any one of the three possible states. The total probability $\psi^* \cdot \psi$ is the sum of the probabilities for finding the particle in each of the three states: $\psi_1^* \cdot \psi_1 = |\psi_1|^2$ is the probability that the particle is found in state 1; $|\psi_2|^2$ is the probability that the particle is found in state 2; and $|\psi_3|^2$ is the probability that the particle is found in state 3. The sum of these three probabilities is equal to 1, since any measurement is guaranteed to find the particle in one of the three possible states. An arbitrary symmetry transformation U maps the wavefunction ψ into a new wavefunction, $\psi \to U\psi$. The dot products of all complex 3-vectors ψ are left invariant under this mapping if U is special (its determinant is equal to one) and unitary (its Hermitian conjugate $U^\dagger$ is equal to its inverse U^{-1}). Thus, SU(3) is the group symmetry transformations of the 3-vector wavefunction ψ that maintain the physical constraint that the total probability for finding the particle in one of the three possible states equals 1.

Any arbitrary SU(3) matrix can be written in the form

$$U(\alpha) = \exp\left(i \sum_{a=1}^{8} \alpha^a T^a\right),$$

where the α^a are arbitrary real numbers, and the eight 3×3 matrices T^a, $a = 1, ..., 8$, are all traceless:

$$\text{Tr } T^a = 0$$

and Hermitian:

$$T^{a\dagger} = T^a.$$

The T^a are called the generators of SU(3) since all SU(3) group transformations can be written as exponentials of linear combinations of these eight generators. Because the number of SU(3) group transformations U is infinite, it is a great simplification to express them in terms of a finite number of group generators.

It is conventional to define the generators of SU(3) in terms of the eight Gell-Mann matrices λ^a:

$$T^a \equiv \frac{1}{2}\,\lambda^a$$

where

$$\lambda^1 = \begin{pmatrix} 0 & 1 & 0 \\ 1 & 0 & 0 \\ 0 & 0 & 0 \end{pmatrix},\ \lambda^2 = \begin{pmatrix} 0 & -i & 0 \\ i & 0 & 0 \\ 0 & 0 & 0 \end{pmatrix},\ \lambda^3 = \begin{pmatrix} 1 & 0 & 0 \\ 0 & -1 & 0 \\ 0 & 0 & 0 \end{pmatrix},$$

$$\lambda^4 = \begin{pmatrix} 0 & 0 & 1 \\ 0 & 0 & 0 \\ 1 & 0 & 0 \end{pmatrix},\ \lambda^5 = \begin{pmatrix} 0 & 0 & -i \\ 0 & 0 & 0 \\ i & 0 & 0 \end{pmatrix},\ \lambda^6 = \begin{pmatrix} 0 & 0 & 0 \\ 0 & 0 & 1 \\ 0 & 1 & 0 \end{pmatrix},$$

$$\lambda^7 = \begin{pmatrix} 0 & 0 & 0 \\ 0 & 0 & -i \\ 0 & i & 0 \end{pmatrix},\ \lambda^8 = \frac{1}{\sqrt{3}}\begin{pmatrix} 1 & 0 & 0 \\ 0 & 1 & 0 \\ 0 & 0 & -2 \end{pmatrix}.$$

The generators T^a of SU(3) satisfy the commutation relations

$$[T^a, T^b] \equiv T^a T^b - T^b T^a = i\sum_{c=1}^{8} f^{abc} T^c,$$

where the f^{abc}, the structure constants of SU(3), are real numbers. The product of any two SU(3) group transformations can be determined from the commutation relations of the generators, so they determine the structure of the group. For SU(3) the maximal set of generators that commute with each other is given by the two diagonal matrices T^3 and T^8. In quantum mechanics, commuting operators correspond to physical quantities that can be known with certainty at the same time. Thus, SU(3) charges T^3 and T^8 of a physical system can be measured simultaneously because the generators T^3 and T^8 commute.

The fundamental representation of SU(3) is the three-dimensional representation, which is referred to as the **3** of SU(3). The generators T^3 and T^8 are both diagonal, so the three states of the **3** each have definite values of the charges T^3 and T^8. The three independent states of the 3-vector ψ correspond to the (T^3, T^8) states

$$\begin{pmatrix} 1 \\ 0 \\ 0 \end{pmatrix} \leftrightarrow \left(\frac{1}{2},\frac{1}{\sqrt{3}}\right),\ \begin{pmatrix} 0 \\ 1 \\ 0 \end{pmatrix} \leftrightarrow \left(-\frac{1}{2},\frac{1}{\sqrt{3}}\right),$$

$$\begin{pmatrix} 0 \\ 0 \\ 1 \end{pmatrix} \leftrightarrow \left(0,-\frac{2}{\sqrt{3}}\right).$$

It is useful to plot the (T^3, T^8) quantum numbers of any given SU(3) representation in a plane with co-ordinate axes labeled by the charges T^3 and T^8. The fundamental representation **3** is plotted in Figure 1. Note the threefold symmetry of the **3**. SU(3) generators acting on the **3** transform the three states into one another. It is interesting to note that the group SU(3) contains three SU(2) subgroups that transform any two states of the **3** into one another, while leaving the third state invariant.

The **3** of SU(3) is not equivalent to its conjugate representation, which is obtained by reversing the signs of all (T^3, T^8) quantum numbers. The conjugate representation of the **3** is called the $\overline{\textbf{3}}$ (3-bar) and is shown in Figure 2.

All higher-dimensional representations of SU(3) can be obtained as products of the fundamental **3** and antifundamental $\overline{\textbf{3}}$ representations. The product of **3** and $\overline{\textbf{3}}$ representations yields the eight-dimensional representation displayed in Figure 3. The **8** of

FIGURE 1

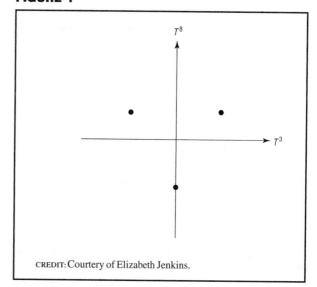

The fundamental representation 3 of SU(3).

FIGURE 2

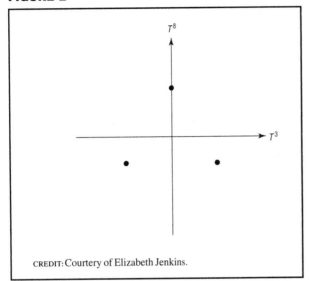

The antifundamental representation $\bar{3}$ of SU(3).

FIGURE 3

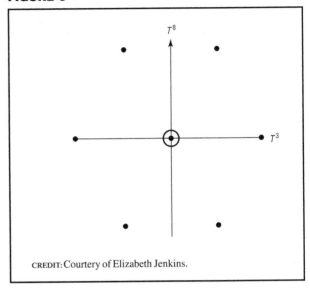

The adjoint representation 8 of SU(3). There are two states in the 8 with (T^3, T^8) equal to (0, 0).

SU(3) is called the adjoint representation of SU(3). The SU(3) generators or charges T^a, $a = 1, \ldots, 8$, form an eight-dimensional adjoint representation of SU(3). In general, every SU(3) representation exhibits threefold symmetry in the (T^3, T^8) plane.

The SU(3) color group is the exact gauge symmetry of the Standard Model, which accounts for the strong interactions of quarks and gluons. The theory of the strong interactions is called quantum chromodynamics (QCD). Quarks occur in the fundamental three-dimensional representation of SU(3) color. The three complex components of the quark color wavefunction

$$\begin{pmatrix} q_{\text{red}} \\ q_{\text{green}} \\ q_{\text{blue}} \end{pmatrix},$$

denote the probability amplitudes for finding a quark with one of three different colors, where color is a charge that comes in three varieties: red, green, and blue. Antiquarks, the antiparticle of quarks, occur in the conjugate $\bar{3}$ representation and carry an anticolor charge. The gauge boson mediators of the strong interactions are massless gluons that occur in the eight-dimensional adjoint representation of SU(3) color. The number of gluons corresponds to the number of SU(3) generators T^a. A different-

colored gluon couples to each of the eight color charges. In QCD color charge is conserved in the interactions of quarks, antiquarks, and gluons. A quark and an antiquark couple to a colored gluon, so a gluon in the eight-dimensional adjoint representation carries both color in the fundamental **3** representation and anticolor in the $\bar{3}$ representation.

The SU(3) flavor group is an approximate symmetry of QCD resulting from the universality of quark-gluon couplings. All quark flavors with a given color couple to gluons in precisely the same manner, that is, gluons are flavor-blind. The light quarks, up, down, and strange, occur in the fundamental three-dimensional representation of SU(3) flavor. The three complex components of the light quark flavor wavefunction

$$\begin{pmatrix} u \\ d \\ s \end{pmatrix}$$

denote the probability amplitudes for finding a light quark with one of the three different flavors, where light quark flavor is a charge that comes in three varieties: up, down, and strange. The antiquarks $\bar{u}$, $\bar{d}$, and $\bar{s}$ occur in the conjugate $\bar{3}$ flavor representation and carry antiflavor charge. SU(3) flavor symmetry is not an exact symmetry because the masses of the u, d, and s quarks are not the same, and so the quark flavors are distinguishable. Nevertheless, the mass difference of the u, d, and s quarks are all small compared to the scale at which the QCD coupling constant becomes large, so neglecting the mass splittings of the three light quarks is a good approximation.

SU(3) flavor symmetry is a useful approximate symmetry in QCD because hadrons containing light quarks and antiquarks of different flavors have similar properties. Colorless hadrons, either mesons or baryons, can be organized into SU(3) flavor multiplets. The lowest-lying meson and baryon multiplets are both in the eight-dimensional representation of SU(3) flavor. For SU(3) flavor multiplets, it is conventional to refer to the charges isospin I^3 and hypercharge Y, which are related to the charges T^3 and T^8 by

$$I^3 = T^3$$
$$Y = \frac{1}{\sqrt{3}}\, T^8.$$

Isospin refers to the SU(2) flavor subgroup for the two lightest quark flavors u and d, whereas hypercharge is proportional to the net number of strange antiquarks minus strange quarks, or strangeness, of a hadron. Isospin and hypercharge are both approximately conserved in decays and scattering processes resulting from strong interactions.

See also: EIGHTFOLD WAY; FAMILY; FLAVOR SYMMETRY; LEPTON; QUARK; STANDARD MODEL

Bibliography

Griffiths, D. *Introduction to Elementary Particles* (Wiley, New York, 1987).

Perkins, D. H. *Introduction to High Energy Physics* (Cambridge University Press, Cambridge, UK, 2000).

Elizabeth Jenkins

SUPERNOVAE

Supernovae are exploding stars. Observations of nearby supernovae are conspicuous entries in the annals of Chinese imperial astrologers dating back to 185 C.E., and supernova observations are among the great works of the Renaissance astronomers Tycho Brahe and Johannes Kepler. However, the true nature of these "guest stars" was not understood until observations in the 1930s revealed the distances to these stellar disasters. Understanding supernova explosions depends on understanding particle physics: the properties of the very smallest components of the universe determine the properties of its most energetic events. Supernovae are important engines in transforming the microscopic properties of the universe. They fuse simple elements such as hydrogen and helium into complex ones such as iron, gold, and uranium, and they blast those products into the gas between stars to enrich the next generation of stars. Supernova debris can help form planets and makes up living things.

Because the brightest supernova explosions are about as bright as 4 billion suns, they can be detected at large distances. Careful measurements of supernovae provide the distance scale of the universe and help establish the 14-billion-year timescale of cosmic expansion. Light from supernovae that has traveled half the span of the observable universe shows that cosmic expansion, surprisingly, has been speeding up. This cosmic acceleration, first glimpsed in 1998, suggests a new property of empty space itself: space has an energy whose outward pressure is revealed only by the supernova data. If this picture is right, supernovae show that two-thirds of the universe resides in an enigmatic dark energy. Explaining this phenomenon in terms of fundamental physics will be an important challenge for the twenty-first century.

The sudden appearance of a new star is a surprise: the lifetime of a short-lived star is 100,000 times that of a very long-lived person, so humans think stars are permanent and unchanging. However, the universe is not constructed on the human scale in space or time. The sudden death of a star in a thermonuclear explosion or a gravitational collapse is rare in any single galaxy, such as the Milky Way, but common throughout the 100 billion galaxies of the observable universe. A single galaxy has a supernova explosion approximately every century, so there should be a billion supernova explosions every year in the observable universe—thirty events per second.

In 2001, 254 supernovae were observed on Earth. Only a small fraction of all these events are actually seen because the entire sky is not observed every night and because the searches are not yet sensitive enough to reveal the most distant supernovae. There is room for improvement in the study of supernovae.

One key element in understanding supernovae is determining their distances: the apparent brightness of a supernova is not very conspicuous, with a few glorious exceptions such as Supernova 1987A in the Large Magellanic Cloud. When the observed flux is coupled with the distance, the intrinsic properties of supernovae become clear. In the early 1900s several new stars ("novae") were noted in spiral nebulae, which today are known as supernovae, in galaxies at distances of millions of light-years. At that time, however, it was more conservative to think that these were ordinary novae as seen in the Milky Way, which would imply that the spiral nebulae were part of the galaxy. In the 1920s the work of Edwin Hubble established that the spiral nebulae were outside the Milky Way, with the nearest of them at distances of millions of light-years. Hubble was puzzled by "that mysterious class of exceptional novae which attain luminosities which are respectable fractions of the systems in which they appear." (Hubble, 103). These were not ordinary novae, they were thousands of times brighter. Fritz Zwicky and Walter Baade dubbed this new class of exploding stars supernovae and set out to study them. Zwicky developed methods to search for supernovae: each time the moon was dark, he repeated a set of photographs of his target galaxies and compared the images by eye to find the new stars. Modern methods use the same approach he developed, except that the telescopes are automated, the detectors are giant electronic cameras with up to 100 million pixels, and the before and after comparison of gigabytes of data is carried out by computer algorithms.

In 1934 Zwicky and Baade proposed that supernova explosions come from the gravitational collapse of a star as it shrivels from 100-million-mile dimensions to "little spheres 14 miles thick." These dense clinkers were neutron stars—objects with the mass of the Sun but made of neutrons. Since neutrons had only been discovered in 1932, this was a remarkable extrapolation. In ordinary matter, electron clouds separate the massive nuclei from one another. Nuclei occupy only about 10^{-15} of the volume of ordinary matter, which is mostly the more or less empty space where the electrons orbit. In a neutron star, electrons and protons are compressed by gravity to form neutrons—a neutron star is a massive object made of 10^{57} particles whose density approaches the nuclear density of 10^{17} kg/m^3.

This brilliant guess has been confirmed by modern work: neutron stars are real objects, and some supernovae do derive their energy from the gravitational collapse to a neutron star. A star with 8 solar masses or more fuses hydrogen to helium during most of its lifetime, which is measured in millions of years. Subsequent stages of nuclear burning, in which the ashes of each burning stage become the fuel for the next, lead to the accumulation of carbon, oxygen, silicon, and finally iron in the core of a massive star. Because iron is the most tightly bound nucleus, no further energy can be extracted by fusion. A star with a hot iron core has huge energy losses from neutrino emission but no energy source, and collapse is inevitable.

The actual moment of collapse is precipitated by energetic gamma rays in the hot, dense interior, which begin to break apart the iron nuclei, leading to a catastrophic loss of pressure. The core of the star collapses from a region about the size of the Earth (10,000 km) to the dimensions of a neutron star (100 km) in less than a second, with the inward velocity approaching one-third of the speed of light. This headlong implosion is halted with a violent snap when the core of the star approaches nuclear density. At that point, repulsive nuclear forces stiffen the forming neutron star, halting its collapse. As the material falling in smashes into the forming neutron star, computer simulations show that a powerful shock wave travels upstream, out through the star. This shock, refreshed by a blast of neutrinos emitted from the hot material raining down on the neutron star, cooks new elements from the iron just outside the forming neutron star and blows the star apart to create the visible explosion seen as a supernova. It is a strange picture: most of the gravitational energy of the collapse is emitted as massless, chargeless, and nearly undetectable neutrinos. Only about 1/10,000 of the energy is converted into the light by which supernova explosions are detected.

Astronomical observations confirm this physical picture as the source of some supernovae—most conspicuously, the observations of supernova 1987A in the Large Magellanic Cloud. There the presupernova star was observed: it was a 20-solar-mass star at an advanced stage of evolution. The supernova was discovered from its optical emission, but subsequent inspection of the records from underground neutrino detectors showed that hours before the supernova began to brighten, there was a brief flash of neutrinos, signaling the formation of a neutron star. Nuclear gamma rays from freshly synthesized

FIGURE 1

Supernova 1994D. This Type Ia supernova is in a galaxy at a distance of about 50 million light years in the Virgo cluster of galaxies. For a month, the light from a single exploding white dwarf is as bright as 4 billion stars like the Sun. CREDIT: COURTESY OF P. CHALLIS, CENTER FOR ASTROPHYSICS/STScI/NASA. REPRODUCED BY PERMISSION radioactive isotopes produced in the shock wave, especially ^{56}Co, were seen in the months after the explosion. Although these observations confirm the basic picture, no neutron star has yet been found in the center of Supernova 1987A.

In the 1940s astronomers discovered there were two basic types of supernovae, as distinguished by their spectra. The original type had no hydrogen lines in the spectrum, but the new type, called Type II, did. Scientists have since discovered that Type II supernovae are core-collapse supernovae. Type Ia supernovae come from different types of stars and erupt by a completely different mechanism, but by coincidence, they emit a similar amount of light. Type Ib and Type Ic supernovae come from massive stars and are powered by gravitation but have no hydrogen in their atmospheres because they have exhaled it in a stellar wind before the explosion.

The modern picture of Type Ia supernovae is that they come from the thermonuclear explosion of white dwarf stars. Stars of less than about 8 solar masses, such as the Sun itself, fuse hydrogen to helium, and helium to carbon and oxygen, but they do not burn all the way up to iron. Nuclear fusion stops in these stars when the core of the star becomes degenerate—when the density becomes high enough so that the quantum mechanical properties of electrons themselves supply the pressure to support the star. For a star like the Sun, a carbon-oxygen white dwarf will be the endpoint of stellar burning about 5 billion years from now. The pressure in a degenerate star does not depend on its temperature, so a cooling white dwarf can be a stable object supported against gravity by its electrons. However, there is an upper limit, called the Chandrasekhar limit, to the mass of a star that can be supported by degeneracy pressure: about 1.4 solar masses. For stars in binary systems, where one star has become a white dwarf, the other star can transfer significant amounts of mass to the white dwarf. As the white dwarf accumulates matter and grows toward the Chandrasekhar limit, computations show that nuclear burning can begin again. In ordinary stars, the heat generated from fusion generates pressure that can make the star expand and cool slightly, decreasing the rate of energy production. This regulating effect ensures that ordinary stars will not explode.

Degenerate matter, on the other hand, is quite different—generating energy by fusing oxygen nuclei increases the rate of energy generation but does not make the star expand and cool. The result is a runaway thermonuclear explosion that rips through the entire white dwarf and destroys it as a Type Ia supernova. The burning wave turns much of a white dwarf into iron and blasts off the outer layers of the star at speeds above 10,000 km/s. Observations show that Type Ia supernovae have the chemistry of exploded white dwarfs: oxygen and carbon on the outside, and radioactive iron ashes inside. Radioactivity powers the light curve of Type Ia supernovae: they take about 20 days to reach their peak brightness, decline by about a factor of 2 in the first 2 weeks after maximum light, and then enter into a long exponential decline powered by the decay of freshly synthesized ^{56}Co.

Because Type Ia supernovae come from a very well-defined physical situation, it is not too surprising to find that they have a well-defined peak energy output. It turns out to be about 4×10^9 solar luminosities. In the 1990s Type Ia supernova became the most powerful tools for measuring cosmic distances. The key improvement was to use the shape of the light curve, which is correlated with intrinsic brightness, to determine which Type Ia supernovae are brighter than the mean and which are dimmer. In 2002 the precision of the distance estimate to a single Type Ia supernova, after taking into account the light curve shape, was determined to be 8 percent, which makes them the best standard candles for judging cosmic distances.

The Hubble Space Telescope has been used to measure the distance to galaxies in which Type Ia supernovae have exploded by observing the brightness of Cepheid variable stars. This establishes the relation between cosmic redshift and cosmic distance, based on Type Ia supernovae. In 2002 the values of the Hubble Constant found from Type Ia supernovae ranged from about 60 to 75 km/s/mpc, with most of the uncertainty associated not with the supernovae but with lower rungs on the cosmic distance ladder.

Because Type Ia supernova are so bright, they can be detected at very large distances about halfway back to the Big Bang. This provides a way to study the history of cosmic expansion. The expansion of the universe during the time the light from a supernova is en route affects its apparent brightness. In a universe that is decelerating due to gravity, the light from a distant supernova travels a slightly shorter path from the explosion to a telescope. It would appear a little brighter than in a universe that is expanding at a constant rate. Observations reported in 1998 show the opposite: distant supernovae are about 25 percent dimmer than they would be in an empty universe. This surprising result, which implies that the expansion of the universe is speeding up over time, points to the presence of a significant amount of dark energy whose pressure produces the observed acceleration. When combined with information obtained from observing the fluctuations in the cosmic microwave background, the supernova results suggest a universe that is one-third dark matter and two-thirds dark energy. The intensive study of supernovae near and far has revealed two-thirds of the universe!

See also: ASTROPHYSICS; BIG BANG; COSMOLOGY

Bibliography

Goldsmith, D. *The Runaway Universe* (Perseus Books, Cambridge, MA, 2000).

Hubble, E. P. "A Spiral Nebula as a Stellar System, Messier 31." *Astrophysical Journal* **69**, 103–158 (1929).

Kirshner, R. P. *The Extravagant Universe: Exploding Stars, Dark Energy, and the Accelerating Cosmos* (Princeton University Press, Princeton, NJ, 2002).

Marschall, L. *The Supernova Story* (Princeton University Press, Princeton, NJ, 1994).

Woosley, S., and Weaver, T. "The Great Supernova of 1987" in *Stars and Galaxies: Citizens of the Universe,* edited by D. E. Osterbrock (W. H. Freeman, New York, 1990).

Robert P. Kirshner

SUPERSYMMETRY

Supersymmetry is a space-time symmetry, an extension of the symmetries of translations, rotations, and boosts. Supersymmetry has played an important role in a broad range of modern developments in physics and mathematics. In particle physics, it is conjectured to be a fundamental symmetry of the elementary particles and provides the framework for many attempts to unify the electromagnetic, weak, strong, and gravitational interactions. In this context, supersymmetry predicts as of yet undiscovered partner particles for each of the known elementary particles. The search for these supersymmetric particles and other evidence for supersymmetry is currently the subject of intense research activity spanning a variety of disciplines in particle physics, astrophysics, and cosmology.

New Space-time Symmetry

Symmetries play an essential role in descriptions of the physical world. Among the most fundamental of these are space-time symmetries. These include translations, rotations, and boosts. Translations shift an object, such as a particle or system of particles, from one place and time to a different place and time. Similarly, rotations transform an object into the same object rotated in three-dimensional space, and boosts transform an object into the identical object with a new velocity. Rotations and boosts together form Lorentz symmetry. When supplemented by translations, the full set of symmetries is called Poincaré symmetry. All known physical laws are invariant under these symmetries. Under general assumptions, stated precisely in the Coleman-Mandula theorem, Poincaré symmetry is the maximal space-time symmetry that transforms particles into identical particles.

Supersymmetry is a new space-time symmetry. It extends Poincaré symmetry without violating the Coleman-Mandula theorem by transforming particles into particles that differ from the original by one-half unit of spin. Spin is an inherently quantum mechanical property of all elementary particles. It has no classical analogue but may be thought of as internal angular momentum. In four dimensions, all particles have integer or half-integer spin. Those with integer spin, such as the photon, are bosons. Those with half-integer spin, such as the electron, are fermions. Supersymmetry therefore transforms bosons into fermions and fermions into bosons. All other particle properties, such as mass and charge, are preserved under supersymmetry transformations.

Rudolph Haag, Jan Lopuszanski, and Martin Sohnius showed in 1975 that supersymmetry is the maximal possible extension of Poincaré symmetry. If supersymmetry is discovered, all mathematically consistent space-time symmetries will have been realized in nature.

Superpartners

Because supersymmetry transforms particles into distinct particles with different spin, it predicts the existence of as of yet undiscovered supersymmetric partners, or superpartners, for all known particles. (The possibility that some of the known particles are the superpartners of other known particles is excluded by comparing basic properties.)

The Standard Model of particle physics describes all known fundamental particles and their interactions. It includes matter fermions, such as the electron, neutrino, and quarks, and interaction bosons, such as the photon, which transmit forces. These and their superpartners are listed in Table 1. Superpartner names are derived from their Standard Model counterparts by appending the suffix "-ino" for supersymmetric fermions and adding the prefix "s-" for supersymmetric bosons. Symbolically, they are denoted by adding tildes ($\sim$) to the symbols for their Standard Model partners.

Exact supersymmetry is not realized in nature—for example, there is no boson with electric charge -1 that has a mass equal to that of the electron. Therefore, if it exists in nature, supersymmetry must be broken. In theories with softly broken supersymmetry, the equality of masses of supersymmetric pairs is broken, but the charges and other quantum numbers of superpartners remain identical. Such theories possess a number of important virtues and are the most widely studied supersymmetric theories. General indirect evidence suggests that superpartners, if they exist, should have masses not far beyond those of Standard Model particles. Various supersymmetric theories make specific predictions for the superpartner masses, but these predictions vary widely from theory to theory.

Unification of Forces

Although there is no direct evidence for supersymmetry, there are a number of indirect motivations. Among these, two are related to the unification of forces and are of special significance.

The first motivation stems from the observed weakness of gravity relative to the other forces. An understanding of this discrepancy is among the most important challenges for those seeking a unified description of the fundamental interactions. That gravity is weak may be understood in several ways. For example, the electromagnetic repulsion between two electrons is roughly 10^{42} times stronger than their gravitational attraction. Alternatively, one may determine the mass required for a hypothetical particle with unit charge to experience gravitational and electromagnetic interactions equally. This is the Planck mass, and it is approximately 10^{19} GeV. From

TABLE 1

		Spin	Mass (GeV)			Spin	Mass (GeV)
Standard Model particles	Matter fermions			Superpartners			
	e electron	1/2	0.0005		$\tilde{e}$ selectron	0	> 95
	ν neutrino	1/2	$< 10^{-7}$		$\tilde{\nu}$ sneutrino	0	> 41
	q quarks	1/2	$0.004 - 174$		$\tilde{q}$ squarks	0	> 200
	Interaction bosons						
	γ photon	1	0		$\tilde{\gamma}$ photino	1/2	> 37
	W	1	80		$\tilde{W}$ Wino	1/2	> 68
	Z	1	91		$\tilde{Z}$ Zino	1/2	> 37
	g gluon	1	0		$\tilde{g}$ gluino	1/2	> 200
	G graviton	2	0		$\tilde{G}$ gravitino	3/2	?
	h Higgs boson	0	> 114		$\tilde{h}$ Higgsino	1/2	> 37

CREDIT: Courtesy of Jonathan L. Feng.

Standard Model particles and their conjectured superpartners. Masses are given for the known particles in units of GeV, approximately the proton mass. For particles not yet discovered, approximate lower bounds on masses are listed.

this point of view, gravity is weak because the masses of even the heaviest elementary particles, such as the *W* and *Z* gauge bosons, are so far below the Planck mass. In this guise, the puzzle of the weakness of gravity is also known as the gauge hierarchy problem.

The gauge hierarchy problem is especially severe in quantum field theories like the Standard Model, where the classical masses of particles are modified by contributions from quantum effects. In the Standard Model, some of these quantum contributions are naturally of the order of the Planck mass, and the weakness of gravity then results from an inexplicable and nearly exact cancellation between enormous classical and quantum contributions to yield relatively tiny observed physical masses.

Supersymmetry solves the gauge hierarchy problem by introducing superpartners that generate additional quantum mass contributions. For exact supersymmetry, these contributions exactly cancel the quantum corrections of the Standard Model. Physical masses are then given solely by their classical values, and no fine-tuned cancellations are required. In softly broken supersymmetry, the quantum corrections do not cancel exactly but are of the order of the superpartner masses. In these theories, large cancellations are therefore also avoided, provided the superpartner masses are not substantially larger than typical masses in the Standard Model.

Supersymmetry is also motivated by the desire to unify the other three forces. Although the strengths of the electromagnetic, weak, and strong forces are roughly similar, especially when compared with gravity, they nevertheless differ, providing another impediment to attempts to unify forces. The observed coupling strengths of these three forces are modified at very short distances, where effects similar to the screening of electromagnetic charge are eliminated. However, in the Standard Model, these interaction strengths differ even at short distances. In supersymmetric models, though, superpartners modify these screening effects. When these are removed, the strengths of the electromagnetic, weak, and strong forces agree with remarkable precision at very short distances. This quantitative result is indirect evidence for supersymmetric grand unified theories (GUTs), in which the electromagnetic, weak, and strong forces are described by one underlying interaction. The sim-

ple elegance of these theories provides further impetus for the study of supersymmetry.

Current Searches and Future Prospects

Although significant, indirect evidence is no substitute for the discovery of superpartners or other supersymmetric effects. The search for supersymmetry is currently an area of intense activity and may be divided into three broad categories.

The first category includes searches for superpartners in high-energy collider experiments. Such searches have been conducted at all major colliders, with the most sensitive searches to date conducted at the Large Electron Positron (LEP) collider at the European Laboratory for Particle Physics (CERN) in Geneva, Switzerland, and the Tevatron proton-antiproton collider at Fermilab in Batavia, Illinois. No evidence for supersymmetry has been found, and these searches have yielded only lower limits on superpartner masses. While these lower bounds depend on the particular supersymmetric theory being considered, characteristic limits are listed in Table 1. Limits for specific theoretical models are updated annually by the Particle Data Group in the *Review of Particle Physics*.

In the near future, the Large Hadron Collider (LHC) at CERN will collide protons with energies far above those accessible at present. The LHC will copiously produce superpartners that interact through the strong interaction, namely, squarks and gluinos, unless they are very heavy. The discovery reach of the LHC is well above 1,000 GeV for squark and gluino masses. For the other superpartners, the reach is somewhat less. Nevertheless, given that a supersymmetric explanation for the weakness of gravity requires superpartner masses not far above those of the Standard Model, the LHC is expected to provide a stringent test of many of the most attractive supersymmetric theories.

Searches for supersymmetry are also underway in a variety of low-energy particle physics experiments. Although such experiments cannot produce superpartners, they may be sensitive to the fleeting effects of short-lived superpartners that may exist as a result of Heisenberg's uncertainty principle. The most promising of these experiments include those that are extremely sensitive to small deviations, such as those measuring the magnetic dipole moment of the muon, and those searching for phenomena absent in the Standard Model, such as searches for electron-muon transitions, electric dipole moments, and rare decays.

Finally, in many supersymmetric theories, the lightest superpartner is stable and interacts weakly with ordinary matter. Such particles are natural candidates for dark matter, the mass responsible for the observed binding together of galaxies and galaxy clusters, which has not yet been identified. Searches for dark matter are also then searches for superpartners, and many current and future dark matter detection experiments are sensitive to supersymmetric dark matter.

If discovered, supersymmetry will drastically modify the understanding of the microscopic world. Measurements of superpartner masses and properties from the LHC and other experiments will favor some supersymmetric theories while excluding others, and provide new insights into attempts to unify the fundamental interactions in grand unified theories or superstring theory.

See also: Standard Model; String Theory

Bibliography

Coleman, S. R., and Mandula, J. "All Possible Symmetries of the *S* Matrix." *Physical Review* **159**, 1251–1256 (1967).

Groom, D. E., et al. "Review of Particle Physics." *The European Physical Journal C* **15**, 817–846 (2000). See also <http://pdg.lbl.gov>.

Haag, R; Lopuszanski, J. T.; and Sohnius, M. "All Possible Generators of Supersymmetries of the *S* Matrix." *Nuclear Physics B* **88**, 257 (1975).

Polonsky, N. "Supersymmetry: Structure and Phenomena. Extensions of the Standard Model." *Lecture Notes in Physics* **M68**, 1–169 (2001).

Weinberg, S. *The Quantum Theory Of Fields*, Vol. 3: *Supersymmetry* (Cambridge University Press, Cambridge, UK, 2000).

Wess, J., and Bagger, J. *Supersymmetry and Supergravity* (Princeton University Press, Princeton, NJ, 1990).

Jonathan L. Feng

SYMMETRY PRINCIPLES

The natural world is a complicated place. Symmetries allow people—and scientists—to discern order in nature. In physics, it has long been understood that symmetries are closely connected to conservation laws. Three of the most familiar conservation laws are the conservation of energy, the conservation of momentum, and the conservation of angular momentum. Conservation of energy is a consequence of the fact that the laws of nature do not change with time. For example, in Newton's law of gravitation,

$$F = G_N \frac{m_1 m_2}{r^2}$$

one could imagine that G_N, the gravitational constant, depended on time. In this case, energy would not be conserved. From experimental searches for violations of energy conservation, one can set strong limits on any such time variation (astronomical observations provide stronger constraints). This principle is quite broad and applies in quantum mechanics as well as classical mechanics. Physicists sometimes call this symmetry—that there is no special time—the "homogeneity of time." Similarly, conservation of momentum is a consequence of the fact that there is no special place. If one describes the world with Cartesian coordinates, the laws of nature don't care what one takes to be the origin. This symmetry is called "translation invariance," or the homogeneity of space. Finally, conservation of angular momentum is related to a familiar symmetry of daily life: the laws of nature are invariant under rotations. For example, not only does it not matter how we choose the origin or our coordinate system, but it doesn't matter how we choose to orient the axes.

The symmetries of time and space translation, and rotations, are called continuous symmetries because one can translate the coordinate axes by any arbitrary amount, and one can rotate through any angle. Another class of symmetries are called discrete symmetries. An example is the symmetry of reflection in a mirror, or "parity." Newton's laws possess this symmetry. Watch the motion of an object falling in a gravitational field, and then examine the same motion in a mirror. While the motion is different, each appears to obey Newton's laws. This is familiar to anyone who has ever stood in front of a clean, well-polished mirror and gotten confused as to what was the object and what the mirror image. Another way to describe this symmetry is as a symmetry between left and right. For example, three-dimensional Cartesian coordinates are usually written according to the "right hand rule," as in Figure 1. The positive direction along the z-axis lies in the direction in which your thumb points if you rotate your right hand about the z-axis, starting at the x-axis and moving to the y-axis. The unconventional coordinate system in Figure 2 is the opposite; here the z-axis points in the direction your left hand would point. The statement that Newton's laws are invariant is the statement that we can use either kind of coordinate system, and the laws of nature look the same. The symmetry of parity is usually denoted by the letter P.

Parity is not the only discrete symmetry of interest in science. Another is called time reversal. In Newtonian mechanics, one can imagine taking a video of an object falling under the influence of gravity. Now consider running the video backward. Both the motion "forward in time" and the motion "backward" will obey Newton's laws (the backward motion may describe a situation which is not very plausible, but

FIGURE 2

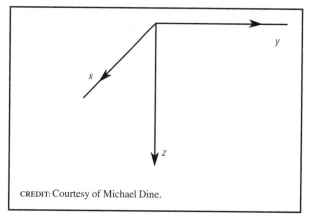

CREDIT: Courtesy of Michael Dine.

An unconventional coordinate system.

it will not violate the laws). Time reversal is usually denoted by the letter T.

A third discrete symmetry is called charge conjugation. For every known particle, (the electron, proton, etc.) there is an antiparticle. The antiparticle has exactly the same mass, but the opposite electric charge. The antiparticle of the electron is called the positron. The antiparticle of the proton is called the antiproton. Recently, antihydrogen has been produced and studied. Charge conjugation is a symmetry between particles and their antiparticles. Clearly particles and antiparticles are not the same. But the symmetry means that, for example, the behavior of an electron in an electric field is identical to that of a positron (antielectron) in the opposite field. Charge conjugation is denoted by the letter C.

These symmetries, however, are not exact symmetries of the laws of nature. In 1956, experiments showed, surprisingly, that in the type of radioactivity called beta decay, there is an asymmetry between left and right. The asymmetry was first studied in decays of atomic nuclei, but it is most easily described in the decay of the negatively charged π^- meson, another strongly interacting particle. The π^- meson decays either to a muon and its antineutrino or an electron and its antineutrino. But the decays to the electron are very rare. This is related (by an argument which uses special relativity) to the fact that the antineutrino always emerges with its spin parallel to its direction of motion. If nature were symmetric between left and right, one would find the neutrino half the

FIGURE 1

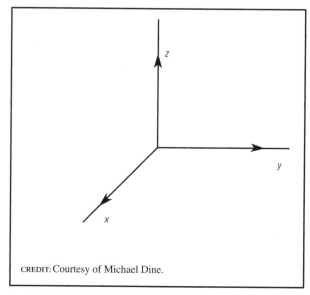

CREDIT: Courtesy of Michael Dine.

Three-dimensional Cartesian coordinates written according to the "right hand rule."

time with its spin parallel and half with its spin antiparallel. This is because, in a mirror, the direction of motion does not change, but the spin or angular momentum flips. Related to this is the positively charged π^+ meson, the antiparticle of the π^-. This particle decays to an electron neutrino, with its spin *parallel* to its momentum. This difference between the behavior of the neutrino and its antiparticle is an example of the violation of charge conjugation invariance.

After these discoveries, the question was raised whether time reversal invariance, T, was violated. By general principles of quantum mechanics and relativity, violation of T is related to violation of $C \times P$, the product of charge conjugation and parity. CP, if it is a good symmetry, would state that the decay of the $\pi^+ \rightarrow e^+ + v$ should proceed at the same rate as $\pi^- \rightarrow e^- + \bar{v}$. In 1964, an example of a process which violates CP was discovered, involving another set of strongly interacting particles, called the K mesons. It turns out that these particles have very special properties which allow the very tiny violation of CP to be measured. Only in the year 2001 was CP violation persuasively measured in the decays of another set of particles, the B mesons.

These results clearly show that absence of symmetry is often as interesting as its presence. Indeed, shortly after the discovery of CP violation, Andrei Sakharov pointed out that violation of CP in the laws of nature is a necessary ingredient to understanding the predominance of matter over antimatter in the universe.

It is still believed that the combination CPT, charge conjugation times parity times time reversal, is conserved. This follows from rather general principles of relativity and quantum mechanics and is, to date, supported by experimental studies. If any violation of this symmetry should be discovered, it would have profound implications.

So far, the symmetries discussed are significant in that they lead to conservation laws, or relations between rates of reaction between particles. There is another class of symmetries which actually determine many of the forces between particles. These symmetries are known as local symmetries or gauge symmetries.

One such symmetry leads to the electromagnetic interactions. Another, in Einstein's theory, leads to gravitation (in the form of Einstein's theory of general relativity). Einstein asserted, in enunciating his principle of general relativity, that it should be possible to write the laws of nature not only so they are invariant, for example, under a rotation of coordinates all at once everywhere in space, but under any change of coordinates. The mathematics for describing this had been developed by Friedrich Riemann and others in the nineteenth century; Einstein in part adapted, and in part reinvented, it for his needs. It turns out that to write equations (laws) which obey the principle, it is necessary to introduce a field, similar in many ways to the electromagnetic field (except that it has spin two). This field couples in just the right way to give Newton's law of gravitation, for things which are not too massive or moving too fast or not too dense. For systems which are very fast (compared to the speed of light) or very dense, general relativity leads to a rich array of exotic phenomena such as black holes and gravitational waves. All of this follows from Einstein's rather innocuous sounding symmetry principle.

The symmetry which leads to electricity and magnetism is another example of a local symmetry. To introduce it requires a bit of mathematics. In quantum mechanics, the properties of the electron are described by a "wave function," $\psi(x)$. It is crucial to the way quantum mechanics works that ψ is a complex number, in other words, it is, in general, the sum of a real number and an imaginary number. A complex number can always be written as the product of a real number, ρ, and a phase, $e^{i\theta}$. For example,

$$\psi(x) = \rho(x) e^{i\theta(x)}.$$

In quantum mechanics, one can multiply the wave function by a constant phase, with no effect. But if one insists on something stronger, that the equations don't depend on the phase (more precisely, if there are many particles with different charges, as there are in nature, a particular combination of phases is not important), one must, as in general relativity, introduce another set of fields. These fields are the electromagnetic fields. Enforcing this symmetry principle (plus the symmetries of special relativity) requires that the electromagnetic field obey Maxwell's equations. Today, all of the interactions of the Standard Model are understood as arising from such local gauge symmetry principles. The existence of the W and Z bosons, as well as their masses, half-lives, and other detailed properties, were successfully predicted as consequences of these principles.

What might lie beyond? A number of other possible symmetry principles have been proposed, for a variety of reasons. One such hypothetical symmetry is known as supersymmetry. This symmetry has been suggested for two reasons. First, it might explain a longstanding puzzle: why are there very small dimensionless numbers in the laws of nature? For example, when Planck introduce his constant h, he realized that one could use this to write a quantity with the dimensions of mass starting with Newton's constant. This quantity is now known as the Planck mass. It is given by

$$M_p = \sqrt{\frac{c^3 h}{2\pi G_N}} \approx 10^{19} m_p$$

where c is the speed of light, G_N is Newton's constant, and m_p is the proton mass. Related to the fact that gravity is a very weak force, M_p is a very large number. In fact, compared to the mass of the W and Z,

$$M_p / M_Z = 10^{17}$$

The great quantum physicist Paul Dirac (who predicted the existence of antimatter) called this the "problem of the large numbers." It turns out that postulating that nature is supersymmetric can help with this problem. Supersymmetry also seems to be an integral part of understanding how the principles of general relativity can be reconciled with the principles of quantum mechanics.

What is supersymmetry? Supersymmetry, if it exists, relates fermions (particles with half integral spin, which obey the Pauli exclusion principle) to bosons (particles with integer spin, which obey what are known as Bose statistics—the statistics which gives rise to the behaviors of lasers and Bose condensates).

At first sight, however, it seems silly to propose such a symmetry, since if it were manifest in nature, one would expect that for every fermion there would be a boson of exactly the same mass, and vice versa.

In other words, in addition to the familiar electron, there should be a particle called the selectron, which has no spin and does not obey the exclusion principle, but is otherwise the same in every way as the electron. Similarly, related to the photon there should be another particle with spin $1/2$ (which obeys the exclusion principle, like the electron) with zero mass and properties in many ways similar to those of photons. No such particles have been seen. It turns out, however, that these facts can be reconciled, and this brings us to one final point about symmetries. Symmetries can be symmetries of the laws of nature but need not be manifest in the world around us. Space around us is not homogeneous. It is filled with all kinds of different stuff, sitting at particular (special) places. Yet from conservation of momentum, we know that the laws of nature are symmetric under translations. In these circumstances, the symmetries are "spontaneously broken." In particle physics, the term is used more narrowly: a symmetry is said to be spontaneously broken if the state of lowest energy is not symmetric. This phenomenon occurs in many instances in nature: in permanent magnets, where the alignment of the spins which gives rise to magnetism in the lowest energy state breaks rotational invariance; in the interactions of the π mesons, which violate a symmetry called chiral symmetry. Whether supersymmetry exists in such a broken state is now a subject of intense experimental investigation.

See also: CP SYMMETRY VIOLATION; SUPERSYMMETRY

Bibliography

Feynman, R. P. *Six Not-So-Easy Pieces: Lectures on Symmetry, Relativity and Space-Time* (Addison-Wesley, Reading, MA, 1997).

Icke, V. *The Force of Symmetry* (Cambridge University Press, Cambridge, UK, 1995).

Michael Dine

T

TAU

See LEPTON

TECHNICOLOR

The discovery of the *W* and *Z* bosons in 1983 at the European Laboratory for Particle Physics (CERN) in Geneva, Switzerland, provided compelling evidence that the electroweak theory proposed by Sheldon Glashow, Abdus Salam, and Stephen Weinberg was an excellent description of both electromagnetic and weak interactions. By that time it had become a central part of what was known as the Standard Model of particle physics, and Glashow, Salam, and Weinberg (GSW) had already been awarded the 1979 Nobel Prize in Physics for their electroweak theory. In spite of this success, particle theorists were eager to determine whether a more fundamental theory could be behind the GSW model.

Electroweak Symmetry Breaking

A key ingredient of the GSW electroweak theory is the mechanism that forces the vacuum to distinguish the *W* and *Z* from the photon (breaking the symmetry of the electroweak gauge interactions), giving masses to the former while keeping the latter exactly massless. This symmetry breaking goes by the name Higgs mechanism. In addition to fields corresponding to the spin-1 gauge bosons (force carriers), it is necessary to add another boson field with zero spin, called the Higgs field, which has the peculiar property of a nonzero probability to measure the Higgs field in a pure vacuum. It may be helpful to recall the situation that arises in superconductors since this provides a relatively simple analogy. In a superconductor, electrons can attract each other (very weakly) by exchanging phonons (quanta of lattice vibrations) forming Cooper pairs. If the charged Cooper pairs undergo Bose condensation, the lowest energy state of the system has an arbitrarily large charge (limited only by the size of the superconductor). Photons moving through this charged medium are effectively massive, as can be seen by the fact that magnetic fields cannot penetrate a superconductor. In the analogy with the GSW electroweak theory, the analog of the Cooper pair is the Higgs field, the analog of the lowest-energy state is the vacuum of space-time, and the analogs of the massive photon are the massive *W* and *Z*.

In the GSW electroweak theory, the Higgs field has an electroweak charge, and the theory is arranged so that in empty space there is a nonzero

467

probability to measure the Higgs field (this is the analog of Bose condensation). Allowing the Higgs field to couple to fermions (particles with spin $\frac{1}{2}$) also gives masses to the quarks and leptons. This relatively simple set of ideas leads to a prediction of the ratio of W and Z masses that was confirmed by experiment. During the 1990s the GSW electroweak theory was subjected to precision tests by experiments at the Stanford Linear Accelerator and at CERN. In order to accurately compare the theory to experiment, quantum corrections to the predictions had to be included. In fact, many particle theorists believed that the GSW electroweak theory was not self-consistent until an understanding of how to perform such quantum corrections had been developed in the early 1970s by Gerardus 't Hooft and Martinus Veltmann. GSW electroweak theory was confirmed at better than the 1 percent level, and 't Hooft and Veltmann won the 1999 Nobel Prize in Physics.

Fine-tuning

In spite of the great successes of the GSW electroweak theory, particle theorists began to suspect as early as the 1970s that it could not be the whole story. The reason was that although 't Hooft and Veltmann had shown how to calculate the quantum corrections to the theory, some of these corrections were extremely sensitive to the highest possible energy scales, and as a consequence, the parameters of the theory had to be fine-tuned to many decimal places. To see how this arises, one can look at the quantum corrections to the propagation of the Higgs boson. For GSW electroweak theory to be consistent, the Higgs boson mass must be below $1{,}000$ GeV/c^2. Consider, for example, how the coupling of the Higgs to the top quark leads to an extremely large correction to the Higgs mass. This coupling allows a single Higgs boson moving through space to turn into (for a short amount of time governed by Heisenberg's uncertainty principle) a top quark and an antitop quark; the top and antitop quickly recombine to again form a single Higgs boson. (A diagram of the trajectory of the quarks in space-time looks like a closed loop, and this type of correction is called a quantum loop correction.) Quantum mechanics requires that if the intermediate state of the system (the top-antitop pair) is not measured, all possibilities must be summed over in order to calculate

that probability amplitude. The sum of the energy and momentum of the top-antitop pair must, of course, add up to the original energy and momentum of the single Higgs, but the differences in energy and momentum of the top and the antitop are unconstrained.

Since energy and momentum are continuous quantities, one has to integrate over these variables. The amplitudes for top-antitop propagation are such that the integrals diverge; that is, if it is assumed there is some maximum amount of energy that the intermediate top and antitop quarks can have, then the integral is proportional to the square of this energy cut-off. This integral gives a direct contribution to the square of the mass of the Higgs boson. If one imagines that the energy cut-off is associated somehow with the scale of gravity (i.e., the Planck scale that is 10^{19} GeV/c^2), then the correction to the square of the Higgs mass is about 10^{32} times larger than the answer needed. One can obtain a reasonable answer for the Higgs mass, but it requires fine-tuning the parameters of the theory to thirty-two decimal places in order to arrange for a tremendous cancellation that allows the answer to be many orders of magnitude smaller than the individual contributions. There are many such quantum loop corrections to the Higgs boson mass, and considering multiloop corrections makes the situation even worse. It is known that in supersymmetric extensions of the GSW theory there is a cancellation of divergent corrections between particles and their superpartners, but until the origin of electroweak symmetry breaking is uncovered experimentally, other explanations must be considered.

New Interactions

In the 1970s Stephen Weinberg and Leonard Susskind independently proposed that composite particles formed by a new strong interaction could replace the Higgs boson in GSW theory. The new interactions were supposed to be similar to those of quantum chromodynamics (QCD), and these theories were hence dubbed technicolor theories. Susskind showed that if the Higgs boson was absent from the Standard Model, QCD would provide electroweak symmetry breaking through quark composites (although it would give masses for the W and Z that are about a factor of $2{,}600$ too small). Technicolor theories thus harkened back to superconductivity where a gauge symmetry is broken by a composite of two fermions, a crucial difference being that the interactions responsible for superconductivity are quite weak, whereas the technicolor interactions must remain strong. Technicolor theories essentially resolve the fine-tuning problem by lowering the effective cut-off scale to 1,000 GeV/c^2. Remarkably, technicolor theories predicted the correct ratio for the W and Z masses; however, producing masses for the quarks and leptons requires several complicated extensions of the model. A further problem with technicolor was revealed by the comparison with precision experiments. Following the idea of scaling up QCD to obtain the correct W and Z masses, it was possible to scale up QCD data (essentially using QCD as an "analog computer") to predict the deviations of a technicolor theory from the GSW theory. These deviations were not seen at SLAC or at CERN. It remains logically possible that there is another version of technicolor that does not behave like QCD, but in the absence of an explicit, workable model interest in technicolor waned during the 1990s.

See also: ELECTROWEAK SYMMETRY BREAKING; HIGGS PHENOMENON; STANDARD MODEL

Bibliography

CERN. "Hands on CERN." <http://hands-on-cern.physto.se/hoc_v1en/index.html>.

Dixon, L. "From Superconductors to Supercolliders." <http://www.slac.stanford.edu/pubs/beamline/26/1/26-1-dixon.pdf>.

Nobel e-Museum. "Nobel Laureates in Physics." <http://particleadventure.org/particleadventure/index_old.html>.

Particle Data Group. "The Particle Adventure." <http://particleadventure.org/particleadventure/index.html>.

SLAC. "SLAC Virtual Visitor Center." <http://www2.slac.stanford.edu/vvc/home.html>.

John Terning

THOMAS JEFFERSON NATIONAL ACCELERATOR FACILITY

Scientists from across the country and around the world visit the U.S. Department of Energy's (DOE)

Thomas Jefferson National Accelerator Facility—Jefferson Lab, in Newport News, Virginia—to further their knowledge of the structure of atomic nuclei. They investigate the boundary between nuclear and particle physics, seeking to determine how the neutrons and protons (more generally, the hadrons) are "constructed" from the more fundamental quarks and gluons of quantum chromodynamics (QCD), and how the forces between the hadrons arise from QCD. They also seek to identify and expand the limits of the understanding of the behavior of nuclei by high-precision studies of their properties.

The Continuous Electron Beam Accelerator Facility (CEBAF) at Jefferson Lab shows the transition region between two views of the nucleus. In the traditional view, the nucleus appears as a cluster of nucleons—protons and neutrons. The more detailed view, which began to emerge around 1970, reveals nucleons as composite objects, made from quarks and gluons. Ultimately, the process of bridging these two views will yield a complete understanding of nuclear matter—99.5 percent of the observable universe. We will learn both how matter itself is constructed and how it obtains its characteristic properties.

The experimenters probe nuclei using continuous-wave (CW) beams of electrons from CEBAF, a 6-GeV (giga electron volt) research instrument that stretches through a racetrack-shaped tunnel nearly a mile long. It delivers beams for simultaneous experiments in three cavernous experimental halls, where advanced particle-detection equipment observes the probing and where ultra-high-speed data-acquisition equipment gathers the resulting data.

The CEBAF accelerator has its roots in the decades-old tradition of electromagnetic nuclear physics, which exploits two fundamental advantages of electrons: pointlike structure and a well understood interaction with other particles. Since the 1960s, it has been recognized that CW, high-energy beams of electrons would constitute a unique, powerful new tool. The 100 percent duty factor of CEBAF's CW electron beams allows the extension of electromagnetic interaction studies to a broad range of reactions in which the probing electron is observed in coincidence with the particles emitted as a consequence of its interaction with the nuclear target. CEBAF's 6-GeV energy is optimized for probing spatial

Thomas Jefferson National Accelerator Facility. CREDIT: COURTESY OF UNITED STATES DEPARTMENT OF ENERGY'S JEFFERSON LAB. REPRODUCED BY PERMISSION.

scales ranging from the size of a large nucleus down to a fraction of the size of a nucleon.

The construction of CEBAF for the U.S. Department of Energy began in 1987 under the inspired leadership of Hermann Grunder, with the Southeastern Universities Research Association (SURA) serving as the prime contractor. The decision to have SURA build the laboratory in the southeast was motivated, in part, by a strong desire to strengthen science in that region. In 1996, with experiments getting under way, the new laboratory was named after Thomas Jefferson (1743–1826), the United States president and statesman of science who fostered American optimism about science, technology, and the future.

The CEBAF accelerator consists of two antiparallel linear accelerators (linacs) interconnected by 180° beam-recirculation arcs, which give the accelerator its racetrack shape. A 1,500-MHz train of polarized electron bunches is preaccelerated before injection into the first linac "straightaway." The beam then makes as many as five acceleration passes around the racetrack, with each pass raising the energy by about 1 GeV. After any pass, every third bunch—constituting a 500-MHz CW beam—can be directed to a particular experimental hall. In each linac, superconducting radio-frequency (SRF) accelerating cavities transfer rf energy to the beam. Liquid helium at 2 K, supplied by the world's largest refrigeration plant for that tem-

perature, cools the SRF components for superconducting operation. Electron bunches at first-pass through fifth-pass energy travel together through the linacs, but the bunches at each of the energies in the linac are separated into individual recirculation arc beam lines for transport through a 180° arc and then recombined in preparation for further acceleration. CEBAF overall incorporates more than 2,200 beam-transport magnets.

The electron beams from CEBAF are sent to three experimental halls. The first hall is equipped with two high-resolution spectrometers for precision electron-scattering measurements; the second has a large-acceptance, lower-resolution spectrometer for studying reactions with many particles in the final state; and the third is a multipurpose hall used mainly for one-of-a-kind experiments. The laboratory has developed intense, laser-driven sources of polarized electrons that produce high-current, highly polarized beams of electrons. The unique, new combination of high-energy, continuous, highly polarized electron beams gives scientists, for the first time, sufficient precision to test key theoretical predictions about nucleon structure.

Research at Jefferson Lab began in earnest in 1997, and although the research program is in its early stages, the laboratory has already made significant contributions to the understanding of nuclear and nucleon structure. Precise measurements have been made on the charge and magnetization distributions of nucleons. Since the proton and neutron are each made up of quarks bound together by the exchange of gluons, and since each quark carries its charge in an extraordinarily small volume, mapping the charge and magnetization of the nucleon tells how the quarks are organized within it. The data for the proton indicate that its magnetization density peaks in the center and falls off rapidly near the edges. The charge density, however, is significantly lower than the magnetization density at the center and drops off somewhat more slowly with increasing distance from the center. Similar data for the neutron show that although its total charge is zero, the distribution of charge is not; it is positive near the center and becomes negative near its outer edge. While this general feature of the neutron has been known for some time, the new data from CEBAF provide a precise map of exactly how the charge is distributed. These data provide a kind of "X-ray" picture of the internal structure of the nucleons that helps to guide theorists who are trying to explain how they are constructed from quarks and gluons.

Considerable progress has also been made in understanding the transition between the classical nuclear physics description of nuclei (in which they are described in terms of nucleons acting as fundamental particles held together by the exchange of mesons) and the modern view, in which the underlying quark structure is accounted for explicitly. Data indicate that the classical description works well down to a scale of about one-half the size of the proton. Quarks are tightly bound inside the nucleon, and only their aggregate properties are observed until one probes more deeply. The quark substructure becomes evident on distance scales smaller than approximately one-tenth the size of the proton. Understanding the transition region where the individual quark properties start to appear is an important goal of nuclear physics.

As a secondary mission, Jefferson Lab has used its SRF electron-accelerating technology to take the lead in developing a powerful, versatile new kind of laser for science, applied research, and industry: the free-electron laser (FEL). All lasers convert electron energy into laser light, although usually with only one fixed choice of wavelength. However, a beam of electrons from a CEBAF-style SRF accelerator can be manipulated to cost-effectively generate immensely powerful laser light. Moreover, the wavelength can be precisely selected over a broad range of wavelengths—a crucial feature for making light perform useful work. The first Jefferson Lab FEL produced infrared (IR) wavelengths at over 2 kilowatts—more than two orders of magnitude higher average power than any predecessor. As of 2002, following FEL's use by over twenty-five research groups in biology, physics, chemistry, and materials science, an upgrade was under way to 10 kilowatts in the IR and 1 kilowatt in the ultraviolet (UV).

The recirculating SRF linac that drives the Jefferson Lab FEL has served as the first proof at significant power of the energy-recovery principle, in which the electron beam is returned through the accelerating structures for deceleration, enabling the recycling of its unspent energy. As of 2002, groups in the United States and Europe were envisioning, proposing, and developing energy-recovery linacs for a variety of accelerators and light sources.

See also: BENEFITS OF PARTICLE PHYSICS TO SOCIETY; FUNDING OF PARTICLE PHYSICS

Bibliography

Leemann, C.; Douglas, D.; and Krafft, G. "The Continuous Electron Beam Accelerator Facility: CEBAF at the Jefferson Laboratory." *Annual Reviews of Nuclear and Particle Science* **51**, 413–450 (2001).

Thomas Jefferson National Accelerator Facility. <http://www.jlab.org/>.

Lawrence S. Cardman

THOMSON, JOSEPH JOHN

Joseph John Thomson is remembered for his recognition of the electron in 1897, which was the basis of elementary particle physics. But this was just one aspect of his prolific study of the relationship between the electromagnetic ether and matter: he was the first to suggest electromagnetic mass; his theory of gaseous discharge by ionization is still broadly accepted; his atomic theories succored both the nuclear atom and ideas of ionic bonding; his work on the structure of light expedited acceptance of the quantum theory of radiation in Britain. Finally Thomson was a leading spokesperson for science and a renowned teacher; his students held influential posts throughout the English-speaking world, and eight of them won Nobel Prizes.

Early Life

Joseph John Thomas was born December 18, 1856, at Cheetham Hill near Manchester, England. Thomson's father, Joseph James Thomson, was a Manchester bookseller; his mother, Emma Swindells, came from a textile manufacturing family. His younger brother, Frederick Vernon Thomson, joined a firm of calico merchants. As a child Thomson developed a lifelong interest in botany. His parents encouraged his scientific interests, and entered him at Owens College, Manchester, at the age of

British physicist Joseph John Thomson (1856–1940) received the Nobel Prize in Physics in 1906 for his theory of gaseous discharge by ionization. CREDIT: COURTESY OF BETTMANN/CORBIS. REPRODUCED BY CORBIS CORPORATION.

fourteen, to begin engineering training. When his father died two years later, Thomson made his way by scholarships in mathematics and physics (taught by Thomas Barker and Balfour Stewart, respectively) at which he excelled.

In 1876 Thomson obtained a scholarship at Trinity College, Cambridge, to study mathematics. Cambridge mathematics at that time was dominated by an emphasis on physical analogies and a mechanical worldview, elucidated by analytical dynamics (the use of Lagrange's equations and Hamilton's principle of least action). Thomson's coach, Edward Routh, gave him a thorough grounding in these methods, and in 1880 he graduated as Second Wrangler (second place).

Thomson remained in Cambridge, working for a College Fellowship. He used analytical dynamics to explore Maxwell's electrodynamics, which he had encountered at Owens College, and then learnt from William Niven at Cambridge. In 1881 he showed that the mass of a charged particle increases as it moves, suggesting that the particle drags some ether with it. In 1882 he won Cambridge's Adams Prize for "A Treatise on Vortex Motion," which investigated the stability of interlocked vortex rings, and developed the then-popular idea that atoms were ethereal vortices into a theory that could account for the periodic table. This work laid the foundations of all his subsequent atomic models.

Thus Thomson was working in the mainstream of Cambridge mathematical physics. In college also he identified himself with Cambridge values and social mores. In 1884, his conventionality and scientific accomplishments established, Thomson was elected Cavendish Professor of Experimental Physics at Cambridge at the age of twenty-eight.

Cavendish Professorship

Thomson became, overnight, a leader of British science. He held an increasing number of positions in scientific administration, was on the Board for Invention and Research during World War 1, President of the Royal Society from 1915 to 1920, and from 1919 to 1927 was an active member of the Advisory Council to the Department for Scientific and Industrial Research. His social position was strengthened by his marriage, in 1890, to Rose Paget, daughter of Cambridge's Professor of Physic (that is, Medicine). He sent his children (George and Joan) to private schools and joined the Athenaeum and Saville clubs. He received a knighthood in 1908, the Order of Merit in 1912, and in 1918 was appointed Master of Trinity College, a Crown appointment.

Under Thomson's leadership the Cavendish Laboratory became a place of lively debate, at the forefront of physics, with a colloquium and a dynamic social life. But it was also a place of financial stringency where space and equipment were bitterly fought over.

Gaseous Discharge

As Cavendish Professor, Thomson had free choice of scientific direction and a duty to undertake experimental physics, which coincided with a realization of the limitations of analytical dynamics. He chose the academically unpopular subject of discharge of electricity through gases that appealed to him for its visual effects. By 1890 he had developed the concept of a discrete electric charge, modeled by the terminus of a vortex tube in the ether, which guided his later work.

The discovery of X rays in 1895 proved crucial for they ionized a gas in a controllable manner, allowing the effects of ionization and secondary radiation to be distinguished. Within a year, working with his student Ernest Rutherford, Thomson had convincing evidence for his theory of discharge by ionization of gas molecules.

X rays also rekindled interest in the cathode rays that caused them. With new confidence in his apparatus and theories, Thomson, in 1897, suggested that the properties of cathode rays could be explained by assuming that they were subatomic charged particles, which were a universal constituent of matter. He called these "corpuscles," but they soon became known as "electrons." Thomson unified his ionization and corpuscle ideas into a widely applicable theory of gaseous discharge for which he won the Nobel Prize in Physics in 1906.

Thomson next investigated the role of corpuscles in matter. His "plum pudding" atomic model, in which thousands of corpuscles orbited in a sphere of positive electrification, was highly sophisticated, giving a qualitative explanation of the periodic table and ionic bonding. In 1906 Thomson pioneered the use of scattering calculations in atomic theory, using scattering of X rays and beta rays to show that the number of corpuscles in the atom was comparable with the atomic weight. His methods proved invaluable, leading Rutherford to the nuclear atom, and promoting analysis of the structure of light. But his first result, that there were only hundreds rather than thousands of corpuscles in the atom, was fatal for his own model which lost both its mass and its stability. Thomson began experiments with positive ions to investigate the mass of the atom. This work led to recognition of the H_3^+ ion and the discovery of the first non-radioactive isotopes, those of neon, in 1913, a discovery which prompted the invention of the

mass spectrograph by Thomson's collaborator, Francis Aston, in 1919.

Later Life

In 1919 Thomson resigned the Cavendish Professorship. As Master of Trinity College, he now had a major social and administrative role. But he continued to experiment until a few years before his death laying, among other things, the foundations of plasma physics. He died on August 30, 1940, and was buried in Westminster Abbey.

See also: ELECTRON, DISCOVERY OF

Bibliography

Buchwald, J., and Warwick, A., eds. *Histories of the Electron* (MIT Press, Cambridge, MA, 2001).

Davis, E. A., and Falconer, I. J. *J.J. Thomson and the Discovery of the Electron* (Taylor and Francis, London, 1997).

Falconer, I. J. "J.J. Thomson's Work on Positive Rays, 1906–1914." *Historical Studies in the Physical Sciences* **18**, 265–310 (1988).

Falconer, I. J. "J.J. Thomson and 'Cavendish' Physics" in *The Development of the Laboratory,* edited by F. James (Macmillan, London, 1989).

Heilbron, J. L. "The Scattering of Alpha and Beta Particles and Rutherford's Atom." *Archive for History of Exact Science* **4**, 247–307 (1968).

Rayleigh, L. *The Life of Sir J.J. Thomson* (Cambridge University Press, Cambridge, England, 1942).

Thomson, J. J. *Recollections and Reflections* (Bell and Sons, London, 1936).

Wheaton, B. *The Tiger and the Shark* (Cambridge University Press, Cambridge, England, 1983).

Isobel Falconer

TOMONAGA, SIN-ITIRO

The Japanese theoretical physicist Sin-itiro Tomonaga, who shared the 1965 Nobel Prize in Physics with Richard P. Feynman and Julian Schwinger, was born in Tokyo on March 31, 1906, the son of Hide and Sanjuro Tomonaga. A professor of philosophy at Shinshu University, Sanjuro moved in 1907 to Kyoto Imperial University, and it was in Kyoto that Sin-itiro was educated.

Undergraduate and Postgraduate Work and Early Research

At Kyoto Imperial University Tomonaga and his classmate Hideki Yukawa (who later won a Nobel Prize in Physics for his meson theory of nuclear forces) studied quantum mechanics from the original physics articles which arrived from abroad during the critical years 1926–1929. After graduation, both budding physicists stayed at Kyoto for several additional years of research. In April 1932, Tomonaga joined the Tokyo group of Yoshio Nishina, Japan's leading nuclear physicist. Nishina, who had worked in Europe for eight years, returned in 1928 to establish a laboratory of nuclear physics at the Institute for Physical and Chemical Research (Riken) in Tokyo, the organization which had financed his stay abroad.

Japanese physicist Sin-itiro Tomonaga (1906–1979) shared the 1965 Nobel Prize in Physcis with Julian Schwinger and Richard P. Feynman for refinement and development of the theory of quantum electrodynamics. CREDIT: COURTESY OF BETTMANN/CORBIS. REPRODUCED BY CORBIS CORPORATION.

Working with Nishina, Tomonaga did theoretical research on the annihilation of positrons, on the nature of the neutron-proton force, and on the probability of collision of a high-energy neutrino with a neutron. At the end of 1937, Tomonaga traveled to Leipzig, Germany, where he worked with Werner Heisenberg until the outbreak of World War II in 1939.

Heisenberg suggested that Tomonaga work on improving Bohr's theory of the compound nucleus. In that theory, a nucleus struck by a high-energy particle behaves like a drop of liquid. The nucleus "heats up" and evaporates one or more nuclear particles. Tomonaga published his study in a German journal and also submitted it as a thesis to Tokyo Imperial University; he received the degree of D.Sc. in 1939.

As his second research project in Leipzig, Tomonaga worked on the properties of recently discovered cosmic-ray particles, which resembled the mesons proposed by Yukawa. The theory predicted that these new particles should decay to an electron and a neutrino, but the observed mean lifetime for decay was about one hundred times too long. Tomonaga's model, using quantum field theory, led to a result that was infinite. Such infinite predictions had also appeared in quantum electrodynamics (QED), and this Leipzig experience eventually led Tomonaga to his Nobel Prize.

Research during the Second World War

In mid-August 1939, Yukawa visited Tomonaga in Leipzig on his first trip abroad, where he was invited to speak at several European conferences. However, on August 25, 1939, both physicists were advised by the Japanese Embassy in Berlin to return to Japan because of the impending outbreak of war in Europe. War broke out on September 1, the same day that they began their homeward voyage via the Panama Canal.

Tomonaga continued his association at Riken and also became professor at the Tokyo University of Science and Literature (which later became Tokyo University of Education and, in 1973, the University of Tsukuba). After Pearl Harbor, he did military research for the Japanese Navy on the theory of microwave circuits and waveguides related to radar. Especially, he worked out the theory, from first principles, of the magnetron, an oscillator to generate powerful microwaves. In this work, he applied mathematical techniques that had originated in the earlier study of nuclear collisions. With Masao Kotani, Tomonaga received the Japan Academy Prize in 1949 for this research.

Aside from his war work, Tomonaga made important contributions during this period to quantum field theories, both mesons and QED. The approach to QED involved an approximation called perturbation theory, which was an expansion of the various probabilities for scattering, absorption, etc., in powers of the so-called fine structure constant, whose value is approximately 1/137. However, expansion terms beyond the first generally gave the absurd result infinity, unless an arbitrary cutoff procedure was adopted.

The situation in meson theory was even worse, as perturbation theory failed, even with cutoffs. The reason was that the analogue of the fine structure constant in meson theory was close to unity. Tomonaga, and some collaborators, developed and applied an intermediate coupling approximation, which worked for both strong and weak coupling as well.

Quantum Electrodynamics

Tomonaga's most important paper, written in 1943, is called "On a Relativistically Invariant Formulation of the Quantum Theory of Wave Fields." It is a generalization of a 1932 work of Dirac, in which each of a set of electrons (or other elementary particles) carries its own time variable, Dirac's many-time theory. Treating time and space on an equal footing, this makes possible a fully relativistic treatment of many particles in interaction. Tomonaga generalized this to quantum field theory, with its infinite number of degrees of freedom—his so-called super-many-time theory. Effectively, the field is described on a succession of arbitrarily chosen space-like surfaces (curved in four dimensions), taking the place of "flat" planes of constant time.

Almost the same point of view was taken independently by Schwinger (several years later). Both theorists used this new approach to treat problems in QED, especially allowing them to carry out a procedure known as renormalization, which gave finite, sensible, and it turned out, extremely accurate answers to outstanding problems in QED.

In April 1947, Willis E. Lamb Jr. and Robert C. Retherford discovered an unexpected feature of the spectrum of the hydrogen atom (now called the Lamb Shift) and reported it at a small private conference organized by J. Robert Oppenheimer. In attendance were Schwinger, Feynman, Hans A. Bethe, and Victor F. Weisskopf, all of whom contributed to calculating the Lamb Shift. Tomonaga heard about this and similar effects that experiment had turned up by reading American news magazines. That was sufficient stimulus for the Tokyo group to apply Tomonaga's methods to calculate these effects.

During 1949–1950, Tomonaga was a Member of the Institute for Advanced Study in Princeton, New Jersey. He became president of the Tokyo University of Education in 1956. Besides the 1965 Nobel Prize in Physics, he received other international honors. He died in Tokyo on July 8, 1979.

See also: FEYMAN, RICHARD; QUANTUM ELECTRODYNAMICS; QUANTUM FIELD THEORY; SCHWINGER, JULIAN

Bibliography

Brown, L. M.; Kawabe, R.; Konuma, M.; and Maki, Z.; eds. "Elementary Particle Theory in Japan. 1930–1960." *Progress of Theoretical Physics Supplement* **105** (1991).

Matsui, M., and Ezawa, H. *Sin-itiro Tomonaga—Life of a Japanese Physicist* (MYU Publishing Company, Tokyo, Japan, 1995).

Schwinger, J. "Two Shakers of Physics: Memorial Lecture for Sin-itiro Tomonaga" in *The Birth of Particle Physics,* edited by L. M. Brown and L. Hoddeson (Cambridge University Press, Cambridge, U.K., 1983).

Tomonaga, S. "Development of Quantum Electrodynamics: Personal Recollections." *Physics Today* **19**(9), 25–32 (1966).

Tomonaga, S. *Quantum Mechanics: Volume 1. Old Quantum Theory,* translated by M. Koshiba (Interscience, New York, 1962).

Tomonaga, S. *Quantum Mechanics: Volume 2. New Quantum Theory,* translated by M. Koshiba (Interscience, New York, 1966).

Tomonaga, S. *The Story of Spin,* translated by T. Oka (University of Chicago Press, Chicago, IL, 1997).

Laurie M. Brown

TOP

See QUARKS

U

UNIFIED THEORIES

The quest for unification has been a perennial theme of modern physics, although it dates back many millennia. The belief that all physical phenomena can be reduced to simple elements and explained by a small number of natural laws is the central tenet of physics, indeed of all science. One of the first unifying scientific principles was the atomic hypothesis, beautifully expressed by Democritus (Presocratics, Fragment 125) in 400 B.C.E.:

By convention there is color,

by convention sweetness,

by convention bitterness,

but in reality there are atoms and space.

Separate laws do not dictate the nature of gases, liquids, and solids; rather, these are different phases of the same matter that obeys the same laws. Both life and superconductivity are consequences of the electric forces between the same kinds of atoms.

The reductionist program of reducing complex phenomena to simpler physical processes, which in turn can be reduced to even simpler and more encompassing laws of nature, is at the heart of the search for unified theories. In this pursuit, physics has been extremely successful. Physicists believe that the turbulent motion of fluids can be understood by the laws of classical mechanics, which are a good approximation to the quantum mechanical laws that govern the interaction of atoms, that the structure of atoms can be explained by the laws of interaction of nuclei and electrons, that the structure of nuclei can explained by the theory of quarks and gluons, and finally that these are consequences of an even more comprehensive unified theory to be developed.

Each stage in this development has required exploring natural phenomena at shorter and shorter distances, explaining complex macroscopic phenomena in terms of simple microscopic constituents. Ordinary matter is made of atoms, atoms are made of nuclei and electrons, and nuclei are made of quarks. At each stage, the laws of physics are more unified, explaining a larger class of phenomena and reducing to the laws of the previous in appropriate circumstances. Typically, the more unified theories exhibit more symmetry, are more predictive, and are less arbitrary.

Isaac Newton's theory of gravity was the first grand success of this saga, unifying in a precise mathematical framework the laws that governed the motion of apples and planets. It spurred the search for similar unifying theories that would explain ordinary matter.

James Clerk Maxwell's theory unifying electricity and magnetism was the next major step in the quest for unification. Maxwell's theory is the very paradigm of a unified theory. Its new mathematical formalism was based on new concepts (local fields) and exhibited new symmetries of nature (Lorenz or relativistic invariance, as well as local gauge symmetry). It explained the many electric and magnetic phenomena that had been discovered over the years as manifestations of a single entity, the electromagnetic field. It also had many new consequences. The most dramatic of these was the prediction of the existence of electromagnetic waves and the demonstration (by calculating the velocity of the waves) that light was such a wave. Thus, optics was unified with electromagnetism.

Albert Einstein, after the successful formulation of his general theory of relativity, which explained gravity as the consequence of the dynamics of the metric field of space-time, and unifying the structure of the geometry of space-time with gravitation, dreamed of a unified theory of all the forces of nature and of all forms of matter. Since general relativity was a nonlinear theory, he hoped that its solutions could behave as localized lumps of matter and that these might even imitate quantum mechanics. Although it is now believed that Einstein's quest to explain quantum mechanics in a classical field theory was in vain, many believe that the goal of unification is achievable. Indeed, Einstein's belief that "nature is constituted so that it is possible to lay down such strongly determined laws that within these laws only rationally completely determined constants appear, not constants therefore that could be changed without completely destroying the theory" (Einstein 1949, p.63) beautifully expresses the view that a theory containing arbitrary parameters that cannot be calculated from first principles is incomplete and is to be superceded by a more unified and predictive theory.

The development of quantum mechanics in the 1920s and throughout the twentieth century enabled the completion of the atomic program, unifying chemistry and atomic physics. All the properties of ordinary matter, in all of its variety of forms, can be explained in terms of atoms and the electromagnetic forces between them, realizing Democritus's vision.

Quantum mechanics also provides for a theory of the structure of atoms in terms of the electromagnetic forces between the atomic nucleus and the electrons orbiting them.

In the latter half of the twentieth century, a comprehensive theory of the constituents of matter and of the forces of nature was completed—the Standard Model of elementary particle physics. This quantum theory of fields identifies the basic constituents of matter. They are the quarks that make up the nuclei at the center of atoms and the leptons (such as the electron) that revolve about the nuclei. The Standard Model also explains the forces that act on the elementary particles (the electromagnetic, the weak, and the strong or nuclear forces) as consequences of local gauge symmetries.

An essential part of the Standard Model is the unification of the electromagnetic and weak forces in a combined electroweak theory. Much as electricity and magnetism were seen as different phenomena before Maxwell's theory, the electromagnetic and weak forces originally were seen as very different in nature. The electromagnetic forces are long-range and the quanta of the electromagnetic fields (photons) are massless particles. The weak forces are short-ranged and their quanta (the W and Z bosons) are massive. In the Standard Model, both are consequences of a unified gauge symmetry, and the differences between them are a consequence of the fact that this symmetry is spontaneously broken. Indeed, the theory predicts that if one were to heat the universe to very high temperatures (a circumstance that did occur at very early cosmological times), the symmetry would be restored, and all apparent differences between the forces would disappear.

The Standard Model provides many hints that further unification is required. First, there is remarkable similarity, at the fundamental level, between the various forces and particles. The electroweak and strong forces are both consequences of local gauge invariance and differ only in the specific group responsible for each ($SU(2) \times U(1)$ and $SU(3)$, respectively). The fundamental particles, the leptons and the quarks, are very similar in their properties. Indeed, it is quite easy and natural to unify these theories in a way whereby the quarks and leptons fit naturally into larger patterns of symmetry.

At first sight it, might appear that the disparity between the strength of the electroweak force (which is rather weakly coupled at low energies) and the strength of the strong nuclear force (that is strongly coupled at low energies) would argue against unification. However, the strength of these forces varies with energy due to the dynamical properties of the fluctuating quantum mechanical vacuum. The strong force decreases at a short distance or for high-energy processes (asymptotic freedom), whereas the electric and weak forces grow stronger. Precise measurements and theoretical calculations enable us to extrapolate these forces to very high energies, where they coincide in strength at the very-high-energy scale of 10^{18} GeV. (This agreement is greatly improved if one also assumes that a new symmetry, supersymmetry, and many new associated particles are present at energies of approximately 10^4 GeV.)

This extrapolation provides a compelling hint that unification of the forces of nature might occur at energies of $\sim 10^{18}$ GeV. Furthermore, since the force of gravity (which is an extraordinarily weak force at low energy) becomes equally strong at this energy, it is suggested that the next stage of unification should include gravity.

Finally, the many unanswered questions raised by the Standard Model, and the many parameters that are incalculable within the Standard Model, suggest that further unification within a more symmetric and predictive theory is necessary.

Currently, the best hope for a unified theory is based on string theory. String theory is largely based on the notion that the elementary constituents of matter are extended one-dimensional objects, not the pointlike particles of the Standard Model. This is a unifying concept. Since a string has infinitely many shapes, it can describe in one entity many elementary particles. Indeed, each vibrational mode of a string will behave as an elementary pointlike particle. One of the most alluring features of string theory is that when certain special solutions of the theory are analyzed, they contain precisely the spectrum of elementary particles found in nature—the quarks and leptons of the Standard Model, as well as the quanta of the gauge theories that provide for all of the observed forces.

String theories are inherently theories of gravity. Unlike the situation in ordinary quantum field theory, one does not have the option in string theory of turning off gravity. The gravitational, or closed string, sector of the theory must always be present for consistency, even if one starts by considering only open strings, since these can join at their ends to form closed strings. One of string theory's greatest successes is that it is a mathematically consistent quantum theory of gravity, free of the infinities that plagued field theoretic attempts to quantize gravity. Thus, string theory appears to provide the framework to unify all the forces of nature including gravity into a single, tightly woven pattern.

String theories, as is appropriate for unified theories of physics, are incredibly unique. In principle, they contain no freely adjustable parameters, and all physical quantities should be calculable in terms of the fundamental dimensional units of nature: the velocity of light c, Planck's constant of action, and Newton's constant of gravity. However physicists' understanding of the structure of string theory is still quite primitive, and thus in practice, they are not yet in the position to exploit such enormous predictive power.

Will the quest for unification ever end? Will a theory of everything, a theory that unifies all the phenomena of physics once and for all, a final theory, ever be formulated? There is no known reason why this is impossible. Experience teaches that each stage of unification leaves many questions unanswered and reveals new mysteries that only find their explanation at the next stage of unification. But, this might end.

Time will tell.

See also: GRAND UNIFICATION; PARTICLE PHYSICS, ELEMENTARY; PLANCK SCALE; STRING THEORY

Bibliography

Einstein, A., and Infeld, L. *The Evolution of Physics* (Simon and Schuster, New York, 1938).

Einstein, A., and Schilpp, P. A., ed. *Albert Einstein: Philosopher-Scientist* (Harper & Brothers, New York, 1949).

Greene, B. *The Elegant Universe: Superstrings, Hidden Dimensions, and the Quest for the Ultimate Theory* (W. W. Norton, New York, 1999).

Weinberg, S. *Dreams of a Final Theory* (Pantheon Books, New York, 1992).

David Gross

UNIVERSE

Today's universe—exceedingly large (10^{28} cm) and frigidly cold (~3K)—has seemingly little to do with the microscopic high-energy world of elementary particle physics. However, because today's universe is expanding, and its temperature is nonzero, the connection between the macroworld of cosmology and the microworld of the elementary particle is unavoidable. The relatively simple equations describing the expanding universe (filled with ubiquitous microwave background radiation) can just as easily be run backwards in time. It is then apparent that the universe evolved from a Big Bang—a hot dense soup consisting of all the particles that inhabit the Standard Model of particle physics. The connection between particle physics and cosmology is so strong that the standard cosmological model (i.e., the Hot Big Bang) assumes the Standard Model of particle physics, along with general relativity, as fundamental components.

There is no evidence for appreciable antimatter anywhere in the universe. For that matter, there are relatively few baryons in the universe: the success of Big Bang nucleosynthesis reveals that today there is roughly one baryon for every 10^9 photons. That is, the baryon asymmetry of the universe (i.e., the net number of baryons over antibaryons relative to the number of photons) is 10^{-9}. This small number turns out to be a big problem for cosmology. If the universe had started out with equal numbers of baryons and antibaryons, they would have annihilated with great efficiency, and today's baryon-to-photon ratio would be 10^{-19}, much too small. A solution to this baryogenesis problem may be derived from particle physics. Generating an excess of baryons over antibaryons in a universe that starts with an equal number of baryons and antibaryons (i.e., one with no net baryon excess) requires physical processes that violate baryon number. Andrey Sakharov pointed out in the 1960s that out-of-equilibrium (e.g., the decay of a massive particle) baryon number violation (along with the violation of CP [the simultaneous conservation of charge and parity]) is the necessary ingredient in any model that explains the observed baryon excess. These Sakharov conditions, including baryon number violation, are a natural feature of many Grand Unified Theory (GUT) extensions of the Standard Model. Although baryon number is almost perfectly conserved at low energies (protons, no matter how long they are observed, do not seem to decay), GUTs, in their quest to unify the strong, weak, and electromagnetic interactions, still possess interactions that violate both baryon and lepton number conservation, and these interactions become strong at large energies. As the universe cooled through the temperature associated with these large energies, the baryon violating interactions allow it to dynamically generate a baryon asymmetry of just the right order of magnitude.

GUT baryogenesis is the fair-haired child resulting from the marriage of particle physics and cosmology. Not nearly so pleasant are the cosmological implications of another GUT-cosmology offspring: magnetic monopoles. As the GUT universe symmetry evolves from the unified to the broken (via spontaneous symmetry breaking), it is impossible, in the standard cosmological model, to avoid the production of very massive relics (so-called topological defects). The pointlike variety of these topological relics are called magnetic monopoles, and the prediction of most GUTs is that they would be very massive (10^{16} GeV) and produced in abundances that would be easily detectable, either by their sheer dynamical mass density or in a variety or other astrophysical environments. It appears that the marriage of GUTs and cosmology is doomed by the monopole problem.

In an effort to save the GUT-cosmology union, particle physicists began to examine the evolution of the early universe during the time that spontaneous symmetry breaking must occur. And there they found not only a solution to the monopole problem but also a significant addition to the standard cosmology: inflation. In the generic theory of inflation, the universe becomes dominated by vacuum energy during one of the GUT symmetry breakings and undergoes a period of exponential growth (in the standard cosmology the universe grows as a fractional power of time), increasing in size by some forty-three orders of magnitude so that a microscopic patch can become larger than the visible universe! In this scenario, there remains only one GUT monopole in the universe, and thus the monopole problem vanishes. More important, inflation helps explain several deficiencies of the standard cosmology: (1) the universe, at least at an age of 100,000 years, appears to be very nearly isothermal even though the standard cosmology predicts that it was not causally connected, (2) there is no origin for the density perturbations that eventually give rise to galaxies and clusters of galaxies, and (3) the universe is remarkably close to, if not, flat (i.e., no curvature), which requires a fine-tuning of at least one part in 10^{60}. Inflation solves these problems by inflating a causally connected patch, along with the quantum fluctuations present in the patch, into a nearly zero-curvature region larger than the observable universe. The solution to the monopole problem has led scientists to a significant revision of their thinking about the evolution of the universe that is so compelling that most cosmologists accept an epoch of inflation as a necessary ingredient of the standard Big Bang cosmology.

What is the particle physicist's next cosmological conquest? Most likely, it will occur within the realm of the dark matter problem. When the rotational velocity of hydrogen gas in galaxies is measured, it is found that there is roughly ten to twenty times more mass present than that tied up in stars. When the amount of gravitating material in large clusters of galaxies is measured, there appears to be roughly thirty to forty times more gravitational mass than the mass associated with light mass. Both of these measurements provide irrefutable evidence of the dark matter problem. Galaxies themselves and groups of galaxies contain much more mass than astronomers can see! One finds similar discrepancies in the mass budget if one compares these estimates of the gravitating mass of the universe to the mass associated with baryons (as derived from Big Bang nucleosynthesis). Not only can most of the universe not be seen, but most of the universe is not made of baryons!

Again, just as in baryogenesis, particle physics was quick to provide two candidates to resolve the dark matter problem. They existed as by-products of extensions of the Standard Model designed to solve significant problems in particle physics. One dark matter candidate is known as the weakly interacting massive particle (WIMP), and it can naturally occur in the supersymmetric models designed to solve the hierarchy problem. In these models, there is a symmetry introduced that pairs each existing fermion with a hypothetical boson and likewise for the existing bosons. The lightest new particle in the theory is stable against decay and can easily survive as a Big Bang relic in numbers large enough to contribute significantly to the mass density at the current epoch. Although this type of supersymmetric dark matter has yet to be discovered, experimentalists are hot on its trail. The other popular dark matter candidate is the axion, a particle that results from the breaking of the symmetry introduced to solve the strong CP problem (namely, where nonperturbative effects in quantum chromodynamics [QCD], unless suppressed, would predict an electric dipole moment for the neutron that is ten orders of magnitude too large). The axion that results would be produced in the Big Bang, and it couples to photons with significant strength so that it may be detected via resonant photon production in a large magnetic field. Again, it is likely that experimentalists will soon be able to determine if the axion is a significant component of dark matter.

Over the past several years, a new dark matter problem has appeared—the dark energy problem—and its solution will almost surely come from the world of particle physics. When astronomers looked at distant supernovae, their apparent brightness implied that they were further away than cosmologists would have predicted based on the amount of gravitating material so far surveyed (dark matter). The universe is accelerating! And, the culprit is smoothly distributed throughout the universe, makes up 70 percent of the mass of the universe, and has a negative pressure (thus, the name dark energy rather than dark matter). Albert Einstein first called this newly discovered dark energy the cosmological constant. The cause for acceleration may, in fact, be Einstein's cosmological constant, or equivalently, vacuum energy. The problem with this solution is that the same symmetry breakings that occur in GUT cosmology would contribute to vacuum energy—a rough estimate would be that the vacuum energy should be 100 orders of magnitude larger than the current data suggest. Particle cosmologists have recognized that they do not know what makes up 90 percent of the universe and that two-thirds of that 90 percent is currently unexplainable by any theory. However, many are confident an answer will be found, and it is very likely that such an explanation will come from the world of particle physics.

See also: ASTROPHYSICS; COSMOLOGICAL CONSTANT AND DARK ENERGY; COSMOLOGY; INFLATION; INFLUENCE ON SCIENCE

Bibliography

Peebles, P. J .E. "Making Sense of Modern Cosmology." *Scientific American* **284**, 54–55 (2001).

Terry P. Walker

UP

See QUARKS

V

VIRTUAL PARTICLE

See VIRTUAL PROCESS

VIRTUAL PROCESSES

Most processes in quantum field theory occur via virtual particles. One can think of them as short-lived imposters of real particles. They act as a kind of currency, allowing real particles to exchange momentum, energy, and charges. They explain how particles decay, scatter off one another, resonantly produce other particles, form bound states, and ultimately explain how long-range forces arise. All these processes can be calculated from Feynman diagrams, invented by Richard Feynman in 1949. External lines in the Feynman diagram of Figure 1(a) represent real particles, and internal lines represent virtual particles. In this diagram, a pair of particles annihilate each other, producing the wavy virtual particle, which then turns into another pair of particles. There are also diagrams where virtual particles produce other virtual particles, which loop around before being reabsorbed, as in Figure 1(b).

FIGURE 1

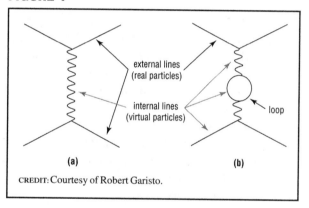

CREDIT: Courtesy of Robert Garisto.

The most obvious difference between virtual and real particles is their energy. In special relativity, the energy of real particles is determined by their mass and momentum, which is called being on mass shell. For massive particles viewed in their rest frame (where their momentum is zero), this relation reduces to $E = mc^2$. Virtual particles have the same mass as their real cousins, but E is not fixed to being mc^2, and they are said to be off mass shell. They get away with having the "wrong energy" because of the uncertainty inherent in quantum mechanics. Still, the further off mass shell the virtual particle, the smaller its contribution to a process, and the more ephemeral its existence.

Scattering

Real particles scatter via the exchange of momentum through virtual particles. Thus, virtual particles are carriers of force. If the real particles just change direction, the scattering is called elastic, but if they change into other particles, it is called inelastic.

Virtual gluons mediate the strong force, virtual photons mediate the electromagnetic force, and virtual W and Z bosons mediate the weak force. Thus, an electron and a positron can scatter electromagnetically via a virtual photon. In Figure 2(a), they change direction after the virtual photon is emitted or absorbed. In Figure 2(b), they annihilate into a virtual photon, which then produces another electron and positron. Since the initial and final particles are the same, the diagrams contribute coherently.

Resonances

Scattering is enhanced if it occurs through a particle nearly on mass shell. This is called a resonance. Suppose an electron and a positron collide, and they happen to have a combined kinetic energy equal to the mass of the Z boson times c^2. Then they will annihilate and produce a real Z, as in Figure 3. The Z is unstable and decays very quickly, here into a quark-antiquark pair. Z bosons can be studied by measuring the increase in electron-positron inelastic scattering at the resonance energy.

FIGURE 3

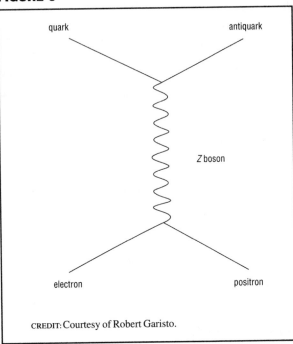

CREDIT: Courtesy of Robert Garisto.

Long-range Forces

Quantum field theory aims to explain all particle interactions. Therefore, although it excels at predicting short-distance scattering, it should apply to long-range forces as well. However, an explanation in terms of virtual particles is tricky. The discussion here will be limited to trying to understand heuristically how virtual carriers of force operate with attractive and repulsive forces.

As previously discussed, virtual particles transfer momentum between scattering particles, so it is easy to see how this can result in a repulsive force. Imagine two players, Alice on the left and Bob on the right, who are exchanging a virtual ball. As Alice throws the ball to Bob, she recoils to the left, away from him. When Bob catches it, he is pushed to the right, away from Alice. Hence, they are pushed apart. How then do attractive forces arise?

Imagine that Alice throws the ball to the *left*. She does get pushed toward Bob, but now the ball has momentum *away* from Bob. However, it is a virtual ball, and where and when it can go defy classical intuition. In virtual processes, both forwards- and backwards-in-time exchanges occur, as ex-

FIGURE 2

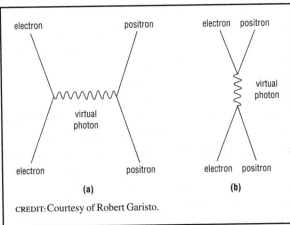

CREDIT: Courtesy of Robert Garisto.

plained in Feynman's famous *QED* published in 1985. The ball can thus have momentum to the left but arrive at Bob on the right by going backward through time. (The change in position is just the momentum divided by the mass times the change in time, so if the change in time is negative, the change in position is to the *right*.) When Bob catches the ball, he receives an impulse *toward* Alice, since it has momentum to the left. In a naive classical view, they can exchange a virtual ball that pulls them closer together because the catch can precede the throw!

Decays

Most elementary particles are unstable and decay in a fraction of a nanosecond. This is also due to virtual particles. For example, in Figure 4, a muon decays via a virtual *W* boson. The muon weighs about 0.1 proton mass, whereas the *W* boson weighs about 80 proton masses. This has two effects. First, the virtual *W* is very far off mass shell and appears for only a trillionth of a trillionth of a second. Second, the overall process is very suppressed by the ratio of these masses, so that the muon lives for about two millionths of a second, which is a long time by unstable particle standards. So, the *W* appears for only a tiny fraction of the muon's lifetime, roughly the same ratio as a centimeter to a light-year!

Virtual Loops

As previously discussed, virtual particles can spawn other virtual particles in loops, as in Figure 1(b). Virtual particles in such a loop can have any energy, and although the contribution from each energy is small, the sum over the infinite range of energies is infinite. An infinity signals a breakdown of a theory. In this case, it shows that the theory is valid only up to some finite energy. This problem can be fixed by a procedure called renormalization. For example, one can cut off the sum at some finite high energy, and then the contributions from loops become finite. Using renormalization group equations, it can be shown that the loops make the strength of

FIGURE 4

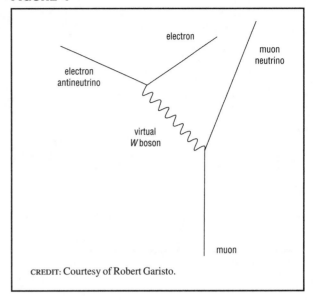

CREDIT: Courtesy of Robert Garisto.

the forces of nature scale with the energy of the process. At an energy of about 10 quadrillion times the proton mass times c^2, the strength of the electromagnetic, strong, and weak forces all reach the same value, leading to the speculation that they can be unified into one theory at that scale.

See also: ANNIHILATION AND CREATION; FEYNMAN DIAGRAMS; QUANTUM FIELD THEORY; QUANTUM MECHANICS; RELATIVITY; RESONANCE; RENORMALIZATION; SCATTERING; QUANTUM STATISTICS; Z FACTORY

Bibliography

Cheng, T. P., and Li, L. F. *Gauge Theory of Elementary Particle Physics* (Oxford Science Publications, Oxford, UK, 1988).

Feynman, R. P. "Space-Time Approach to Quantum Electrodynamics. *Physical Review* **76**, 769–789 (1949).

Feynman, R. P. *QED: The Strange Theory of Light and Matter* (Princeton Science Library, Princeton, NJ, 1985).

Halzen, F., and Martin, A. D. *Quarks and Leptons: An Introductory Course in Modern Particle Physics* (Wiley, New York, 1984).

Robert Garisto
Rashmi Ray

W, Z

See BOSON, GAUGE

WEAK INTERACTION

See BASIC INTERACTIONS AND FUNDAMENTAL FORCES

WIGNER, EUGENE

Eugene Wigner was one of the dominant theoretical physicists of the twentieth century. His life and work spanned much of that century, in particular, the exciting era when quantum mechanics burst onto the physics scene. He was born in Budapest, Hungary, on November 17, 1902, and died in Princeton, New Jersey, on January 1, 1995. His association with Princeton University exceeded fifty years.

His early education was in Budapest, and he received much stimulation from an excellent secondary school and from a classmate, the great mathematician, John von Neumann. He was also strongly influenced all of his life by a group of Hungarian physicists and chemists who were roughly his contemporaries: Michael Polanyi, Leo Szilard, and Edward Teller. To satisfy his father's wishes he obtained his doctorate in chemical engineering—at the Technical University of Berlin, in 1925—but dramatic events in science soon turned his interests to his beloved physics.

Much of twentieth-century physics—indeed the heart of modern physics—is connected to the invention and application of quantum mechanics, which governs all science at small distance scales. The need for the new framework arose from puzzles pertaining to the light spectra of atoms. The ideas that brought the breakthrough for quantum mechanics came in the middle of the 1920s. Wigner's work began just a year or two later, and he was a central member of that brilliant group of physicists who appreciated what could be done with the new framework of quantum mechanics. It opened the door for the description of almost all of science, first the structure of atoms, then of atomic nuclei, and eventually of particle physics, and it led to the creation of entirely new fields of science, such as microelectronics and microbiology.

Wigner led. He seems to have grasped intuitively, right from the beginning, how important symmetry principles would be for the description of quantum

Hungarian physicist Eugene Wigner (1902–1995) shared the 1963 Nobel Prize in Physics with Maria Goeppert-Mayer and J. Hans D. Jensen. Wigner received the Nobel Prize for his discovery of fundamental symmetry principles as well as his contributions to quantum mechanics. CREDIT: COURTESY OF ARCHIVE PHOTOS, INC. REPRODUCED BY PERMISSION.

systems and therefore also the importance of the field of mathematics, group theory, which lent itself naturally to the articulation of symmetry ideas. His early book *Group Theory and its Application to the Quantum Mechanics of Atomic Spectra* was epoch making. When nuclear physics was born in 1932 with the discovery of the neutron, Wigner at once became one of the leaders in the early development of this important new field out of which particle physics eventually grew. Also, with the first three of his more than forty Princeton Ph.D. students—Frederick Seitz, John Bardeen, and Conyers Herring—he built the foundations of modern solid state physics.

World War II and the Manhattan Project brought Wigner his greatest challenge and with it the greatest flourishing of his scientific genius. He was head of theoretical physics in Enrico Fermi's team at Chicago that built the first nuclear reactor. But even before the Stagg Field reactor went critical, Wigner's small team had, within the span of a few weeks, fully designed the whole set of Hanford reactors for the production of plutonium, which was so critically important for the success of the Manhattan Project. His expertise in chemical engineering was a great help in the detailed design of the Hanford facility. He spent the rest of the war at Chicago literally creating the field of reactor physics and thus laying the foundations for the entire world effort in nuclear power.

Wigner's contributions to particle physics pertain to the foundations of the subject but also to many quantum mechanical concepts—for example, line breadth, resonance analysis, etc.—which he originally devised for atomic or nuclear physics but which later became essential for articulating particle physics. In applying symmetry principles to quantum systems he made a remarkable progression of ideas from the use of compact groups (the permutation group, the rotation group, etc.) for the classification of the symmetries of atomic states to the noncompact crystallographic groups for solid-state physics and finally to the noncompact Poincaré group for the invariance principles of particle physics. The work on the Poincaré group also led to important contributions to the understanding of relativistic wave equations by Theodor Newton and Eugene Wigner and to the concept of localization of the position of elementary systems. Although his contribution to symmetries and quantum mechanics was immense, he focused on global symmetries and had no fondness for the local symmetries of the electromagnetic theory of Maxwell in which the physics remains invariant to adjustment of a gauge at any point in space. Such local symmetries underlie almost all of the important developments in particle physics of the last few decades of the twentieth century. Nonetheless, Wigner's contribution to symmetry and quantum mechanics was immense and earned him the Nobel Prize in Physics in 1963.

Looking at Wigner's long career, what is most impressive is the depth and great breadth of his contributions to science. There is scarcely any area of physics which was not deeply affected by his work and in which some important phenomenon is not named after him. His work continues to have an impact on entirely new fields such as quantum chaos. In particle physics his influence affects the whole field.

The breakup of the Austro-Hungarian empire after the Great War and the chaotic socialist dictatorship of Bela Kun that briefly followed had a profound effect on Wigner, as did the advent of Hitler in Germany. Like most of his great Hungarian scientific contemporaries, Wigner's roots were Jewish. He tended to be profoundly pessimistic about the development of international affairs, and his magnificent contributions during World War II were carried out under great personal stress. He greatly valued stability in governance and became a very proud American. As a person he was very kind and considerate of his colleagues although somewhat formal, consistent with his European (Hungarian) roots. In fact, his formality was almost legendary. But his many students and colleagues revered him as a person with an unparalleled combination of strengths in science which he imparted freely and joyously to those with whom he interacted.

See also: Conservation Laws; Symmetry Principles

Bibliography

Wigner, E. P. *Group Theory and Its Application to the Quantum Mechanics of Atomic Spectra,* translated by J. J. Griffin (Academic Press, New York, 1959).

Wigner, E. P. *Symmetries and Reflections* (Ox Bow Press, Woodbridge, CT, 1979).

Wigner, E. P. *The Collected Works of Eugene Paul Wigner Part A: The Scientific Papers,* edited by A. S. Wightman (Springer-Verlag, New York, 1993).

Wigner, E. P. *The Collected Works of Eugene Paul Wigner Part B: Historical and Biographical Reflections and Syntheses Volume 7,* edited by A. Wightman and J. Mehra (Springer-Verlag, New York, 2001).

Erich Vogt

WILSON, ROBERT R.

Robert Rathbun Wilson has been a central figure in accelerator design and development since the birth of the cyclotron in 1932. High-energy particle accelerators are the essential tool of physicists for the discovery and investigation of the properties of elementary particles the fundamental building blocks of matter. Wilson was the driving force in the cre-

ation of two of the four world-class high-energy physics laboratories in the United States: the Cornell Laboratory of Elementary Particle Physics (at that time called the Cornell Laboratory for Nuclear Studies) and the Fermi National Accelerator Laboratory (Fermilab) in Batavia, Illinois, which houses the world's highest-energy accelerator (initially the five hundred billion electron volt [GeV] energy proton synchrotron and, since 1990, the trillion electron volt [TeV] Tevatron).

A brief review of his career cannot begin to describe his central role in high-energy experimental physics. His insistence on bolder, more compact, and economical design, seen clearly in the accelerators he built at Cornell, influenced the design of most

American physicist Robert R. Wilson (1914–2000) was the driving force in the creation of the Cornell Laboratory of Elementary Particle Physics and the Fermi National Accelerator Laboratory. He was also a central figure in accelerator design, developing the first superconducting magnet accelerator. CREDIT: COURTESY OF AIP EMILIO SEGRE VISUAL ARCHIVES, PHYSICS TODAY COLLECTION. REPRODUCED BY PERMISSION.

modern accelerators. His development of the first superconducting magnet accelerator at Fermilab made possible, both technically and economically, the very-high-energy accelerators that were later constructed.

Early Years

Wilson was born March 4, 1914, in Frontier, Wyoming, the son of Platt and Edith Rathbun Wilson, a pioneering ranching family. He was admitted to the University of California at Berkeley in 1932 and received the A.B. degree cum laude in 1936. During his junior year, he began research under E. O. Lawrence. His first work, in which he developed a new method of studying time lag in gaseous discharges, a work of considerable importance, was published in the *Physical Review* during his senior year. Wilson continued his studies under Lawrence as a graduate student. Among the four papers he published as a graduate student were the first theoretical analysis of the stability of cyclotron orbits, which he verified experimentally, and a paper on the theory of the cyclotron. During his graduate career, he made major contributions to the development of the cyclotron as an important tool in the study of the atomic nucleus. He received his Ph.D. in 1940.

The War Years

In 1940 he married Jane Inez Scheyer of San Francisco, accepted an appointment as instructor at Princeton, and very soon became involved in the scientific war effort. He collaborated with Enrico Fermi in some preliminary experiments on the production of a chain reaction. In the fall of 1941 he invented a new method of separating uranium isotopes and led a group of about fifty scientists and engineers in developing this technique.

Early in 1943 the work on the separation of uranium isotopes was limited to methods that were ready for production. The work at Princeton was terminated, and Wilson was asked to set up a cyclotron at the new Los Alamos laboratory. He and some of his Princeton staff moved the Harvard cyclotron to Los Alamos and began to study the fission process. At Los Alamos, he directed the Cyclotron Group, and in the summer of 1944 he was appointed head of the Physics Research Division that was responsible for experimental nuclear

research and later for nuclear measurements made during the test of the first atomic bomb.

Wilson was greatly troubled by the bomb. After witnessing the first explosion, he wrote: "I determined that having played even a small role in bringing it about, I would go all out in helping to make it become a positive factor for humanity" (Wilson 1970b). He played a leading role in the formation of the Federation of Atomic Scientists, became its Chairman in 1946, and worked effectively for civilian control of atomic energy.

In the fall of 1946 Wilson accepted an associate professorship at Harvard where he designed a one hundred fifty million electron volt (MeV) cyclotron. His stay at Harvard was short, for in the winter of 1947 he went to Ithaca to become the director of the Laboratory of Nuclear Studies and Professor of Physics at Cornell University. He remained in that position until 1967, when he left Cornell to assume the directorship of Fermilab, in Batavia, Illinois.

The Cornell Years

During Wilson's tenure at Cornell, he and his colleagues built four successively more energetic electron synchrotrons, each with some unique capability. Wilson built accelerators because they were the best instruments for doing the physics he wanted to do. No one was more aware of the technical subtlety of accelerators, no one more ingenious in practical design, but it was the physics potential that came first, and Wilson had very clear ideas about that physics. During his twenty years at Cornell, he remained deeply embedded in the experimental program, both as mentor and experimenter. Among his many researches at Cornell, his work on the structure of the proton and neutron stand out.

Fermilab

In 1967, after completing the 10-GeV synchrotron, Wilson left Ithaca to become the director of Fermilab. Starting on a "greenfield" site with no staff, he began the job of building the most ambitious accelerator project ever undertaken. In addition to the challenge of building at a virgin site a cascade of large accelerators in six years, Wilson promised to double the energy of the accelerator over the original proposal without any increase in cost, and he did. He was able to do that primarily by redesigning the magnets and their arrangement. The new magnet was smaller with twice the magnetic field, thereby doubling the energy of the protons circulating in the same size tunnel. The achievement of higher energy and more physics capability at the same cost are hallmarks of Wilson's work. Many of Wilson's ideas, often considered risky and unrealistic when proposed, were later adopted in subsequent accelerators, a further tribute to Wilson's vision and courage.

In 1980 the accelerator's energy was doubled again, to 1,000 GeV, by the installation of superconducting magnets in the same tunnel. The Tevatron, the name given the new machine, was vintage Wilson. To guide its circulating beams, the accelerator required about one thousand very accurate and reliable superconducting magnets. Nothing of this scale and refinement had ever been undertaken before. Wilson provided the project's vision and leadership and was devotedly and personally involved in the difficult research and development (R&D) to establish the mass production technology required to bring the project to a successful and low cost conclusion. Without the superconducting technology, the capital and operating cost for multi-TeV accelerators, such as the Tevatron, would be prohibitive. Since 1980, the Tevatron has been the world's highest energy accelerator.

Though the demands of the directorship prevented Wilson's personal involvement in any particular experiment, his influence was crucial to the physics program. Two of the most important physics results at Fermilab have been the discovery of the family of heaviest quarks: the bottom quark in 1977 and the top quark in 1995. It was Wilson's decision to double the energy of the initial design and his insistence on running the accelerator at the highest energy that made the discovery of the bottom quark possible. The heavier top quark required the full energy of the Tevatron.

Fermilab was an architectural, as well as a scientific, triumph. It was designed with a grace and beauty unique among such facilities; and several of Wilson's own sculptures are installed on the grounds at Fermilab. Here the other side of Wilson is seen: the artist who believed art and science form a harmonious whole that advances both the science and culture of society. Wilson's concern is eloquently expressed in testimony before the Congressional Committee on Atomic Energy, April 1, 1947.

> Senator John Pastore: Is there anything connected with the hopes of this accelerator that in any way involves the security of this country?
>
> Robert Wilson: No sir, I don't believe so.
>
> Pastore: Nothing at all?
>
> Wilson: Nothing at all.
>
> Pastore: It has no value in that respect?...
>
> Wilson: It has only to do with the respect with which we regard one another, the dignity of men, our love of culture. It has to do with are we good painters, good sculptors, great poets? I mean all the things we venerate in our country and are patriotic about. It has nothing to do with defending our country except to make it worth defending.

Hadron Cancer Therapy

A paper of Wilson's published in the *Journal of Radiology* in 1946 entitled "Radiological Use of Fast Protons" has assumed great importance. In 1941, Wilson made an accurate measurement of how energetic protons lost energy as they traversed matter. He observed that protons deposit most of their energy at the end of their path. There was nothing unexpected in this measurement, but it led him to an exciting and far-reaching idea—to use protons for cancer therapy. Wilson noted that by carefully controlling the energy of a proton beam most of its energy could be deposited in a cancerous tumor inside the body, leaving other cells undamaged. This is in stark contrast to radiation treatment with electron or photon beams, which lose energy more or less uniformly in traversing matter and so attack healthy and cancerous tissues indiscriminately.

The first facility for proton therapy was at the Harvard Cyclotron that Wilson designed after the war. The last decade has seen an explosion of interest in this therapy. At present, proton cancer treatment facilities have been installed in hospitals in many different countries. Wilson was honored for his pioneering work in proton therapy at an international conference held at the European Laboratory for Particle Physics (CERN) in 1996.

Awards and Honors

Wilson was awarded honorary degrees from Notre Dame University, Harvard University, the University of Bonn in Germany, and Wesleyan University. Among many other honors he has received are the Elliot Cresson Medal from the Franklin Institute, the National Medal of Science, the Enrico Fermi Award, the Wright Prize, and the del Regato Medal. He was elected to the National Academy of Sciences, the American Academy of Arts and Sciences, and the American Philosophical Society. In 1985 he was elected president of the American Physical Society.

Wilson lived a very rich life. He was, in the words of the citation for the American Institute of Physics (AIP) Andrew Gemant Award granted him in 1995, "an outstanding experimenter, master-builder and designer, sculptor of stone, wood, metal," to which one could add architect, humanist, and, above all, physicist.

Robert Wilson died on January 16, 2000, in Ithaca, NY. He is survived by his wife Jane; sons Daniel, Jonathan, and Rand; and four grandchildren.

See also: CORNELL LABORATORY FOR ELEMENTARY PARTICLE PHYSICS; FERMILAB

Bibliography

Wilson, R. R. "Radiological Use of Fast Protons." *Radiology* **47**, 487–91 (1946).

Wilson, R. R. "An Anecdotal Account of Accelerators at Cornell" in *Perspectives in Modern Physics* (Interscience Publishers, New York, 1966, 225–46).

Wilson, R. R. "My Fight Against Team Research." *Daedalus* **99**, 1076–1087 (1970a).

Wilson, R. R. "The Conscience of a Physicist." *Bulletin of the Atomic Scientist* **26**, 30–34 (1970b).

Wilson, R. R. "The Batavia Accelerator." *Scientific American* **230**(2), 72–83 (1974).

Wilson, R. R. "A Recruit for Los Alamos." *Bulletin of the Atomic Scientist* **31**, 41–47 (1975).

Wilson, R. R. "High Energy Physics at Fermilab." *Bulletin of the American Academy of Arts and Sciences* **29**, 18–24 (1975).

Wilson, R. R. "The Tevatron." *Physics Today* **30**(10), 23–30 (1977).

Wilson, R. R. "Fantasies of Future Fermilab Facilities." *Reviews of Modern Physics* **51**, 259–73 (1979).

Albert Silverman
Boyce D. McDaniel

WU, CHIEN-SHIUNG

Chien-Shiung Wu ranks as one of the foremost physicists of the twentieth century. Her pioneering experiments in beta decay and weak interactions were the preeminent tests of new paradigms and models of subatomic physics.

Chien-Shiung Wu was born to Wu Zhongyi and Fan Funhua on May 31, 1912, in Liu He, a small town near Shanghai. Her father, an engineer, was the headmaster of the School for Girls, one of the first schools to admit girls in China. Wu graduated from this school in 1922 and continued her studies at the Soochow School for Girls in Nanjing. As was generally expected for women at the time, she enrolled in

Chinese physicist Chien-Shiung Wu (1912–1997) experimentally confirmed the Fermi theory of beta decay. In her major accomplishment she showed that reflection symmetry, or parity, is a fundamental symmetry not conserved in electron decay of radioactive nuclei. She was the first women to serve as President of the American Physical Society. CREDIT: COURTESY OF AP/WIDE WORLD PHOTOS. REPRODUCED BY PERMISSION.

the Normal School program, which led to a teaching career. Pursuing her interest in physics and mathematics, she enrolled in 1930 at the National Central University in Nanjing, from which she graduated in 1934 at the head of her class. She taught for a year at the National Chekiang University and started research in X-ray crystallography at the National Academy of Sciences (Academia Sinica) in Shanghai in 1935 and 1936. Wu immigrated to the United States in 1936 to pursue graduate studies at the University of Michigan. However, she first visited the University of California at Berkeley where she met another Chinese physics student, Luke Chia Yuan, who introduced her to Professor Ernest Lawrence. Lawrence immediately recognized Wu's intelligence and potential and convinced her to stay at Berkeley. This encounter with Luke Yuan was the beginning of a long and warm relationship, which was further strengthened by marriage in 1942.

Wu worked under the direct supervision of Professor Emilio Segrè. Her success hinged on her appreciation of the importance of careful and accurate measurements. Wu's Ph.D. thesis (1940) involved studies of fission products of uranium, a topic of major interest at the time. In particular, the identification of two Xe isotopes earned her wide recognition a few years later, during World War II, when the development of nuclear piles depended critically on the knowledge and avoidance of materials that would poison and shut down the reactors. Her very careful work enabled her to identify ^{135}Xe as the culprit in reactor malfunction and made it possible to devise techniques to control the operation of reactors.

After graduation, Wu taught at Smith College and Princeton University. By then the nation was at war, and she was invited to join the Manhattan District Project at Columbia University. She first worked on gaseous diffusion of uranium and later on measurements by time-of-flight of the energy dependence of neutron reaction cross sections.

The end of the war in 1945 allowed Wu to finally take control of her career and focus on a problem that was to make significant advances in the understanding of nature. Enrico Fermi had developed a mathematical theory to explain the radioactive process of beta decay. This approach involved a new force called the weak interaction. In this process a neu-

tral particle, named neutrino by Fermi, and postulated simultaneously by Enrico Fermi and Wolfgang Pauli, was emitted. This particle was assumed to have remarkable properties, namely, it was massless; had spin just like protons, neutrons, and electrons; and barely interacted with matter. Understanding the weak interaction and the nature of the neutrino became Wu's life-long commitment. Existing experiments disagreed with Fermi's theory, but Wu quickly understood the experimental problems and assembled the most suitable apparatus to measure, with exquisite precision, the shapes of the electron spectra resulting from these decays. She investigated different types of beta decay, the so-called "allowed" and "forbidden" transitions, and showed unambiguously that the Fermi theory of beta decay was correct in all its details.

These experiments brought Wu worldwide recognition. She was now poised to handle the next challenge. It came in the form of a puzzle in particle physics. Two of the newly discovered particles, the τ and θ, had the same mass, spin, and lifetime, and yet one decayed into two pions while the other decayed into three pions. These two decay modes were of opposite parity, meaning that reflection symmetry was not obeyed. This symmetry, a property of physical systems obeyed in all observed interactions studied up to that time, requires that a mirror image of a process obtained by reversing all directions and velocities be identical to the original process. The $\tau - \theta$ puzzle led Professors Tsung Dao Lee and Chen-Ning Yang to question the accumulated evidence for conservation of parity in various decay processes. They realized, after extensive discussions with Wu, the undisputed experimentalist in beta decay and weak interaction physics, that there was no evidence for parity conservation or nonconservation in weak interactions. Wu designed the experiment that would test directly this symmetry principle. She formed collaboration with the experts in low-temperature spin polarization at the National Bureau of Standards in Washington, D. C., to measure the forward-backward asymmetry of electron emission of spin-polarized ^{60}Co nuclei. Again, in this instance, as in many cases in her work, she contributed a crucial element to the experiment, namely, the large crystals of paramagnetic salts necessary for the polarization of the ^{60}Co nuclei. The historic paper describing this work, "Experimental Test of Parity Conservation in Beta Decay," has become a classic.

Wu's beautiful and definitive work on beta decay established the Fermi theory of weak interactions. The elegant and momentous experiment, which established the nonconservation of parity and the violation of particle-antiparticle charge conjugation symmetry in physics, altered forever the view of the universe. As T. D. Lee described her and her work, "C. S. Wu was one of the giants of physics. In the field of beta-decay, she has no equal" (p. 7).

Wu continued research on fundamental problems. In 1963 she observed the phenomenon of weak magnetism, which confirmed the symmetry between the weak and the electromagnetic currents and set the cornerstone for the unification of these two basic forces into the electroweak force. Wu's interest in the nature of the neutrino led her to studies of double beta decay in which either two neutrinos or none are emitted, depending on their characteristics. She examined the radiations of "exotic" atoms with muons or pions replacing electrons in order to determine nuclear charge radii with higher accuracy than previously measured. She used new techniques such as the Mössbauer effect to study another fundamental property, time-reversal invariance. Wu also conducted research in condensed matter physics through the examination of magnetic transitions and relaxation effects in materials, and she explored a current problem in biology, the structure of sickle cell hemoglobin. In later years she devoted much effort to educational programs in both the Republic of China and Taiwan, as well as to the development of new facilities such as synchrotron radiation light sources.

Wu was frequently honored. She was promoted to a full professorship at Columbia in 1958, the first woman to hold a tenured faculty position in the physics department. She was appointed the first Michael I. Pupin Professor of Physics in 1973 and retired in 1981. She was elected to the National Academy of Sciences (1958) and received the National Medal of Science (1958), the Research Corporation Award (1958), and the John Price Wetherill Medal of the Franklin Institute (1962). Many distinguished awards followed, most notably the Cyrus B. Comstock Award (1964), the Tom Bonner Prize of the American Physical Society (1975), and the Wolf Prize from the State of Israel (1978). She was inducted in 1998 into the American National Women's Hall of Fame. She was the first woman to receive an honorary doctorate from Princeton University, and the first woman to serve as President of the American Physical Society (1975).

Beauty and aesthetics defined her work, her demeanor, and her relationships with family and friends. She was proud of the intellectual development of her son, Vincent Yuan, who as a physicist also worked in parity nonconservation in compound nuclei, and of the academic achievements of her grand daughter. She nurtured about thirty-three graduate students and many visiting scientists and postdoctoral fellows.

Wu died in New York, following a stroke, on February 16, 1997. Her remarkable life can be portrayed by an ancient Chinese poem by Qu Yuan ($\sim$340 B.C.E.): "Although the road is long and arduous, I am determined to explore its entire length."

See also: PARITY, NONCONSERVATION OF; PAULI, WOLFGANG

Bibliography

Lee, T. D. "Chien-Shiung Wu, 84, Dies; Top Experimental Physicist." *The New York Times* (February 18, 1997) p. 7.

McGrayne, S. B. *Nobel Prize Women in Science* (Carol Publishing Group, Secaucus, NJ, 1992).

Wu, C. "Recent Investigations of the Shapes of Beta-Ray Spectra." *Reviews of Modern Physics* **22**, 386–398 (1950).

Wu, C. Ambler, E.; Hayward, R. W.; Hoppes, D. D.; and Hudson, R. P. "Experimental Test of Parity Conservation in Beta Decay." *Physical Review* **105**, 1413–1415 (1957).

Wu, C. and Moszkowski, S A. *Beta Decay* (Wiley, New York, 1966).

Noemie Benczer Koller

Y

YUKAWA, HIDEKI

Hideki Yukawa, the first Japanese Nobel Laureate in physics and the originator of the meson theory of nuclear forces, was born in Tokyo on January 23, 1907. During most of his life, he lived in Kyoto, Japan's ancient capital, and he died there on September 8, 1981. He was the fifth of seven children of Koyuki and Takuji Ogawa. Both of his parents stemmed from scholarly families of the samurai tradition. The family moved from Tokyo to Kyoto in 1908 when Yukawa's father, a geologist, was appointed Professor of Geography at Kyoto Imperial University. Besides Hideki (who changed his name to Yukawa when he married Sumi Yukawa and was adopted by her family in 1932), three other Ogawa sons became university professors and renowned scholars in their respective fields.

Early Education and Research

As a youth, Yukawa's interests were mainly literary, and he read widely in world literature and philosophy. In 1923 he entered the Third High School, actually a junior college, one of only eight then existing in Japan, and found himself strongly attracted to physics. One of his classmates (from elementary school through university) was Sin-itiro Tomonaga, who shared a Nobel Prize in Physics in 1965 for his contributions to quantum electrodynamics. Both young men became acquainted with modern atomic and nuclear physics from their high school physics teacher.

Yukawa and Tomonaga enrolled in the physics program at Kyoto Imperial University in 1926 and graduated in 1929. They stayed on as unpaid assistants at the university due mainly to the worldwide depression. As a postgraduate, Yukawa first tried to solve the so-called divergence problems of quantum electrodynamics (QED): the theory predicted infinite electron charge and infinite mass, both predictions being absurd. Failing in this attempt (in which Tomonaga and others succeeded in the 1940s), Yukawa turned to what seemed to be an "easier" problem, that of nuclear forces.

The Meson Theory of Nuclear Forces

Beginning in 1933, Yukawa tried to improve a theory that had been proposed in 1932 by the German physicist Werner Heisenberg to explain the forces holding the nucleus together. In this theory, the main attractive force arose through the exchange of an electron between a neutron and a proton in the nucleus. That is, a neutron would emit an electron and become a proton. Normally the electron would be absorbed by a proton in the same nucleus, turning into a neutron. Heisenberg's theory attributed the

Japanese physicist Hideki Yukawa (1907–1981) received the Nobel Prize in Physics in 1949 for his meson theory of nuclear forces.
CREDIT: COURTESY OF BETTMANN/CORBIS. REPRODUCED BY CORBIS CORPORATION.

radioactive process known as beta decay, in which a nucleus emits an electron, to the occasional escape of an electron from the nucleus.

However, Heisenberg's theory had a number of serious difficulties. As a theory of the forces holding the nucleus together, it violated several principles of the theory of quantum mechanics. For one thing, Heisenberg's own uncertainty principle would not allow an electron to be confined in a space as small as the nucleus. Other laws were violated as well, including the conservation law of angular momentum. As for radioactive beta decay, while the nucleus lost a definite amount of energy, the emitted electron had lesser, and variable, amounts of energy, and Heisenberg's theory did not account for this missing energy. Accounting for this energy would require that a new particle be emitted with the electron; this particle was called the neutrino, proposed in 1930 by Wolfgang Pauli. One way out of these contradictions was to assume that entirely new laws of physics held in the nucleus.

Yukawa sought to retain the principles of quantum theory and the conservation laws and to eliminate the difficulties of Heisenberg's theory by relating the exchange force to a new field that he called the U-field. He took for a model the electromagnetic field that, according to quantum field theory, results from the exchange of massless light quanta (photons). Allowing his new U-quanta to be massive, Yukawa discovered in 1934 an important relation between the mass of the exchanged heavy quanta and the range of the force in question. From this relation he inferred that the quanta of the nuclear force field must be about two hundred times as massive as the electron, based upon the known short range of the strong nuclear forces. He assumed that the new heavy quanta have charges of the same size as the electron's, either positive or negative

To account for beta decay, Yukawa assumed that his heavy quanta, interacting with the nuclear particles with high probability (that is, they were strongly coupled), could also decay with small probability (weak coupling) into an electron and a neutrino. Yukawa and his students later calculated that the decay of the free heavy quantum should take about 10^{-8} seconds. Yukawa was the first physicist to make a clear distinction between the two kinds of nuclear force: the strong force to give binding and the weak force for beta decay.

Yukawa's theory involving the new heavy charged quanta (now called mesons) was published in an English-language Japanese journal at the beginning of 1935. However, it attracted worldwide notice only in 1937, after American physicists Carl Anderson and Seth Neddermeyer, studying the cosmic rays, found particles that appeared to fit Yukawa's requirements.

Beginning in 1937 many theorists, Yukawa and his students, and several groups working in Europe and America worked out versions of the meson theory to account for phenomena occurring in nuclear physics as well as in the cosmic rays. However, there was only rough qualitative agreement with the experiments which were simultaneously undertaken. For example, the absorption of the new cosmic ray particles by nuclei was much weaker than expected.

In 1947 cosmic ray workers in England demonstrated, with a new technique that used sensitive photographic emulsion, that the particles that had been observed, mainly at sea level, were not Yukawa mesons as was believed, but rather they were decay products of Yukawa mesons. These decay products were later called muons, and the Yukawa mesons were called pions. Pions are produced by collisions of cosmic rays with air molecules higher in the atmosphere, at mountain altitude or higher. The pion is an example of a class of particles having strong nuclear interaction called hadrons (including the proton and the neutron), while the muon belongs to the weakly interacting class called leptons (including the electron). After 1947, new particles of both classes were found in experimental searches in the cosmic rays and at new laboratories where large particle accelerators were constructed to create and study these particles. Among the new particles forming part of the particle explosion was a neutral pion. Its existence had been predicted in the 1930s by the Yukawa group in Japan and by Nicholas Kemmer in England, who had claimed that it was required by the known charge independence of nuclear forces, that states that the nuclear forces between any pair of nuclear particles is the same.

Yukawa's Academic Career

In 1933, Yukawa taught at Osaka Imperial University, and in 1939 he returned to Kyoto as a full professor. From 1948 to 1949 he was a member of the Institute for Advanced Study in Princeton, New Jersey. He was then a visiting professor at Columbia University in New York City until 1951, the year in which he was awarded the Nobel Prize in Physics, when he was made a full professor. In 1953, he returned to Kyoto to become the director of a new interuniversity research institute founded (and later named) in his honor. He wrote many essays on cultural and scientific subjects and participated in international movements for world federation and peace, such as the Pugwash Movement.

See also: ANDERSON, CARL D.; MUON, DISCOVERY OF

Bibliography

Brown, L. M. "Yukawa's Prediction of the Meson." *Centaurus* **25**, 71–132 (1981).

Hayakawa, S. "The Development of Meson Physics in Japan" in *Birth of Particle Physics*, edited by L.M. Brown and L. Hoddeson (Cambridge University Press, Cambridge, UK, 1983).

Kemmer, N. "Hideki Yukawa." *Biographical Memoirs of Fellows of the Royal Society* **29**, 661–676 (1983).

Mukherji, V. "History of the Meson Theory of Nuclear Forces from 1935 to 1952." *Archive for the History of Exact Sciences* **13**, 27–102 (1974).

Yukawa, H. *Creativity and Intuition*, translated by J. Bester (Kodansha, New York, 1973).

Yukawa, H. *Tabibito (The Traveler)*, translated by L.M. Brown and R.Y Yoshida (World Scientific, Singapore, 1982).

Laurie M. Brown

Z

Z FACTORY

The highest-energy electron-positron colliders ever built were LEP and SLC. Both projects were initiated around 1980 as Z factories to study the properties of the Z boson at a center-of-mass energy of approximately 91 GeV. The Large Electron Positron project (LEP) was a conventional circular storage ring built at the European Laboratory for Particle Physics (CERN) in Geneva, Switzerland. The Stanford Linear Collider (SLC), built at the Stanford Linear Accelerator Center (SLAC) in California, was a prototype for an entirely new approach to electron-positron colliders. The switch to a new technology was motivated by the fact that the circumference of an electron storage ring has to increase as the square of the desired energy. To reach an eventual energy of 100 GeV per beam, LEP required a 27-km tunnel bored underneath the Jura Mountains in France and Switzerland. To reach ten times higher energy in a storage ring, the tunnel would have to be more than 2,000 km long, too big to be practical. The ring has to be so long because electrons radiate energy in the form of light whenever they bend out of a straight path. The sharper the bending angle, the more energy is lost. This energy must be replaced on each turn the electrons make around the ring, so a large ring is necessary to limit the energy loss.

The new technique pioneered at the SLC was to use a linear accelerator, the 3-km-long SLAC linac, to raise the electrons and positrons to the desired energy. Because the electrons are not bent in a circle, the tunnel length only grows linearly with energy, and the cost can be kept reasonable. Present designs for linear colliders to reach 1,000 GeV are about 30 km long, only slightly longer than LEP. The disadvantage of the linear collider approach is that the bunches only collide at the machine repetition rate, about 100 times a second. In a storage ring the bunches collide tens of thousands of times a second. In a collider the rate of events per second, or luminosity, depends on the bunch crossing frequency and the density of particles in the bunch. Since a linear collider has many fewer crossings per second, it must use bunches that have a much smaller size. The challenges for a linear collider are to produce bunches of electrons cooled to a very small phase space, or emittance; to preserve that emittance as the bunches are accelerated; and then to focus the bunches to a very small size at the interaction point.

LEP built on the experience from a long series of electron-positron storage rings, and many problems were already familiar and their solutions well understood. LEP began operation in 1989, was commissioned rather quickly, and reached design performance in 1993. In 1992 the number of bunches was doubled from four to eight by using a pretzel scheme to separate the orbit of the two beams at unwanted extra collision points. In 1995 for the last Z physics run, four trains of up to three bunches were used. LEP was then converted to LEP2, and the energy increased to eventually 104.4 GeV per beam. To reach these energies required the largest superconducting rf system ever built, with more than 3 GV of accelerating voltage. One of the most interesting challenges at LEP was determining the exact energy of the beams for precision measurements of the Z mass. Because the LEP ring is so large, the circumference changes by ± 1 mm due to the tidal stretching of the Earth's surface. Seasonal changes in the local water table have a measurable effect on the tunnel. It was also found that the field of the bending dipoles increased steadily during each fill as a result of the thermal expansion of the concrete cores on which the magnet coils were wound and of leakage currents from the rails of nearby high-speed electric trains. To accurately predict the beam energy, the temperature and stray currents as well as numerous other quantities had to be carefully monitored.

The SLC was the first linear collider ever attempted, and it took many years to develop the understanding and techniques required to collide very small beams. New diagnostics and new procedures were needed to center the beams through the focusing magnets and accelerating structures to minimize effects that could increase the beam size. An innovative algorithm for finding the best beam trajectory, called dispersion-free steering, was used first at SLC and later at LEP to improve beam quality. A key breakthrough was moving micrometer-sized wires through the beams to measure the beam size noninvasively during routine operation. Feedback systems were required to stabilize the energy and position of the beams throughout the machine. Specialized feedback was used to bring the tiny beams into collision and even to optimize the final tuning of the beam size. The SLC eventually had more than forty wire scanners and fifty feedback systems. The beam size at the collision point was 1.4 micrometers horizontally and 0.7 micrometers vertically, much smaller than a human hair and about a factor of 100 smaller than in storage rings. With such small intense beams, the interaction of the two beams

causes them to shrink even further in size. This effect is called pinch enhancement and was measured for the first time at the SLC, where it increased the luminosity (event rate) by as much as a factor of 2.

LEP had four experimental detectors: Aleph, Delphi, L3, and Opal. SLC had only one interaction point and began with the Mark-II, which was replaced by the SLD experiment in 1991. Because LEP reached high luminosity more quickly, the LEP experiments had the advantage of a much larger data sample. The SLD exploited two unique features of a linear collider: the very small interaction point and a highly polarized electron beam, where the electron spins point predominantly in the same direction. These advantages allowed SLD to make measurements that were complementary to the high-precision LEP measurements, including the world's most precise measurement of the critical electroweak mixing angle.

Building on the success of the SLC, many groups around the world are actively developing proposals for a new linear collider project to start at an energy of 500 GeV and eventually reach 1,000 GeV or higher. High-energy physics advisory groups in the United States, Europe, and Asia have all recently endorsed such a collider as the next major project needed to advance their research. Several different technologies for accelerating the beams are under study. The TESLA proposal led by the Deutsches Elektronen-Synchroton Laboratory (DESY) in Hamburg, Germany, uses superconducting cavities. The NLC proposal led by SLAC and the JLC proposal led by the Japanese High-Energy Accelerator Research Organization (KEK) in Japan use accelerating structures similar to those at the SLC, but at a higher frequency. The CLIC proposal led by CERN uses a novel two-beam design, which could form the basis for a later linear collider of even higher energy.

See also: ACCELERATORS, COLLIDING BEAMS: ELECTRON-POSITRON

Bibliography

Assmann, R., et al. "Calibration of Center-of-Mass Energies at LEP-1 for Precise Measurements of Z Properties." *European Physical Journal* **C6**, 187–223 (1999).

Barklow, T., et al. "Experimental Evidence for Beam-Beam Disruption at the SLC." <http://accelconf.web.cern.ch/AccelConf/p99/PAPERS/WEBR3.PDF>.

LD Collaboration (Kenji Abe, et al.) "A High Precision Measurement of the Left-Right Z Boson Cross-section Asymmetry." *Physical Review Letters* **84**, 5945–5949 (2000).

Sally Dawson

c. 600 B.C.E: Thales of Miletus speculates that water is the common principle out of which all matter is made.

c. 540 B.C.E: Pythagoras and his pupils claim that all of nature is bound together by numbers.

c. 540 B.C.E: Anaximenes suggests air as the primary matter of all things, whereas Heraclitus (c. 480 B.C.E) favors fire.

c. 450 B.C.E: Democritus develops an atomic theory that may have been first suggested by his teacher Leucippus. According to Democritus, everything consists of atoms—hard, indivisible and indestructible particles—within a vacuum. At about the same time, Empedocles proposes his theory of the four elements, water, air, earth, and fire.

c. 350 B.C.E: Aristotle rejects atomism and adopts the four-element theory. He maintains a strict separation between terrestrial matter and the substance of the heavenly bodies, later known as *quinta essentia* or the fifth element.

c. 50 B.C.E: Lucretius, in his *De Rerum Natura* (On the Nature of Things), gives a full exposition of philosophical atomism based on Democritus and Epicurus.

c. 1200: Following the Arabian philosopher Averroës, medieval thinkers develop the Aristotelian notion of *minima naturalia*, a corpuscular theory of matter.

c. 1530: In Paracelsus's chemical philosophy, all matter and changes are explainable in terms of three principles called salt, sulfur, and mercury.

1661: In *The Sceptical Chymist*, Robert Boyle bases chemistry on the theory of atoms. He rejects the Aristotelian four-element theory and Paracelsian ideas.

1704: Isaac Newton, in his *Opticks*, conceives light as a stream of tiny particles and suggests that matter is explained by forces acting between primary particles.

1718: Georg Stahl proposes a theory according to which all combustible bodies include *phlogiston*, a hypothetical principle.

1758: Rudjer Boscovich develops a dynamic conception of matter based on point-atoms.

1789: Antoine-Laurent Lavoisier's *Treatise of Chemistry* signals the coming of a new, antiphlogistic chemistry based on the principle of mass conservation. In his table of chemical elements, Lavoisier includes imponderable (weightless) bodies such as light and heat.

1808: In his *New System of Chemical Philosophy*, John Dalton revises the atomic theory. Dalton's atoms are characterized by their weights, with one kind of atom corresponding to one kind of chemical element.

1811: Amedeo Avogadro introduces compound atoms, or molecules.

1815: William Prout suggests that all atomic weights are multiples of hydrogen's atomic weight and that hydrogen may be the common element, or prototype, of all matter.

1847: The law of energy conservation receives its full formulation by Hermann von Helmholtz, who draws upon earlier work by Robert Mayer, James Joule, and others.

1859: Gustav Kirchhoff introduces the concept of a black body. Together with Robert Bunsen, he invents the spectroscope for chemical analysis and discovers two new elements, cesium and rubidium.

c. 1860: The ether becomes an important concept in many physical theories, in particular in electrodynamics where it serves as the medium for transmission of signals. The ether is only abandoned around 1910, after the acceptance of Albert Einstein's special theory of relativity.

1864: James Clerk Maxwell publishes his *Dynamical Theory of the Electromagnetic Field* in which the laws of electromagnetism are presented in terms of fields, but with no room for electrified particles.

1865: Johann Loschmidt makes the first calculation of the number of molecules in a volume of gas, corresponding to Avogadro's number.

1867: William Thomson (later, Lord Kelvin) suggests that atoms are vortical modes of motion in a primitive, perfect, and all-pervading fluid. The vortex atomic theory, a truly unified theory of all matter, is developed during the following decades but abandoned in the 1890s.

1868: Based on a few unidentified lines in the solar spectrum, Norman Lockyer suggests the existence of a new element, helium, in the sun.

1869: Dmitri Mendeleev publishes his periodic system in which the chemical elements are ordered according to their atomic weights and arranged in periods and groups.

1886: William Crookes suggests that the periodic system reflects a compound structure of atoms and that the basic unity of matter may be helium, supposed to have an atomic weight of one-half.

1891: George Johnstone Stoney introduces the term electron as a measure of an atomic unit charge. During the following years, electron theories are developed by Hendrik Lorentz, Joseph Larmor, and others.

1895: Wilhelm Röntgen discovers a new kind of rays, called X rays or Röntgen rays.

1896: Henri Becquerel announces the discovery of radioactivity, a radiation produced by uranium.

1896: Pieter Zeeman discovers the effect of magnetism on spectral lines, and Hendrik Lorentz deduces from it that atoms contain oscillating negative charges of a mass that is perhaps 1,000 times as small as that of a hydrogen atom.

1897: Joseph John Thomson demonstrates that cathode rays are corpuscular in nature. He calls the particles *corpuscles* and determines their charge-to-mass ratio. Thomson's corpuscles are soon known as electrons.

1898: Marie and Pierre Curie discover two strongly radioactive elements, radium and polonium. It is gradually recognized that radioactivity is a spontaneous process, a disintegration of atoms.

1900: Max Planck suggests that radiation energy is quantized. Planck's law of blackbody radiation is announced.

1904: Joseph John Thomson develops a quantitative electron theory of atomic structure. In the Thomson model, a large number of electrons move in a hypothetical fluid of positive electricity.

1905: Albert Einstein proposes that light and other electromagnetic radiation is not only emitted and absorbed in quanta, but consists of such quanta, later (1926) called photons. In his theory of relativity, he shows that energy and matter are equivalent according to $E = mc^2$.

1908: Ernest Rutherford and Thomas Royds show conclusively that alpha particles are doubly charged helium ions. During the next two decades, alpha particles are often considered to be elementary particles.

1909: Pieter Zeeman identifies the photon as a genuine elementary particle with both energy and momentum.

1911: Based on experiments with scattering of alpha particles by gold foils, Ernest Rutherford suggests that atoms possess a tiny, positively charged nucleus in which almost all the mass is concentrated.

1912: Victor Franz Hess discovers cosmic rays; their existence is confirmed in 1913.

1913: Niels Bohr proposes his quantum theory of atomic structure, according to which the electrons can move around the nucleus only in certain stationary orbits allowed by quantum conditions.

1919: Ernest Rutherford studies the action of alpha particles on gases and observes the first artificial disintegration of atomic nuclei.

1919: Ernest Rutherford discovers the proton, which he identifies with the nucleus of the hydrogen atom.

1922: Niels Bohr's theory of the periodic system promises a reduction of chemistry to physics.

1922: Arthur Compton provides direct experimental confirmation of the particle nature of light.

1923: Louis de Broglie proposes that particles have a wave character.

1924–25: Satyendranath Bose and Albert Einstein discover new statistics for particles of integral spin.

1925: Wolfgang Pauli formulates the exclusion principle for electrons in an atom.

1925: The hypothesis of electron spin is introduced from spectroscopic evidence by Samuel Goudsmit and George Uhlenbeck.

1925: Werner Heisenberg suggests a new quantum mechanics, where electrons do not move in definite orbits. The theory is quickly developed by Max Born, Pascual Jordan, Wolfgang Pauli, Paul Dirac, and Heisenberg himself.

1926: Erwin Schrödinger develops an alternative form of quantum mechanics, wave mechanics, based on a wave function governed by an equation known as the Schrödinger equation.

1926: Enrico Fermi and Paul Dirac develop the statistics for particles of half-odd-integral spin.

1927: Werner Heisenberg proposes the uncertainty principle.

1927–28: Clinton Joseph Davisson and George Paget Thomson provide confirmation of the wave nature of electrons.

1928: Paul Dirac finds an equation that combines quantum mechanics and the special theory of relativity. The new quantum equation accounts for the electron's spin.

1928: George Gamow applies quantum mechanics to the atomic nucleus and explains the mechanism of alpha radioactivity.

1929: Dimitry V. Skobelzyn observes energetic cosmic electrons and a shower produced by a cosmic ray. This is considered the birth of cosmic-ray particle physics.

1930: Wolfgang Pauli suggests the existence of the neutrino, a very light neutral particle emitted together with the electron in beta decay.

1930: Based on his relativistic equation, Paul Dirac introduces the notion of the (positively charged) antielectron that he identifies with the proton. Dirac's hypothesis reduces all matter to electrons.

1931: Paul Dirac revises his 1930 theory and identifies the antielectron with a hypothetical particle of electronic mass. He also suggests that negatively charged antiprotons and magnetic monopoles may exist.

1931: Georges Lemaître proposes, for the first time, a Big Bang model of the universe. He likens the initial state of the universe to a huge radioactive atom.

1931: Harold Urey reports the discovery of deuterium, the heavy hydrogen isotope.

1931: Robert Jemison Van de Graaff invents the electrostatic accelerator.

1931: Ernest Orlando Lawrence tests the first cyclotron.

1932: Called the "miraculous year" of nuclear physics, this year also marks the beginning of modern particle physics. James Chadwick discovers a new nuclear constituent, the neutron, initially believed to be a proton-electron composite. In the cosmic radiation, Carl David Anderson finds positively charged electrons, or positrons, that are soon identified with Paul Dirac's antielectrons.

1933: Enrico Fermi's theory of beta decay initiates weak-interaction physics and justifies the neutrino.

1935: Hideki Yukawa suggests the existence of a "heavy quantum," or meson, a particle that mediates nuclear forces. At first Yukawa's particle is mistakenly identified with the cosmic-ray meson (muon) and only in the late 1940s is it discovered experimentally and then called a π meson or, later, pion.

1937: Carl David Anderson and Seth Neddermeyer discover a "heavy electron," what will later be identified as and called a muon.

1939: Hans Bethe applies nuclear physics to the interior of stars and proposes mechanisms (the CNO cycle) that explain stellar energy production.

1941: Donald William Kerst builds the first betatron and accelerates electrons to an energy of 2.3 MeV.

1944: Louis Leprince-Ringuet and Michel Lhéritier produce the first evidence of the positive kaon (K^+).

1945: Edwin McMillan invents the principle of phase stability for particle accelerators.

1946: George Gamow links the expansion of the universe and the abundance of the chemical elements with an early time of high density and temperature. This idea calls attention to Big Bang cosmology.

1947: Marcello Conversi, Ettore Pancini, and Oreste Piccioni publish evidence showing that the muon is not the mediator of the nuclear force.

1947: The two-meson theory is vindicated by detection in the cosmic radiation of heavy mesons (pions) decaying into muons. The first indication of pions observed in nuclear emulsions is reported by Donald H. Perkins. This is confirmed by Giuseppe Paolo Stanislao Occhialini and Cecil Frank Powell.

1947: The Lamb shift in the hydrogen spectrum, measured by and named for Willis Lamb, is explained by the new quantum electrodynamics (QED) pioneered by Julian Schwinger, Richard Feynman, and Sin-itiro Tomonaga.

1947: The first strange particles are reported by George Dixon Rochester and Clifford Charles Butler.

1948: Polykarp Kusch and Henry Foley establish the anomalous magnetic moment of the electron. Julian Schwinger calculates the anomalous magnetic moment of the electron.

1949: Jack Warren Keuffel creates the spark chamber method for particle tracking.

1950: Physicists Russell Foster Bjorklund, Walter Ellis Crandall, Burton Jones Moyer, and Herbert Frank York at the Lawrence Radiation Laboratory discover evidence for the neutral pion, B^0. The existence of the B^0 is confirmed in 1951.

1953: Murray Gell-Mann introduces the strangeness quantum number to account for the new, strongly interacting particles.

1953: In an extension of George Gamow's earlier work, Ralph Alpher, Robert Hermann, and James Follin give the first detailed calculations of the early universe.

1953: Donald Glaser, creator of the bubble chamber, observes the first particle tracks in his device.

1954: Chen Ning Yang and Robert Mills construct a locally gauge-invariant field theory of non-abelian interactions in analogy with QED, as does Ronald Shaw in his Ph.D. dissertation at Cambridge University.

1955: Using the Bevatron particle accelerator, Owen Chamberlain, Emilio Segrè, Clyde Wiegand, and Thomas Ypsilantis produce the antiprotons that Paul Dirac predicted in 1931.

1955: Murray Gell-Mann and Abraham Pais predict the long-lived kaon K_L.

1955: Abraham Pais and Oreste Piccioni propose an experiment for K_L to K_S regeneration.

1956: Experiments carried out by Clyde Cowan and Frederick Reines prove the existence of neutrinos from beta decays.

1956: Chen Ning Yang and Tsung Dao Lee suggest that parity is not conserved in weak interactions, which is experimentally confirmed the following year.

1956: With a cloud chamber as a detector, physicists at Brookhaven National Laboratory find evidence for the long-lived kaon K_L.

1957: Two groups independently demonstrate experimentally the nonconservation of parity in weak interactions. One group includes Chien-Shiung Wu, Ernest Ambler, Raymond Webster Hayward, Dale D. Hoppes, and Ralph P. Hudson and the other includes Richard L. Garwin, Leon M. Lederman, and Marcel Weinrich. This is confirmed by Jerome I. Friedman and Valentine Lories Telegdi.

1958: Richard Feynman and Murray Gell-Mann develop V-A form of weak interaction.

1961: Murray Gell-Mann and Yuval Ne'eman introduce the Eightfold Way, a particle classification system based on the SU(3) symmetry group.

1961: The staff of the Brookhaven National Laboratory develop the first strong focusing proton synchrotron—the Alternating Gradient Synchrotron (AGS).

1961: The hypothesis of an intermediate boson as the mediator of the electromagnetic and weak interactions is introduced by Sheldon Glashow.

1961: A Johns Hopkins/Northwestern collaboration led by Aihud Pevsner discovers the eta, the last member of the first octet of the Eightfold Way.

1962: Experiments at Brookhaven National Laboratory by Leon Lederman, Melvin Schwartz, and Jack Steinberger demonstrate the existence of a new kind of neutrino, the muon neutrino.

1964: In experiments with neutral kaons, James Cronin, Jim Christensen, Val Fitch, and Rene Turlay prove that CP (the combination of charge conjugation and parity) is not absolutely conserved.

1964: At the Brookhaven National Laboratory, a group led by Nicholas P. Samios discovers the omega-minus particle that vindicates the Eightfold Way and leads to Murray Gell-Mann's and George Zweig's introduction of fractionally charged quarks as constituents of hadrons.

1964: Peter Ware Higgs proposes a mechanism for mass generation of gauge bosons.

1965: Moo-Young Han and Yoichiro Nambu introduce the color quantum number for quarks and gluons.

1966: An article by Yoichiro Nambu marks the beginnings of quantum chromodynamics (QCD).

1967: Steven Weinberg and Abdus Salam develop a formalism that gives a unified description of electromagnetic and weak interactions. With Gerardus 't Hooft's 1971 proof that the theory is renormalizable, and the 1973 detection of neutral currents, the Weinberg-Salam electroweak theory is established.

1967: Andrei Sakharov suggests that the particle-antiparticle asymmetry in the universe is based on three physical conditions: the baryon number not being exactly conserved, the CP symmetry being violated, and the lack of thermal equilibrium during the early moments of the universe.

1967: Gersh Itskovich Budker proposes particle cooling in accelerator storage rings.

1968: George Charpak invents the multiwire proportional chamber for particle detection.

1969: The scattering experiments of Jerome I. Friedman, Henry Kendall, and Richard Taylor give birth to the idea that protons and neutrons consist of smaller particles.

1970: Sheldon Glashow, John Iliopoulos, and Luciano Maiani propose a fourth charmed quark.

1971: The first evidence of charmed particles is seen in emulsions.

1973: David Gross and Frank Wilczek, as well as Hugh David Politzer, demonstrate the asymptotic freedom of Yang-Mills theories and pave the way for QCD as a theory of strong interactions.

1974: The J/ψ particle discovery, made independently by groups led by Samuel Ting and Burton Richter, confirms that constituent quarks must be taken seriously.

1974: Sheldon Glashow and Howard Georgi, building on ideas of Jogesh C. Pati and Abdus Salam, suggest the first grand unified theory of strong, weak, and electromagnetic forces.

1974: John Schwarz and Joel Scherk propose a ten-dimensional string theory that includes gravitation.

1975: Experiments in Stanford by Martin Perl and collaborators show the existence of a superheavy (tau) lepton.

1975: The first evidence of quark jets in electron-positron annihilation is obtained at SLAC.

1977: Leon Lederman and his group at Fermilab discover the Upsilon and two of its excited states, thereby providing the first evidence for the b (bottom) quark.

1979: The gluon is discovered by the JADE, MARK J, PLUTO, and TASSO experiments at DESY with PETRA.

1981: Alan Guth applies particle physics to the very early universe and produces an inflation model based on the idea of a false vacuum.

1981: At the Cornell accelerator facility with the CLEO detector, the first evidence for the B meson is obtained.

1983: The intermediate vector bosons (W and Z particles) predicted by the electroweak theory are detected in experiments conducted on the SPS Proton-antiproton collider at CERN.

1984–85: With the work of Michael Green, John Schwartz, Edward Witten, and others, super-string theory becomes a strong candidate for a "theory of everything."

1987: The first observation of the neutrino bursts from supernova SN1987A, opening the era of neutrino astronomy.

1989: Evidence supporting the conclusion that there are three types of neutrinos comes from experimental data gathered at two laboratories: the SLC accelerator at SLAC using the Mark-II detector and the LEP accelerator at CERN using the ALEPH, DELPHI, L3, and OPAL detectors.

1989–1998: Precision measurements at CERN, SLAC, and Fermilab confirm the Standard Model to 0.1 percent accuracy and point to a Higgs boson with a mass of less than approximately 200 GeV.

1995: The last type of quark, the heavy top quark, is discovered at Fermilab with the CDF and D0 detectors.

1998: Data from cosmic-ray neutrinos indicate that the neutrino may possess mass.

1998: Astronomical observations suggest that the cosmological constant, first introduced by Einstein in 1917, is not zero. The constant corresponds to a vacuum energy that may be responsible for the main part of the energy-mass density of the universe.

1998: Neutrino oscillations are demonstrated at the Super-Kamiokande laboratory in Japan.

1999: Direct violation of the CP symmetry appears to emerge from experiments at Fermilab and CERN.

2000: By smashing lead ions into fixed targets of lead or gold atoms, a new form of nuclear matter, a quark-gluon plasma, may have been observed.

2000: Direct evidence for the tau neutrino is reported by physicists at Fermilab.

2001: The mass and charge of the antiproton is measured. To within 60 parts per billion, the values agree with those of the proton.

2001: CP violation in the decay of B mesons is observed at SLAC and KEK.

2001: The Sudbury Neutrino Observatory confirms solar neutrino oscillation; namely, that neutrinos change from one type to another.

Helge Kragh
John S. Rigden

GLOSSARY

Alpha Particle: The nucleus of ordinary helium, consisting of two protons and two neutrons bound together by the strong nuclear force.

Amplitude: A representation of an interaction between particles. A Feynman diagram represents a particular amplitude. Often a particular process can occur by several different paths, each represented by a particular Feynman diagram. The amplitude is the sum of the Feynman diagrams for a given process. The probability that a process will occur is represented by the absolute square of the sum of all contributing amplitudes.

Angstrom (Å): A unit of length equal to 10^{-10} m, commonly used in atomic and molecular physics where typical distances and sizes can be expressed in a few angstroms. Visible light has wavelengths of 4,000 to 7,000 Å, whereas the wavelengths of X rays are about 1 Å.

Angular Velocity (ω): The rate at which an object is rotating. Angular velocity can be expressed in units of radians per second.

Asymptotic Freedom: The term that is used to refer to the fact that the strong force between quarks becomes weaker as quarks get closer together.

***B* Meson:** A particle consisting of a *b* (bottom) antiquark and a light quark. It is similar to the neutral kaon (*s* quark and one *u* or *d* antiquark) but more massive and more sensitive to violations of CP invariance. Various accelerators have been built to produce *B* mesons to study the predictions of the Standard Model regarding CP violation in *B* decays.

Baryon: A particle with half-odd-integer spin that feels the strong force. A proton and a neutron, for example, are baryons.

Beam: A stream of a large number of identical particles, moving together in the same direction along the same well-defined path, produced by a particle accelerator. Each particle in a beam has approximately the same energy. The accelerator operator directs the beam against a stationary target or head-on into another beam (a colliding beam accelerator) in order to probe very short distances.

Boson: A particle with a spin quantum number equal to an integer, usually shortened by saying that the particle has "integral spin." The alpha particle and photon are bosons because their spins are 0 and 1, respectively. Bosons do not obey the Pauli exclusion principle. Thus, more than one boson of a given type can occupy the same quantum state, as in the case of a Bose-Einstein condensate.

Charge, Electric: One of the basic properties of particles along with mass, spin, baryon number, etc. These properties together uniquely define each type of particle. The charge of a particle can be positive, negative, or zero. Charged particles either attract or repel each other through the electromagnetic force or, expressed in quantum mechanical terms, through the exchange of photons. Quarks have fractional charges; all other particles have integral charges.

Collision: A process in which a moving particle passes close enough to a second particle that the two particles interact with each other, thereby causing changes in their motions or, perhaps, creating new and different particles.

Collision, Elastic: A collision between two particles in which the total kinetic energy of the particles before and after the collision is the same. This means the colliding particles are not changed into other particles.

Collision, Inelastic: A collision between two particles in which the total kinetic energy of the particles is different before and after the collision. The kinetic energy might decrease because some is converted into internal energy of one or both of the particles, some energy might be carried away as radiation, or new particles may be created.

Coulomb Interaction: The interaction of electrons, and other charged particles, with each other due to the electric force acting between them. This interaction is named after the French physicist Charles Augustin Coulomb, who discovered the law that described the electric force.

D Meson: The lightest particle that contains a charm quark and either a u, d, or s antiquark. The D meson cannot decay by the strong interaction: there is no lighter particle with charm. Therefore, it must decay by the weak interaction, which usually changes the charm quark into a strange quark, thereby permitting the decay to a K meson.

Dark Energy: The name given to the cause of the universe's accelerating expansion. This surprising feature of the universe was discovered in 1998 from observations of supernova. Since the known gravitation attraction would slow the expansion, something is apparently overriding the gravitation force, which was called dark energy. No one knows what dark energy really is.

Deuterium: A naturally occurring isotope of hydrogen (1 atom in about 6,700 atoms of hydrogen) with a nucleus consisting of one proton and one neutron. Deuterium is often referred to as heavy hydrogen because its mass is about twice that of ordinary hydrogen.

Duty Factor: The ratio of the time period when a system is actually operating to the total time for a complete cycle of the system. One operating cycle of a physical system consists of two parts: the time the system is actually in operation, and the time the energy sources of the system are reenergized so that the system can operate once again.

Electric Dipole Moment: The measure of two equal but opposite charges, $+q$ and $-q$, separated a distance d, $(q)(d)$.

Electron Volt (eV): A unit of energy equal to the kinetic energy acquired by an electron when it is accelerated through a potential difference of 1 volt, equal to 1.6×10^{-19} J.

Event: Something that happens at a particular location in space and at a particular time. For example, when an electron and a positron collide at a particular spatial point at a particular time, it can be referred to as an event.

Fermion: A particle with a spin quantum number equal to half of an odd integer, usually shortened by saying that the particle has half-integral spin. The electron, with spin $\frac{1}{2}$, is a fermion.

Gamma Ray: Photons with energies in the MeV range or higher. Gamma rays can be created in many ways, including the decay of a nucleus from a high-energy state to a lower-energy state or when particle-antiparticle pairs annihilate each other. For example, in the decay of ^{60}Co into ^{60}Ni, an electron and two gamma rays (1.173 MeV and 1.332 MeV) are emitted.

Gauss: An older unit of magnetic field equal to 10^{-4} Tesla (the SI unit of magnetic field defined such that a particle of charge 1 Coulomb moving at 1 m/s at right angles to a 1-Tesla magnetic field will experience a force of 1 Newton).

iga Electron Volt (GeV): A unit of energy—equal to 10^9 eV—that is useful in describing high-energy elementary particle processes.

adron: An elementary particle that interacts by the strong nuclear interaction.

-bar ($\hbar$): The combination h divided by 2π. This combination appears in many applications and so is shortened to $\hbar$. Planck's constant is symbolized by h.

ertz (Hz): The unit of frequency for repetitive phenomena, equal to one cycle per second. The frequency of the lowest audible musical tone is about 20 Hz; the frequency of green light is approximately 5×10^{14} Hz.

yperons: Any baryon that is more massive than the neutron and proton.

nvariance: A joint property of the laws of physics and certain transformations such that the laws do not change form under the transformations. For example, in his theory of special relativity, Einstein proposed that the laws of physics are the same for different observers in different states of uniform motion (Lorentz invariance); that is, the laws have the same mathematical form when transformed between coordinate systems for the two observers. Similarly, the laws of electromagnetism and the strong nuclear force, but not the weak nuclear force, are unchanged when space is inverted (parity invariance).

sospin: A mathematical spin that treats groups of particles as different states of the same particle. For example, the neutron and proton are treated as two isospin-$\frac{1}{2}$ states of a nucleon, with the proton being the isospin-up state and the neutron the isospin-down state. This treatment is valid to the extent that the different electromagnetic properties of the proton and neutron can be ignored. Similarly, the $\pi^{\pm}$ and π^0 pions are treated as the three isospin states of an isospin-1 particle: spin-up, spin-sidewise, and spin-down.

aon (or K meson): A particle made of a strange antiquark and either an up or down quark. The kaons are symbolized by K^+, K^-, K^0, and anti-K^0. The neutral kaons have played an important role in studies of invariance of CP (charge con-

jugation and parity) and time-reversal invariance. They form a pure two-state quantum system that can be treated as K^0 and anti-K^0, or alternatively as K_S and K_L, the short-lived and long-lived states of the system.

Kilo Electron Volt (KeV): A unit of energy equal to 1,000 electron volts.

Lifetime: The time from the formation of an unstable particle to its decay. The average lifetime τ of all similar particles is related to the half-life $t_{\frac{1}{2}}$ by $\tau = t_{\frac{1}{2}} \ln 2$.

Magnetic Monopole: The magnetic equivalent of an isolated electric charge. Magnetic monopoles have been predicted to exist, but none has ever been found.

Mass: A measure of the rest energy of a particle or system expressed in terms of Einstein's famous equation $m = E/c^2$.

Mega Electron Volt (MeV): A unit of energy equal to one million electron volts (10^6 eV).

Meson: A strongly interacting boson composed of a quark-antiquark combination.

Moment of Inertia (I): An expression of the way mass is distributed around a rotational axis. It plays the same role in rotational motion that mass plays in linear motion.

Neutron: A neutral, spin-$\frac{1}{2}$ baryon with a mass of 1.008665 u, which is equivalent to a mass-energy of 939.6 MeV. Nuclei are composed of neutrons and protons, which are collectively known as nucleons.

Nucleon: One of the constituents of nuclei, that is, a proton or neutron.

Nucleus: The center of an atom, where more than 99.9 percent of the atom's mass resides. Nuclei consist of approximately equal numbers of protons and neutrons, although ordinary hydrogen nuclei consist of a single proton and no neutrons. For heavy atoms the number of neutrons exceeds the number of protons. For example, whereas ordinary oxygen (a light atom) consists of eight protons and eight neutrons, the most common form of uranium (^{238}U) consists of 92 protons and 146 neutrons.

Order of Magnitude: Calculations done when an approximate answer is good enough for the purpose at hand. For example, if one is asked to multiply 7,863 by 145, one can approximate 7,863 by 8×10^3 (8,000) and 145 by 1×10^2 (100), which gives the answer 8×10^5 (800,000). The accurate answer is $7,863 \times 145 = 1,140,135$. The order-of-magnitude answer is often close enough for discussion purposes.

Particle: An elementary constituent of matter. The most elementary of particles are those without any known structure. According to the Standard Model of elementary particles, these include the quarks and leptons, their antiparticles, and the gauge particles (gluons, the photon, the W and Z particles, and the graviton). Baryons and mesons are also called elementary particles even though they are composed of the more elementary quarks and antiquarks.

Partons: Pointlike, "hard" components of neutrons and protons. They were detected during interactions with energetic electrons and are now known to be the quarks making up those particles.

Pion: A spin-0 meson that comes in three states, π^+, π^-, and π^0, of charges ^+e, ^-e, and zero, respectively. Pions are the lightest mesons. The π^+ and π^- are antiparticles of each other, and the π^0 is its own antiparticle, decaying into two photons with a half-life of 8×10^{-17} s.

Planck's Constant: The constant h, which establishes the scale where quantum effects become significant. Planck's constant h appears in all descriptions of atomic and subatomic physics. Its value is: $h = 6.6260755 \times 10^{-34}$ J/s.

Plasma: A state of matter in which atoms are highly ionized. A plasma consists of free electrons, charged ions, and atoms. The behavior of a plasma is characterized by the electric forces acting between the charged entities.

Precession: The motion of a rotating body when a torque changes the direction of its axis of rotation. An example is the familiar motion of a spinning top that is not spinning around the vertical axis, in which case the axis of the top sweeps out a cone as the top itself spins around its own axis.

Proton: A spin-$\frac{1}{2}$ baryon with a mass of 1.00728 u, which is equivalent to a mass-energy of 938.3 MeV, and a positive charge equal in magnitude to the electron charge. Nuclei are composed of neutrons and protons, which are collectively known as nucleons.

Quantum Number: Labels given to the quantum states of atoms and elementary particles.

Radian: A unit used to express the magnitude of angles. 2π radians $= 360°$; 1 radian $= 57.3°$.

Radiation: Either particles emitted by radioactive nuclei or electromagnetic waves. Electromagnetic radiation includes ordinary light, radar, radio waves, microwaves, infrared (thermal) radiation, ultraviolet radiation, X rays, and gamma rays.

Speed of Light (c): A basic constant of physics that appears in many equations. Its value is 3.0×10^8 m/s or, more exactly, 2.99792458×10^8 m/s.

Spin: The intrinsic angular momentum of a particle. It is analogous to the rotational angular momentum of a top but is not produced by actual spinning motion. It is one of the basic properties of a particle, such as mass and charge, that distinguishes one particle from another. The spin quantum number of any particle is an integer (0, 1, 2,...) or half an odd integer (1/2, 3/2,...).

Statistical Error: The uncertainty in a measured quantity originating from random sources. The measured value will lie on a bell-shaped curve called a Gaussian distribution whose width is related to the standard deviation σ.

Systematic Error: The uncertainty of a measured quantity originating from limitations in the experimental apparatus or experimental technique.

Tera Electron Volt (TeV): A unit of energy—equal to 10^{12} eV—that is used in describing very high-energy elementary particle processes.

Torque: The product of a force and a distance that causes things to rotate (as a force causes things to accelerate). Specifically, it is the magnitude of the force multiplied by the distance between the point of application of the applied force and the axis of rotation.

Upsilon: A meson that consists of a bottom quark and an antibottom quark. It was discovered in 1977, and its predicted discovery was an important confirmation of the Standard Model of elementary particles.

Lawrence A. Coleman
John S. Rigden

Page numbers in **boldface** refer to the main entry on the subject. Page numbers in *italics* refer to illustrations. References to figures and tables are denoted by *f* and *t*.